To my wife Judy

PREFACE

Applied mechanics is more than the teaching of the principles of physics. It is an important instrument in developing a method of stripping a problem to essentials and solving it in a logical, organized manner. This method of working can be applied to many other areas. This book therefore shows a consistent pattern of problem solving. The physics principles are presented in small elementary steps, the mathematics is kept at a reasonable level, and the problems are as practical as possible without becoming too involved with many extraneous details.

To accommodate the transition from the U.S. Customary system to the SI metric system, each chapter is a random mix of both systems but predominantly SI metric. There are more than 175 worked examples and 1030 graded problems, of which nearly two-thirds are in the SI metric system.

I would now like to address the student directly. You will no doubt discover problems that defy solution, no matter how well you understand all previous examples or problems. At the end of each chapter there is a list of hints for problem solving. It is not necessarily a summary of the chapter material, but is instead similar to a troubleshooting list. It is a list of the common areas where students have had difficulty or made errors in the past. Hopefully, by using this checklist to go back over your diagrams and calculations, you will find self-study and problem solving not only easier but certainly less frustrating.

I would like to acknowledge the reviewers of this edition: Kenneth Belanus, Oklahoma State University; Samuel W. Coates, Michigan Tech University; and Robert Willison, Sinclair Community College.

Keith M. Walker

BRIEF CONTENTS

PART ONE: STATICS

PART TWO: DYNAMICS

CONTENTS

PART ONE: STATICS

3 Moments and Couples 69

4 Equilibrium 91

5 Structures and Members 157

6 Three-Dimensional Equilibrium 215

CHAPTER 1

Introduction

OBJECTIVES

Upon completion of this chapter the student will be able to:

1. Explain why applied mechanics is necessary in engineering.
2. Submit a problem solution complete in all aspects of organized layout, units, and significant figures.
3. Solve for lengths or angles of a right-angle triangle using the trigonometric functions of sine, cosine, and tangent.
4. Apply the sine law to an appropriate triangle.
5. Apply the cosine law to an appropriate triangle.
6. Solve algebraic equations with one unknown, simultaneous equations, and quadratic equations.
7. Solve for lengths or angles in problems combining all previous trigonometry in addition to basic geometry principles such as opposite angle, supplementary angle, and sum of the included angles of a triangle.

1–1 WHAT AND WHY OF APPLIED MECHANICS

To someone who has never been exposed to applied mechanics, the subject may seem on first examination to be closely akin to a formal physics course, since physics is what it most closely relates to in the high school curriculum. But applied mechanics is basically an engineering science with practical applications. In this book we do not emphasize the purely theoretical approach but endeavor to show the practical applications of new theory.

Basic mechanics is composed of two principal areas—statics and dynamics. In this book, *statics* will be dealt with first; it is the study of forces on and in structures that are at rest or moving at a uniform velocity. A motionless body may have several forces acting on it, for example, gravitational force and a force opposing that gravity. Such a body is therefore *static,* or motionless, and has forces in balance, or *equilibrium.* Statics is the analyzing and determining of such forces. *Dynamics,* which will be studied later, is the next logical step in the study of forces, since it is concerned with *dynamic equilibrium,* or the forces acting on a moving body.

Because applied mechanics deals with the very basic concept of force, it is the origin for all calculations in areas such as stress analysis, machine design, hydraulics, and structural design. The design of an aircraft landing gear would require knowledge in all of these areas.

There are reasons other than those just mentioned for learning mechanics: The discipline is invaluable in developing one's logic or reasoning ability; one also learns a method of applying a little theory in a logical, neatly organized manner to arrive at a solution to a practical problem. The key to success is the method of attacking problems rather than the learning of massive quantities of theory. For those who prefer to memorize equations and to look for a "plug into the formula" solution, a change in method will be required.

The "why" of applied mechanics is therefore twofold: to lay the groundwork of theory for future engineering calculations and to train a person to organize and present his or her work in a logical manner. Such theory and the logical thought processes that must accompany it are the groundwork for many future engineering subjects.

1–2 UNITS AND BASIC TERMS

The units in this book will be predominantly SI metric; the remainder will be the U.S. Customary system. A metric system was standardized in June 1966 when the International Organization for Standardization approved a metric system called *Le Système International d'Unités.* The abbreviation is SI. This supplanted the old MKS metric system.

There are some changes required in the new metric system, but the most marked change will be for countries converting from the U.S. Customary system to the SI metric system. Due to the current phase of conversion to the SI system, the units used in examples and problems are mixed randomly throughout the book.

In the SI system, there are only seven basic units (Table 1–1). These basic units must measure quantities that could vary considerably in magnitude. To avoid awkwardly large or small figures, prefixes representing multiples and submultiples will be used (Table 1–2). You will notice that the multiples and submultiples are in increments of three digits. There are others that do not follow this pattern and therefore are not part of the SI system. Their use is permitted for convenience in certain cases.

TABLE 1–1

Physical Quantity	Name	Symbol
Length	meter	m
Mass	kilogram	kg
Time	second	s
Electric current	ampere	A
Temperature	kelvin	K
Luminous intensity	candela	cd
Amount of substance	mole	mol

TABLE 1–2

Name	Symbol	Multiply By
	Multiples	
kilo	k	10^3
mega	M	10^6
giga	G	10^9
tera	T	10^{12}
	Submultiples	
milli	m	10^{-3}
micro	μ	10^{-6}
nano	n	10^{-9}
pico	p	10^{-12}

Other well-accepted prefixes with more limited application are:

hecto	h	multiply by 10^2
deka	da	multiply by 10
deci	d	multiply by 10^{-1}
centi	c	multiply by 10^{-2}

For those who may not be familiar with the metric prefixes, the following equivalent values will demonstrate their use:

$$1 \text{ kilometer} = 1000 \text{ meters}$$
$$1 \text{ km} = 1000 \text{ m}$$
$$1 \text{ millimeter} = 10^{-3} \text{ meter}$$
$$1 \text{ mm} = 10^{-3} \text{ m}$$
$$10^3 \text{ mm} = 1 \text{ m}$$

Some of the principal SI-derived units are shown in Table 1–3. The units and terms in this table will be discussed when each specific area is covered. The following discussion of each will serve as an introductory explanation and as a central reference.

Length

A base unit of 1 meter (m) is used. The popular multiples are kilometers (km) and millimeters (mm). The centimeter (cm) is used for calculations to avoid unwieldy numbers and for convenience in other cases. The predominant unit in the U.S. Customary system is the foot (ft). Inches (in.) and miles are also used $(1 \text{ ft} = 12 \text{ in.}; 1 \text{ mile} = 5280 \text{ ft})$.

Mass

The *mass* of an object is a measure of the amount of material in the object. A base unit of 1 kilogram (kg) is used (1 tonne = 1000 kg). In the U.S. Customary system, it is the slug. Mass, weight, and force of gravity are discussed further under the heading "Force."

TABLE 1–3

Quantity	Unit	Symbol	Description
Acceleration	meter per second squared	—	m/s^2
Angle	radian	rad	—
Angular acceleration	radian per second squared	—	rad/s^2
Angular momentum	kilogram meter squared per second	—	$kg{\cdot}m^2/s$
Angular velocity	radian per second	—	rad/s
Area	square meter	—	m^2
Density	kilogram per cubic meter	—	kg/m^3
Energy	joule	J	N·m
Force	newton	N	$kg{\cdot}m/s^2$
Frequency	hertz	Hz	s^{-1}
Length	meter	m	—
Mass	kilogram	kg	—
Moment (torque)	newton-meter	—	N·m
Momentum	kilogram meter per second	—	kg·m/s
Power	watt	W	J/s
Pressure	pascal	Pa	N/m^2
Strain	—	—	mm/mm
Stress	pascal	Pa	N/m^2
Time	second	s	—
Velocity	meter per second	—	m/s
Volume			
Solids	cubic meter	—	m^3
Liquids	liter	l	10^{-3} m^3
Work	joule	J	N·m

Time

The base unit of time is 1 second (s). For consistency we will use the abbreviation "s" rather than "sec" as in the U.S. Customary system. Because of universal acceptance, other permitted units are minute (min), hour (h), and day (d).

Area

Area is measured in square meters (m^2) or multiples such as square millimeters (mm^2), square centimeters (cm^2), and square kilometers (km^2). Area in the U.S. Customary system is often in ft^2 or yd^2.

Volume

The base unit for solids is 1 cubic meter (m^3), and for liquids it is 1 liter (l), which is equal to 1 cubic decimeter (dm^3). Another common relationship between solids and liquids is 1 milliliter (ml) = 1 cubic centimeter (cm^3). The U.S. Customary system commonly uses ft^3 and yd^3 for solids and gallons for liquid.

Force

The unit of force is the newton (N). One newton is the force that when applied to a mass of 1 kg gives it an acceleration of 1 m/s^2 (1 N = 1 kg·m/s^2). Similarly, a mass of 1 kg with the standard acceleration of gravity of 9.81 m/s^2 will have a force of gravity of $1 \times 9.81 = 9.81$ N.

In the U.S. Customary system, a mass of 1 slug has a weight or force of gravity = mass × acceleration of gravity = 1 slug × 32.2 ft/s^2 = 32.2 lb. Note that in the SI system the term *weight* of an object is not usually used, but rather the force of gravity expressed in newtons. The newton is a relatively small unit of force; therefore, common multiples are kN and MN. To handle large forces in the U.S. Customary system, the kilopound (kip) is used (1 kip = 1000 lb).

Angle

The *radian* (rad) is used for measuring plane angles. In common practice, plane angles will continue to be measured in degrees although the use of minutes and seconds is discontinued (e.g., 38.2° rather than 38°12′). The radian is the angle between two radii of a circle that cut off on the circumference, an arc equal in length to the radius. Because the circumference equals $2\pi r$, there are 2π radians in 360°.

Pressure

Pressure is force per unit area, and the derived unit used is 1 pascal = 1 newton per square meter (1 Pa = 1 N/m^2). Again, this is a relatively small unit; therefore, kPa and MPa are often used. Units of psi (lb/$in.^2$) are common in the U.S. Customary system.

Stress

Stress is an internal load per unit area; therefore, it is expressed in units of pascals, as is pressure. The U.S. Customary system uses lb/$in.^2$

Strain

Strain is a measure of length per length, and the SI system requires the use of millimeters/millimeter (mm/mm). Units of cm/cm may be encountered since they are still in common usage in some countries. The U.S. Customary system uses units of in./in.

Energy

The *joule* is the work done when a force of 1 newton acts through a distance of 1 meter. Because

$$\text{work} = \text{force} \times \text{distance}$$

$$1\ \text{J} = 1\ \text{N}\cdot\text{m}$$

Similar to the pascal, the joule is a relatively small unit and may often be preceded by the larger prefixes of *kilo* and *mega.* In the U.S. Customary system, a force of 1 pound acts through a distance of 1 foot, giving units of ft-lb.

Work

Because work is energy, it has units of joules.

Power

Power is the rate of doing work. One *watt* of power is the rate of 1 joule of work per second.

$$\text{power} = \frac{\text{work}}{\text{time}}$$

$$1 \text{ W} = 1 \text{ J/s} = 1 \text{ N}\cdot\text{m/s}$$

All forms of power are expressed in watts. An electric motor has an electrical power input of watts and a mechanical power output of watts. In the U.S. Customary system, power is in horsepower.

$$\text{power} = \frac{\text{work}}{\text{time}} = \frac{\text{ft-lb}}{\text{s}}$$

but

$$1 \text{ horsepower (hp)} = 550 \text{ ft-lb/s}$$

Therefore,

$$\text{hp} = \frac{\text{ft-lb/s}}{550}$$

Moment

Moment is equal to a force times a perpendicular distance and is expressed in units of N·m or multiples thereof. In the U.S. Customary system, units of lb-ft are used to make a distinction from work units expressed in ft-lb.

Velocity

Because *velocity* is a rate of change of displacement with respect to time, the units are m/s. Usual U.S. Customary system units are ft/s, ft/min, and miles per hour (mph).

Angular Velocity

Angular velocity is the rate of change of rotational displacement with respect to time and is expressed in units of radians per second (rad/s). *Revolutions per minute* (rpm) is a permitted term but is converted to rad/s for calculations. This applies to both SI and the U.S. Customary systems.

Acceleration

Acceleration is the rate at which velocity changes; hence, the units are (m/s)/s or m/s^2. The commonly used unit in the U.S. Customary system is ft/s^2. The acceleration of gravity in each system is 9.81 m/s^2 and 32.2 ft/s^2.

Angular Acceleration

Angular acceleration is the rate at which angular velocity changes; hence, the units are (rad/s)/s or rad/s^2. The U.S. Customary system is similar.

Density

Because *density* is mass per unit volume, the units are kilogram per cubic meter (kg/m^3). The U.S. Customary system usually uses lb/in.^{3} or lb/ft^3 for *specific weight.*

Frequency

One hertz is the frequency of a periodic occurrence that has a period of 1 second. Formerly used units were cycles per second (1 Hz = 1 s^{-1}). This also applies to the U.S. Customary system.

Momentum

Momentum is mass times velocity; it is expressed in units of kg $\times$ m/s = kg·m/s. In the U.S. Customary system, we have mass in slugs or (lb-s^2)/ft times velocity in ft/s, giving units of lb-s.

Angular Momentum

Angular momentum is mass moment of inertia times angular velocity; it is expressed in units of kg·m^2 $\times$ rad/s = kg·m^2/s. In the U.S. Customary system, we have ft-lb-s^2 $\times$ rad/s = ft-lb-s.

For those who are not familiar with writing the various SI symbols, the following rules may be useful.

1. Symbols of units named after historic persons are written in capital letters, for instance, Hertz, Joule, Newton, and Watt. Exceptions are the multiples Mega (M), Giga (G), and Tera (T) (Table 1–2). All others are written in lowercase letters.
2. Symbols are not written with a plural "s."
3. Symbols are never followed by a period.

TABLE 1–4

Length	1 in.	= 25.4 mm
	1 ft	= 0.3048 m
	1 mile	= 1609 m
Area	1 in.^{2}	= 6.45 cm^2
	1 ft^2	= 0.093 m^2
	1 sq mile	= 2.59 km^2
Volume	1 in.^{3}	= 16.39 cm^3
	1 ft^3	= 0.0283 m^3
Capacity	1 qt	= 1.136 l
	1 gal	= 4.546 l
Mass	1 lb	= 0.454 kg
	1 slug	= 14.6 kg
Velocity	1 in./s	= 0.0254 m/s
	1 ft/s	= 0.3048 m/s
	1 ft/min	= 0.00508 m/s
	1 mph	= 0.447 m/s = 1.61 km/h
Acceleration	1 in./s^2	= 0.0254 m/s^2
	1 ft/s^2	= 0.3048 m/s^2
Force	1 lb	= 4.448 N
	1 poundal	= 0.138 N
Pressure	1 lb/in.^{2}	= 6.895 kPa
	1 lb/ft^2	= 47.88 Pa
Energy	1 ft-lb	= 1.356 J
	1 Btu	= 1.055 kJ
	1 hp-hr	= 2.685 MJ
	1 watt-hr	= 3.6 kJ
Power	1 hp	= 0.746 kW

4. Compound prefixes cannot be used; for example, mμm should be written nm.
5. Avoid the use of a prefix in the denominator of a composite unit; for example, do not use N/mm, but rather kN/m.

Table 1–4 provides some conversion factors that you may need while using this book in order to make conversions between the U.S. Customary system and the SI system. For those converting from U.S. Customary to SI, let me again emphasize that the key to learning the SI metric units is to think and work in the quantities of the system and not to have to convert continually from the U.S. Customary system: *Think metric!*

1–3 METHOD OF PROBLEM SOLUTION AND WORKMANSHIP

When attempting initially to understand and analyze a problem, the student should attempt to utilize logic, experience, and visualization. The use of common sense can be quite helpful here. Once the solution is begun—with the use of free-body diagrams (Section 4–2), for example—there are general rules and equations to be used for the remainder of the solution. In this way, nothing is left to chance. (A free-body diagram is a diagram of the object showing various forces acting on it.)

Upon obtaining an answer, the student may know from experience that a check is required. In many cases, an alternative method may be used as a reliable check. The checking of calculations by the same method can often lead to the same mathematical error; therefore, the use of an alternative method, where possible, is advisable.

A typical problem solution should begin with a reproduction of the given or known information in a concise form. Although this may seem wasteful of time, it serves several useful purposes:

1. It thoroughly acquaints you with the problem.
2. It is typical preliminary organization practiced every day in engineering.
3. The problem can be easily reviewed later without referring to the text or other sources.

The suggested method of solution consists of diagrams on the left side of the page and calculations on the right side. Do not crowd the calculations or double back with them. The format of Example 1–1 demonstrates what has just been described.

You should adhere to any symbols or methods of presentation used in this text. Any deviation or shortcut method might be successful at the early stages but could lead to complications in more sophisticated problems later. It is better to do this now than to wish later that you had. It cannot be overemphasized that through all phases of problem solution, neatness is an important factor contributing to organization. This becomes apparent when one refers back for intermediate answers or checks the final answer.

Although at this point it is premature to expect that the student will understand all the conventions and equations used, Example 1–1 will be solved on page 10. A file of problems solved in this manner might be all that a student would require for review or studying purposes at the end of the course.

EXAMPLE 1–1

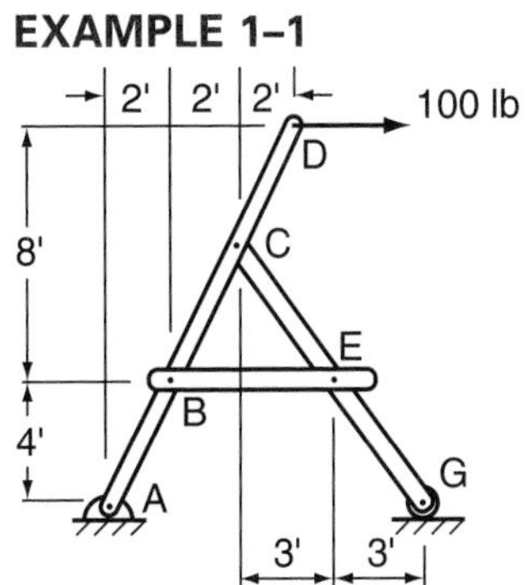

FIGURE 1–1a

A force of 100 lb is applied to the frame shown (Figure 1–1a). Calculate the load in member BE.

At this time, do not worry about how the problem is solved, but instead observe the format of the standard presentation. The 14 steps of this presentation are listed and explained on page 11.

By using this pattern of problem presentation, every future problem using this format could simply be viewed as filling in the blanks of a standard procedure.

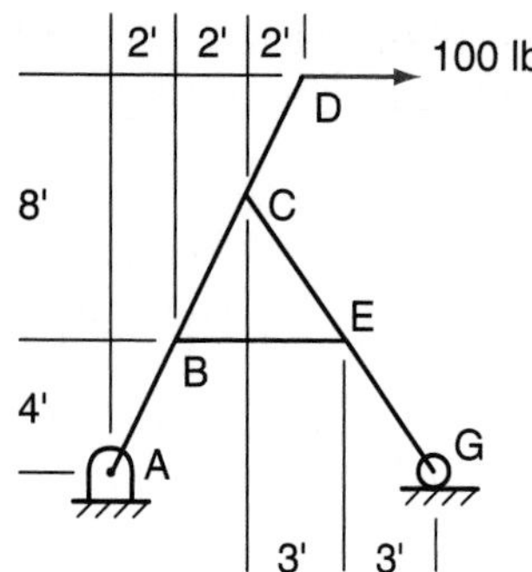

FIGURE 1–1b

Given: As shown
Required: BE
(See the following page for a description of each step.)

Free-Body Diagram of Frame

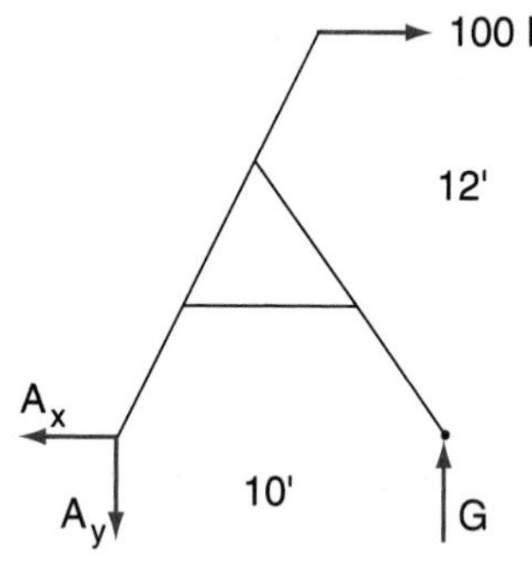

FIGURE 1–1c

$\Sigma M_A = 0$

$$G(10) = 100(12)$$
$$G = 120 \text{ lb}\uparrow$$

Free-Body Diagram of CG

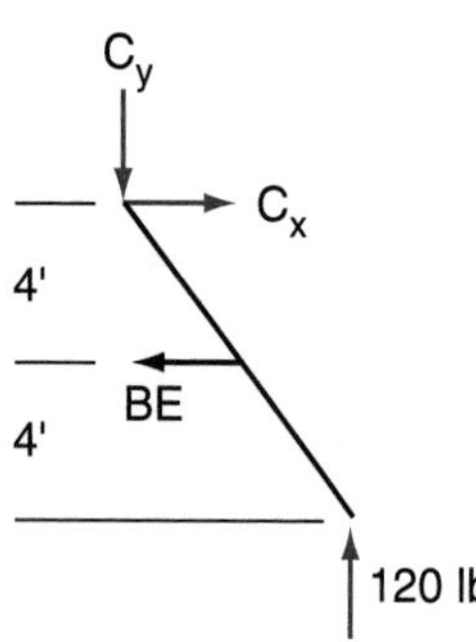

FIGURE 1–1d

$\Sigma M_C = 0$

$$BE(4) = 120(6)$$
$$BE = \frac{120(6)}{4} \leftarrow$$
$$= 180 \text{ lb}$$
$$BE = 180 \text{ lb tension}$$

Do not be distracted by the details of the diagrams or calculations but focus only on problem layout and presentation of Example 1–1's solved solution. Comments on each step of the solution are as follows:

(1) Problem number.

(2) Sketch of object, frame, etc., reproduced in a concise form suitable for reference during problem solution.

(3) State given information not already labeled on sketch.

(4) State required information.

(5) Name of free-body diagram.

(6) First free-body diagram or sketch.

(7) Statement of type of calculation to be used.

(8) Calculations.

(9) Underlined answer to calculations.

(10) Name of free-body diagram.

(11) Next free-body diagram, if necessary.

(12) Statement of type of calculation to be used.

(13) Calculations.

(14) Final answer underlined.

1–4 NUMERICAL ACCURACY AND SIGNIFICANT FIGURES

A value stated with more significant figures indicates a greater *degree of accuracy.* For example, 6.100 is more accurate than 6.1 (four significant figures versus two significant figures).

The number of significant figures is determined by counting the number of digits from left to right starting at the first nonzero and continuing to the last nonzero.

The following values will help explain:

503 — 3 significant figures
5.03 — 3 significant figures
0.0503 — 3 significant figures
503,000 — 3 significant figures

If the last nonzero to the right of the decimal has a further zero, then it is counted.

5.030 — 4 significant figures
503.000 — 6 significant figures
0.05030 — 4 significant figures

In a practical situation, if you measured a distance of 45 ft and recorded it as 45.0 ft, then you would be implying an accuracy of plus or minus 0.05 ft. You must be careful not to indicate any greater accuracy than is intended or possible. The final answer of a calculation should not indicate any greater accuracy than that of the most inaccurate figure of the original data.

Suppose that we perform the calculation $27.1 \times 89.2/203$. The accuracy indicated for each figure is as follows:

$$27.1 \pm 0.05 = \pm 0.18\%$$
$$89.2 \pm 0.05 = \pm 0.06\%$$
$$203 \pm 0.5 = \pm 0.25\%$$

The most inaccurate figure is 203 plus or minus 0.5, giving an accuracy of approximately 0.25%. Performing the complete calculation, we get 11.90798. Stating the answer as 11.9 implies an accuracy of $0.05/11.9 \times 100 = 0.42\%$. Thus, the degree of accuracy of the final answer is no greater than that of the original data. If the answer had been rounded off to the second decimal place and stated as 11.91, the accuracy indicated would have been $0.005/11.91 \times 100 = 0.042\%$. This would indicate an accuracy greater than that of the original data.

The location of the decimal does not affect the number of significant figures. In many engineering areas, the accuracy of three significant figures is adequate. For example, a calculation in stress analysis has to be based on at least two factors:

1. The strength of the material being used in the structure
2. The actual forces being applied to the structure

Neither of these factors can be determined with a high degree of certainty. First, small as they might be, variations occur in both homogeneity of material and consistency of quality. Second, the maximum load to which the structure will be subjected may have to be estimated. For these reasons, calculations to three significant figures are usually adequate. With the use of the electronic calculator, care must be taken not to indicate a misleading degree of accuracy.

Having said all of the above, there are many problems in this book with values given to only one significant figure, but as a rule of thumb (to check your calculations) all answers are given to three significant figures.

Note that carrying intermediate answers to four significant figures will give more accurate final answers. Prudent use of the calculator allows for storage and reuse of intermediate calculations with more significant figures.

1–5 MATHEMATICS REQUIRED

Although addition, subtraction, multiplication, and division are rather elementary, they are mentioned here to emphasize their importance in obtaining correct answers. Carelessness in such elementary mathematical operations is often the source of error as well as much frustration and wasted time in problem solutions.

Incorrect number entry into a calculator also can waste considerable time, so always check your entered value before proceeding to the next operation on the calculator.

Additional math skills will be needed for the following topics:

- Algebraic equations with one unknown
- Simultaneous equations with two unknowns
- Quadratic equations
- Trigonometry functions of a right-angle triangle
- Sine law and cosine law as applied to non-right-angle triangles
- Geometry

Complete coverage of these topics is not possible, but a brief review of each follows.

1–6 ALGEBRAIC EQUATION—ONE UNKNOWN

The first essential skill in solving algebraic equations is that of transposing values while solving for a single unknown. The basic rule of thumb is to treat each side of the equation exactly the same. Whatever you do to one side of the equation, do the same to the other side.

EXAMPLE 1–2

Solve for x in the equation.

$$4 + \frac{3(6 + x)}{2} = 16$$

Subtract 4 from each side.

$$4 + \frac{3(6 + x)}{2} - 4 = 16 - 4$$

$$\frac{3(6 + x)}{2} = 12$$

Multiply each side by $\frac{2}{3}$.

$$\left(\frac{2}{3}\right)\frac{3(6 + x)}{2} = 12\left(\frac{2}{3}\right)$$

$$6 + x = 8$$

Subtract 6 from each side.

$$6 + x - 6 = 8 - 6$$

$$\underline{x = 2}$$

Check by substituting $x = 2$ into the original equation.

$$4 + \frac{3(6 + 2)}{2} = 16$$

$$4 + \frac{3(8)}{2} = 16$$

$$4 + 12 = 16$$

$$16 = 16 \textit{ check}$$

EXAMPLE 1–3

Solve for x in the equation.

$$3.45 + \frac{3x}{5} = \frac{7x}{4}$$

Multiply each side by 20 (20 is a common denominator of $\frac{3}{5}x$ and $\frac{7}{4}x$).

$$(20)(3.45) + (20)\left(\frac{3}{5}x\right) = (20)\left(\frac{7}{4}x\right)$$

$$69 + 12x = 35x$$

Subtract $12x$ from each side.

$$69 + 12x - 12x = 35x - 12x$$

$$69 = 23x$$

Divide each side by 23.

$$\frac{69}{23} = \frac{23x}{23}$$

$$3 = x$$

Transpose the terms into a more appropriate form of stating the variable first and the value second.

$$\underline{x = 3}$$

Check by substituting $x = 3$.

$$3.45 + \frac{(3)(3)}{5} = \frac{(7)(3)}{4}$$

$$3.45 + 1.8 = 5.25$$

$$5.25 = 5.25 \textit{ check}$$

1–7 SIMULTANEOUS EQUATIONS—TWO UNKNOWNS

EXAMPLE 1–4

Solve the simultaneous equations.

$$3x + 4y = 8 \qquad (1)$$

$$6x + 2y = 10 \qquad (2)$$

Either of two methods will work.

Method A

Multiply Equation (1) by -2.

$$-6x - 8y = -16$$

Add Equation (2) $\quad 6x + 2y = 10$

Divide by -6. $\quad 0 - 6y = -6$

$$\underline{y = 1}$$

Substitute $y = 1$ into either Equation (1) or (2). Using Equation (1) in this case

$$3x + (4)(1) = 8$$
$$3x = 8 - 4$$
$$\underline{x = 1.33}$$

Method B

Isolate variable x of Equation (1) as follows.

$$3x + 4y = 8$$
$$3x + 4y - 4y = 8 - 4y$$
$$\frac{3x}{3} = \frac{8 - 4y}{3}$$
$$x = \frac{8 - 4y}{3}$$

Substitute $x = \dfrac{8 - 4y}{3}$ into Equation (2).

$$6x + 2y = 10$$
$$6\left(\frac{8 - 4y}{3}\right) + 2y = 10$$
$$2(8 - 4y) + 2y = 10$$
$$16 - 8y + 2y = 10$$
$$-6y = -6$$
$$\underline{y = +1}$$

Substitute $y = 1$ into

$$x = \frac{8 - 4y}{3}$$
$$x = \frac{8 - (4)(1)}{3}$$
$$x = \frac{4}{3}$$
$$\underline{x = 1.33}$$

1–8 QUADRATIC EQUATIONS

EXAMPLE 1–5

Solve for x in the equation.

$$3x(4 + 2x) - 10 = x^2 - 8$$

Simplify the equation, in preparation for the quadratic equation.

$$12x + 6x^2 - 10 = x^2 - 8$$

Subtract x^2 on each side.

$$12x + 6x^2 - 10 - x^2 = x^2 - 8 - x^2$$
$$5x^2 + 12x - 10 = -8$$

Add 8 to each side.

$$5x^2 + 12x - 10 + 8 = -8 + 8$$
$$5x^2 + 12x - 2 = 0$$

For an equation in the form

$$(a)x^2 + (b)x + c = 0$$

the quadratic equation is

$$x = \frac{-b \pm \sqrt{b^2 - 4ac}}{2a}$$

Substitute the values from our equation $a = 5$, $b = 12$, $c = -2$.

$$x = \frac{-12 \pm \sqrt{(12)^2 - (4)(5)(-2)}}{2 \times 5}$$
$$= \frac{-12 \pm \sqrt{144 + 40}}{10}$$
$$= \frac{-12 \pm \sqrt{184}}{10}$$
$$= \frac{-12 \pm 13.56}{10}$$
$$\underline{x = -2.56 \text{ or } + 0.156}$$

EXAMPLE 1–6

Solve for x in the equation.

$$3x^2 - 5x + 1 = 0$$

Substituting into the quadratic formula

$$x = \frac{-b \pm \sqrt{b^2 - 4ac}}{2a}$$

$$x = \frac{-(-5) \pm \sqrt{(-5)^2 - (4)(3)(1)}}{(2)(3)}$$

$$= \frac{+5 \pm \sqrt{25 - 12}}{6}$$

(Notice the resulting sign when b is negative.)

$$= \frac{+5 \pm \sqrt{13}}{6}$$

$$= \frac{5 + 3.606}{6} \quad \text{or} \quad = \frac{5 - 3.606}{6}$$

$$\underline{x = +1.43 \text{ or } 0.232}$$

1–9 TRIGONOMETRY: RIGHT-ANGLE TRIANGLES

Trigonometry is the study of the relationships among the sides and interior angles of triangles. The angles are usually given in Greek lowercase letters, some popular ones being alpha (α), beta (β), gamma (γ), theta (θ), and phi (ϕ).

The basic trigonometric functions apply only to right-angle triangles. The right- angle triangle can be in any one of four quadrants of an x-y axis system (Figure 1–2). For each triangle, the functions are

$$\text{sine } \theta = \sin \theta = \frac{\text{side opposite}}{\text{hypotenuse}}$$

$$\text{cosine } \theta = \cos \theta = \frac{\text{side adjacent}}{\text{hypotenuse}}$$

$$\text{tangent } \theta = \tan \theta = \frac{\text{side opposite}}{\text{side adjacent}}$$

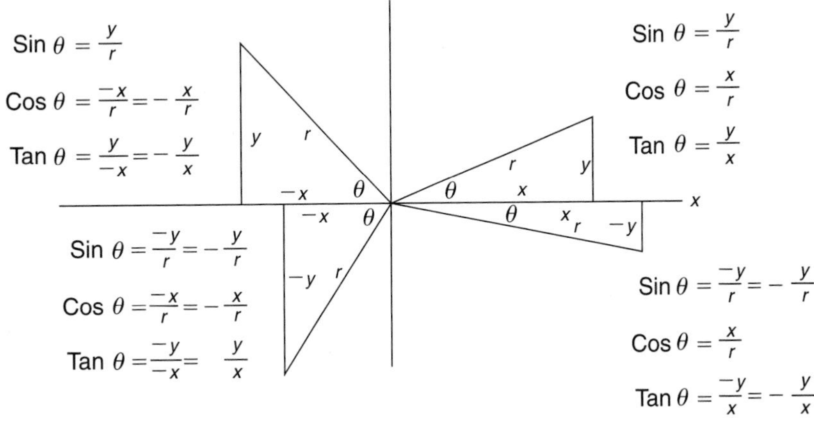

FIGURE 1–2

The signs of these trigonometric functions depend on the quadrant in which the triangle lies (Figure 1–2).

EXAMPLE 1–7

Determine the value of length A for each triangle shown (Figure 1–3).

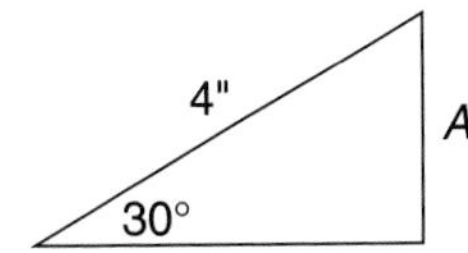

FIGURE 1–3a

$$\sin 30^\circ = \frac{A}{4 \text{ in.}}$$

$$A = \sin 30^\circ(4)$$
$$= 0.5(4)$$
$$\underline{A = 2 \text{ in.}}$$

8'

70°

A

FIGURE 1–3b

$$\cos 70^\circ = \frac{A}{8 \text{ ft}}$$

$$A = \cos 70^\circ(8)$$
$$= 0.342(8)$$
$$\underline{A = 2.74 \text{ ft}}$$

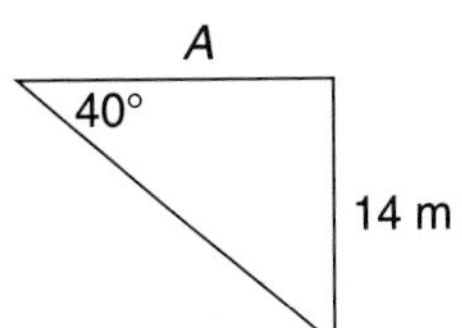

FIGURE 1–3c

$$\tan 40^\circ = \frac{14 \text{ m}}{A}$$

$$A = \frac{14}{\tan 40^\circ}$$
$$\underline{A = 16.7 \text{ m}}$$

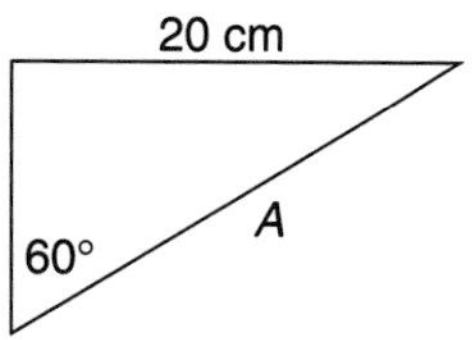

FIGURE 1–3d

$$\sin 60^\circ = \frac{20\text{ cm}}{A}$$

$$A = \frac{20}{\sin 60^\circ} = \frac{20}{0.866}$$

$$\underline{A = 23.1\text{ cm}}$$

1–10 SINE AND COSINE LAWS: NON-RIGHT-ANGLE TRIANGLES

For triangles that are not right-angle triangles, such as the ones in Figures 1–4 and 1–5, either the *sine law* or the *cosine law* is used. The sine law is given by

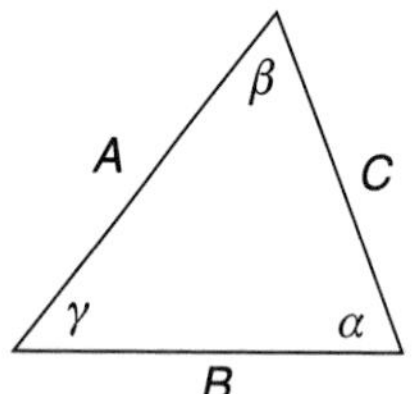

FIGURE 1–4

$$\frac{A}{\sin \alpha} = \frac{B}{\sin \beta} = \frac{C}{\sin \gamma} \tag{1–1}$$

In each case, the side is divided by the sine of the angle opposite the side.

The cosine law is given by

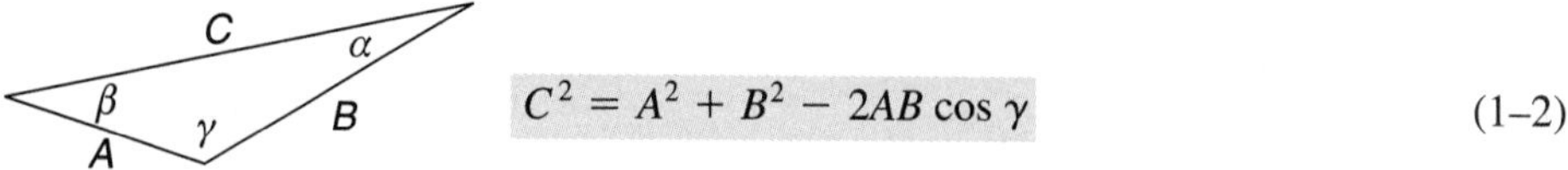

$$C^2 = A^2 + B^2 - 2AB \cos \gamma \tag{1–2}$$

FIGURE 1–5

If the cosine law is applied to a right-angle triangle where $\gamma = 90^\circ$ (Figure 1–6) and $\cos \gamma = \cos 90 = 0$, the equation $C^2 = A^2 + B^2 - 2\,AB \cos \gamma$ becomes

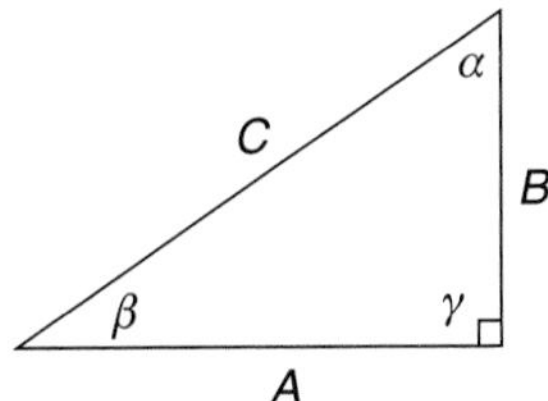

FIGURE 1–6

$$C^2 = A^2 + B^2 \tag{1–3}$$

Equation 1–3 is the *Pythagorean theorem;* it states that the square of the hypotenuse of a right-angle triangle equals the sum of the squares of the two remaining sides.

EXAMPLE 1–8

Find the length of the unknown side and the angle θ (Figure 1–7).

Designating this side C and using the cosine law gives

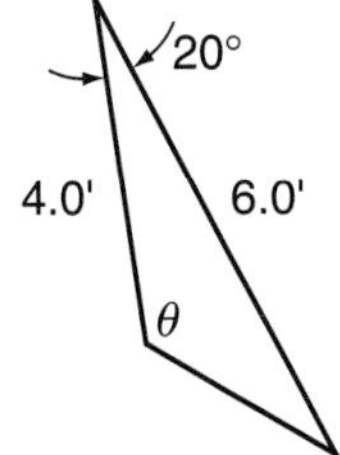

FIGURE 1–7

$$\begin{aligned} C^2 &= A^2 + B^2 - 2AB\cos\gamma \\ &= (4\text{ ft})^2 + (6\text{ ft})^2 - 2(4\text{ ft})(6\text{ ft})\cos 20° \\ &= 16 + 36 - 48(0.94) \\ &= 6.9 \\ \underline{C} &\underline{= 2.6\text{ ft}} \end{aligned}$$

Using the sine law, we get

$$\begin{aligned} \frac{6\text{ ft}}{\sin\theta} &= \frac{2.6\text{ ft}}{\sin 20°} \\ \sin\theta &= \frac{6(0.342)}{2.6} \\ &= 0.789 \\ \theta &= 52.1° \end{aligned}$$

But we know this to be in the second quadrant, so

$$\begin{aligned} \theta &= 180 - 52.1 \\ \underline{\theta} &\underline{= 127.9°} \end{aligned}$$

EXAMPLE 1–9

Find the length of side A (Figure 1–8).

The sum of all the angles of a triangle is 180°. Therefore,

$$\text{angle } \alpha = 180° - (60 + 80) = 40°$$

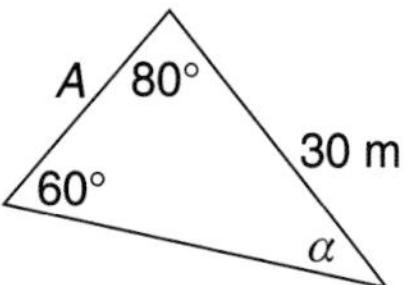

FIGURE 1–8

Using the sine law, we obtain

$$\frac{A}{\sin \alpha} = \frac{B}{\sin \beta}$$

$$\frac{A}{\sin 40°} = \frac{30 \text{ m}}{\sin 60°}$$

$$A = 30\left(\frac{\sin 40°}{\sin 60°}\right)$$

$$= 30\left(\frac{0.643}{0.866}\right)$$

$$\underline{A = 22.3 \text{ m}}$$

EXAMPLE 1–10

Find the length of side A (Figure 1–9).

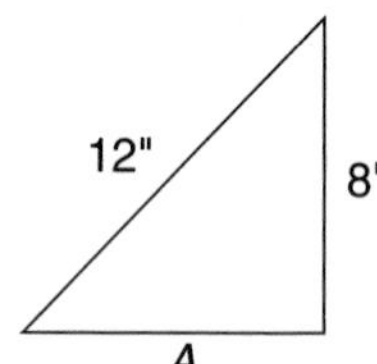

FIGURE 1–9

$$(12)^2 = A^2 + (8)^2$$

$$A^2 = (12 \text{ in.})^2 - (8 \text{ in.})^2$$

$$= 144 - 64$$

$$= 80$$

$$\underline{A = 8.9 \text{ in.}}$$

EXAMPLE 1–11

Determine the interior angles of the triangles shown (Figure 1–10).

Using the cosine law gives

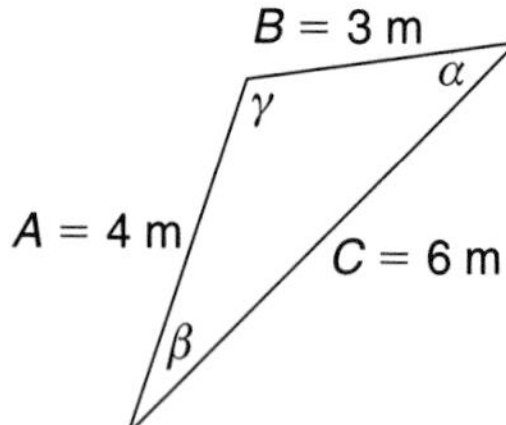

FIGURE 1–10

$$C^2 = A^2 + B^2 - 2\,AB \cos \gamma$$

$$(6 \text{ m})^2 = (4 \text{ m})^2 + (3 \text{ m})^2 - 2(4 \text{ m})(3 \text{ m})(\cos \gamma)$$

$$36 - 16 - 9 = -24 \cos \gamma$$

$$\cos \gamma = -0.4583$$

Be careful to pay attention to signs here.

$$\begin{aligned} \text{If } \cos\gamma &= +0.4583 \\ \text{then } \gamma &= 62.7^\circ \\ \text{But } \cos\gamma &= -0.4583 \\ \text{Therefore } \gamma &= 180 - 62.7 \\ \underline{\gamma} &\underline{= 117.3^\circ} \end{aligned}$$

Using the sine law now yields

$$\begin{aligned} \frac{A}{\sin\alpha} &= \frac{C}{\sin\gamma} \\ \frac{4\text{ m}}{\sin\alpha} &= \frac{6\text{ m}}{\sin 117.3^\circ} \\ \sin\alpha &= \frac{4(\sin 117.3^\circ)}{6} \\ &= 0.5924 \\ \underline{\alpha} &\underline{= 36.3^\circ} \end{aligned}$$

Knowing that the sum of the interior angles is 180° gives

$$\begin{aligned} \beta &= 180 - 117.3 - 36.3 \\ \underline{\beta} &\underline{= 26.4^\circ} \end{aligned}$$

1–11 GEOMETRY

Some basic rules of geometry are as follows:

(a) Opposite angles are equal when two straight lines intersect.

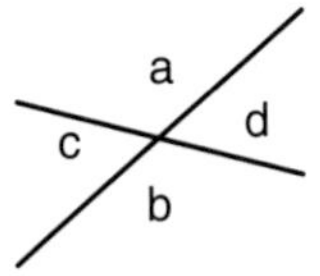

$$\begin{aligned} a &= b \\ c &= d \end{aligned}$$

FIGURE 1–11

(b) Supplementary angles total 180°.

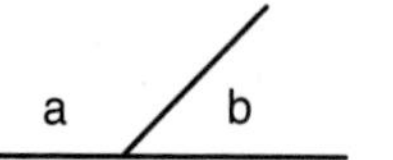

$$a + b = 180°$$

FIGURE 1–12

(c) Complementary angles total 90°.

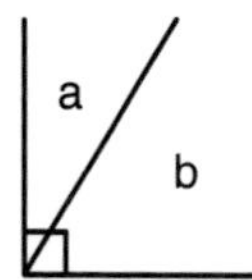

$$a + b = 90°$$

FIGURE 1–13

(d) A straight line intersecting two parallel lines produces the following equal angles:

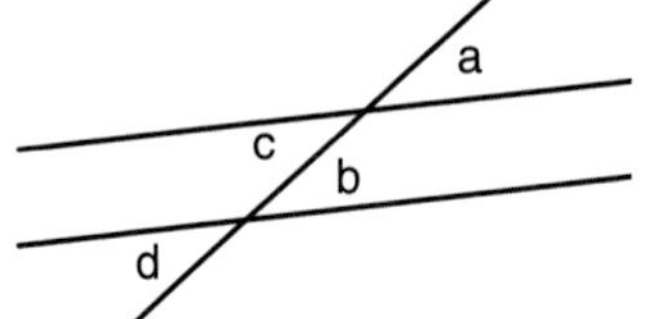

$$a = b$$
$$c = d$$
or
$$a = b = c = d$$

FIGURE 1–14

(e) The sum of the interior angles of any triangle equals 180°.

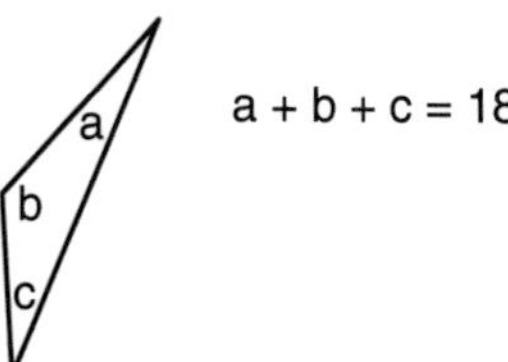

$$a + b + c = 180$$

FIGURE 1–15

(f) Similar triangles have the same shape.

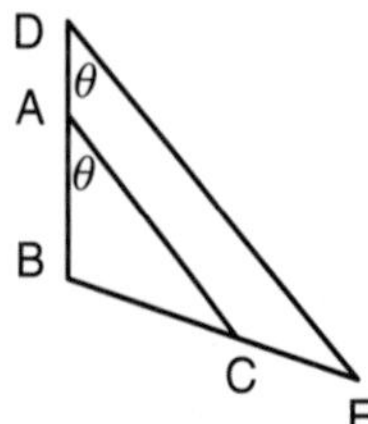

FIGURE 1–16a

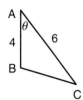

FIGURE 1–16b

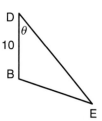

FIGURE 1–16c

If AB = 4, AC = 6, and DB = 10, then by proportion

$$DE = \frac{6}{4} \times 10 = 15$$

(g) Circle equations:

$$\text{Circumference} = \pi D \text{ or } 2\pi r$$

$$\text{Area} = \frac{\pi D^2}{4} \text{ or } \pi r^2$$

Angle θ is defined as one radian when a length of 1 radius is measured on the circumference.

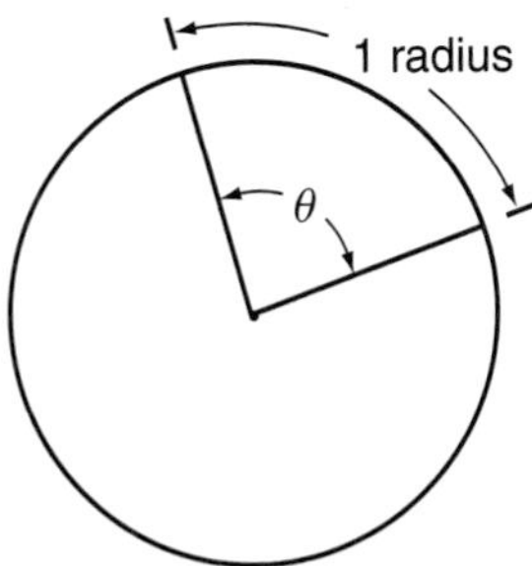

FIGURE 1–17

HINTS FOR PROBLEM SOLVING

1. Check basic geometry. Two intersecting lines have opposite angles that are equal. The sum of the internal angles of a triangle is 180°.
2. Right-angle-triangle trigonometry functions should be checked by writing them out fully. If you have trouble remembering them, try following the arrows shown in Figure 1–18.

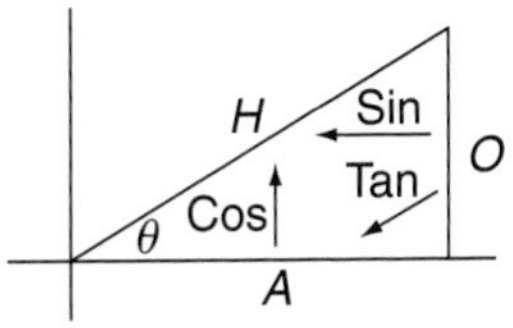

$$\sin\theta = \frac{O}{H}$$

$$\cos\theta = \frac{A}{H}$$

$$\tan\theta = \frac{O}{A}$$

FIGURE 1–18

3. Watch the sign of the cosine function when using the cosine law. Cos 120°, for example, is in the second quadrant and therefore negative.
4. Too many significant figures can give misleading accuracy.
5. If the geometry of several connecting triangles seems confusing, draw a new large sketch, labeling all possible lengths and angles.

PROBLEMS

APPLIED PROBLEMS FOR SECTIONS 1–1 TO 1–9

1–1. Solve for x in the equation

$$(3 + x)8 - 12x = 8x$$

1–2. Solve for x in the equation

$$(2 + x)6 + 3 = 27$$

1–3. Solve for x in the equation

$$28 - \frac{5}{12}x = \frac{3}{4}x$$

1–4. Solve for y in the simultaneous equations

$$2x + 8y = 20$$
$$5x - 3y = 10$$

1–5. Solve for x in the simultaneous equations

$$22x + 3y = 121$$
$$13x - 8y = 56$$

1–6. Solve for x in the equation

$$13x^2 - 2x - 8 = 0$$

1–7. Solve for x in the equation

$$3x + \frac{5}{x} = 8$$

1–8. Determine angles a, b, and c (Figure P1–8).

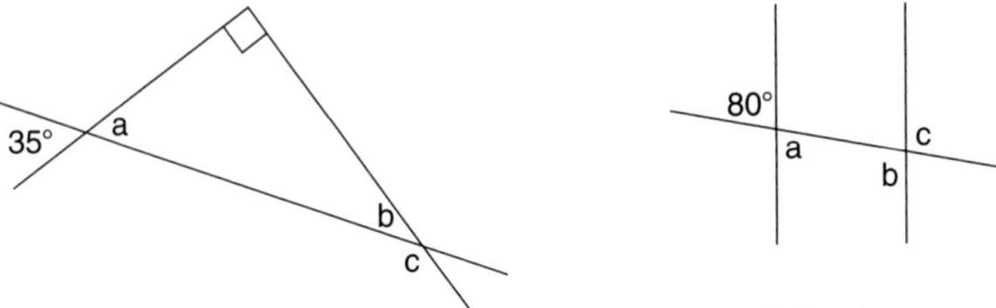

FIGURE P1–9

FIGURE P1–8

1–9. Determine angles a, b, and c (Figure P1–9).
1–10. Determine angles a, b, c, d, and e (Figure P1–10).

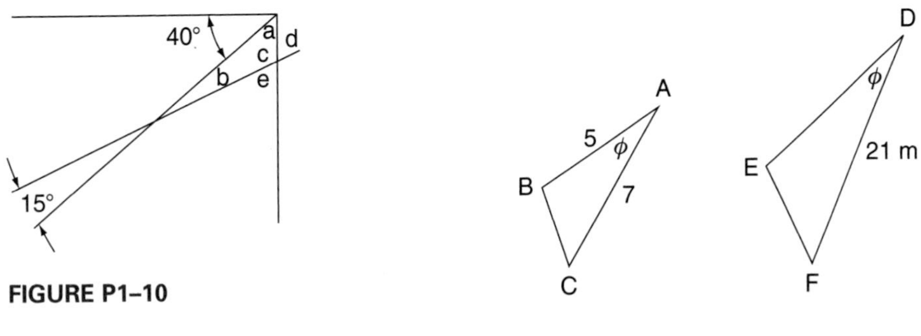

FIGURE P1–10

FIGURE P1–11

1–11. Determine length ED for the similar triangles shown in Figure P1–11.
1–12. Determine length CE for the similar triangles shown in Figure P1–12.
1–13. Find the length of side *A* of the triangle shown in Figure P1–13.

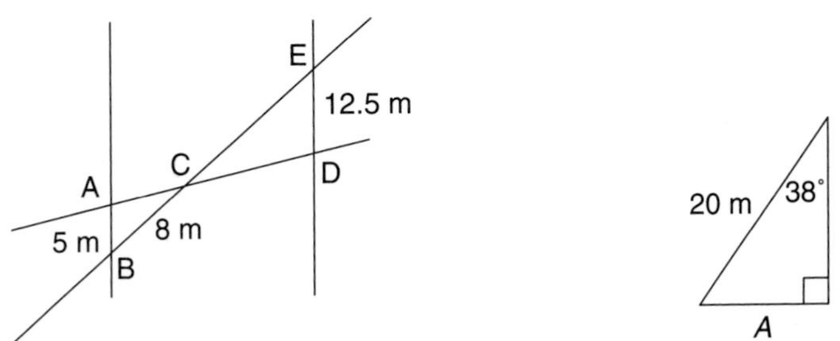

FIGURE P1–12

FIGURE P1–13

1–14. Find angle θ for the triangle shown in Figure P1–14.

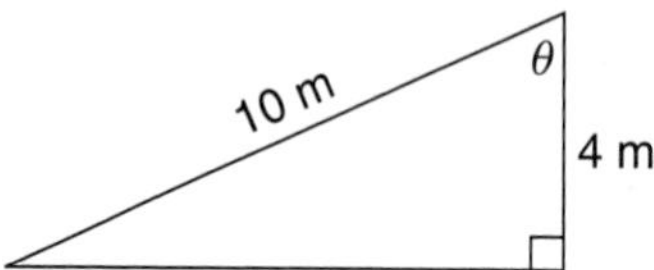

FIGURE P1–14

1–15. Find length *A* of the triangle shown in Figure P1–15.

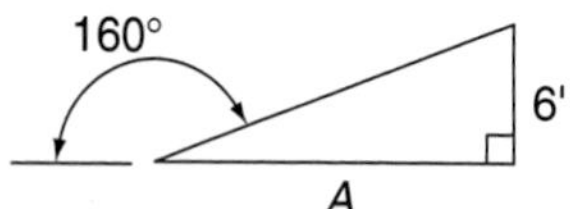

FIGURE P1–15

1–16. Find *y* of the triangle shown in Figure P1–16.

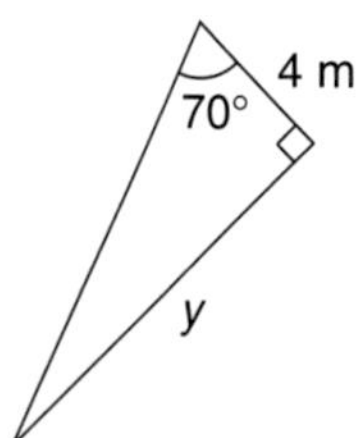

FIGURE P1–16

1–17. Calculate the angle θ and the hypotenuse *R* for each triangle shown in Figure P1–17.

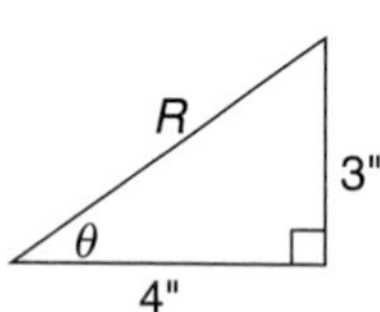

FIGURE P1–17a

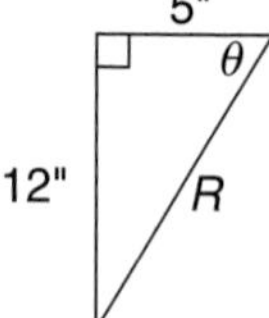

FIGURE P1–17b

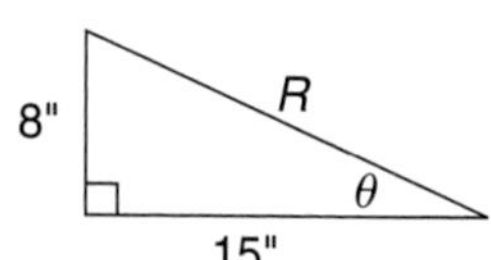

FIGURE P1–17c

1–18. Determine the length of side A for each triangle shown in Figure P1–18.

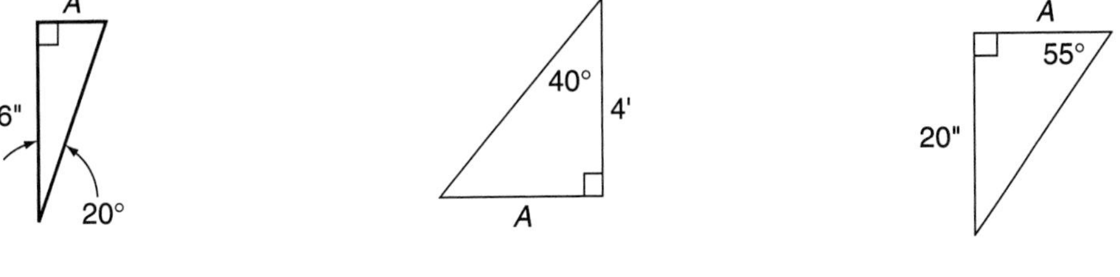

FIGURE P1–18a FIGURE P1–18b FIGURE P1–18c

1–19. The top end of a 40-m conveyor can reach a height of 25 m. What is the angle between the conveyor and the ground?

1–20. Find y of the triangle shown in Figure P1–20.

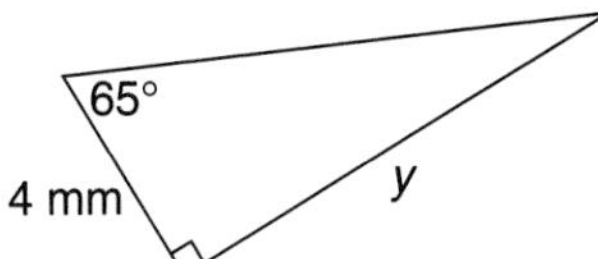

FIGURE P1–20

1–21. Determine angle θ for the triangle shown in Figure P1–21.

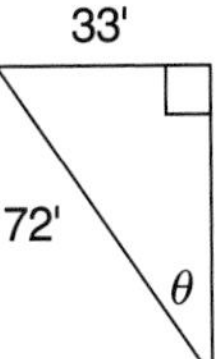

FIGURE P1–21

APPLIED PROBLEMS FOR SECTIONS 1–10 AND 1–11

1–22. A formed piece of sheet metal has a cross section as shown in Figure P1–22. Determine distance c.

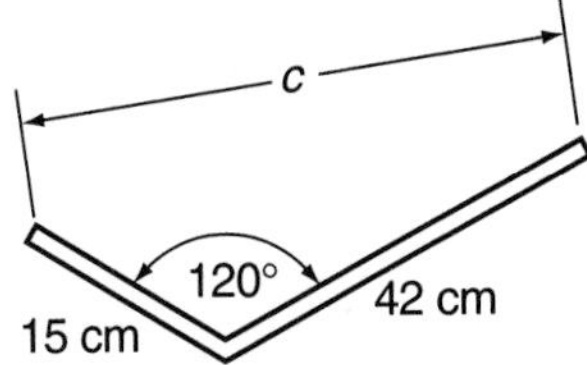

FIGURE P1–22

1–23. Determine the length of side C of the triangle shown in Figure P1–23.

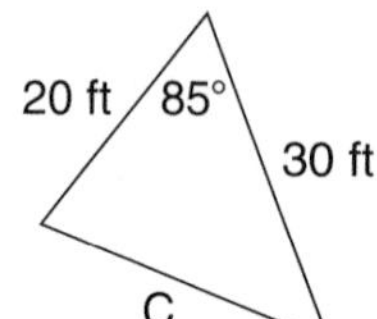

FIGURE P1–23

1–24. An electric winch lifts a weight on the end of a 3-m jib pole. Determine angle θ for the position shown (Figure P1–24).

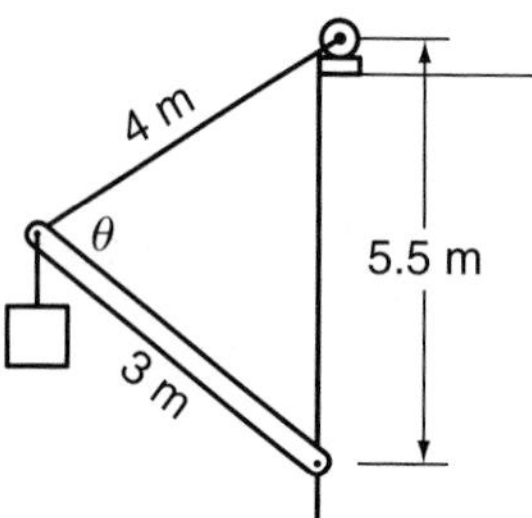

FIGURE P1–24

1–25. Determine the length of cylinder CB in Figure P1–25.

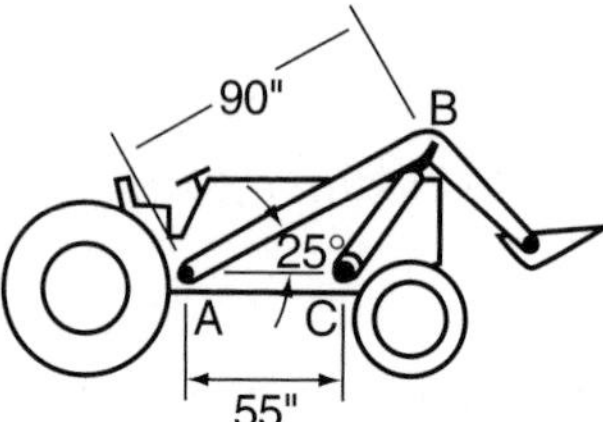

FIGURE P1–25

1–26. Determine the length of side *d* of the triangle shown in Figure P1–26.

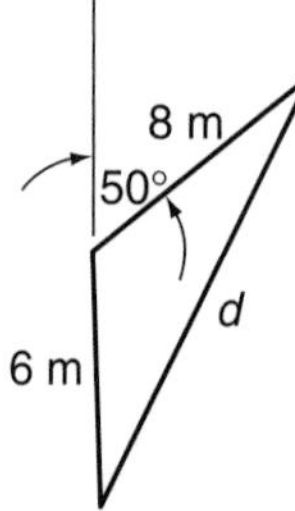

FIGURE P1–26

1–27. Arm AB rotates clockwise to the new position shown in Figure P1–27. Determine the displacement CD. (Hint: Draw triangle EDC.)

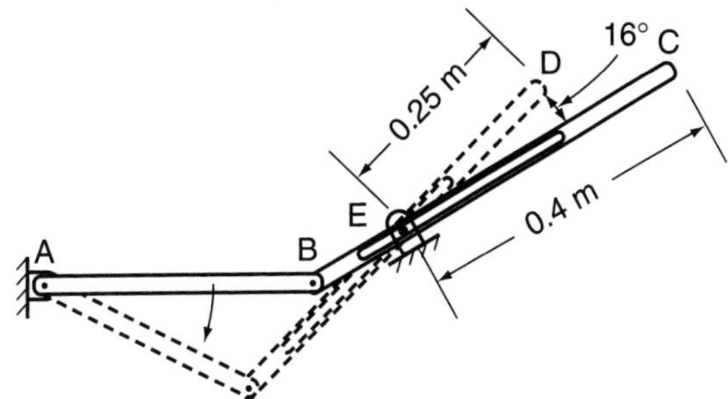

FIGURE P1–27

1–28. A building marquee is supported by cables (Figure P1–28). Determine the length of cable A.

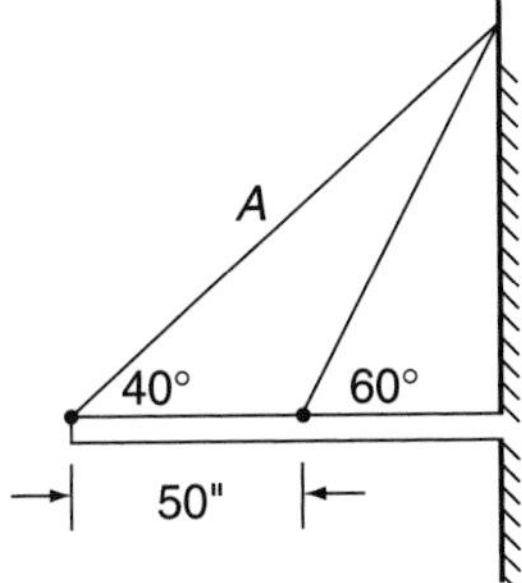

FIGURE P1–28

1–29. A surveying method to find the distance between two points A and B, between which is an obstacle, is shown in Figure P1–29. The length of CD is 640 ft. As an intermediate step of the surveying method, calculate the lengths of AC and AD.

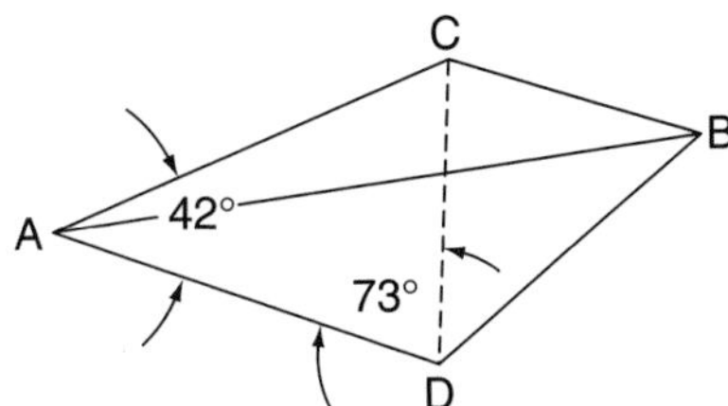

FIGURE P1–29

1–30. Determine the distance d as shown in Figure P1–30.

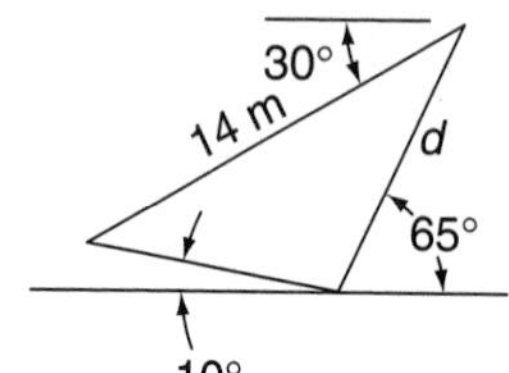

FIGURE P1–30

1–31. When the ramps of the trailer shown in Figure P1–31 are lowered, the support section welded to the hinge end of the ramp is vertical. If the hinge height is 14 inches above the ground and the ramp length is 48 inches, determine the angle between the ramp and its vertical support.

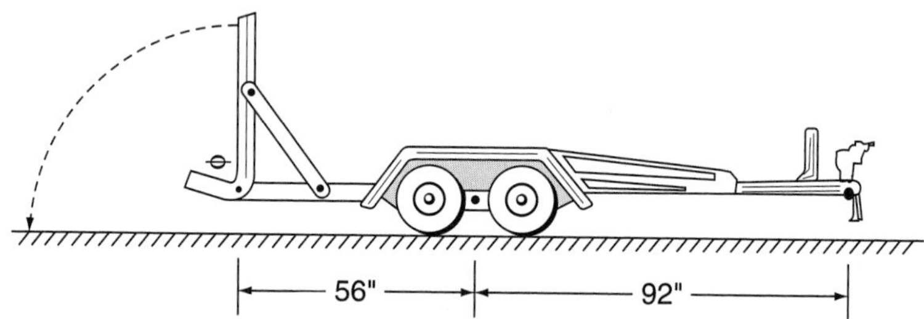

FIGURE P1–31

1–32. Find distance x in Figure P1–32.

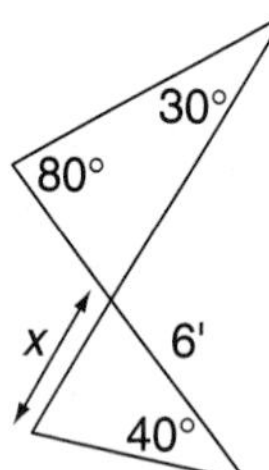

FIGURE P1–32

1–33. A cone-shaped tube is 10 cm long and has diameters of 6 cm and 8 cm. Calculate the angle of taper.

1–34. A rotating shaft level gauge rotates from position A to B (Figure P1–34). What is the measured height of liquid?

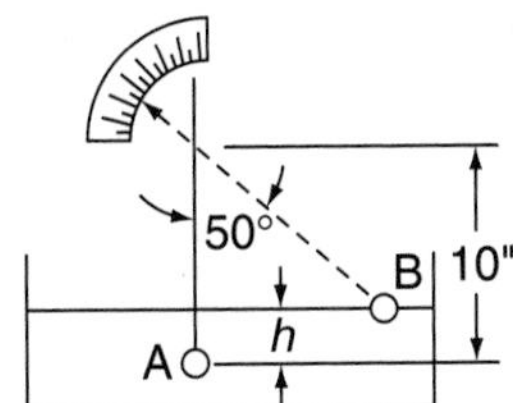

FIGURE P1–34

1–35. A part's dimensions may be determined in either of two ways as shown in Figure P1–35. Determine *x* and *y*.

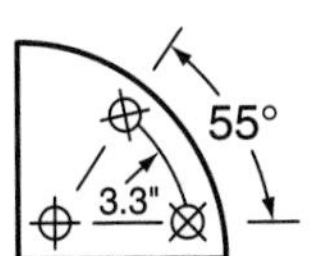

FIGURE P1–35a

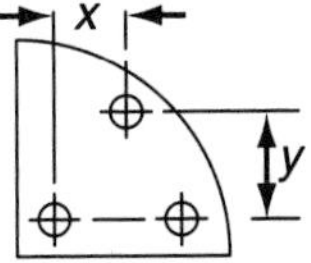

FIGURE P1–35b

1–36. The width across the flats of a hexagonal nut is 1.875 in. What is the width across the corners?

1–37. Find the distance between any two holes of the plate shown in Figure P1–37.

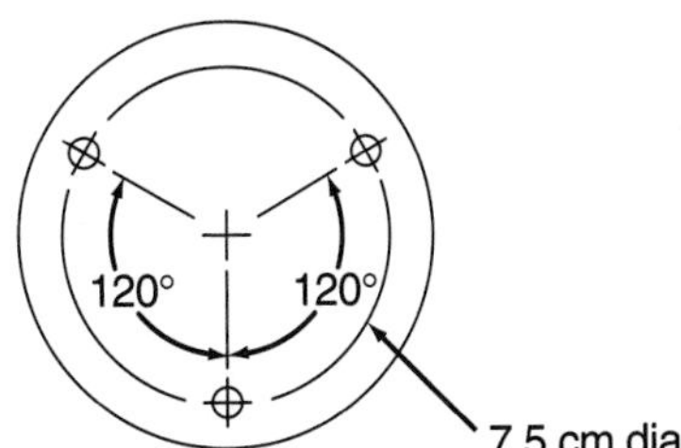

FIGURE P1–37

1–38. For the system shown in Figure P1–38, determine the length *c* and angle θ.

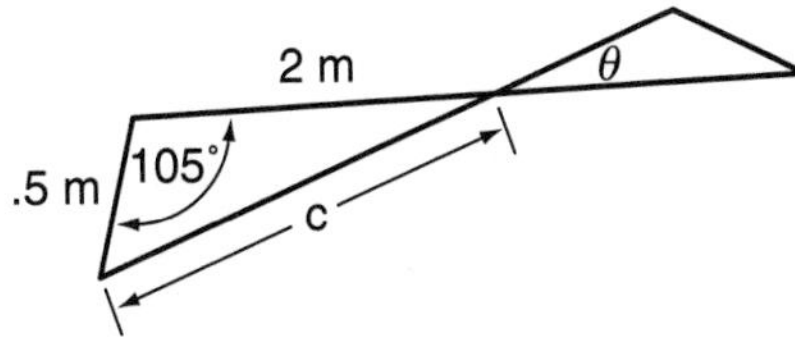

FIGURE P1–38

1–39. A surveying technique for measuring the height of a building on sloping ground has data as shown in Figure P1–39. Determine the height *h* of the building.

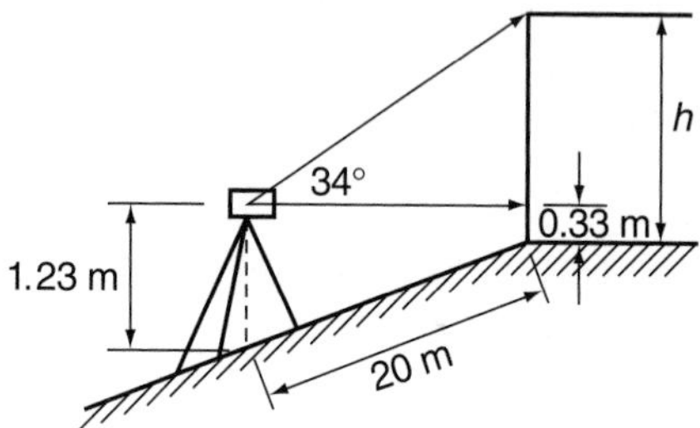

FIGURE P1–39

1–40. Neglecting the pulley diameter at D, determine how far weight A drops as θ changes from 120° to 50° (Figure P1–40).

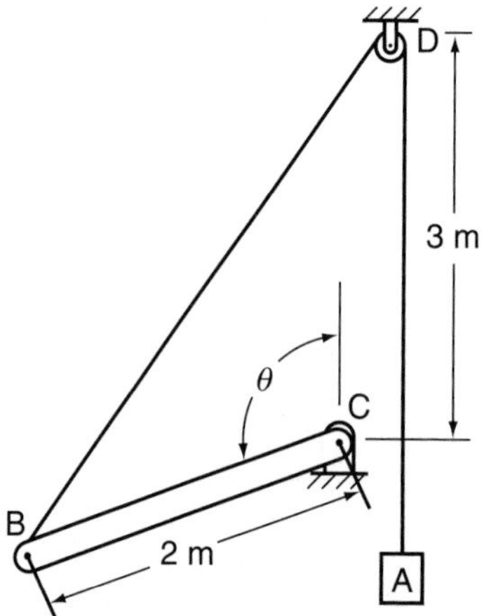

FIGURE P1–40

1–41. A roller and lever mechanism is in the position shown in Figure P1–41. Find the horizontal distance between A and B. (Point B is at the same level as the horizontal surface.)

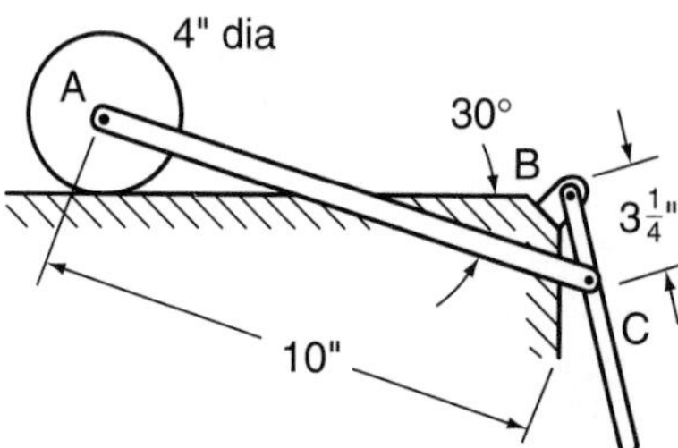

FIGURE P1–41

1–42. The bottom end of a 6-m ladder is placed 2.5 m from point A (Figure P1–42). Determine (a) the distance *d* and (b) the shortest distance from A to the ladder.

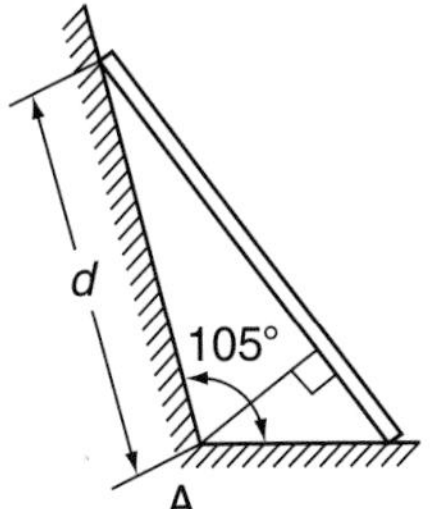

FIGURE P1–42

1–43. The center slines of the arms of the tree pruning lopper intersect at E as shown in Figure P1–43. Determine angles θ_1, θ_2, θ_3, θ_4, θ_5, and θ_6.

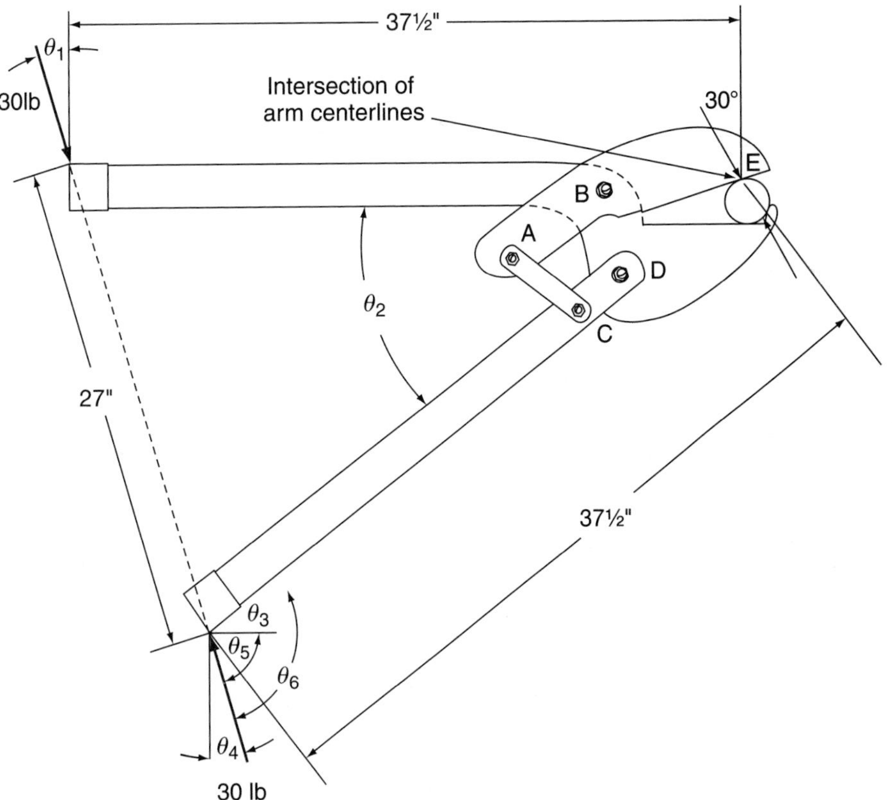

FIGURE P1–43

REVIEW PROBLEMS

R1–1. A sloped surface 15 m long is elevated at 25° to the horizontal. Determine the vertical rise and the horizontal run of the sloped surface.

R1–2. Determine the lengths *A* and *B* of the triangle shown in Figure RP1–2.

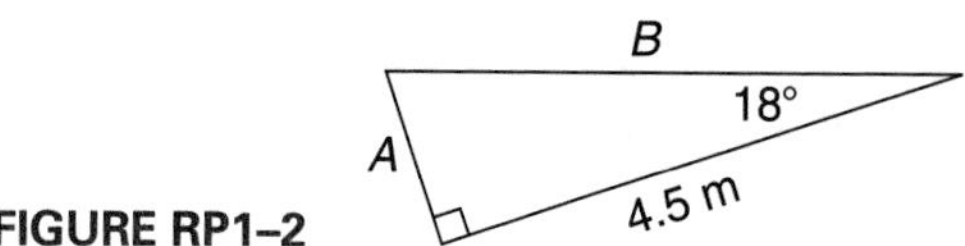

FIGURE RP1–2

R1–3. Determine angle θ of the triangle shown in Figure RP1–3.

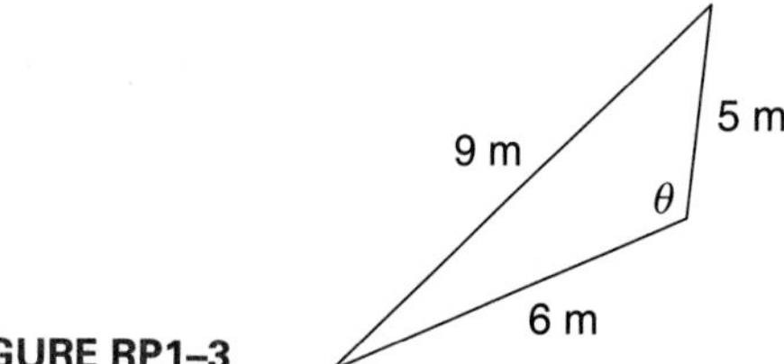

FIGURE RP1–3

R1–4. Determine the included angles ϕ and θ of the triangle shown in Figure RP1–4.

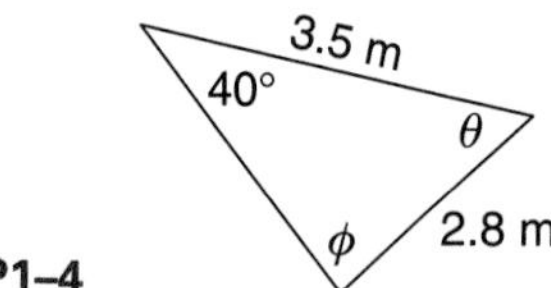

FIGURE RP1–4

R1–5. Determine the angle θ of the triangle shown in Figure RP1–5.

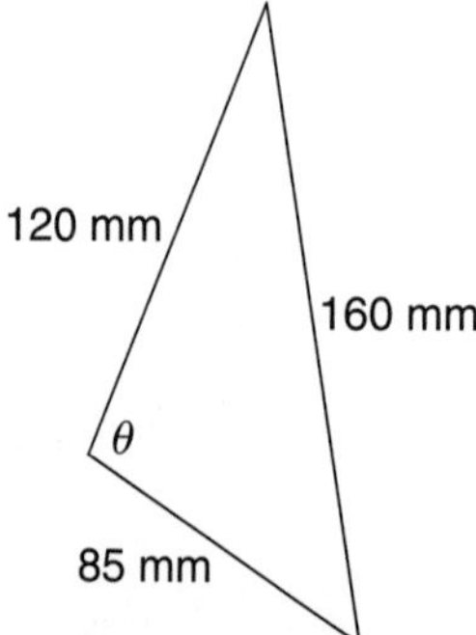

FIGURE RP1–5

R1–6. Using the sine law, determine length R in Figure RP1–6.

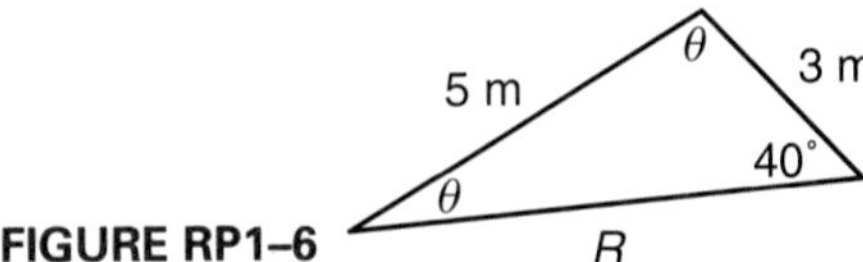

FIGURE RP1–6

R1–7. Lever AB is 2 m long and initially horizontal as shown in Figure RP1–7. Determine the angle θ that lever CD must rotate to cause lever AB to rotate 30° upward.

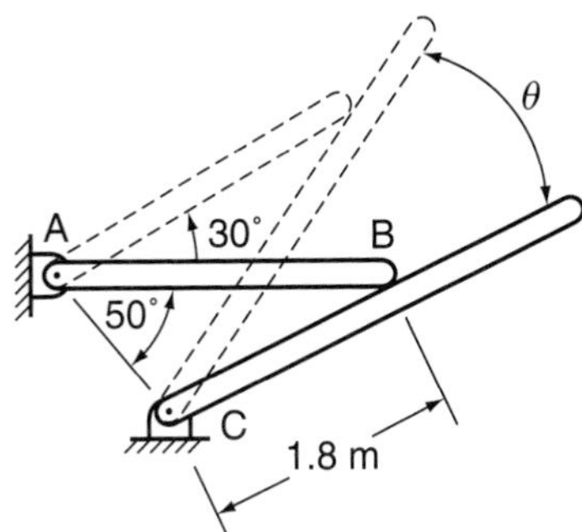

FIGURE RP1–7

R1–8. Member AB rotates 150° clockwise from the original position shown in Figure RP1–8. Determine the length CB'. (*Hint:* Use length AC as a common length to the two triangles formed on either side of it.)

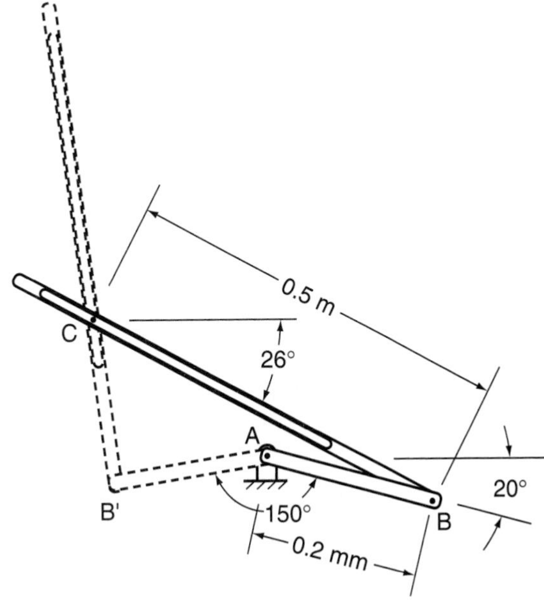

FIGURE RP1–8

R1–9. The length of member BC in Figure RP1–9 is 60 mm. Determine angle ϕ.

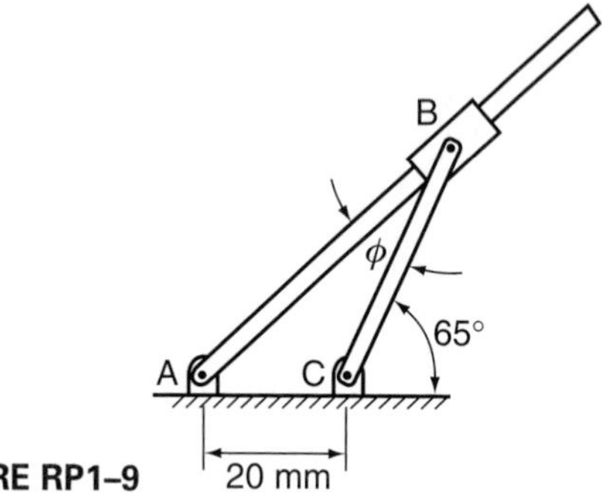

FIGURE RP1–9

R1–10. The toggle mechanism shown (Figure RP1–10) moves from an initial angle $\theta = 30°$ to a final $\theta = 20°$. Calculate the horizontal distance moved by point C.

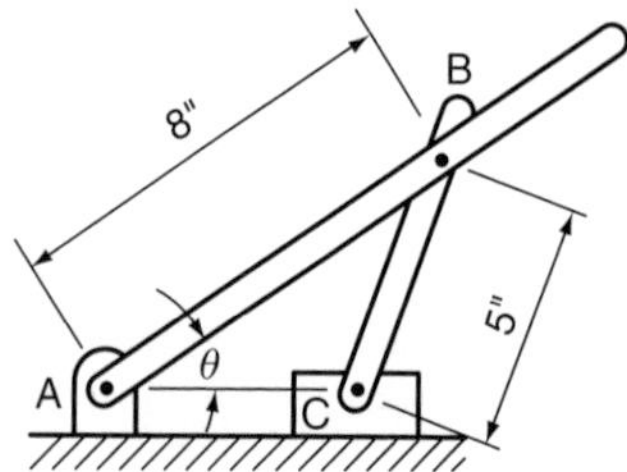

FIGURE RP1–10

R1–11. A surveyor is to determine the internal angles of triangle ACB (Figure RP1–11). Since she cannot set up at point C, she selects point D, 12 m from C, and obtains the data shown. Calculate the internal angle ACB.

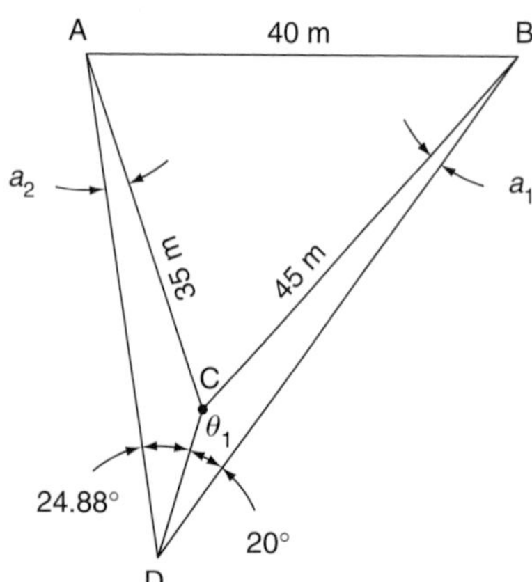

FIGURE RP1–11

R1–12. Block B slides to the left until member BC is vertical (Figure RP1–12). Determine the change in length of distance *AC*.

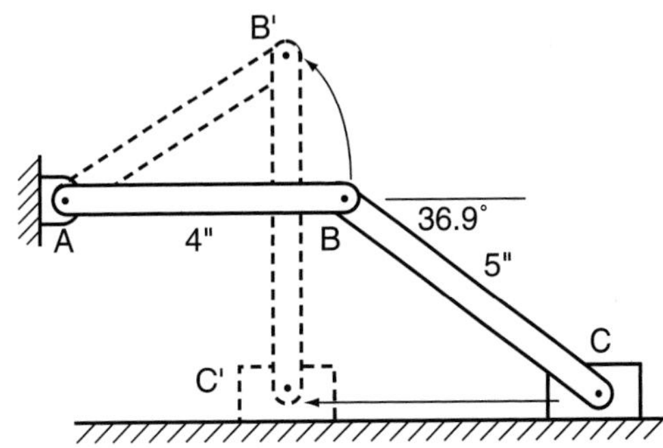

FIGURE RP1–12

CHAPTER 2

Forces, Vectors, and Resultants

OBJECTIVES

Upon completion of this chapter the student will be able to:

1. Determine the resultant of two vectors at right angles to each other by drawing a vector triangle and using the trigonometric functions of sine, cosine, and tangent.
2. Determine the resultant of any two vectors by drawing a vector triangle and using either the sine law or cosine law.
3. Resolve any vector quantity into components in the horizontal and vertical directions.
4. Resolve any vector quantity into components along any two axes.
5. Determine the resultant of several vectors by the method of components.

2–1 VECTORS

Everyone feels that he or she knows what a force is and would probably define it as a push or pull. Although this is true, there are further classifications. However, all forces do have one property in common: They can be represented by vectors.

We must first distinguish between a *vector* quantity and a *scalar* quantity. You are no doubt familiar with scalar quantities. A board 10 ft long, a 2-hour time interval, a floor area of 20 m^2, and a 60-W light bulb all tell us "how much." These are scalar quantities; they indicate size or magnitude.

Vector quantities have the additional property of direction. Some vector quantities are a force of 15 N vertically downward, a distance of 20 km north, a velocity of 20 km/h east, and an acceleration of 7 ft/sec^2 upward. A vector quantity is therefore represented by an arrow; the arrowhead indicates the direction, and the length of the arrow indicates the magnitude.

If we arbitrarily choose a scale of 1 cm = 2 N, a 15-N vertical force would be that shown in Figure 2–1. Similarly, the other vector quantities referred to previously would be those shown in Figures 2–2, 2–3, and 2–4. Drawing vectors to scale is only used for

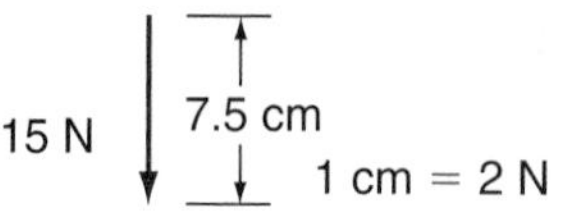

FIGURE 2–1

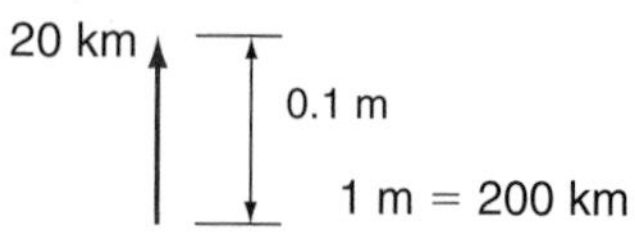

FIGURE 2–2

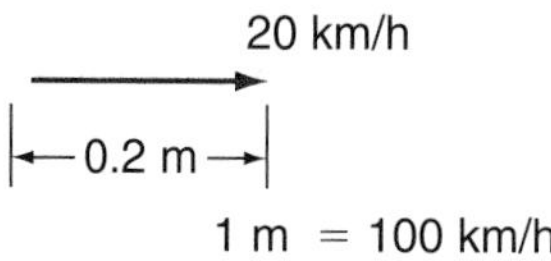

FIGURE 2–3

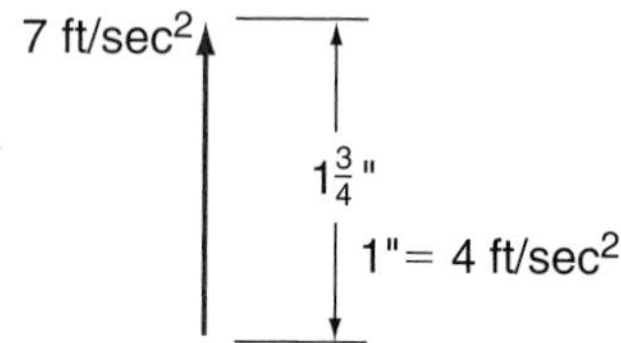

FIGURE 2–4

graphical solutions, but when drawing vectors for analytical solutions, draw them approximately to scale for easier visualization of the problem solution.

To be complete, both direction and magnitude must be labeled for each vector quantity. A vector representing a force has a point of application and a line of action. In Figure 2–5, a force of 20 N is applied to a cart. The vector shown indicates magnitude (20 N), a point of application (A), and the direction along the line of action. If the 20-N force is not sufficient to move the cart, the cart has a balance of external forces acting on it and is said to be in *static equilibrium.*

The principle of *transmissibility* states that a force acting on a body can be applied anywhere along the force's line of action without changing its effect on the body. Thus, the 20-N force can also be applied at point B, as shown in Figure 2–6. Whether the point of application is A or B, the 20-N force has the same effect on the cart. Later in the book, this principle will be used quite frequently in dealing with the subject of *moments.*

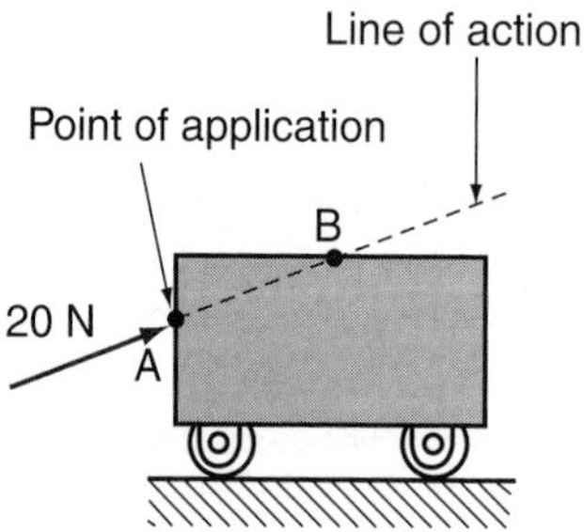

FIGURE 2–5

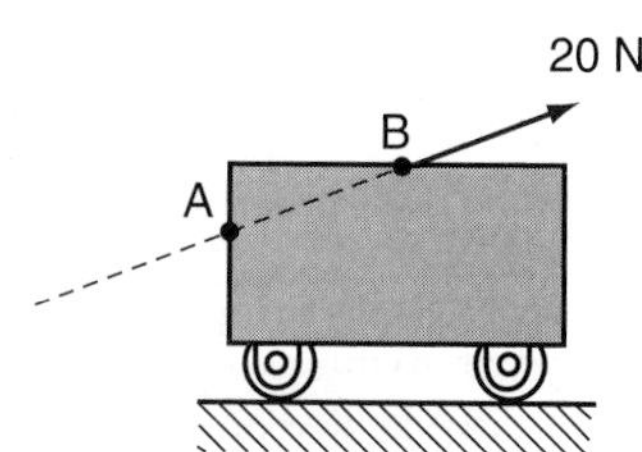

FIGURE 2–6

Vector quantities are not always vertical or horizontal. They may be at some angle or slope. The slope is indicated by reference to a horizontal line. This is done by giving either the angle in degrees (Figure 2–7a) or the rise and run of the slope (Figure 2–7b).

FIGURE 2–7a

FIGURE 2–7b

2–2 FORCE TYPES, CHARACTERISTICS, AND UNITS

Realizing that forces can be represented by vectors, consider now the various types of forces. For the sake of easier discussion in a subject such as mechanics, several classifications are used. One such classification is that of *applied* and *nonapplied forces.* An applied force is a very real and noticeable force applied directly to an object. The force that you would apply to a book (Figure 2–8) to slide it across a table is an applied force. The nonapplied force acting on this same book (Figure 2–9) may not be as readily apparent since it is the force of gravity, or the weight of the book. In Figure 2–9a the weight is shown as a concentrated force acting at the center of gravity of the book. It could also have been shown as a *distributed* load (Figure 2–9b), consisting of many smaller forces distributed over the

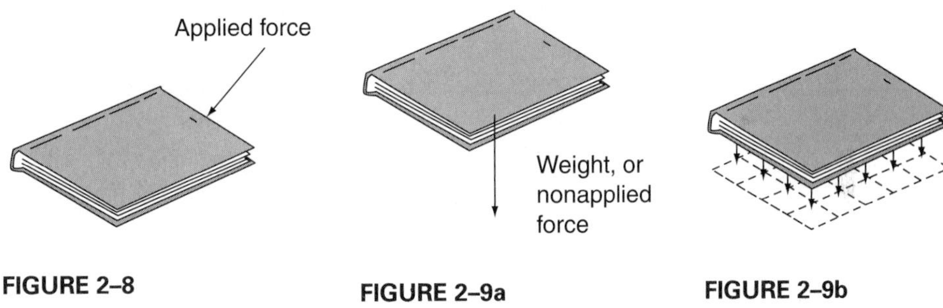

FIGURE 2–8

FIGURE 2–9a

FIGURE 2–9b

entire surface of the book. Other examples of nonapplied forces are the force of magnetic attraction or repulsion and the force due to inertia. In analyzing various force systems later, keep in mind that forces such as weight and inertia are always present and may have to be included in your calculations.

Another classification categorizes forces as *internal* and *external.* The distinction here is very important, since it is often the source of incorrectly drawn free-body diagrams (Section 4–2). Internal forces are often included where they should not be.

An internal force is a force inside a structure, and an external force is a force outside the structure. The pin-connected structure in Figure 2–10 has an external force of 30 lb. Knowing that connection C is on rollers and free to move horizontally, one can visualize that the horizontal member AC is *in tension;* that is, there is a force tending to stretch it. The tensile force in AC is an internal force. There are also internal forces of compression in members AB and BC.

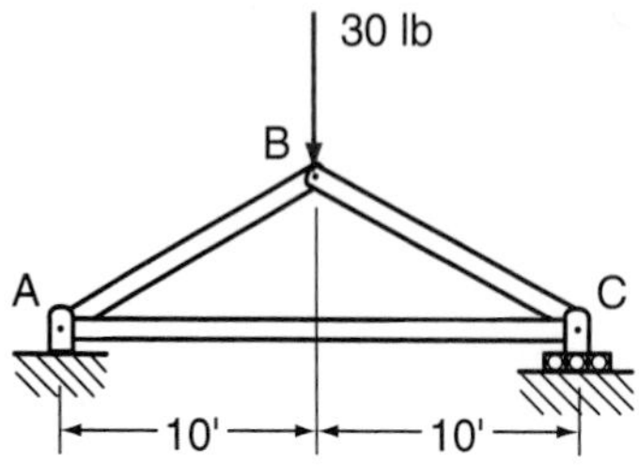

FIGURE 2–10

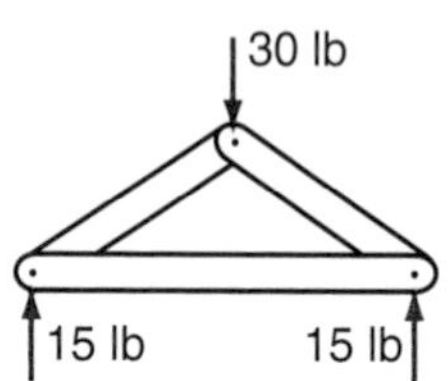

FIGURE 2–11

If we replace the actual supports at points A and C with equivalent supporting forces of 15 lb at each end (Figure 2–11), we have a total of three external forces. These external forces can be further subdivided into *acting* and *reacting forces.* The 15-lb forces are present because of the 30-lb force; thus, they are a reaction to the application of the 30-lb force. Therefore, the 30-lb force is an *acting* force, and the 15-lb forces are *reacting* forces. You will be asked to solve for the reactions on various structures. *Reactions* are simply the reacting forces that are necessary to support the structure when its given method of support is removed. As shown in Figure 2–11, the reaction at A is 15 lb vertically upward or $R_A = 15 \text{ lb} \uparrow$, and similarly, $R_B = 15 \text{ lb} \uparrow$. (Expressing a quantity in italic type will indicate that it is a vector quantity.)

Unless otherwise stated, the weight of all members or structures will be neglected to simplify problem solution. This can be done without appreciable error when the supported load is much greater than the structure's weight. When weights are significant in later problems, they will be included.

2–3 RESULTANTS

Scalar quantities such as 4 m^2 and 3 m^2 can be added to equal 7 m^2. But if we add vector quantities of 4 km and 3 km, their directions must be considered. This is known as *adding vectorially* or *vector addition.* The answer obtained is the *resultant;* it is a single vector giving the result of the addition of the original two or more vectors.

What is the result of walking 4 km east and then 3 km west? You have walked a total distance of 7 km, but how far are you from your original location? Let each distance be represented by a vector (Figure 2–12a). When adding 4 + 3 vectorially, place the vectors tip to tail. The resultant is a vector from the original point to the final point. In this case, $R = 1$ km in an easterly direction.

Suppose that you had walked 4 km east and 3 km north. Again, the distances would be represented by vectors and added by being placed tip to tail (Figure 2–12b). The resultant vector drawn from the original point to the final point is the hypotenuse of a right-angle triangle. For this case, $R = 5$ km in a northeast direction.

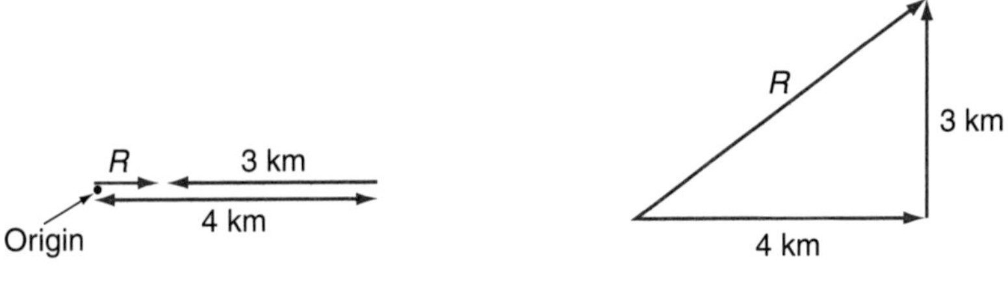

FIGURE 2–12a

FIGURE 2–12b

2–4 VECTOR ADDITION: GRAPHICAL (TIP TO TAIL)

As was described in Section 2–3, the sum of two or more vectors is a resultant. The vector quantity used there was *distance;* other vector quantities often encountered are *force, velocity,* and *acceleration.* Force vectors are going to be our main concern in static mechanics.

Graphical vector addition requires the drawing of the vectors to some scale in their given direction. The resultant can then be measured or scaled from the drawing. This method of scale drawings will not be used to any extent in this book since drafting equipment is required and the accuracy of the solution is somewhat dependent on the scale used. However, graphical vector addition is still important, because the same drawings or sketches are required for the analytical method. When employing the analytical method, one uses the same sketches drawn roughly to scale and then solves mathematically.

The three methods or rules of vector addition are the *triangle, parallelogram,* and *vector polygon* methods. The triangle and parallelogram methods are simplified versions of the vector polygon method. The triangle method of vector addition can be used for a right-angle triangle (Figure 2–13a) or any other triangle (Figure 2–13b). Constructing A and B (Figure 2–13a) to a scale of 1 cm = 1 N, we can calculate the length of R to be 13 cm. R is therefore equal to 13 N. (Vectors are drawn tip to tail in any sequence.) Using a scale of 1 cm = 10 N (Figure 2–13b) and drawing B at the correct slope, we find the length of R to be 5.2 cm or R = 52 N. (Vector triangles are not always right-angle triangles and will therefore require different mathematics when being solved analytically.)

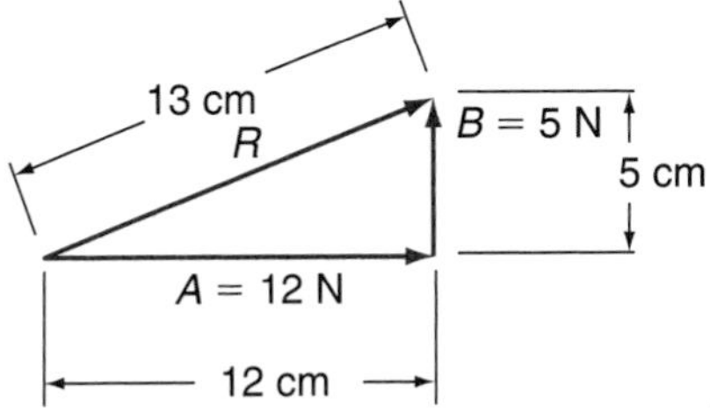

FIGURE 2–13a

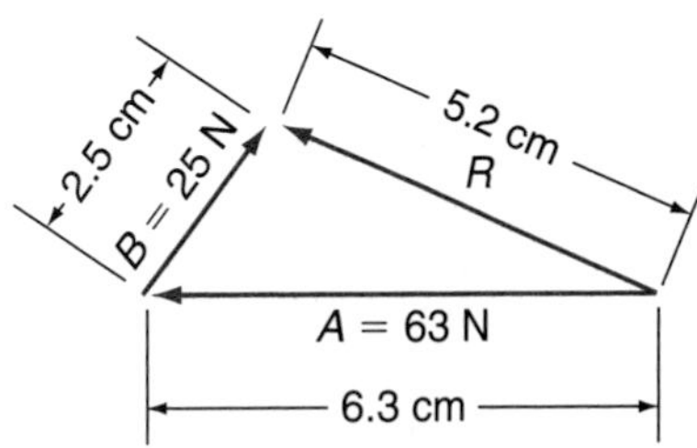

FIGURE 2–13b

EXAMPLE 2–1

Forces of 20 N and 30 N are pulling on a ring (Figure 2–14a). Determine the resultant using the triangle rule.

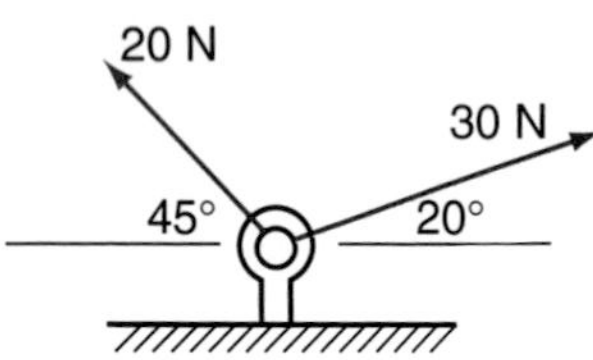

FIGURE 2–14a

Employing an appropriate scale such as 1 cm = 2 N and the angles of 20° and 45° as given, one can construct a vector triangle (Figure 2–14b or c), giving an answer of R = 28.2 N $\angle 60^\circ$. **(Note that it is immaterial whether 20 N is added to 30 N or 30 N is added to 20 N.)**

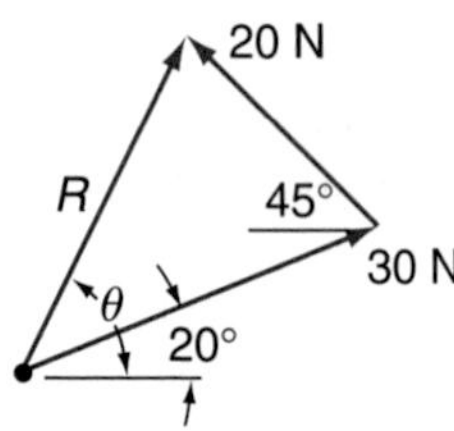
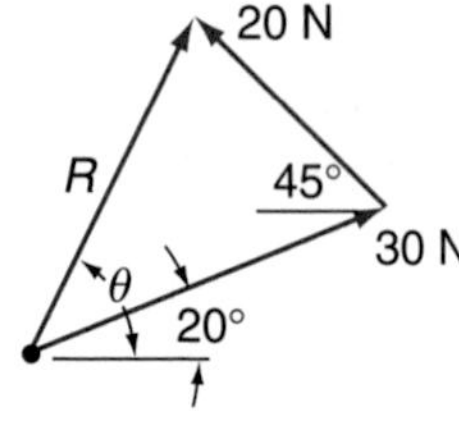

FIGURE 2–14b

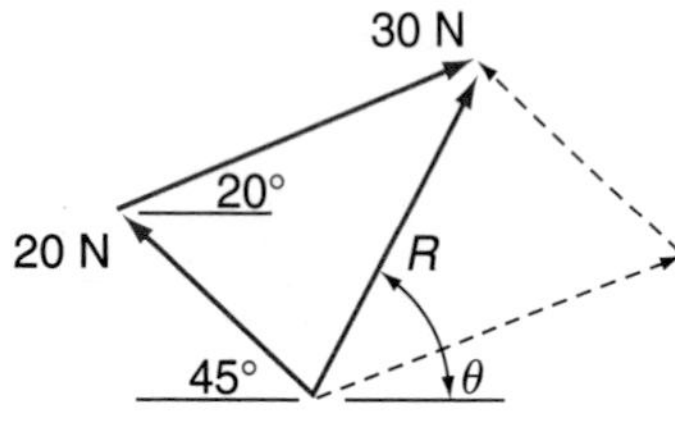

FIGURE 2–14c

Applying the parallelogram rule to the forces in Example 2–1, we get the parallelogram shown in Figure 2–15. From the top of the 20-N force, a line is drawn parallel to the 30-N force. Similarly, a line is drawn parallel to the 20-N force. The diagonal of the parallelogram formed represents the resultant. The parallelogram is simply the two triangles of the triangle method.

The vector polygon is a continuation of the triangle rule to accommodate more than two forces. Several vectors are added tip to tail—the sequence of the addition is not important. The resultant *R* is a vector from the origin of the polygon to the tip of the last vector (Figure 2–16).

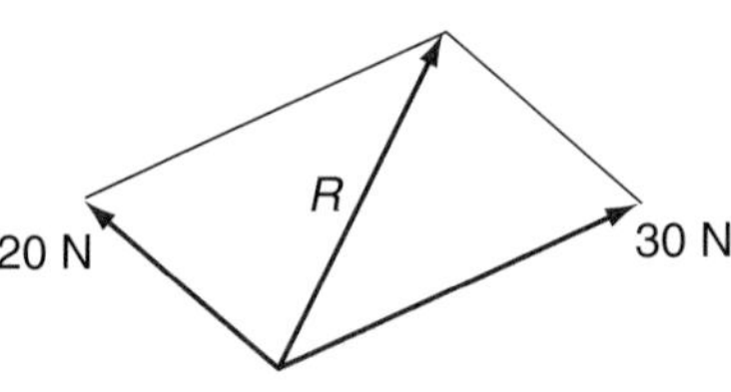

FIGURE 2–15

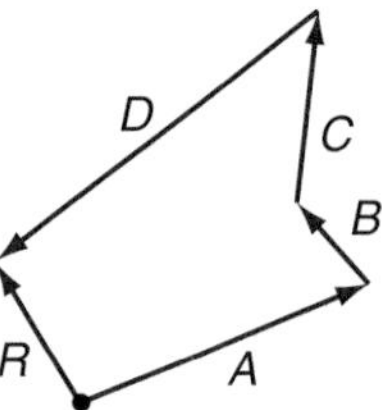

FIGURE 2–16

2–5 VECTOR ADDITION: ANALYTICAL

The solutions to mechanics problems are no exception to the old saying, "A picture is worth a thousand words." Sketches are vitally important in many cases; calculations should be accompanied by a sketch drawn as closely to scale as a little care will allow. A calculated answer can be visually checked for an obvious error in direction or magnitude. Analytical vector addition consists of two main methods:

1. Construction of a triangle and use of the cosine law or other simple trigonometric functions.
2. Addition of the components of vectors (Section 2–7).

EXAMPLE 2–2

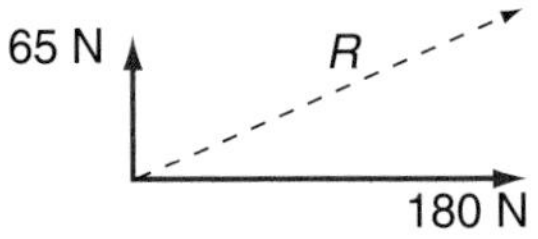

FIGURE 2–17a

Determine the resultant of the vectors shown in Figure 2–17a. The resultant is easily obtained since the vectors, when added, form a right-angle triangle, as shown in Figure 2–17b.

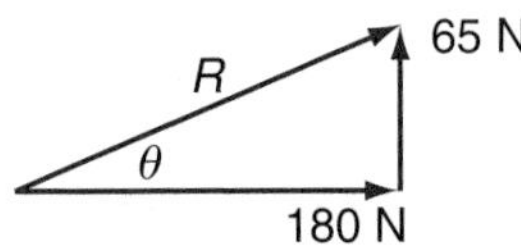

FIGURE 2–17b

$$R^2 = (180\text{ N})^2 + (65\text{ N})^2$$

$$R = \sqrt{32{,}400 + 4225}$$

$$= \sqrt{36{,}625}$$

$$R = 191\text{ N}$$

$$\tan\theta = \frac{65}{180} = 0.361$$

$$\theta = 19.8^\circ$$

The final answer is expressed as

$$\underline{R = 191\text{ N} \nearrow 19.8^\circ}$$

An alternative method would be

$$\tan\theta = 0.361$$
$$\theta = 19.8°$$
$$\sin\theta = \frac{65\text{ N}}{R}$$
$$R = \frac{65}{\sin 19.8°}$$
$$\underline{R = 191\text{ N} \nearrow 19.8°}$$

Note that with this method, the value of R depends on a correct initial calculation of the value of θ. In the first solution, neither R nor θ was dependent on the other. **Thus, the first method is preferred so that one mistake at the beginning does not make remaining calculations incorrect.**

EXAMPLE 2–3

Solve for the resultant of the force system shown in Figure 2–18a acting on point A.

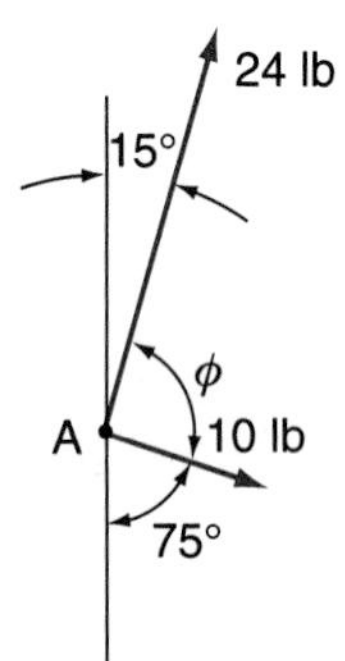

FIGURE 2–18a

$$\phi = 180 - 75 - 15 = 90°$$

Therefore, the vector triangle is a right-angle triangle and

$$R = \sqrt{(24\text{ lb})^2 + (10\text{ lb})^2}$$
$$R = 26\text{ lb}$$
$$\tan\theta = \frac{10}{24} = 0.416$$
$$\theta = 22.6°$$

Now find the angle between R and the horizontal plane.

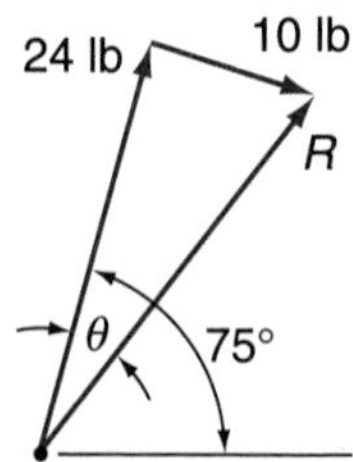

FIGURE 2–18b

$$75 - \theta = 75 - 22.6 \quad \text{(Figure 2–18b)}$$
$$= 52.4°$$
$$\underline{R = 26\text{ lb}\angle 52.4°}$$

EXAMPLE 2–4

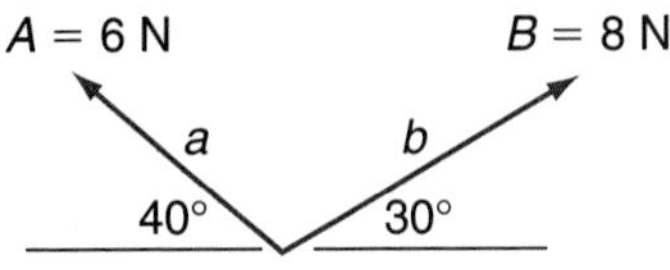

FIGURE 2–19a

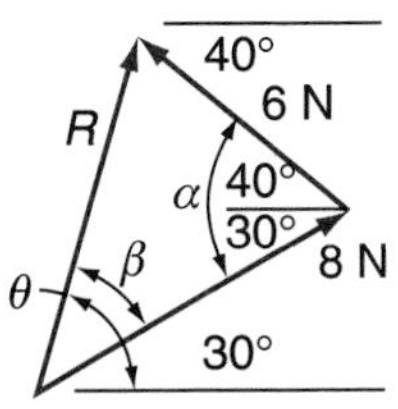

FIGURE 2–19b

Find the resultant in Figure 2–19a.

Sketch a vector triangle (Figure 2–19b), adding the vectors tip to tail and labeling all forces and angles as you construct the triangle. Use the cosine law.

$$\begin{aligned} R^2 &= A^2 + B^2 - 2AB\cos\alpha \\ &= (6\text{ N})^2 + (8\text{ N})^2 - 2(6\text{ N})(8\text{ N})(\cos 70°) \\ &= 36 + 64 - 96(0.342) \\ &= 100 - 32.8 \\ R &= \sqrt{67.2} \\ R &= 8.2\text{ N} \end{aligned}$$

To show a direction for R, we must solve for $\theta = \beta + 30°$. By the sine law

$$\begin{aligned} \frac{6\text{ N}}{\sin\beta} &= \frac{8.2\text{ N}}{\sin 70°} \\ \sin\beta &= \frac{6(0.94)}{8.2} \\ &= 0.688 \\ \beta &= 43.5° \\ \theta &= 43.5 + 30 = 73.5° \end{aligned}$$

Our final answer is

$$\underline{R = 8.2\text{ N} \nearrow 73.5°}$$

2–6 COMPONENTS

Previously, our main concern was the addition of two or more vectors to obtain a single vector, the resultant. *Resolution* of a vector into its components is the reverse of adding to get the resultant. A single force can be broken up into two separate forces. This is known as resolution of a force into its components. It is often convenient in problem solutions to be concerned only with forces in either the vertical or the horizontal direction. Therefore, the *x-y* axis system is used: A component in the horizontal direction has a subscript *x*, and a component in the vertical direction has a subscript *y*.

EXAMPLE 2–5

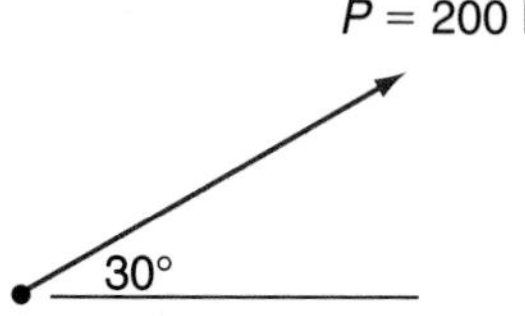

FIGURE 2–20a

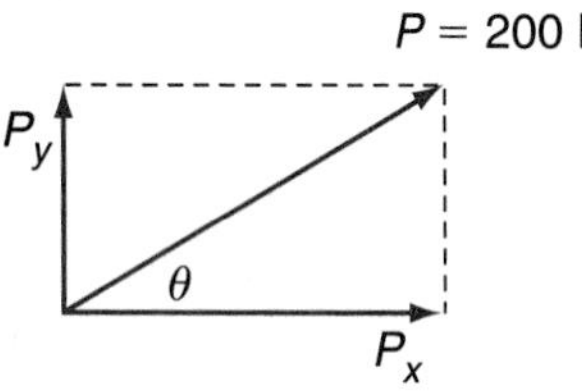

FIGURE 2–20b

Determine the horizontal and vertical components of P (Figure 2–20a).

By constructing a parallelogram or rectangle, we are left with a right-angle triangle in which

$$\sin\theta = \frac{P_y}{P} \qquad \text{(Figure 2–20b)}$$

$$P_y = (200\text{ N})(\sin 30^\circ)$$
$$= 200(0.5)$$
$$\underline{P_y = 100\text{ N}\uparrow}$$

$$\cos\theta = \frac{P_x}{P}$$

$$P_x = (200\text{ N})(\cos 30^\circ)$$
$$= 200(0.866)$$
$$\underline{P_x = 173\text{ N}\rightarrow}$$

EXAMPLE 2–6

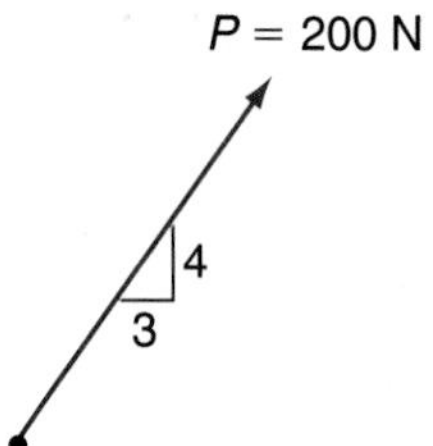

FIGURE 2–21a

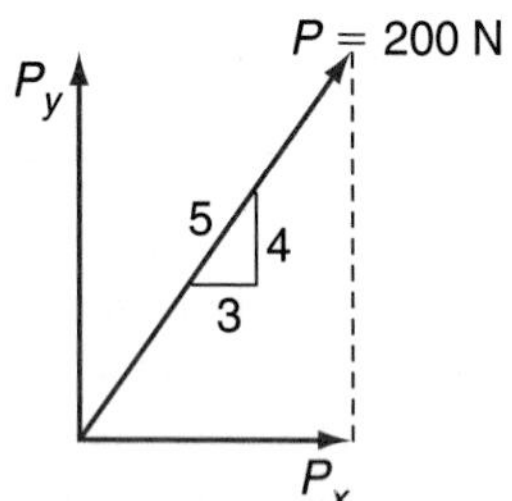

FIGURE 2–21b

Determine the horizontal and vertical components when the direction of P is shown as a slope (Figure 2–21a).

Construct the vector triangle (Figure 2–21b). Notice that we have two similar triangles. The hypotenuse of the small triangle $= \sqrt{(4)^2 + (3)^2} = 5$. Therefore,

$$\frac{P_y}{4} = \frac{P}{5} \quad \text{and} \quad \frac{P_x}{3} = \frac{P}{5}$$

or

$$P_y = \frac{4}{5}P \qquad P_x = \frac{3}{5}P$$

$$= \frac{4}{5}(200\text{ N}) \qquad = \frac{3}{5}(200\text{ N})$$

$$\underline{P_y = 160\text{ N}\uparrow} \qquad \underline{P_x = 120\text{ N}\rightarrow}$$

There are three main combinations of slope numbers that produce a whole number for the hypotenuse of a right-angle triangle. These combinations are 3, 4, 5; 5, 12, 13; and 8, 15, 17, or multiples thereof.

The following figures (Figure 2–22a, 2–22b, and 2–22c) illustrate the use of these combinations in calculating components. Each component is a fraction or ratio of the total as given by the slope numbers.

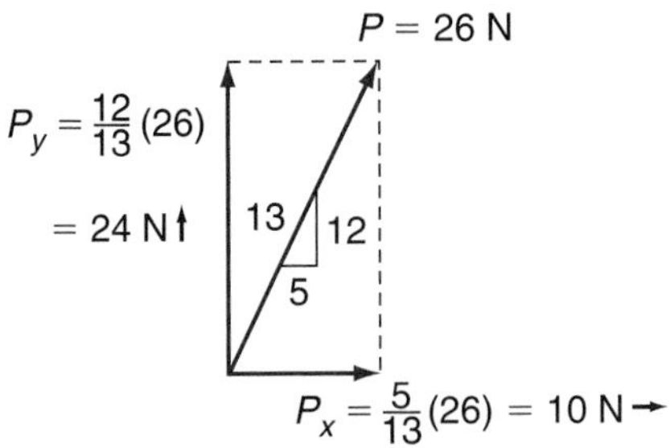

FIGURE 2–22a

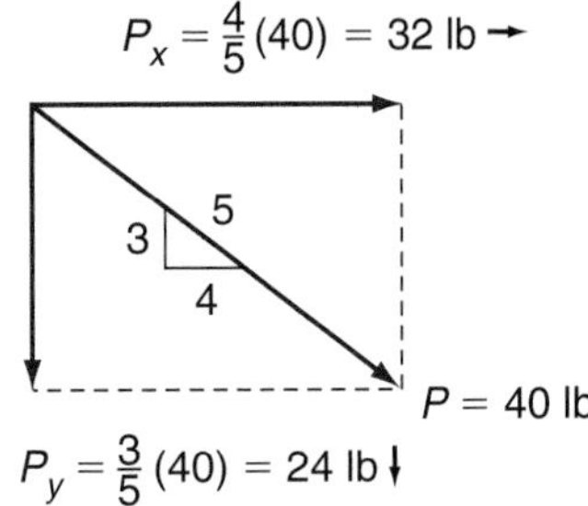

FIGURE 2–22b

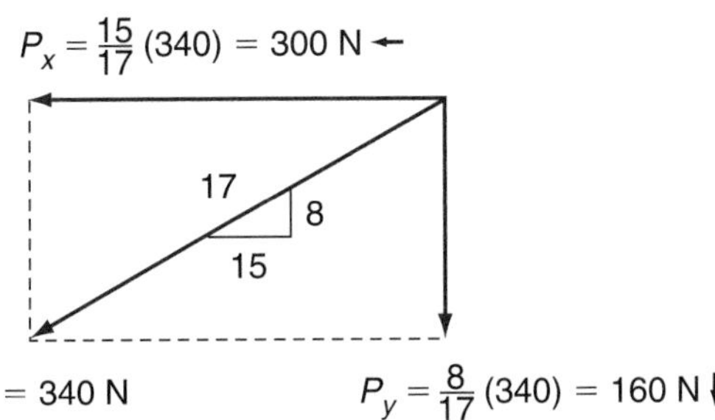

FIGURE 2–22c

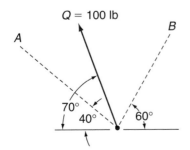

The components are not always in the horizontal and vertical directions, nor are they always at right angles to one another—as illustrated in the following example.

EXAMPLE 2–7

Find the components of force Q for the axis system of A and B as shown in Figure 2–23a.

Construct a vector parallelogram by drawing lines, parallel to axes A and B, from the tip of Q (Figure 2–23b). Applying the sine law to the left half of the parallelogram, we obtain

$Q = 100$ lb
A
B
70°
40°
60°

FIGURE 2–23a

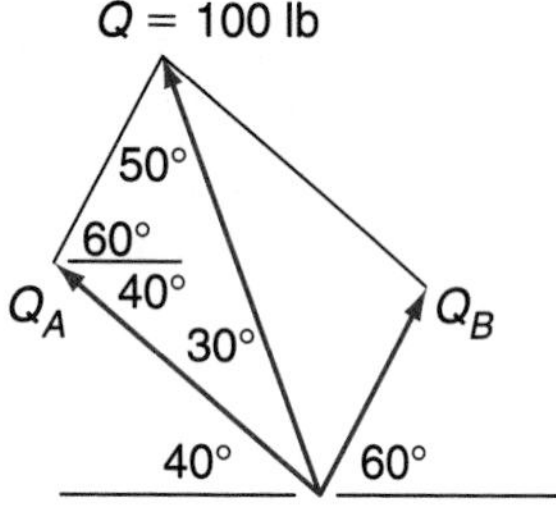

FIGURE 2–23b

$$\frac{Q_A}{\sin 50^\circ} = \frac{Q}{\sin 100^\circ}$$

$$Q_A = \frac{Q \sin 50^\circ}{\sin 100^\circ}$$

$$= \frac{Q \sin 50^\circ}{\sin 80^\circ}$$

$$= (100 \text{ lb})\left(\frac{0.766}{0.985}\right)$$

$$\underline{Q_A = 77.8 \text{ lb } 40^\circ}$$

$$\frac{Q_B}{\sin 30^\circ} = \frac{Q}{\sin 100^\circ}$$

$$Q_B = \frac{Q \sin 30^\circ}{\sin 80^\circ}$$

$$= (100 \text{ lb})\left(\frac{0.5}{0.985}\right)$$

$$\underline{Q_B = 50.8 \text{ lb } \nearrow 60^\circ}$$

2–7 VECTOR ADDITION: COMPONENTS

In Section 2–4 a graphical solution was shown where several vectors were added by the use of a vector polygon. To add these vectors analytically using the method of components, proceed according to the following steps:

1. Resolve each vector into a horizontal and vertical component.
2. Add the vertical components, $R_y = \Sigma F_y$.
3. Add the horizontal components, $R_x = \Sigma F_x$.
4. Combine the horizontal and vertical components to obtain a single resultant vector.

$$R = \sqrt{(R_x)^2 + (R_y)^2}$$

Note: The Greek capital letter Σ (sigma) means "the sum of." When writing $R_y = \Sigma F_y$, it is recommended that you say to yourself, "The resultant in the y-direction equals the sum of the forces in the y-direction."

EXAMPLE 2–8

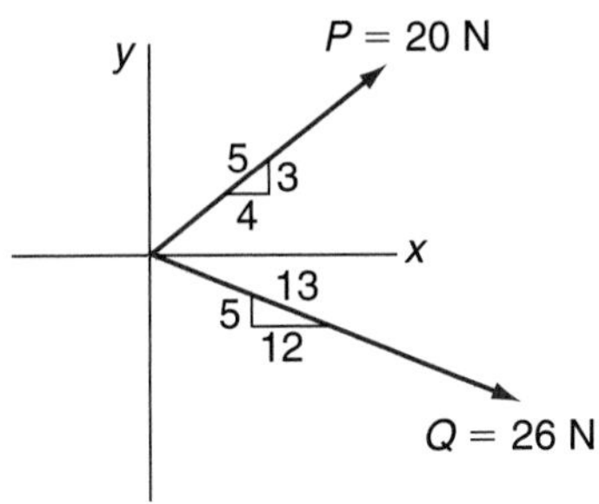

FIGURE 2–24

Find the resultant of forces P and Q as shown in Figure 2–24. Using the x- and y-axes for the algebraic signs of the components, we have

$$P_y = \frac{3}{5}(20\text{ N}) \qquad P_x = \frac{4}{5}(20\text{ N})$$

$$P_y = +12\text{ N} \qquad P_x = +16\text{ N}$$

$$Q_y = -\frac{5}{13}(26\text{ N}) \qquad Q_x = \frac{12}{13}(26\text{ N})$$

$$Q_y = -10\text{ N} \qquad Q_x = +24\text{ N}$$

Now sum the y and x components

$$R_y = 12\text{ N} - 10\text{ N} \qquad R_x = 16\text{ N} + 24\text{ N}$$

$$R_y = 2\text{ N} \qquad R_x = 40\text{ N}$$

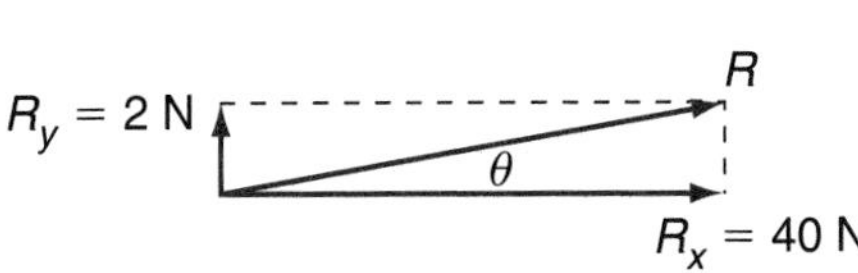

FIGURE 2–25

$$R = \sqrt{(40\text{ N})^2 + (2\text{N})^2} \quad \text{(Figure 2–25)}$$

$$R = 40.1\text{ N}$$

$$\tan\theta = \frac{2}{40} = 0.05$$

$$\theta = 2.9^\circ$$

$$\underline{R = 40.1\text{ N} \nearrow 2.9^\circ}$$

When determining the resultant of several forces, one may find it more convenient to tabulate all the components as follows:

Force	X	Y
P	+16 N	+12 N
Q	+24 N	−10 N
	R_x = +40 N	R_y = +2 N

EXAMPLE 2–9

Determine the resultant of the forces shown in Figure 2–26.

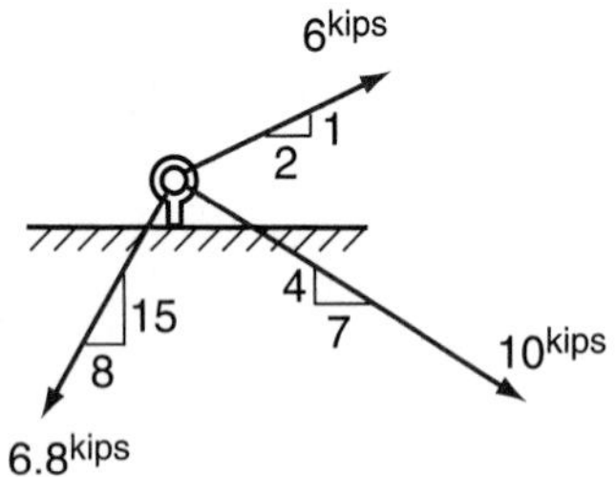

FIGURE 2–26

For easier visualization, the components of each force can be drawn as in Figure 2–27. Tabulate the components as follows:

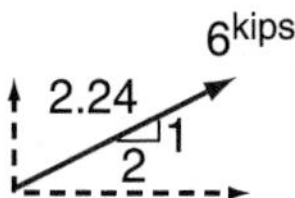

FIGURE 2–27a

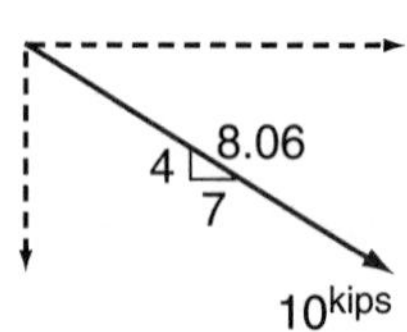

FIGURE 2–27b

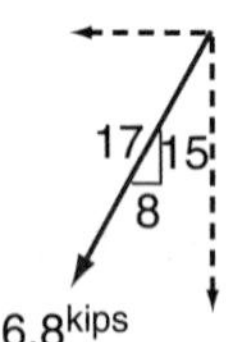

FIGURE 2–27c

Force (kips)	X	Y
6^{kips}	$\frac{2}{2.24} \times 6^{kips} = +5.35^{kips}$	$\frac{1}{2.24} \times 6^{kips} = +2.68^{kips}$
10^{kips}	$\frac{7}{8.06} \times 10^{kips} = 8.69^{kips}$	$\frac{4}{8.06} \times 10^{kips} = -4.97^{kips}$
6.8^{kips}	$\frac{8}{17} \times 6.8^{kips} = \underline{-3.2^{kips}}$	$\frac{15}{17} \times 6.8^{kips} = \underline{-6^{kips}}$
	$R_x = +10.84^{kips}$	$R_y = -8.29^{kips}$

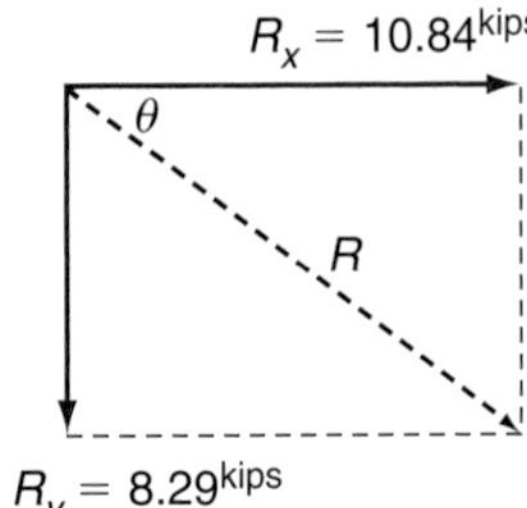

FIGURE 2–28

From Figure 2–28,

$$\tan\theta = \frac{8.29}{10.84}$$

$$\theta = 37.4°$$

$$R = \sqrt{(10.84^{kips})^2 + (8.29^{kips})^2}$$

$$R = 13.6^{kips} \searrow 37.4°$$

HINTS FOR PROBLEM SOLVING

1. For a *complete* answer of a vector quantity you must have magnitude, units, and direction (including slope or angle).
2. When drawing a vector triangle
   (a) Add the vectors in any sequence.
   (b) Show the direction of each vector.
   (c) Label all possible angles as you construct the triangle.
3. When solving for components of a vector whose slope is given as rise and run such as 3↘4, use the ratios 3:5 and 4:5. Converting to an angle in degrees and using trigonometry may only increase the chance of error due to increased calculations.
4. Watch that you do not mistakenly switch given slope numbers.
5. Lines perpendicular to each other have opposite slope numbers. For example, 5↘12 is perpendicular to 12↗5.
6. Pay special attention to the algebraic signs when summing the horizontal or vertical components of a force system.

PROBLEMS

APPLIED PROBLEMS FOR SECTIONS 2–1 TO 2–3

2–1. Determine the resultant and indicate the angle or slope for the right-angle force system shown in Figure P2–1.

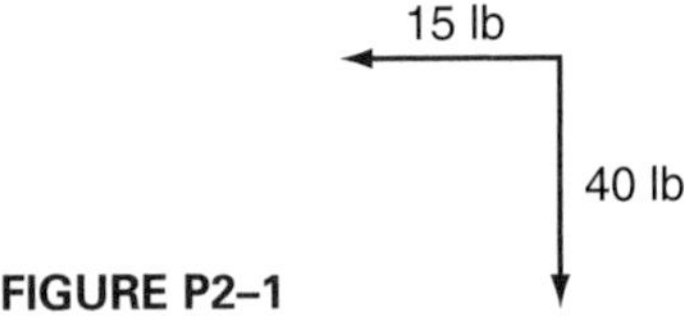

FIGURE P2–1

2–2. Determine the resultant and indicate the angle or slope for each right-angle force system shown in Figure P2–2.

FIGURE P2–2a FIGURE P2–2b FIGURE P2–2c

2–3. Determine the resultant and indicate the angle or slope for each right-angle force system in Figure P2–3.

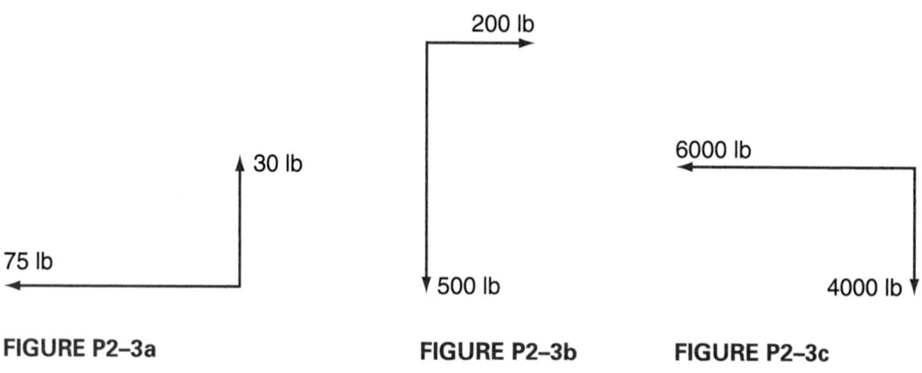

FIGURE P2–3a FIGURE P2–3b FIGURE P2–3c

2–4. Determine the resultant and indicate the angle or slope for each right-angle force system shown in Figure P2–4.

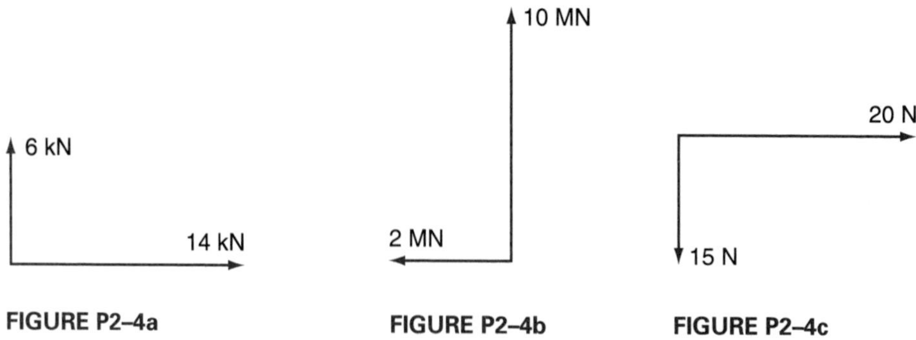

FIGURE P2–4a FIGURE P2–4b FIGURE P2–4c

2–5. A horizontal shaft exerts a sideways thrust of 200 N to the right and a downward load of 600 N on a bearing. What is the resultant?

2–6. A ladder is supported at the floor by forces as shown in Figure P2–6. Determine the resultant of these forces.

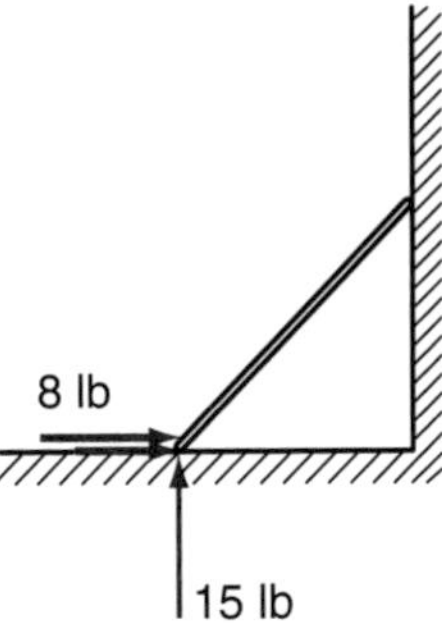

FIGURE P2–6

2–7. Member AB of the frame shown has two forces acting on it. Find the resultant force (Figure P2–7).

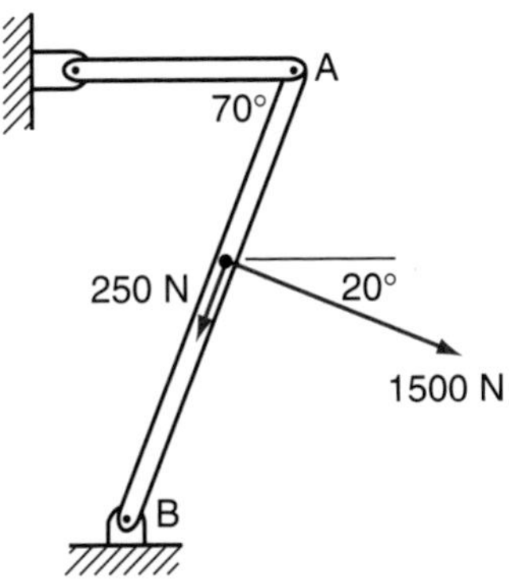

FIGURE P2–7

2–8. Each drive wheel of a car accelerating up a 12° slope has the forces shown (Figure P2–8) acting on it. Determine the resultant.

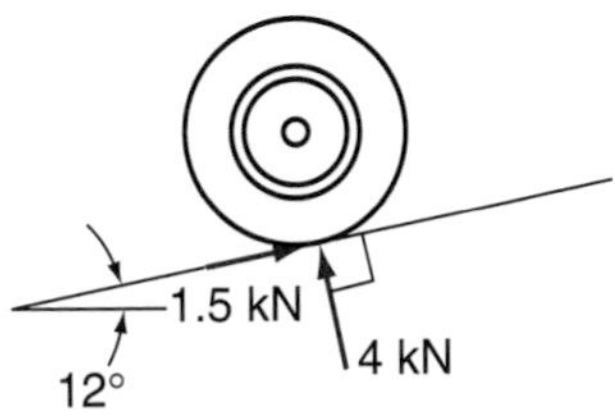

FIGURE P2–8

APPLIED PROBLEMS FOR SECTIONS 2–4 AND 2–5

2–9. Determine the resultant force on the lever shown in Figure P2–9.

FIGURE P2–9

2–10. Determine the resultant of the forces shown in Figure P2–10 using a vector triangle, cosine law, and sine law.

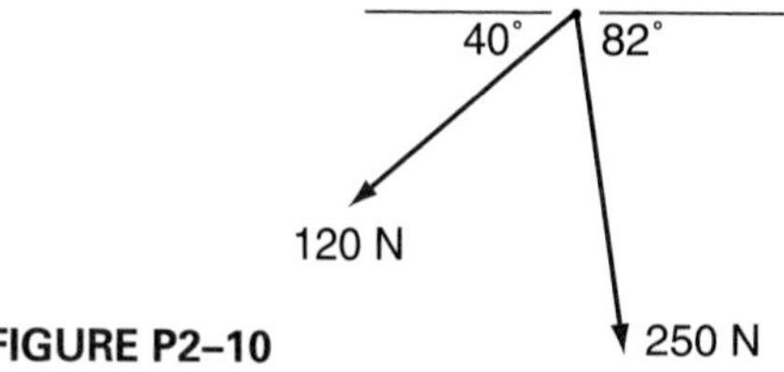

FIGURE P2–10

2–11. Using the cosine law, and then the sine law, solve for the resultant of the forces shown in Figure P2–11.

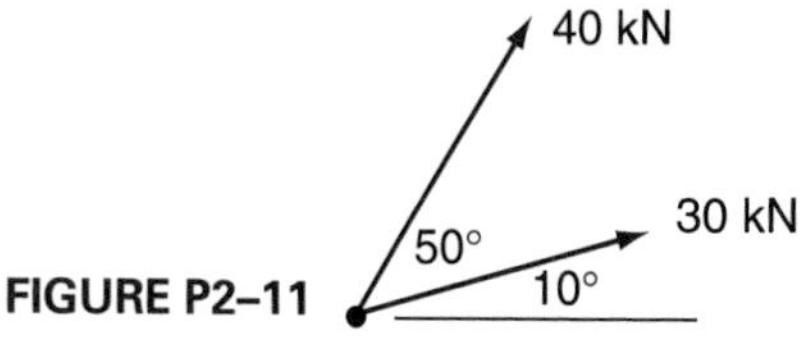

FIGURE P2–11

2–12. Using the cosine law, solve for the resultant of the forces shown in Figure P2–12.

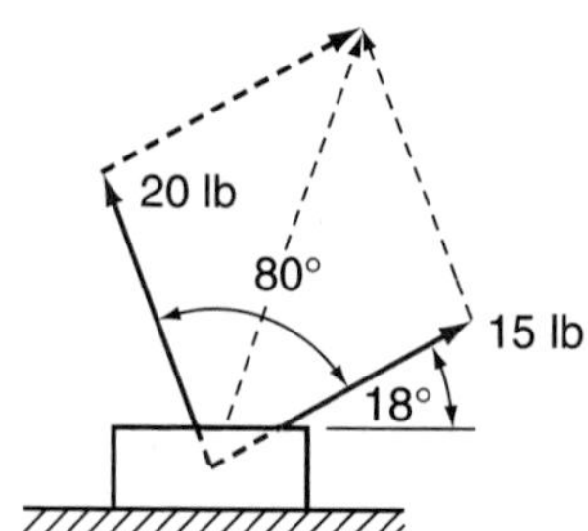

FIGURE P2–12

2–13. Two tow trucks are pulling a car from a ditch as shown in Figure P2–13. If truck A is pulling with a force of 8 kN and truck B is pulling with a force of 6.5 kN, determine the resultant force on the car.

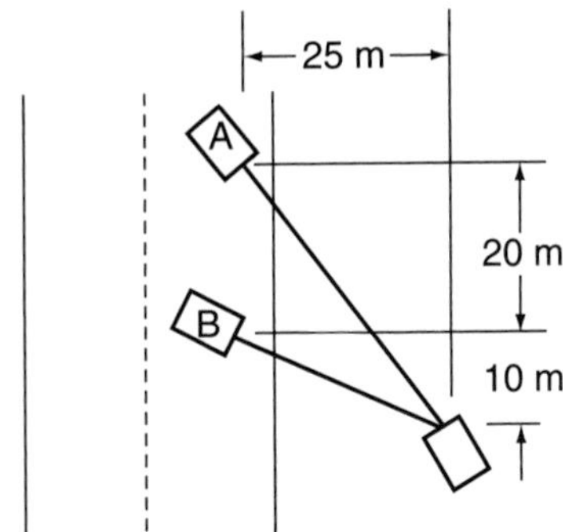

FIGURE P2–13

2–14. Using a vector triangle, cosine law, and sine law, determine the resultant of the forces shown in Figure P2–14.

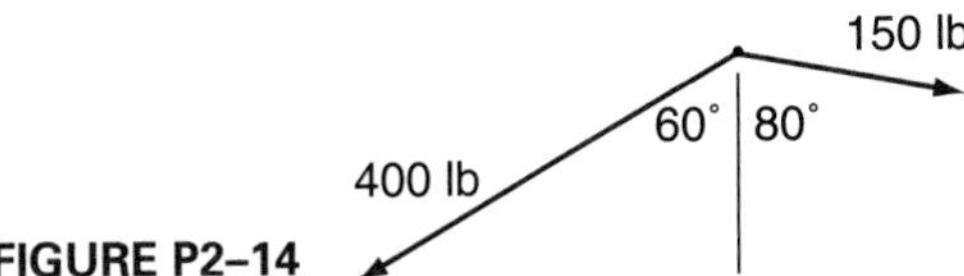

FIGURE P2–14

2–15. Two people pull on ropes as shown in Figure P2–15. Person B then moves left to pull from a center location. Determine the resultant force on the crate in each case.

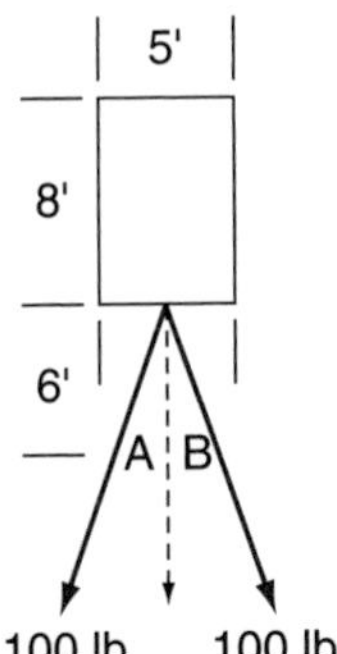

FIGURE P2–15

2–16. One method of clearing bush from land is to use a large steel ball with chains attached pulled by two caterpillar tractors (Figure P2–16). The chain tensions are AB = 4 kips and CD = 3 kips. Find the resultant force on the steel ball.

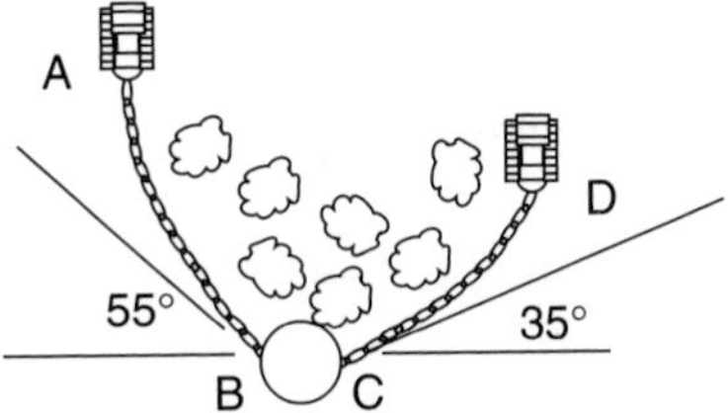

FIGURE P2–16

APPLIED PROBLEMS FOR SECTION 2–6

2–17. Determine the horizontal and vertical components of each force shown in Figure P2–17.

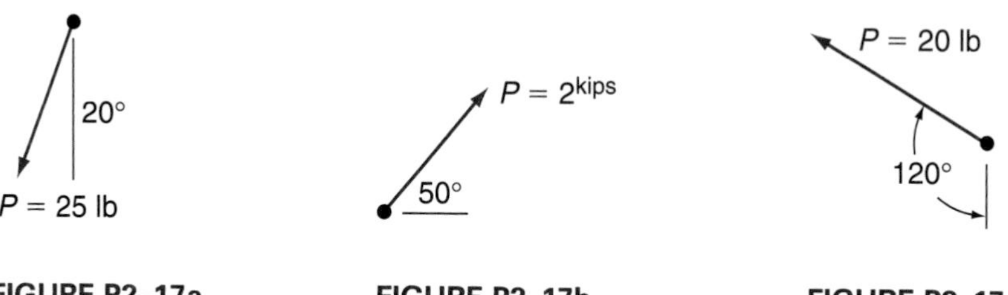

FIGURE P2–17a **FIGURE P2–17b** **FIGURE P2–17c**

2–18. Determine the horizontal and vertical components of each force shown in Figure P2–18.

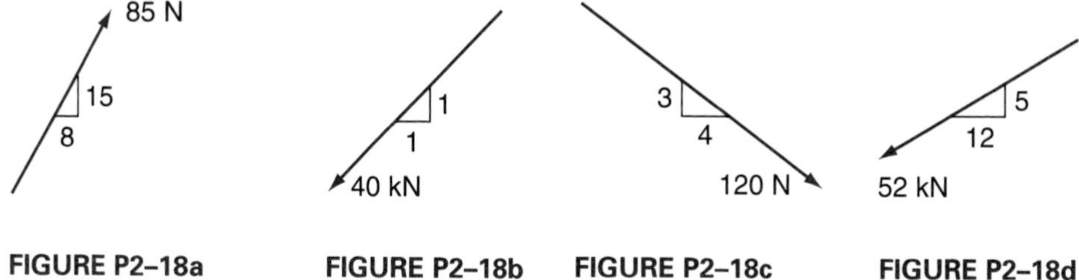

FIGURE P2–18a **FIGURE P2–18b** **FIGURE P2–18c** **FIGURE P2–18d**

2–19. Find the horizontal and vertical components of each vector in Figure P2–19.

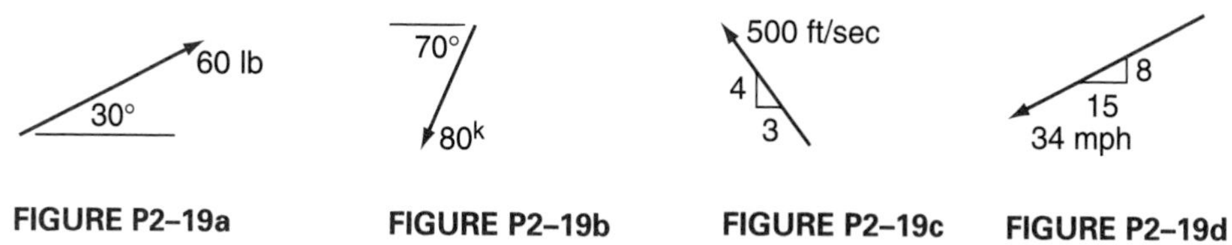

FIGURE P2–19a **FIGURE P2–19b** **FIGURE P2–19c** **FIGURE P2–19d**

2–20. Determine the horizontal and vertical components of each vector in Figure P2–20.

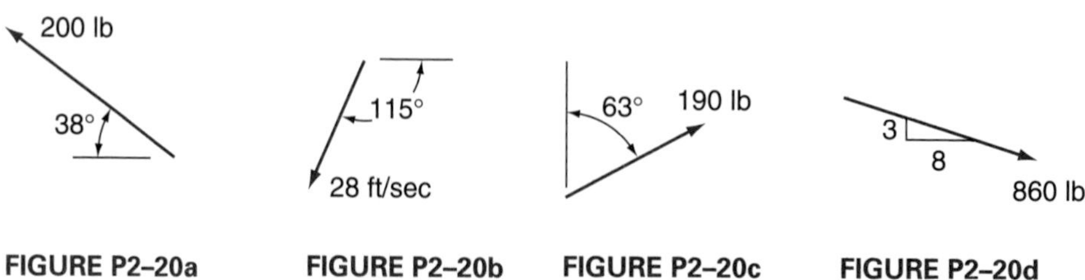

FIGURE P2–20a **FIGURE P2–20b** **FIGURE P2–20c** **FIGURE P2–20d**

2–21. A purlin hanger is used to join timbers A and B (Figure P2–21). Timber B exerts a resultant force on the purlin hanger of 1.8 kN$\searrow$80°. Calculate the horizontal and vertical components.

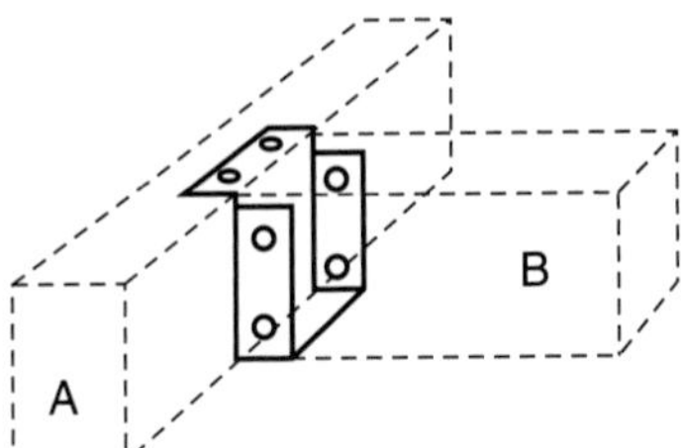

FIGURE P2–21

2–22. A frame supports a 65-kg block by means of a pulley and rope. Determine the horizontal and vertical components of the rope tension acting on A (convert kg to a force in newtons). (See Figure P2–22.)

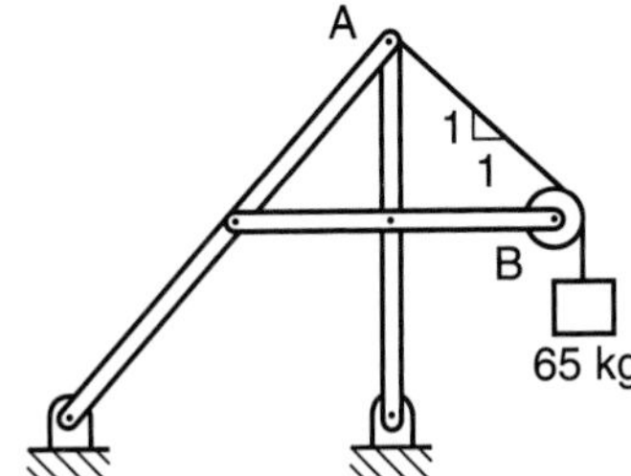

FIGURE P2–22

2–23. Resolve each force shown in Figure P2–23 into components in the *x*- and *y*-directions for the given *x*-*y*-axes orientation (Figure P2–23a).

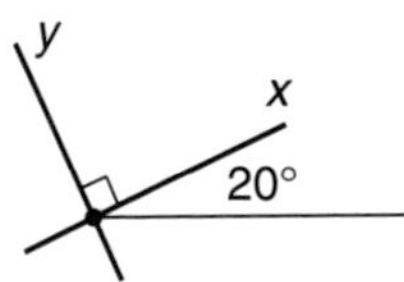

FIGURE P2–23a

FIGURE P2–23b

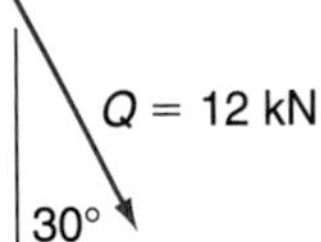

FIGURE P2–23c

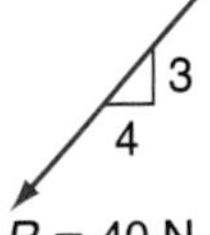

FIGURE P2–23d

2–24. The 7-m A-frame ladder shown in Figure P2–24 is rated to carry 1.25 kN but is being tested at 5 kN. Determine the component of the 5-kN force that is acting perpendicular to the ladder at A, causing it to bend.

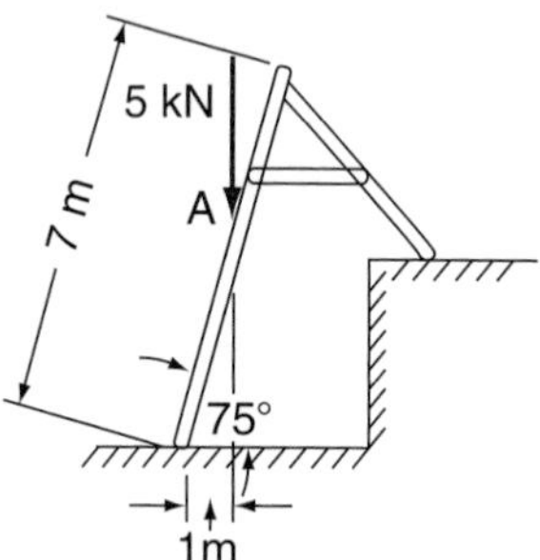

FIGURE P2–24

2–25. A lawn mower is pushed up a 15° slope by a 20-lb force. Find the component acting parallel to the slope (Figure P2–25).

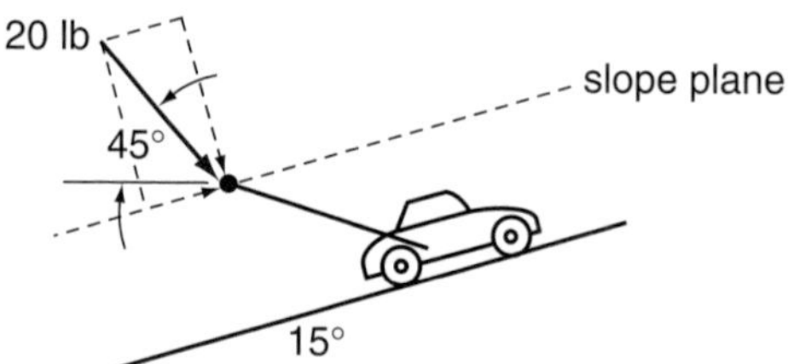

FIGURE P2–25

2–26. A crate is pushed up a slope by a force of 80 N. Find the component parallel to the slope and the component perpendicular to the slope (Figure P2–26).

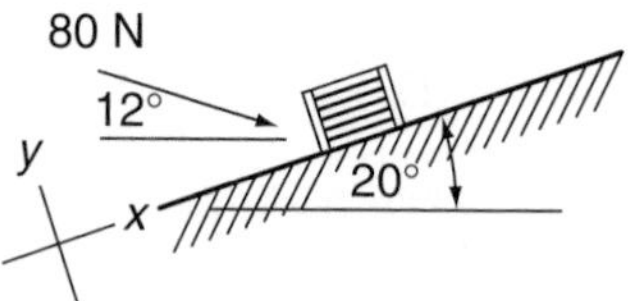

FIGURE P2–26

2–27. A person on a swing seat is pushed by a 25-lb force at the position shown (Figure P2–27). Determine the component perpendicular to the swing chain.

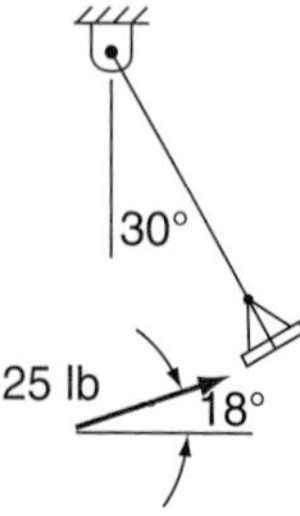

FIGURE P2–27

2–28. A crate is winched up the slope shown in Figure P2–28. If the rope tension is 400 N, determine the components acting on the crate, parallel to the slope and perpendicular to the slope.

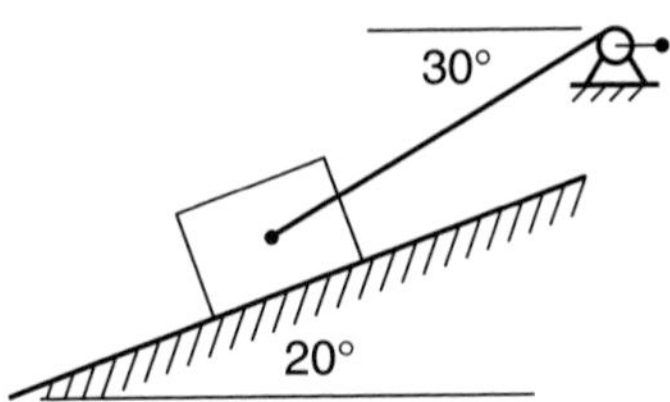

FIGURE P2–28

APPLIED PROBLEMS FOR SECTION 2–7

2–29. Find the resultant of the forces shown in Figure P2–29.

FIGURE P2–29

2–30. Find the resultant of the force system shown in Figure P2–30.

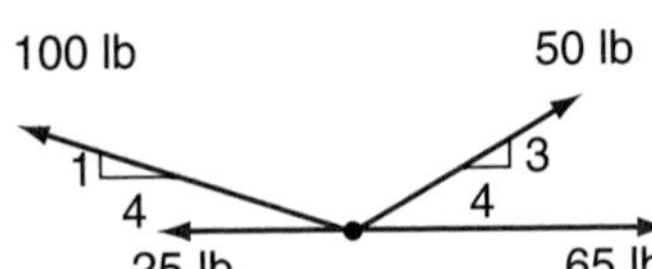

FIGURE P2–30

2–31. Determine the resultant of the force system shown in Figure P2–31.

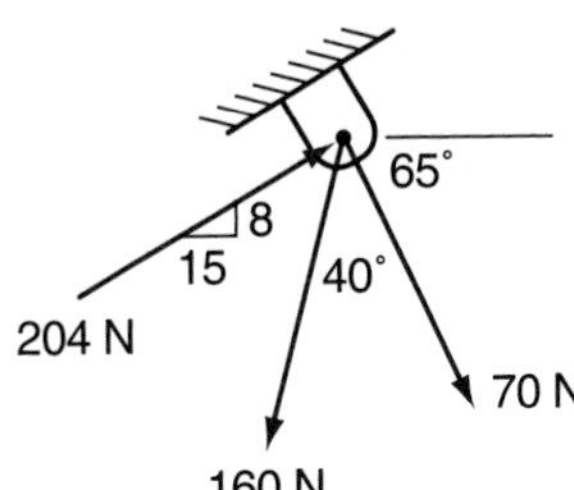

FIGURE P2–31

2–32. Determine the resultant of the force system shown in Figure P2–32.

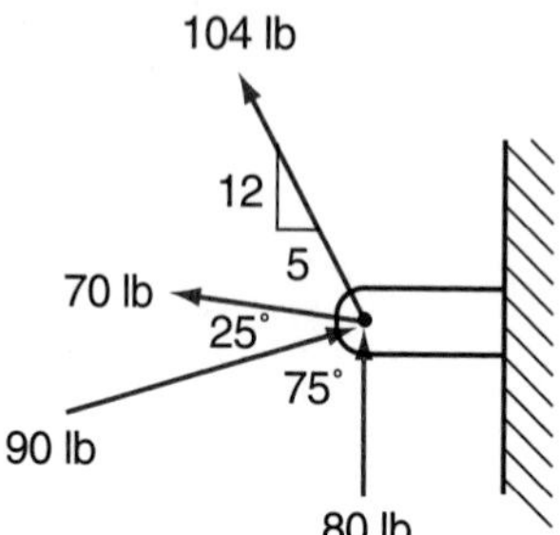

FIGURE P2–32

2–33. Find the resultant of the forces shown in Figure P2–33.

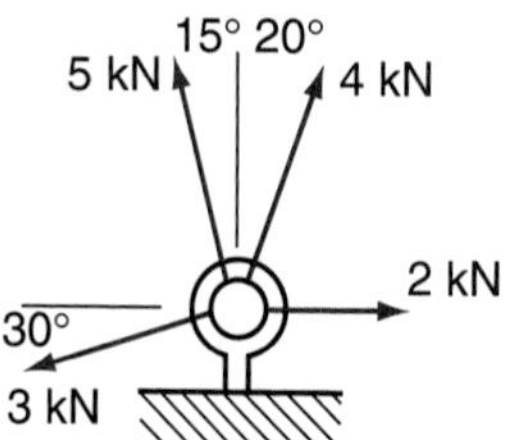

FIGURE P2–33

2–34. Determine the resultant of the force system shown in Figure P2–34.

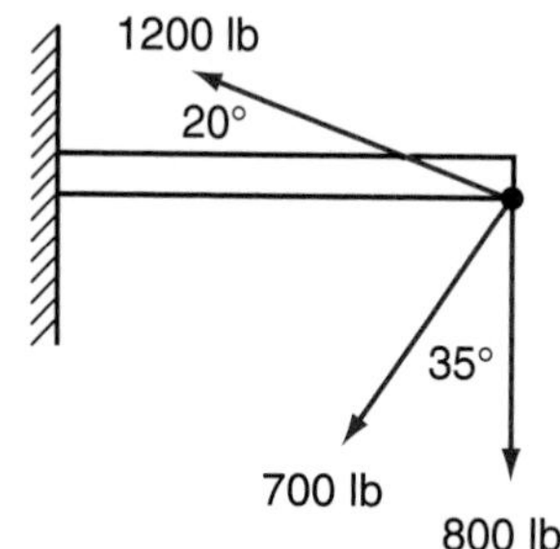

FIGURE P2–34

2–35. Determine the resultant of the force system shown in Figure P2–35.

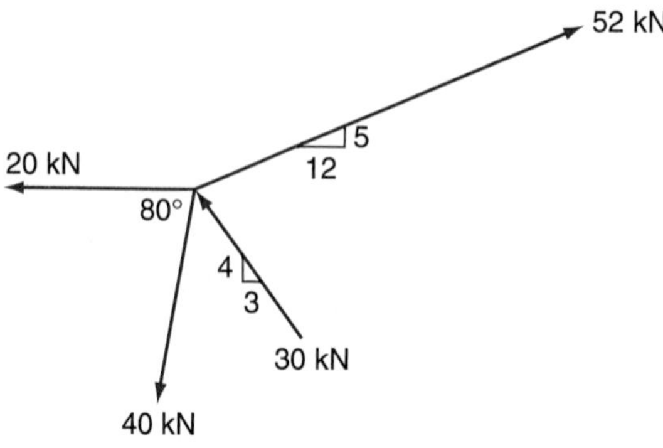

FIGURE P2–35

2–36. The anchor point for several cables has forces acting as shown in Figure P2–36. Determine the resultant.

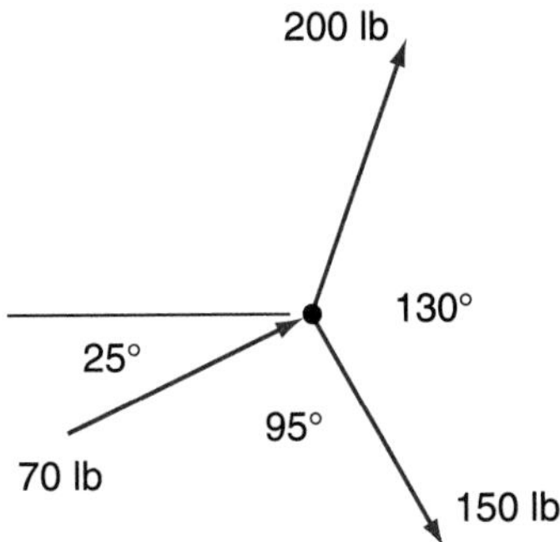

FIGURE P2–36

2–37. Find the resultant of the force system shown in Figure P2–37.

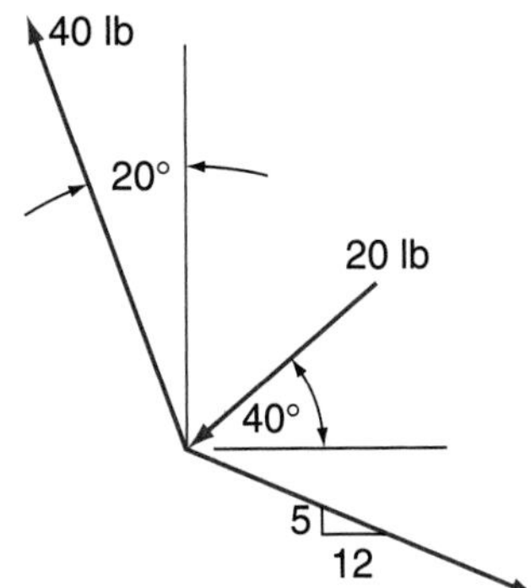

FIGURE P2–37

REVIEW PROBLEMS

R2–1. Determine the resultant and indicate the angle or slope for each right-angle force system in Figure RP2–1.

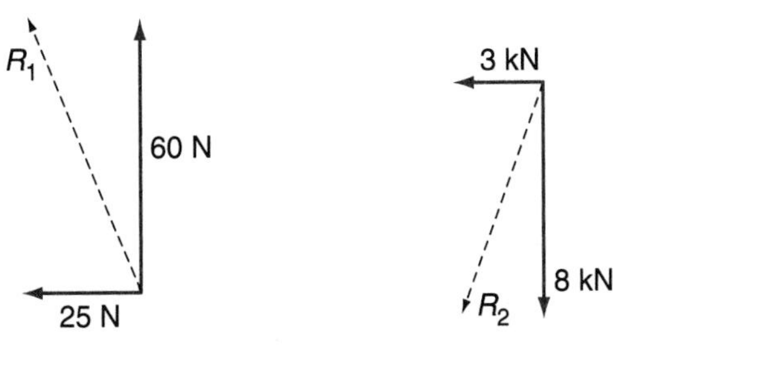

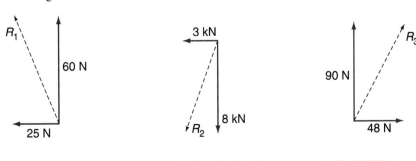

FIGURE RP2–1a FIGURE RP2–1b FIGURE RP2–1c

R2–2. Find the resultant force acting on point A (Figure RP2–2).

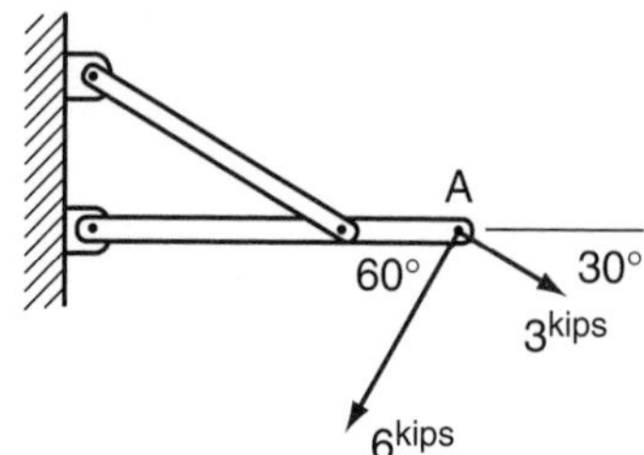

FIGURE RP2–2

R2–3. A carton is pushed onto a platform by the forces shown in Figure RP2–3. Find the resultant force.

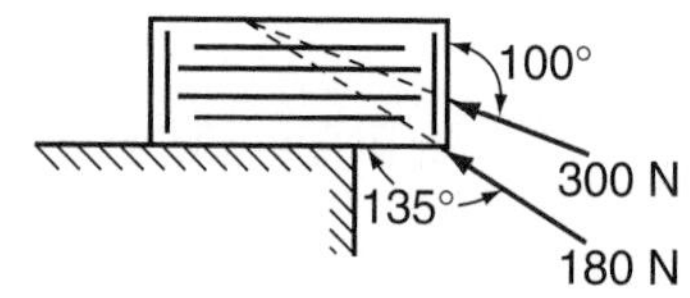

FIGURE RP2–3

R2–4. Determine the horizontal and vertical components of each force shown in Figure RP2–4.

FIGURE RP2–4a **FIGURE RP2–4b** **FIGURE RP2–4c** **FIGURE RP2–4d**

R2–5. Find the horizontal and vertical components of each vector in Figure RP2–5.

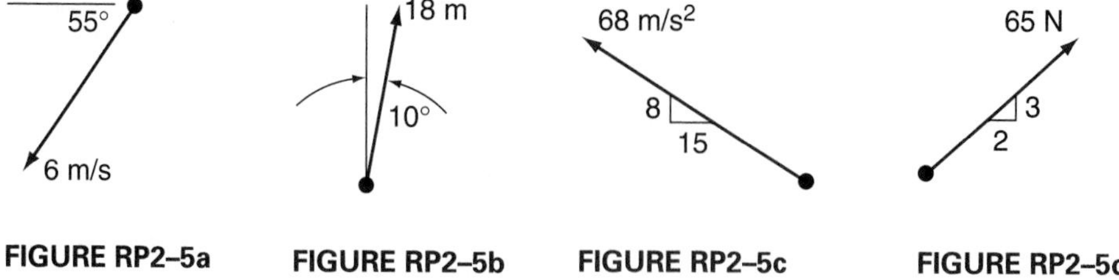

FIGURE RP2–5a **FIGURE RP2–5b** **FIGURE RP2–5c** **FIGURE RP2–5d**

R2–6. A block is winched from A to B by a constant cable tension of 3 kN. Determine both the initial and final components of the cable tension that is pulling the block parallel to the slope (Figure RP2–6).

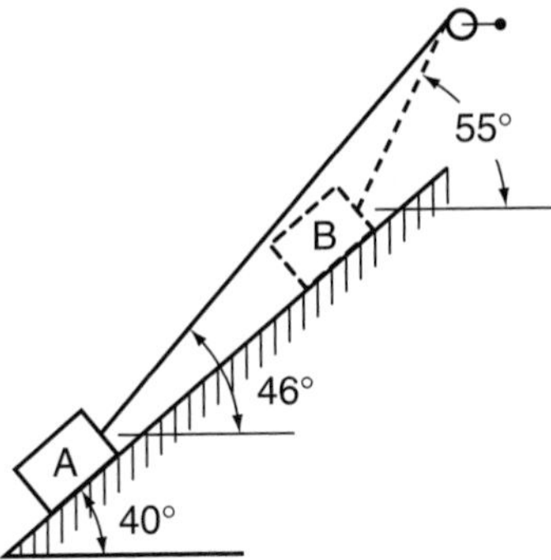

FIGURE RP2–6

R2–7. Determine the resultant of the force system shown in Figure RP2–7.

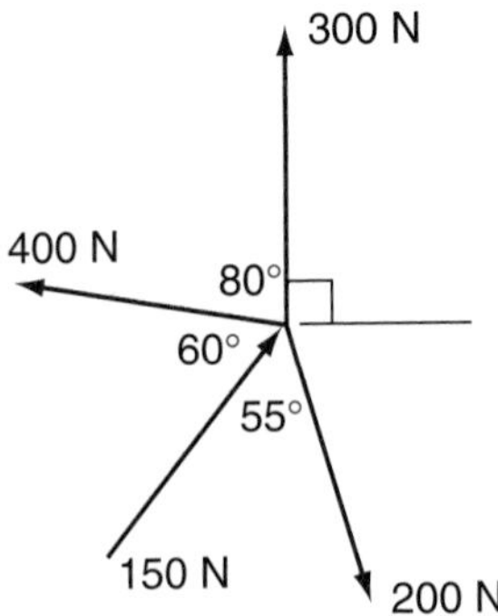

FIGURE RP2–7

R2–8. A block on an inclined plane has forces shown in Figure RP2–8 acting on it. Find the resultant force.

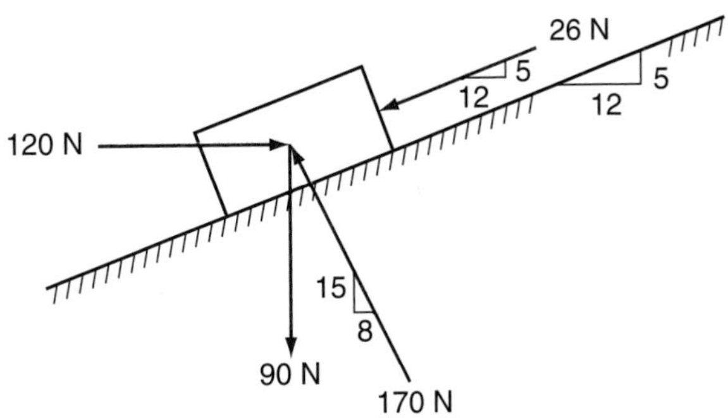

FIGURE RP2–8

CHAPTER 3

Moments and Couples

OBJECTIVES

Upon completion of this chapter the student will be able to:

1. Calculate the moment of a force by
   (a) multiplying the total force by its perpendicular distance.
   (b) multiplying each of a force's components by their respective perpendicular distances.
2. Name the three factors that constitute a couple, and calculate the moment.
3. Replace a given couple with an equivalent couple at a different location.

3–1 MOMENT OF A FORCE

Moment is merely another term meaning *torque,* which is something producing or tending to produce rotation or torsion. Common examples of moment or torque are numerous. Pushing on a revolving door while walking through it, tightening a nut with a wrench, and turning the steering wheel of a car all involve moments. The moment is present whether there is actual rotation or only a tendency to rotate. If a force is acting some distance away from a point, such as the fulcrum of a lever, it causes a twisting action about the point. This twisting action, or torque, is called a moment. The magnitude of the moment depends upon both the size of the force and the perpendicular distance from the force to the point.

FIGURE 3–1

In Figure 3–1, the moment $M = (F)(d)$ is in a clockwise direction. Recall from Section 2–1 the principle of transmissibility states that a force can act anywhere along its line of action. Force F can therefore act at any of the three locations shown in Figure 3–1 and still produce the same moment.

moment = force × the perpendicular distance between the axis and the line of action of the force

The units of moment are pound-feet (lb-ft), pound-inches (lb-in.), or kip-feet (kip-ft). When expressed this way, moment is easily distinguished from work, which is stated in

FIGURE 3–2

units of ft-lb and in.-lb. In the SI metric system, force is expressed in newtons, and distance in meters. The units of moment or torque are newton-meter (N·m).

In two-dimensional drawings—which are frequently used—the axis appears as a point. Moments taken about a point are indicated as being clockwise (↻) or counterclockwise (↺) (Figure 3–2). For the sake of uniformity in calculations, we will assume clockwise to be negative and counterclockwise to be positive. Moments are vector quantities, and their direction must be indicated in one of three ways. For example, the same moment can be expressed as 10 lb-ft (↻), −10 lb-ft, or 10 lb-ft clockwise.

EXAMPLE 3–1

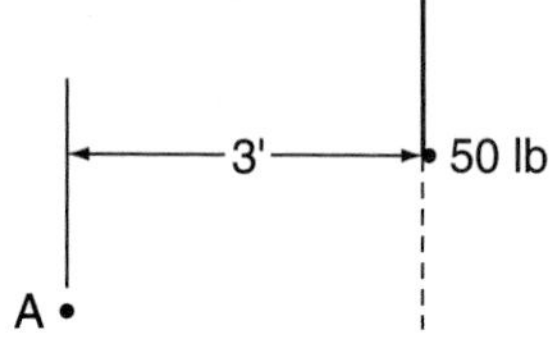

FIGURE 3–3

Calculate the moment about point A in Figure 3–3.

Notice that the perpendicular distance can be measured to the line of action of the force.

$$\begin{aligned} M &= (F)(d) \\ &= -(50\ \text{lb})(3\ \text{ft}) \\ \underline{M} &\underline{= 150\ \text{lb-ft}\ \circlearrowright} \end{aligned}$$

EXAMPLE 3–2

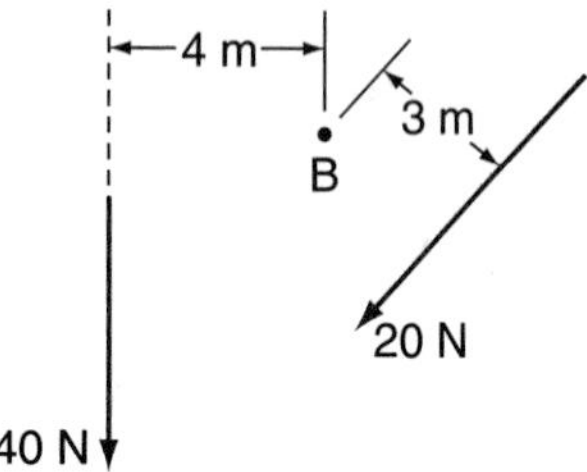

FIGURE 3–4

Calculate the moment about point B due to the forces shown in Figure 3–4.

$$\begin{aligned} M_B &= \text{the sum of moments due to each force} \\ M_B &= -(20\ \text{N})(3\ \text{m}) + (40\ \text{N})(4\ \text{m}) \\ &= -60 + 160 \\ \underline{M_B} &\underline{= 100\ \text{N}\cdot\text{m}\ \circlearrowleft} \end{aligned}$$

It is frequently easier to calculate moments by using a distance and a perpendicular force instead of a force and a perpendicular distance. Example 3–3 is solved by two methods; the second method illustrates the breaking down of a 100-lb force into its components. The total moment is equal to the sum of each component times its perpendicular distance.

EXAMPLE 3–3

Solve for the moment about A due to the 100-lb force (Figure 3–5).

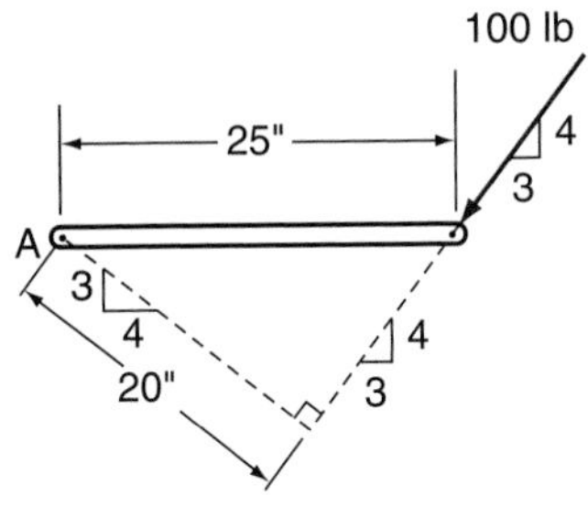

FIGURE 3–5

Method 1

$$\text{moment} = (\text{force})(\text{perpendicular distance})$$
$$= -(100\text{ lb})(20\text{ in.})$$
$$\underline{M_A = 2000\text{ lb-in.} \circlearrowright}$$

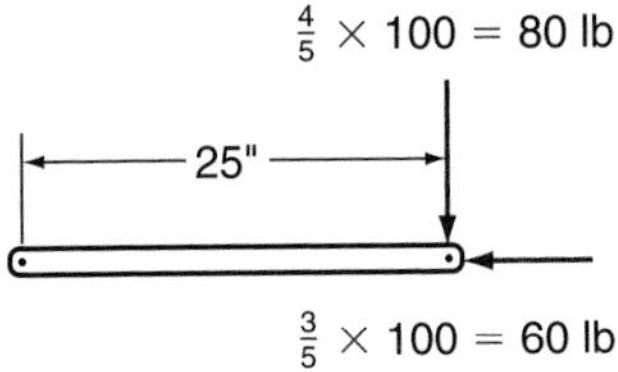

FIGURE 3–6

Method 2

Resolve the 100-lb force into horizontal and vertical components (Figure 3–6).

$$M_A = -(80\text{ lb})(25\text{ in.}) + (60\text{ lb})(0)$$
$$= -2000 + 0$$
$$\underline{M_A = 2000\text{ lb-in.} \circlearrowright}$$

If the perpendicular distance (20 in.) **is not given, it is often easier to calculate the force** (80 lb) **perpendicular to a given distance** (25 in.).

EXAMPLE 3–4

Calculate the moment about the center of the nut (Figure 3–7a).

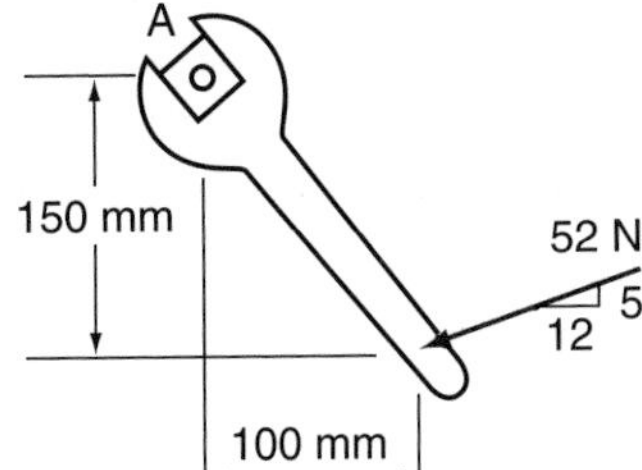

FIGURE 3–7a

Resolving the 52-N force into horizontal and vertical components (Figure 3–7b), we have

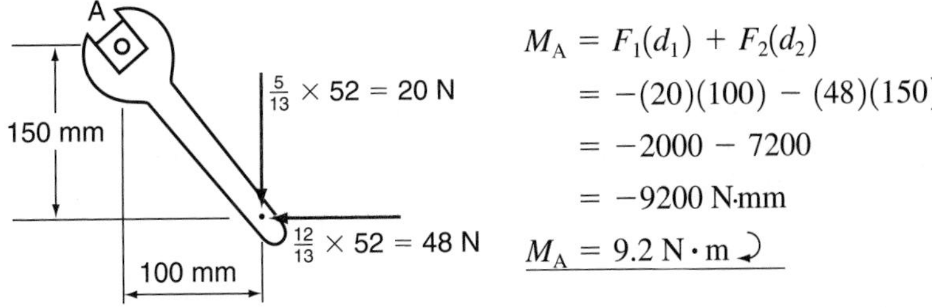

FIGURE 3–7b

$$M_A = F_1(d_1) + F_2(d_2)$$
$$= -(20)(100) - (48)(150)$$
$$= -2000 - 7200$$
$$= -9200\text{ N·mm}$$
$$\underline{M_A = 9.2\text{ N}\cdot\text{m} \circlearrowright}$$

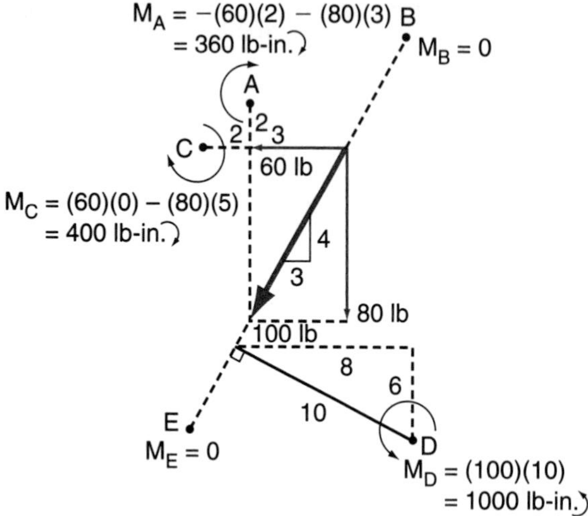

FIGURE 3–8

Figure 3–8 shows how a force of 100 lb can produce moments about various points (A, B, C, D, E).

The 100-lb force can be resolved into components anywhere along the line of action, to produce the simplest moment equation, as done for moments about A and C.

Leaving the 100-lb force intact and calculating a perpendicular distance is easiest for moments about D.

3–2 COUPLES

A couple consists of two equal forces, acting in opposite directions and separated by a perpendicular distance. Let us look further at an application of moments by considering the rotation of a steering wheel (Figure 3–9). It is pulled down on the left side and pushed up on the right side with equal forces of 5 lb. The moment about the center due to each of the two forces is (5 lb)(10 in.), or 50 lb-in. ↻.

$$\text{total moment} = 50 + 50$$
$$= 100 \text{ lb-in.} \circlearrowright$$

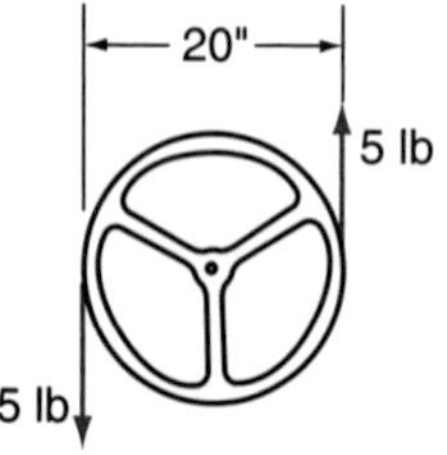

FIGURE 3–9

These forces could have been treated as a *couple,* which consists of two forces that are:

1. Equal
2. Acting in opposite directions
3. Separated by some perpendicular distance d

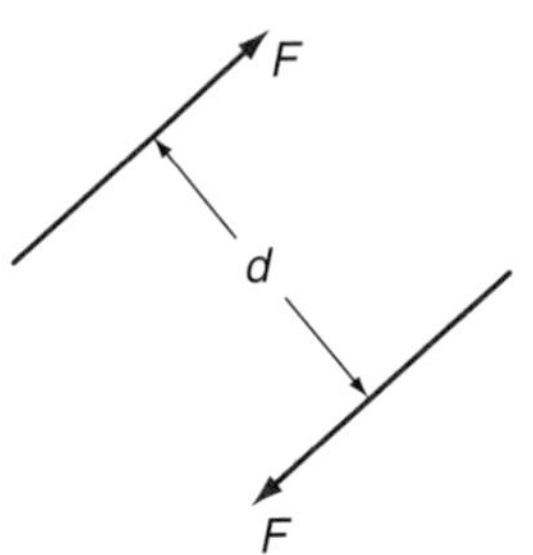

FIGURE 3–10

These three requirements of a couple are shown in Figure 3–10. Referring back to the steering wheel in Figure 3–9, we have

$$\text{couple moment} = (F)(d) \quad \circlearrowleft$$
$$= (5\text{ lb})(20\text{ in.})$$
$$\underline{M = 100\text{ lb-in.}}$$

This is the same answer that we obtained when we multiplied the individual forces by their distances from the pivot.

To illustrate how easily you can be mistaken and assume that two forces are a couple when they really are *not,* not one of the systems of Figure 3–11 is a couple for the following reasons:

1. The forces are not equal (Figure 3–11a).
2. The forces are not in opposite directions (Figure 3–11b).
3. The forces are neither parallel nor in opposite directions (Figure 3–11c).
4. The forces are not separated by a distance d (Figure 3–11d).

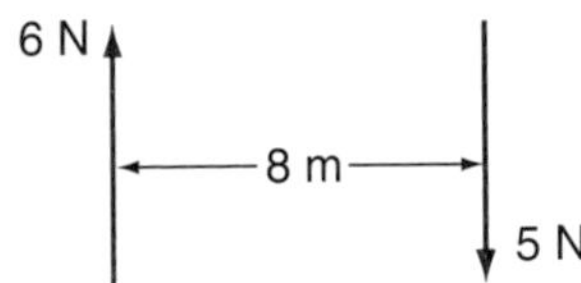

FIGURE 3–11a

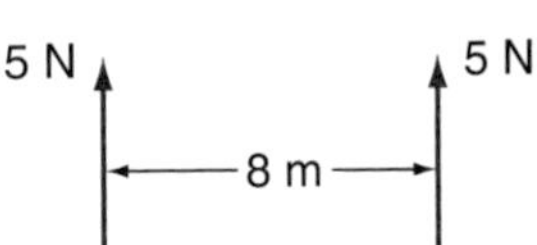

FIGURE 3–11b

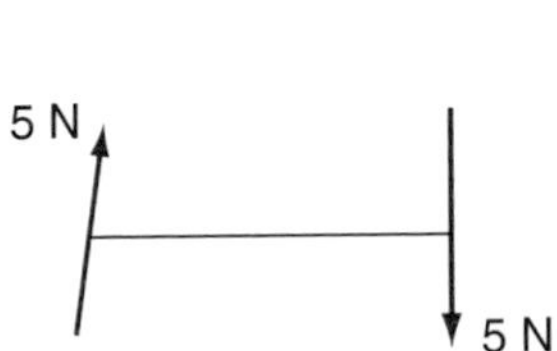

FIGURE 3–11c

FIGURE 3–11d

You will have noticed that when we calculated moments (Section 3–1), we specified the point or moment center about which the moments were calculated. It does not matter where the moment center is located when we deal with couples. **A couple has the same moment about all points on a body.** To illustrate this, consider a lever (Figure 3–12a) loaded as shown with its moment center located at A. Neglecting the weight of the lever and considering the forces as a couple, we can calculate the moment of the couple:

$$(10\text{ N})(6\text{ m}) = 60\text{ N}\cdot\text{m} ⤺$$

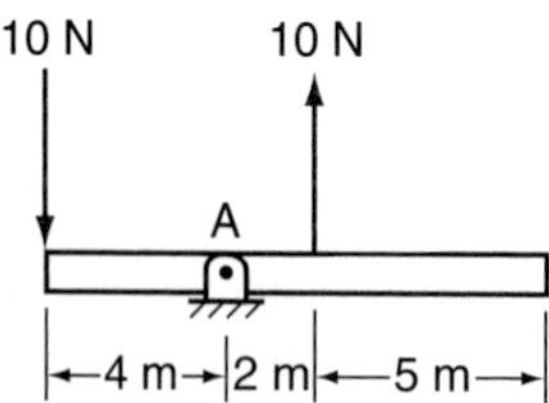

FIGURE 3–12a

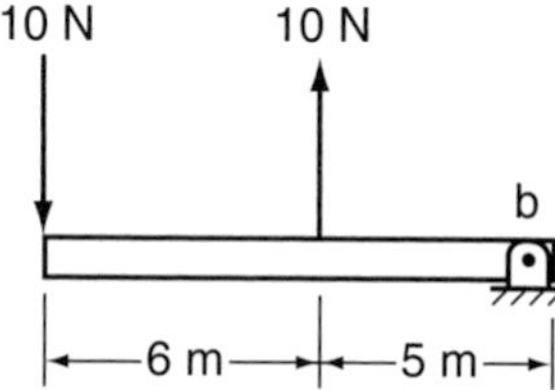

FIGURE 3–12b

Checking this value by considering the moment of each force about A, we have

$$\begin{aligned} M_A &= (10\text{ N})(4\text{ m}) + (10\text{ N})(2\text{ m}) \\ &= 40 + 20 \\ \underline{M_A} &\underline{= 60\text{ N}\cdot\text{m} ⤺} \end{aligned}$$

Taking the same lever with the same forces and moving the moment center to B (Figure 3–12b), we now have

$$\begin{aligned} M_B &= (10\text{ N})(11\text{ m}) - (10\text{ N})(5\text{ m}) \\ &= 110 - 50 \\ \underline{M_B} &\underline{= 60\text{ N}\cdot\text{m} ⤺} \end{aligned}$$

Thus, regardless of the moment center location or point about which we take moments, we still have a couple of 60 N·m in a counterclockwise direction.

Suppose now that we have a fixed moment center and move the couple. In each part of Figure 3–13, a couple of 60 N·m is located on the lever. The moment calculation for each is as follows:

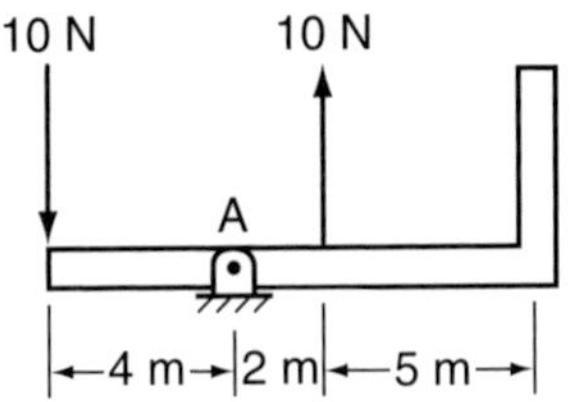

FIGURE 3–13a

1. Figure 3–13a:

$$M_A = (10\text{ N})(4\text{ m}) + (10\text{ N})(2\text{ m})$$
$$= 40 + 20$$
$$\underline{M_A = 60\text{ N}\cdot\text{m}} \circlearrowleft$$

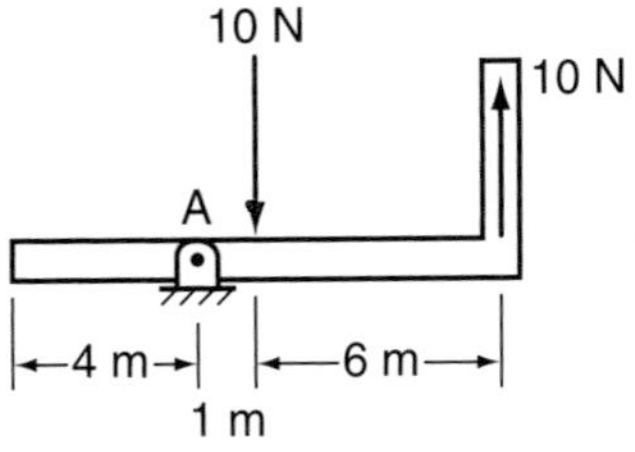

FIGURE 3–13b

2. Figure 3–13b:

$$M_A = -(10\text{ N})(1\text{ m}) + (10\text{ N})(7\text{ m})$$
$$= -10 + 70$$
$$\underline{M_A = 60\text{ N}\cdot\text{m}} \circlearrowleft$$

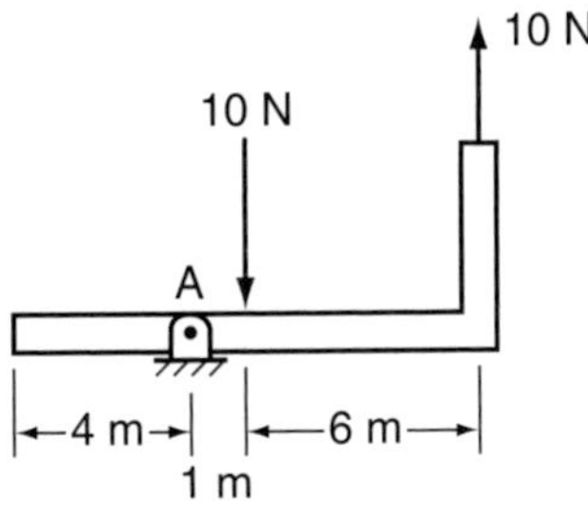

FIGURE 3–13c

3. The moment is the same as in step 2 since a force acts anywhere along its line of action (Figure 3–13c).

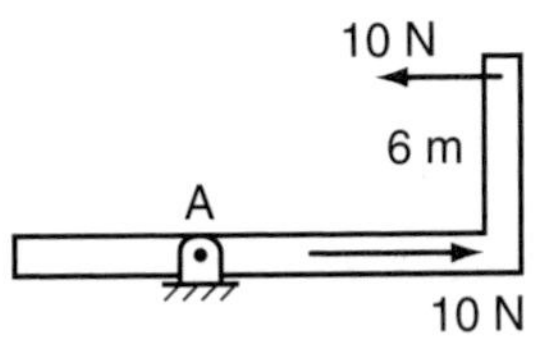

FIGURE 3–13d

4. Note that in Figure 3–13d the couple has been rotated 90° and the bottom 10-N force is acting at a distance of zero from point A since its line of action passes through A.

$$M_A = (10\text{ N})(0) + (10\text{ N})(6\text{ m})$$
$$\underline{M_A = 60\text{ N}\cdot\text{m}} \circlearrowleft$$

We can also have *equivalent couples.* In this case, couples with a combination of forces and perpendicular distances that multiply to equal 60 N·m ↺ are equivalent couples. Figure 3–13e, Figure 3–13f, and Figure 3–13g illustrate three equivalent couples with moments of 60 N·m ↺.

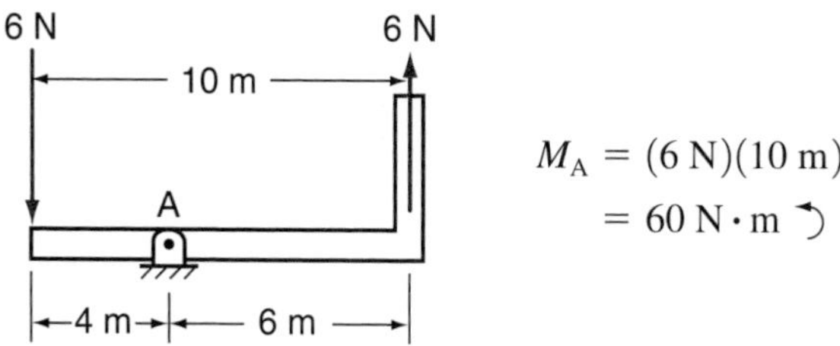

$$M_A = (6\ \text{N})(10\ \text{m})$$
$$= 60\ \text{N} \cdot \text{m} \circlearrowleft$$

FIGURE 3–13e

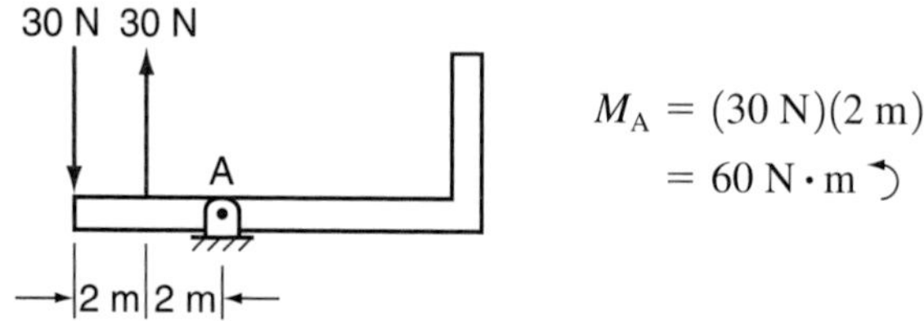

$$M_A = (30\ \text{N})(2\ \text{m})$$
$$= 60\ \text{N} \cdot \text{m} \circlearrowleft$$

FIGURE 3–13f

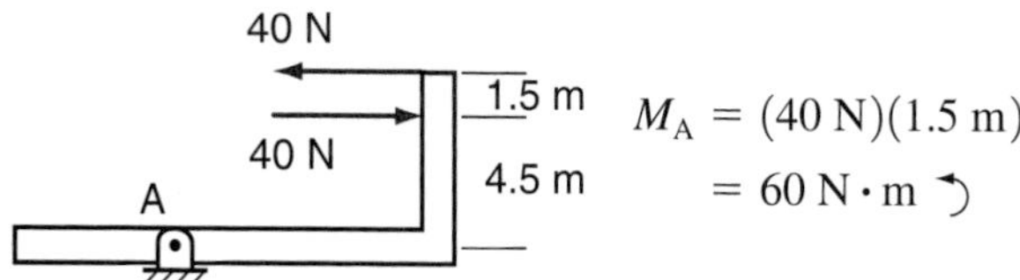

$$M_A = (40\ \text{N})(1.5\ \text{m})$$
$$= 60\ \text{N} \cdot \text{m} \circlearrowleft$$

FIGURE 3–13g

The highlights to remember about couples are:

1. A couple is always characterized by two *equal* and *opposite* forces separated by a *perpendicular distance.*
2. The couple moment is unaffected by the pivot location.
3. A couple can be shifted and still have the same moment about a given point.
4. Equivalent couples will have different forces and perpendicular distances.

EXAMPLE 3–5

A *truss* is a structure composed of bars or members joined to form one or more connected triangles. Calculate the moment about point A of the truss shown in Figure 3–14.

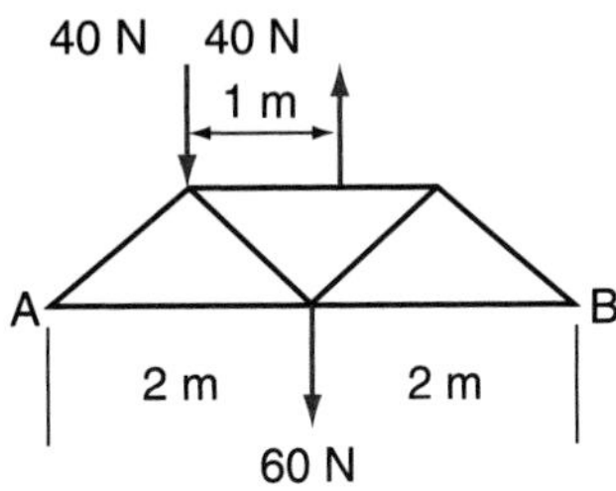

FIGURE 3–14

$$\text{moment about A} = \text{moment of a couple} + \text{moment of 60-N force}$$

$$M_A = (40\text{ N})(1\text{ m}) - (60\text{ N})(2\text{ m})$$
$$= 40 - 120$$
$$= -80$$
$$\underline{M_A = 80\text{ N}\cdot\text{m} \curvearrowright}$$

EXAMPLE 3–6

Calculate the moment about point A due to the forces on the truss shown in Figure 3–15a.

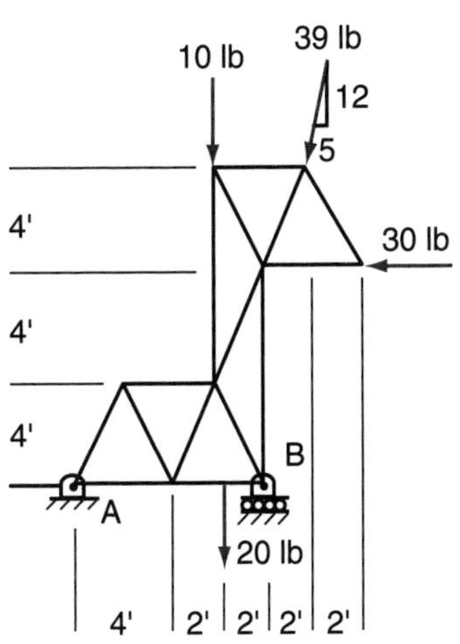

FIGURE 3–15a

Because we are concerned only with the complete truss and the external forces shown, we can break the 39-lb force into horizontal and vertical components; Figure 3–15b will be the result.

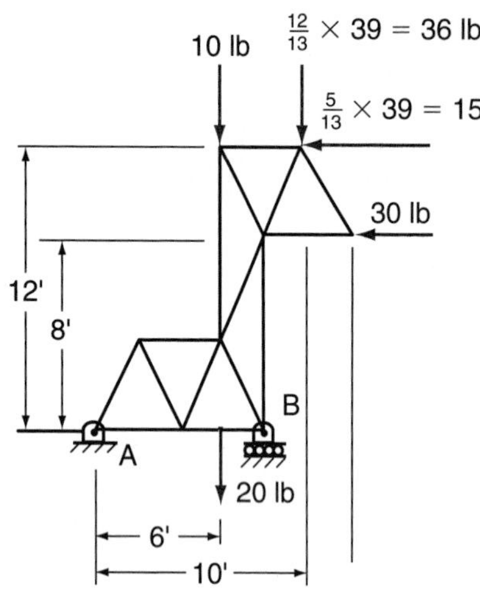

FIGURE 3–15b

Taking moments about point A, we get

$$M_A = -(10\text{ lb})(6\text{ ft}) - (20\text{ lb})(6\text{ ft}) - (36\text{ lb})(10\text{ ft}) + (15\text{ lb})(12\text{ ft}) + (30\text{ lb})(8\text{ ft})$$
$$= -60 - 120 - 360 + 180 + 240$$
$$= -540 + 420$$
$$\underline{M_A = 120\text{ lb-ft} \circlearrowright}$$

HINTS FOR PROBLEM SOLVING

1. Moment = (force) (*perpendicular* distance), where perpendicular distance is the *shortest possible distance* between the line of action of the force and the point about which you are taking moments.
2. The final answer for a moment must have the direction shown.
3. When writing moment equations, clockwise is negative and counterclockwise is positive.
4. There should be no force alone in a moment equation. Check that each force has been multiplied by its perpendicular distance.
5. A couple has the same moment about *any* point.

PROBLEMS

APPLIED PROBLEMS FOR SECTION 3–1

3–1. Calculate the moment about point A (Figure P3–1). (Each section represents 1 ft^2.)

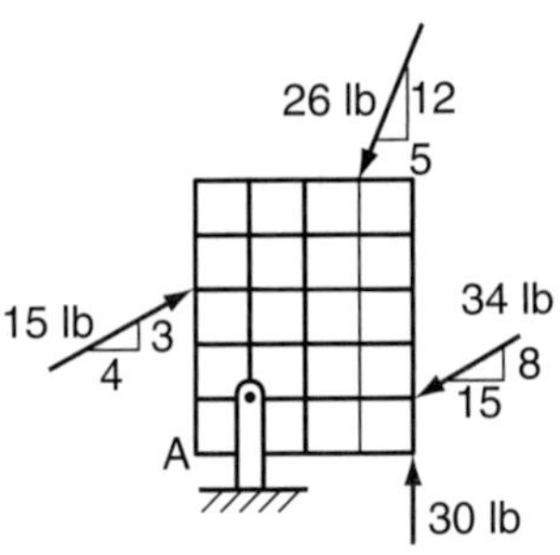

FIGURE P3–1

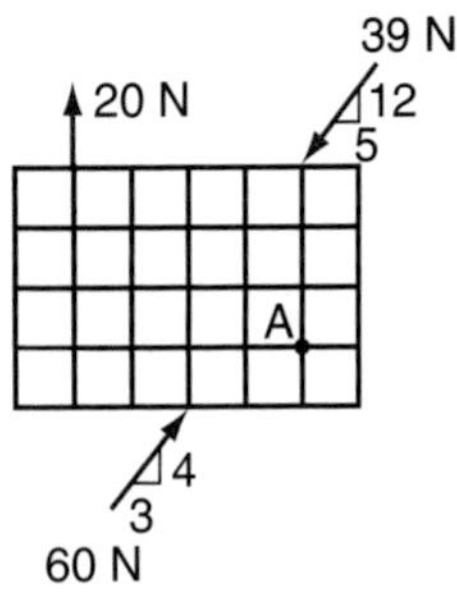

FIGURE P3–2

3–2. Determine the moment about point A for the force system shown in Figure P3–2 if each square represents 1 m.

3–3. When viewed from above, a worker slides a cabinet by applying the forces shown in Figure P3–3. Determine the moment about corner A.

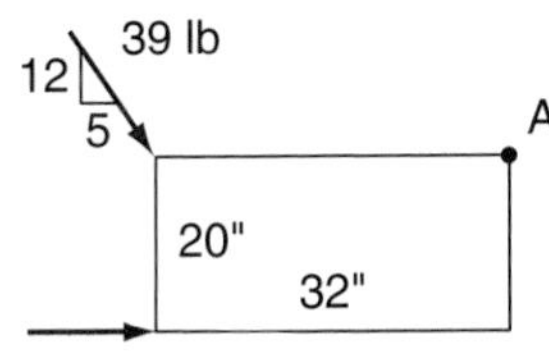

FIGURE P3–3

3–4. In an effort to tip a crate about the edge shown as point A (Figure P3–4), three forces are applied. Determine the moment about A due to these forces.

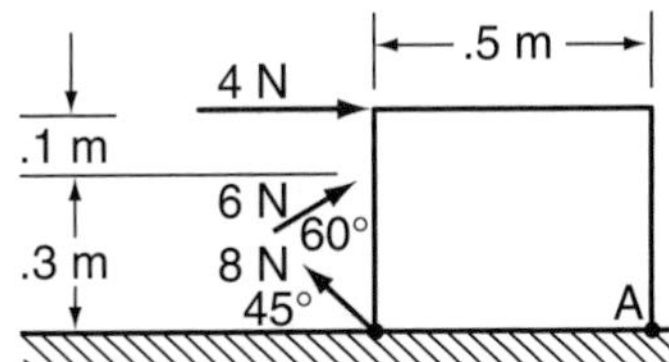

FIGURE P3–4

3–5. Calculate the moment about point A in Figure P3–5.

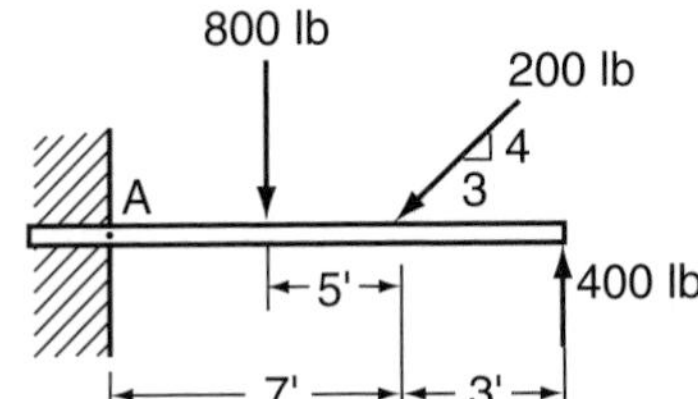

FIGURE P3–5

3–6. The bending moment in the beam at point A is equal to the moment of the 800-N force about point A. Calculate the moment about point A (Figure P3–6). What force P could be applied at the same location, perpendicular to the beam, and still produce the same moment?

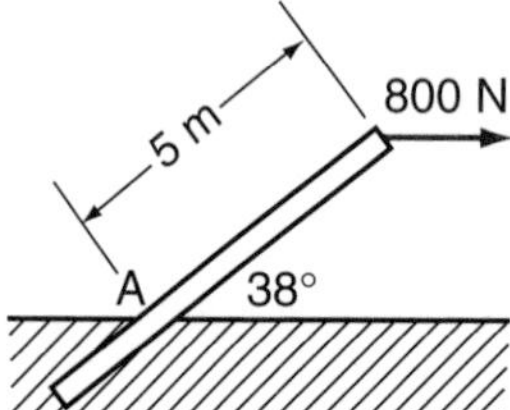

FIGURE P3–6

3–7. The box of a dump truck pivots at A and has a force of 850 lb applied by a hydraulic cylinder as shown in Figure P3–7. Determine the moment about A due to this force.

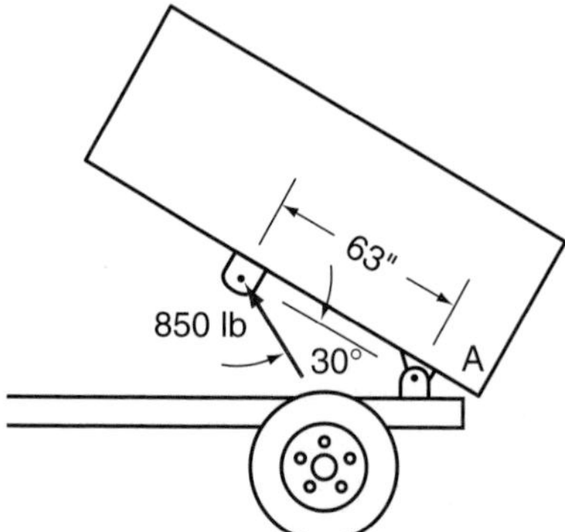

FIGURE P3–7

3–8. Determine the moment of the 1800-N force about (a) point C and (b) point B of the engine hoist shown in Figure P3–8.

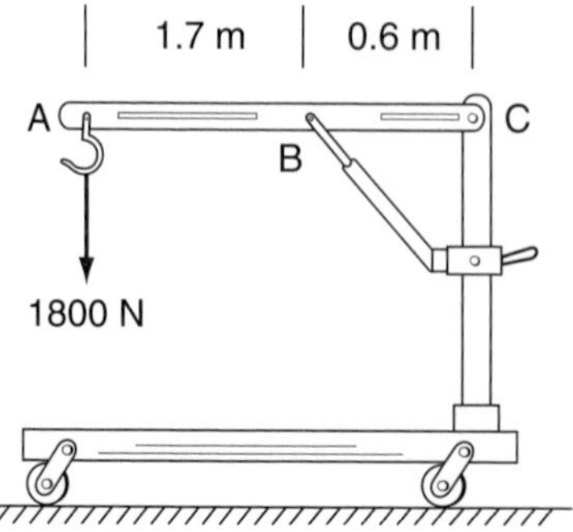

FIGURE P3–8

3–9. Determine the moment about point A of the lever shown in Figure P3–9.

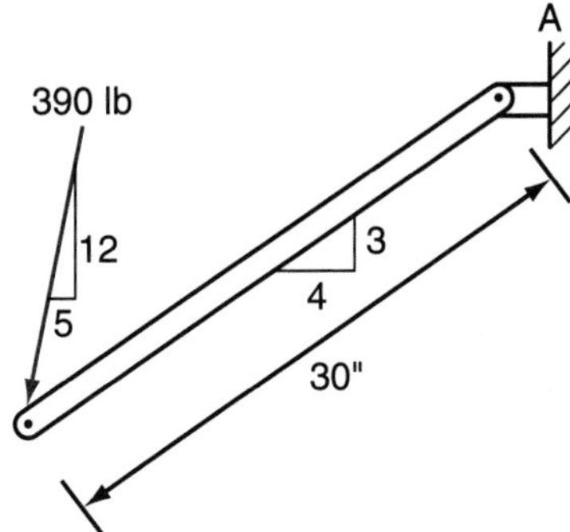

FIGURE P3–9

3–10. Calculate the moment or torque tightening the pipe in Figure P3–10.

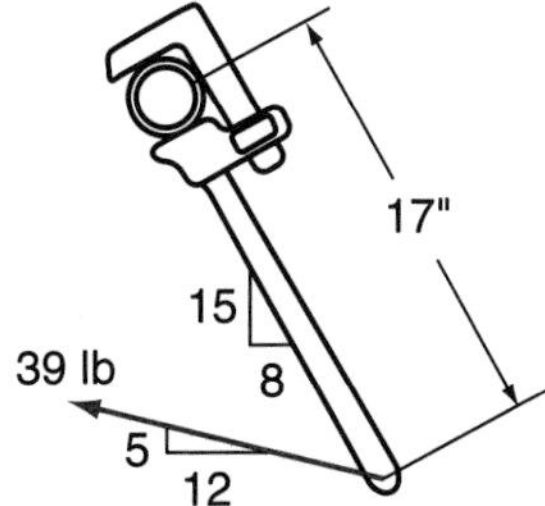

FIGURE P3–10

3–11. Force *P* (Figure P3–11) causes a moment of 500 N·m about point A. Determine force *P*.

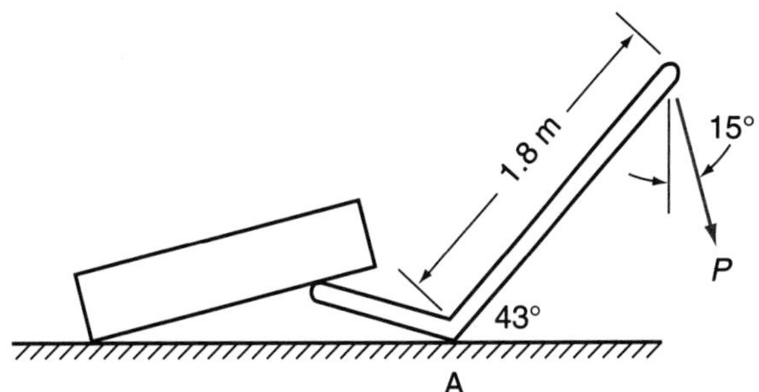

FIGURE P3–11

3–12. Calculate the moment about point A in Figure P3–12.

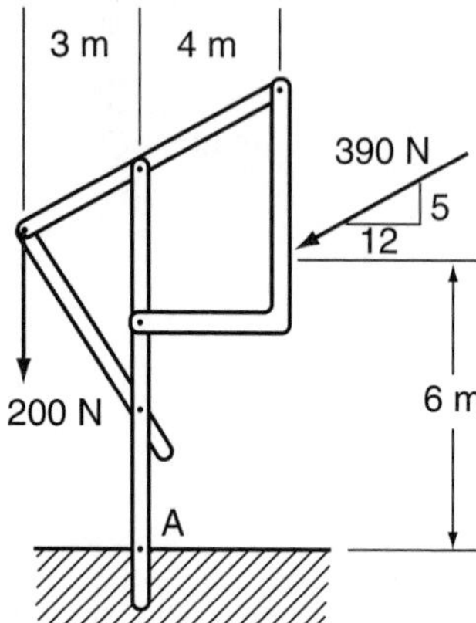

FIGURE P3–12

3–13. Calculate the moment about point A due to the 150-N force shown in Figure P3–13.

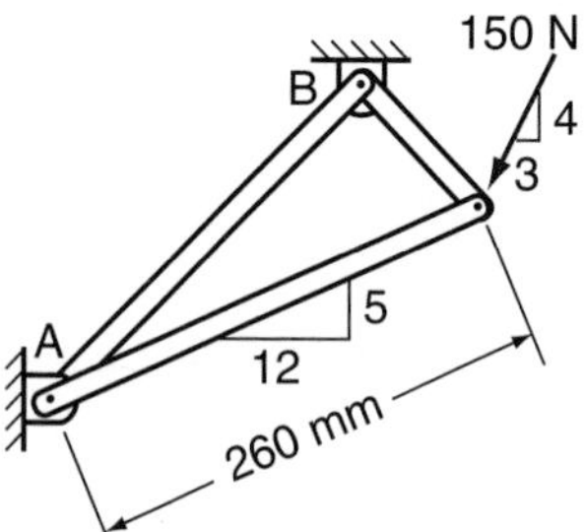

FIGURE P3–13

3–14. Determine the moment about point A for the forces shown in Figure P3–14. (*Hint:* The missing dimension is not required if you show all possible couples.)

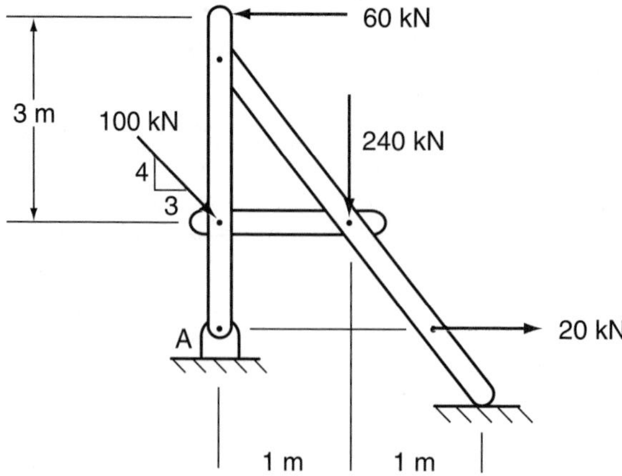

FIGURE P3–14

3–15. A large rock has forces acting on it as shown in Figure P3–15. Determine the moment about point A.

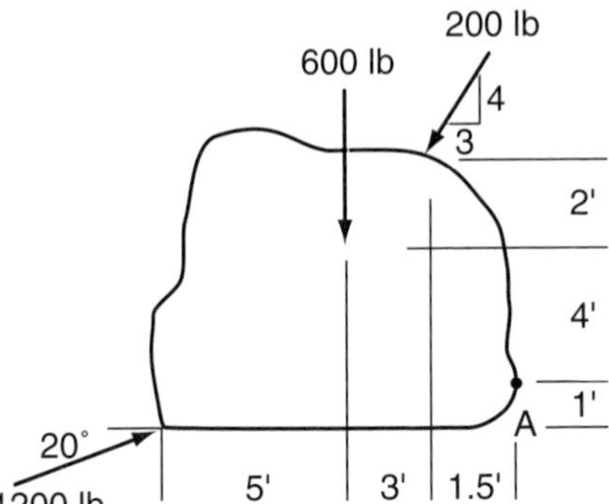

FIGURE P3–15

3–16. The spring at A in Figure P3–16 opposes the moment about B due to the 500-N force. Determine the moment about B due to the 500-N force. If the spring has an equal and opposite moment about B, determine the spring tension.

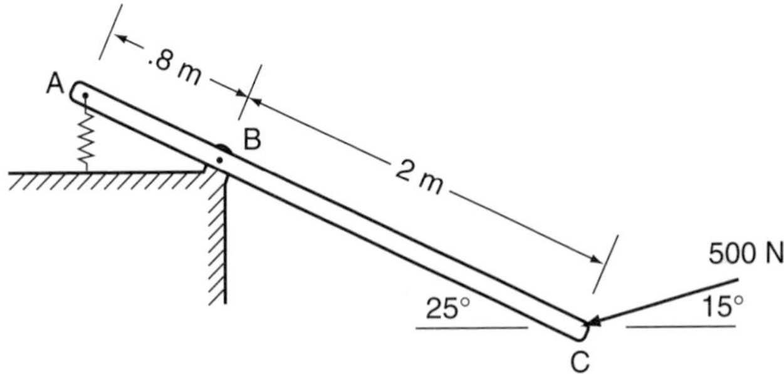

FIGURE P3–16

3–17. For the forklift truck shown in Figure P3–17 determine (a) the moment about A due to the two weights shown and (b) the moment about B due to the two weights shown.

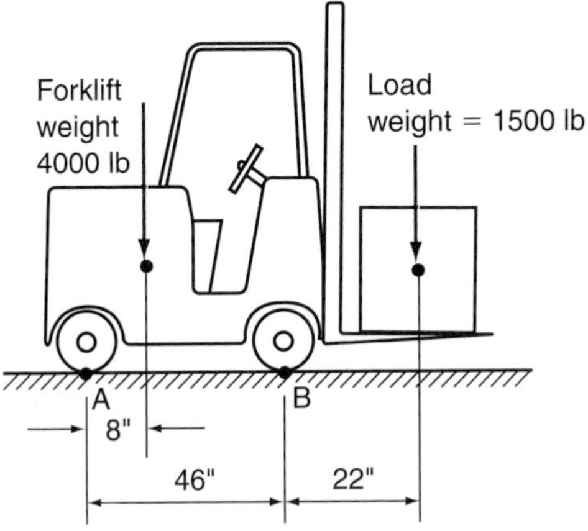

FIGURE P3–17

3–18. The load carried by the hitch of the trailer shown in Figure P3–18 is 45 N. Determine the load on the jack when it is cranked down to unhitch the trailer.

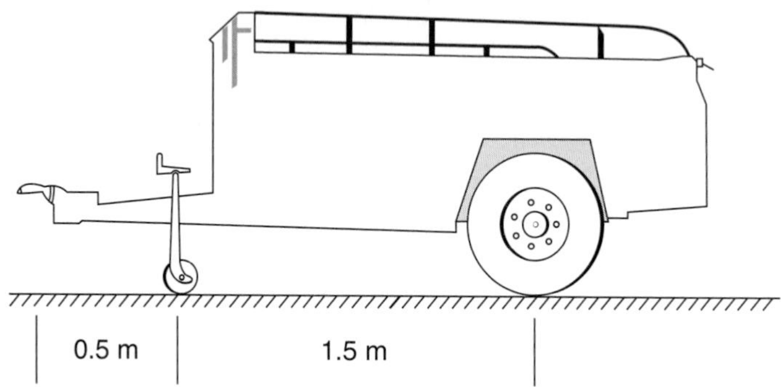

FIGURE P3–18

3–19. Calculate the moments about pins B and C due to the 30-lb force shown in Figure P3–19. Which arm has the highest moment tending to bend it?

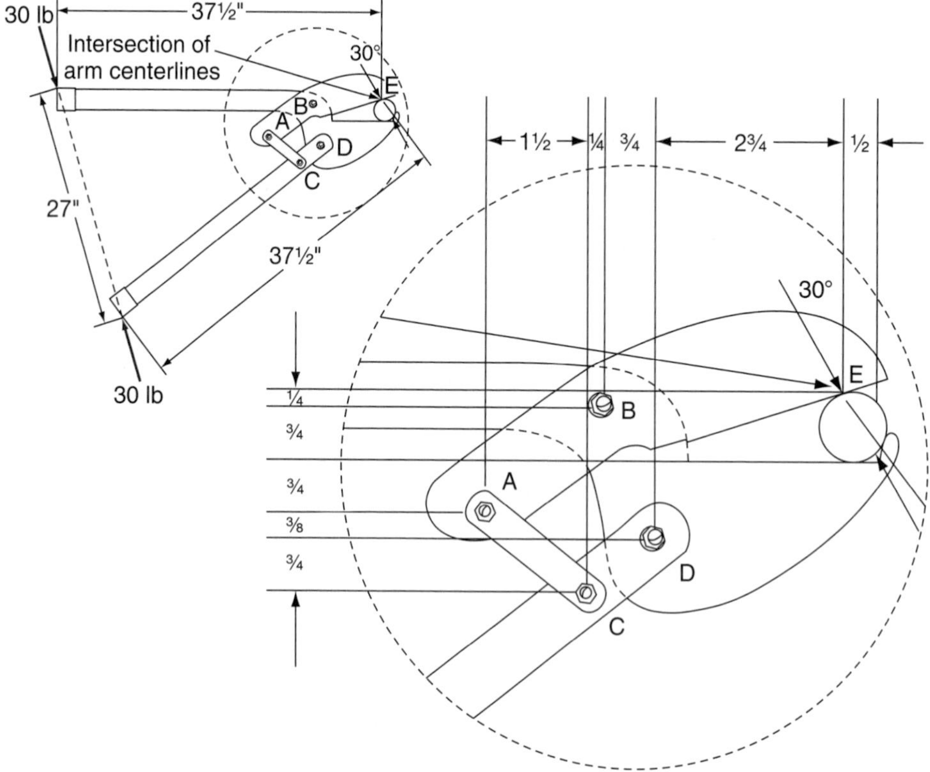

FIGURE P3–19

APPLIED PROBLEMS FOR SECTION 3–2

3–20. For the system shown in Figure P3–20 determine the moment about (a) point A and (b) point B.

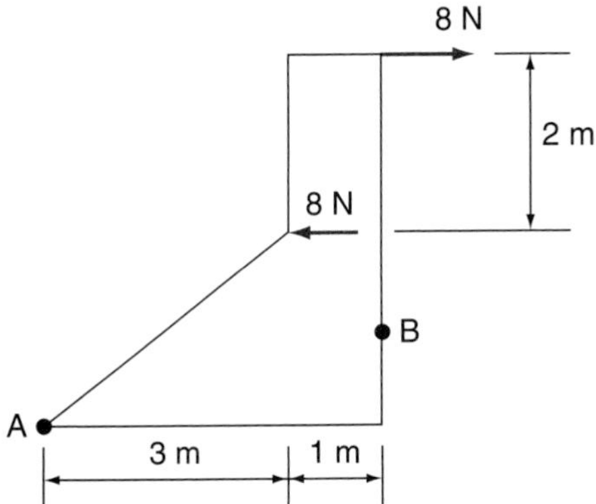

FIGURE P3–20

3–21. Determine the moment about point A for the forces shown in Figure P3–21.

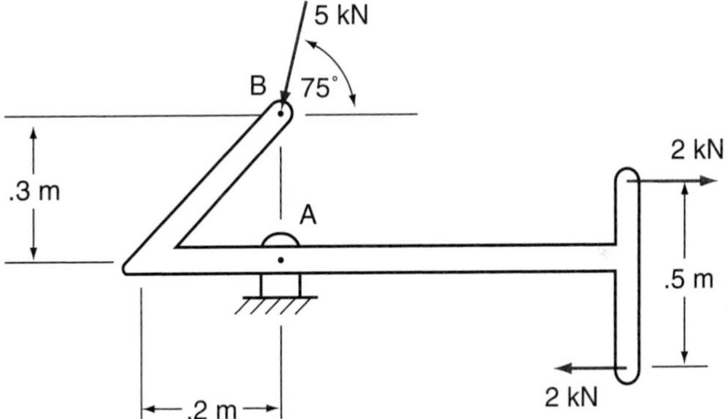

FIGURE P3–21

3–22. Determine the moment about point A for the system shown in Figure P3–22.

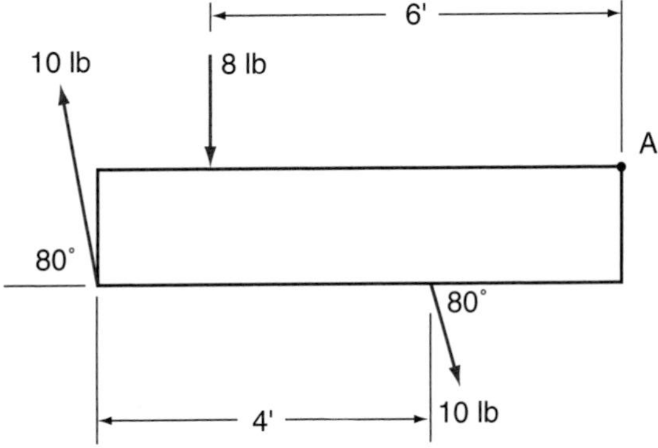

FIGURE P3–22

3–23. Determine the moment about point A due to the forces shown in Figure P3–23.

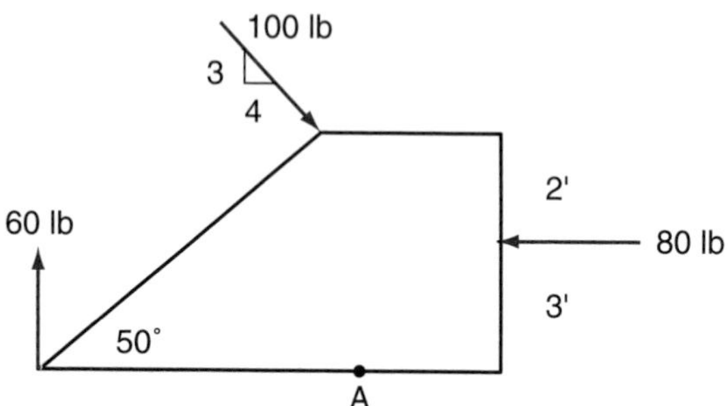

FIGURE P3–23

3–24. Find the moment about point A for the lever shown in (a) Figure P3–24a and (b) Figure P3–24b.

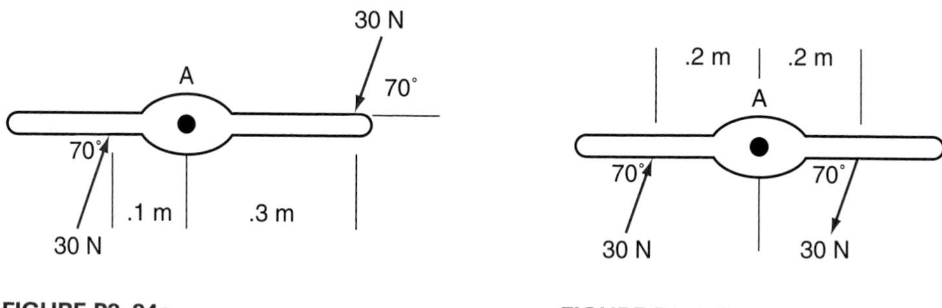

FIGURE P3–24a

FIGURE P3–24b

3–25. The bucket shown in Figure P3–25 pivots at A and has the forces shown acting on it. Determine the moment about A.

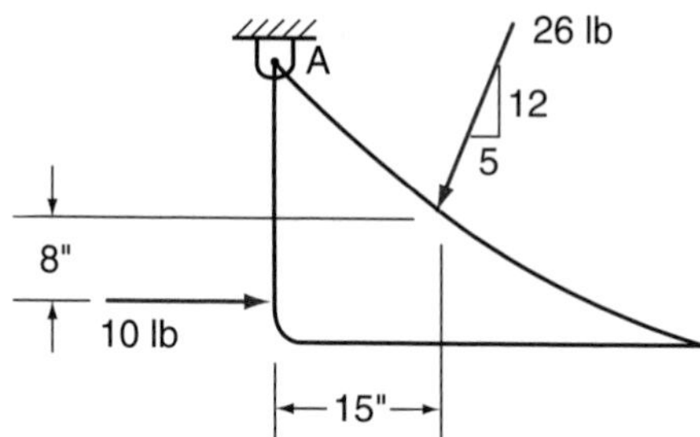

FIGURE P3–25

3–26. A portion of a beam has forces applied as shown in Figure P3–26. Forces A and B form a couple that opposes the couple made up by the vertical forces. Determine forces A and B.

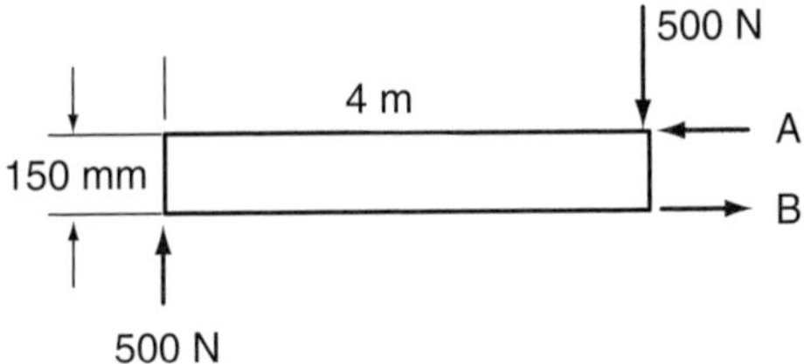

FIGURE P3–26

3–27—3–31. The forces shown form a couple. Replace the shown couple with an equivalent couple acting at points A and B.

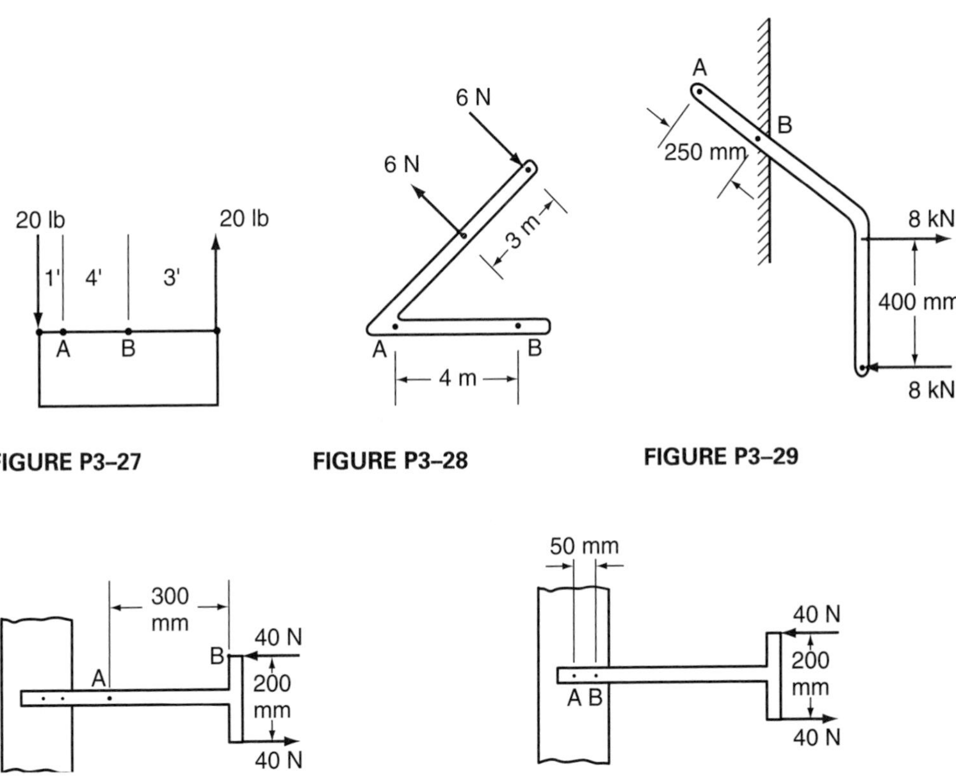

FIGURE P3–27

FIGURE P3–28

FIGURE P3–29

FIGURE P3–30

FIGURE P3–31

REVIEW PROBLEMS

R3–1. Each square in Figure RP3–1 represents 1 m. Determine the moment about point A.

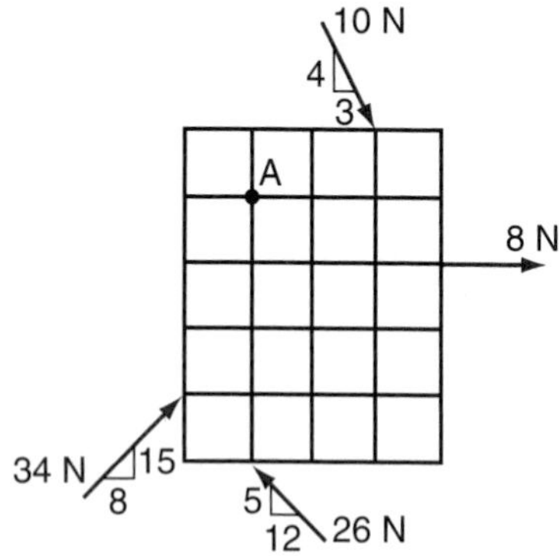

FIGURE RP3–1

R3–2. Boom AC is extended as shown in Figure RP3–2. Determine (a) the moment about A due to the 1200-lb force and (b) the cylinder force pushing in the same direction as the cylinder at B, if its moment is to be equal to the moment of the 1200-lb force about A.

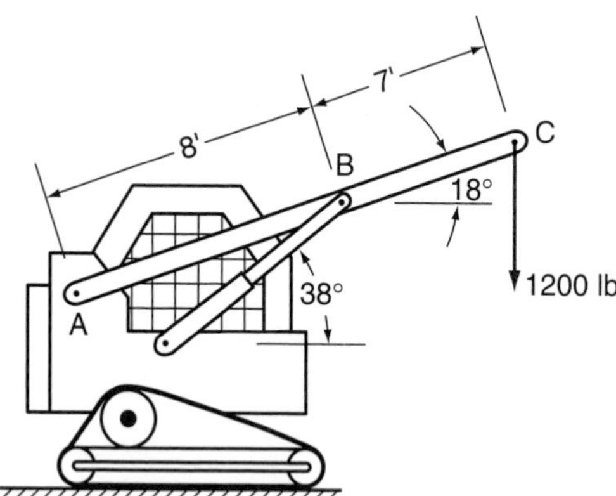

FIGURE RP3–2

R3–3. A flywheel must be held from rotating about shaft A while a nut on it is tightened as shown (Figures RP3–3a and RP3–3b). Which position should the wrench be in to have minimum moment about A? Calculate the moment in each case.

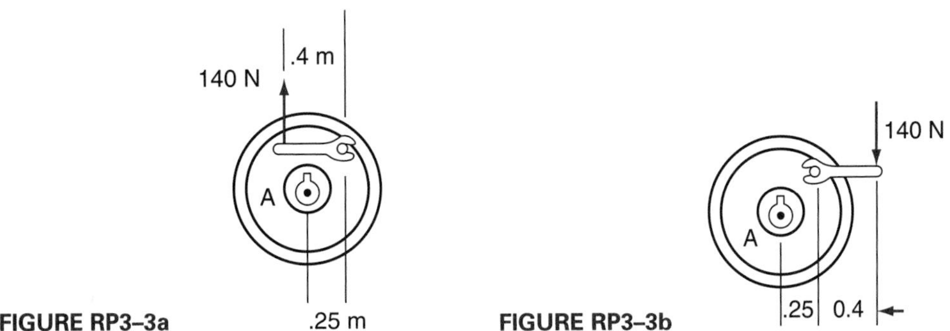

FIGURE RP3–3a

FIGURE RP3–3b

R3–4. Find the moments about points A and B due to the forces shown in Figure RP3–4.

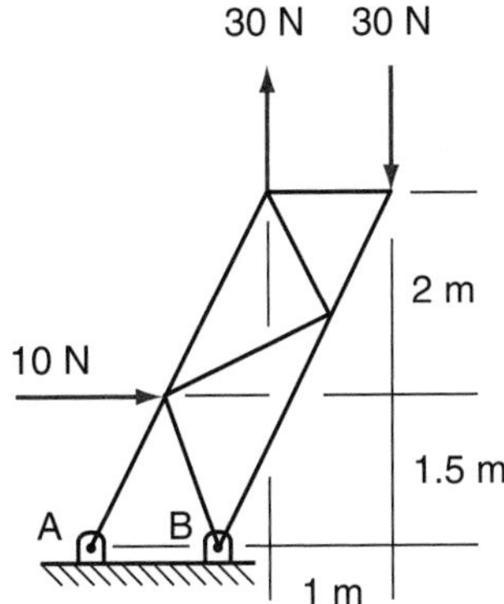

FIGURE RP3–4

R3–5. The forces shown in Figure RP3–5 form a couple. Replace the couple shown with an equivalent couple acting at points A and B.

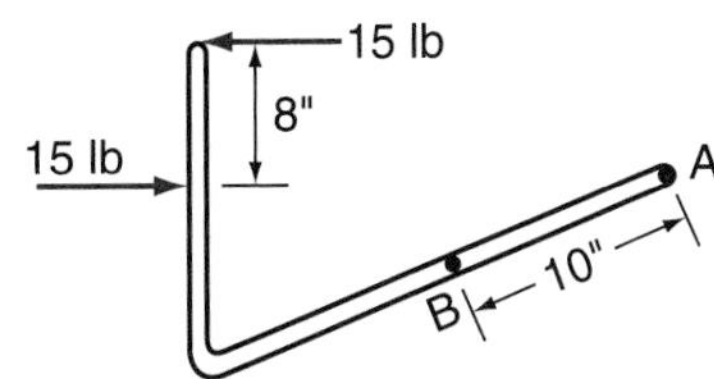

FIGURE RP3–5

CHAPTER 4

Equilibrium

OBJECTIVES

Upon completion of this chapter the student will be able to:

1. Draw complete free-body diagrams of whole or part mechanisms.
2. Apply the three equations of equilibrium, $\Sigma F_x = 0$, $\Sigma F_y = 0$, and $\Sigma M = 0$, to free-body diagrams of the following coplanar systems:
   (a) Concurrent
   (b) Parallel (including uniform and nonuniform beam loading)
   (c) Nonconcurrent

4–1 FREE-BODY DIAGRAMS

Free-body diagrams are diagrams of objects in static equilibrium.

"Static" means the object is not moving but is "at rest." "Equilibrium" means the forces acting on the object are in equilibrium or balanced against each other so that the object does not move.

A *diagram* that shows an object or *body* with all supports removed and replaced by forces in balance appears to be *floating freely* in space. It is called a *free-body diagram* (FBD). Drawing it is a necessary first step in calculating the forces acting on the object.

The crate in Figure 4–1 is drawn as a free-body diagram in Figure 4–2 and is acted upon by the following forces:

1. The rope tension is replaced by a force of 60 lb pulling upward on the crate.
2. The pull of gravity is 100 lb on the crate.
3. The floor had been partially supporting or pushing up on the crate. The support of the floor on the crate is shown as *N*.

A free-body diagram of an object must show all the forces acting *on* the object.

A final check is made to ensure that no forces have been omitted and that our free-body diagram is complete. (Each removed support has been replaced by a force vector[s].)

If, when you are drawing a free-body diagram, you are confused as to what forces are acting on the object and in which direction each is acting, visualize yourself in the place of

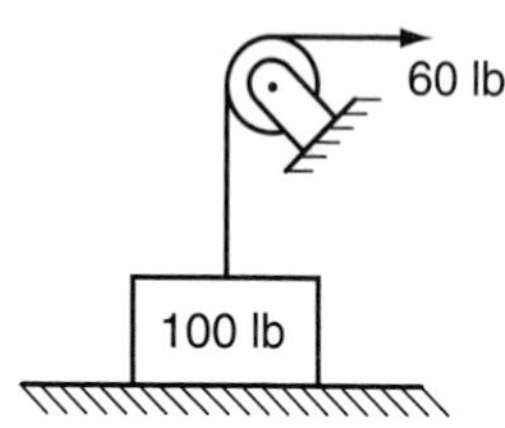

FIGURE 4–1

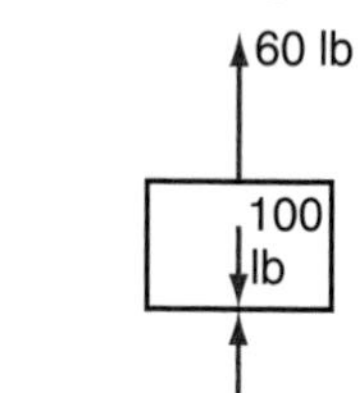

FIGURE 4–2

the object. Ask yourself the following questions: What forces would be acting on me? Where would they be acting? Would they be pushing or pulling?

A free-body diagram of a member is a picture showing how the rest of the world is acting *on* the member, not what the member is doing *to* anything else.

The forces are of three categories:

1. Applied forces
2. Nonapplied forces such as weight
3. Forces replacing a support or sectioned member

Forces can also be called "acting" or "reacting." Acting forces would be applied forces and weight. Due to these acting forces, the forces at a support would react accordingly and are therefore known as *reactions*.

The FBD Riddle

WHO uses them?
– every organized problem solver

WHAT are they?
– a picture upon which calculations are based

WHEN are they used?
– always

WHERE are they used?
– as step one of all problems with forces

WHY are they used?
– to show clear documentation of a problem solution

HOW accurate must they be?
– 100% (except for assumed directions)

A final check must be made to ensure that the correct number of forces are shown. Show too many and the problem may appear unsolvable. Show too few and incorrect calculations will occur. An incorrect direction can often be corrected during the calculations, with some degree of inconvenience.

The importance of a complete and correct free-body diagram cannot be overemphasized. Checking it is time well spent—check that each removed support has been replaced by a force vector(s).

4–2 FREE-BODY DIAGRAM CONVENTIONS

When drawing free-body diagrams and replacing supports with equivalent supporting forces, you must employ some definite assumptions or conventions. Figure 4–3 shows the forces drawn to replace various supports or connections on the main members.

1. *Roller:* The roller cannot exert a horizontal force; therefore, only a force perpendicular to the surface is present (Figure 4–3a).

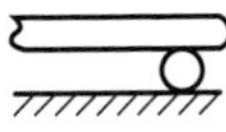
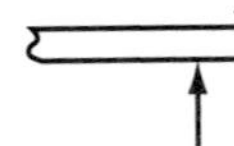

FIGURE 4–3a

2. *Roller:* The only force present is that perpendicular to the roller surface (Figure 4–3b).

FIGURE 4–3b

3. *Smooth surface:* Zero friction is assumed; therefore, only one force, that perpendicular to the surface, is present (Figure 4–3c).

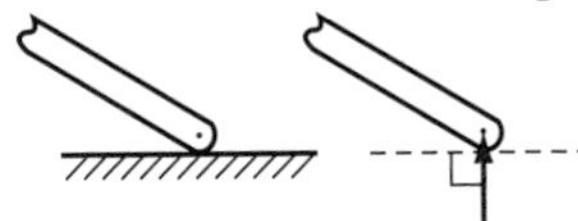

FIGURE 4–3c

4. *Slot:* The same principle applies as for a smooth surface: There is only one force present, that perpendicular to the slot (Figure 4–3d).

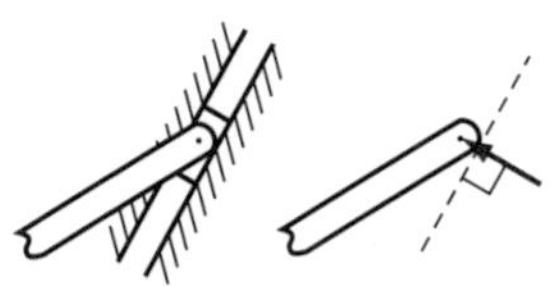

FIGURE 4–3d

5. *Pinned:* Both horizontal and vertical components must be assumed at a pinned connection unless it is on a roller or smooth surface (Figure 4–3e).

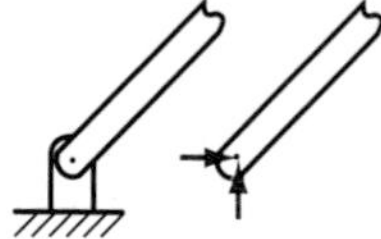

FIGURE 4–3e

6. The orientation of the support is immaterial; a horizontal and vertical force must still be assumed (Figure 4–3f).

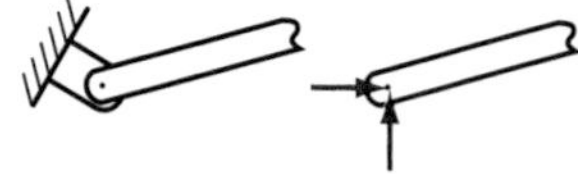

FIGURE 4–3f

7. *Cable:* There is always a single force pulling in the direction of the cable (Figure 4–3g).

FIGURE 4–3g

8. *Fixed support* (Figure 4–3h): The beam is embedded in the wall or support and therefore has three possible reactions: a moment, vertical force, and horizontal force.

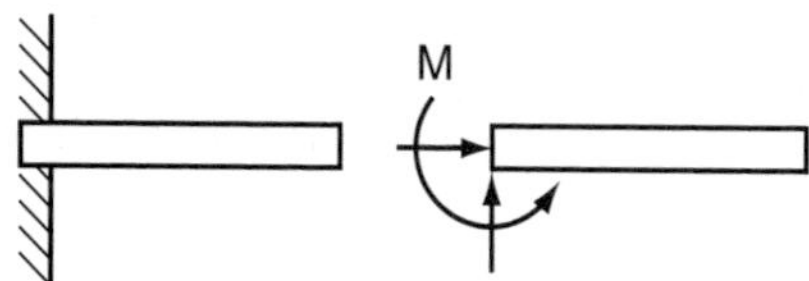

FIGURE 4–3h

As indicated earlier (Section 2–2), all objects or members will be assumed to be weightless unless the weight is specifically stated. When considering weight later on, one can easily include it in the calculation by adding another force passing through the center of gravity of the object or member. Another assumption is that of zero friction on smooth surfaces. This simplifies calculations for our initial problems. When friction is to be included later, the coefficient of friction or some other clear indication that it must be considered will be given.

Because complete free-body diagrams are so crucial to correct problem solution, the following examples show free-body diagrams of various members or structures. Be certain to dimension your free-body diagrams. This will help avoid errors in later work, such as moment equations and slopes of forces.

If you experience some difficulty in deciding whether a vertical force is acting up or down, do not worry. If you assume an incorrect direction, the calculated answer will be negative; if this answer must be used in further calculations, substitute it into equations as a negative value and do not change any vector directions until the calculations are complete. The calculated answer can then be restated as positive and in the opposite direction of the original assumed vector. All calculations are unchanged. Your other choice is to immediately change the direction of the force vector and change the sign in your most recent calculation (usually the most comfortable option).

Note that the horizontal bar in Figure 4–4 appears to have pinned connections at both A and B. Pin B has both horizontal and vertical forces but pin A has only a vertical force due the rollers above it.

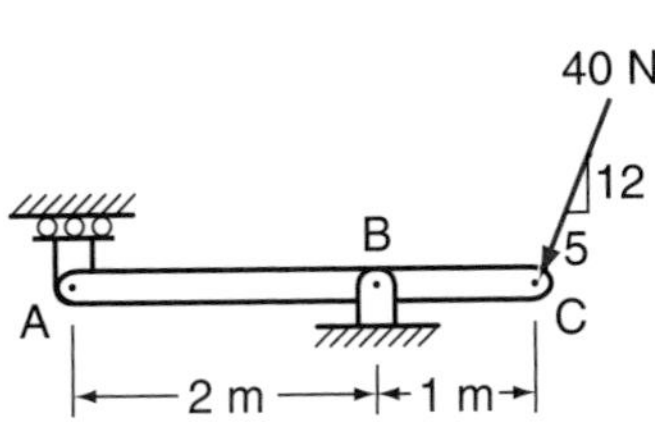

FIGURE 4–4a

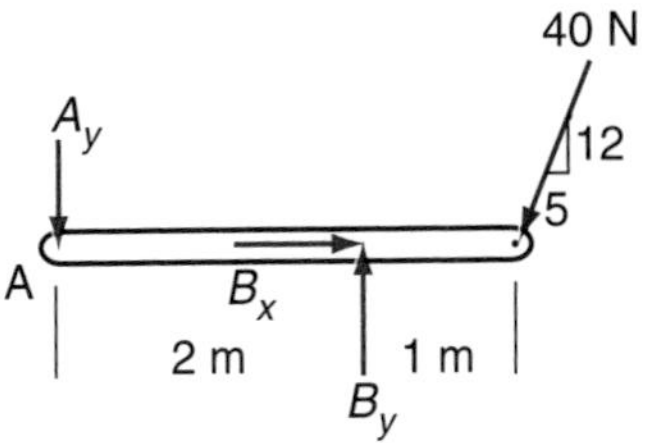

FIGURE 4–4b

In Figure 4–5, a pin fastened to BE is free to slide in the slot of AC. Note the difference in the direction of B in the free-body diagrams of BE and AC. The slot of AC acts downward to the right *on BE,* and BE acts upward to the left *on AC.* This is the case of equal and opposite forces, with their direction depending on the object for which a free-body diagram is drawn. Note that force *B* has a slope of 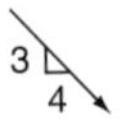 3 4 and is perpendicular to member AC, which has a slope of 4 3.

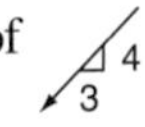

This illustrates that whenever you need the slope of a line that is perpendicular to another line you simply reverse the slope numbers.

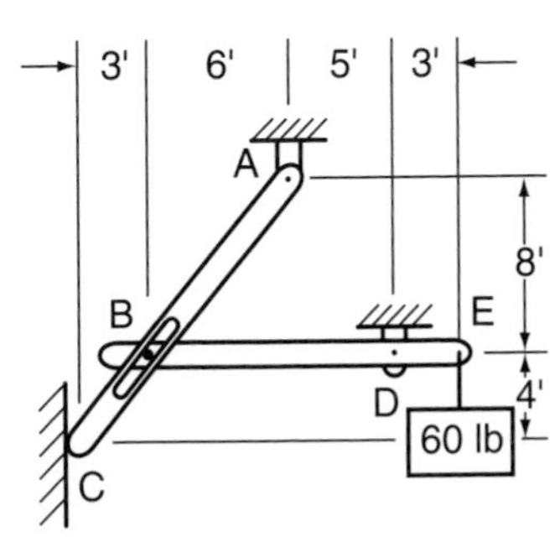

FIGURE 4–5a

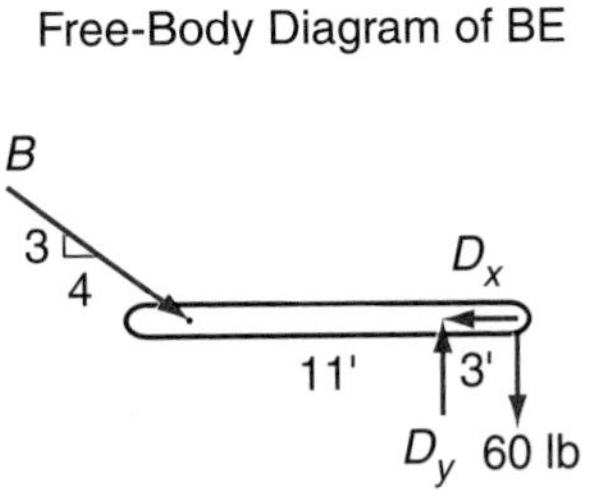

Free-Body Diagram of BE

B
3
4
D_x
11'
3'
D_y 60 lb

FIGURE 4–5b

Free-Body Diagram of AC

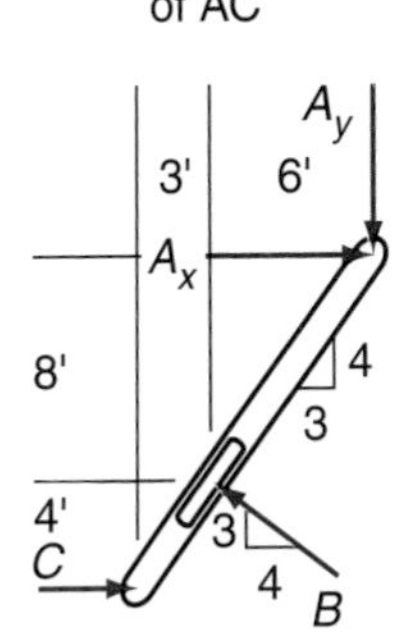

FIGURE 4–5c

Because cables can only be in tension (not compression), the free-body diagram of ring B (Figure 4–6) shows three forces pulling on B.

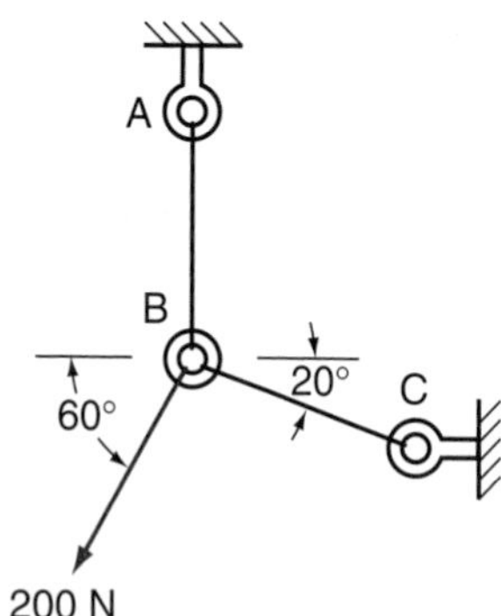

FIGURE 4–6a

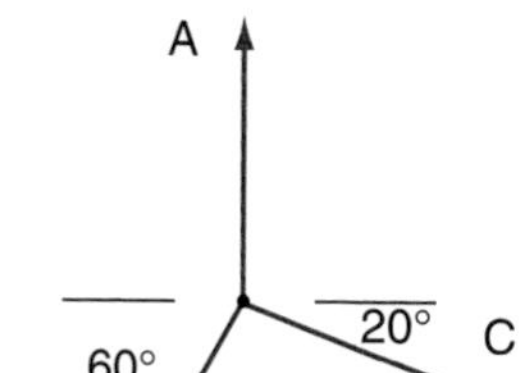

FIGURE 4–6b

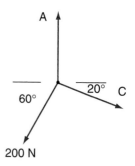

Note that while there may be loads in each member of the pin-connected structure shown in Figure 4–7, they are internal loads and are thus of no concern in a free-body diagram of the complete frame; only external forces must be accounted for. Because there are rollers at B, there cannot be any horizontal force at B.

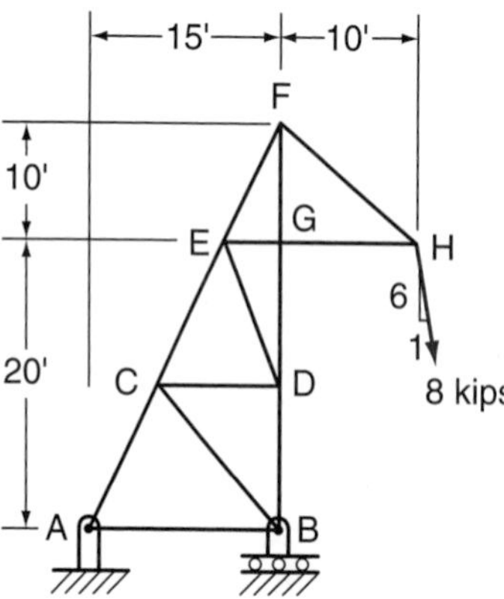

FIGURE 4–7a

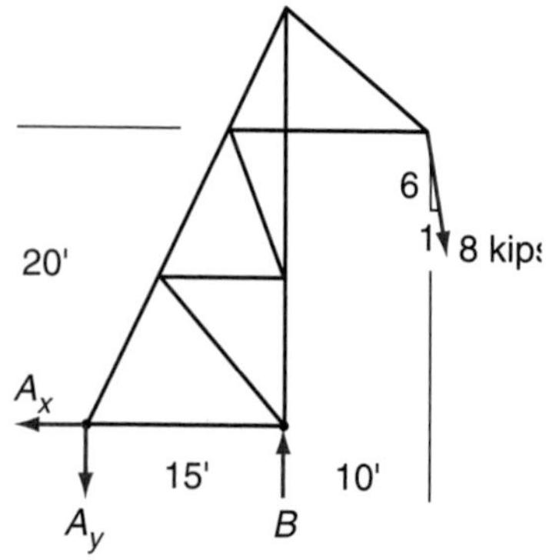

FIGURE 4–7b

EXAMPLE 4–1

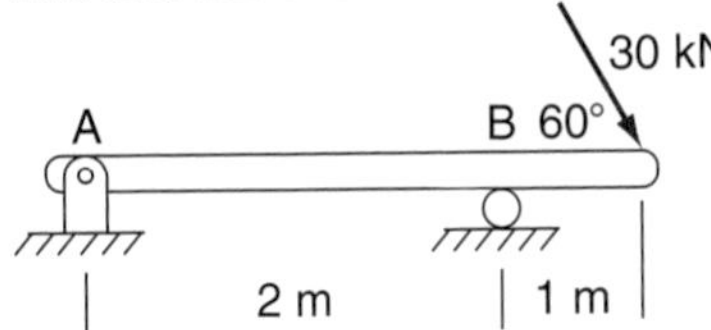

FIGURE 4–8

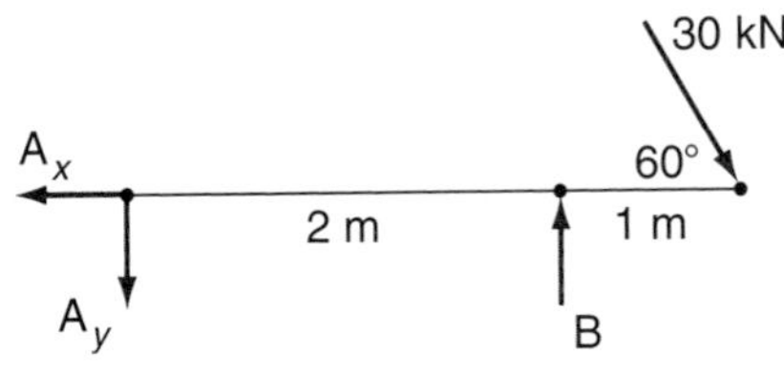

FIGURE 4–9

Draw a free-body diagram of the beam shown in Figure 4–8 showing the forces at points A and B.

- Title the diagram (Figure 4–9).
- Identify the three locations for the forces.
- Show the 30-kN force at 60°. Always begin by showing the known force, in this case, 30 kN, as this force will often help you assume the directions of unknown forces.
- Show the force at B pushing upward perpendicular to AB. We know that AB is pushing down on the roller but remember that this a free-body diagram of AB and we need to know what the roller is doing to AB, that is, pushing upward. Since it is free to roll, it cannot exert a horizontal force. If you mistakenly assume a horizontal force, you will have too many unknowns.
- Show A_x acting to the left since the 30-kN force acts partially to the right.
- Show A_y acting down to provide rotational balance about B.
- Show all dimensions.

EXAMPLE 4–2

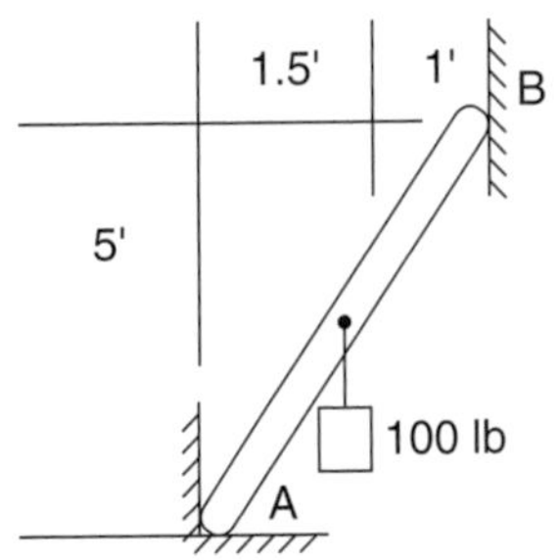

FIGURE 4–10

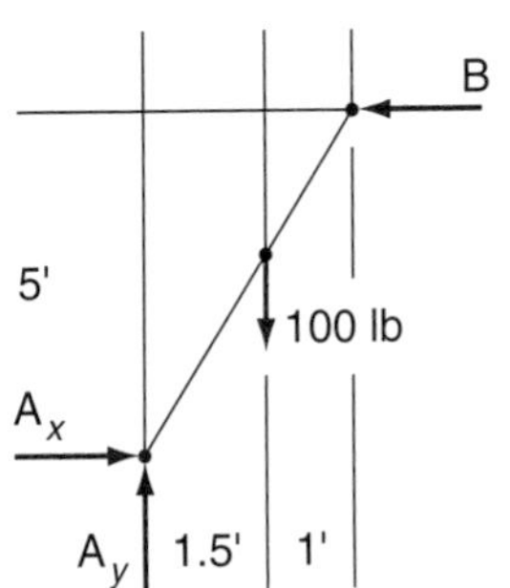

FIGURE 4–11

Draw a free-body diagram of member AB (Figure 4–10).

- Title the diagram (Figure 4–11).
- Identify the three locations for the forces.
- Show the 100-1b force acting downward.
- The vertical wall at B can only act horizontally to the left on AB (our assumption of zero friction means that there cannot be a vertical force at B).
- At point A, the vertical wall provides a horizontal force A_x and the horizontal floor provides a vertical force A_y (both must be pushing on AB).
- Complete the free-body diagram by showing all dimensions.

EXAMPLE 4–3

Draw complete free-body diagrams of members AC and DE (Figure 4–12).

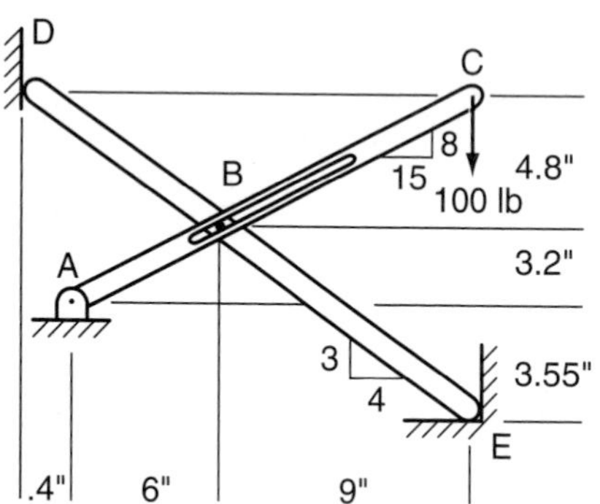

FIGURE 4–12

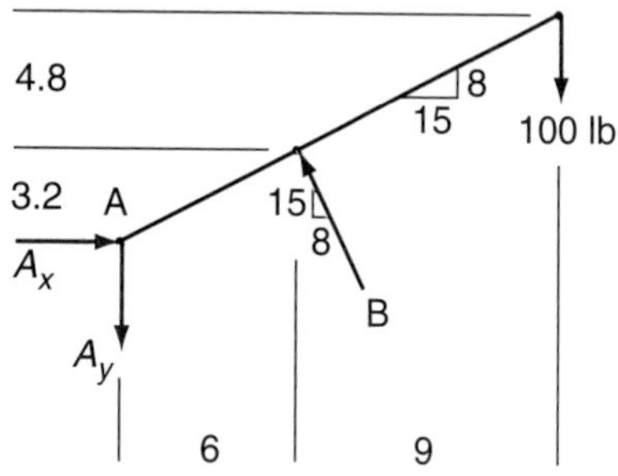

FIGURE 4–13

A good rule of thumb is to begin with a member where a force is known—in this case, member AC (Figure 4–13). The suggested steps are:

- Title the diagram.
- Identify the three locations for the forces.
- Show the 100-lb force.
- Show force *B* pushing at right angles to AC since there is no friction in the slot. If the pin at B in DE were to shear off, member AC would drop; therefore, member DE must be pushing upward on AC.
- Show A_x acting to the right, since *B* has a component to the left.
- Show A_y acting downward to counteract the clockwise moment of the 100-lb force about point B.
- Show all necessary dimensions.

The suggested steps for member DE (Figure 4–14) are:

- Title the diagram.
- Identify the three locations for the forces.
- Show the force at *D*, since the vertical surface can only push to the right on DE.
- Show the force at *B* (an internal force) as equal and opposite to what we just assumed in the free-body diagram of AC.
- Show E_x pushing to the left (at right angles to the smooth vertical surface).
- Show E_y pushing upward (at right angles to the smooth horizontal surface).
- Show the necessary dimensions.

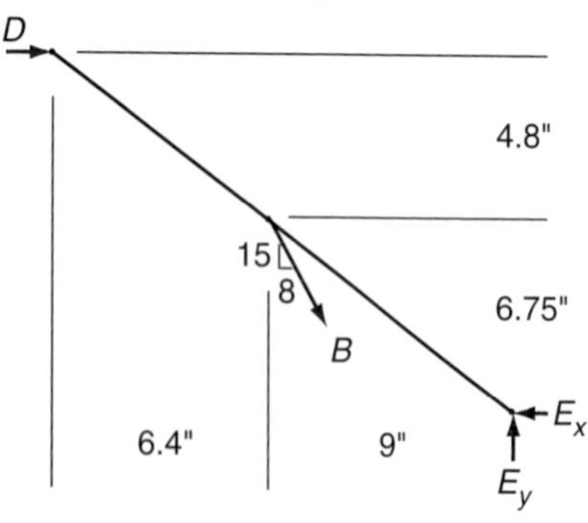

FIGURE 4–14

At this point you should get some experience by drawing free-body diagrams for many of the problems P4–1 to P4–24.

If you are still having difficulty, glance ahead to any example in Chapters 4 and 5, referring only to the free-body diagrams and not the calculations.

As a final step of full understanding, slowly go through the following detailed discussion that explains why and where each force is shown.

Drawing a free-body diagram requires a proper sequence of steps and a good understanding of the concept of a free-body diagram. It is a *diagram* of a *body* appearing to float *freely* in space because visible supports have been replaced by forces. Both the number and types of forces must be correct since this diagram is the basis of all future calculations.

To fully describe the sequence of steps, including suggested easier first steps, consider now the structure of Figure 4–15.

There are three possible free-body diagrams to draw: the entire frame, member BG, and member AD.

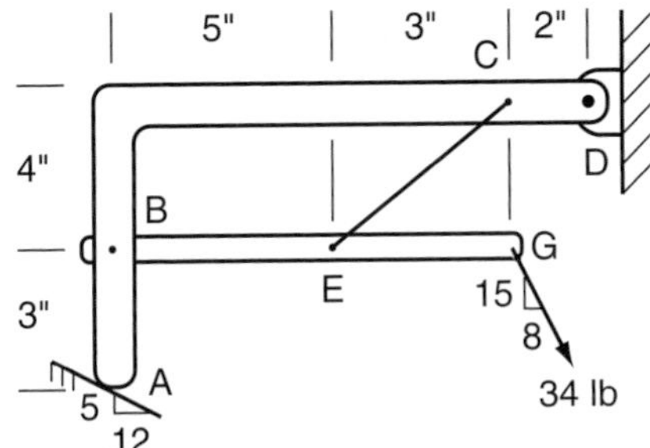

FIGURE 4–15

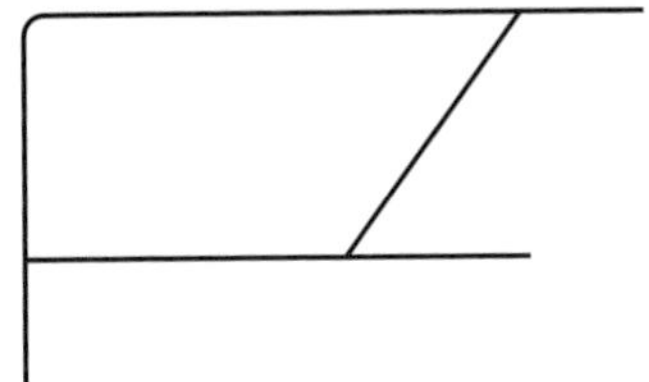

FIGURE 4–16a

We will now draw them in that sequence, itemizing each step.

Step 1. Title the diagram "FBD of Frame" as in Figure 4–16a (FBD is an abbreviation of free-body diagram).
Step 2. Redraw the entire frame without showing the supports. A "stick diagram" is easiest and quite suitable.

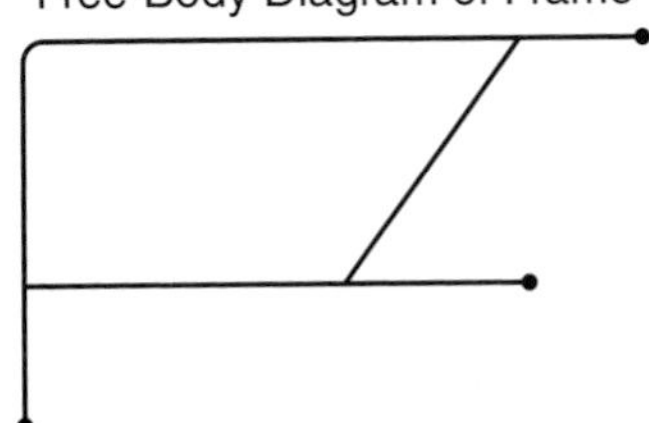

FIGURE 4–16b

Step 3. Identify all locations where there were external forces or supports at points A, D, and G (Figure 4–16b).

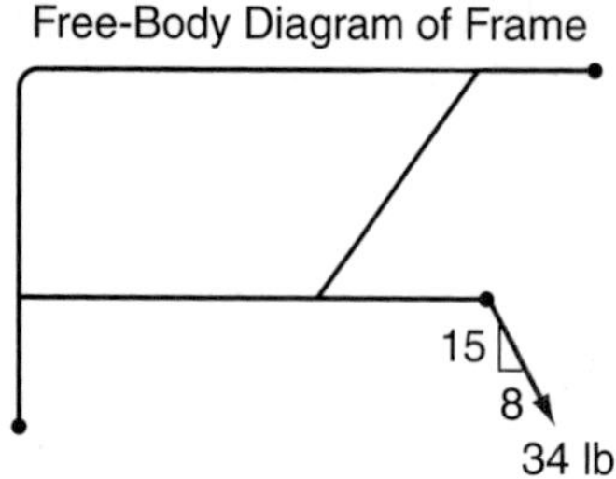

FIGURE 4–16c

Step 4. Draw a given force such as the 34-lb force at G. Be sure to indicate slope, direction, and magnitude (Figure 4–16c).

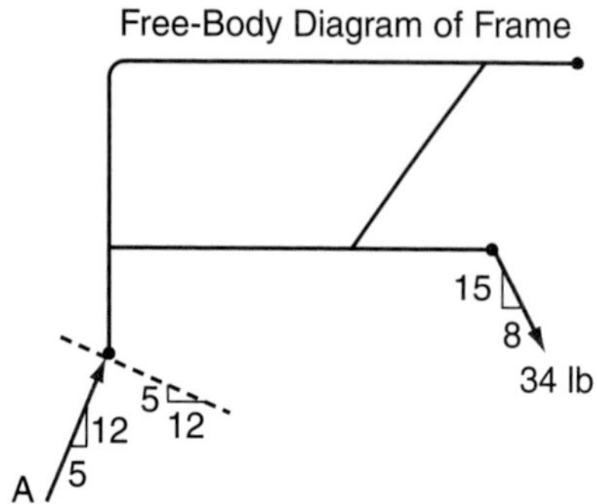

FIGURE 4–16d

Step 5. Because we assume no friction at A, the force at A must be perpendicular to the sloped surface (Figure 4–16d). Note the reversing of the slope numbers as you move from the sloped surface to the force. The sloped surface had been holding the frame up or pushing upward, so the replacement force is shown *pushing on* the frame at A. This force is now labeled A and is complete, with both direction and slope labeled.

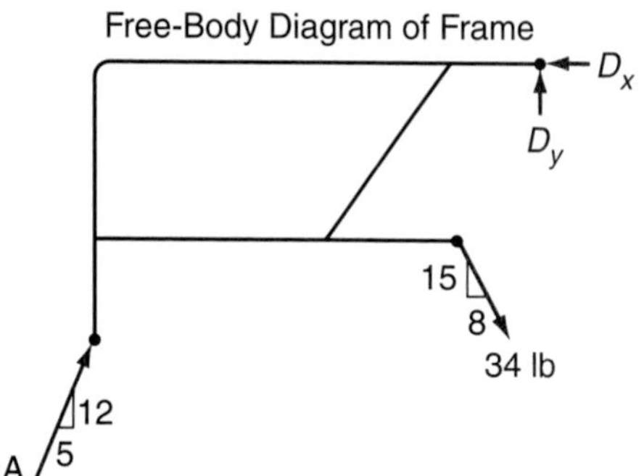

FIGURE 4–16e

Step 6. Both horizontal and vertical components must be shown at point D (Figure 4–16e). This point has purposely been left for last since assuming correct directions is often easier when all the other force directions are shown. Both force A and the 34-lb force have horizontal components acting to the right. Forces acting to the left must balance with forces to the right; therefore, you can safely assume force D_x to be acting to the left. Draw this component and label it D_x.

The direction to assume for force D_y is not as obvious since the vertical components of A and the 34-lb force are in opposite directions. As you become more familiar with estimating moments, the direction of D_y may be more apparent, but for now assume it is acting upward.

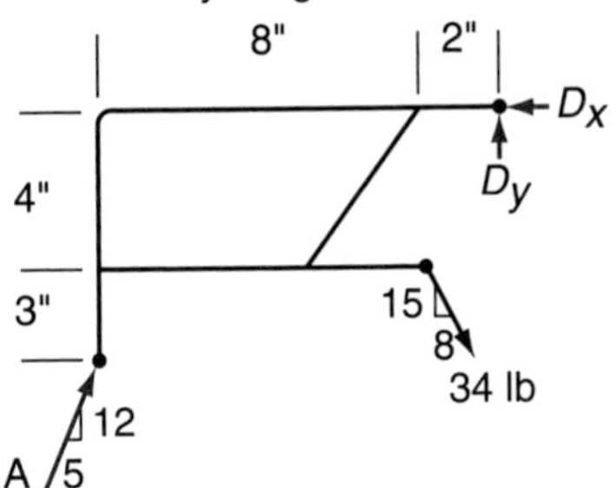

FIGURE 4–16f

Step 7. The FBD of the frame (Figure 4–16f) now appears complete, but your future calculations would be much easier and less prone to error if all distances to force locations were labeled on the diagram.

The self-contained information on this diagram will now allow for calculation of all the forces without reference to any of the original problem.

Consider member BG now and follow a similar sequence of steps.

Free-Body Diagram of BG

FIGURE 4–17a

Step 1. The free-body diagram of BG can begin with the title "FBD of BG" and a stick diagram with the three locations identified where forces will be drawn (Figure 4–17a).

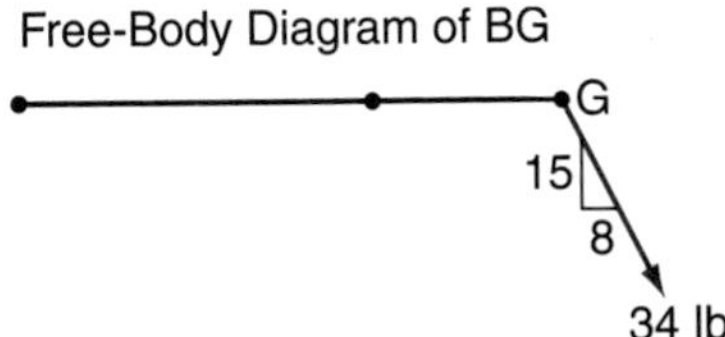

FIGURE 4–17b

Step 2. Show the given force of 34 lb acting at point G (Figure 4–17b).

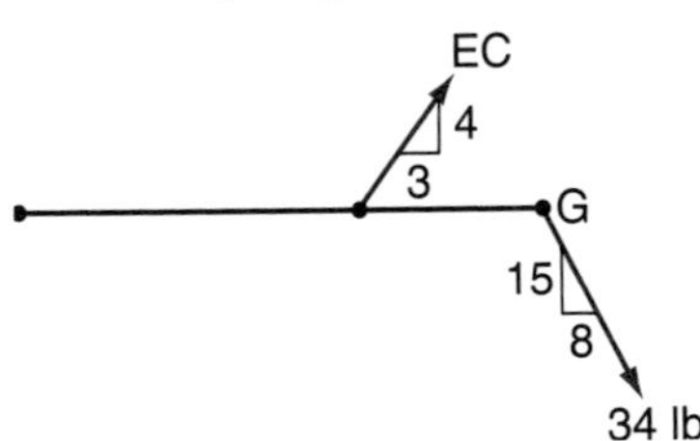

FIGURE 4–17c

Step 3. Cable EC is in tension and therefore pulling on point E. Show all three details: direction, slope, and label EC (Figure 4–17c).

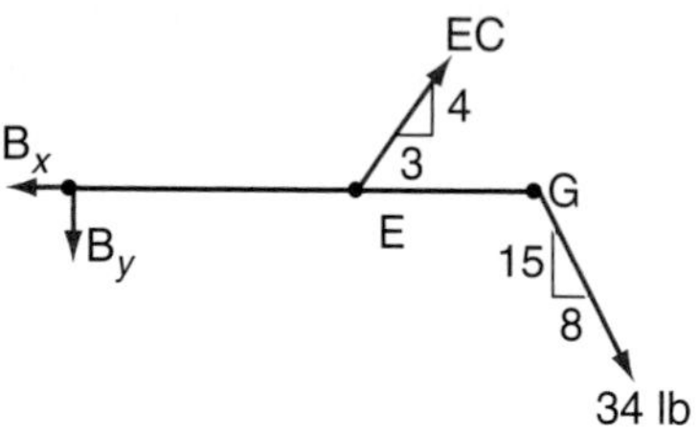

FIGURE 4–17d

Step 4. Point B (Figure 4–17d) is a pinned connection with both horizontal and vertical components. Force EC and the 34-lb force both have horizontal components to the right. Since forces left must equal forces right, assume B_x to be acting to the left.

If you visualize member BG pivoting at E, the 34-lb force causes clockwise rotation so we need a vertical force, B_y, acting downward to balance this rotation. Draw and label B_y.

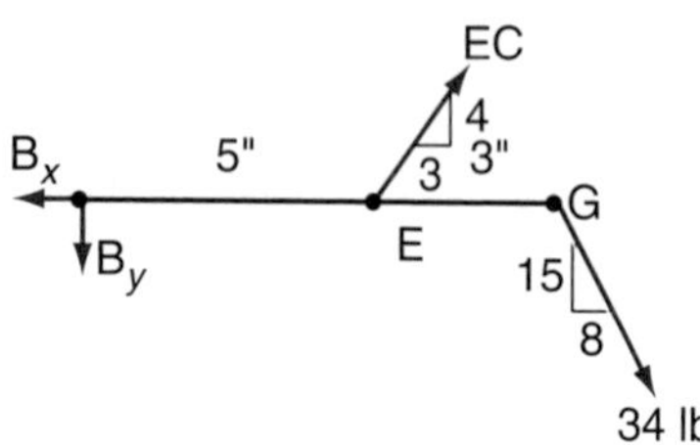

FIGURE 4–17e

Step 5. The free-body diagram (Figure 4–17e) is completed by labeling the given distances.

The final free-body diagram, member AD, can now be considered.

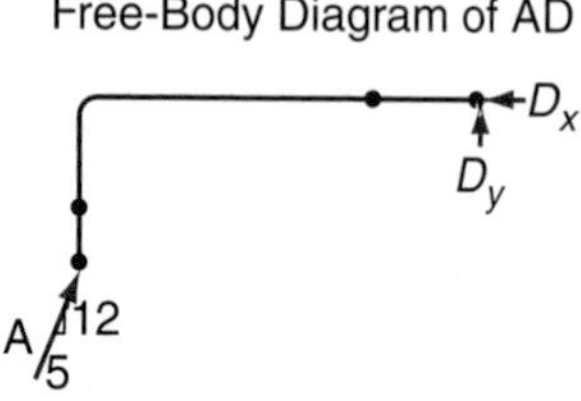

FIGURE 4–18a

Step 1. Title the diagram and identify the four locations at which forces will be shown (Figure 4–18a).

Care must be taken when showing any of the forces on this diagram because we have already assumed directions for all of them and we must be consistent. The forces at points A and D are *external* forces and must be repeated exactly the same as on the free-body diagram of the frame (Figure 4–16f).

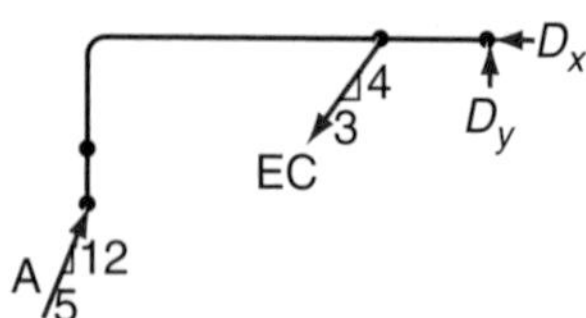

FIGURE 4–18b

Step 2. Cable EC is in tension and therefore pulling on point C (Figure 4–18b). Label this force EC and show both direction and slope. This may appear to be in the opposite direction to our previous assumption, but since a cable can only pull (not push), it therefore pulls on both E and C.

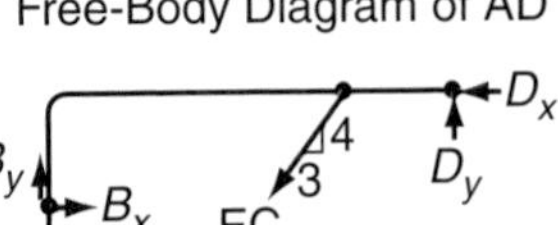

FIGURE 4–18c

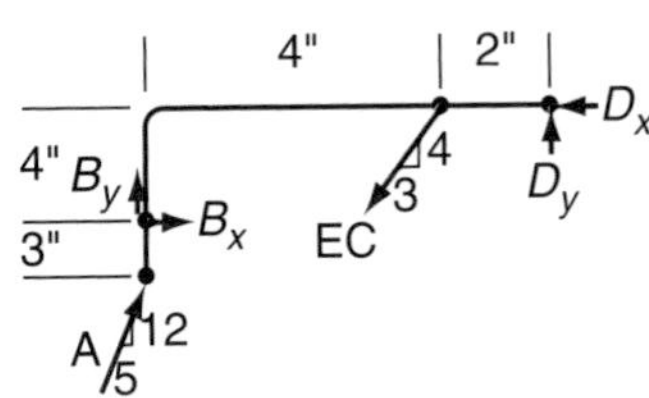

FIGURE 4–18d

Step 3. In the previous free-body diagram of BG (Figure 4–17e) we showed B_x acting to the left and B_y acting downward. These were the forces of AD acting on *BG*. The forces of BG on *AD* are equal and opposite. To therefore agree with our previously assumed directions for B_x and B_y, show B_x acting to the right and B_y acting upward (Figure 4–18c).

Whenever you have internal forces at the same point on two different free-body diagrams, always reverse directions.

Step 4. Showing the given dimensions completes the free-body diagram of AD (Figure 4–18d).

4–3 THREE EQUATIONS OF EQUILIBRIUM

Initial understanding of this section will be easier if we consider only coplanar force systems. *Coplanar forces* are those that act only in one plane, such as the sheet of paper that they are drawn on. All the force systems that we have used so far have been coplanar.

The chair in which you sit is in static equilibrium. It is not moving up or down; therefore, total forces up equal total forces down. In this case, your weight and that of the chair are equal to the force of the floor pushing up on the chair.

The sideways forces are also equal. An externally applied sideways force that fails to move the chair is equal to the force of friction that the floor exerts on the chair. A stationary chair is not twisting or turning either. This is because any clockwise (cw) moment has an equal and opposite counterclockwise (ccw) moment. If the chair begins to move, it does so because there is an imbalance of forces on it, and static equilibrium no longer exists.

For complete static equilibrium, three requirements must be met:

1. vertical forces balance.
2. horizontal forces balance.
3. moments balance; cw = ccw (about any point).

A concise form of stating the same points is:

1. $\Sigma F_y = 0$
2. $\Sigma F_x = 0$
3. $\Sigma M = 0$

If we take upward vertical forces as positive and downward vertical forces as negative, the algebraic sum of all vertical forces is zero, and a body is in balance or static equilibrium. $\Sigma F_y = 0$, stated in words, is "summation of forces in the y-direction equals zero," or "forces up minus forces down equals zero." If horizontal forces are positive to the right and negative to the left, then we can also say, "summation of forces in the x-direction equals zero," or "forces to the right minus forces to the left equals zero" ($\Sigma F_x = 0$).

The same principle applies to moments. The sum of counterclockwise moments equals the sum of clockwise moments about *any* point on the object ($\Sigma M = 0$). Taking clockwise moments as negative and counterclockwise as positive, we find that the algebraic sum of moments is zero.

The free-body diagram of the create that was drawn earlier when introducing the concept of free-body diagrams (Figure 4–2) is complete and can have any or all of the three equilibrium equations applied to it. The equation to be used for Figure 4–2 would be

$$\Sigma F_y = 0$$

or

$$\text{forces up} - \text{forces down} = 0$$
$$N + 60\text{ lb} - 100\text{ lb} = 0$$
$$N = +40$$
$$N = 40\text{ lb}\uparrow$$

EXAMPLE 4–4

The free-body diagram of AB in Example 4–1 (Figure 4–19) is repeated here so that we can now solve for the forces at A and B.

There are only two horizontal forces shown, so we can solve for A_x by stating:

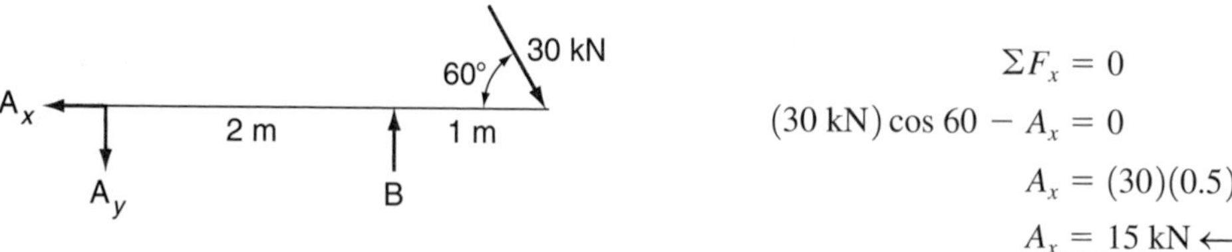

$$\Sigma F_x = 0$$
$$(30\text{ kN})\cos 60 - A_x = 0$$
$$A_x = (30)(0.5)$$
$$\underline{A_x = 15\text{ kN}\leftarrow}$$

FIGURE 4–19

There are three vertical forces, of which two are unknown, so we cannot use $\Sigma F_y = 0$ yet.

If we take moments about A then A_x and A_y will not appear in the equation and we can solve for B.

$$\Sigma M_A = 0$$

$$B(2\text{ m}) - (30\text{ kN})(\sin 60)(3\text{ m}) = 0$$

or

$$2B = (30 \sin 60)(3)$$

$$B = \frac{78}{2}$$

$$\underline{B = 39\text{ kN}\uparrow}$$

Taking moments about B will give us A_y.

$$\Sigma M_B = 0$$

$$(A_y)(2\text{m}) - (30\text{ kN})(\sin 60)(1\text{ m}) = 0$$

or

$$2A_y = (30 \sin 60)(1)$$

$$A_y = \frac{26}{2}$$

$$\underline{A_y = 13\text{ kN}\downarrow}$$

EXAMPLE 4–5

Solve for the forces at A and B of the beam shown in Figure 4–20.

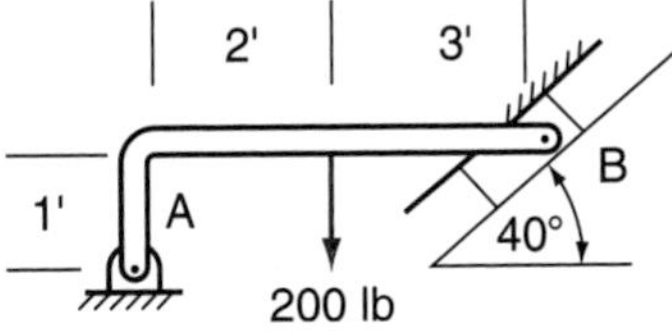

FIGURE 4–20

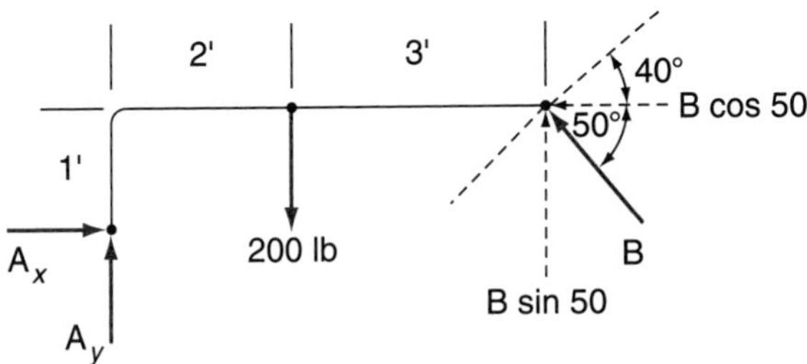

FIGURE 4–21

Note the direction of force B. (Figure 4–21). The slot holds the beam up so B is shown acting upward. The force is perpendicular to the slot, which is 40° from horizontal, so the force is 50° from horizontal.

The horizontal force A_x can be shown acting to the right to balance the horizontal component of B.

Assume A_y acting upward. Our calculations will confirm if this direction is correct or incorrect.

You must begin by taking moments about A because considering horizontal forces, vertical forces, or moments elsewhere would give you two unknowns.

$$\Sigma M_A = 0$$

$$(\text{B} \cos 50)(1 \text{ ft}) + (\text{B} \sin 50)(5 \text{ ft}) - (200 \text{ lb})(2 \text{ ft}) = 0$$

$$0.643 \text{ B} + 3.83 \text{ B} = 400$$

$$\underline{\text{B} = 89.4 \text{ lb } 50°}$$

$$\Sigma F_x = 0$$

$$A_x = (89.4 \text{ lb})\cos 50$$

$$\underline{A_x = 57.5 \text{ lb} \rightarrow}$$

$$\Sigma F_y = 0$$

$$A_y + (89.4 \text{ lb})\sin 50 = 200 \text{ lb}$$

$$\underline{A_y = 132 \text{ lb} \uparrow}$$

EXAMPLE 4–6

Determine the horizontal and vertical reactions at C for the beam shown in Figure 4–22.

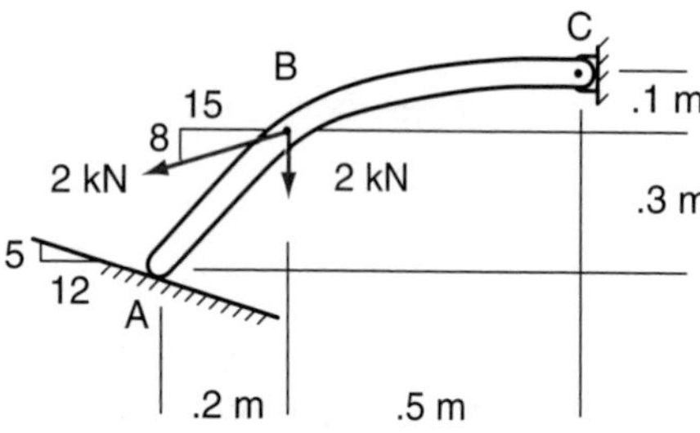

FIGURE 4–22

Taking moments about C will eliminate two unknowns, C_x and C_y from the equation.

$\Sigma M_C = 0$ (Figure 4–23)

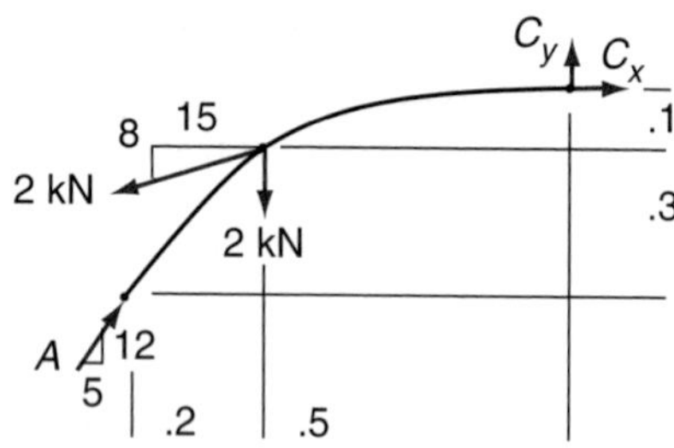

FIGURE 4–23

$$\frac{12}{13}A(0.7\text{ m}) + \frac{15}{17}(2\text{ kN})(0.1\text{ m}) = \frac{5}{13}A(0.4\text{ m}) + (2\text{ kN})(0.5\text{ m}) + \frac{8}{17}(2\text{ kN})(0.5\text{ m})$$

$$0.646\,A + 0.176 = 0.154\,A + 1 + 0.471$$

$$A = 2.63\text{ kN} \quad \angle\,12/5$$

You can now choose either vertical or horizontal forces, as there will only be one unknown in each case.

$\Sigma F_y = 0$

$$C_y + \frac{12}{13}(2.63\text{ kN}) = 2\text{ kN} + \frac{8}{17}(2\text{ kN})$$

$$C_y = -2.43 + 2 + 0.941$$

$$\underline{C_y = 0.513\text{ kN}\uparrow}$$

$\Sigma F_x = 0$

$$C_x + \frac{5}{13}(2.63\text{ kN}) = \frac{15}{17}(2\text{ kN})$$

$$\underline{C_x = 0.753\text{ kN}\rightarrow}$$

4–4 TWO-FORCE MEMBERS

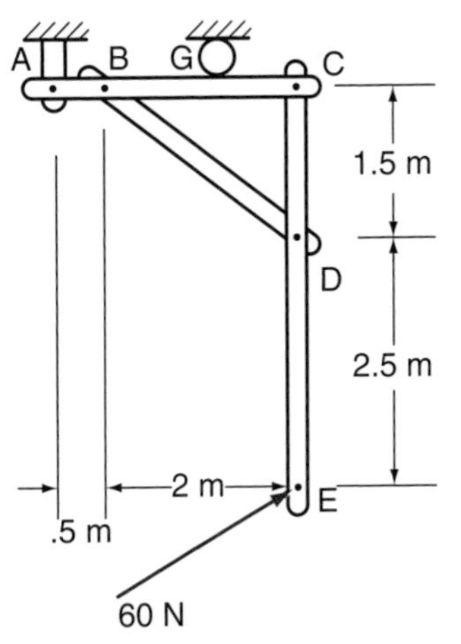

FIGURE 4–24

A member that is acted upon by two forces—for example, one at each end—is known as a *two-force member.* A two-force member will always be in either tension or compression. When a member is acted upon by at least three forces at several locations, there will likely be not only compression or tension but also bending. In Figure 4–24, member BD is a two-force member, and members AC and CE are three-force members.

As was stated earlier, when drawing a free-body diagram of a pinned connection, you must assume both horizontal and vertical components. This rule still applies to two-force members, which are in tension or compression. In the case of the two-force member, the direction of the resultant of the horizontal and vertical components is directly in line with the member. If the member is curved the forces would be acting along the straight line connecting the two points.

The free-body diagram of BD can be drawn as in either Figure 4–25a or 4–25b. For BD to be in static equilibrium and not to rotate, forces *B* and *D* must be equal and opposite and must have the same line of action as the member. Components will also be equal and opposite.

$$B_y = D_y$$
$$B_x = D_x$$
$$B = B_x + B_y \text{ (vectorially)}$$
$$D = D_x + D_y \text{ (vectorially)}$$

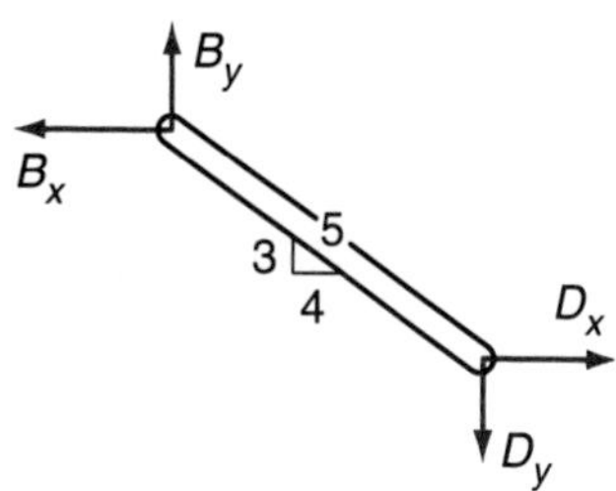

FIGURE 4–25a

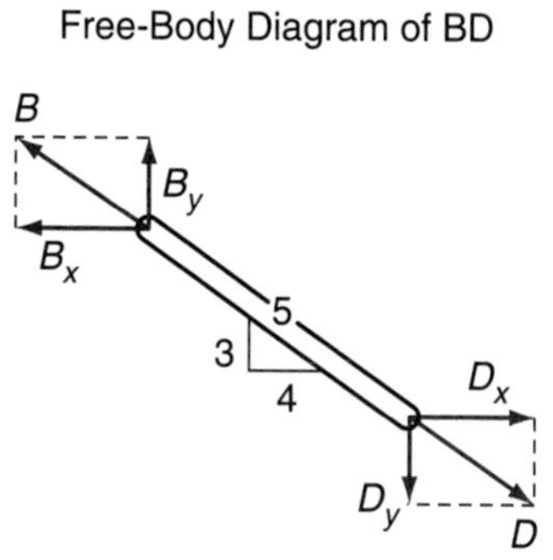

Free-Body Diagram of BD

B, B_y, B_x, 3, 5, 4, D_x, D_y, D

FIGURE 4–25b

Suppose that $B_x = 80$ N. Knowing the slope of the member and by similar triangles, we get

$$\frac{B_y}{B_x} = \frac{3}{4}$$
$$B_y = \frac{3}{4}(B_x)$$
$$= \frac{3}{4}(80 \text{ N})$$
$$\underline{B_y = 60 \text{ N}}$$

Similarly,

$$B = \frac{5}{4}(B_x)$$
$$= \frac{5}{4}(80\text{ N})$$
$$\underline{B = 100\text{ N}}$$

Thus, for two-force members, the one component can be used to solve for the other component or the total force.

From the frame of Figure 4–24, free-body diagrams of BD (Figure 4–26a), CE (Figure 4–26b), and AC (Figure 4–26c) are drawn.

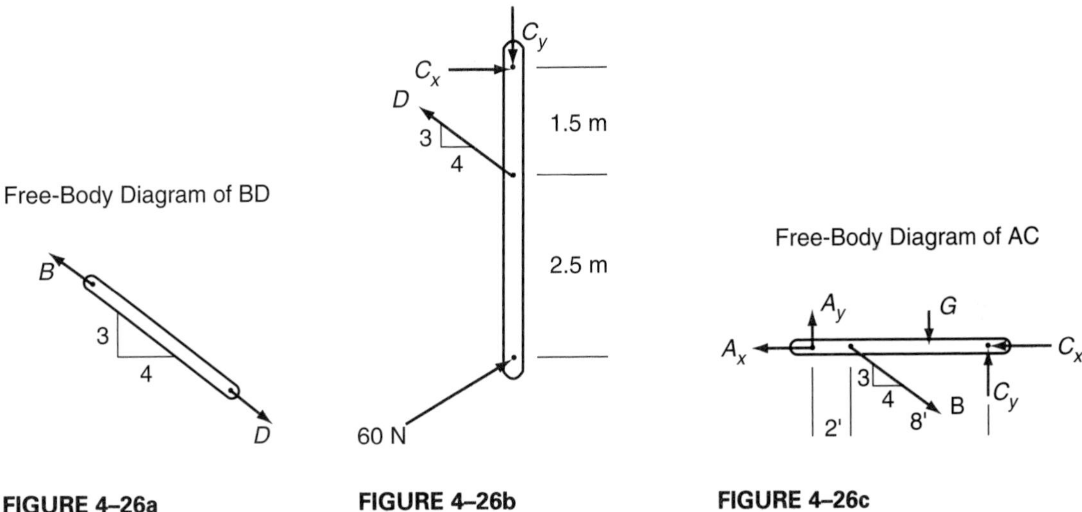

FIGURE 4–26a

FIGURE 4–26b

FIGURE 4–26c

In the free-body diagram of BD (Figure 4–26a), it is not necessary to show horizontal and vertical components at pinned connections *B* and *D*. Since member BD is in tension only, the total force at *B* must have the same direction or line of action as member BD. If the line of action of force *B* had not coincided with member BD, an unbalanced moment would have been present, and we would not have had static equilibrium. Notice how the forces at *B, C,* and *D* change direction depending on the member for which the free-body diagram is drawn. This is because they are internal forces with equal and opposite reactions.

4–5 PULLEYS

Free-body diagrams of pulleys or structures that have pulleys may require careful analysis. Consider the pulley in Figure 4–27. Cable tension due to the 20.4-kg mass is 20.4 (9.81) = 200 N. Since the cable has the same tension throughout its length, the free-body diagram of the pulley would have two forces of 200 N and horizontal and vertical components at D (Figure 4–28).

Adding the 200-N forces vectorially, we obtain a resultant, $R = 179$ N. For moment equilibrium about the center of the pulley, the resultant R must pass through the center (Figure 4–29). Since R can be applied anywhere along its line of action, let us suppose that it acts at D and resolve it into its original components, that is, two 200-N forces (Figure 4–30).

The free-body diagram of member AD without the pulley (Figure 4–31) can show two 200-N forces acting on the pin D.

The free-body diagram of the complete structure can be drawn either without the pulley (Figure 4–32) or with the pulley (Figure 4–33).

In Figure 4–32, there are three 200-N external forces, with the vertical one acting 0.6 m from point *G*. In Figure 4–33, there is only one 200-N external force (the remaining rope tension is an internal force). Notice, however, that the 200-N vertical force is now acting 0.7 m from point *G*. In this case, the diameter of the pulley is significant.

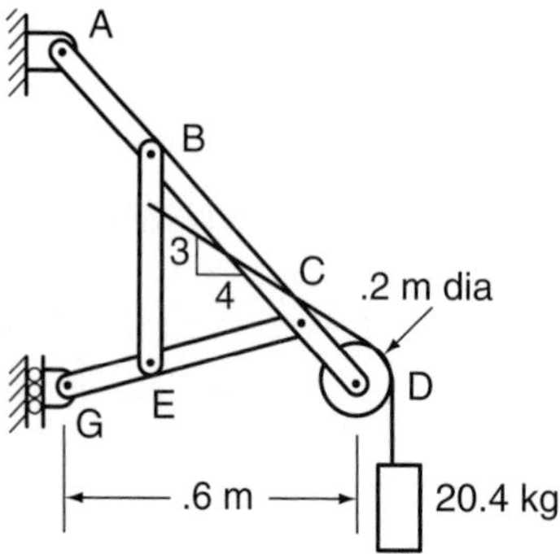

FIGURE 4–27

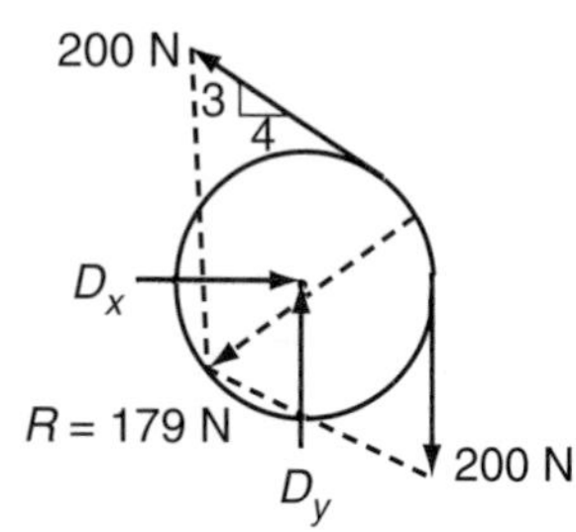

FIGURE 4–28

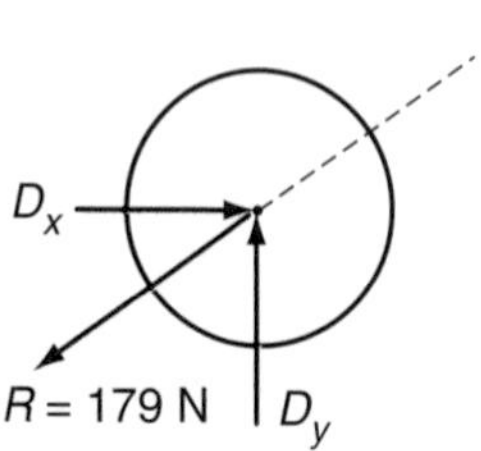

FIGURE 4–29

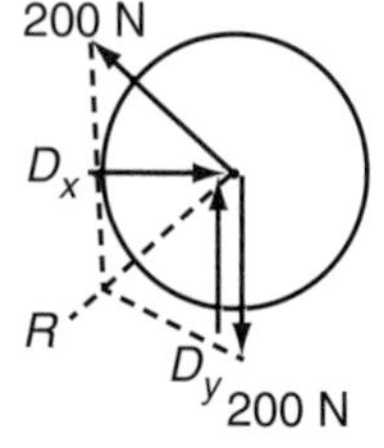

FIGURE 4–30

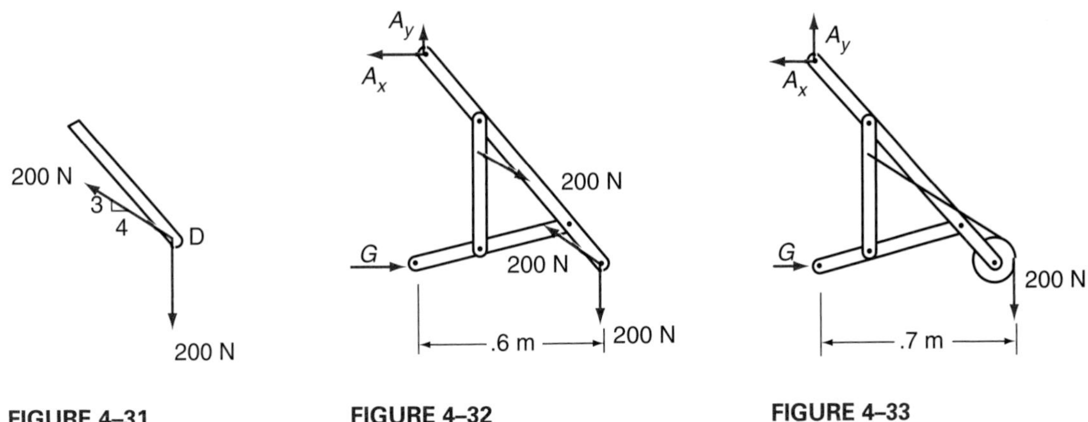

Either figure is correct, and the solution of either will yield the same answers for A_y, A_x, and G. (Figure 4–33 would be easier to solve.) Remember to analyze each problem individually—do not blindly ignore pulley diameters.

EXAMPLE 4–7

Determine the tension T in the cable of the pulley system shown in Figure 4–34.

Hint: Follow the cable labeled T throughout its length, labeling the same tension T between various pulleys. If we now cut through the cables just below pulley A and above pulley C, we can draw a free-body diagram of the lower half (Figure 4–35).

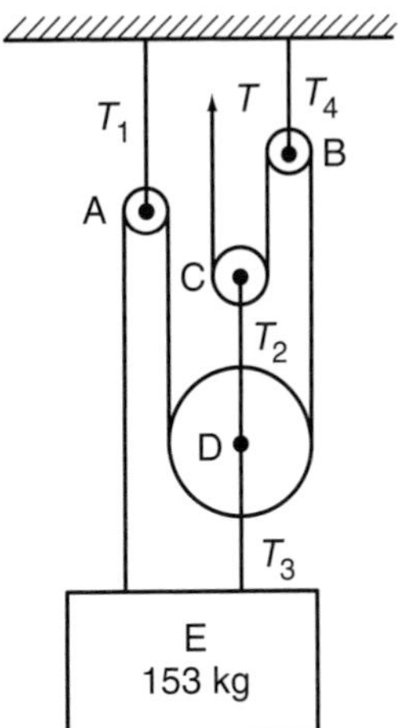

FIGURE 4–34

Free-Body Diagram of Lower Half

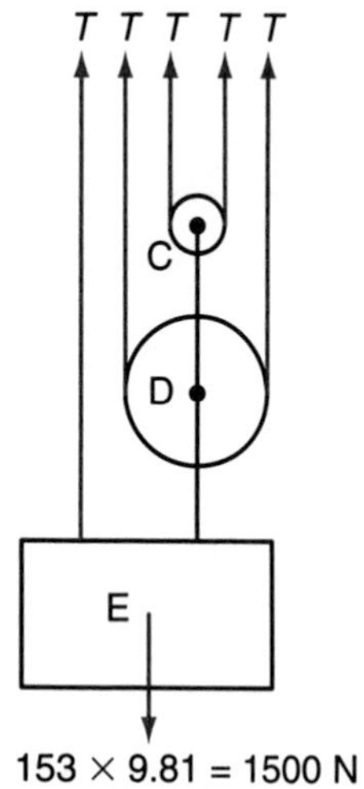

FIGURE 4–35

The most direct solution would be as follows:

$\Sigma F_y = 0$ (Figure 4–35)

$$5T = 1500 \text{ N}$$
$$\underline{T = 300 \text{ N}}$$

A more detailed approach would provide other cable tensions along with *T*.

Free-Body Diagram of C

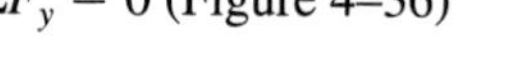

$\Sigma F_y = 0$ (Figure 4–36)

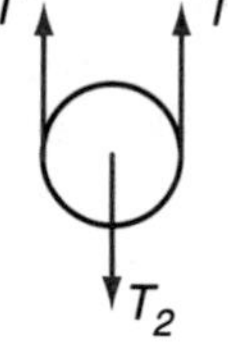

FIGURE 4–36

$$T_2 = 2T$$
$$\underline{T_2 = 600 \text{ N}}$$

Free-Body Diagram of D

$\Sigma F_y = 0$ (Figure 4–37)

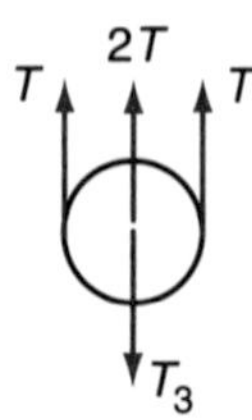

FIGURE 4–37

$$T_3 = 4T$$
$$\underline{T_3 = 1200 \text{ N}}$$

Free-Body Diagram of E

$\Sigma F_y = 0$ (Figure 4–38)

$$T + 4T = 1500 \text{ N}$$

$$\underline{T = 300 \text{ N}}$$

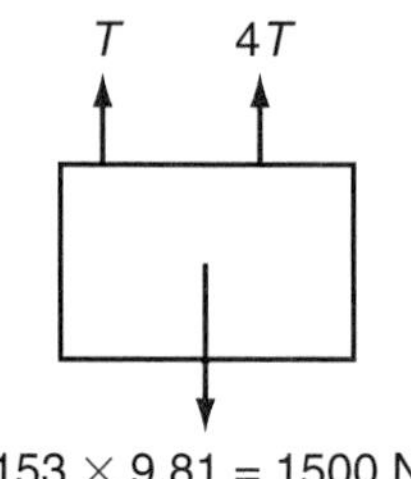

FIGURE 4–38

4–6 COPLANAR CONCURRENT FORCE SYSTEMS

With a *coplanar concurrent* force system, which has all forces in one plane and intersecting at one point, we usually have a free-body diagram of a point or pin. There is only one point on the free-body diagram. All the forces have zero moment about this point, so using $\Sigma M = 0$ yields no equation. We can only apply the two equations of equilibrium, $\Sigma F_x = 0$ and $\Sigma F_y = 0$. Two unknowns and simultaneous equations may result. An alternative solution consists of drawing a vector polygon. If there are only three forces, the result is a vector triangle to which the sine law can be applied.

In Section 2–4, the vector triangle was used to obtain a resultant. The resultant was the third and unknown force required to complete the triangle; it began at the origin and was directed away from the origin. For a system to be in equilibrium, the equilibrant force must be equal and opposite to the resultant of all other forces. The equilibrant force vector then closes the triangle or polygon as the resultant force vector did, but it points *toward* the origin rather than *away* from it (Method 2, Example 4–9).

EXAMPLE 4–8

Find the load in each section of the cable system in Figure 4–39.

$$1223 \text{ kg } (9.81 \text{ m/s}^2) = 12 \text{ kN}$$

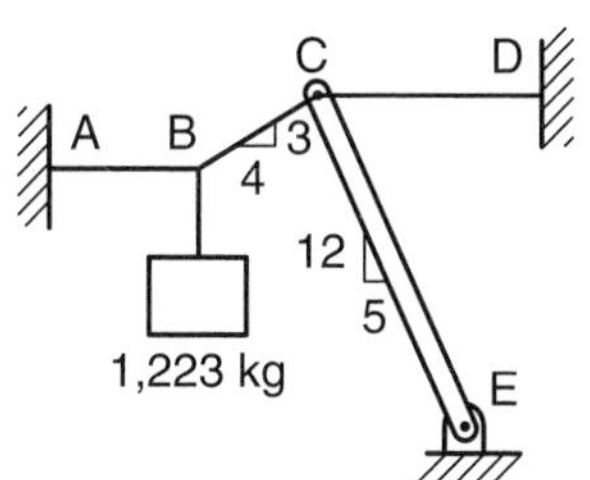

FIGURE 4–39

Free-Body Diagram of B

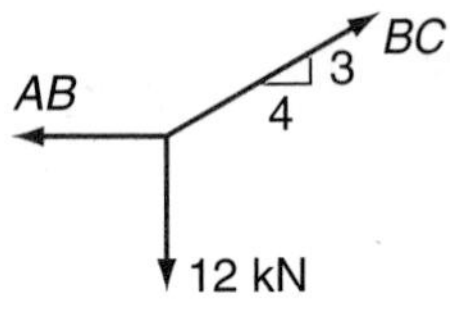

FIGURE 4–40

Starting with a free-body diagram that has a known force (Figure 4–40), we have

$\Sigma F_y = 0$

$$\frac{3}{5}BC - 12\text{ kN} = 0$$

$$BC = \frac{5(12)}{3}$$

$$\underline{BC = 20\text{ kN } T}$$

(T will be used to represent tension; C will represent compression.)

$\Sigma F_x = 0$

$$-AB + \frac{4}{5}(20\text{ kN}) = 0$$

$$\underline{AB = 16\text{ kN } T}$$

$\Sigma F_y = 0$ (Figure 4–41)

Free-Body Diagram of C

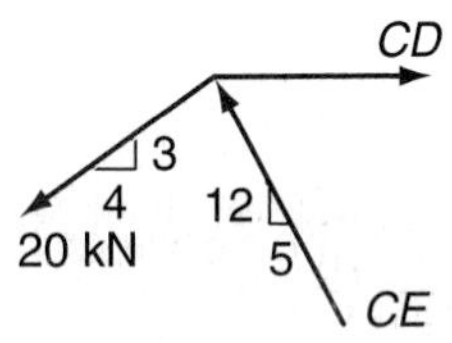

FIGURE 4–41

$$\frac{12}{13}CE - \frac{3}{5}(20\text{ kN}) = 0$$

$$CE = \frac{12(13)}{12}$$

$$\underline{CE = 13\text{ kN } C}$$

$\Sigma F_x = 0$

$$CD - \frac{4}{5}(20\text{ kN}) - \frac{5}{13}(13\text{ kN}) = 0$$

$$CD = 16 + 5$$

$$\underline{CD = 21\text{ kN } T}$$

EXAMPLE 4–9

Determine the loads in members AB and CB of Figure 4–42.

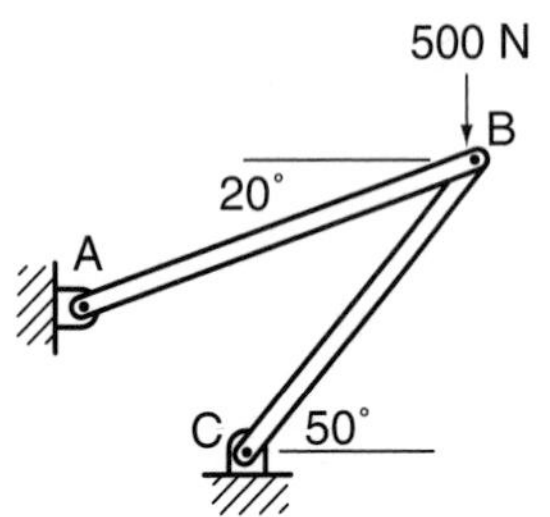

FIGURE 4–42

Free-Body Diagram of B

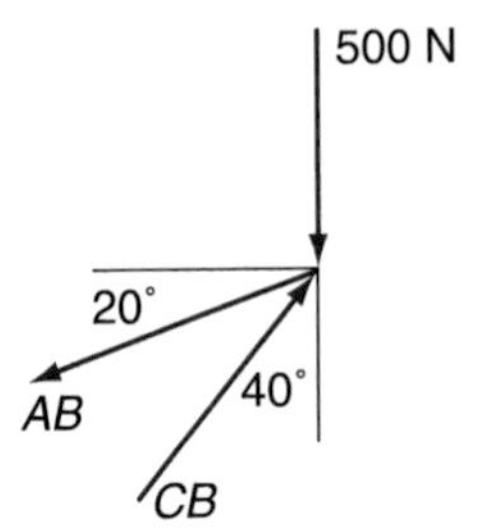

FIGURE 4–43

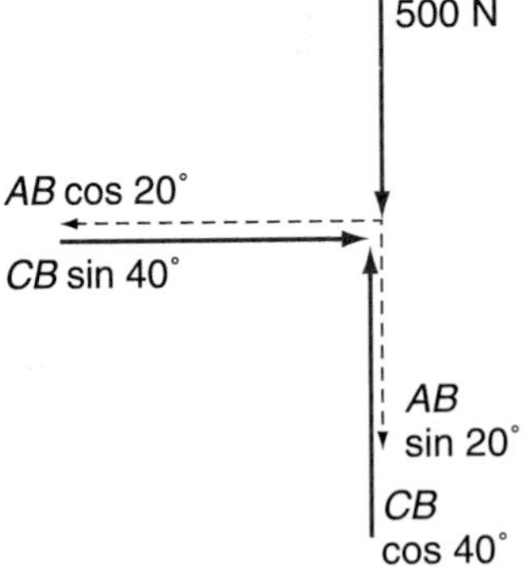

FIGURE 4–44

Vector Triangle

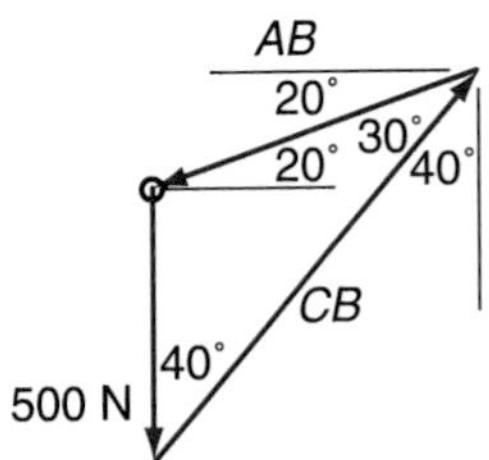

FIGURE 4–45

Method 1: Simultaneous Equations

$\Sigma F_y = 0$ (Figures 4–43 and 4–44)

$$CB \cos 40° - AB \sin 20° - 500 \text{ N} = 0$$
$$0.766CB = 0.342AB + 500$$
$$CB = 0.447AB + 653 \qquad (1)$$

$\Sigma F_x = 0$

$$-AB \cos 20° + CB \sin 40° = 0$$
$$AB(0.94) = CB\,(0.643)$$
$$AB = 0.684\,CB \qquad (2)$$

Substituting Equation (2) into (1), we get

$$CB = 0.447(0.684CB) + 653$$
$$= 0.306CB + 653$$
$$CB - 0.306\,CB = 653$$
$$CB = \frac{653}{0.694}$$
$$\underline{CB = 940 \text{ N } C}$$

From Equation (2),

$$AB = 0.684(940 \text{ N})$$
$$\underline{AB = 643 \text{ N } T}$$

Method 2: Vector Triangle and Sine Law

Construct a vector triangle (Figure 4–45) from Figure 4–43, adding the vectors tip to tail until they close at the origin.

$$\frac{AB}{\sin 40°} = \frac{500 \text{ N}}{\sin 30°}$$
$$AB = 500\left(\frac{0.643}{0.5}\right)$$
$$\underline{AB = 643 \text{ N } T}$$
$$\frac{CB}{\sin 110°} = \frac{500 \text{ N}}{\sin 30°}$$

$$CB = 500\frac{\sin 110^\circ}{\sin 30^\circ}$$

$$= 500\left(\frac{0.94}{0.5}\right)$$

$$\underline{CB = 940 \text{ N } C}$$

As you can see, the easier solution is usually the sine-law method. The important step in this method is the proper construction of the vector triangle. All vectors must be added tip to tail. Label all angles as you construct the triangle. (The angles of 20° and 40° are each shown in two locations as you draw the respective vectors.)

EXAMPLE 4–10

Weight A is pulled to the right by means of the ropes, a pulley, and a 100-lb force (Figure 4–46). Calculate the tension in ropes AB and BD.

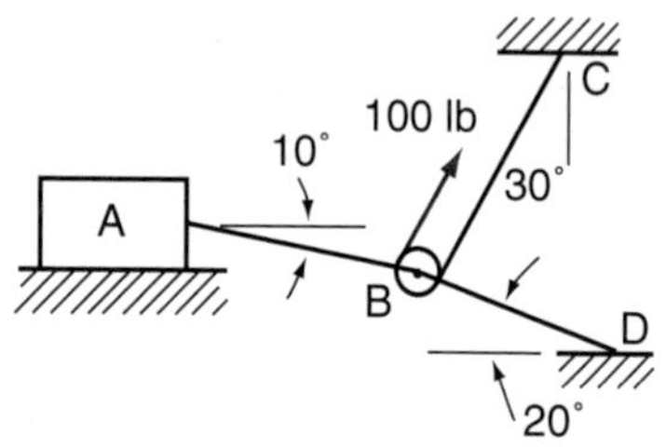

FIGURE 4–46

Free-Body Diagram of B

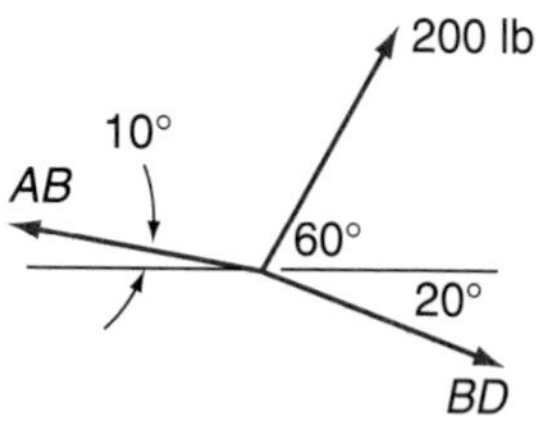

FIGURE 4–47

Since rope BC has a tension of 100 lb, there is a total applied force of 200 lb at B (Figure 4–47).

There are two 100-1b forces pulling on pulley B, and both of them can be shown acting through the center of the pulley.

Constructing the vector triangle (Figure 4–48) and applying the sine law, we have

$$\frac{BD}{\sin 70^\circ} = \frac{200 \text{ lb}}{\sin 10^\circ}$$

$$BD = 200\left(\frac{0.94}{0.174}\right)$$

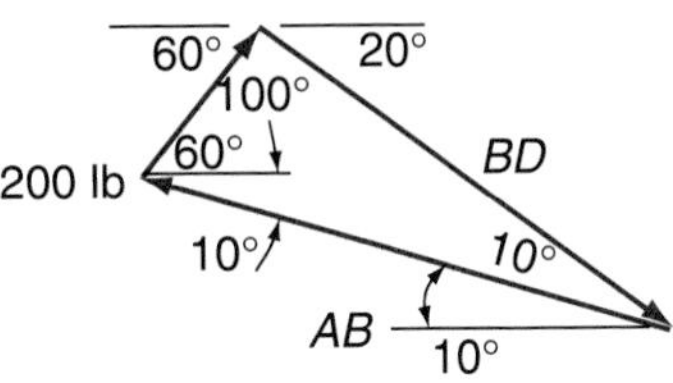

FIGURE 4–48

$$\underline{BD = 1080 \text{ lb } T}$$

$$\frac{AB}{\sin 100°} = \frac{200 \text{ lb}}{\sin 10°}$$

$$AB = 200\left(\frac{\sin 100°}{\sin 10°}\right)$$

$$= 200\left(\frac{0.985}{0.174}\right)$$

$$\underline{AB = 1130 \text{ lb } T}$$

4–7 COPLANAR PARALLEL FORCE SYSTEMS

A horizontal beam with only vertical loading on it is an example of a force system in which all the forces are parallel and in the same plane—a *coplanar parallel system.* The beam is supported at two points. The free-body diagram replaces these supports with equivalent forces or reactions. One of the reaction forces is determined by taking moments about the other point of support.

EXAMPLE 4–11

A beam has concentrated loads applied as shown in Figure 4–49. Calculate the reactions at A and B. (Ignore the weight of the beam.)

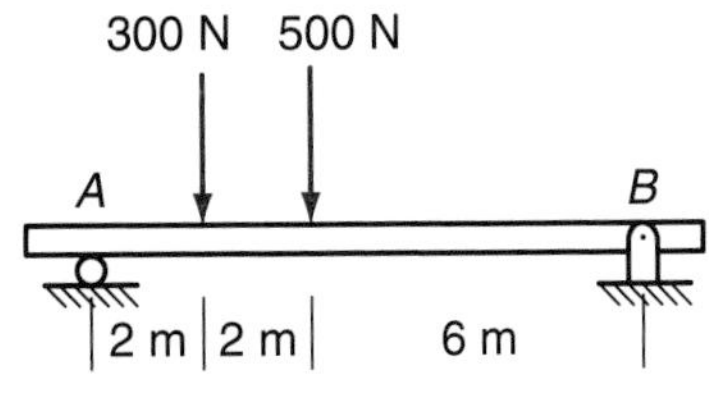

FIGURE 4–49

Free-Body Diagram of AB

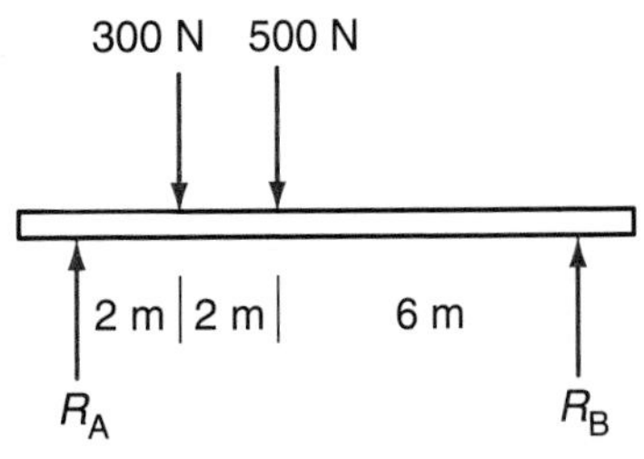

FIGURE 4–50

The reaction at A, R_A, and the reaction at B, R_B, are indicated on the free-body diagram (Figure 4–50). Point B is pin connected and should also have a horizontal component, but since there is no other horizontal force acting on the beam, there cannot be any horizontal component at B.

We now have a free-body diagram to which three equations may be applied:

$$\Sigma F_x = 0$$

$$\Sigma F_y = 0$$

$$\Sigma M = 0$$

If moments are taken about point A, R_A acts at a distance of zero from the center of moments and therefore drops out of the moment equation. Stating the point about which moments are to be taken, we have

$$\Sigma M_A = 0$$

or

$$\text{counterclockwise moment} - \text{clockwise moment} = 0$$

$$(R_B)(10 \text{ m}) - (300 \text{ N})(2 \text{ m}) - (500 \text{ N})(4 \text{ m}) = 0$$

$$10R_B = 60 + 2000$$

$$\underline{R_B = 260 \text{ N} \uparrow}$$

Equating vertical forces, we obtain

$$\Sigma F_y = 0$$

$$R_A + 260 \text{ N} - 300 \text{ N} - 500 \text{ N} = 0$$

$$\underline{R_A = 540 \text{ N} \uparrow}$$

Since the value of R_A depends on the correct calculation of R_B, R_A can be checked by taking moments about point B.

$$\Sigma M_B = 0$$

$$-10R_A + (500 \text{ N})(6 \text{ m}) + (300 \text{ N})(8 \text{ m}) = 0$$

$$10R_A = 3000 + 2400$$

$$\underline{R_A = 540 \text{ N} \uparrow \textit{ check}}$$

EXAMPLE 4–12

Solve for the reactions at points A and B (Figure 4–51).

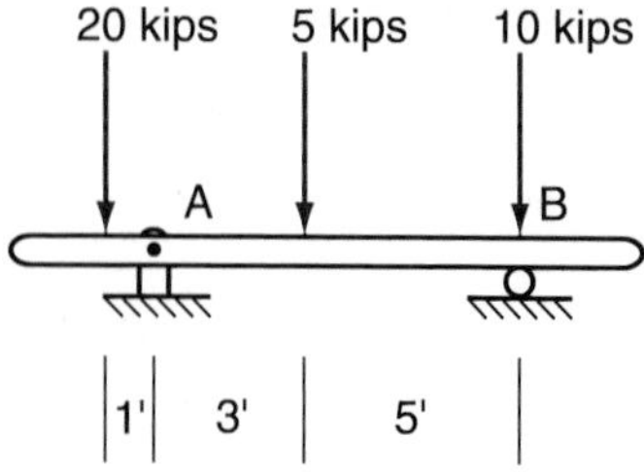

FIGURE 4–51

$$\Sigma M_A = 0 \text{ (Figure 4–52)}$$

$$(R_B)(8 \text{ ft}) + (20 \text{ kips})(1 \text{ ft}) - (5 \text{ kips})(3 \text{ ft}) - (10 \text{ kips})(8 \text{ ft}) = 0$$

$$8R_B + 20 - 15 - 80 = 0$$

$$8R_B = 75$$

$$\underline{R_B = 9.4 \text{ kips} \uparrow}$$

Free-Body Diagram of AB

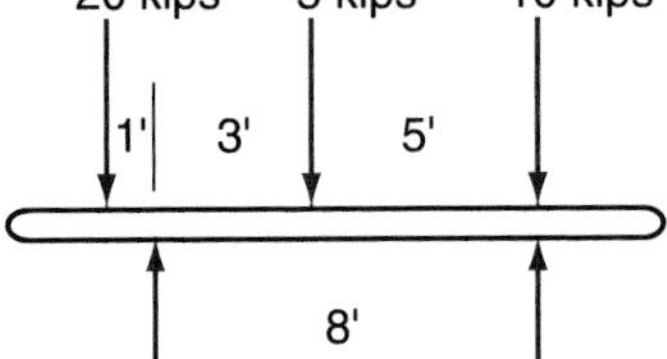

FIGURE 4–52

$$\Sigma M_B = 0$$

$$(20 \text{ kips})(9 \text{ ft}) + (5 \text{ kips})(5 \text{ ft}) - 8R_A = 0$$

$$180 + 25 = 8R_A$$

$$\underline{R_A = 25.6 \text{ kips} \uparrow}$$

Check $\Sigma F_y = 0$

$$R_A + 9.4 \text{ kips} - 20 \text{ kips} - 5 \text{ kips} - 10 \text{ kips} = 0$$

$$R_A = 35 - 9.4$$

$$\underline{R_A = 25.6 \text{ kips} \uparrow \textit{ check}}$$

Instead of being subject to concentrated loads, a beam may have a *distributed load* applied along its length. The distributed load may be *uniform* or nonuniform. A distributed load can be visualized as the piling of concrete blocks along a beam's length either uniformly or to varying depths. The loading of a beam, whether it is uniform or nonuniform, is specified as weight per unit of length. In the U.S. Customary system, it may be lb/ft or kips/ft, and in the SI metric system it will be newtons per meter (N/m) or kN/m.

To determine the reactions where a beam is supported, the distributed load will be converted to a single force acting at the center of gravity of the distributed load. This assumption does not give the true bending effect on the beam, but it does allow us to solve for the reactions. The following examples show how to locate the center of gravity for various distributed loads.

EXAMPLE 4–13

Suppose that a beam is loaded at 20 kN/m over its first 2 m and at 30 kN/m over its remaining 8 m (Figure 4–53). Calculate R_A and R_B.

A concentrated force is assumed to be acting through the center of gravity of a distributed load. The first distributed load

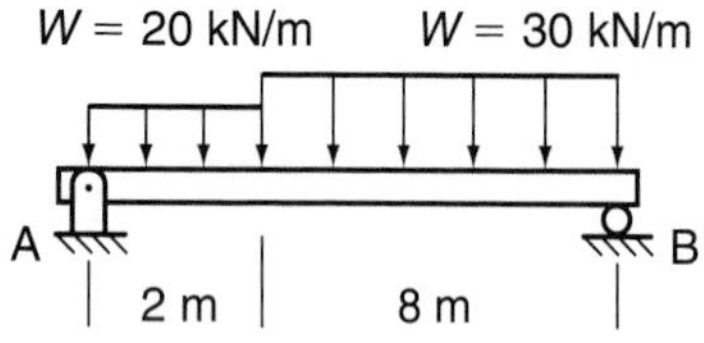

FIGURE 4–53

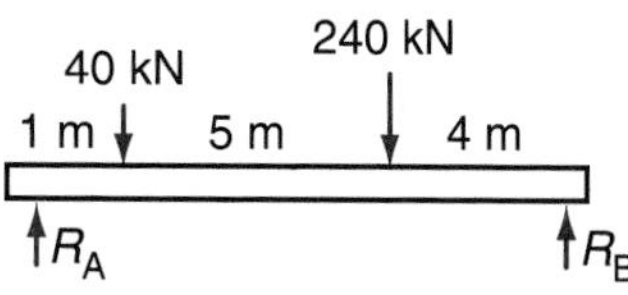

FIGURE 4–54

we could convert would be 20 kN/m for 2 m, or $20 \times 2 = 40$ kN, acting at a distance of 1 m from A. Similarly, 30 kN/m for 8 m gives a force of $30 \times 8 = 240$ kN acting 4 m from B.

$\Sigma M_A = 0$ (Figure 4–54)

$$10R_B - (40\text{ kN})(1\text{ m}) - (240\text{ kN})(6\text{ m}) = 0$$
$$10R_B = 40 + 1440$$
$$\underline{R_B = 148\text{ kN}\uparrow}$$

$\Sigma M_B = 0$

$$(40\text{ kN})(9\text{ m}) + (240\text{ kN})(4\text{ m}) - (10\text{ m})R_A = 0$$
$$360 + 960 = 10R_A$$
$$\underline{R_A = 132\text{ kN}\uparrow}$$

Check $\Sigma F_y = 0$

$$R_A + 148\text{ kN} - 40\text{ kN} - 240\text{ kN} = 0$$
$$\underline{R_A = 132\text{ kN}\uparrow\ \textit{check}}$$

To solve for a more complex beam loading, we break it into simple loadings of either uniform or nonuniform loads. In this way, the center of gravity and concentrated force of each are more easily found. A nonuniform load, triangular in shape, has a center of gravity located as shown in Figure 4–55. The concentrated force replacing this nonuniform load is essentially equal to the area of the triangle where we have units of N/m × m = N.

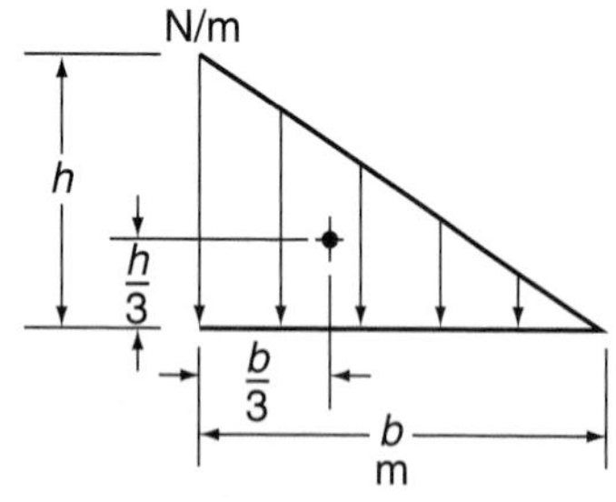

FIGURE 4–55

EXAMPLE 4–14

Solve for the reactions at A and B for the beam loaded as shown in Figure 4–56.

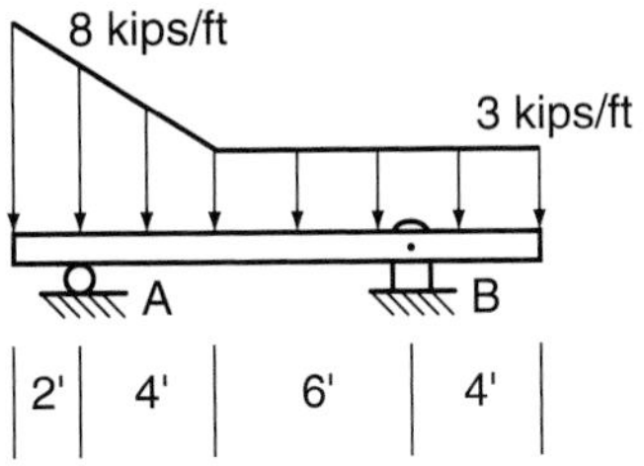

FIGURE 4–56

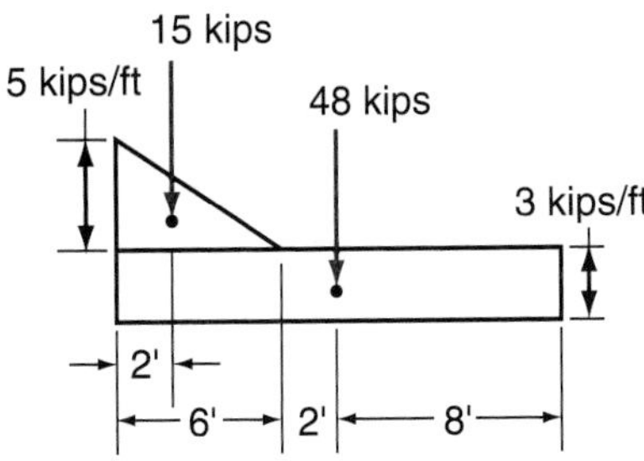

FIGURE 4–57

When converting the beam loading to equivalent concentrated forces, use as few forces as possible (Figure 4–57).

Each area in Figure 4–57 represents a force. The area of the triangle is 1/2 (6)(5) or 15 kips. The rectangular area is 3(16) = 48 kips.

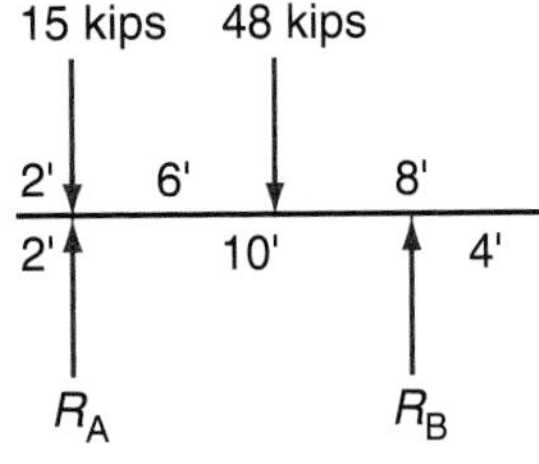

FIGURE 4–58

$\Sigma M_A = 0$ (Figure 4–58)

$$R_B(10\text{ ft}) - (48\text{ kips})(6\text{ ft}) = 0$$

$$\underline{R_B = 28.8\text{ kips}\uparrow}$$

$\Sigma M_B = 0$

$$(15\text{ kips})(10\text{ ft}) + (48\text{ kips})(4\text{ ft}) - (10\text{ ft})\,R_A = 0$$

$$\underline{R_A = 34.2\text{ kips}\uparrow}$$

Check $\Sigma F_y = 0$

$$R_A + 28.8\text{ kips} - 15\text{ kips} - 48\text{ kips} = 0$$

$$R_A = 63 - 28.8$$

$$\underline{R_A = 34.2\text{ kips}\uparrow\ \textit{check}}$$

EXAMPLE 4–15

The two beams in Figure 4–59 are fastened together by means of a bolt at D and B. Calculate the tension in the bolt and the reactions at A and C.

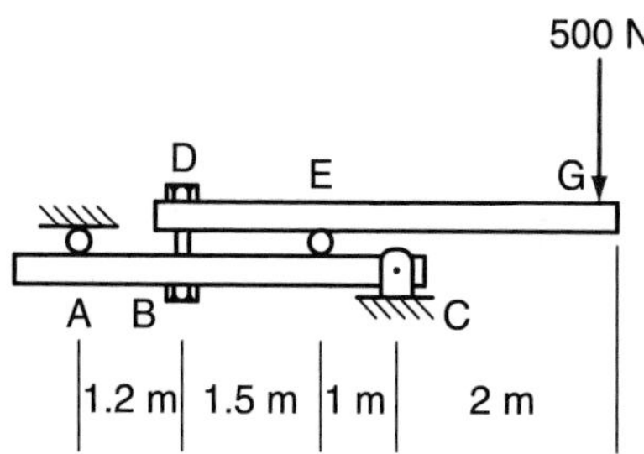

FIGURE 4–59

You have the choice of drawing free-body diagrams of DG or AC, or of both beams combined. In this case, we will start with beam DG because the known force is applied to this member.

Free-Body Diagram of DG

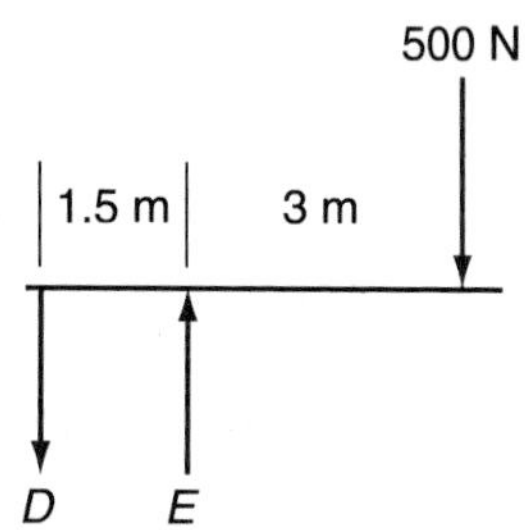

FIGURE 4–60

$\Sigma M_E = 0$ (Figure 4–60)

$$D(1.5\text{ m}) - (500\text{ N})(3\text{ m}) = 0$$

$$\underline{D = 1000\text{ N}\downarrow}$$

The bolt tension is therefore 1 kN.

Free-Body Diagram of Beams

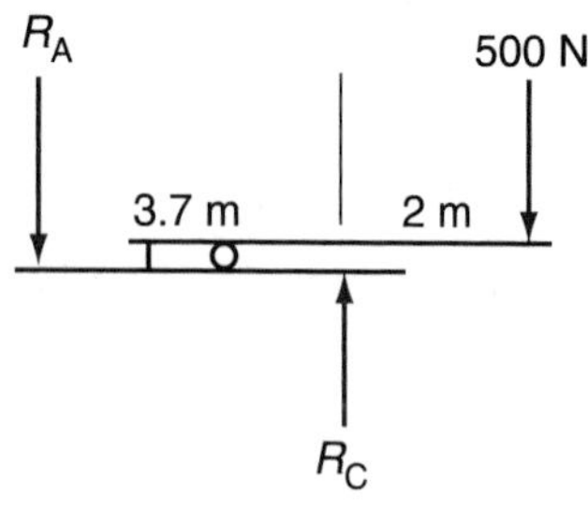

FIGURE 4–61

$\Sigma M_A = 0$ (Figure 4–61)

$$R_C(3.7\text{ m}) - (500\text{ N})(5.7\text{ m}) = 0$$

$$\underline{R_C = 770\text{ N}\uparrow}$$

$\Sigma M_C = 0$

$$R_A(3.7\text{ m}) - (500\text{ N})(2\text{ m}) = 0$$

$$\underline{R_A = 270\text{ N}\uparrow}$$

Check $\Sigma F_y = 0$

$$770\text{ N} - R_A - 500\text{ N} = 0$$

$$\underline{R_A = 270\text{ N}\uparrow \textit{ check}}$$

4–8 COPLANAR NONCONCURRENT FORCE SYSTEMS

When dealing with *coplanar nonconcurrent force systems,* we are faced not only with vertical forces and moment equations but also with horizontal forces since the applied forces are no longer parallel. Pinned connections will now have horizontal components.

EXAMPLE 4–16

Solve for reactions R_A and R_B in Figure 4–62.

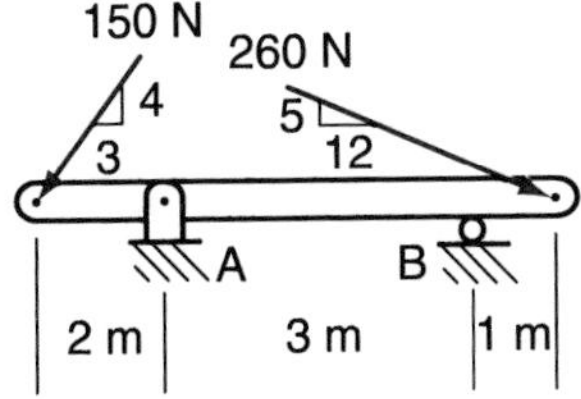

FIGURE 4–62

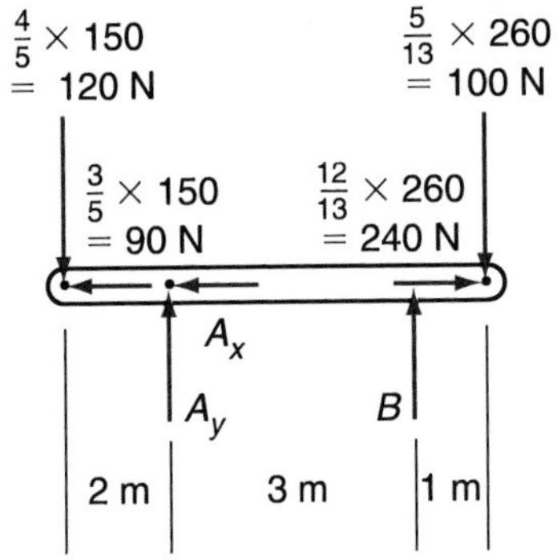

$\frac{4}{5} \times 150 = 120$ N

$\frac{5}{13} \times 260 = 100$ N

$\frac{3}{5} \times 150 = 90$ N

$\frac{12}{13} \times 260 = 240$ N

A_x

A_y

B

2 m 3 m 1 m

FIGURE 4–63

Resolve the 150- and 260-N forces into their horizontal and vertical components (Figure 4–63). Using $\Sigma F_y = 0$ would give an equation with two unknowns, A_y and B; instead, use either ΣM or $\Sigma F_x = 0$.

$\Sigma F_x = 0$

$$240 \text{ N} - 90 \text{ N} - A_x = 0$$

$$\underline{A_x = 150 \text{ N} \leftarrow}$$

Moments can be taken about A or B. Use point A because all horizontal forces and A_y have zero moments about A.

$\Sigma M_A = 0$

$$(120 \text{ N})(2 \text{ m}) + B(3 \text{ m}) - (100 \text{ N})(4 \text{ m}) = 0$$

$$3B = 400 - 240$$

$$\underline{B = 53.3 \text{ N} \uparrow}$$

$\Sigma M_B = 0$

$$(120 \text{ N})(5 \text{ m}) - (100 \text{ N})(1 \text{ m}) - (3 \text{ m})A_y = 0$$

$$600 - 100 = 3A_y$$

$$\underline{A_y = 167 \text{ N} \uparrow}$$

Check $\Sigma F_y = 0$

$$A_y + 53.3\text{ N} - 120\text{ N} - 100\text{ N} = 0$$

$$\underline{A_y = 167\text{ N}\uparrow \textit{ check}}$$

EXAMPLE 4–17

Find the reactions at A and C for the pin-connected structure shown in Figure 4–64.

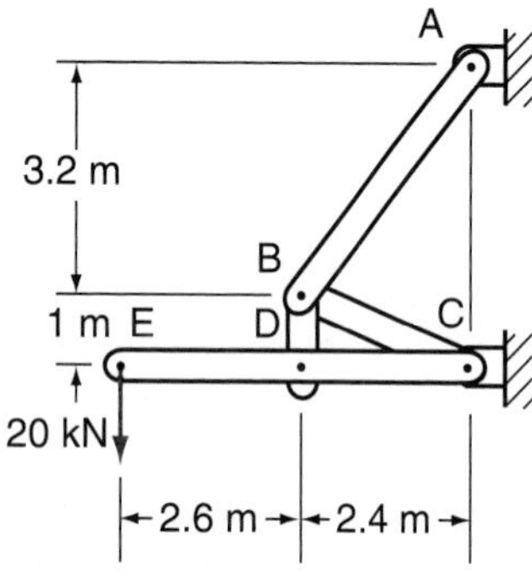

FIGURE 4–64

$\Sigma M_C = 0$ (Figure 4–65)

$$-(A_x)(4.2\text{ m}) + (20\text{ kN})(5\text{ m}) = 0$$

$$\underline{A_x = 23.8\text{ kN} \rightarrow}$$

Free-Body Diagram of Frame

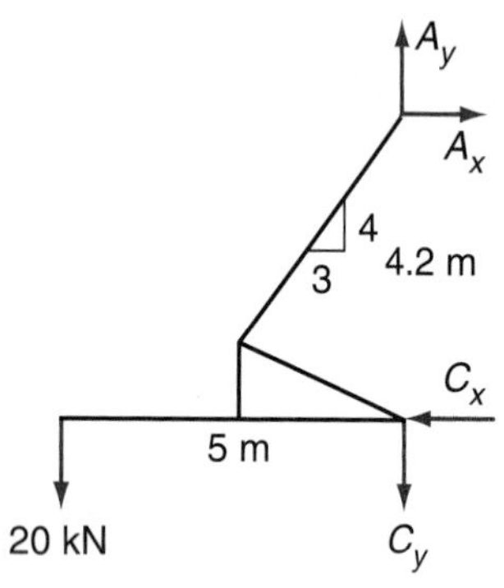

FIGURE 4–65

Because AB is a two-force member, $A_x + A_y$ must be in the direction (slope 4 vertical to 3 horizontal). Therefore,

$$A_y = \frac{4}{3}(A_x)$$

$$= \frac{4}{3}(23.8\text{ kN})$$

$$\underline{A_y = 31.7\text{ kN}\uparrow}$$

$\Sigma F_x = 0$

$$C_x = A_x$$

$$\underline{C_x = 23.8\text{ kN} \leftarrow}$$

$\Sigma F_y = 0$

$$31.7\text{ kN} - C_y - 20\text{ kN} = 0$$

$$\underline{C_y = 11.7\text{ kN}\downarrow}$$

EXAMPLE 4–18

For the lever shown in Figure 4–66, solve for the pin reactions at B. Assume a smooth surface at C.

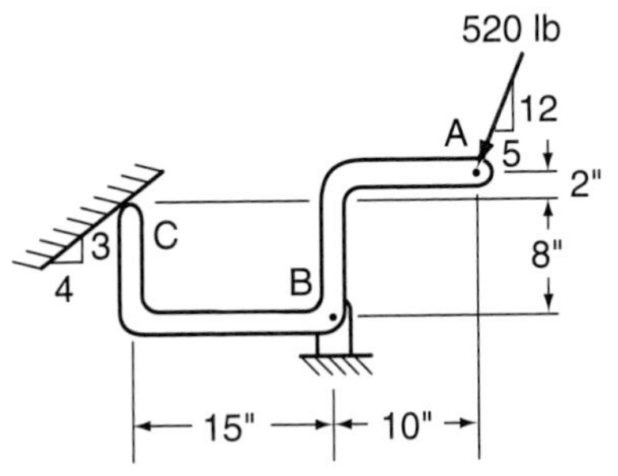

FIGURE 4–66

The free-body diagram may be constructed as in either Figures 4–67 or 4–68. Notice the slope of force *C*. We were given a slope of 3 to 4 for the smooth surface at *C*. The reaction force *C* is perpendicular to this surface and therefore has a slope of 4 to 3. This is a rule that may be applied to all lines that are perpendicular to one another; that is, when the slope of one is known, merely reverse the numbers to find the slope of the other.

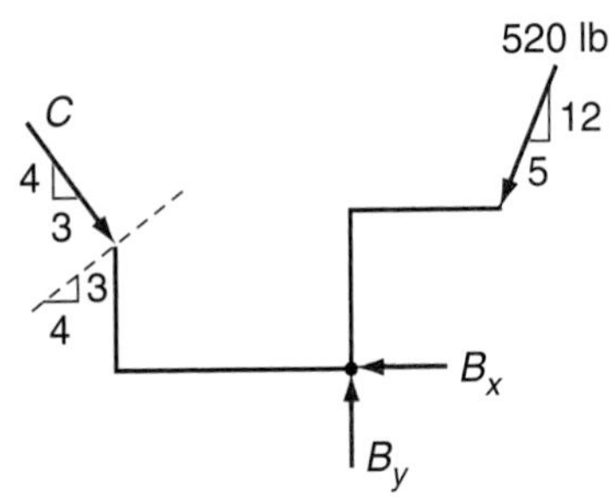

FIGURE 4–67

The free-body diagram in Figure 4–68 shows all forces as horizontal and vertical components. The value of *C* must be found in order to obtain B_x and B_y; therefore, taking moments about point B (Figure 4–68), we get

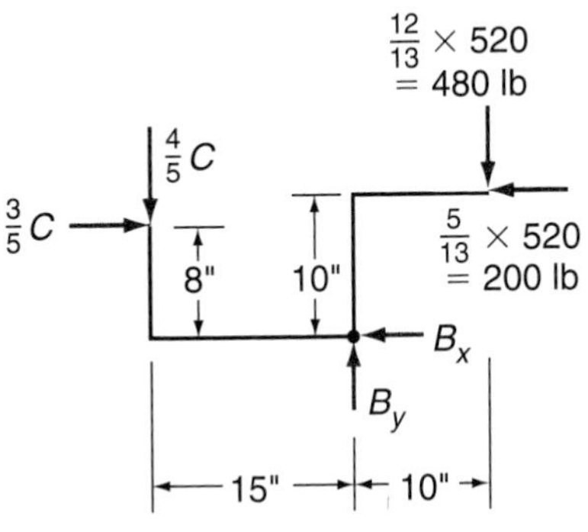

FIGURE 4–68

$\Sigma M_B = 0$

or counterclockwise moment = clockwise moment

$$\frac{4}{5}C(15\text{ in.}) + (200\text{ lb})(10\text{ in.}) = \frac{3}{5}C(8\text{ in.}) + (480\text{ lb})(10\text{ in.})$$

$$12C + 2000 = 4.8C + 4800$$

$$7.2C = 2800$$

$$C = 389\text{ lb}$$ (slope 4 to 3, directed down and to the right)

$\Sigma F_x = 0$

or forces right = forces left

$$\frac{3}{5}(389 \text{ lb}) = B_x + 200 \text{ lb}$$

$$\underline{B_x = 33 \text{ lb} \leftarrow}$$

$\Sigma F_y = 0$
or forces up = forces down

$$B_y = \frac{4}{5}(389 \text{ lb}) + 480 \text{ lb}$$

$$\underline{B_y = 791 \text{ lb} \uparrow}$$

EXAMPLE 4–19

Calculate the horizontal and vertical reactions at point E of the frame shown in Figure 4–69. AB is a cable.

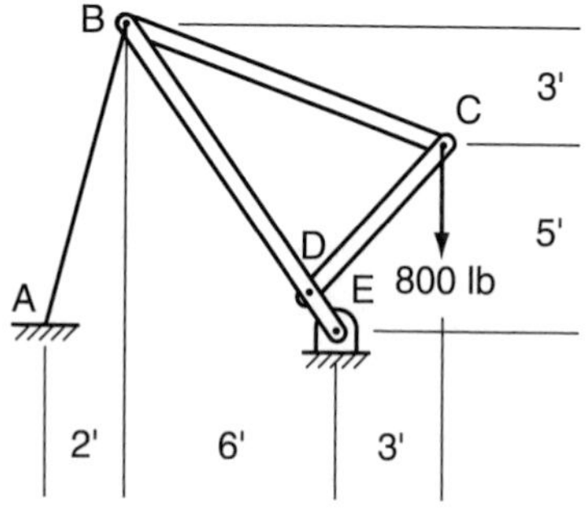

FIGURE 4–69

Free-Body Diagram of Frame

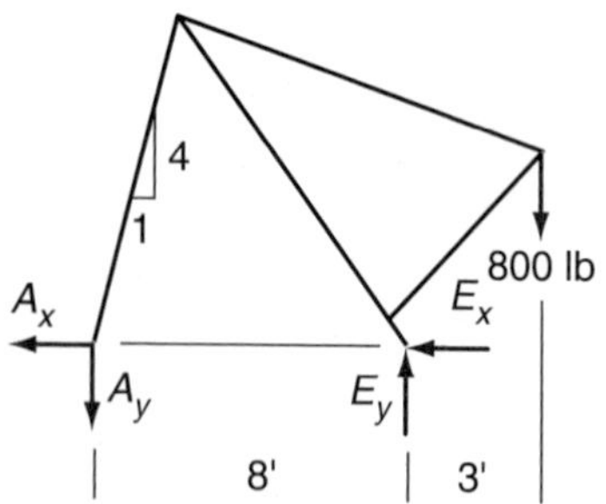

FIGURE 4–70

Note that it is preferable to show the cable tension components at A rather than at B since, when moments are taken about E, A_x passes through the pivot (Figure 4–70).

$\Sigma M_E = 0$

$$A_y(8 \text{ ft}) - (800 \text{ lb})(3 \text{ ft}) = 0$$

$$\underline{A_y = 300 \text{ lb} \downarrow}$$

$$A_x = \frac{1}{4}(A_y)$$

$$= \frac{1}{4}(300 \text{ lb})$$

$$\underline{A_x = 75 \text{ lb} \leftarrow}$$

$$\Sigma M_A = 0$$

$$+(8 \text{ ft})E_y - (800 \text{ lb})(11 \text{ ft}) = 0$$

$$\underline{E_y = 1100 \text{ lb} \uparrow}$$

$$\Sigma F_x = 0$$

$$-75 \text{ lb} - E_x = 0$$

$$E_x = -75 \text{ lb} \leftarrow$$

$$\underline{E_x = +75 \text{ lb} \rightarrow}$$

Our calculation is still correct—only the direction of E_x must be changed.

HINTS FOR PROBLEM SOLVING

1. When drawing a free-body diagram, remember:
   (a) A FBD *of* a member shows forces acting *on* the member.
   (b) Replace a support with an equivalent force acting *on* the member as shown in the conventions of Section 4–2. (Do not show internal forces if you have not changed or removed anything.)
   (c) Locate all the points where forces should be acting.
   (d) Show forces and label each.
   (e) Show slopes of forces if possible.
   (f) Label distances.
   (g) Cables cannot push and rollers cannot pull on a member.
   (h) Internal forces, between connecting members, switch directions when you switch free-body diagrams.
2. For concurrent force systems, draw only free-body diagrams of pins or points, not whole members.
3. For free-body diagrams that result in simultaneous equations, the alternate solution is to use the vector triangle and sine law (often a shorter solution).
4. If a pulley has been removed, the cable forces can be shown acting at the center of the pulley.
5. A two-force member that is in *compression* pushes on the pins at each end and, if in *tension,* pulls on each pin. The slope of the force is the same as the slope of the member.
6. Assuming an incorrect force direction on an FBD will give an answer of the correct magnitude, but will be negative. If you choose not to go back and change your solution to the correct direction, you must use the negative answer as a negative value in any later calculations.

7. Check to be sure you have not mistakenly switched the given slope numbers.
8. Thoroughly check that the FBD is *complete* before beginning calculations.
9. When writing equations using $\Sigma F_x = 0$, $\Sigma F_y = 0$, and $\Sigma M = 0$, you may *equate* opposite forces or opposite moments rather than use the sign convention previously used.

PROBLEMS

APPLIED PROBLEMS FOR SECTIONS 4–1 TO 4–4

4–1. Draw a free-body diagram of pump handle AC in Figure P4–1.

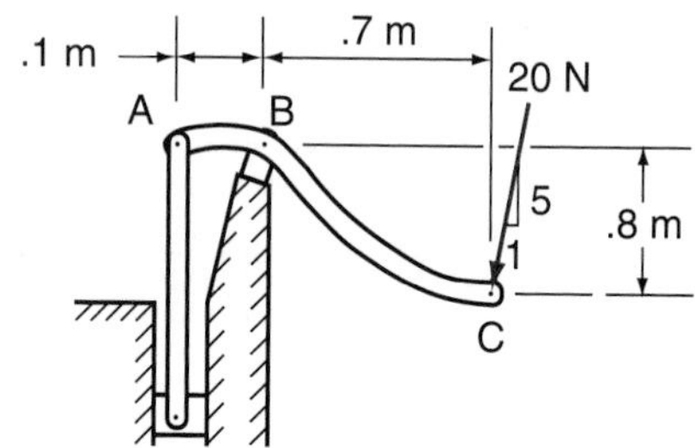

FIGURE P4–1

4–2. Draw a free-body diagram of member AC in Figure P4–2.

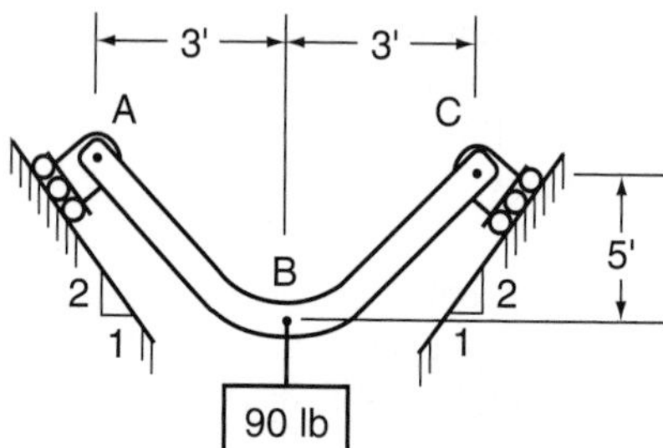

FIGURE P4–2

4–3. Draw a free-body diagram of member BD in Figure P4–3.

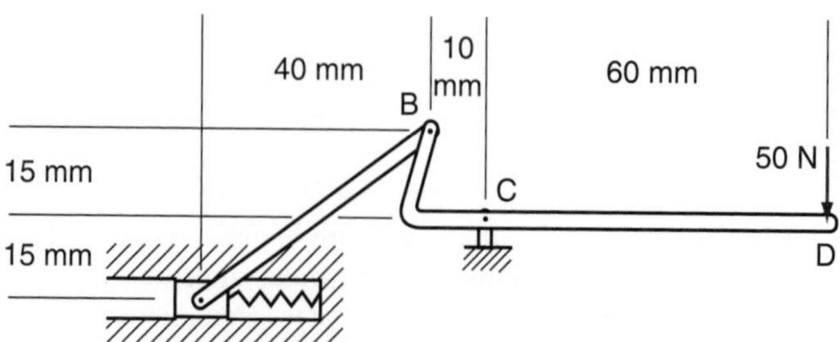

FIGURE P4–3

4–4. Draw a free-body diagram of member AC in Figure P4–4.

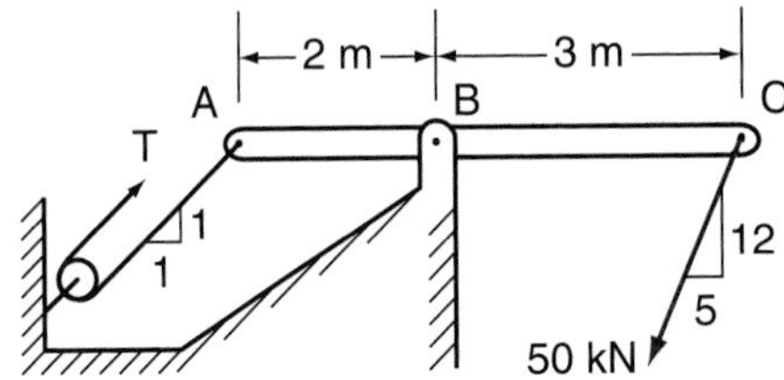

FIGURE P4–4

4–5. Rollers A and B each weigh 200 N (Figure P4–5). Assume smooth surfaces and draw a free-body diagram of roller A.

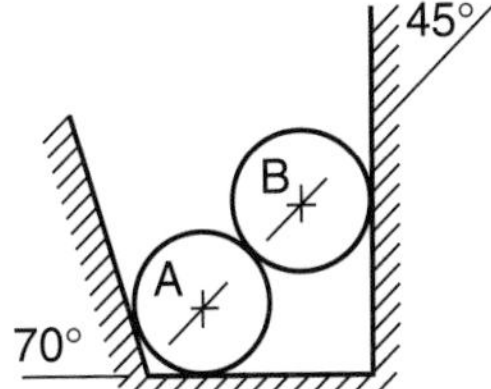

FIGURE P4–5

4–6. Draw a free-body diagram of the frame shown in Figure P4–6.

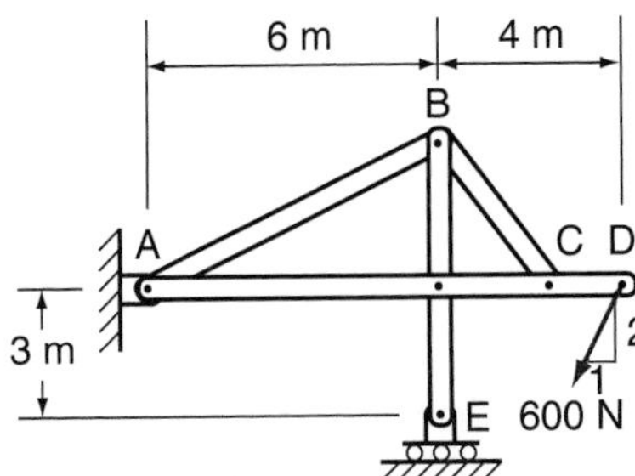

FIGURE P4–6

4–7. The two-force member CD has a compressive load of 2 kips when the frame in Figure P4–7 is loaded as shown. Draw a free-body diagram of member BE, and label all horizontal and vertical components of forces acting on it. (Do not calculate actual values. Surface at A is smooth.)

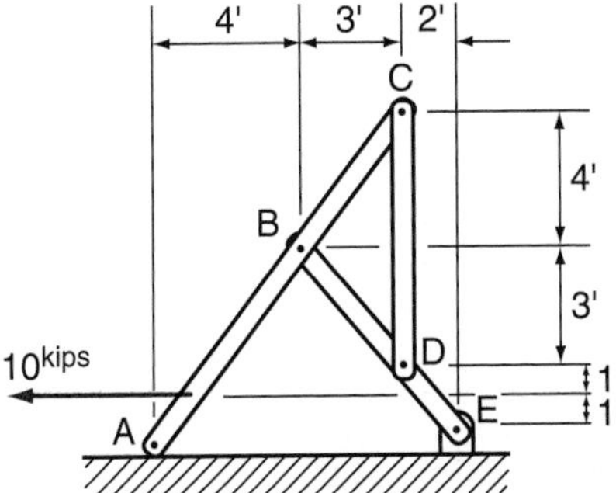

FIGURE P4–7

4–8. Draw a free-body diagram of member AC in Figure P4–8, and label all the horizontal and vertical components of the forces acting on it. Show the forces at B as fractions of BD.

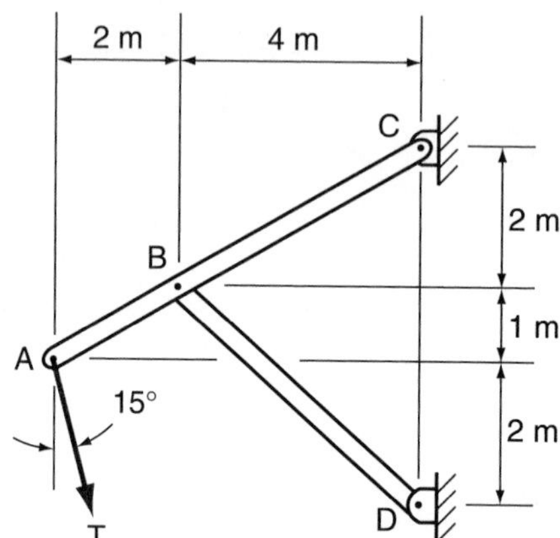

FIGURE P4–8

4–9. A jib crane supports an 80-lb load (Figure P4–9). Assume that all surfaces are smooth. Draw a free-body diagram of the crane frame and label all forces.

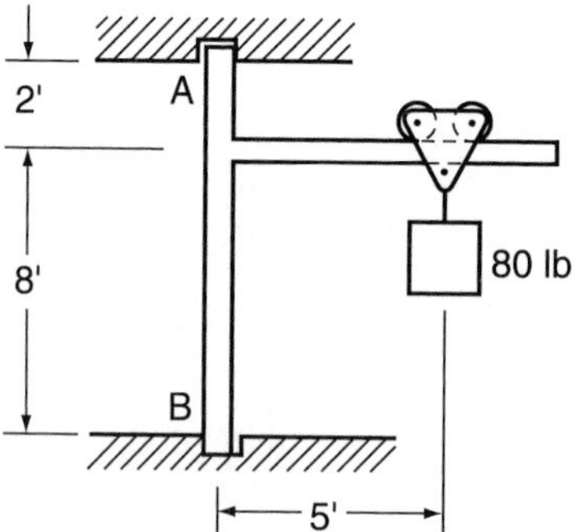

FIGURE P4–9

4–10. Draw a free-body diagram of member CD in Figure P4–10.

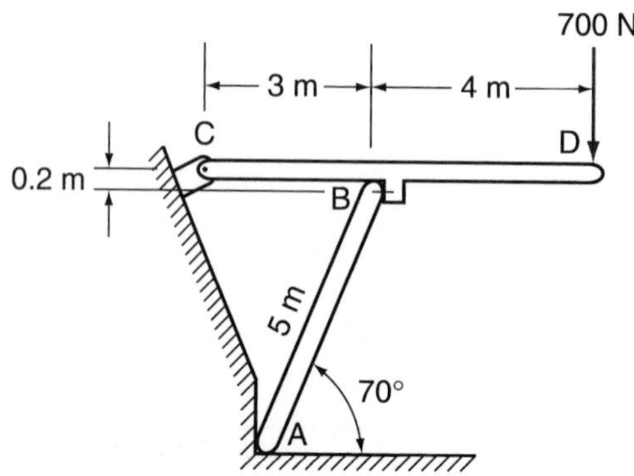

FIGURE P4–10

4–11. Draw a free-body diagram of member DE in Figure P4–11. Do not calculate values but label all forces acting on DE as horizontal and vertical components.

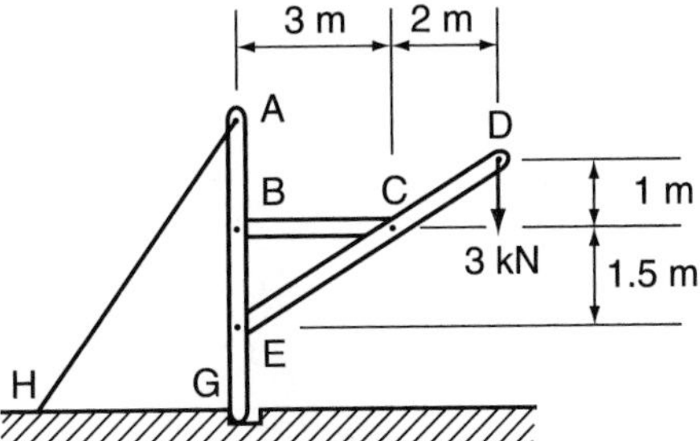

FIGURE P4–11

4–12. Draw a free-body diagram of member AE in Figure P4–12, and label all horizontal and vertical components. Show the components of BC and DC as fractions of the total load in each. (Do not calculate actual values.)

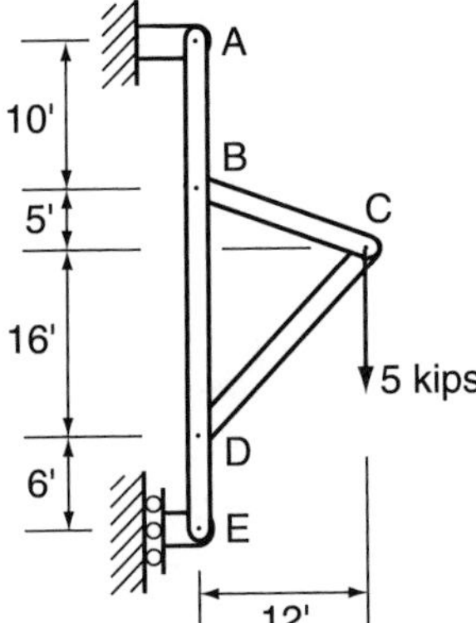

FIGURE P4–12

4–13. A truck-mounted articulating crane has a lifting force of 1800 N at the lifting hook when in the position shown (Figure P4–13). Draw a free-body diagram of arm ABC.

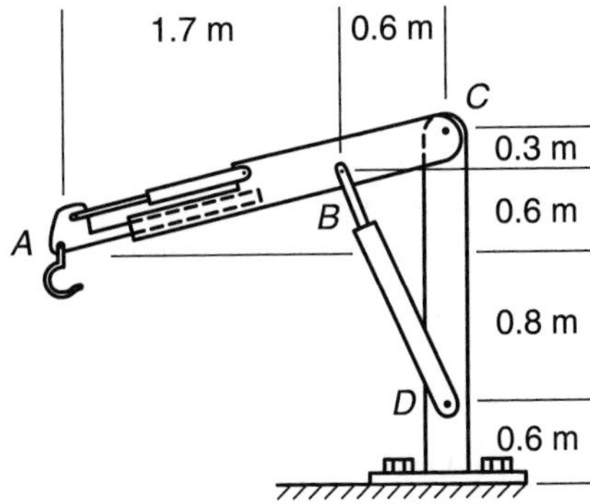

FIGURE P4–13

4–14. A grain auger is supported by a frame and cylinder as shown in Figure P4–14. For the position shown, draw free-body diagrams of the auger tube EGH and frame member EDC. Neglect the weight of the frame and show the 900-lb weight of the auger tube as shown.

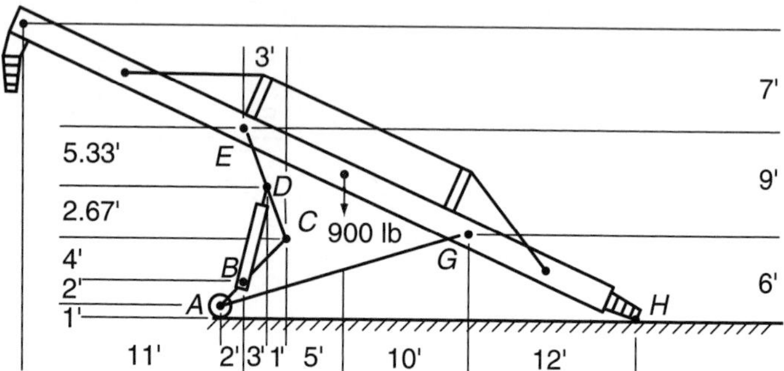

FIGURE P4–14

4–15. The bucket of a "bobcat" is lifted and tilted by means of hydraulic cylinders EL and HJ (Figure P4–15). Draw a free-body diagram of the bucket.

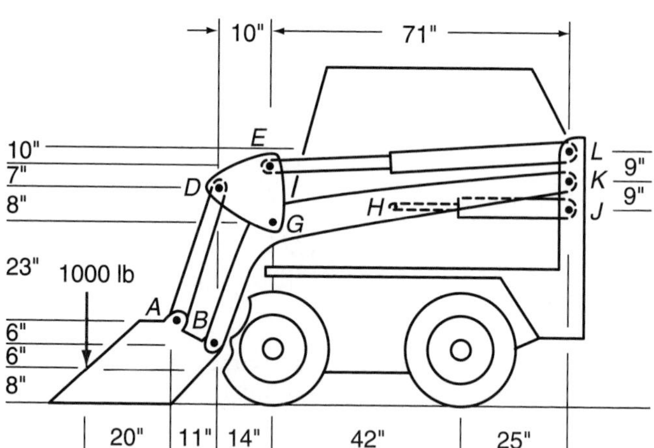

FIGURE P4–15

4–16. Draw a free-body diagram of plate DEG (Figure P4–15).

4–17. Draw a free-body diagram of loader arm BGK (Figure P4–15).

4–18. For the clamping wrench shown in Figure P4–18, draw free-body diagrams of members ABC, BDG, and ACEH.

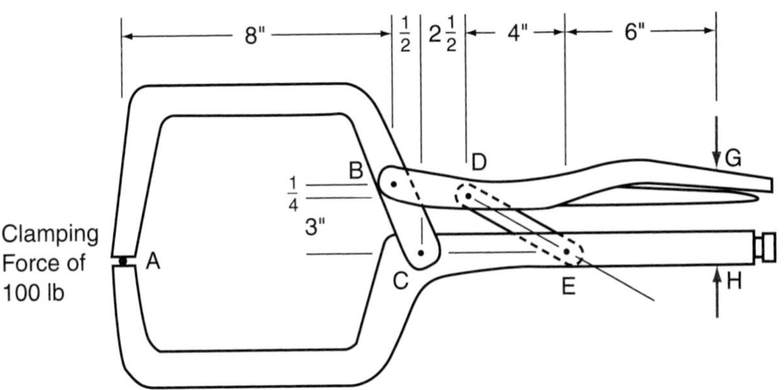

FIGURE P4–18

APPLIED PROBLEMS FOR SECTION 4–5

4–19. Draw a free-body diagram of beam AC in Figure P4–19.

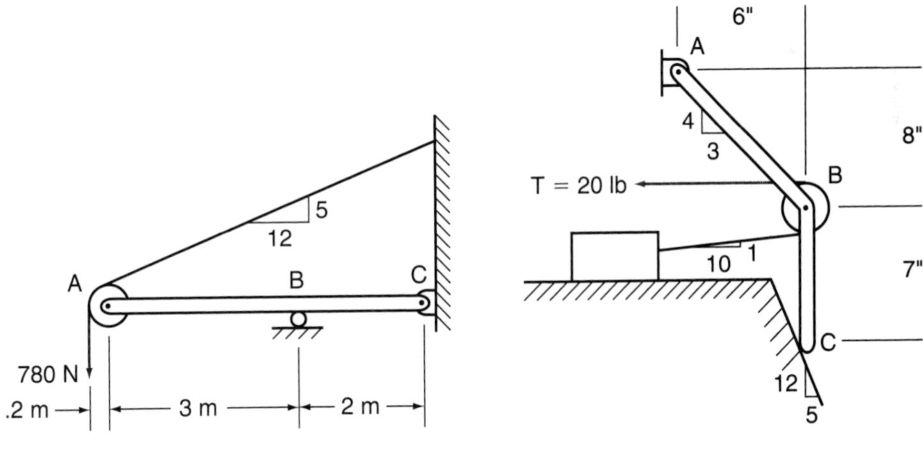

FIGURE P4–19

FIGURE P4–20

4–20. Draw a free-body diagram of member ABC (Figure P4–20).

4–21. Draw a free-body diagram of member ABDC (Figure P4–21).

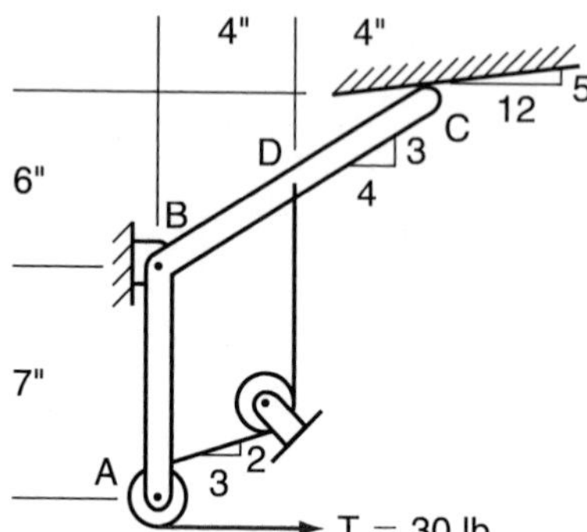

FIGURE P4–21

4–22. Draw free-body diagrams of members AD and EG shown in Figure P4–22.

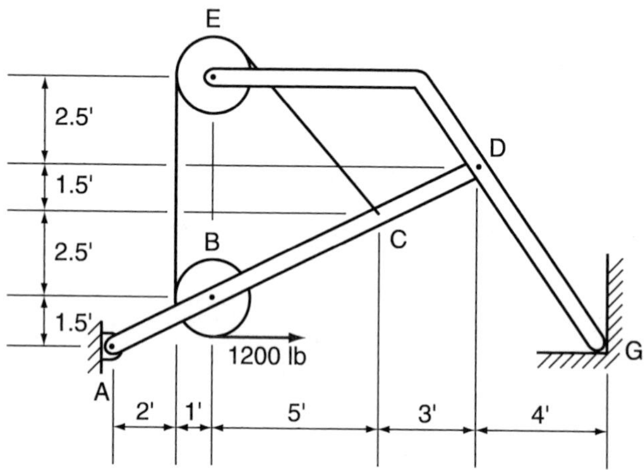

FIGURE P4–22

4–23. Draw free-body diagrams of members AG and CE (Figure P4–23).

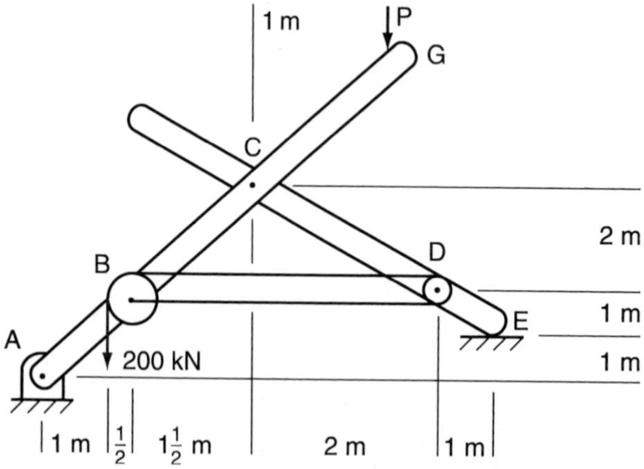

FIGURE P4–23

4–24. Draw free-body diagrams of members AC and BG of Figure P4–24.

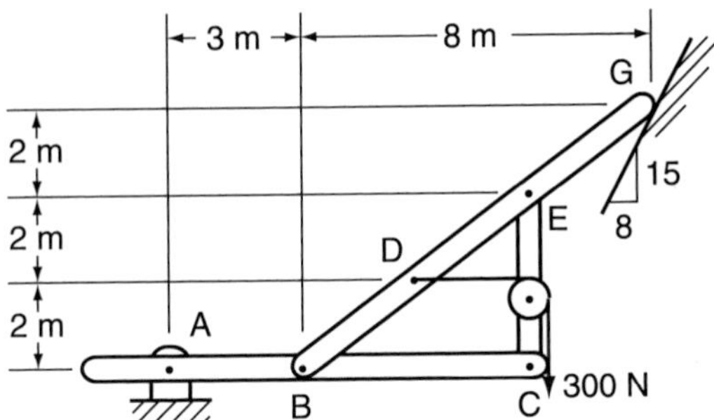

FIGURE P4–24

4–25. Determine the tension T in the cable for each pulley system in Figure P4–25.

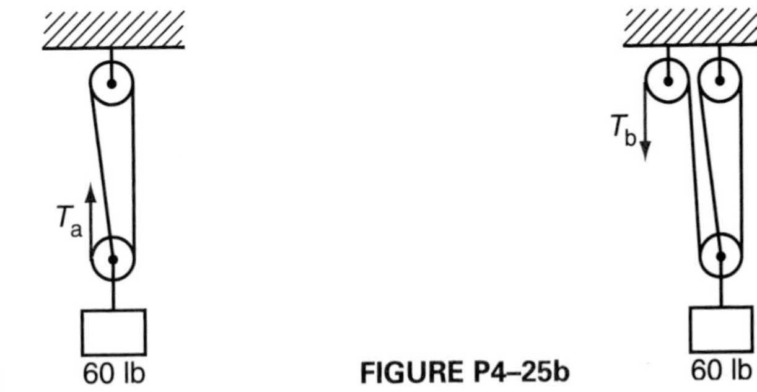

FIGURE P4–25a **FIGURE P4–25b**

4–26. Determine tension T for the pulley system shown in Figure P4–26. Show appropriate free-body diagrams to support your calculations.

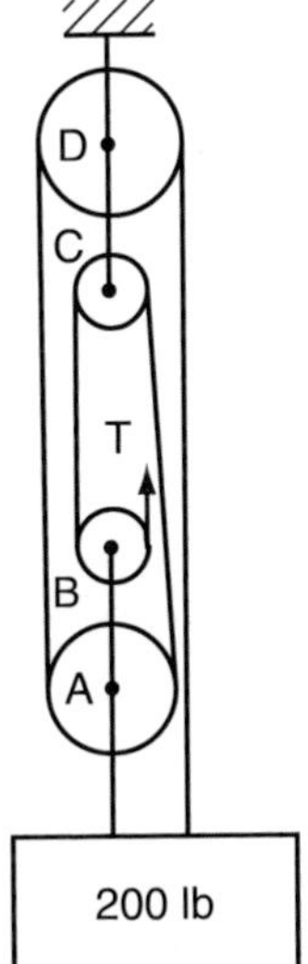

FIGURE P4–26

4–27. Determine the tension T in the cable for each pulley system in Figure P4–27.

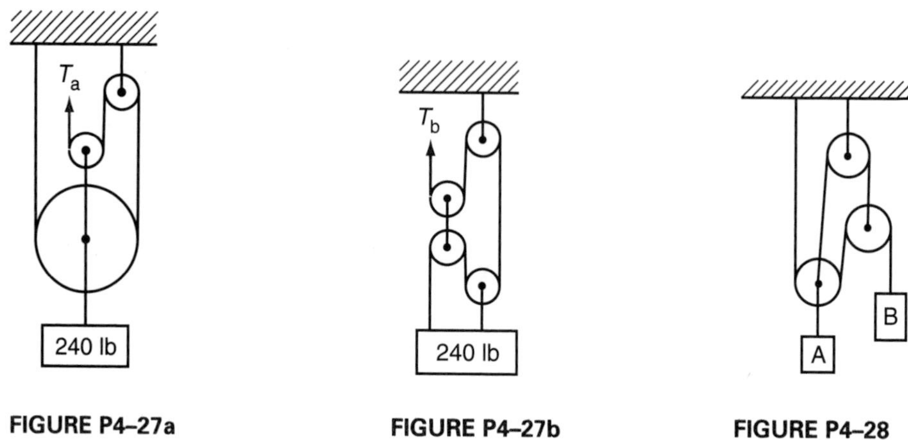

FIGURE P4–27a

FIGURE P4–27b

FIGURE P4–28

4–28. If A has a mass of 300 kg, determine the mass of B if the system in Figure P4–28 is to be in balance.

4–29. Determine tension T_B of Figure P4–29.

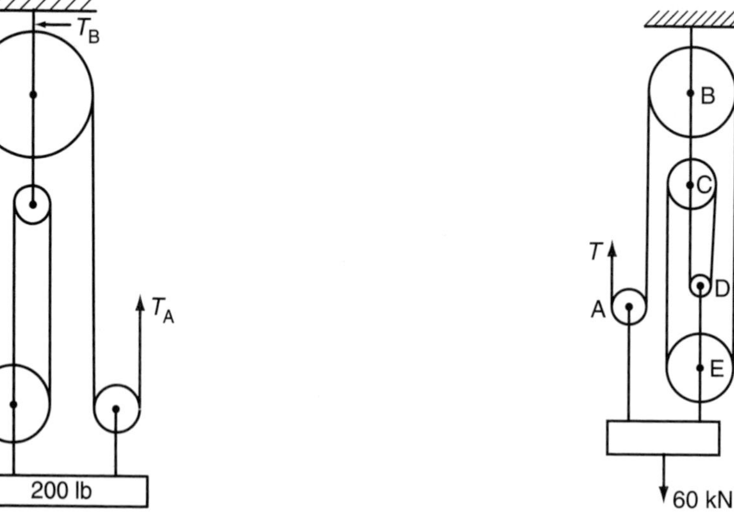

FIGURE P4–29

FIGURE P4–30

4–30. Determine tension T of the pulley system of Figure P4–30.

4–31. Determine tension T for static equilibrium of the system shown in Figure P4–31.

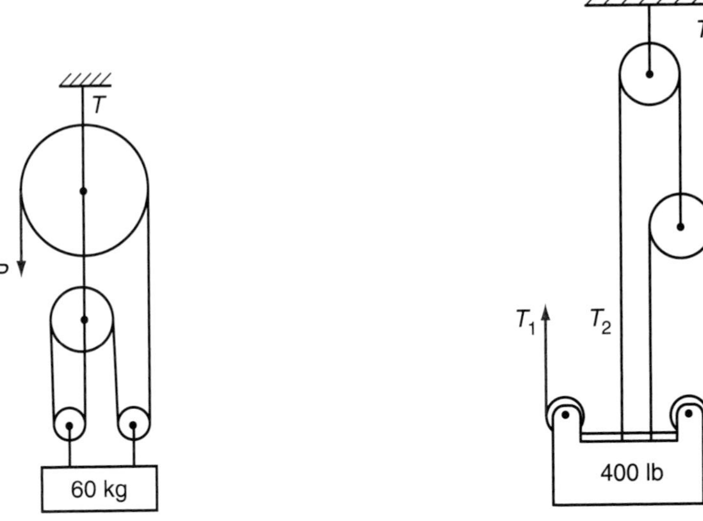

FIGURE P4–31

FIGURE P4–32

4–32. Determine tensions T, T_2, and T_3 for the system shown in Figure P4–32.

4–33. Determine the force T required to start pulling the crate up the slope if it starts to move when the 40 kN is applied (Figure P4–33).

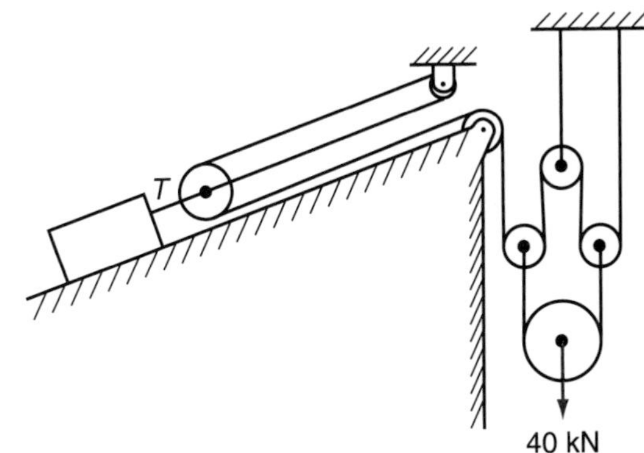

FIGURE P4–33

APPLIED PROBLEMS FOR SECTION 4–6

4–34. Determine the load in members AC and BC in Figure P4–34.

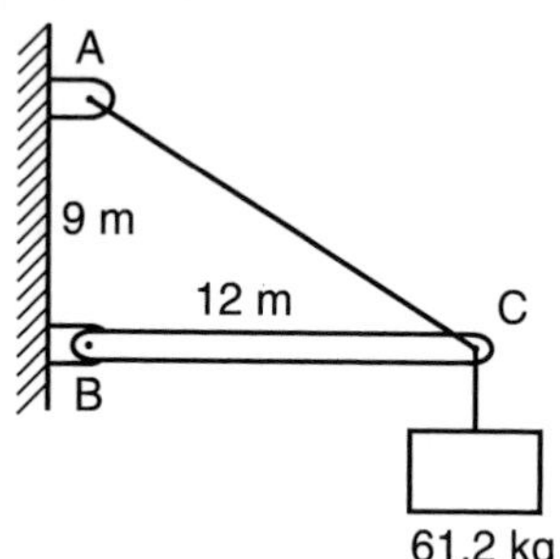

FIGURE P4–34

4–35. Determine the compressive loads in members AB and BC in Figure P4–35.

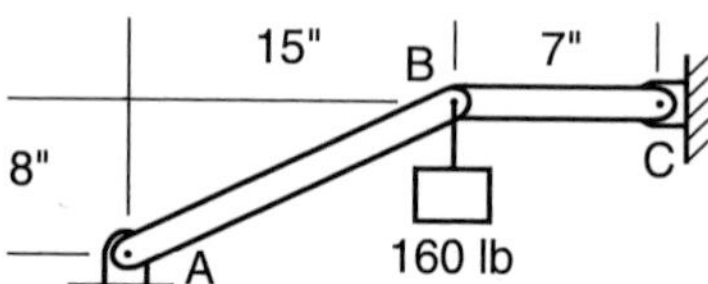

FIGURE P4–35

4–36. Determine the cable tension in each length of cable shown in Figure P4–36.

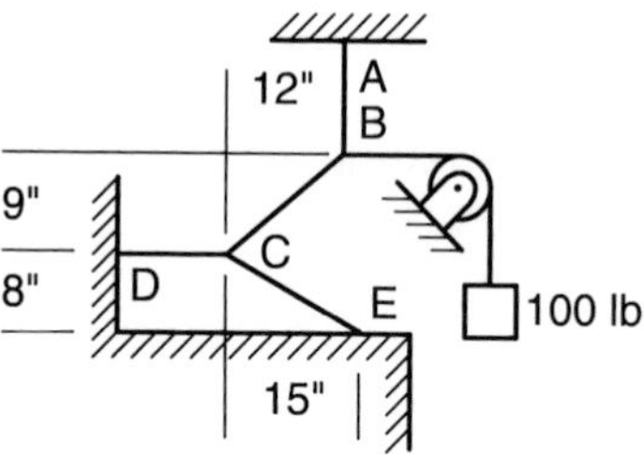

FIGURE P4–36

4–37. Determine the load in members AB and BC in Figure P4–37.

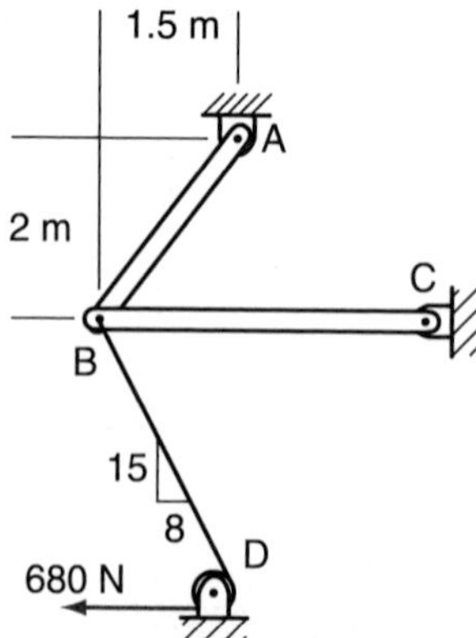

FIGURE P4–37

4–38. The system shown in Figure P4–38 is in static equilibrium and member AB has a load of 100 lb. Determine the weight *W*.

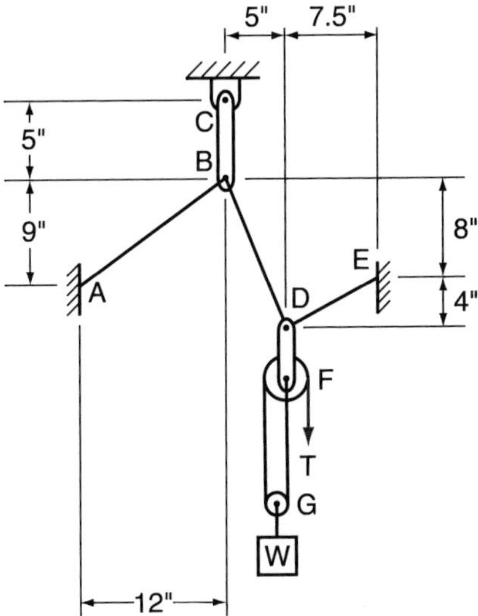

FIGURE P4–38

4–39. Determine the loads in members BC and BD of the system shown in Figure P4–39.

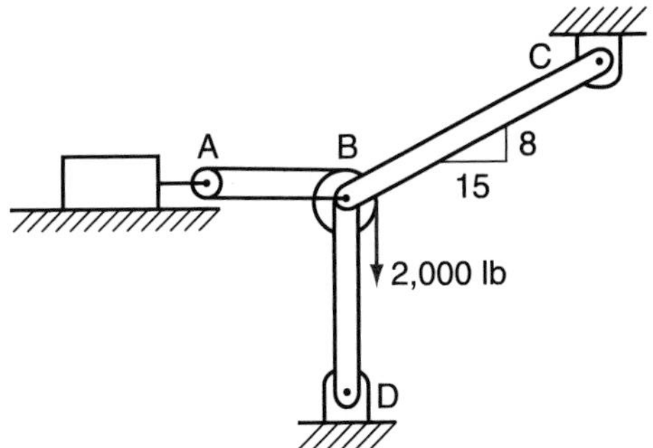

FIGURE P4–39

4–40. Determine the load in each member of the system shown in Figure P4–40.

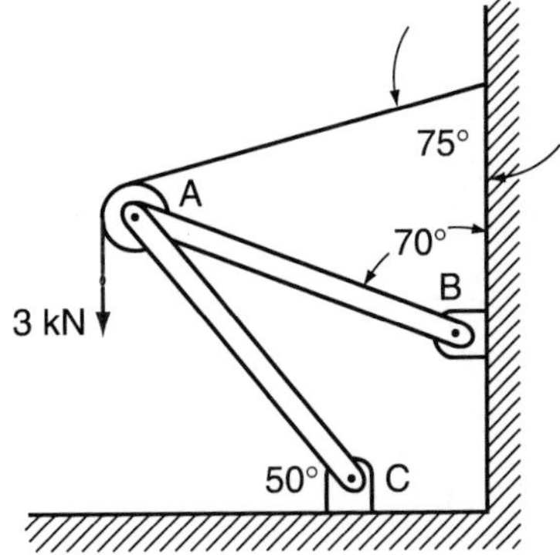

FIGURE P4–40

4–41. Determine the cylinder force required to lift the weight of 100 1b shown in Figure P4–41.

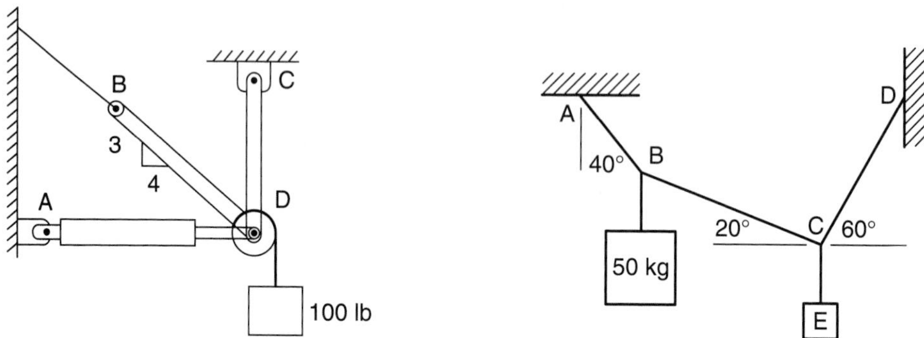

FIGURE P4–41

FIGURE P4–42

4–42. Determine all the cable tensions and the mass of block E of the system shown in Figure P4–42. Use the vector triangle method.

4–43. Determine the load in cable CD of the system shown in Figure P4–43.

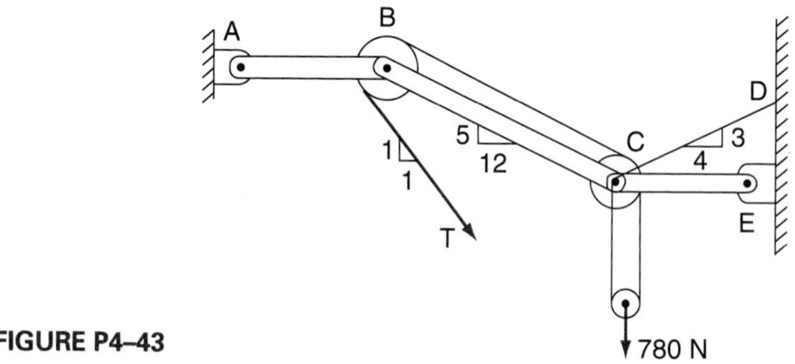

FIGURE P4–43

4–44. Blocks A and B each require 30-N cable tension to start them sliding (Figure P4–44). Determine force P and angle θ to cause them to start sliding simultaneously.

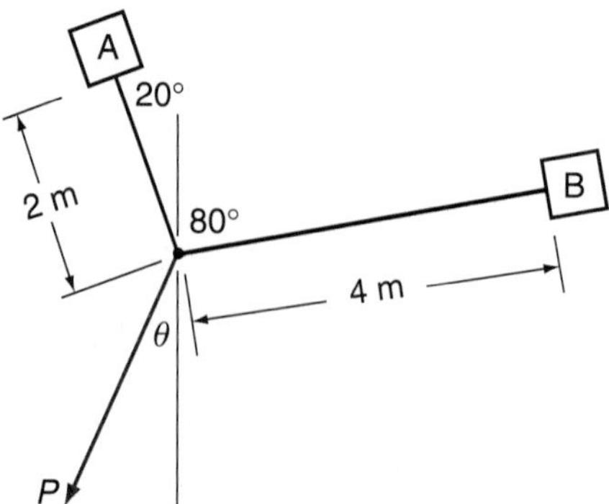

FIGURE P4–44

4–45. Each smooth roller in Figure P4–45 weighs 50 lb and has a diameter of 20 in. Calculate the reaction forces on the cylinder at points A and B.

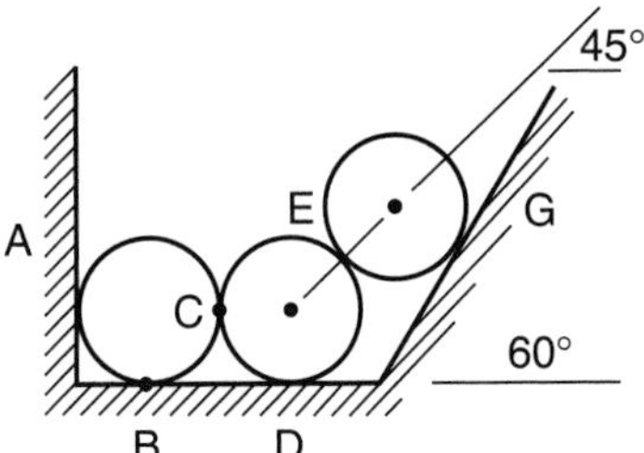

FIGURE P4–45

4–46. The cable and beam construction in Figure P4–46 supports a load of 2 kN. Determine the load in each beam and cable.

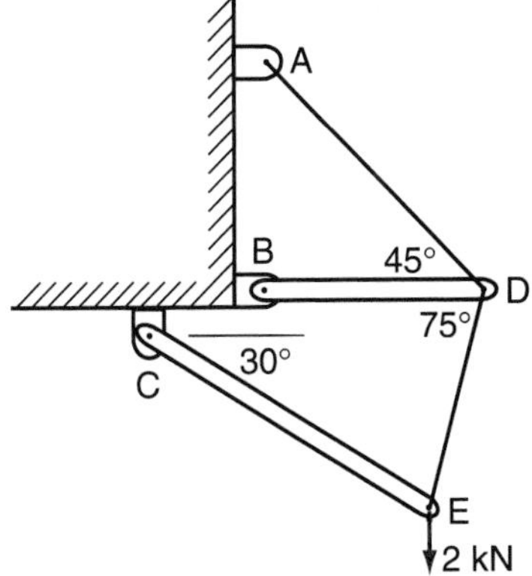

FIGURE P4–46

4–47. A belt is passed over two pulleys of equal diameter and then is tightened. Center-to-center distance between pulleys is 30 in. In checking the belt tension, the belt is pushed inward $\frac{1}{4}$ in. by a force of 10 lb at a point midway between the pulleys. Calculate the belt tension.

4–48. The belt tension is adjusted by means of idler pulley A in Figure P4–48. At the position shown, the belt tension is 65 N. What is the moment about B due to this tension?

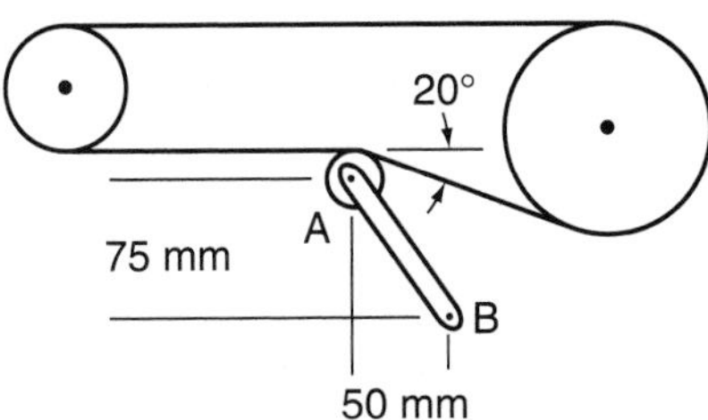

FIGURE 4–48

APPLIED PROBLEMS FOR SECTION 4–7

4–49—4–61. Determine the reactions at A and B for the beams loaded as shown in Figures P4–49 to P4–61. Beam weight may be neglected in all cases.

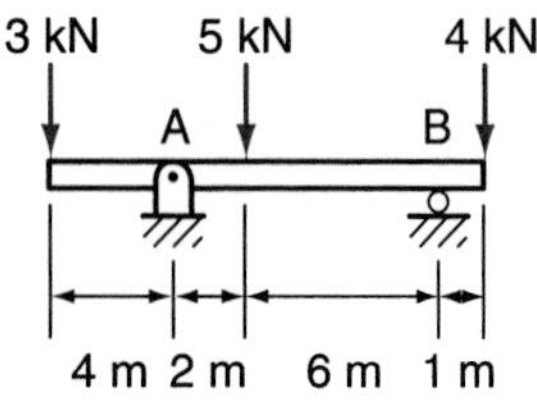

FIGURE P4–49

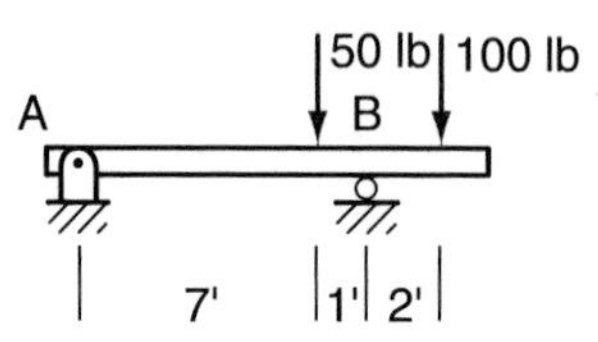

FIGURE P4–50

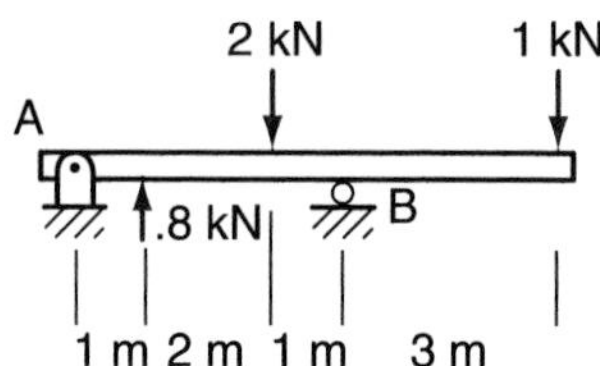

FIGURE P4–51

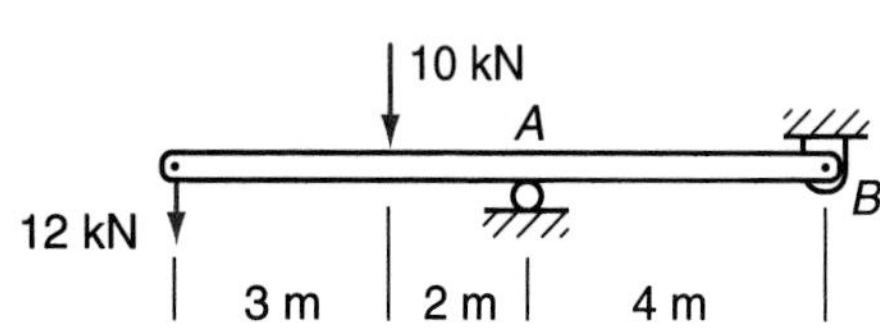

FIGURE P4–52

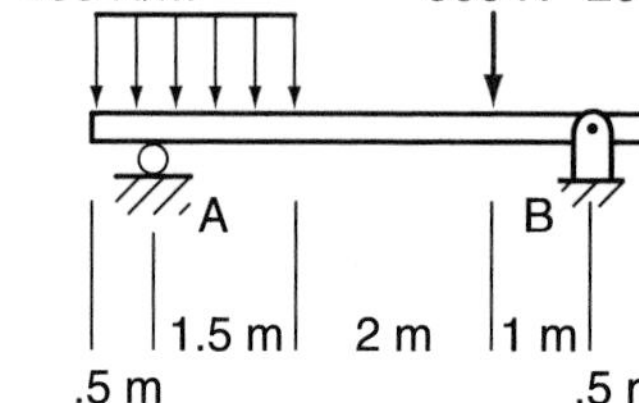

FIGURE P4–53

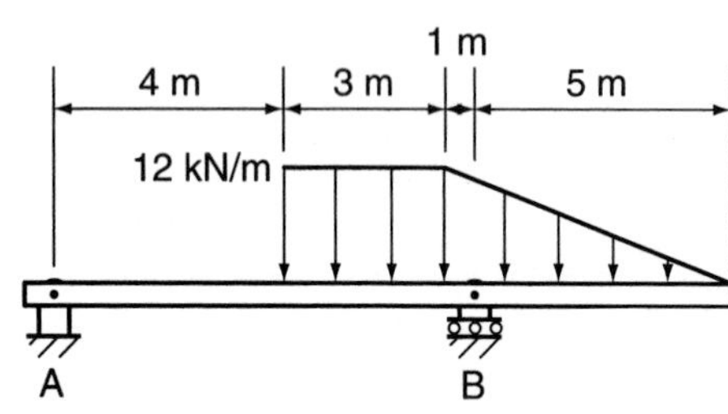

FIGURE P4–54

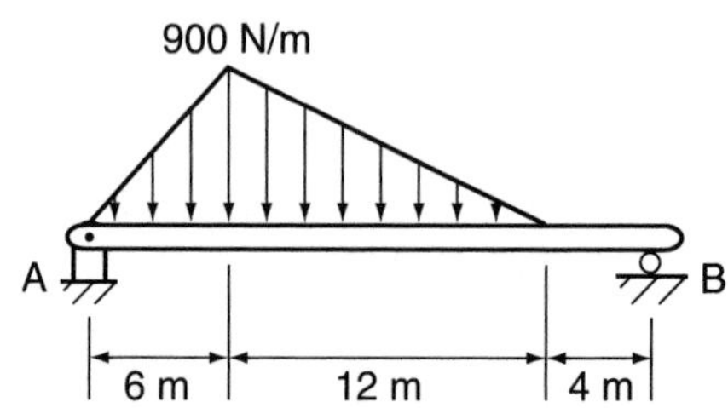

FIGURE P4–55

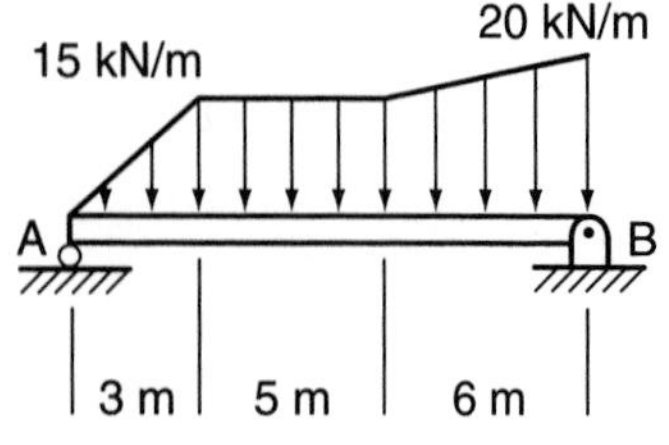

FIGURE P4–56

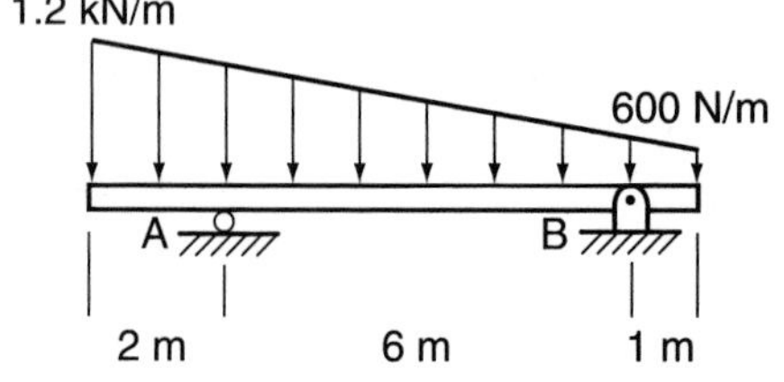

FIGURE P4–57

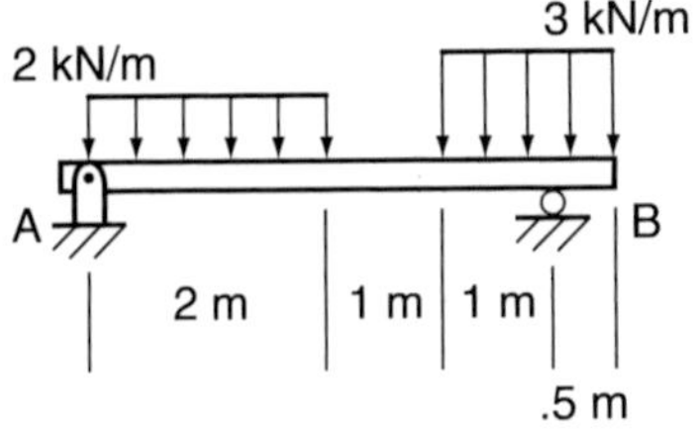

FIGURE P4–58

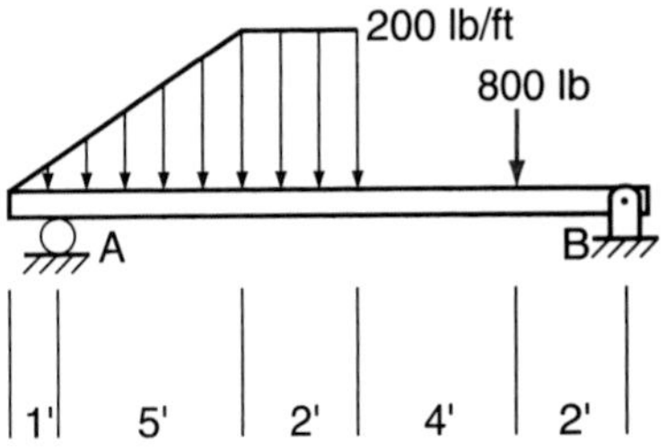

200 lb/ft
800 lb
A
B
1'
5'
2'
4'
2'

FIGURE P4–59

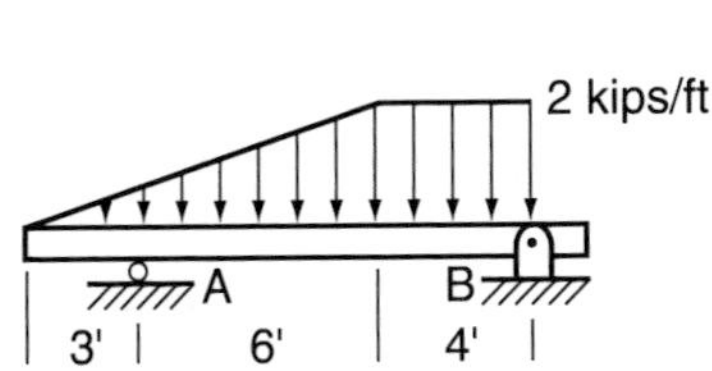

FIGURE P4–60

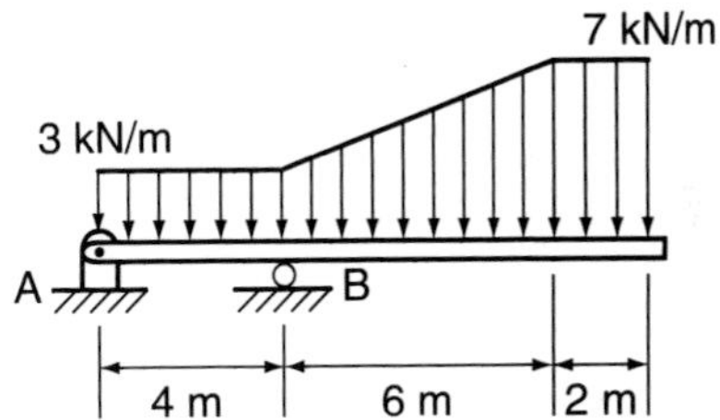

FIGURE P4–61

4–62. Determine the bolt tension at B for the combined beams shown in Figure P4–62.

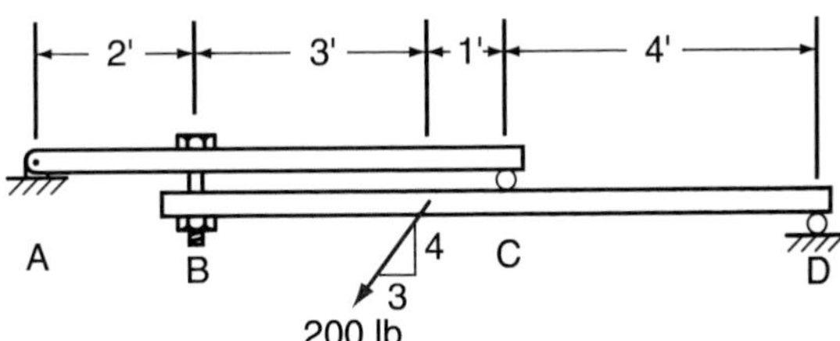

FIGURE P4–62

4–63. Determine the load in cable BE of the beam system shown in Figure P4–63.

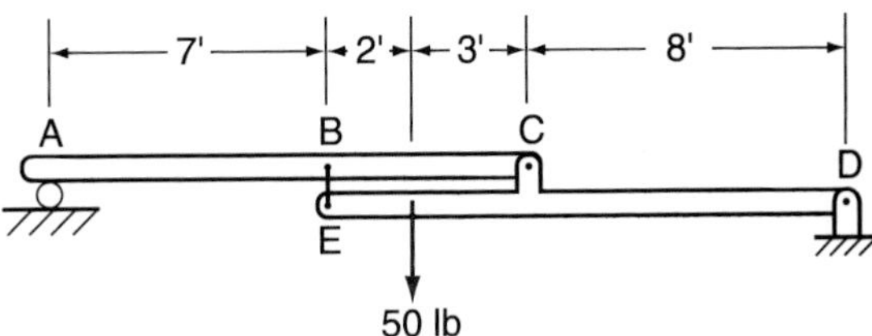

FIGURE P4–63

4–64. A centrifugal trash pump has individual weights and centers of gravity as shown in Figure P4–64. Calculate distance *d* at which the trailer wheels should be located so that the hitch weight is 200 1b.

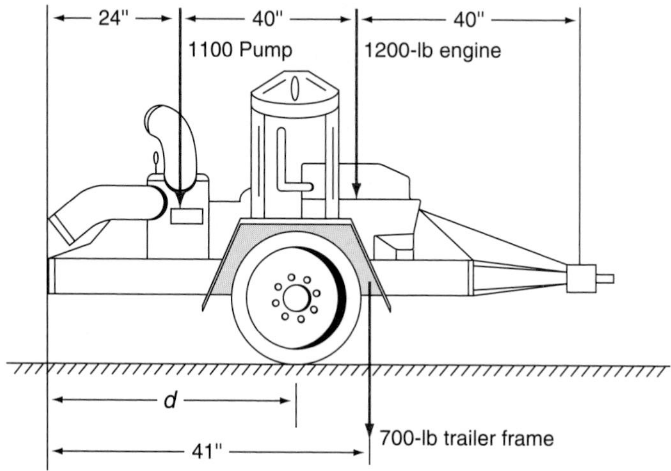

FIGURE P4–64

4–65. The roller shown in Figure P4–65 is driven onto the trailer (Figure P4–31) and the ramps are raised for transport. The trailer bed weighs 800 1b with its center of gravity located 18" in front of the tandem axle. How far ahead of the tandem axle should the front axle of the roller be located so that the hitch weight is 250 1b?

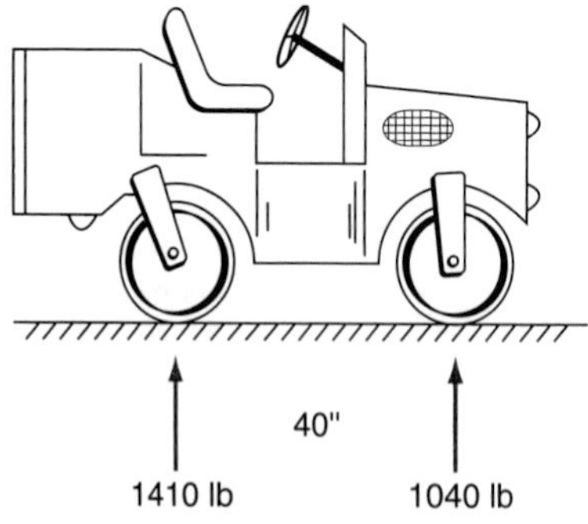

FIGURE P4–65

4–66. A 3000-lb truck when empty has 60 percent of its weight on the front wheels. An 800-lb tractor that has 65% of its weight on the rear wheels is placed in the truck as shown in Figure P4–66. Determine the load carried by each set of truck wheels.

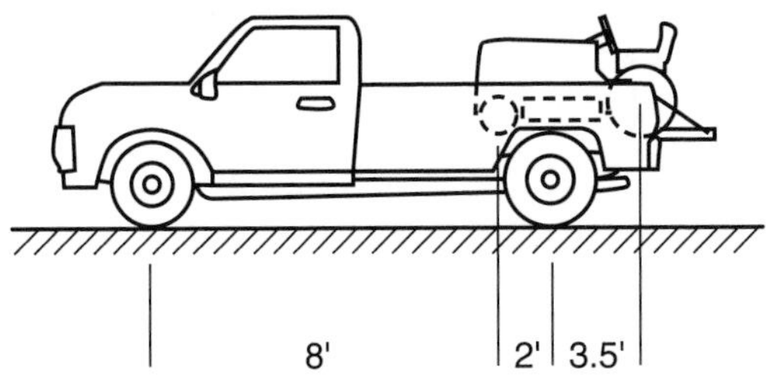

FIGURE P4–66

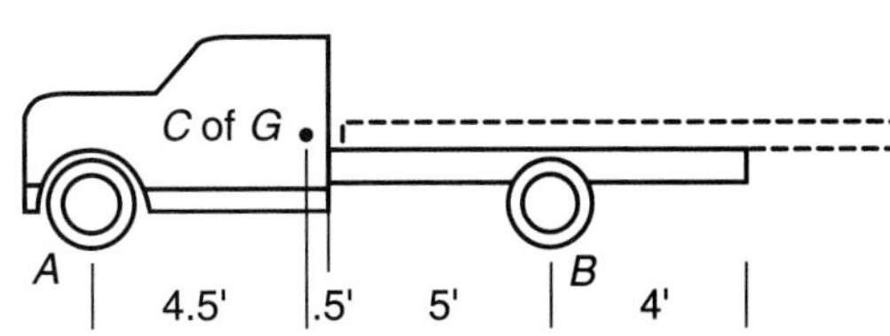

FIGURE P4–67

4–67. The flatbed truck shown in Figure P4–67 weighs 3000 lb and has a center of gravity as shown. How many 14-ft-long beams, each weighing 800 lb, could be placed on the flatbed, resulting in no weight on the truck's front wheels?

4–68. The tractor in Figure P4–68 weighs 5000 lb, the trailor weighs 4000 lb, and the load weighs 2000 lb. Determine the load on each set of wheels at A, B, and C.

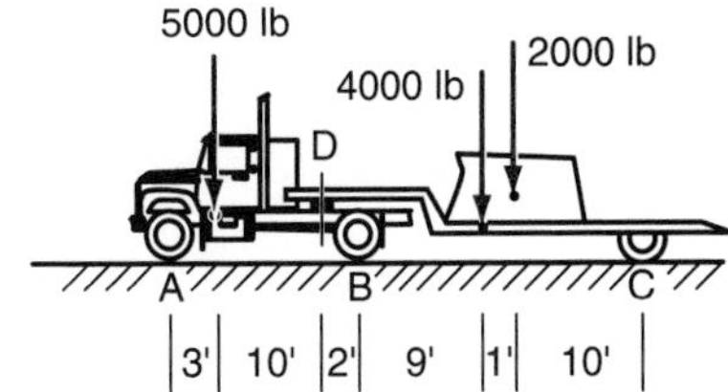

FIGURE P4–68

4–69. Determine the forces at B and C for the system shown (Figure P4–69).

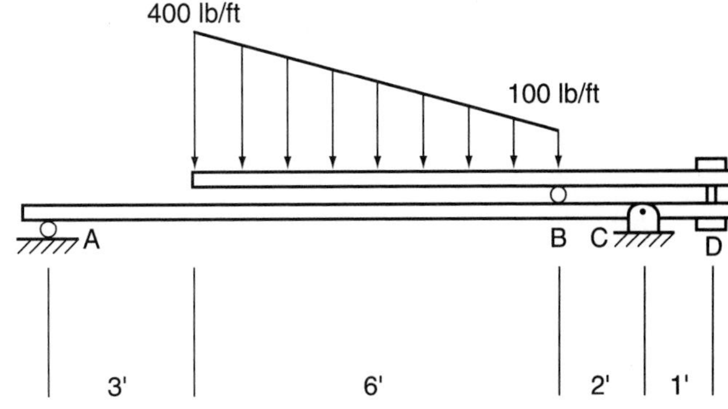

FIGURE P4–69

4–70. The two beams shown in Figure P4–70 are bolted together with spacers between them at B and C. Determine which bolt is in tension and what tensile load it is carrying.

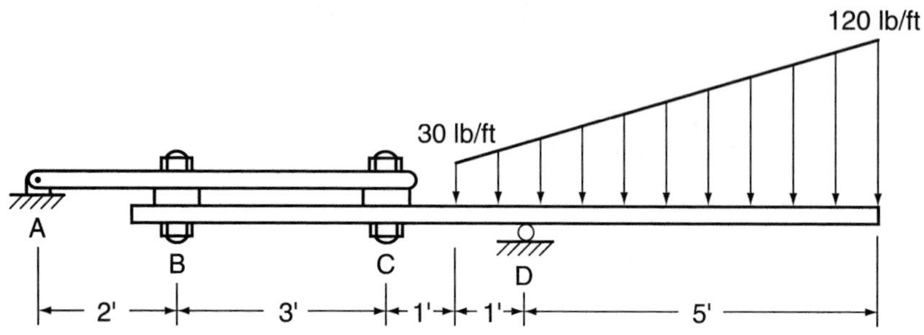

FIGURE P4–70

4–71. Determine the reactions at A, B, C, and D of the system shown in Figure P4–71.

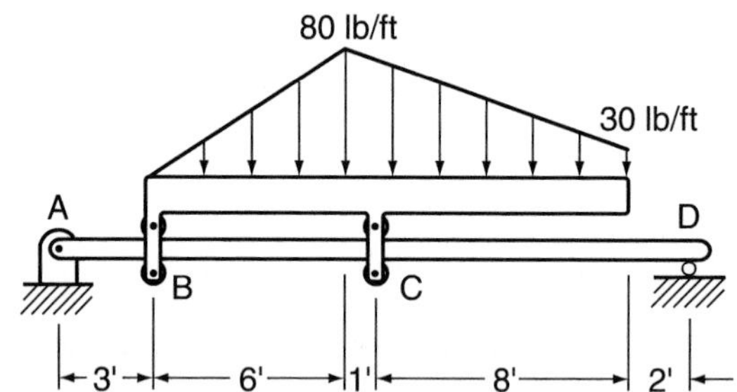

FIGURE P4–71

4–72. The top roller assembly is moved left until the reactions at A and D are equal (Figure P4–71). Determine the new horizontal dimension between A and B.

4–73. A 4-m-long beam is extended 2 m by bolting at points B and C (Figure P4–73). Neglecting friction, the maximum design load on bolt C is 8 kN. Determine the distance "d" between the bolts.

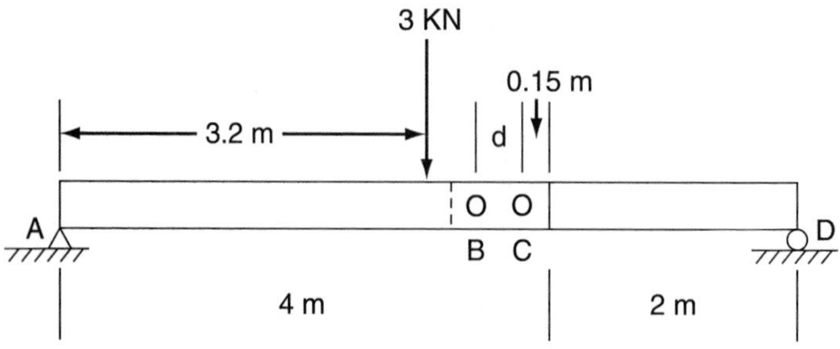

FIGURE P4–73

APPLIED PROBLEMS FOR SECTION 4–8

4–74. Find the reactions at A and B for the beam shown in Figure P4–74.

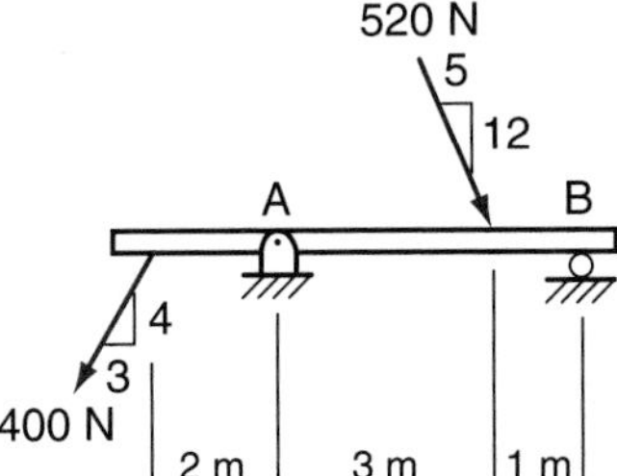

FIGURE P4–74

4–75. Determine the reactions at A and B for the beam shown in Figure P4–75.

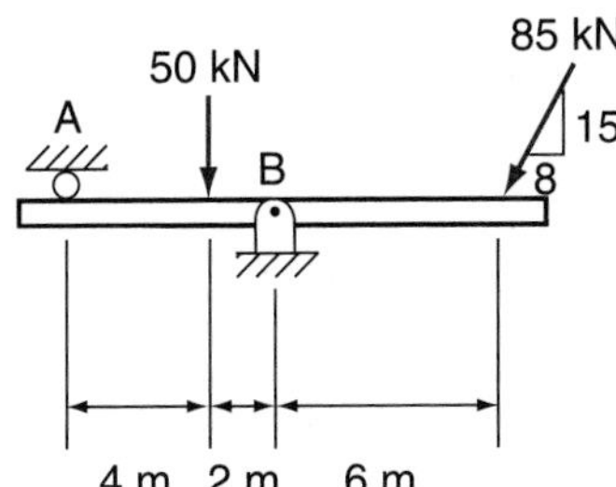

FIGURE P4–75

4–76. Find the cable tension and the reactions at A for the beam shown in Figure P4–76.

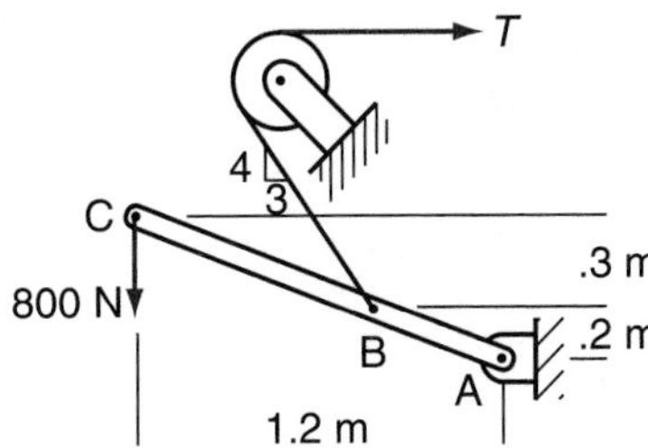

FIGURE P4–76

4–77. If the spring tension is 680 N in Figure P4–77, determine the reactions at B and C on member AC.

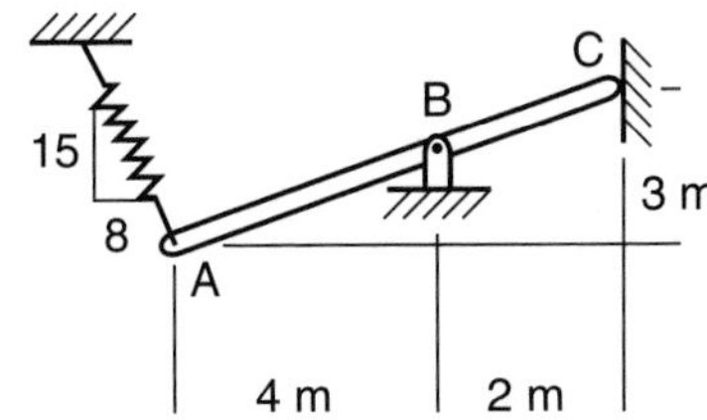

FIGURE P4–77

4–78. Determine the load in members AB and AC if force P is sufficient to hold the 3-kN force as shown in Figure P4–78.

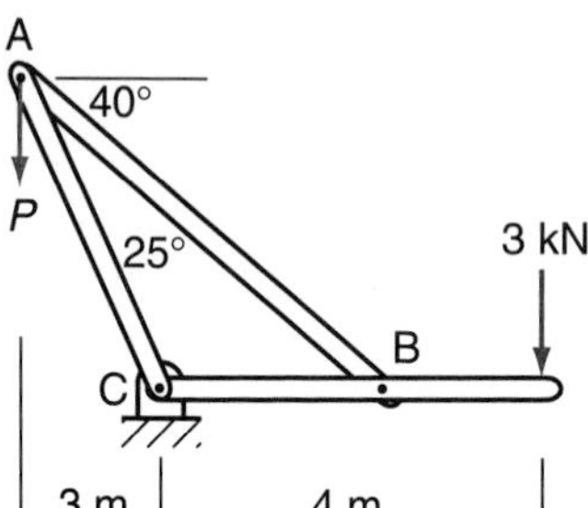

FIGURE P4–78

4–79. Assume smooth surfaces at points A and E in Figure P4–79. Determine the load in each member supporting the 3000-lb load.

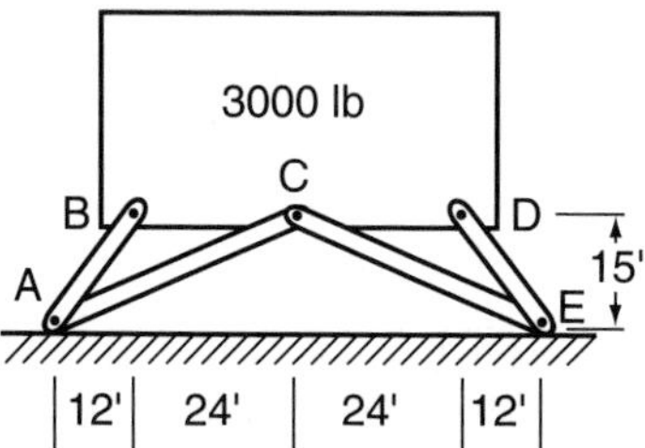

FIGURE P4–79

4–80. A storage rack is bolted to the wall at A and rests on the floor at B in Figure P4–80. Determine the reactions at A and B if the rack carries a total load of 900 lb and the load has a center of gravity 2 ft from the wall.

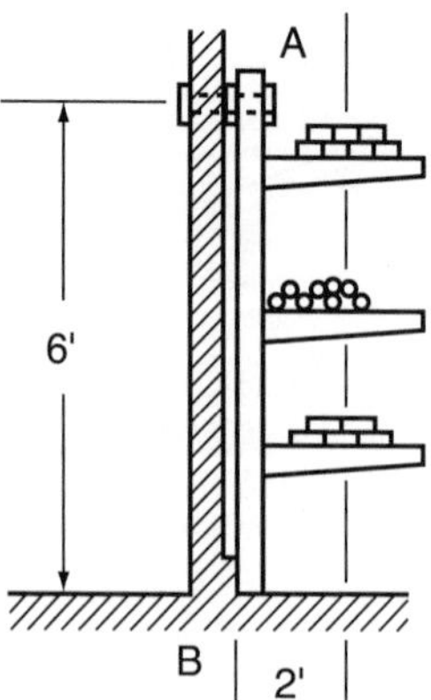

FIGURE P4–80

4–81. The frame shown in Figure P4–81 has a pulley at C that has a diameter of 2 ft. Determine the reaction on the frame at A and B. (The cable is parallel to AB.)

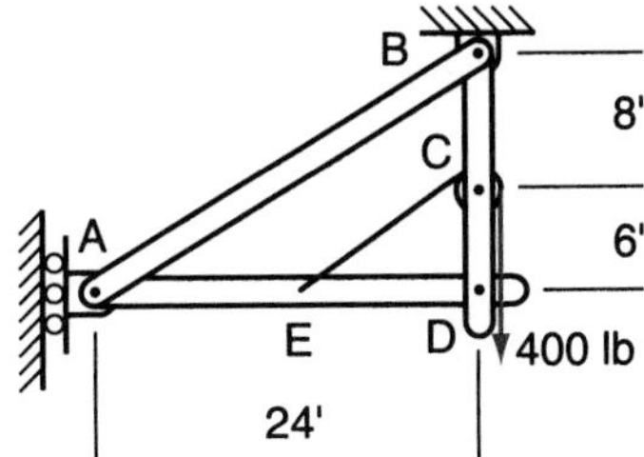

FIGURE P4–81

4–82. Determine the reactions at A and B for the frame shown in Figure P4–82.

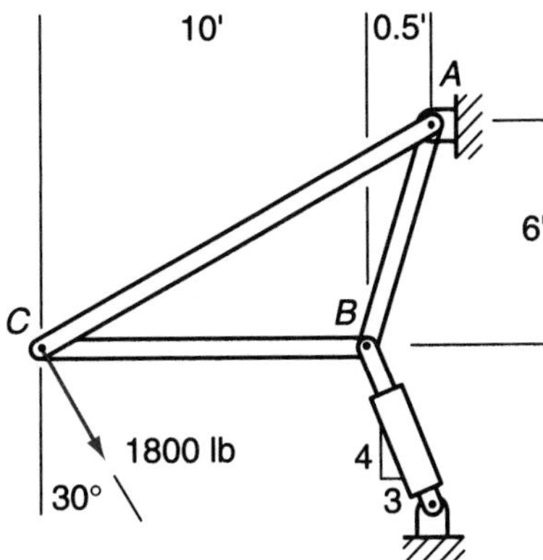

FIGURE P4–82

4–83. Determine the reactions at A for the beam shown in Figure P4–83.

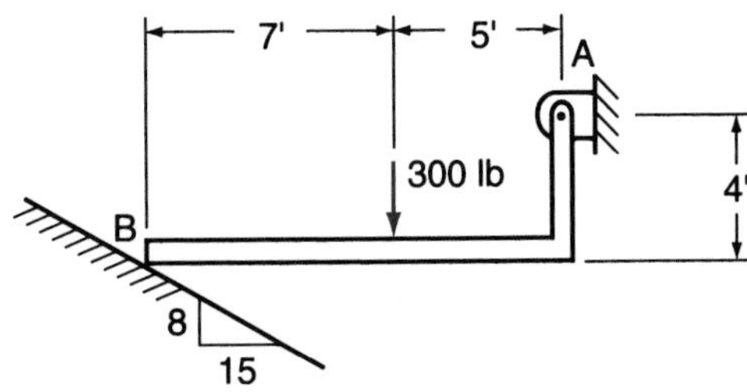

FIGURE P4–83

4–84. Determine the reactions at A and C for the beams shown in Figure P4–84.

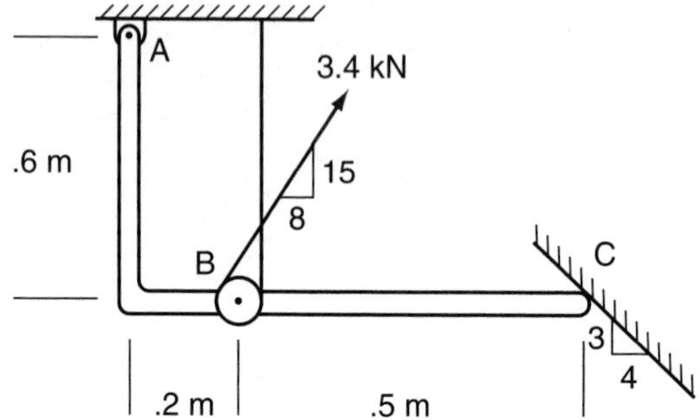

FIGURE P4–84

4–85. Determine the reactions at A and B for the system shown in Figure P4–85.

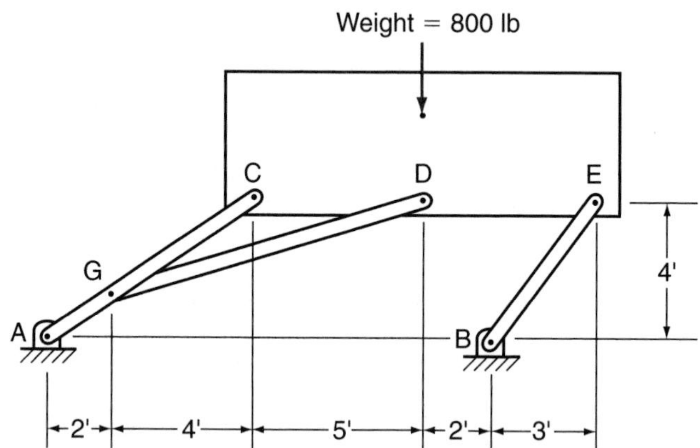

FIGURE P4–85

4–86. The tower shown in Figure P4–86 supports two loads of 2000 lb each due to wire weight. A side wind from the right causes the insulator cables AB and JK to form a 10° angle with the vertical. Assume zero tension initially in the guy wires and calculate the tension in guy wire DE.

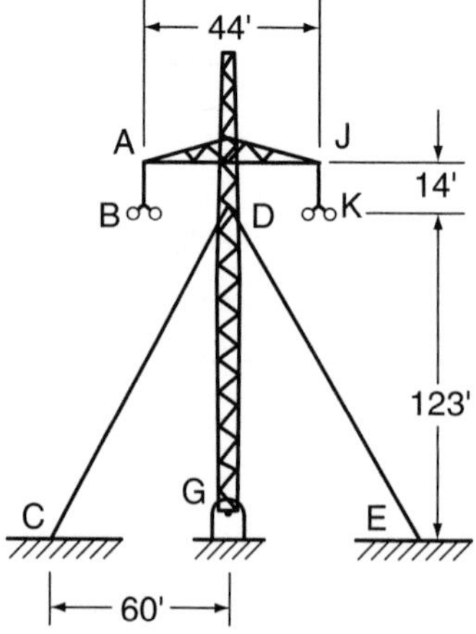

FIGURE P4–86

REVIEW PROBLEMS

R4–1. Draw free-body diagrams of members AE and DB as shown in Figure P5–53.

R4–2. Draw a free-body diagram of member BF in Figure RP4–2 and label all horizontal and vertical components. Show the components of BC as fractions of the total compressive force *BC*.

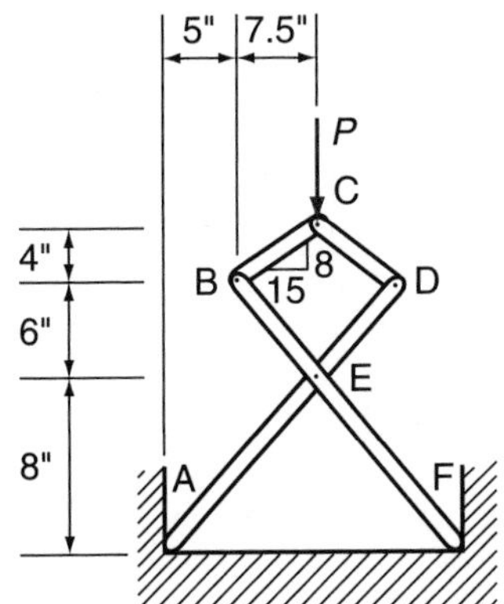

FIGURE RP4–2

R4–3. Draw a free-body diagram of member AC for each of the systems shown in Figure RP4–3.

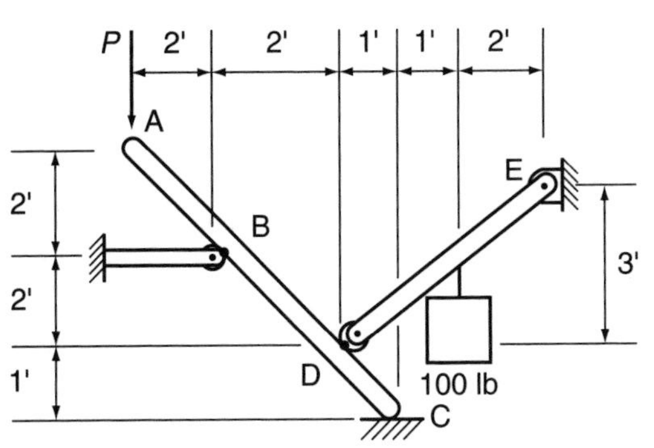

FIGURE RP4–3a

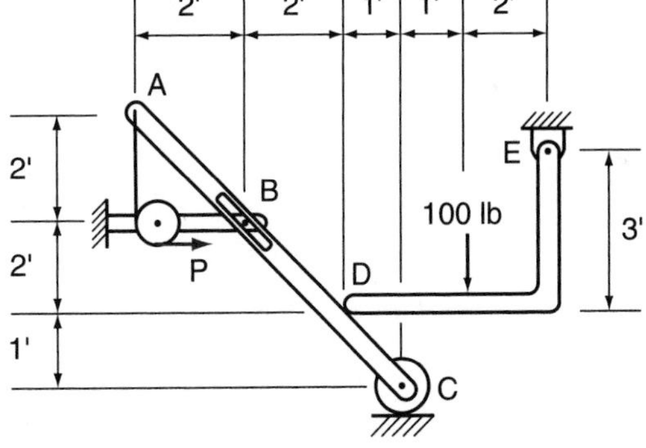

FIGURE RP4–3b

R4–4. Determine tensions T_1, T_2, and T_3 for the system shown in Figure RP4–4.

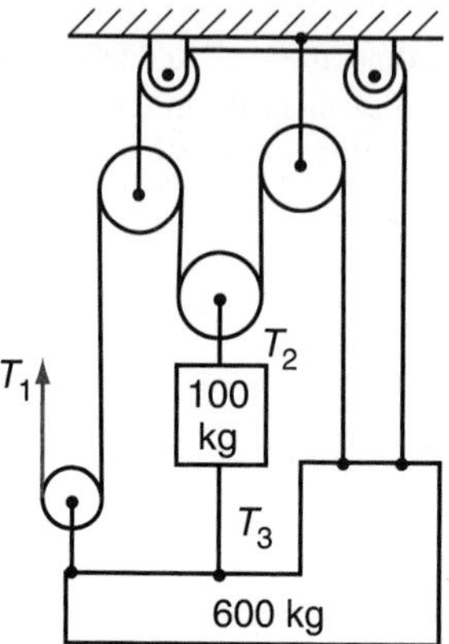

FIGURE RP4–4

R4–5. Draw free-body diagrams of members AC and DE shown in Figure RP4–5.

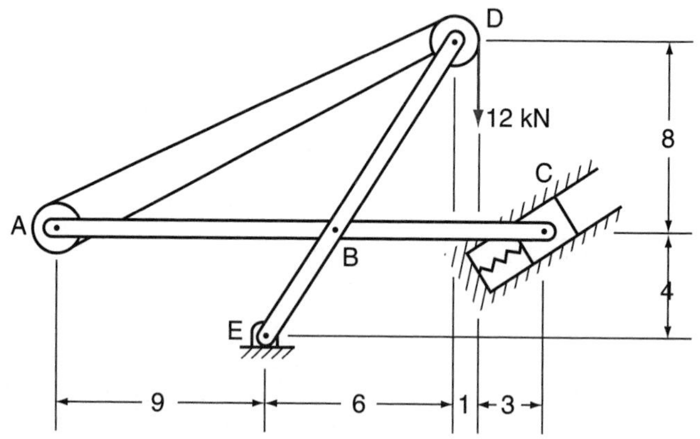

FIGURE RP4–5

R4–6. Determine the tension T required for static equilibrium in Figure RP4–6.

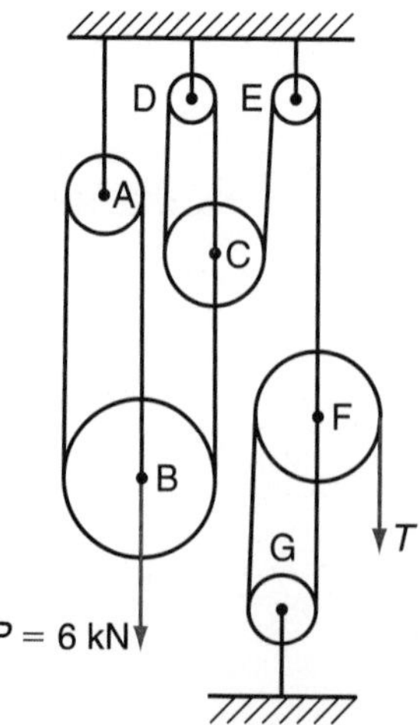

FIGURE RP4–6

R4–7. Determine the load in each pin-connected member shown in Figure RP4–7.

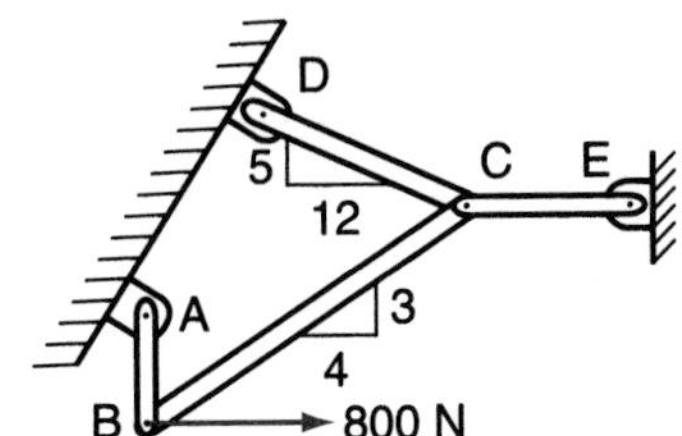

FIGURE RP4–7

R4–8. Determine the load in member BC of Figure RP4–8.

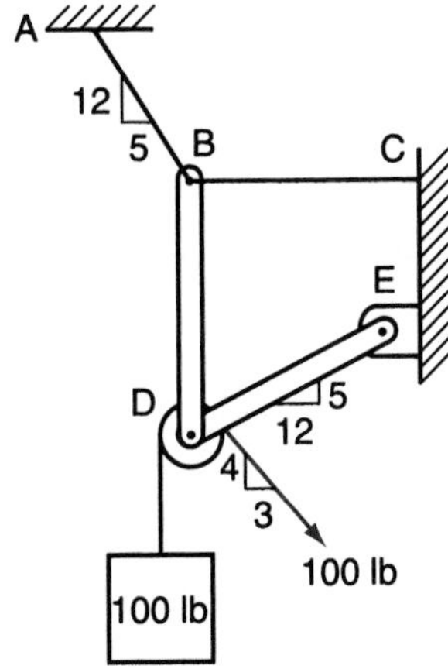

FIGURE RP4–8

R4–9. In Figure RP4–9, beam AB is rotated from position A to position B by lengthening cable CB. Determine the cable tension *CB* for each position.

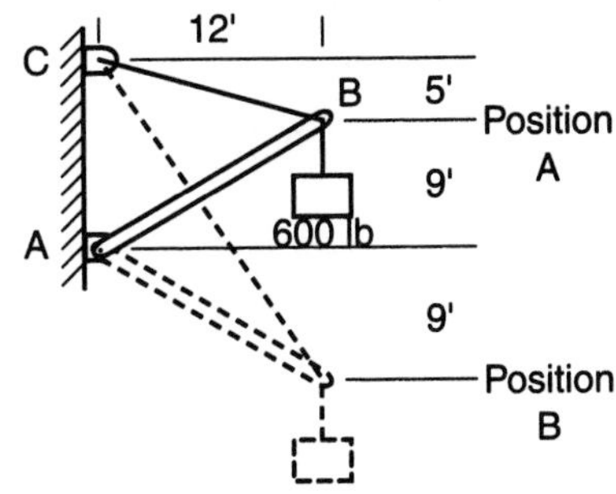

FIGURE RP4–9

R4–10. A 2-kN load is lifted by the cable system in Figure RP4–10. Determine the tension T in the cable that passes over pulley A.

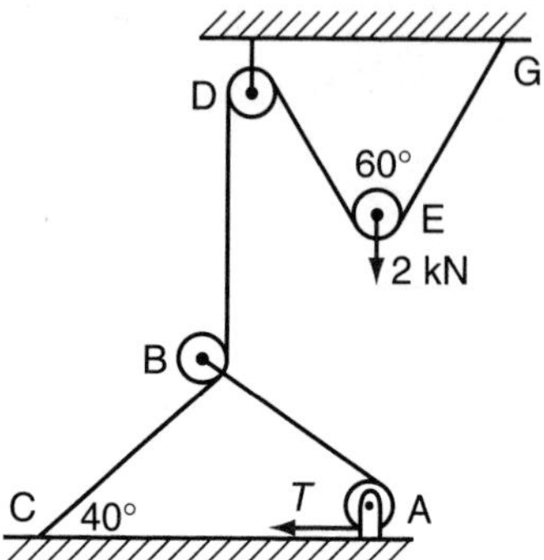

FIGURE RP4–10

R4–11. Determine the reactions at A and B for the beam loaded as shown in Figure RP4–11.

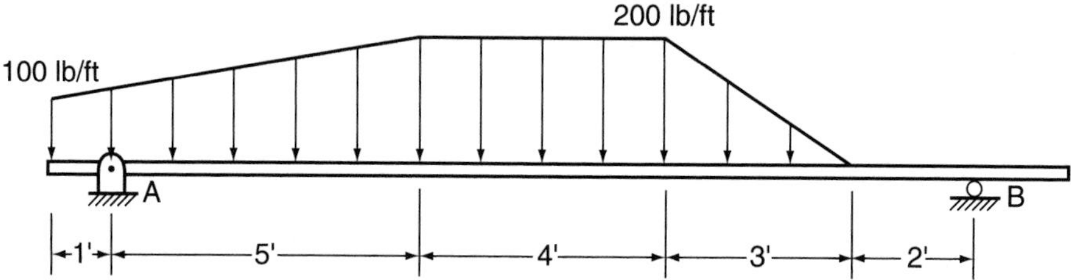

FIGURE RP4–11

R4–12. Determine the load in cable DC in Figure RP4–12.

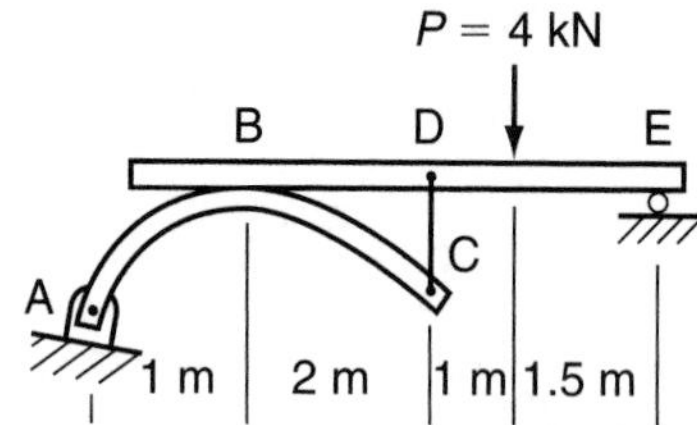

FIGURE RP4–12

R4–13. Determine the force P required to hold the 800-lb weight A by means of the system shown in Figure RP4–13.

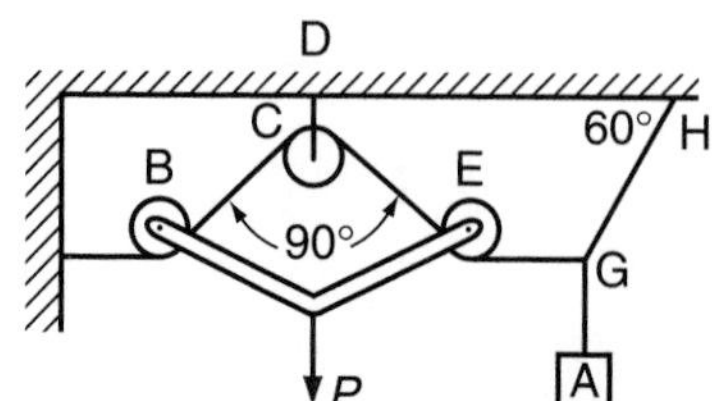

FIGURE RP4–13

R4–14. The light standard in Figure RP4–14 is bolted to a concrete pedestal. The weight of the light, 10 lb, can be assumed to be 12 ft from the standard. The beam supporting the traffic lights weighs 150 lb (the weight can be assumed to be at the beam center), and each traffic light weighs 40 lb. Calculate the tension in bolts A and B if they are the only two installed.

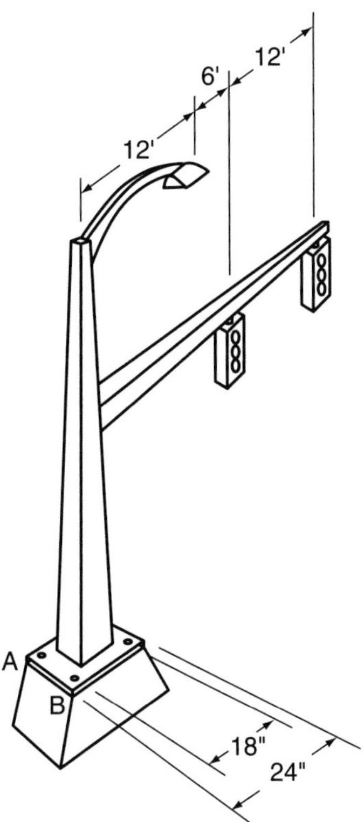

FIGURE RP4–14

CHAPTER 5

Structures and Members

OBJECTIVES

Upon completion of this chapter the student will be able to apply previous knowledge of coplanar concurrent force systems in:

1. Determining the loads in individual members of a pin-connected truss.
2. Calculating truss loads by the method of sections, which requires the drawing of a free-body diagram of a partial truss.
3. Calculating the pin reactions for various mechanisms, using the method of members.

5–1 METHOD OF JOINTS

As shown in Section 4–4, a two-force member has forces acting on each end. These forces line up with the member and cause either tension or compression. A truss is formed if several two-force members are joined in one or more connected triangles. Each of the members is pinned at each end and, if carrying a load, is in either tension or compression. The direction of the member indicates the direction of the tensile or compressive force in the member acting on the joint or pin. The ends of the members are pinned together to form a joint. A member in tension and pinned at a certain joint exerts a pull on the pin. A free-body diagram of this joint shows a vector acting away from the joint in the direction of the member. Conversely, a compression member pushes on the pin. Each truss member is a two-force member if we neglect the weight of the member. This is a relatively safe assumption since the member weight is often small compared to the loads carried by the truss.

The *method of joints* consists of a number of free-body diagrams of adjacent joints. The first joint selected must have only two unknown forces and one or more known forces. The unknown forces are determined by using $\Sigma F_x = 0$ and $\Sigma F_y = 0$. These newly found forces are used in the free-body diagram of an adjacent joint. The load in each truss member is found by taking consecutive free-body diagrams of joints throughout the complete truss.

An alternative to the method of joints is the method of sections. A brief introductory comparison of the two methods can be seen in Figure 5–1 and Figure 5–2. More details on each method are covered in future examples.

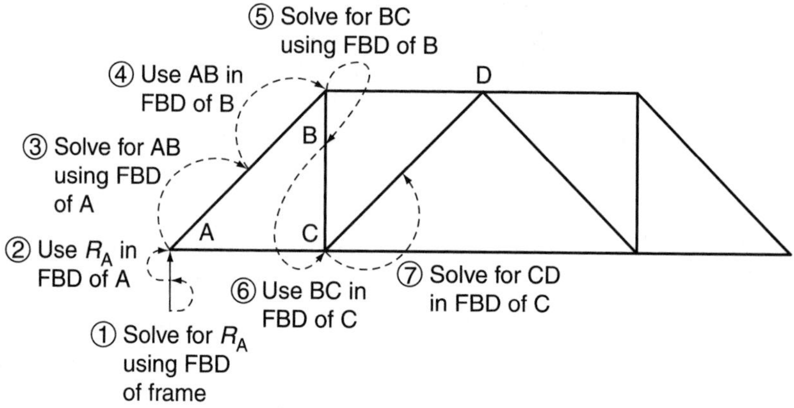

FIGURE 5–1

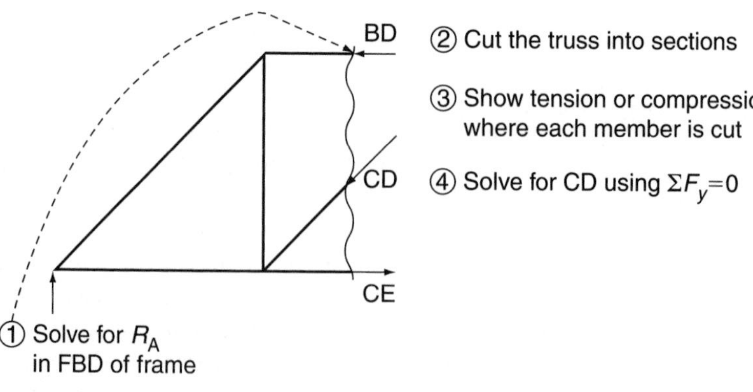

FIGURE 5–2

EXAMPLE 5–1

Determine the load in each member of the truss shown in Figure 5–3.

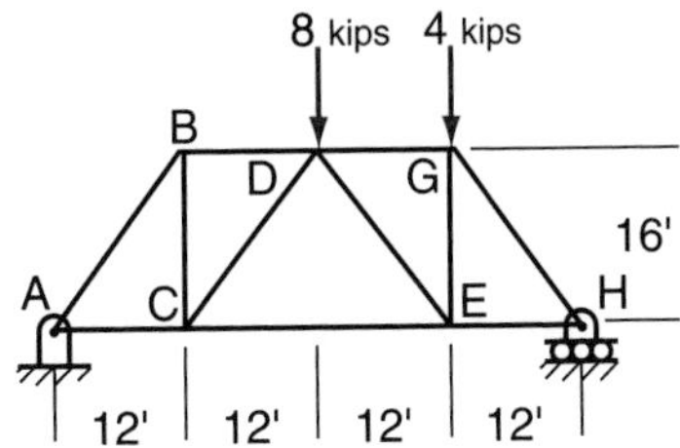

FIGURE 5–3

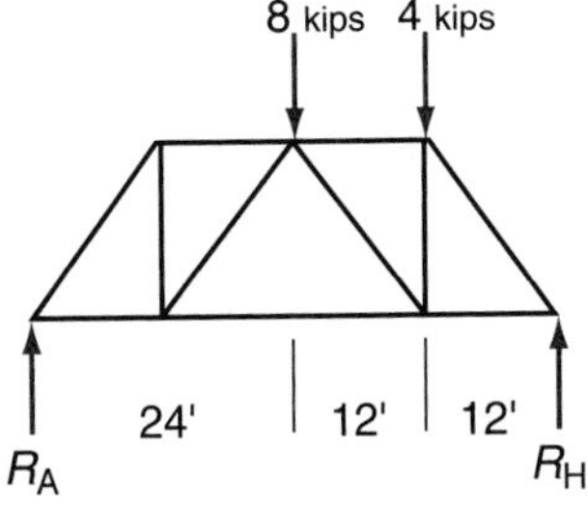

FIGURE 5–4

Step 1. Solve for the external reactions R_A and R_H.

The first step of the solution is one with which you are already familiar: Solve for the external reactions at points A and H (Figure 5–4). Taking moments about point H and solving for R_A, we obtain:

$$\Sigma M_H = 0$$

$$(8 \text{ kips})(24 \text{ ft}) + (4 \text{ kips})(12 \text{ ft}) - R_A(48 \text{ ft}) = 0$$

$$48R_A = 192 + 48$$

$$R_A = \frac{240}{48}$$

$$\underline{R_A = 5 \text{ kips} \uparrow}$$

$$\Sigma F_y = 0$$

$$5 \text{ kips} + R_H - 8 \text{ kips} - 4 \text{ kips} = 0$$

$$\underline{R_H = 7 \text{ kips} \uparrow}$$

Step 2. Choose a pin or joint for the first free-body diagram.

In choosing the first joint of which you will draw a free-body diagram, notice that only four joints—A, D, G, and H—have known forces. Joint D has four unknowns, G has three unknowns, and joints A and H each have two unknowns. Thus, the first free-body diagram could be of joint A or H. Let us arbitrarily choose joint A.

Step 3. Draw a free-body diagram of A.

Considering member AB with its dimensions of 16 ft and 12 ft, we find the slope to be 4 to 3 (Figure 5–5).

Free-Body Diagram of A

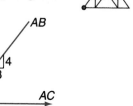

FIGURE 5–5

Assume member AB to be in compression and member AC to be in tension. (Experience will make you more confident of these assumptions later.) The vectors are drawn so that member AB is pushing on the joint (compression) and member AC is pulling on it (tension).

In considering vertical forces first, we find that the equation contains only one unknown, *AB*. The vertical component of *AB* is equal to 5 kips since there are no other vertical forces.

Step 4. Solve for *AB*, using

$$\Sigma F_y = 0$$

$$5\text{ kips} - \frac{4}{5}AB = 0$$

$$\underline{AB = 6.25\text{ kips } C}$$

Step 5. Solve for *AC*, using

$$\Sigma F_x = 0$$

$$AC - \frac{3}{5}AB = 0$$

$$AC = \frac{3}{5}(6.25\text{ kips})$$

$$\underline{AC = 3.75\text{ kips } T}$$

The compression or tension of a member should be indicated following the value with *C* or *T*, respectively.

Free-Body Diagram of B

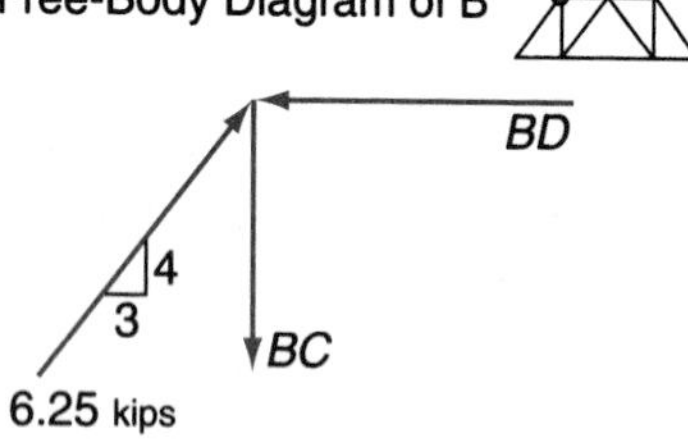

FIGURE 5–6

Step 6. Follow the value of a new known load, AB = 6.25, from pin A to an adjacent pin, B.

We draw a free-body diagram of joint B next, since it has only two unknowns (Figure 5–6). AB was found to have a compression of 6.25 kips; therefore, 6.25 kips must be pushing on joint B even though vector direction is opposite to that in the free-body diagram of A.

Step 7. Solve for *BD*, using

$$\Sigma F_x = 0$$

$$\frac{3}{5}(6.25\text{ kips}) - BD = 0$$

$$\underline{BD = 3.75\text{ kips } C}$$

Step 8. Solve for *BC*, using

$\Sigma F_y = 0$

$$\frac{4}{5}(6.25 \text{ kips}) - BC = 0$$

$$\underline{BC = 5 \text{ kips } T}$$

Step 9. Knowing the value of BC = 5, move from pin B to pin C and draw a free-body diagram (Figure 5–7).

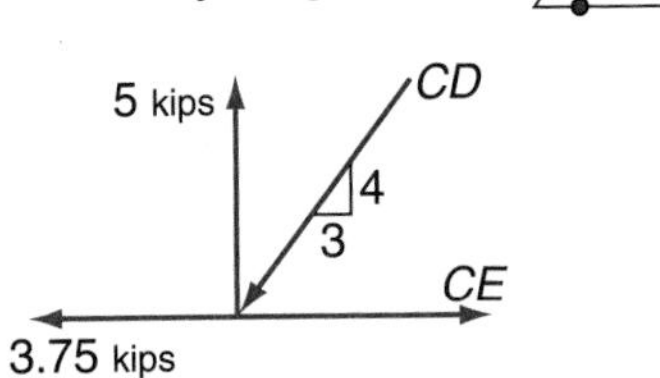

FIGURE 5–7

Step 10. Solve for *CD*, using

$\Sigma F_y = 0$

$$5 \text{ kips} - \frac{4}{5}CD = 0$$

$$\underline{CD = 6.25 \text{ kips } C}$$

Step 11. Solve for *CE*, using

$\Sigma F_x = 0$

$$CE - \frac{3}{5}CD - 3.75 \text{ kips} = 0$$

$$CE = \frac{3}{5}(6.25 \text{ kips}) + 3.75 \text{ kips}$$

$$\underline{CE = 7.5 \text{ kips } T}$$

Step 12. Reduce your possibility of error.

At this point, approximately one-half of the truss member loads have been determined. Each calculated value depends on our having calculated the previous value correctly. If we now go to joint H and work back toward the center of the truss, the possibility of an error being perpetuated through the complete calculation is lessened. The problem solution is, in essence, broken into two halves, and we can conclude with a final check, using a free-body diagram of a joint near the center of the truss.

Step 13. Show $R_H = 7$ kips on a free-body diagram of H (Figure 5–8).

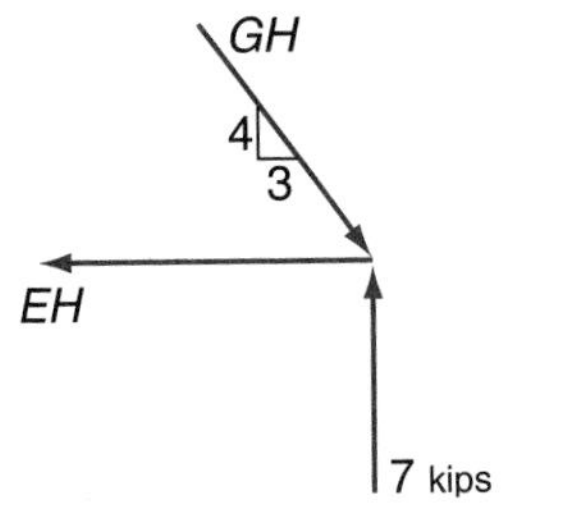

FIGURE 5–8

Step 14. Solve for *GH*, using

$\Sigma F_y = 0$

$$\frac{4}{5}GH = 7 \text{ kips}$$

$$\underline{GH = 8.75 \text{ kips } C}$$

Step 15. Solve for *EH,* using

$\Sigma F_x = 0$

$$\frac{3}{5}GH - EH = 0$$

$$\frac{3}{5}(8.75\text{ kips}) - EH = 0$$

$$\underline{EH = 5.25\text{ kips } T}$$

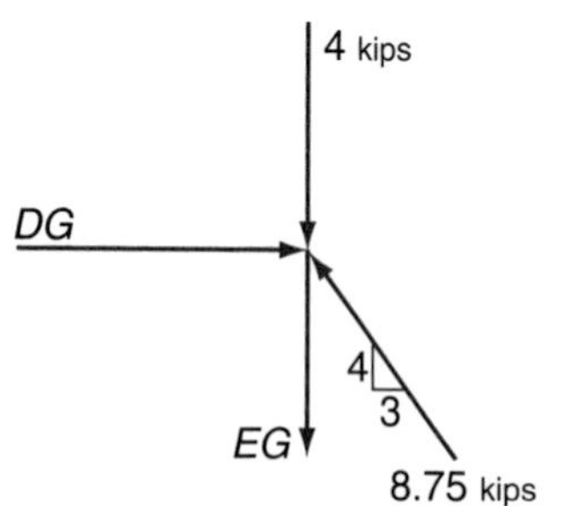

FIGURE 5–9

Step 16. Knowing GH = 8.75 kips C, move from pin H to a free-body diagram of G (Figure 5–9).

Do not forget to include the external 4-kip load in the free-body diagram of G (Figure 5–9).

Step 17. Solve for *EG,* using

$\Sigma F_y = 0$

$$\frac{4}{5}(8.75\text{ kips}) - EG - 4\text{ kips} = 0$$

$$EG = 7 - 4$$

$$\underline{EG = 3\text{ kips } T}$$

Step 18. Solve for *DG,* using

$\Sigma F_x = 0$

$$DG - \frac{3}{5}(8.75\text{ kips}) = 0$$

$$\underline{DG = 5.25\text{ kips } C}$$

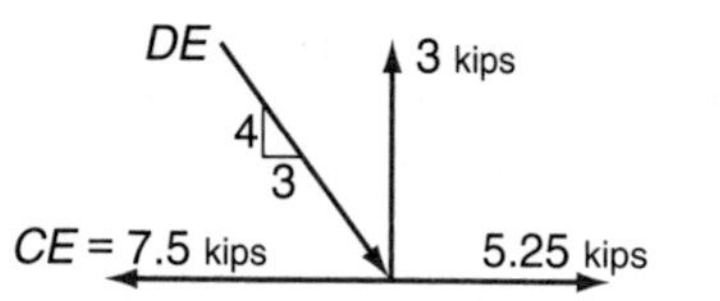

FIGURE 5–10

Step 19. Draw a free-body diagram of E showing CE = 7.5 kips, EG = 3 kips, and EH = 5.25 kips (Figure 5–10).

Step 20. Solve for *DE,* using

$\Sigma F_y = 0$

$$3\text{ kips} - \frac{4}{5}DE = 0$$

$$\underline{DE = 3.75\text{ kips } C}$$

Step 21. Check accuracy of calculations, using

$\Sigma F_x = 0$

$$\frac{3}{5}(3.75\text{ kips}) + 5.25\text{ kips} - 7.5\text{ kips} = 0$$

$$2.25 + 5.25 - 7.5 = 0$$

$$\underline{7.5 = 7.5\ \textit{check}}$$

Since $CE = 7.5$ kips was calculated in the first half of the solution and now gives us a balance of horizontal forces at joint E, it would seem to indicate that all values calculated are correct.

Step 22. Label forces for all truss members.

It is suggested that you conclude the solution with a sketch of the truss, labeling all member loads. Such a sketch is shown in Figure 5–11. This is useful not only as a final summary of all answers but, if you fill it in as you progress through the problem solution, it also serves as a quick and easy way to find formerly calculated values when required for a new free-body diagram.

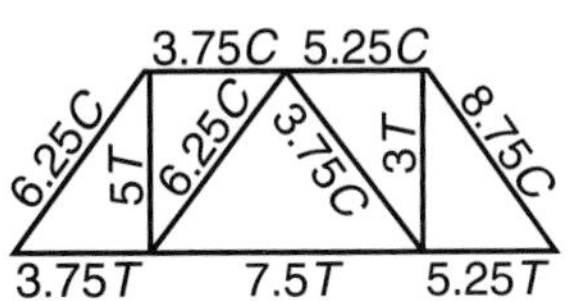

FIGURE 5–11

EXAMPLE 5–2

Use the method of joints to solve for the loads in members AB and CE of the truss shown in Figure 5–12.

This truss will illustrate that some truss members are needed to maintain alignment of other members but may not carry a load, depending upon how the external forces act on the truss.

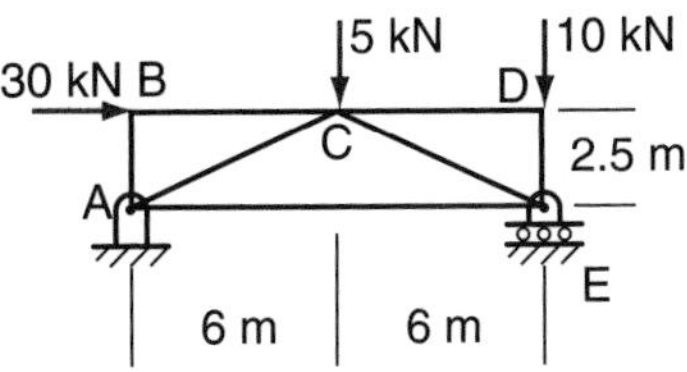

FIGURE 5–12

Step 1. Draw a free-body diagram of B (Figure 5–13).

A free-body diagram of joint B shows the load of member AB. Tension has been assumed but, since there is no other vertical force present, the load in member AB is zero.

$$\underline{AB = 0}$$

Free-Body Diagram of B

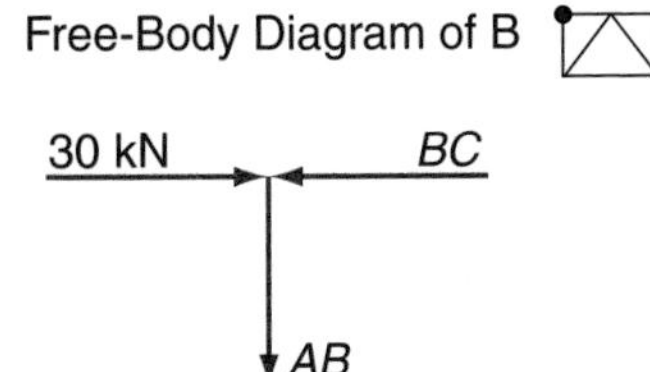

FIGURE 5–13

Step 2. Analyze the truss to select the next free-body diagram.

The load in member CE can be found from a free-body diagram of either joint C or E. The solution of joint C may be more difficult because of the joint's two members being sloped and unknown. Joint E will be easier to solve, but the load in member DE and the reaction at E must be found first.

Perhaps the best method of solution will not be readily apparent to you until you have solved several problems; so until that time, you may simply have to draw many free-body diagrams and use some trial-and-error methods. There is usually more than one method of solution, and experience will teach you the shortest one.

Step 3. Draw a free-body diagram of D (Figure 5–14).

Free-Body Diagram of D

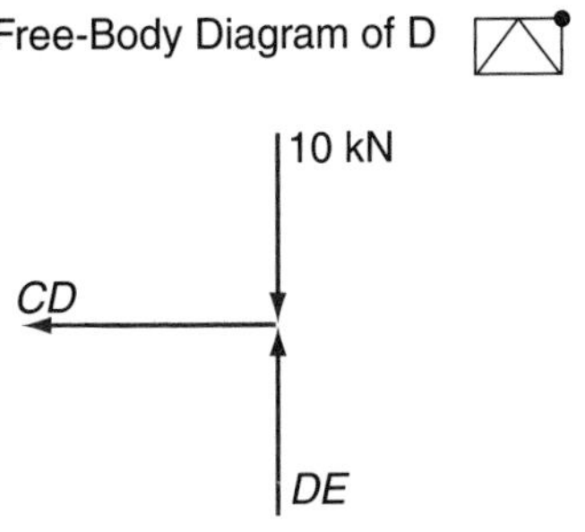

FIGURE 5–14

Step 4. Solve for member *DE*, using

$\Sigma F_y = 0$

$$\underline{DE = 10 \text{ kN } C}$$

(Also note that the load in member CD equals zero.)

Step 5. Draw a free-body diagram of the truss (Figure 5–15).

Step 6. Solve for reaction *E*, using

$\Sigma M_A = 0$

$$E(12 \text{ m}) - (30 \text{ kN})(2.5 \text{ m}) - (5 \text{ kN})(6 \text{ m}) - (10 \text{ kN})(12 \text{ m}) = 0$$

$$12E - 75 - 30 - 120 = 0$$

$$\underline{E = 18.75 \text{ kN} \uparrow}$$

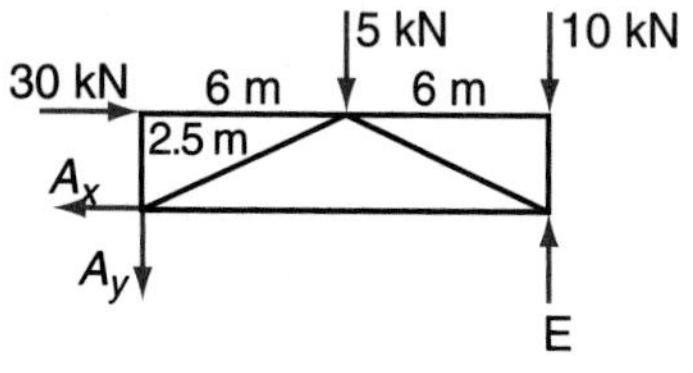

FIGURE 5–15

Step 7. Draw a free-body diagram of E (Figure 5–16).

Step 8. Solve for *CE*, using

$\Sigma F_y = 0$

$$18.75 \text{ kN} + \frac{5}{13}CE - 10 \text{ kN} = 0$$

$$\frac{5}{13}CE = 10 - 18.75$$

$$CE = -8.75 \times \frac{13}{5}$$

$$\underline{CE = -22.8 \text{ kN } T}$$

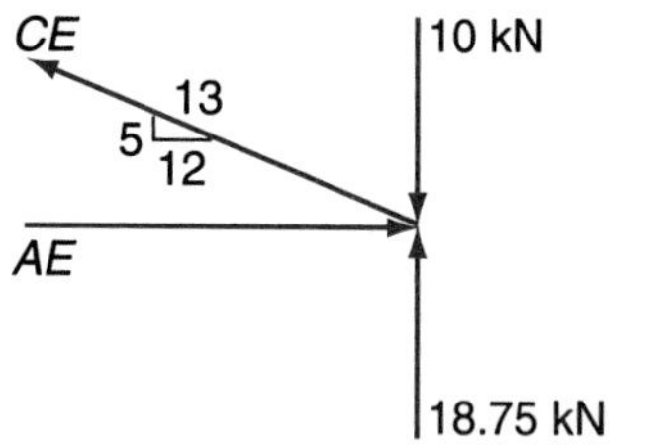

FIGURE 5–16

Since this is a negative value, the direction assumed was incorrect; therefore, $\underline{CE = +22.8 \text{ kN } C}$.

EXAMPLE 5–3

Solve for the load in each member of the pin-connected truss shown in Figure 5–17.

FIGURE 5–17

Free-Body Diagram of A

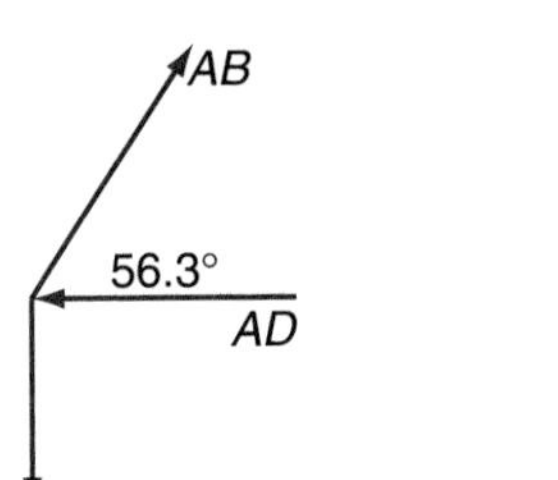

FIGURE 5–18

Step 1. Since the only known force is at joint A, start with a free-body diagram of A (Figure 5–18).

Step 2. Solve for AB, using

$\Sigma F_y = 0$

$$\sin 56.3°(AB) - 200 \text{ lb} = 0$$

$$\underline{AB = 240 \text{ lb } T}$$

Step 3. Solve for *AD*, using

$\Sigma F_x = 0$

$$\cos 56.3°(AB) - AD = 0$$

$$0.555(240 \text{ lb}) - AD = 0$$

$$\underline{AD = 133 \text{ lb } C}$$

Free-Body Diagram of B

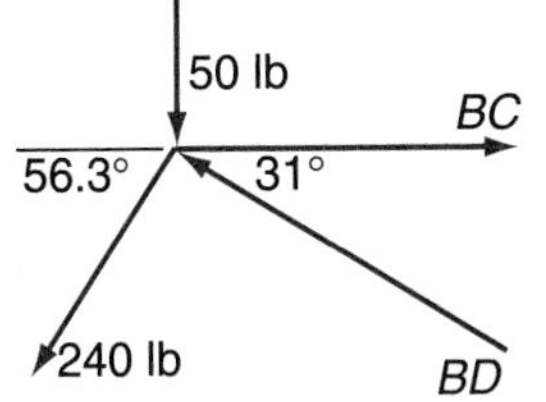

FIGURE 5–19

Step 4. Knowing AB = 240 lb *T*, move to a free-body diagram of joint B (Figure 5–19).

Step 5. Solve for *BD*, using

$\Sigma F_y = 0$

$$\sin 31°(BD) - 50 \text{ lb} - \sin 56.3°(240 \text{ lb}) = 0$$

$$\sin 31°(BD) - 50 - 200 = 0$$

$$BD = \frac{250}{\sin 31°}$$

$$\underline{BD = 486 \text{ lb } C}$$

Step 6. Solve for *BC*, using

$\Sigma F_x = 0$

$$BC - \cos 31°(486 \text{ lb}) - \cos 56.3°(240 \text{ lb}) = 0$$

$$BC - 417 - 133 = 0$$

$$\underline{BC = 550 \text{ lb } T}$$

EXAMPLE 5–4

Solve for the load in each member of the system in Figure 5–20.

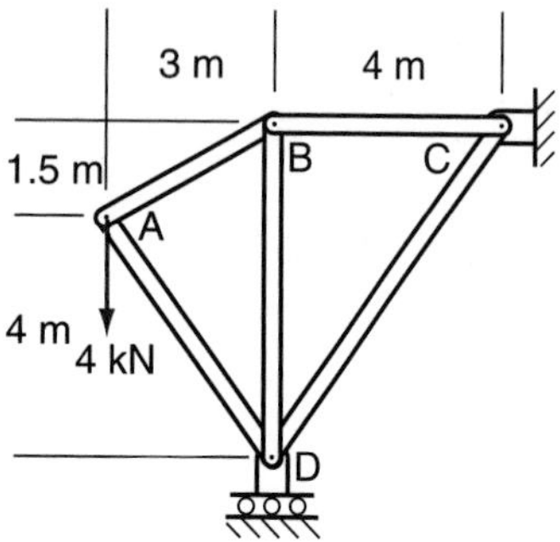

FIGURE 5–20

The solution generally begins with a free-body diagram of a joint on which a known force is acting. For this example, such a solution is not the easiest one, because we will encounter simultaneous equations, but we will first solve it in this fashion; an alternative solution will follow.

The frame is statically determinate. There are three unknowns, not four, because the reaction at D is a single vertical force perpendicular to the horizontal rollers (Figure 5–24).

Free-Body Diagram of A

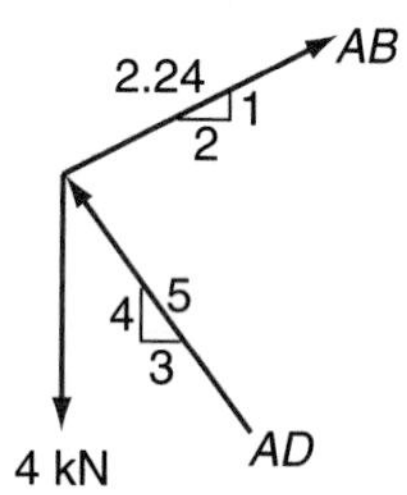

FIGURE 5–21

Step 1. Draw a free-body diagram of A (Figure 5–21).
Step 2. Write an equation for the vertical forces.

$\Sigma F_y = 0$

$$\frac{1}{2.24}(AB) + \frac{4}{5}(AD) - 4 \text{ kN} = 0$$

$$0.447AB + 0.8AD = 4 \qquad (1)$$

Step 3. Write an equation for the horizontal forces.

$\Sigma F_x = 0$

$$\frac{2}{2.24}AB - \frac{3}{5}AD = 0$$

$$1.49AB = AD \qquad (2)$$

Step 4. Substitute Equation (2) into Equation (1):

$$0.447AB + 0.8(1.49AB) = 4 \text{ kN}$$

$$0.447AB + 1.19AB = 4$$

$$AB = \frac{4}{1.637}$$

$$\underline{AB = 2.44 \text{ kN } T}$$

Step 5. Substitute $AB = 2.44$ into Equation (2):

$$AD = 1.49(2.44 \text{ kN})$$
$$\underline{AD = 3.64 \text{ kN } C}$$

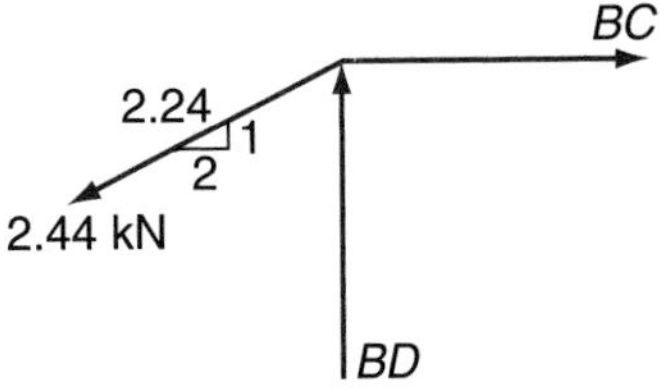

FIGURE 5–22

Step 6. Knowing AB = 2.44 kN T, move to a free-body diagram of B (Figure 5–22).

Step 7. Solve for BC, using

$\Sigma F_x = 0$

$$BC - \frac{2}{2.24}(2.44 \text{ kN}) = 0$$
$$\underline{BC = 2.18 \text{ kN } T}$$

Step 8. Solve for BD, using

$\Sigma F_y = 0$

$$BD - \frac{1}{2.24}(2.44 \text{ kN}) = 0$$
$$\underline{BD = 1.09 \text{ kN } C}$$

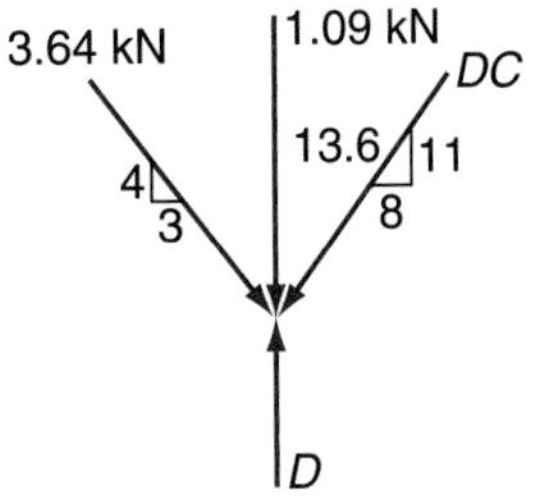

FIGURE 5–23

Step 9. The final unknown, DC, can be found from a free-body diagram of joint D (Figure 5–23).

$\Sigma F_x = 0$

$$\frac{3}{5}(3.64) - \frac{8}{13.6}DC = 0$$
$$\underline{DC = 3.71 \text{ kips } C}$$

Simultaneous equations were used in this first solution; therefore, the alternative solution (which follows) may be preferred. The alternative solution begins with a free-body diagram of the frame to find the other external reactions, such as at C. The sequence of free-body diagrams of joints is then C, B, and A.

EXAMPLE 5–4 (ALTERNATIVE SOLUTION)

Free-Body Diagram of Frame

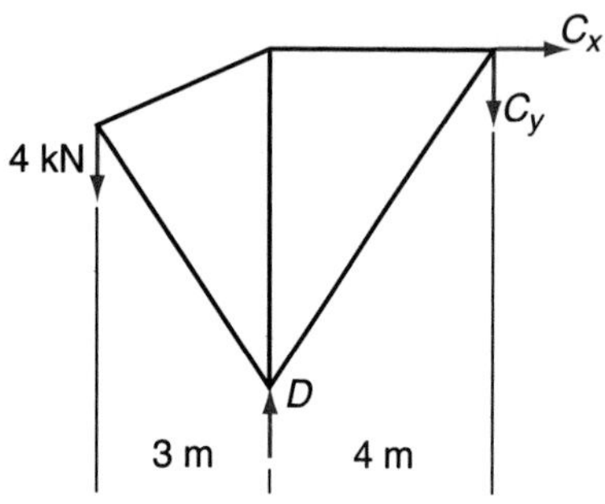

FIGURE 5–24

Step 1. Draw a free-body diagram of the frame (Figure 5–24) to get other external reactions.

Step 2. Solve for C_x using

$\Sigma F_x = 0$

$$C_x = 0$$

Step 3. Solve for C_y, using

$\Sigma M_D = 0$

$$(4\text{ kN})(3\text{ m}) - C_y(4\text{ m}) = 0$$

$$\underline{C_y = 3\text{ kN}\downarrow}$$

Step 4. Knowing $C_y = 3$ kN we can draw a free-body diagram of joint C (Figure 5–25).

Free-Body Diagram of C

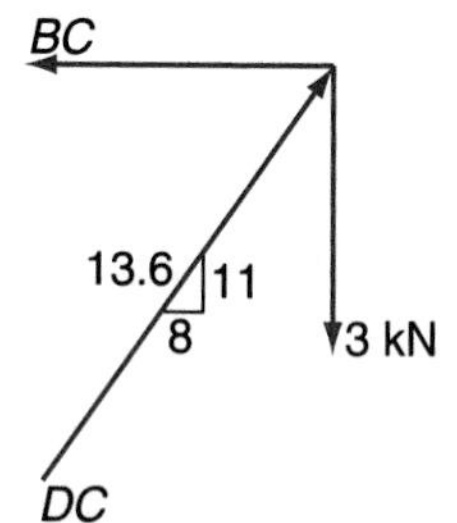

FIGURE 5–25

Step 5. Solve for *DC*, using

$\Sigma F_y = 0$

$$\frac{11}{13.6}DC - 3\text{ kN} = 0$$

$$DC = 3\left(\frac{13.6}{11}\right)$$

$$\underline{DC = 3.71\text{ kN } C}$$

Step 6. Solve for *BC*, using

$\Sigma F_x = 0$

$$\frac{8}{13.6}(3.71\text{ kN}) - BC = 0$$

$$\underline{BC = 2.18\text{ kN } T}$$

Step 7. Knowing $BC = 2.18$ kN T, draw a free-body diagram of joint B (Figure 5–26).

Free-Body Diagram of B

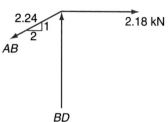

FIGURE 5–26

Step 8. Solve for *AB*, using

$\Sigma F_x = 0$

$$2.18\text{ kN} - \frac{2}{2.24}AB = 0$$

$$\underline{AB = 2.44\text{ kN } T}$$

Step 9. Solve for BD, using

$\Sigma F_y = 0$

$$BD - \frac{1}{2.24}(2.44 \text{ kN}) = 0$$

$$\underline{BD = 1.09 \text{ kN } C}$$

Free-Body Diagram of A

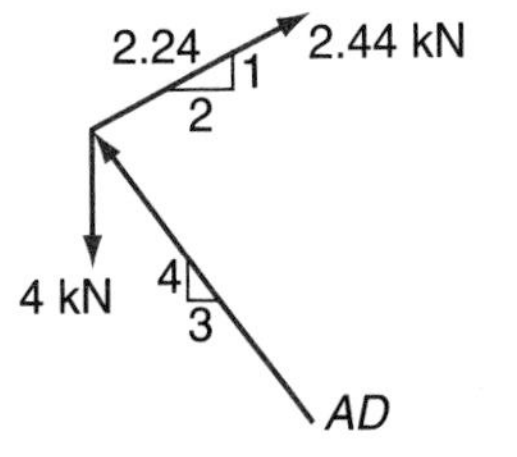

FIGURE 5–27

Step 10. Knowing $AB = 2.44$ kN T, draw a free-body diagram of joint A (Figure 5–27).

Step 11. Solve for AD, using

$\Sigma F_x = 0$

$$\frac{2}{2.24}(2.44 \text{ kN}) - \frac{3}{5} AD = 0$$

$$\underline{AD = 3.64 \text{ kN } C}$$

Step 12. Now check by taking:

$\Sigma F_y = 0$ (Figure 5–27)

$$\frac{1}{2.24}(2.44 \text{ kN}) + \frac{4}{5}(3.64 \text{ kN}) - 4 \text{ kN} = 0$$

$$1.09 + 2.91 - 4 = 0$$

$$\underline{0 = 0 \textit{ check}}$$

This second method avoided simultaneous equations but was slightly longer. The choice will be yours in future problems. The load summary is shown in Figure 5–28.

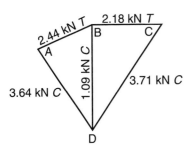

FIGURE 5–28

5–2 METHOD OF SECTIONS

The *method of sections* is used to solve for the force in a member near the middle of a truss. The time-consuming method of joints is avoided. The method of sections consists of cutting a truss into two sections by cutting through the truss where a member force is required; one section is discarded. A free-body diagram of the remaining section is drawn. On this free-body diagram, we show a tensile or compressive force where each member was cut. These are equivalent forces that have the same effect as the discarded section had. Suppose that a truss member, cut in two by the method of sections, had been in compression. The free-body diagram would show a force pushing on the remaining half of the member. Only three members are usually cut at one time, although a partial solution is possible when four or more members are cut.

EXAMPLE 5–5

Solve for the load in members CB, AB, and JK of the truss shown in Figure 5–29.

A cutting plane is drawn through the truss, cutting members AB, CB, and CD. A free-body diagram of the left half of the truss is drawn since the external force of 900 N is acting on this section. The right half could be used, but then the reactions at E and H would have to be solved for first.

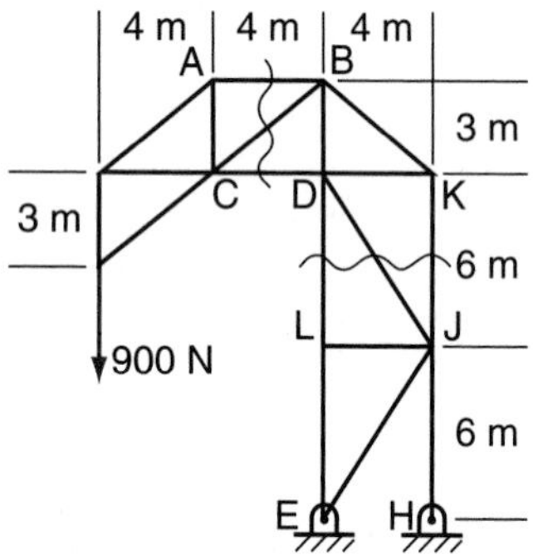

FIGURE 5–29

Free-Body Diagram of Left Half

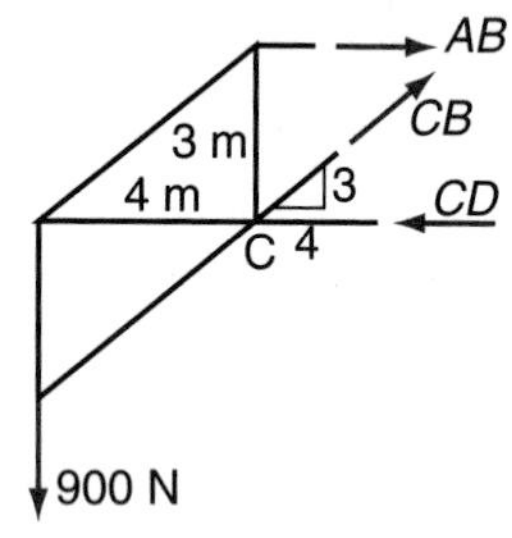

FIGURE 5–30

Member AB is assumed to be in tension, and vector *AB* is shown as a pull in the free-body diagram (Figure 5–30). Similarly, vector *CD* is pushing due to assumed compression. The nature of the load in member CB may not be so obvious, but it can be assumed to be in tension. The free-body diagram is now complete, and any one of the three equations of equilibrium can be applied.

$\Sigma F_y = 0$ (Figure 5–30)

$$\frac{3}{5}CB = 900 \text{ N}$$

$$\underline{CB = 1500 \text{ N } T}$$

$\Sigma M_C = 0$

$$(900 \text{ N})(4 \text{ m}) - AB(3 \text{ m}) = 0$$

$$\underline{AB = 1200 \text{ N } T}$$

Free-Body Diagram of Top Half

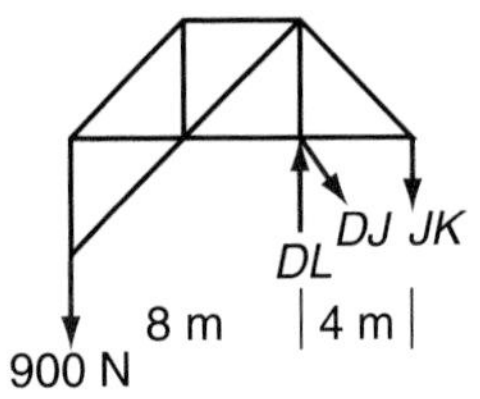

FIGURE 5–31

A second cutting plane is needed to solve for JK.

$\Sigma M_D = 0$ (Figure 5–31)

$$-JK(4\text{ m}) + (900\text{ N})(8\text{ m}) = 0$$

$$\underline{JK = 1800\text{ N } T}$$

EXAMPLE 5–6

The truss of a canopy is shown in Figure 5–32. Determine the load in member CD.

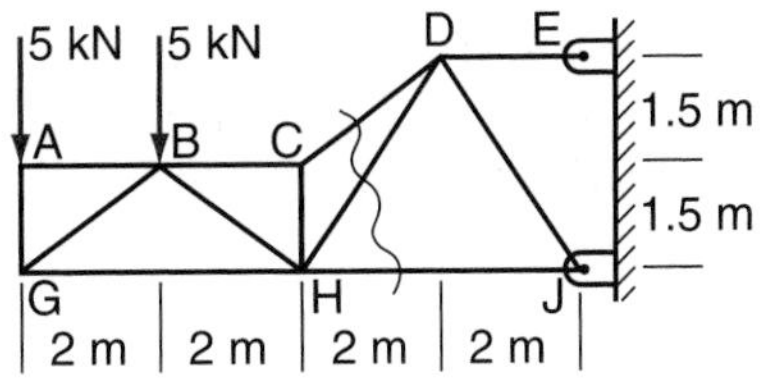

FIGURE 5–32

The cutting plane is chosen such that it cuts only three members. Moments are taken about point H since the two other unknowns, HD and HJ, pass through this point (Figure 5–33). Rather than calculate the perpendicular distance between CD and point H, we can take horizontal and vertical components of CD at point C. Taking moments about point H, we find that the only unknown is $\frac{4}{5}$ CD.

Free-Body Diagram of Left Half

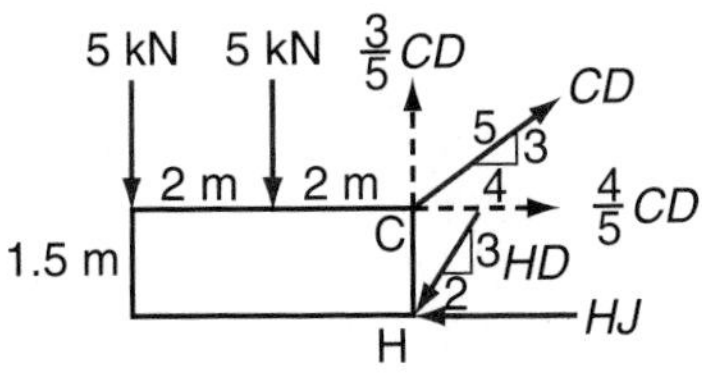

FIGURE 5–33

$\Sigma M_H = 0$ (Figure 5–33)

$$(5\text{ kN})(2\text{ m}) + (5\text{ kN})(4\text{ m}) - \frac{4}{5}CD(1.5\text{ m}) = 0$$

$$1.5\left(\frac{4}{5}CD\right) = 10 + 20$$

$$\underline{CD = 25\text{ kN } T}$$

EXAMPLE 5–7

For the truss loaded as shown in Figure 5–34, determine the load in member GE.

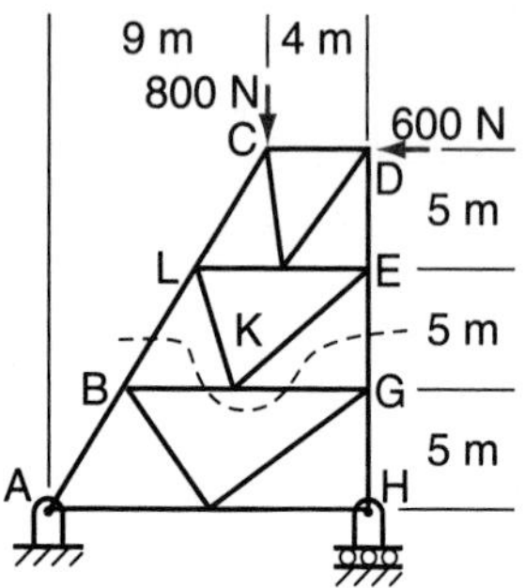

FIGURE 5–34

Free-Body Diagram of Top Half

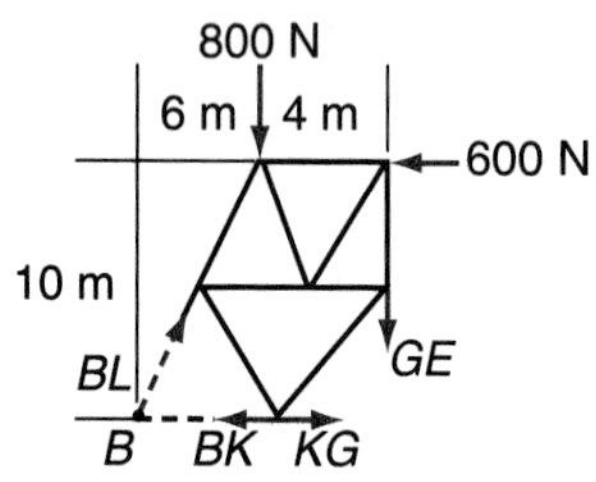

FIGURE 5–35

The cutting plane chosen (Figure 5–35) cuts four members. We cannot solve for all the members that are cut, but you will see that the load in member GE is quite readily found since three of the unknowns have zero moment about B.

$\Sigma M_B = 0$ (Figure 5–35)

$$-(800\text{ N})(6\text{ m}) + (600\text{ N})(10\text{ m}) - GE(10\text{ m}) = 0$$

$$\underline{GE = 120\text{ N } T}$$

EXAMPLE 5–8

A crane has the truss framework shown in Figure 5–36. Determine the loads in members CE, DE, and DG.

A free-body diagram of the right half (Figure 5–37) is used to avoid solving for the reactions at A and B. Moments must be used, and point J is chosen since the lines of action of two of the unknown forces pass through this point.

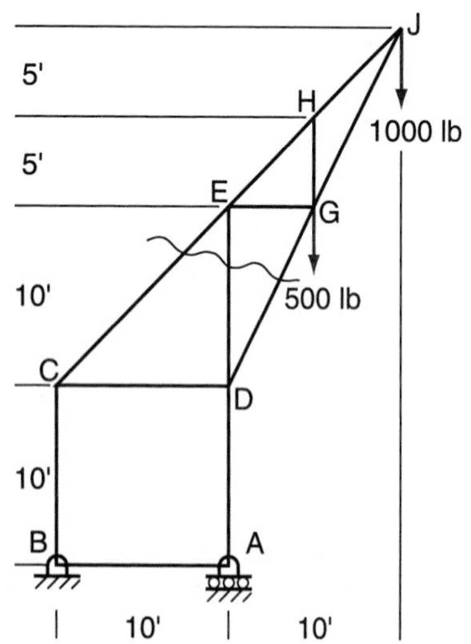

FIGURE 5–36

$\Sigma M_J = 0$ (Figure 5–37)

$$DE(10\text{ ft}) = (500\text{ lb})(5\text{ ft})$$

$$\underline{DE = 250\text{ lb } C}$$

The use of $\Sigma F_x = 0$ or $\Sigma F_y = 0$ forces would give simultaneous equations with two unknowns. Moments can be taken about point D, thus eliminating DE and DG from the moment equation and allowing us to find CE. The distance d, between CE and point D, will have to be calculated.

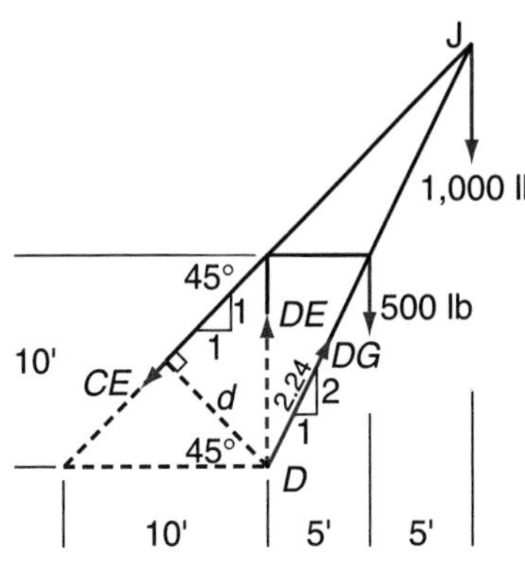

FIGURE 5–37

$$\cos 45° = \frac{d}{10 \text{ ft}}$$

$$d = 0.707(10)$$

$$d = 7.07 \text{ ft}$$

$$\Sigma M_D = 0$$

$$CE(7.07 \text{ ft}) - (500 \text{ lb})(5 \text{ ft}) - (1000 \text{ lb})(10 \text{ ft}) = 0$$

$$7.07\ CE = 2500 + 10{,}000$$

$$CE = \frac{12{,}500}{7.07}$$

$$\underline{CE = 1770 \text{ lb } T}$$

$$\Sigma F_x = 0$$

$$CE_x = DG_x$$

$$\cos 45°(1770 \text{ lb}) = \frac{1}{2.24}(DG)$$

$$1250 = \frac{DG}{2.24}$$

$$\underline{DG = 2800 \text{ lb } C}$$

5–3 METHOD OF MEMBERS

In the previous truss problems, the truss members were two-force members; there was a force acting at each end of the member; each member was in either tension or compression. The force of a member on a joint had the same direction as the slope of the member. For this reason, we could draw free-body diagrams of individual joints.

Free-body diagrams of joints cannot be used when the members have three or more forces acting on them. A three-force member may be subject to bending, and the force that it exerts on a joint no longer has the same slope or direction as the member. Therefore, a free-body diagram is drawn of the member, not of the joint. A *frame* consists of a number of members fastened together so that each member has two or more forces acting on it. Determining these forces consists of drawing free-body diagrams of individual members or of the complete frame.

EXAMPLE 5–9

Solve for the horizontal and vertical components of the force at pin B (Figure 5–38).

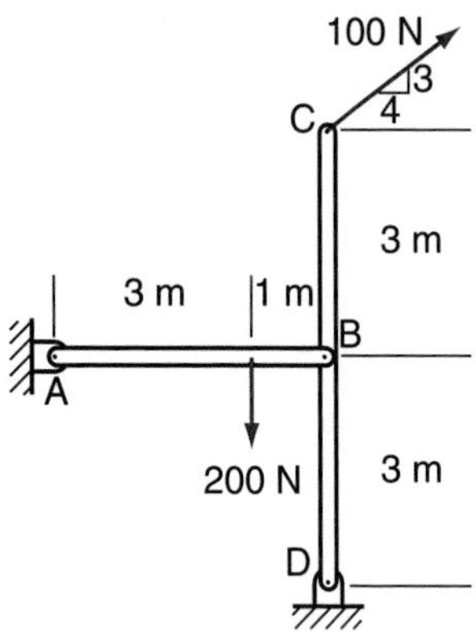

FIGURE 5–38

Each member has loads that both stretch and bend them.

Step 1. Sketch free-body diagrams of both members (Figure 5–39 and Figure 5–40).

Step 2. Note that the *internal* forces at B are in opposite directions in the two diagrams. If AB pushes down on CD, then CD pushes up on AB.

Step 3. Use the free-body diagram of AB (Figure 5–39) to solve for B_y (B_x cannot be found at this time because A_x is unknown).

Free-Body Diagram of AB

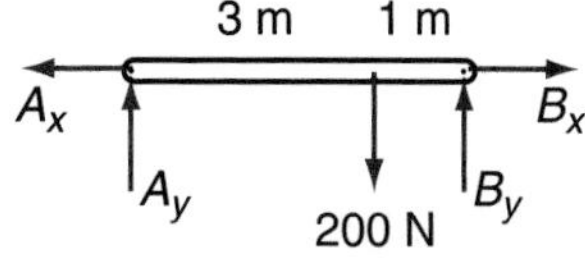

FIGURE 5–39

$$\Sigma M_A = 0$$

$$B_y(4\text{ m}) - (200\text{ N})(3\text{ m}) = 0$$

$$\underline{B_y = 150\text{ N}\uparrow\text{ on AB}}$$

Step 4. Use the free-body diagram of CD (Figure 5–40) to solve for B_x.

Free-Body Diagram of CD

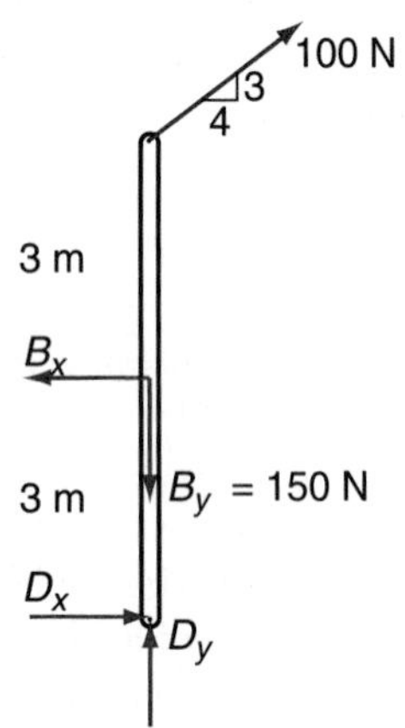

FIGURE 5–40

$$\Sigma M_{\text{D}} = 0$$

$$B_x(3\text{ m}) - \frac{4}{5}(100\text{ lb})(6\text{ m}) = 0$$

$$B_x = 160\text{ N}\leftarrow\text{ on CD}$$

or

$$\underline{B_x = 160\text{ N}\rightarrow\text{on AB}}$$

Remember the general rule for *internal* forces: Whenever a force direction is assumed at a point on a member, the opposite force direction must be shown at that point on the other connected member.

EXAMPLE 5–10

Determine the horizontal and vertical reactions at D (Figure 5–41).

Note that member BC is a two-force member in tension and will pull *horizontally only* on pin B.

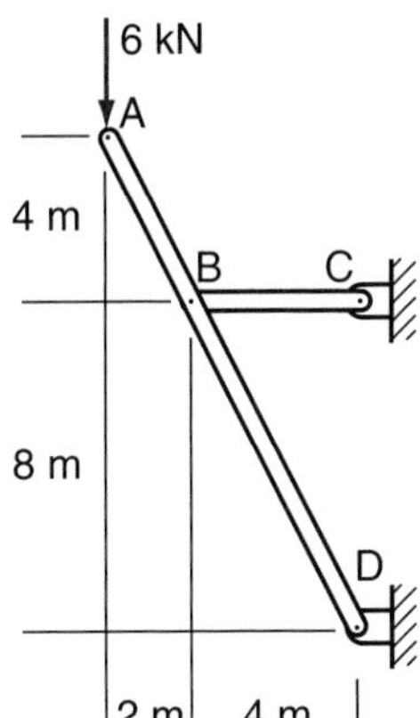

FIGURE 5–41

Free-Body Diagram of AD

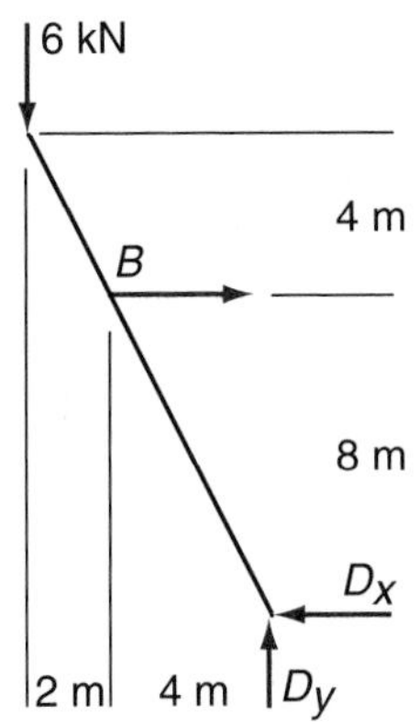

FIGURE 5–42

$\Sigma F_y = 0$ (Figure 5–42)

$$\underline{D_y = 6 \text{ kN}\uparrow}$$

$\Sigma M_B = 0$

$$(6 \text{ kN})(4 \text{ m}) + (6 \text{ kN})(2 \text{ m}) - D_x(8 \text{ m}) = 0$$

$$8D_x = 36$$

$$\underline{D_x = 4.5 \text{ kN} \leftarrow}$$

An alternative solution would have been $\Sigma M_D = 0$ (to get B), then $\Sigma F_x = 0$ (to get D_x). By using $\Sigma M_B = 0$ the solution was shorter, and we avoided unnecessarily solving for B.

EXAMPLE 5–11

A load of 175 lb applied to the members shown in Figure 5–43 causes a tension of 400 lb in cable DC. Determine the reaction at B; assume a smooth surface. Determine the horizontal and vertical reactions at A and E.

Note that cable DC is a two-force member in tension and will be pulling *vertically only* on pins D and C.

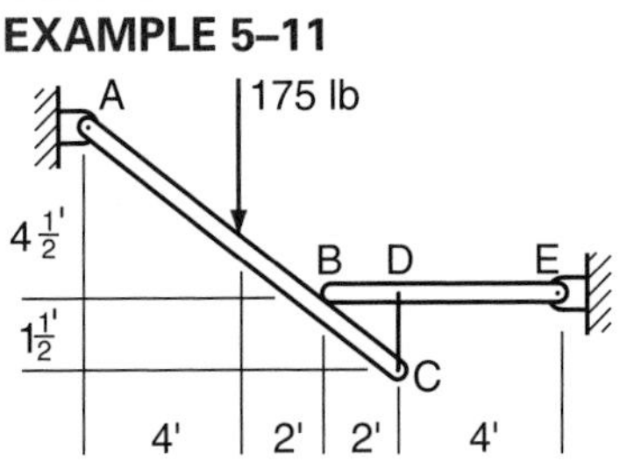

FIGURE 5–43

Free-Body Diagram of Frame

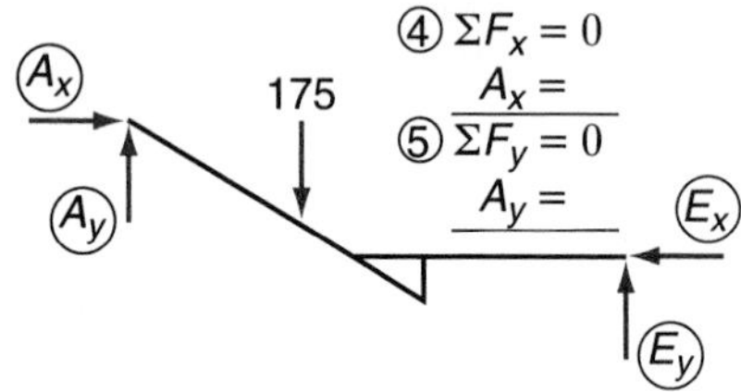

FIGURE 5–44

Free-Body Diagram of BE

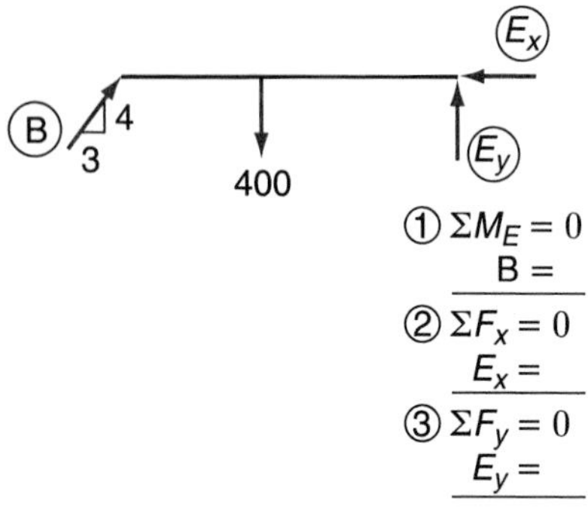

FIGURE 5–45

As the problems get more difficult, it is often useful to sketch several free-body diagrams and plot your strategy or sequence of steps as follows:

1. Sketch all possible free-body diagrams (Figures 5–44, 5–45, 5–46). (Dimensions can be omitted at this stage.)
2. Circle the required forces.
3. Select a diagram that has known forces and some of the required unknowns. The diagram of the frame has too many unknowns. Member BE appears simpler to work with than member AC.
4. Number the sequence of calculations that you could perform on BE (Figure 5–45). Taking moments where two unknowns intersect is often useful.
5. Transfer proposed solved values to other free-body diagrams to see if new calculations are possible. In this case transferring values of E_x and E_y to the free-body diagram of the frame (Figure 5–44) will allow you to solve for A_x and A_y.
6. With your sketches and calculation sequences labeled, your strategy, or "game plan," is complete, and you can now proceed to draw *complete* free-body diagrams, write the appropriate equations, and calculate the required values.

A neat, sequential, organized solution without distractions will result as follows.

Free-Body Diagram of AC

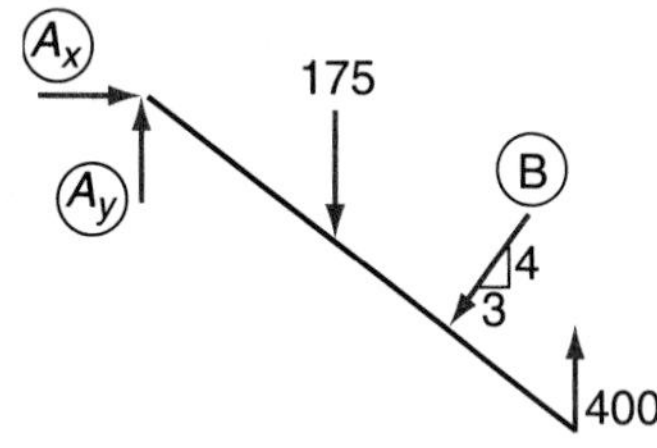

FIGURE 5–46

Since member AC has a slope of 3 to 4, force B perpendicular to it has a slope of 4 to 3 (Figure 5–47).

Free-Body Diagram of BE

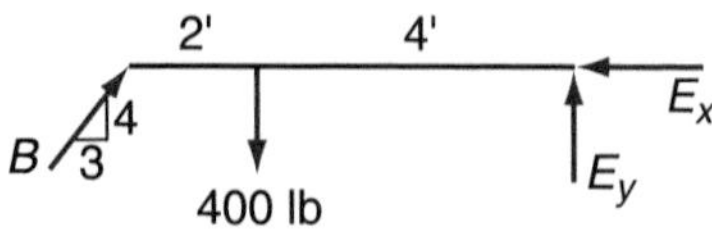

FIGURE 5–47

$\Sigma M_{\mathrm{E}} = 0$

$$(400\ \text{lb})(4\ \text{ft}) - \frac{4}{5}B(6\ \text{ft}) = 0$$

$$\underline{B = 333\ \text{lb}}\ \angle^{4}_{3}$$

$\Sigma F_x = 0$

$$\frac{3}{5}(333\ \text{lb}) - E_x = 0$$

$$\underline{E_x = 200\ \text{lb} \leftarrow}$$

$$\Sigma F_y = 0$$

$$\frac{4}{5}B + E_y - 400 \text{ lb} = 0$$

$$\frac{4}{5}(333 \text{ lb}) + E_y = 400 \text{ lb}$$

$$E_y = 400 - 266$$

$$\underline{E_y = 134 \text{ lb} \uparrow}$$

The reactions at A can be found by using a free-body diagram either of AC or of the complete frame. The latter is easier (Figure 5–48).

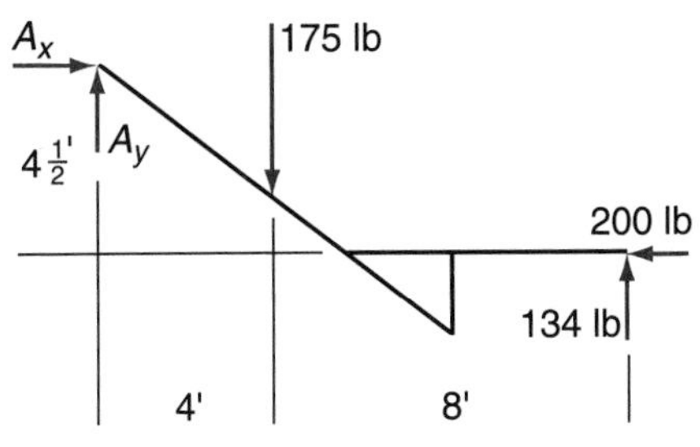

FIGURE 5–48

$$\Sigma F_x = 0$$

$$A_x = 200 \text{ lb} \rightarrow$$

$$\Sigma F_y = 0$$

$$A_y + 134 \text{ lb} - 175 \text{ lb} = 0$$

$$\underline{A_y = 41 \text{ lb} \uparrow}$$

EXAMPLE 5–12

The frame shown in Figure 5–49 has a pulley 0.5 m in diameter at B. Determine the horizontal and vertical pin reactions at C.

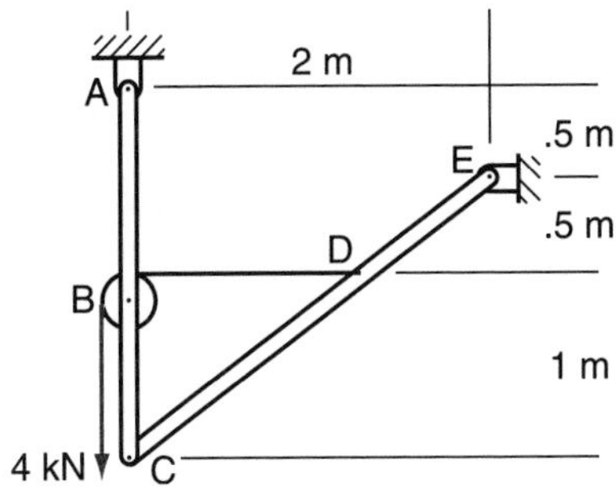

FIGURE 5–49

Free-Body Diagram of AC

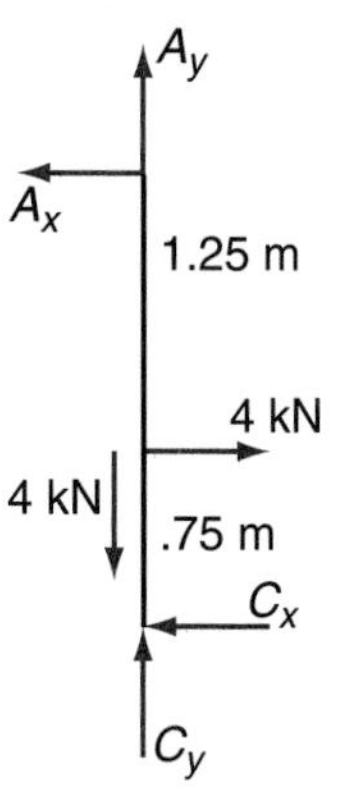

FIGURE 5–50

Free-Body Diagram of CE

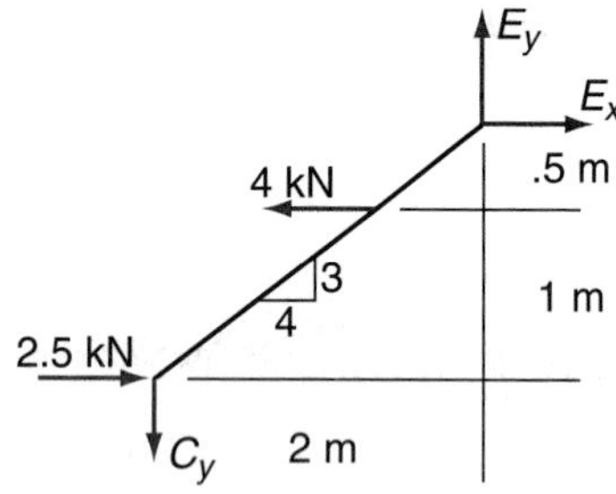

FIGURE 5–51

The cable tension of 4 kN is shown acting at the center of the pulley in Figure 5–50. Note the distances of 1.25 m and 0.75 m on AC. There are too many unknowns in the horizontal and vertical directions; therefore, a moment equation must be used. Since C_x must be determined, moments are taken about A.

$$\Sigma M_A = 0$$

$$(4\text{ kN})(1.25\text{ m}) - (2\text{ m})C_x = 0$$

$$\underline{C_x = 2.5\text{ kN} \leftarrow}$$

C_y cannot be determined from Figure 5–50, so a free-body diagram of CE (Figure 5–51) must be drawn. C_x and C_y are shown in a direction opposite to that assumed in the free-body diagram of AC.

$$\Sigma M_E = 0$$

$$(2\text{ m})C_y + (2.5\text{ kN})(1.5\text{ m}) - (4\text{ kN})(0.5\text{ m}) = 0$$

$$2C_y = 2 - 3.75$$

$$C_y = -0.87\text{ kN} \downarrow$$

$$\underline{C_y = 0.87\text{ kN} \uparrow \text{ on CE}}$$

The magnitude of this answer is correct, but the direction of C_y was incorrectly assumed in both free-body diagrams. If C_y were to be used in further calculations, the incorrectly assumed direction could remain the same but the value of C_y would be used as -0.87 kN.

EXAMPLE 5–13

For the pin-connected frame shown in Figure 5–52, determine the horizontal and vertical reactions at B on AC.

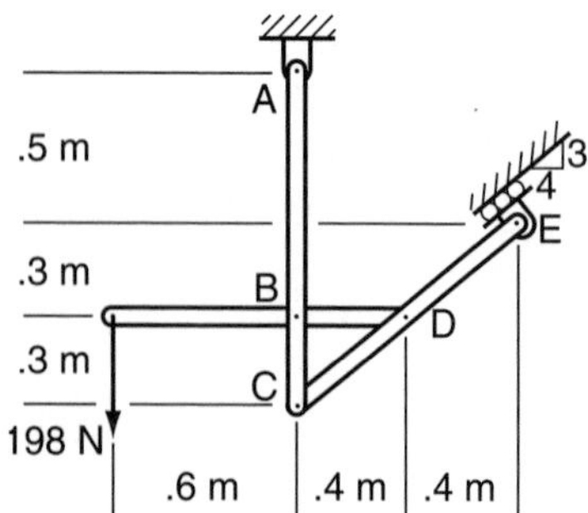

FIGURE 5–52

Free-Body Diagram of BD

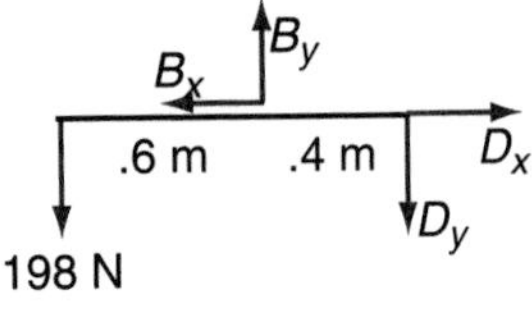

FIGURE 5–53

A free-body diagram of BD (Figure 5–53) contains the only known force and the horizontal and vertical reactions at B.

$\Sigma M_D = 0$

$$(198\text{ N})(1\text{ m}) - (0.4\text{ m})B_y = 0$$

$$B_y = 495\text{ N} \uparrow \text{ on BD}$$

or

$$\underline{B_y = 495\text{ N} \downarrow \text{ on AC}}$$

Free-Body Diagram of Frame

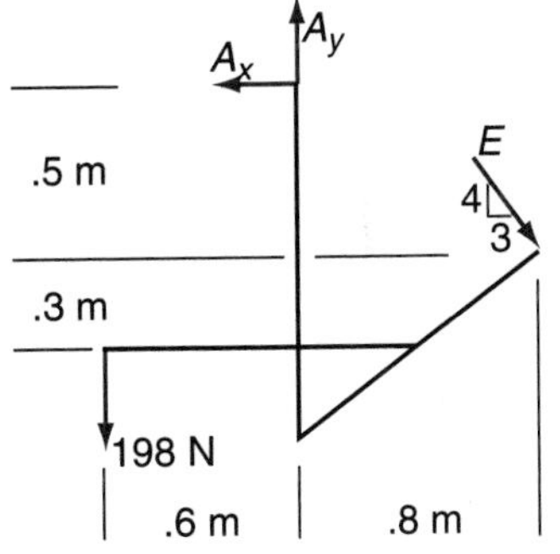

FIGURE 5–54

A free-body diagram of AC would seem to be next, but it is found to have too many unknowns. Some of these unknowns, particularly those at A, can be found with a free-body diagram of the frame (Figure 5–54).

$\Sigma M_A = 0$

$$-\frac{4}{5}E(0.8\text{ m}) + (198\text{ N})(0.6\text{ m}) + \frac{3}{5}E(0.5\text{ m}) = 0$$

$$-0.64E + 0.3E = -119$$

$$\underline{E = 350\text{ N}} \quad (4, 3 \text{ slope})$$

Free-Body Diagram of AC

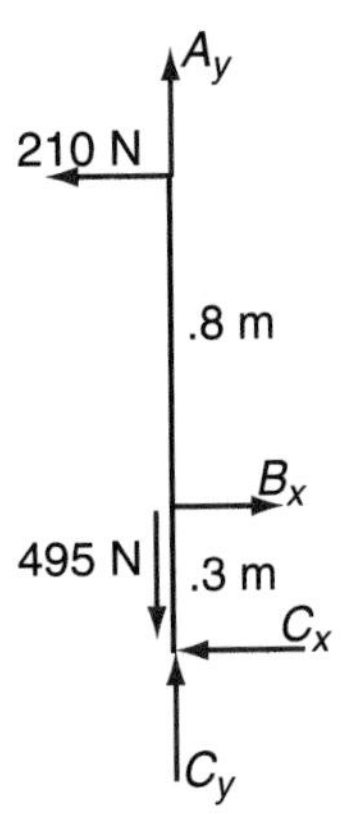

FIGURE 5–55

$\Sigma F_x = 0$

$$A_x = \frac{3}{5}E$$

$$A_x = \frac{3}{5}(350\text{ N})$$

$$\underline{A_x = 210\text{ N} \leftarrow}$$

The value of $A_x = 210$ N is sufficient information to allow us to go to the free-body diagram of AC (Figure 5–55).

$\Sigma M_C = 0$

$$(210\text{ N})(1.1\text{ m}) - (0.3\text{ m})B_x = 0$$

$$\underline{B_x = 770\text{ N} \rightarrow \text{on AC}}$$

EXAMPLE 5–14

Solve for the horizontal and vertical pin reactions at A and C of the frame shown in Figure 5–56.

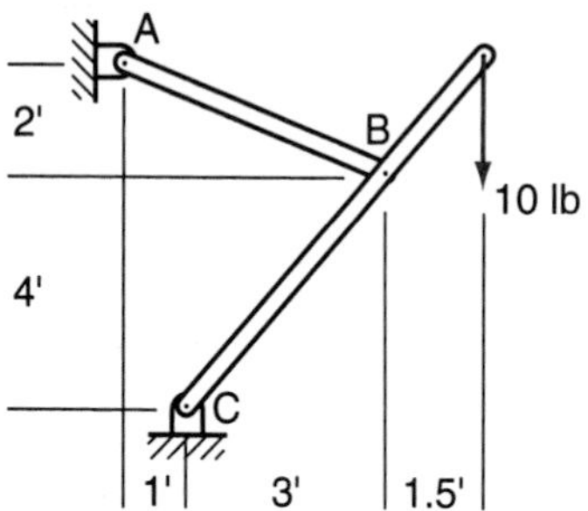

FIGURE 5–56

Although the free-body diagram of the frame (Figure 5–57) appears to have four unknowns, the key to the solution of this problem is to notice that member AB is a two-force member in tension. We note that the vector sum of A_x and A_y must have the same slope as member AB. Therefore,

$$A_x = 2A_y$$

Free-Body Diagram of Frame

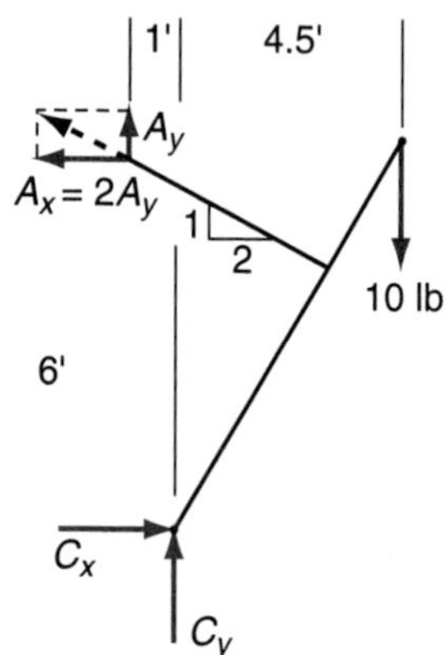

FIGURE 5–57

If moments are taken about point C, there is only one unknown, A_y.

$\Sigma M_C = 0$

$$-A_y(1\text{ ft}) - (10\text{ lb})(4.5\text{ ft}) + 2A_y(6\text{ ft}) = 0$$

$$-A_y - 45 + 12A_y = 0$$

$$A_y = \frac{45}{11}$$

$$\underline{A_y = 4.1\text{ lb}\uparrow}$$

$$A_x = 2A_y$$

$$= 2(4.1\text{ lb})$$

$$\underline{A_x = 8.2\text{ lb}\leftarrow}$$

We can now calculate C_x and C_y by considering forces in the horizontal direction and forces in the vertical direction.

$\Sigma F_x = 0$

$$C_x = A_x$$

$$\underline{C_x = 8.2\text{ lb}\rightarrow}$$

$\Sigma F_y = 0$

$$C_y + 4.1\text{ lb} = 10\text{ lb}$$

$$\underline{C_y = 5.9\text{ lb}\uparrow}$$

EXAMPLE 5–15

Suppose that the frame in Figure 5–56 has an additional load of 20 lb applied as shown in Figure 5–58. Solve for the horizontal and vertical pin reactions at A and C.

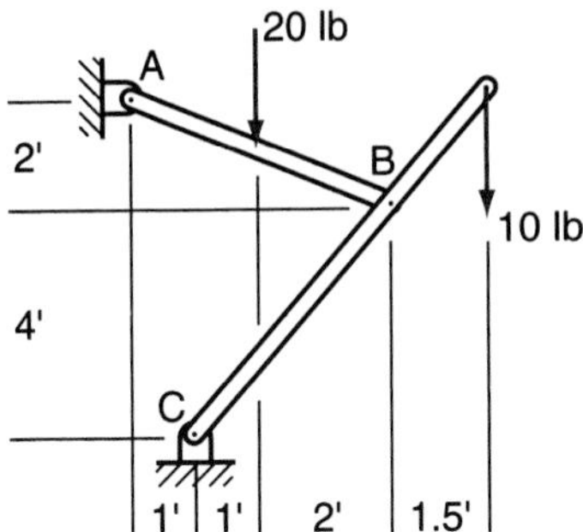

FIGURE 5–58

Both members of this frame are three-force members, and we do not know the relationship between the horizontal and vertical components at any point. In the free-body diagram of the frame (Figure 5–59) or that of AB (Figure 5–60), there are two unknowns regardless of which one of the three equilibrium equations is used. The solution requires your finding a point in each diagram that will give you moment equations with the same two unknowns. The two equations are then solved simultaneously.

In Figure 5–59, moments about C will give an equation with A_x and A_y.

$\Sigma M_C = 0$ (Figure 5–59)

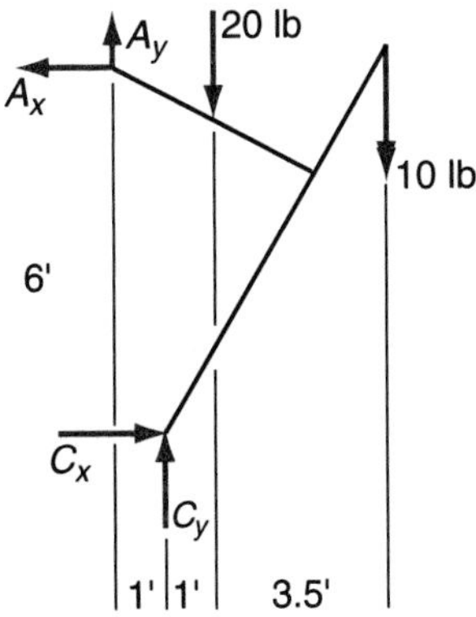

FIGURE 5–59

$$+(6\text{ ft})A_x - (1\text{ ft})A_y - (20\text{ lb})(1\text{ ft}) - (10\text{ lb})(4.5\text{ ft}) = 0$$

$$6A_x - A_y = 65 \quad (1)$$

Considering the free-body diagram of AB (Figure 5–60), moments about B will give another equation with A_x and A_y as unknowns.

$\Sigma M_B = 0$ (Figure 5–60)

$$(2\text{ ft})A_x + (20\text{ lb})(2\text{ ft}) - (4\text{ ft})A_y = 0$$

$$2A_x - 4A_y = -40 \quad (2)$$

Multiplying Equation (2) by -3 and adding Equation (1), we have

$$-6A_x + 12A_y = 120$$

$$6A_x - A_y = 65$$

$$11A_y = 185$$

$$A_y = 16.8\text{ lb}\uparrow$$

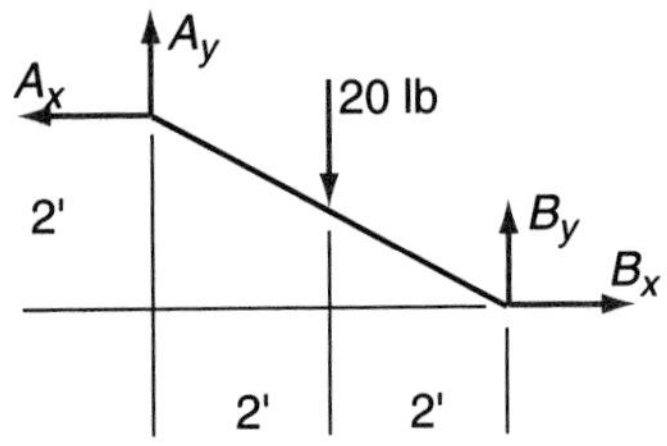

FIGURE 5–60

Substituting $A_y = 16.8$ into Equation (1), we get

$$6A_x - 16.8\text{ lb} = 65$$

$$A_x = 13.6\text{ lb}\leftarrow$$

Referring now to the free-body diagram of the frame and solving for C_x and C_y, we have

$$\Sigma F_x = 0 \quad \text{(Figure 5–59)}$$

$$C_x = A_x$$

$$\underline{C_x = 13.6 \text{ lb} \rightarrow}$$

$$\Sigma F_y = 0$$

$$C_y + 16.8 \text{ lb} - 20 \text{ lb} - 10 \text{ lb} = 0$$

$$\underline{C_y = 13.2 \text{ lb} \uparrow}$$

HINTS FOR PROBLEM SOLVING

1. You usually have two choices as to where to begin a truss problem:
   (a) An FBD of a joint where a force is given.
   (b) An FBD of the complete truss, to get more external forces.
2. Be ready to spot zero-force members since they can simplify and shorten a complicated-looking problem.
3. Keep your free-body diagrams large and well labeled, with the calculations beginning opposite and to the right of them.
4. When using the method of joints to solve for all the loads in the members of a truss, try to work into the truss from several external points so that a single initial error will not be perpetrated throughout the entire solution.
5. As in Chapter 4, a tension member pulls on both pins and a compression member pushes.
6. A section line cutting through a truss:
   (a) Usually cuts only three members, thereby avoiding too many unknowns.
   (b) Does not have to be straight but may take any path through the truss.
7. After sectioning a truss, draw an FBD of the simplest portion with known forces. Try taking moments about a point where two unknown forces intersect. This point can be on or off the FBD.
8. The shortest solution in the method of members occurs when the first FBD incorporates both given information and what is required. If this fails to give a solution, then draw several or all possible free-body diagrams to see in which sequence they may be used. (An FBD of the complete frame is often helpful.)
9. When drawing the FBD of a member that has a two-force member pinned to it, I suggest not showing the x and y component but rather a single force with the same slope as the two-force member. This will point out that there is only one unknown at that point and not two.
10. Remember, for internal forces between connecting members, if you switch free-body diagrams, you switch the directions of the forces for the new FBD.

PROBLEMS

APPLIED PROBLEMS FOR SECTION 5–1

5–1. Using the method of joints, determine the load in member AC of Figure P5–1.

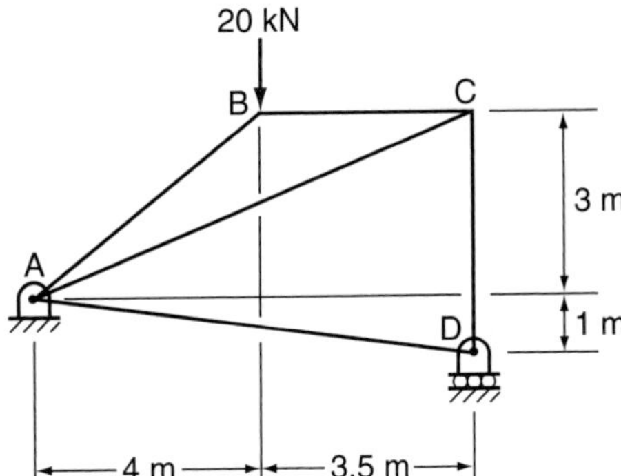

FIGURE P5–1

5–2. Determine the load in each member of the truss shown in Figure P5–2.

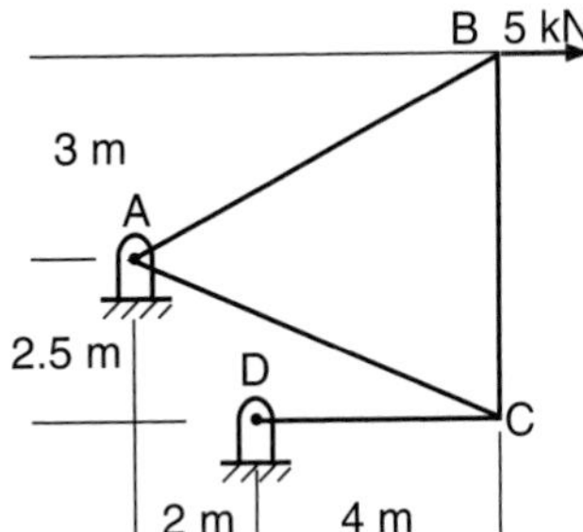

FIGURE P5–2

5–3. Determine the load in member BC of the pin-connected truss shown in Figure P5–3.

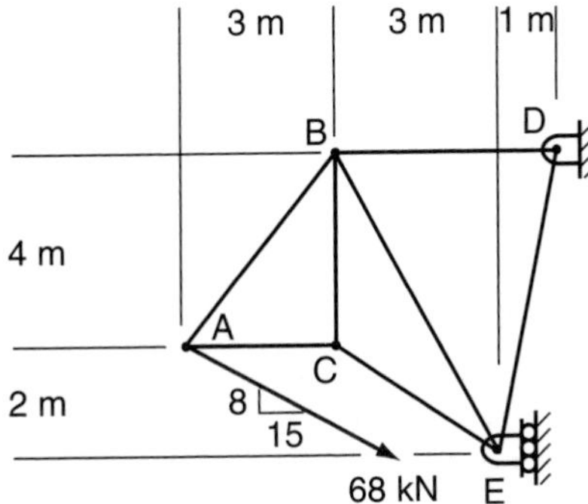

FIGURE P5–3

5–4. Determine the load in member CH of the truss shown in Figure P5–4.

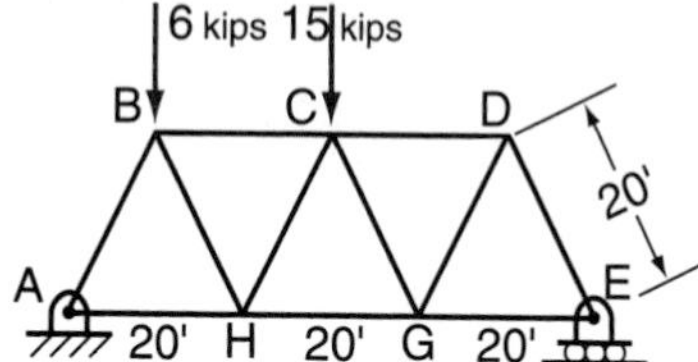

FIGURE P5–4

5–5. Determine the load in member BD of the truss shown in Figure P5–5.

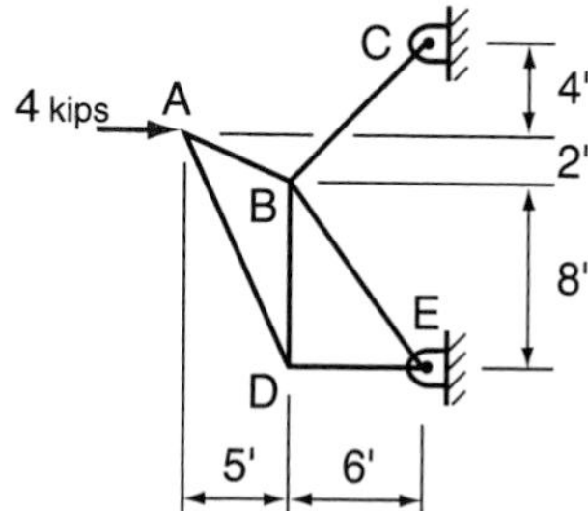

FIGURE P5–5

5–6. Determine the force in each member of the truss shown in Figure P5–6.

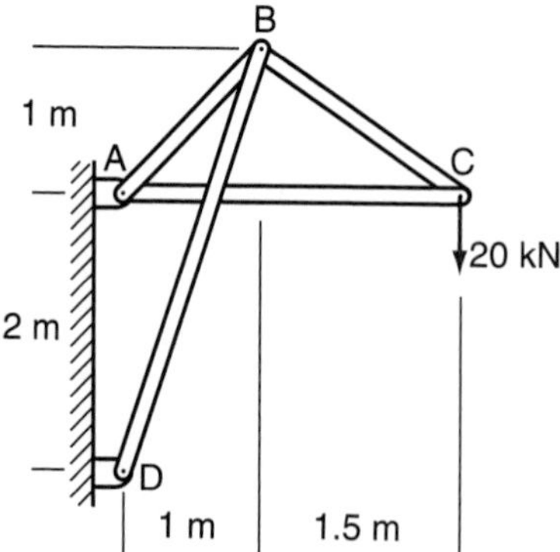

FIGURE P5–6

5–7. Determine the force in each member of the truss shown in Figure P5–7.

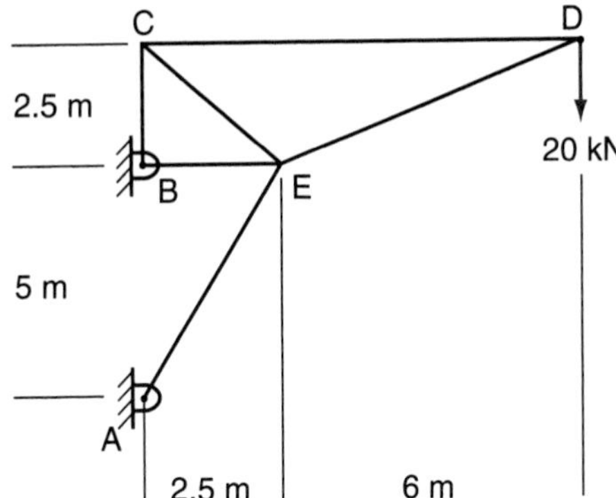

FIGURE P5–7

5–8. Determine the loads in members CE and BE of the truss shown in Figure P5–8.

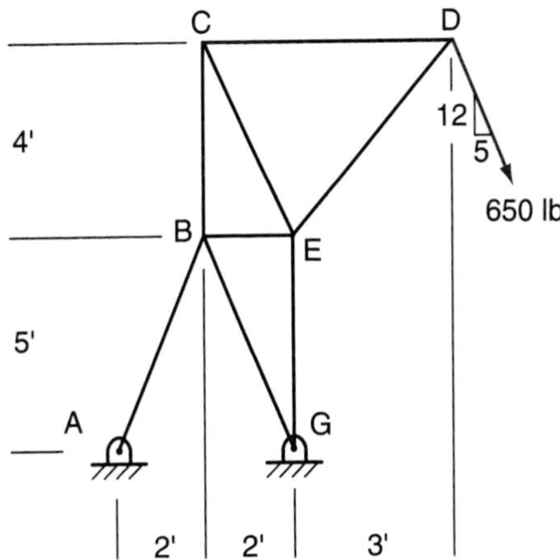

FIGURE P5–8

5–9. Determine the force in each member of the truss shown in Figure P5–9.

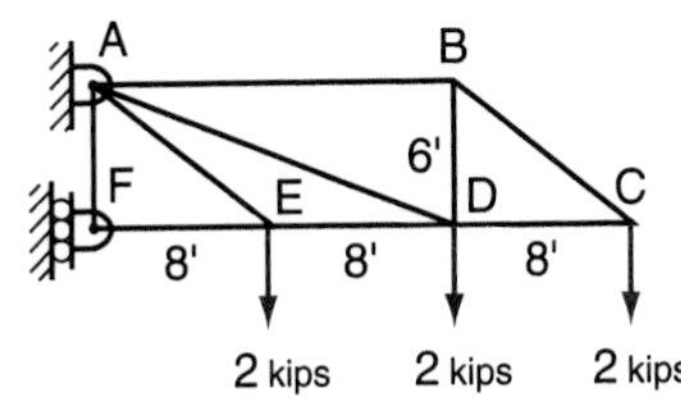

FIGURE P5–9

5–10. Determine the load in each member of the truss shown in Figure P5–10.

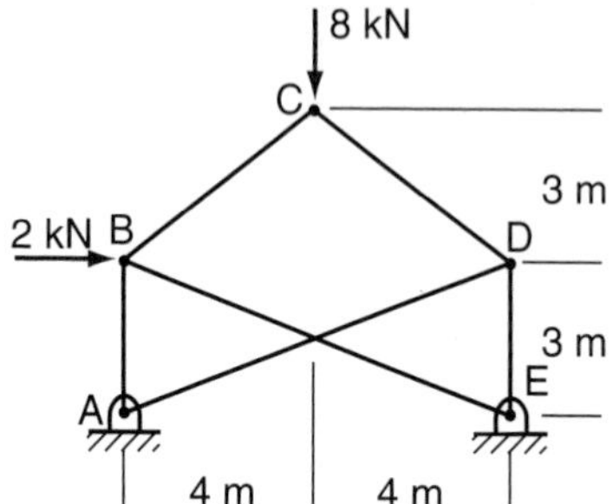

FIGURE P5–10

5–11. Determine the loads in members EB and BD of the pin-connected truss shown in Figure P5–11.

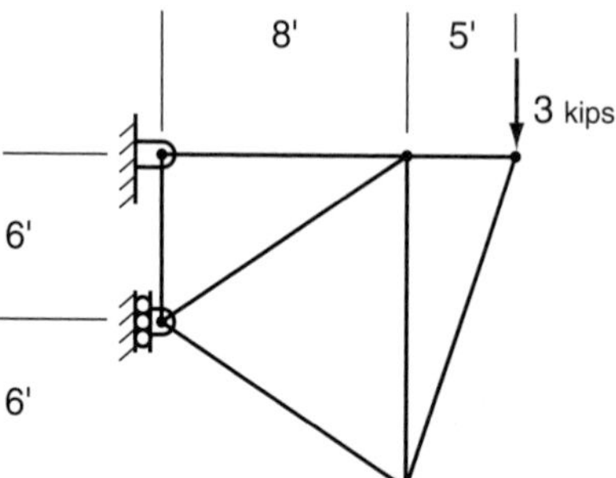

FIGURE P5–11

5–12. The scissor linkage shown in Figure P5–12 is controlled by cylinder CD and is used to compress material in a container below with a vertical force of 1000 lb. What cylinder force is required if the force of 1000 lb is being applied at the position shown?

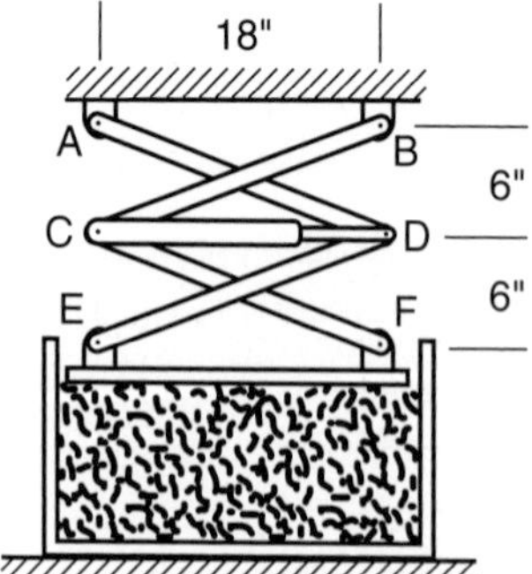

FIGURE P5–12

5–13. Determine the load in member CG of the truss shown in Figure P5–13.

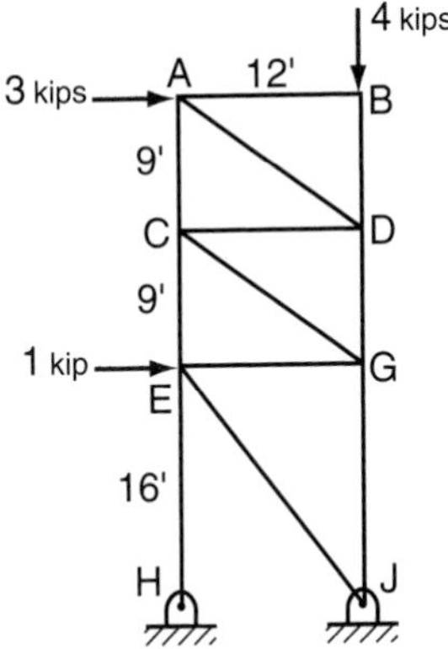

FIGURE P5–13

5–14. Cylinder BD acts as a truss member to support the load of 600 lb (Figure P5–14). Determine the loads in the cylinder and member CA.

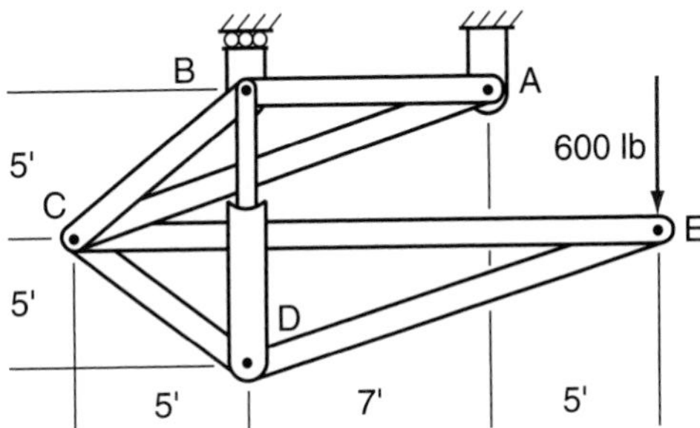

FIGURE P5–14

5–15. Using the method of joints, determine the loads in members AC and CE of Figure P5–15.

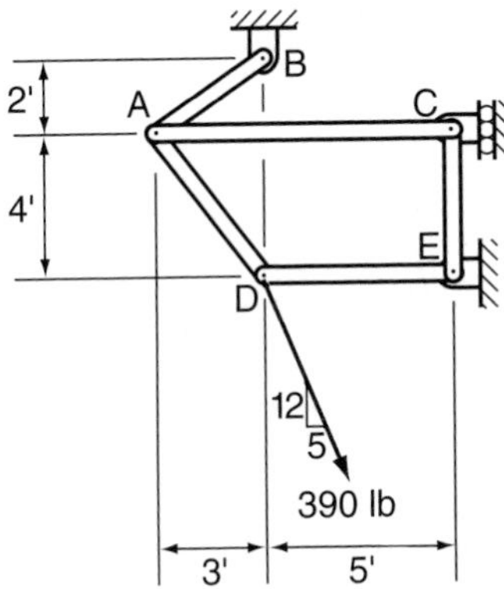

FIGURE P5–15

5–16. Using the method of joints, determine the loads in members BC and CE of the pin-connected truss shown in Figure P5–16.

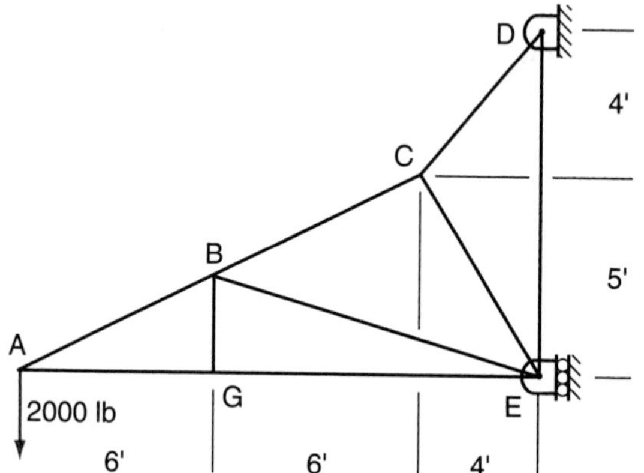

FIGURE P5–16

5–17. Using the method of joints, determine the load in member AC of the truss shown in Figure P5–17.

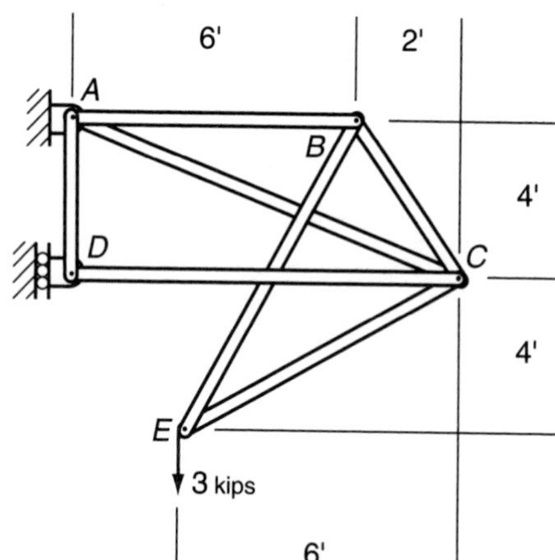

FIGURE P5–17

5–18. Determine the load in each member of the truss shown in Figure P5–18.

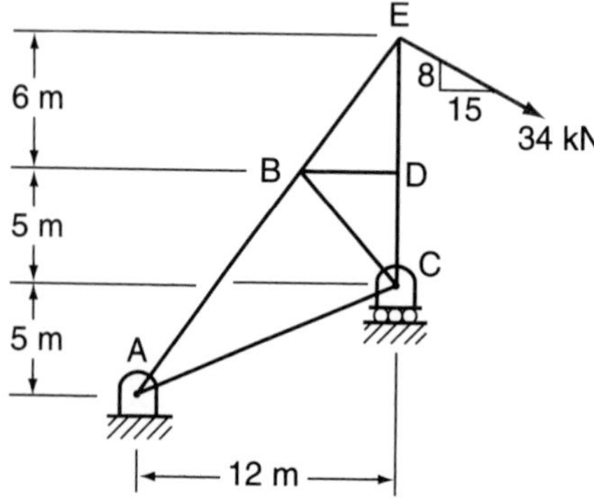

FIGURE P5–18

5–19. Determine the load in each member of the truss shown in Figure P5–19.

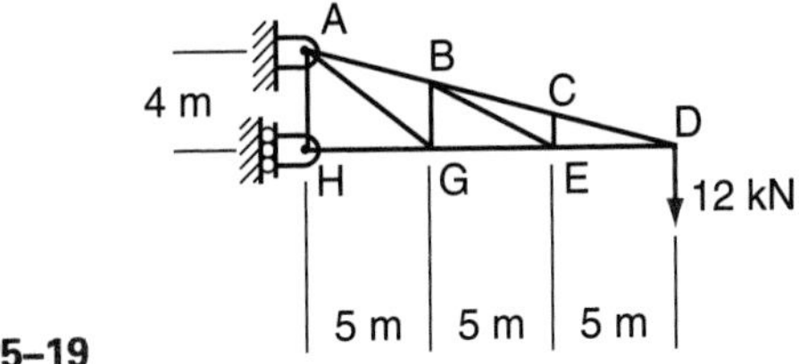

FIGURE P5–19

5–20. Determine the load in each member of the pin-connected truss shown in Figure P5–20.

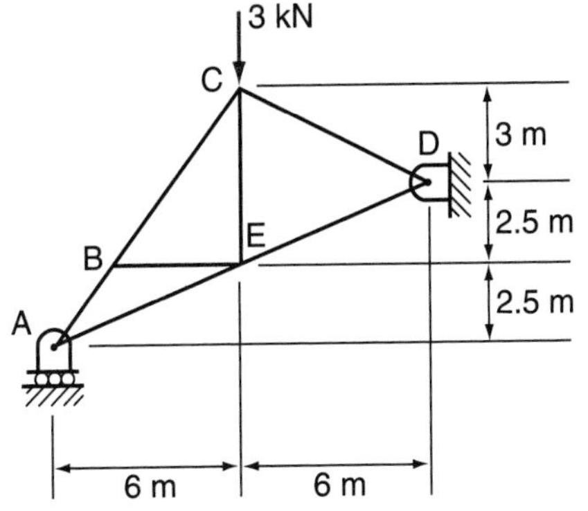

FIGURE P5–20

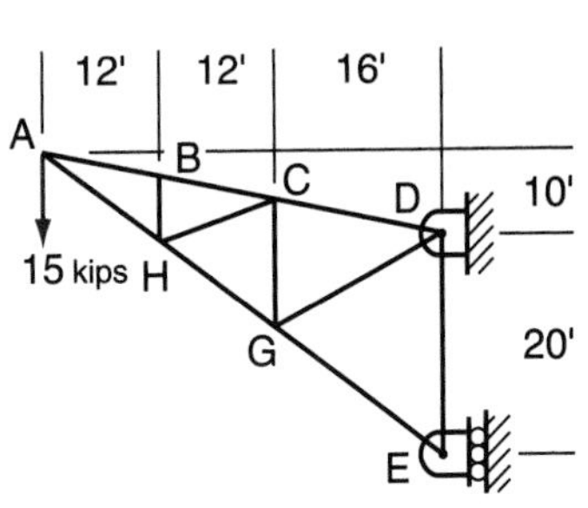

FIGURE P5–21

5–21. Determine the force in members BC and CG of the pin-connected truss shown in Figure P5–21.

5–22. Determine the loads in members GE and CE of the pin-connected truss shown in Figure P5–22.

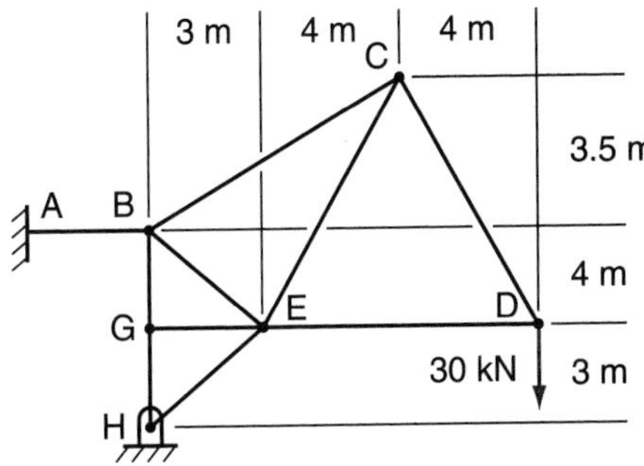

FIGURE P5–22

5–23. Use the method of joints to solve for the force in member BC of the truss shown in Figure P5–23.

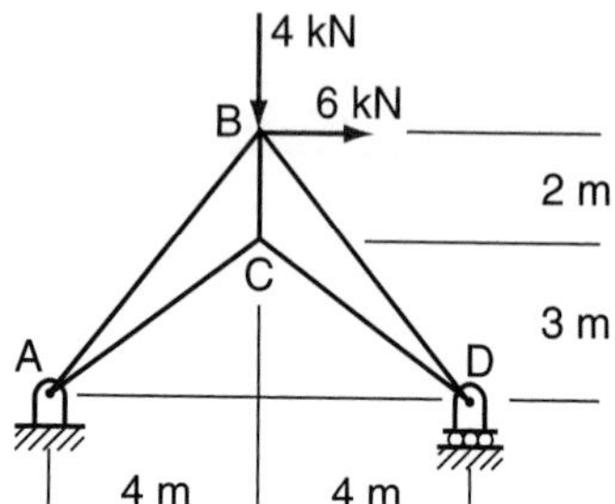

FIGURE P5–23

5–24. Use the method of joints to solve for the load in member AC of the system shown in Figure P5–24.

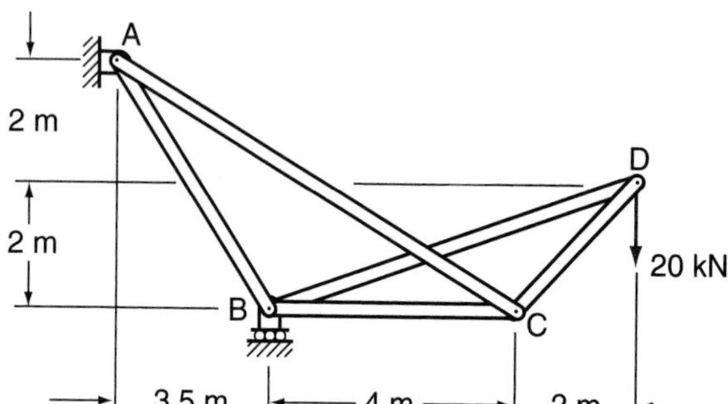

FIGURE P5–24

APPLIED PROBLEMS FOR SECTION 5–2

5–25. Determine the force in members BC, BG, and EG of the truss loaded as shown in Figure P5–25.

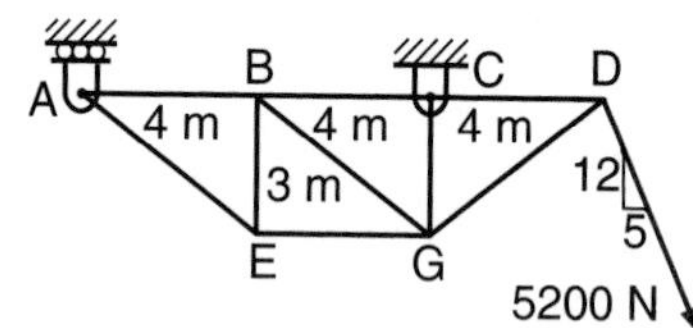

FIGURE P5–25

5–26. Using the method of sections, determine the load in members BD, CD, and CE of the truss shown in Figure P5–26.

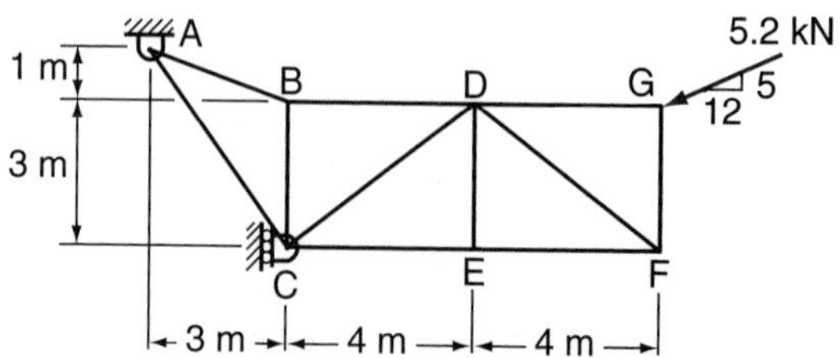

FIGURE P5–26

5–27. Determine the force in members BC, BE, BD, and DE of the truss shown in Figure P5–27.

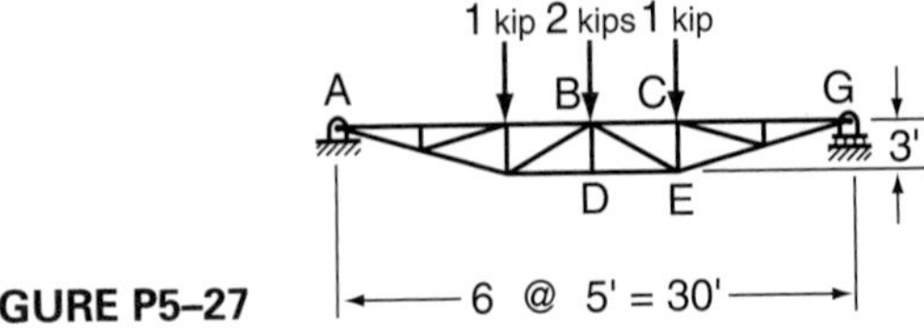

FIGURE P5–27

5–28. Using the method of sections, determine the loads in members BD, CD, and CE of the truss shown in Figure P5–28.

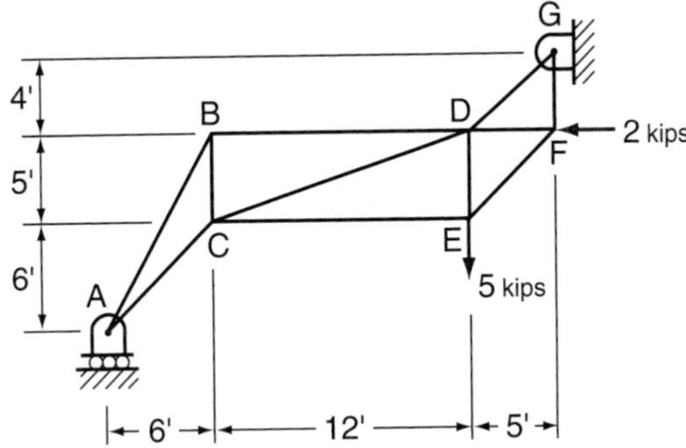

FIGURE P5–28

5–29. Using the method of sections, determine the loads in member BG (Figure P5–29).

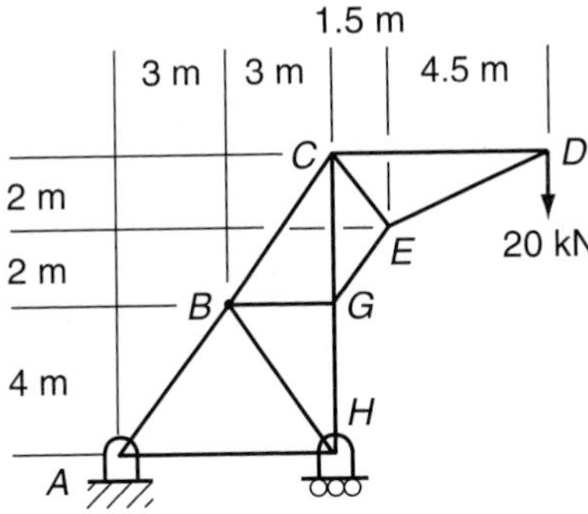

FIGURE P5–29

5–30. Determine the force in member EG of the K-truss shown in Figure P5–30.

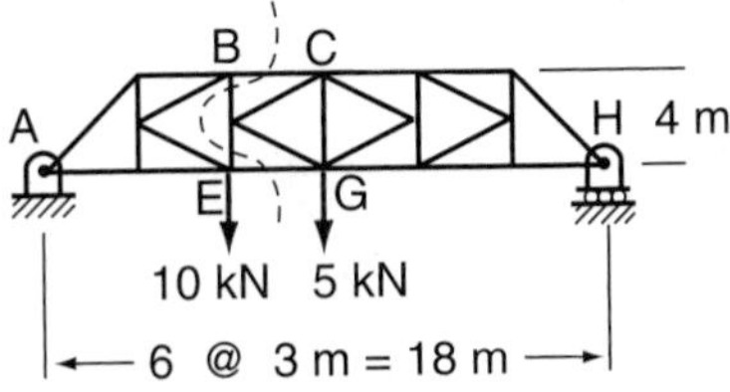

FIGURE P5–30

5–31. Using the method of sections, determine the load in member CD of the truss shown in Figure P5–31.

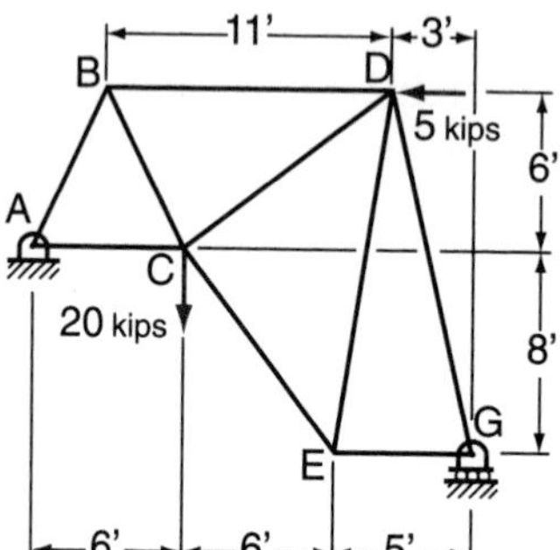

FIGURE P5–31

5–32. Use the method of sections to determine the loads in members CD and ED of the truss shown in Figure P5–32.

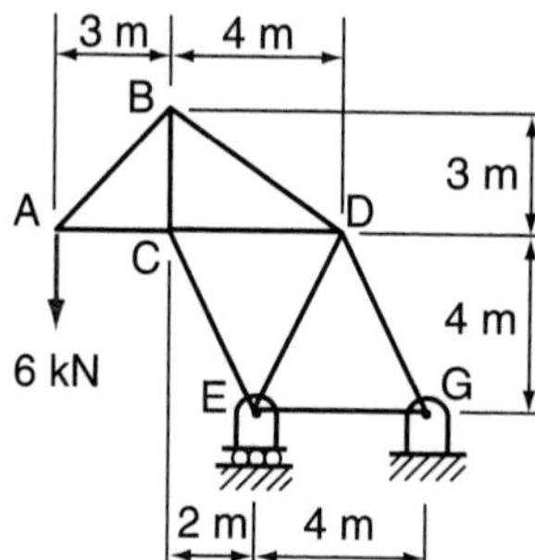

FIGURE P5–32

5–33. Determine the load in members BC, BH, and JH of the truss shown in Figure P5–33.

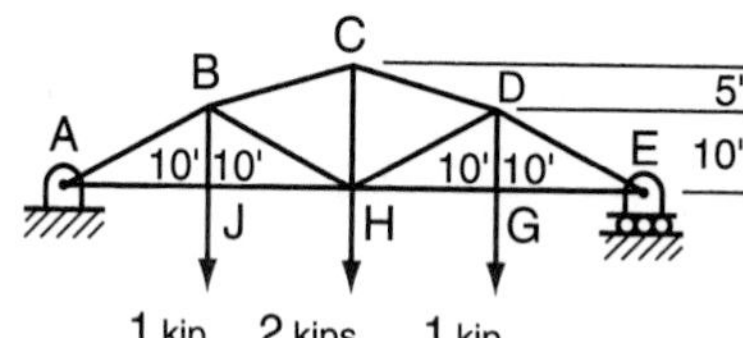

FIGURE P5–33

5–34. Determine the loads in members DE and DG of the pin-connected truss shown in Figure P5–34. All triangles are equilateral with 4-m sides.

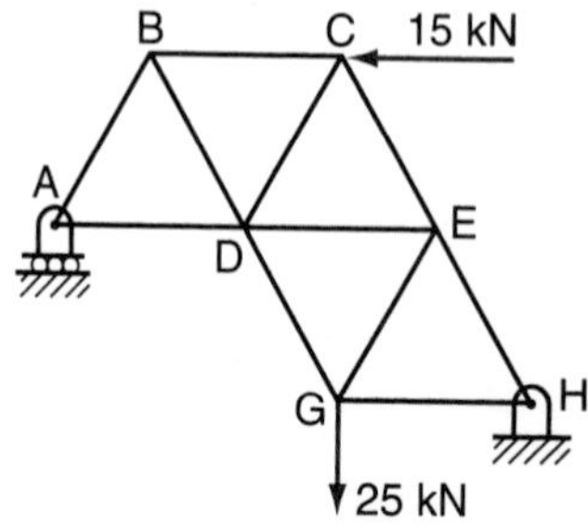

FIGURE P5–34

5–35. The truss framework of a billboard is subjected to the forces due to wind and the billboard weight as shown in Figure P5–35. Use the method of sections to determine the force in members CB, BE, and BG.

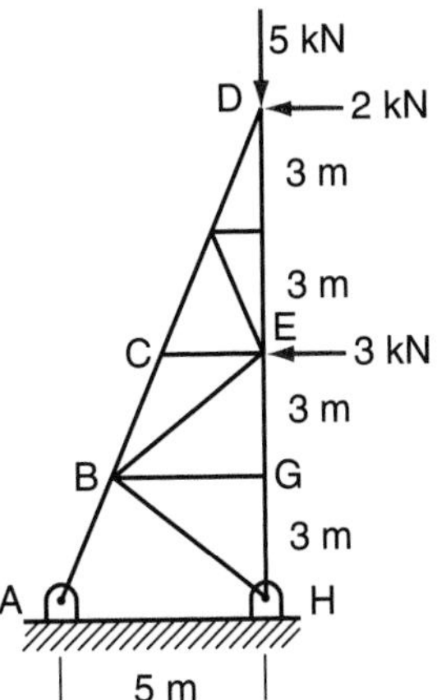

FIGURE P5–35

5–36. Using the method of sections, determine the loads in members CE, ED, and DG (Figure P5–36).

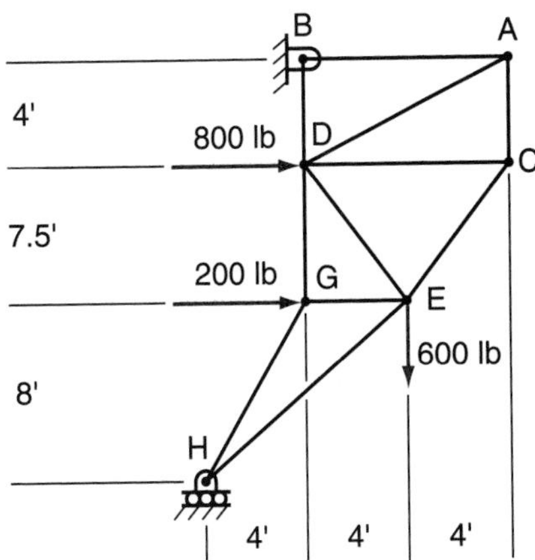

FIGURE P5–36

5–37. Using the method of sections, determine the loads in members BD, CD, and CE of the truss shown in Figure P5–37.

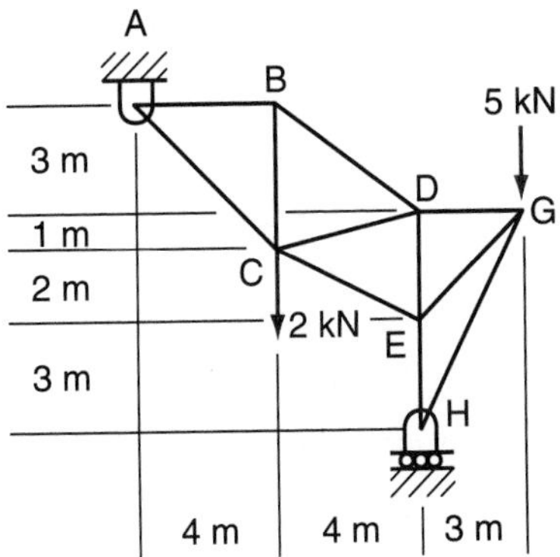

FIGURE P5–37

5–38. Determine the loads in members CE, ED, and BD of the pin-connected truss shown in Figure P5–38. Use the method of sections.

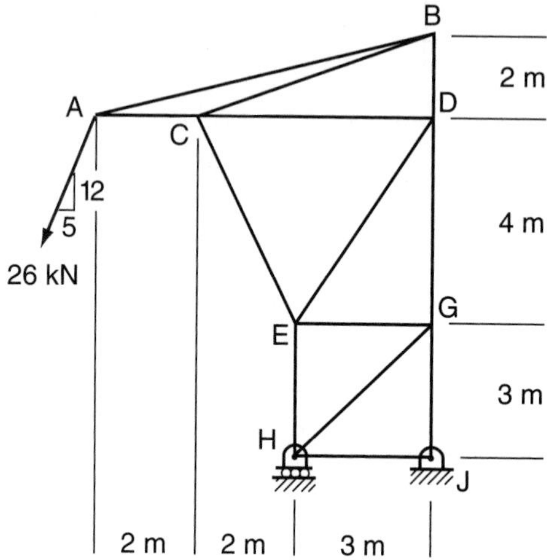

FIGURE P5–38

5–39. Using the method of sections, determine the loads in members CD, HG, and JG of the truss shown in Figure P5–39.

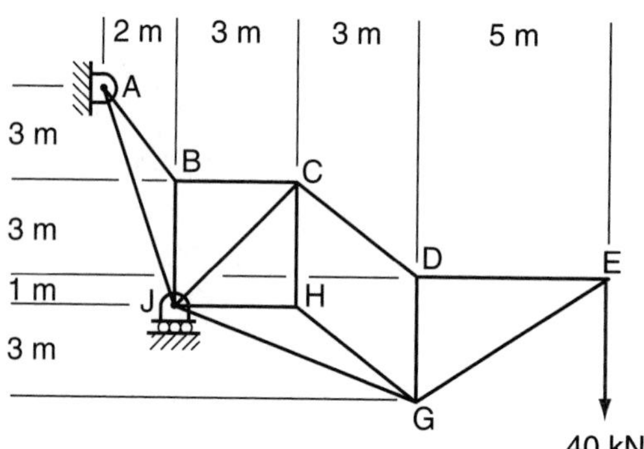

FIGURE P5–39

5–40. Use the method of sections to determine the load in members CD and CG of the sawtooth truss shown in Figure P5–40.

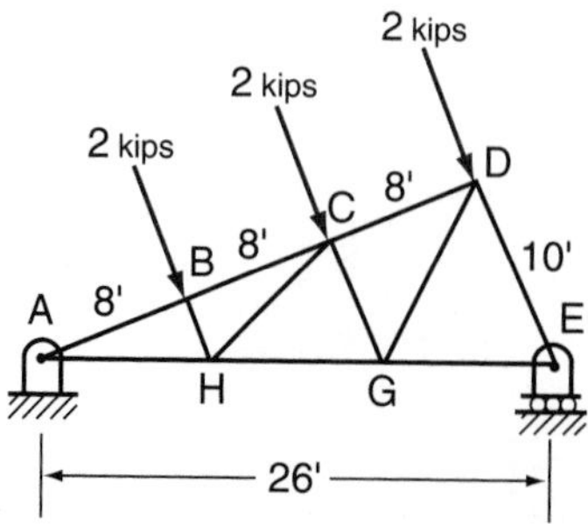

FIGURE P5–40

5–41. For the Parker truss shown in Figure P5–41, determine the force in members BC and BG.

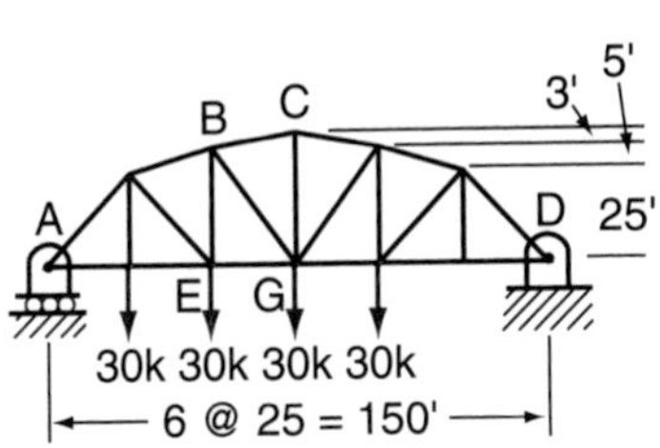

FIGURE P5–41

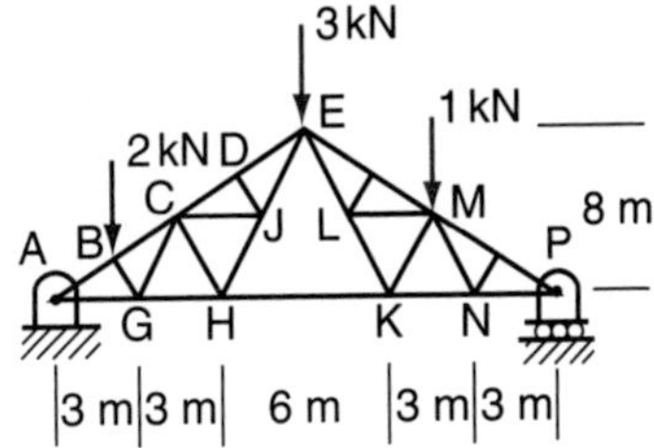

FIGURE P5–42

5–42. For the Fink truss shown in Figure P5–42, determine the load in members DE, JE, and KH. What is the load in members LM and MN?

APPLIED PROBLEMS FOR SECTION 5–3

5–43. Determine the horizontal and vertical components of the pin reactions at B and D of the frame shown in Figure P5–43.

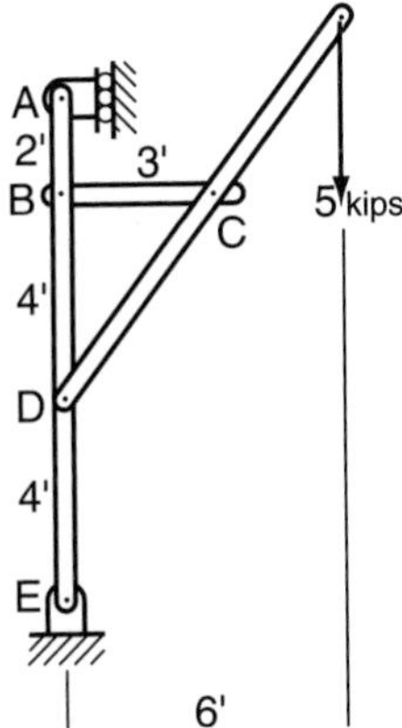

FIGURE P5–43

5–44. Determine the horizontal and vertical components of the pin reaction at D of the system shown in Figure P5–44.

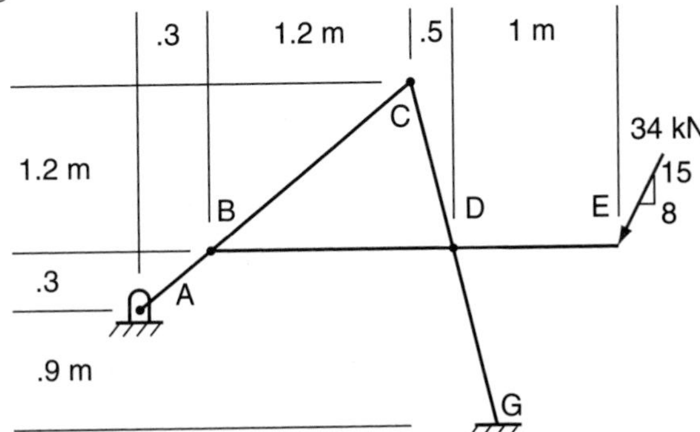

FIGURE P5–44

5–45. Determine the pin reactions at B and C of the structure shown in Figure P5–45.

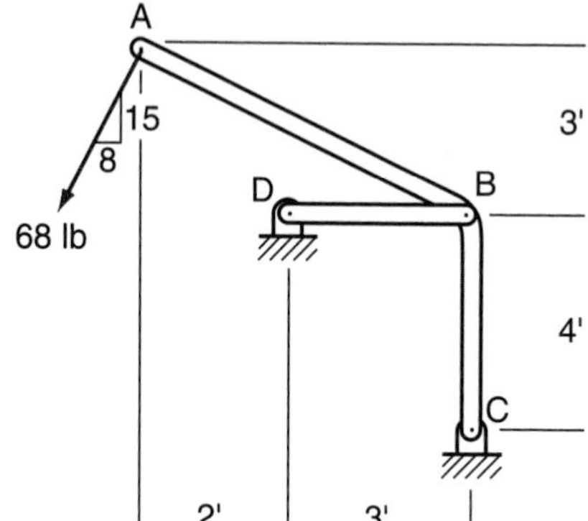

FIGURE P5–45

5–46. Determine the horizontal and vertical components of the pin reactions at B and C on member AC in Figure P5–46.

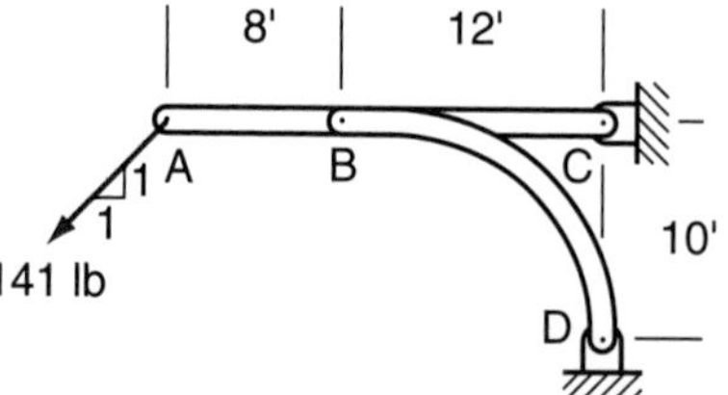

FIGURE P5–46

5–47. The frame in Figure P5–47 is held by cable HG when loaded as shown. Determine the horizontal and vertical components of the pin reactions at B and C.

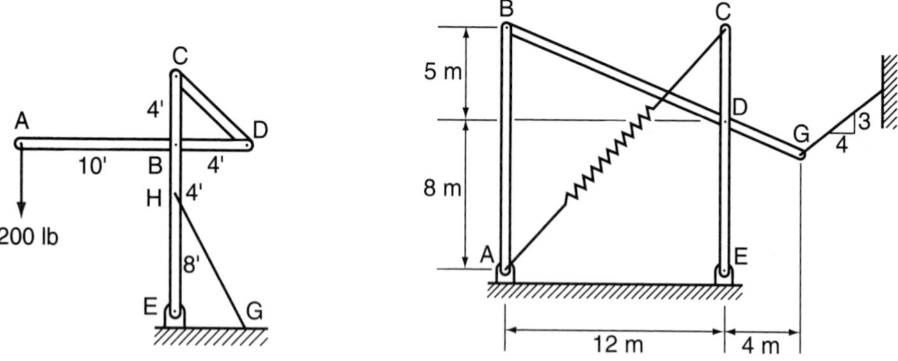

FIGURE P5–47

FIGURE P5–48

5–48. Determine the force in the spring shown in Figure P5–48 if the cable tension at G is 50 kN.

5–49. Determine the horizontal and vertical components of the pin reactions at A and B (Figure P5–49).

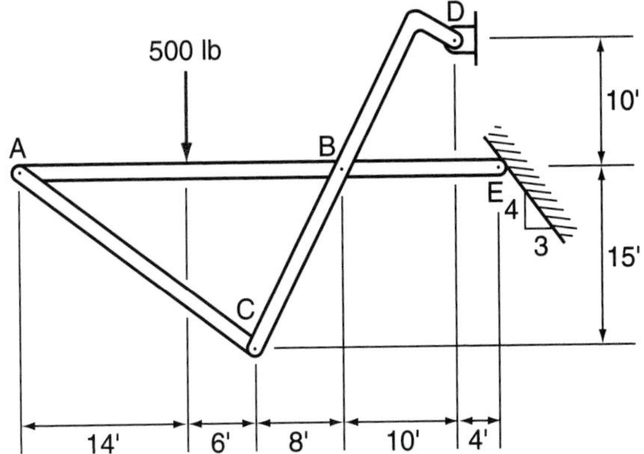

FIGURE P5–49

5–50. Determine the horizontal and vertical components of the pin reaction at D of the system shown in Figure P5–50.

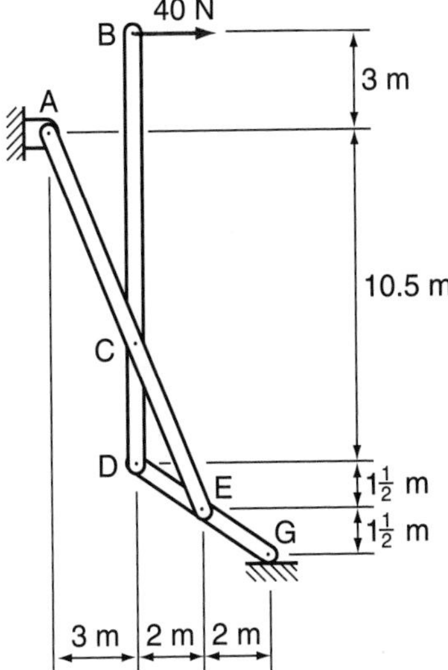

FIGURE P5–50

5–51. The platform supporting a 4.2-kip load in Figure P5–51 can be leveled by means of a cable and a winch at C. Neglect the weight of the platform and the drum diameter of the winch. Calculate the tension in the cable and the pin reactions at B on the platform.

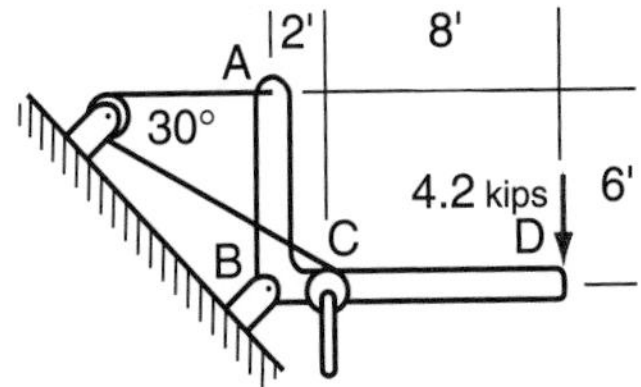

FIGURE P5–51

5–52. Determine cylinder force BD, and the horizontal and vertical components of the pin reaction at C, for the truck-mounted articulating crane shown in Figure P4–13.

5–53. Determine the pin reactions at C of the system shown in Figure P5–53.

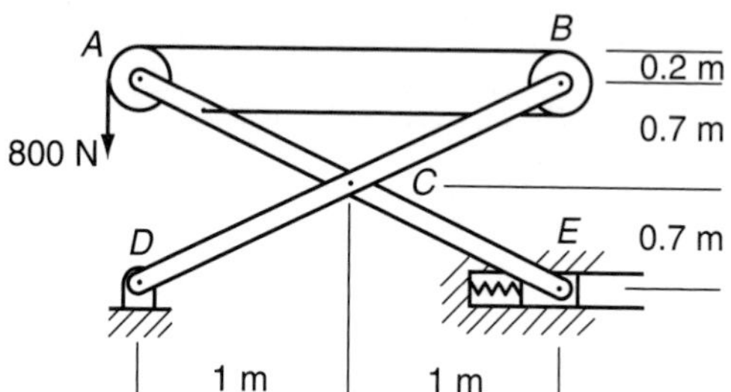

FIGURE P5–53

5–54. Pulley B is belt-driven by pulley A in Figure P5–54. By means of adjusting the turnbuckle DE, there is a belt tension of 400 N at pulley C. What is the tensile force in turnbuckle DE?

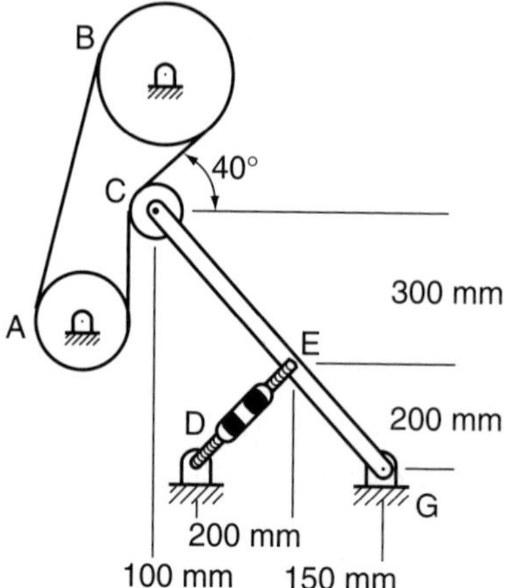

FIGURE P5–54

5–55. An automobile wheel assembly supports 3.5 kN. Determine the force compressing the spring and the components of the forces acting on the frame at points A and E (Figure P5–55).

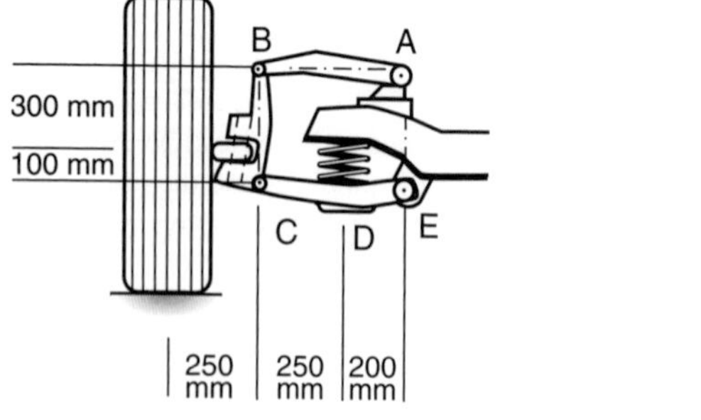

FIGURE P5–55

FIGURE P5–56

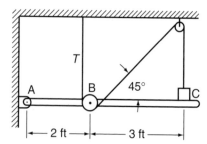

5–56. The horizontal beam shown in Figure P5–56 weighs 40 lb and block C weighs 200 lb. If block C is partially resting on the beam, determine the cable tension *T*.

5–57. The system shown in Figure P5–57 is in static equilibrium. Member CB is 12 in. long. Determine force *P*.

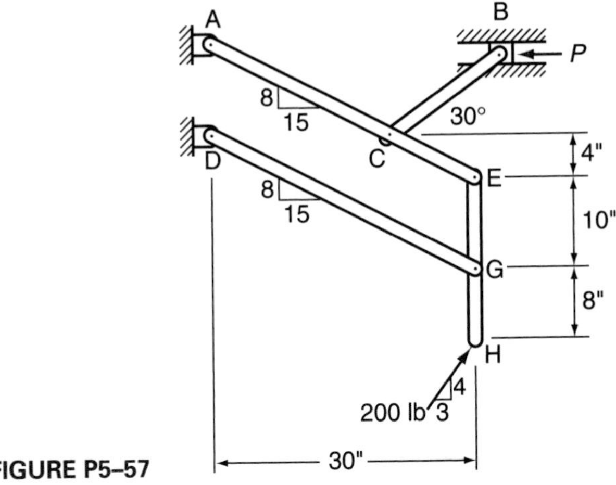

FIGURE P5–57

5–58. Bar AC of the frame shown in Figure P5–58 has a loading of 100 lb/ft. Determine the horizontal and vertical components of the pin reactions at A and D.

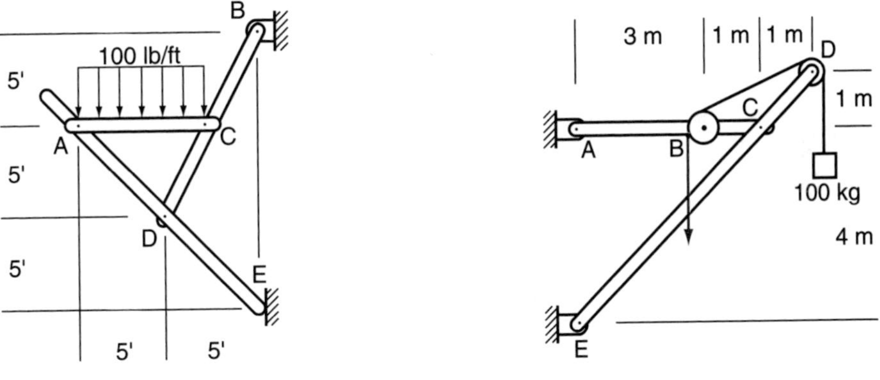

FIGURE P5–58

FIGURE P5–59

5–59. A 100-kg mass is lifted by means of the mechanism shown in Figure P5–59. Neglect the diameters of the pulleys and determine the horizontal and vertical components of the pin reaction at C on member ED.

5–60. Determine the force P for the system shown in Figure P5–60 to be in static equilibrium. What are the horizontal and vertical components of the pin reactions at A and B?

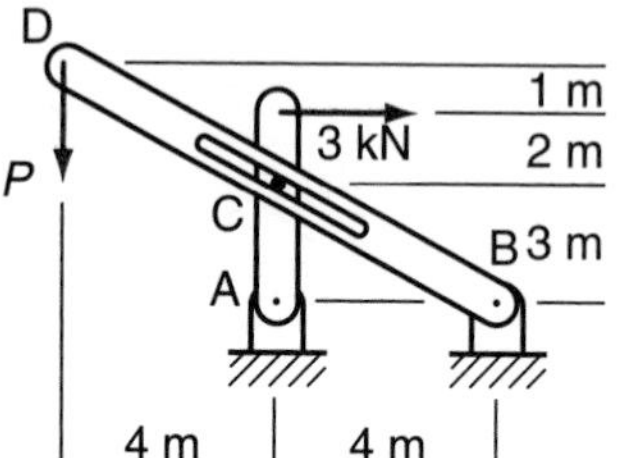

FIGURE P5–60

5–61. Determine the force P for equilibrium of the system shown in Figure P5–61.

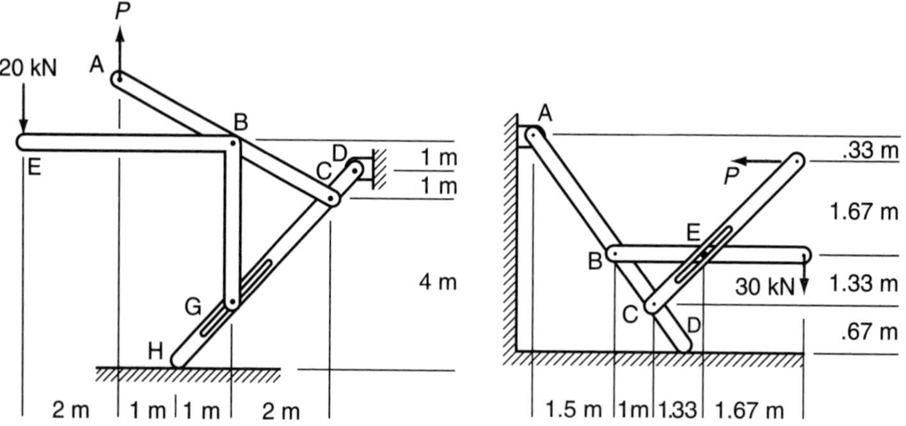

FIGURE P5–61

FIGURE P5–62

5–62. Force P produces equilibrium of the system shown in Figure P5–62. Determine force P and the pin reactions at B.

5–63. Determine the load in cylinder BD of the grain auger shown in Figure P4–14.

5–64. Determine the force P for equilibrium of the system shown in Figure P5–64.

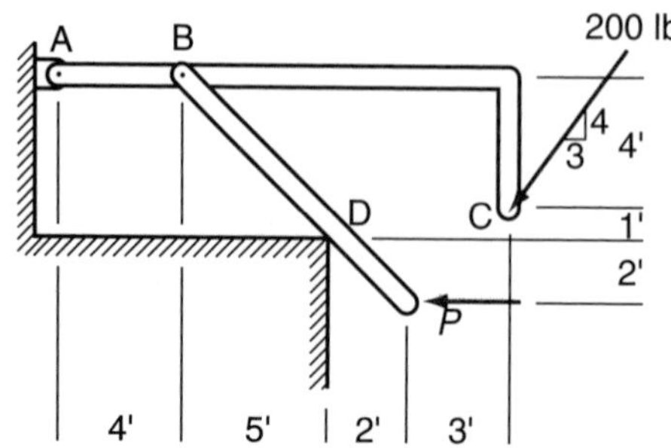

FIGURE P5–64

5–65. Determine the pin reactions at G for the "bobcat" loader shown in Figure P4–15.

5–66. Solve for the force *P* and tension *T* for the system shown (Figure P5–66) to be in equilibrium. Assume the pulley diameters to be negligible.

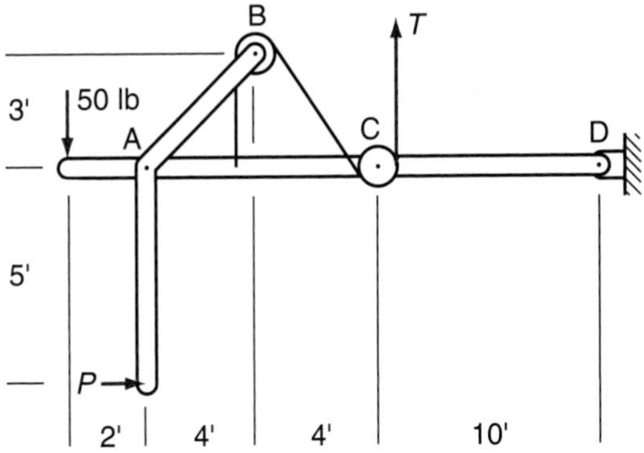

FIGURE P5–66

5–67. A concealed door closer has a spring load of 5 lb in the position shown (Figure P5–67). Determine the torque tending to close the door.

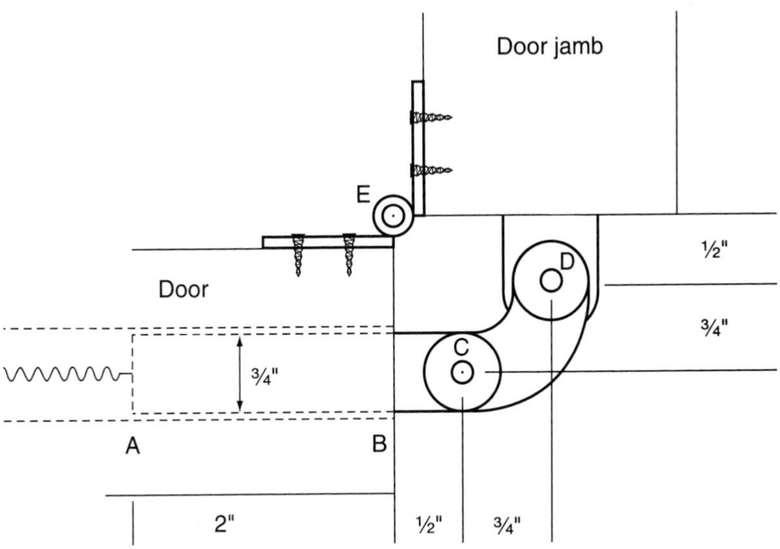

FIGURE P5–67

5–68. When the hand clamping force of 300 N is applied, the spring-loaded toggle bar locks on the horizontal bar of the clamp shown (Figure P5–68). This produces motion of the sliding portion to the right and a clamping force *P*.

Determine the clamping force *P* and the reactions at points C and D.

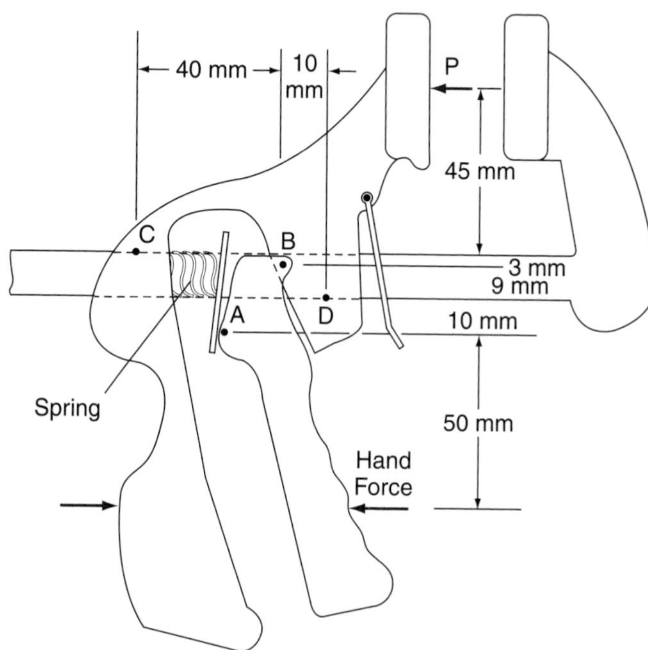

FIGURE P5–68

5–69. For the clamping wrench shown in (Figure P5–69), calculate the force required to squeeze the handles together at GH.

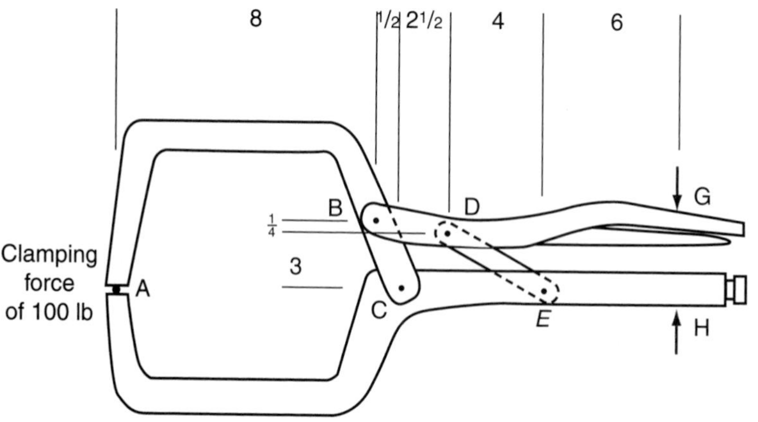

FIGURE P5–69

5–70. When two plates are spot welded together as in Figure P5–70, electrodes A and B squeeze the plates together with a force of 150 lb. What force must air cylinder EH apply in order to accomplish this?

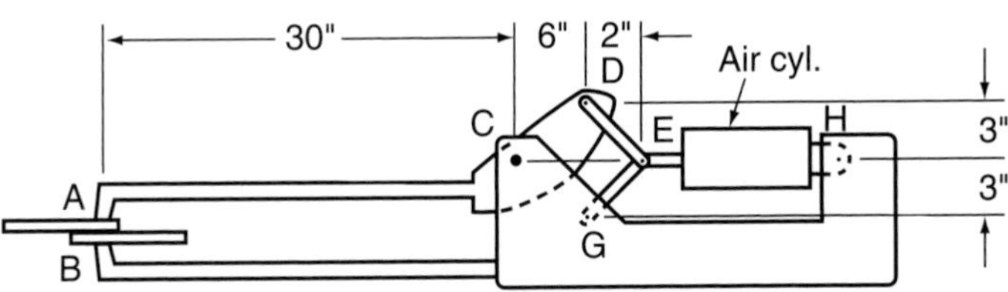

FIGURE P5–70

5–71. The side view of an evenly loaded clamshell bucket is shown in Figure P5–71. The holding line and closing line have loads of 6 kN and 800 N, respectively. If the crosshead casting has a mass of 40 kg and the mass of arms AB and CD is neglected, determine the load in arms AB and CD.

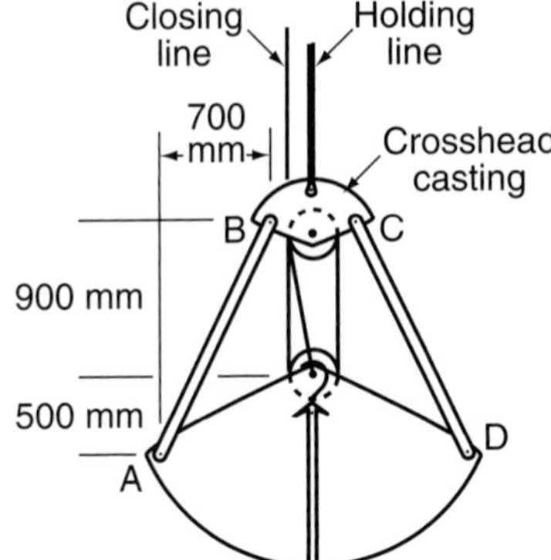

FIGURE P5–71

5–72. Determine the spring force required for static equilibrium to exist in the mechanism shown in Figure P5–72.

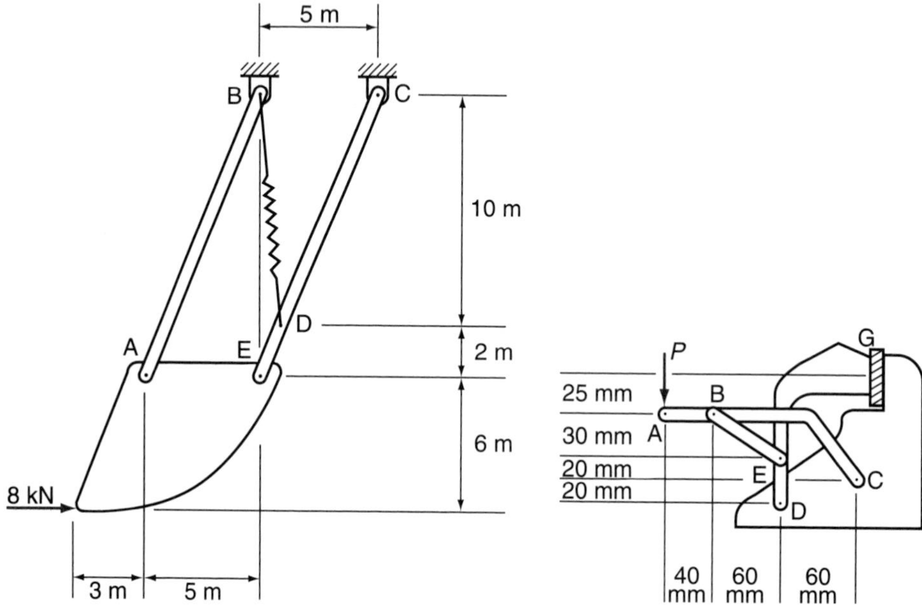

FIGURE P5–72

FIGURE P5–73

5–73. The toggle linkage shown in Figure P5–73 is used to clamp a workpiece at G with a clamping force of 200 N. Determine (a) the pin reactions at D and C and (b) the applied force *P*.

5–74. Find the force applied by each pair of cylinders of the loader when the bucket is positioned and loaded as shown in Figure P5–74. The lengths of CD and DF are 11.5 in. and 7.7 in., respectively. Member BC is parallel to AG.

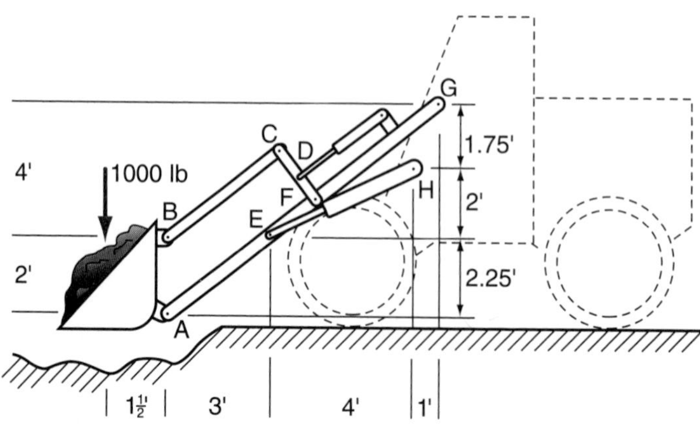

FIGURE P5–74

5–75. Cylinder BG is used to adjust the height of the scissor lift table in Figure P5–75. If a weight of 300 lb is placed on the table, what cylinder force is required? (Neglect the weight of the table and that of all other members.)

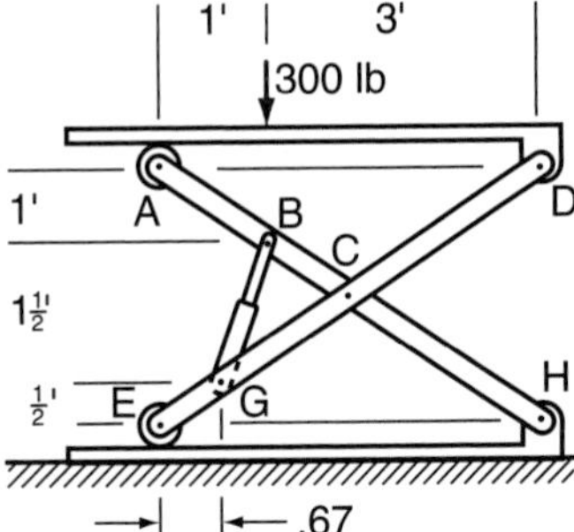

FIGURE P5–75

5–76. The gear pulley shown in Figure P5–76 is used to pull gears and pulleys from a shaft by tightening the vertical screw. The screw pushes on the shaft with a vertical force of 800 N to remove the pulley shown. Assume smooth surfaces and determine the force at B and the force in member AC.

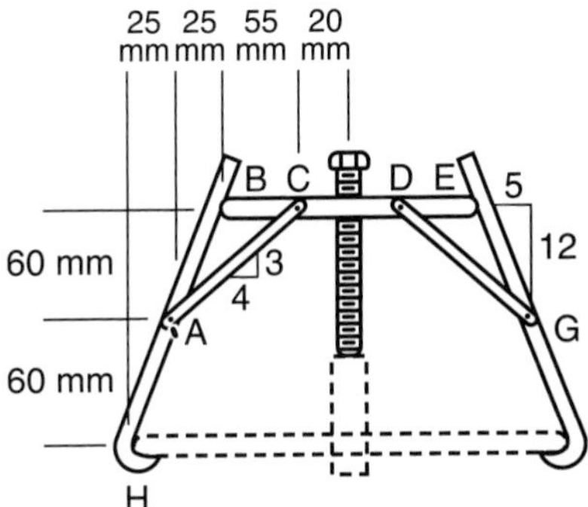

FIGURE P5–76

5–77. A tube may be bent around a roller as shown in Figure P5–77. Forces of 39 lb are applied at the positions indicated. Find the normal force at A and the pin reaction components at B on member BC. (Neglect the force due to friction at A.)

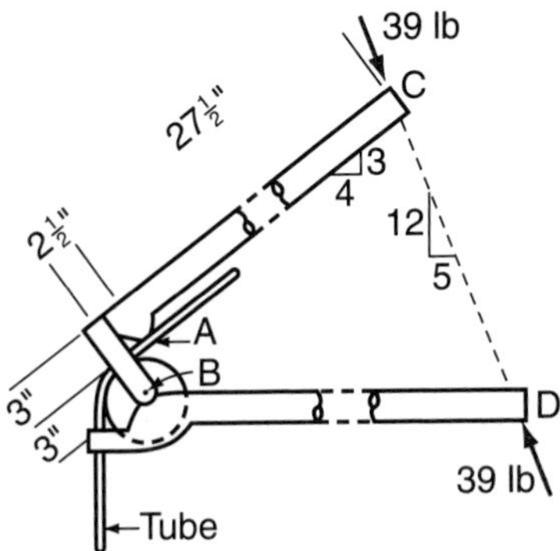

FIGURE P5–77

5–78. Using data from problem 3–19, solve for the load in member AC and the branch shearing force at E for the preening lopper shown in Figure P3–18.

5–79. Using data from problems 3–19 and 5–78, solve for the total shearing force loads in bolts D and B.

5–80. A refuse truck lifts a dumpster from the lowered position (Figure P5–80a) to the raised position (Figure P5–80b). If the dumpster weight of 2400 lb is shared equally by the two front forks (Figure P5–80c), determine:

(a) the pin reactions at B,
(b) cylinder force CD,
(c) the pin reactions at G, and
(d) the cylinder force EH.

FIGURE P5–80a

FIGURE P5–80b

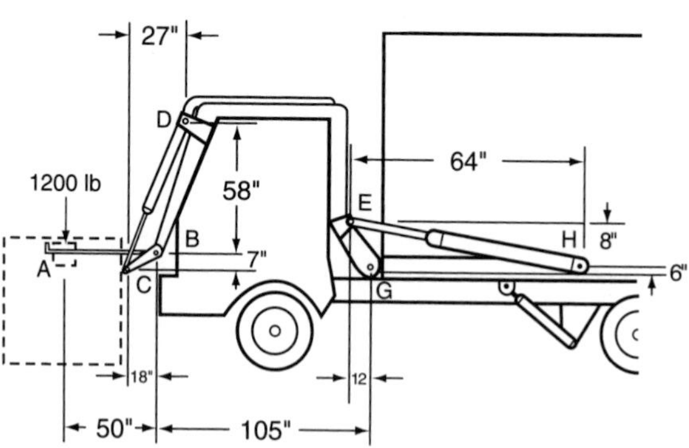

FIGURE P5–80c

5–81. Members AC and BC each weigh 10 lb/ft and support a 500-lb load as shown in Figure P5–81. Determine the horizontal and vertical components of the pin reactions at A and B.

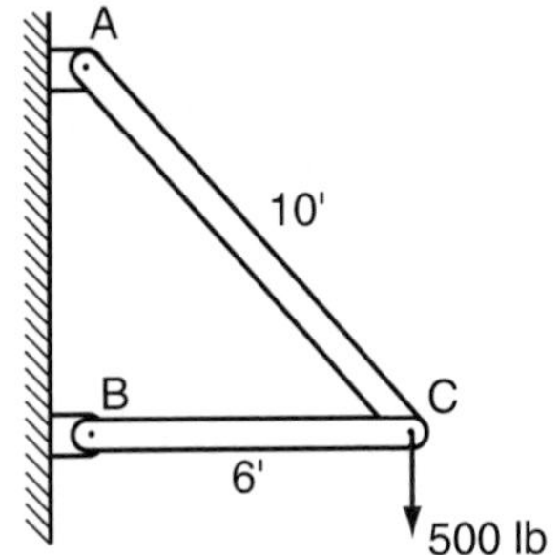

FIGURE P5–81

5–82. Determine the horizontal and vertical components of the pin reactions at A and B on the frame shown in Figure P5–82.

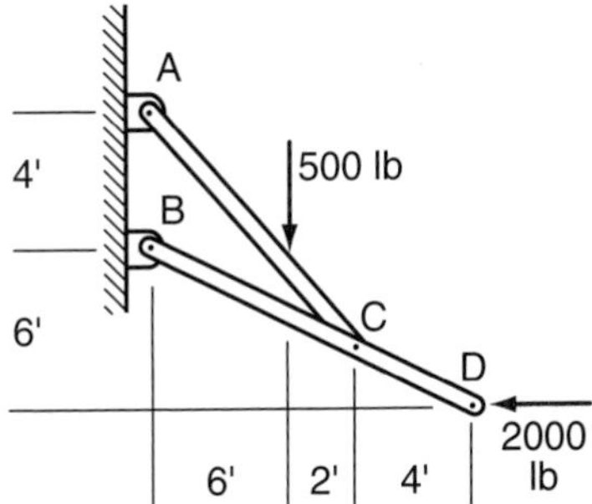

FIGURE P5–82

5–83. A cable is fastened at C and passes over a pulley at A. The cable tension is 6 kN. Neglect the pulley diameter and determine the pin reaction components at D and B (Figure P5–83).

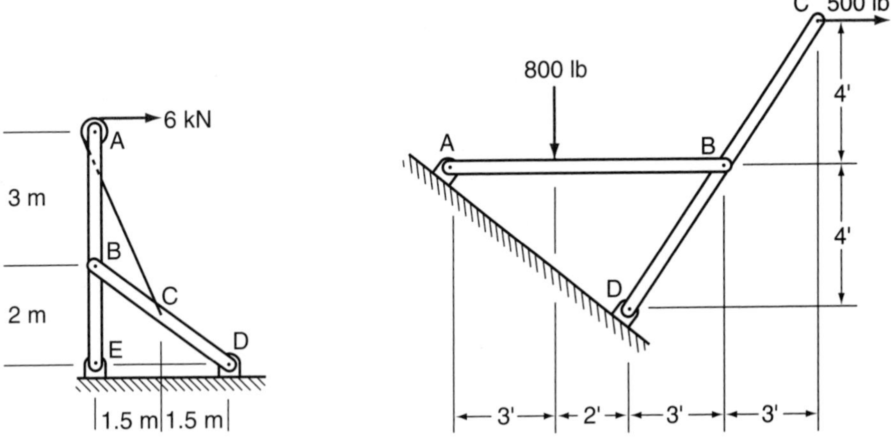

FIGURE P5–83

FIGURE P5–84

5–84. Determine the horizontal and vertical components of the pin reaction at D of the system shown in Figure P5–84.

5–85. Determine the horizontal and vertical components of the pin reactions at A and D of the system shown in Figure P5–85.

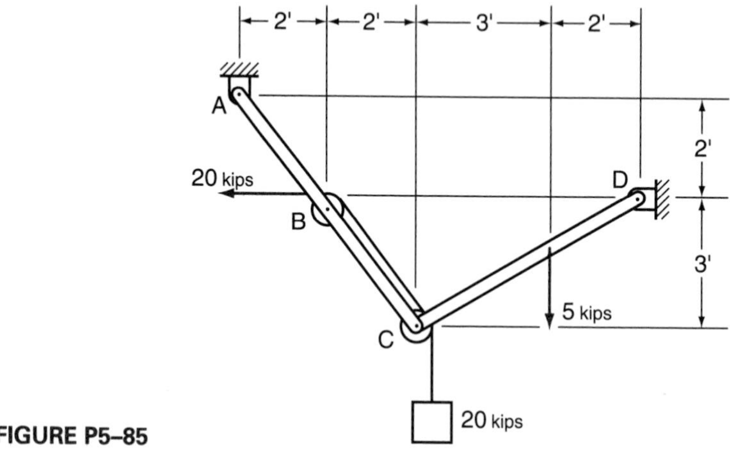

FIGURE P5–85

5–86. Determine the horizontal and vertical components of the pin reactions at A and C of the system shown in Figure P5–86.

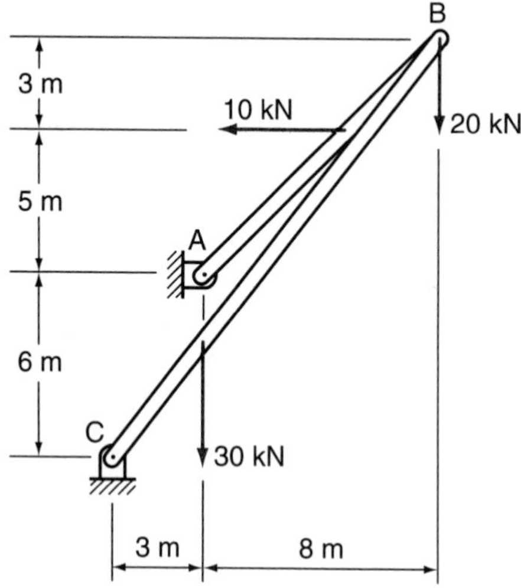

FIGURE P5–86

REVIEW PROBLEMS

R5–1. Determine the load in each member of the truss shown in Figure RP5–1.

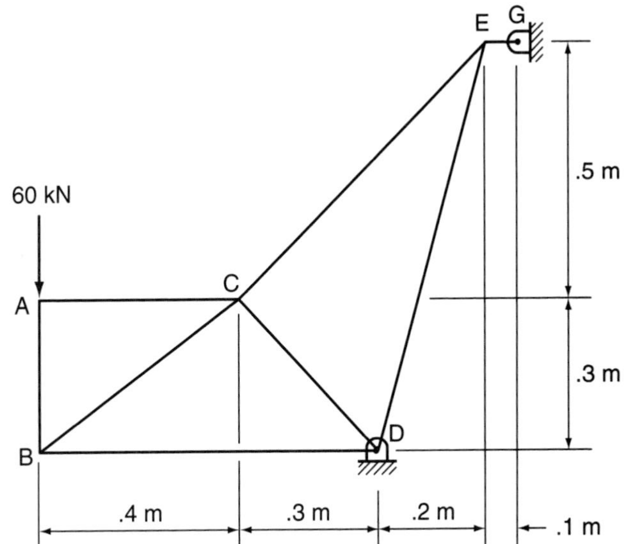

FIGURE RP5–1

R5–2. Determine the load in member GD of the pin-connected truss shown in Figure RP5–2.

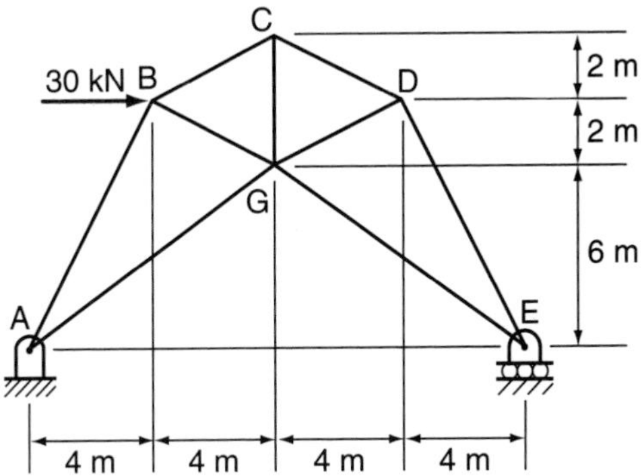

FIGURE RP5–2

R5–3. Determine the load in each member of the pin-connected truss shown in Figure RP5–3.

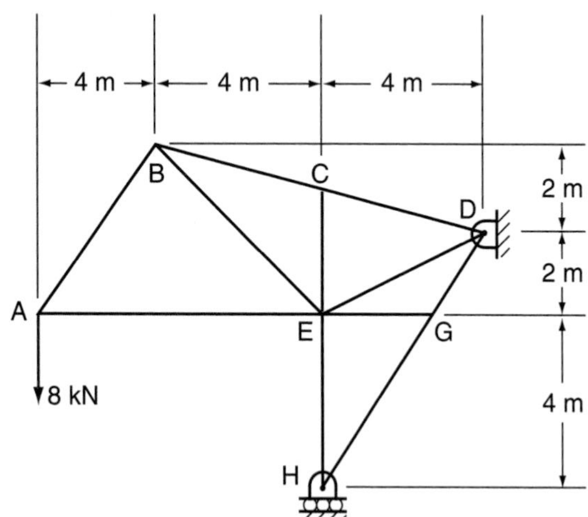

FIGURE RP5–3

R5–4. For the truss loaded as shown in Figure RP5–4, determine the loads in members CD, EG, and EH.

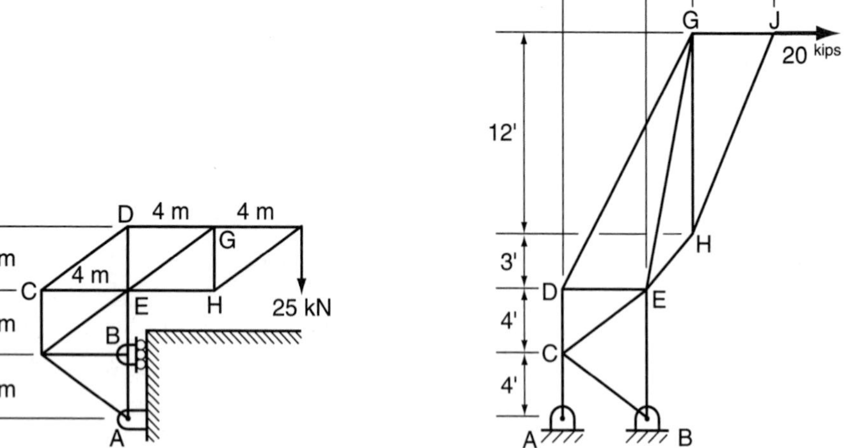

FIGURE RP5–4

FIGURE RP5–5

R5–5. Using the method of sections, determine the load in member GE of the truss shown in Figure RP5–5.

R5–6. Using the method of sections, determine the loads in members CD, CH, and GH of the truss shown in Figure RP5–6.

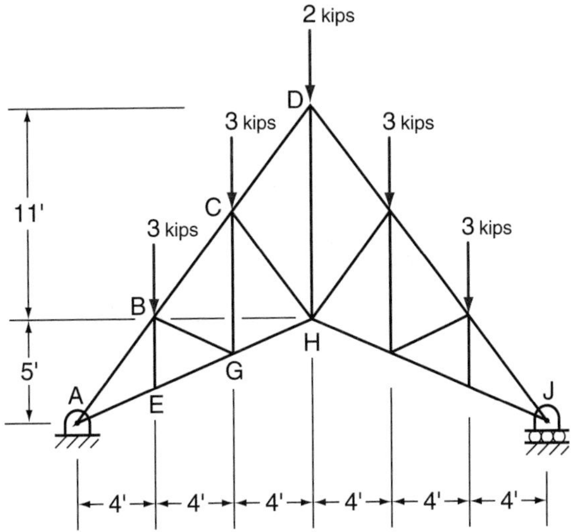

FIGURE RP5–6

R5–7. Determine the horizontal and vertical components of the pin reactions at C of the system shown in Figure RP5–7.

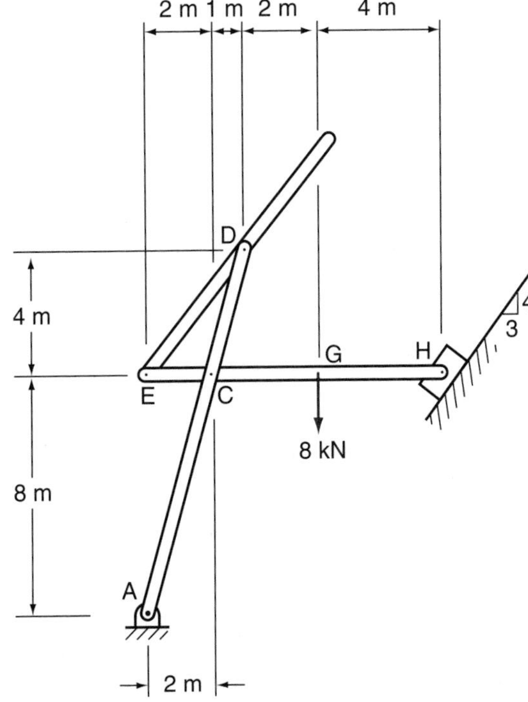

FIGURE RP5–7

R5–8. Determine the horizontal and vertical components of the pin reaction at B of the mechanism shown in Figure RP5–8.

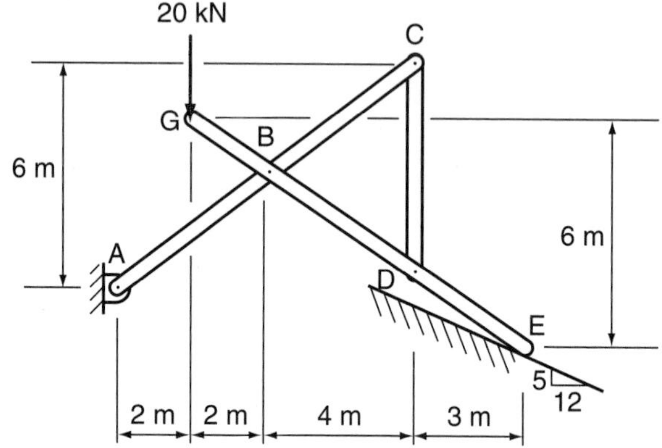

FIGURE RP5–8

R5–9. Determine the tension T and the reaction at E as shown in Figure RP5–9.

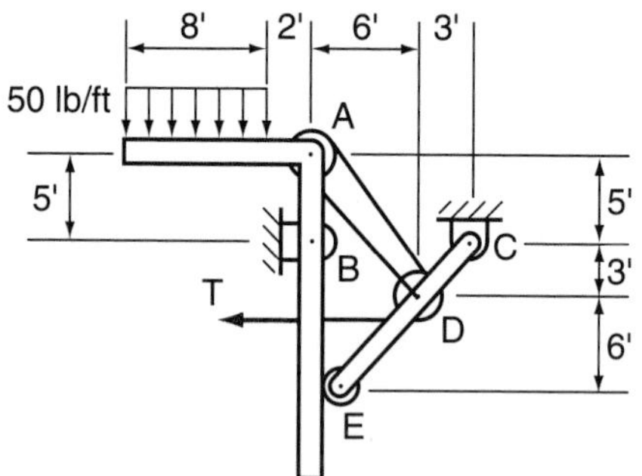

FIGURE RP5–9

R5–10. Material slid under the blade of the gap shear shown in Figure RP5–10 requires 700 lb of vertical force to be sheared. Calculate the cylinder force required at D. Member AC pivots at point B.

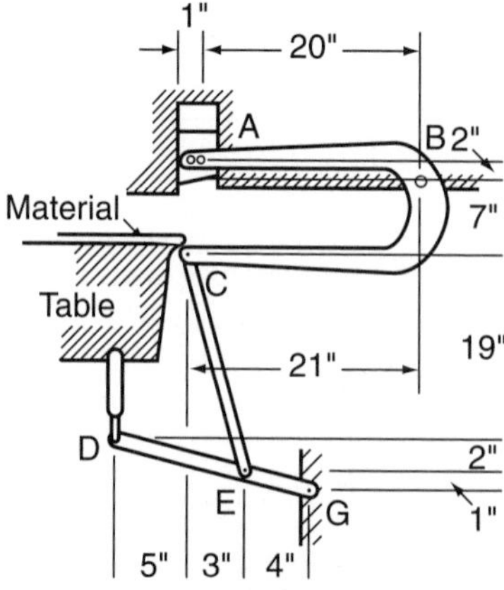

FIGURE RP5–10

R5–11. A bicycle with a spring shock absorber has a load of 300 lb applied as shown in Figure RP5–11. Determine the load in the spring and the pin reactions at pin C. (The spokes and sprocket have been partially cut away to reveal details such as point C on the Frame.)

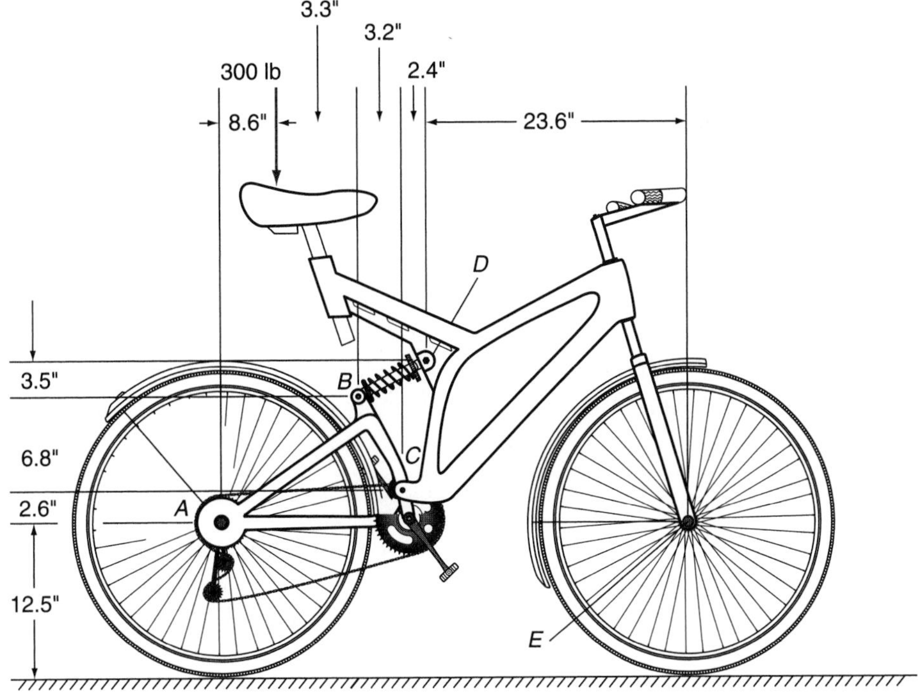

FIGURE RP5–11

R5–12. As shown in Figure RP5–12, a new pipe wrench design has part B sliding on part A when gear C rotates. Gear C is rotated by the handle fastened to it. If 15 lb is applied as shown, determine (a) the torque applied to the pipe and (b) the gripping force acting perpendicular to the jaw gripping surface. Assume that the center of C and the center of the pipe are vertically aligned.

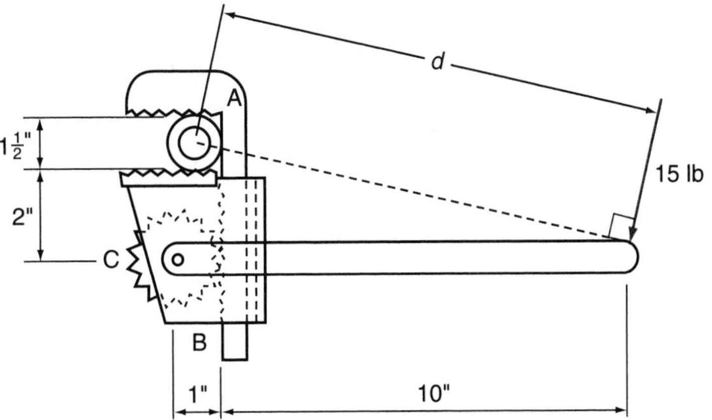

FIGURE RP5–12

R5–13. Determine the horizontal and vertical components of the pin reactions at C and D on the frame shown in Figure RP5–13.

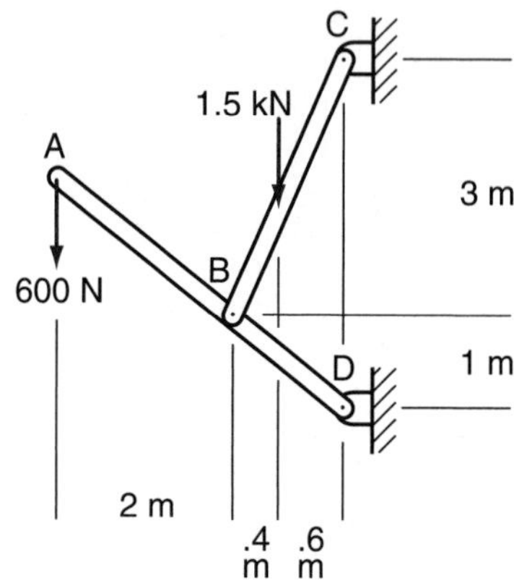

FIGURE RP5–13

R5–14. Determine the horizontal and vertical components of the reaction at A in Figure RP5–14.

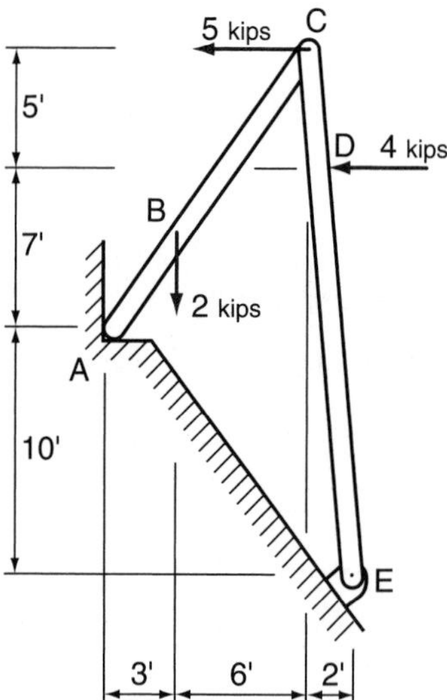

FIGURE RP5–14

CHAPTER 6

Three-Dimensional Equilibrium

OBJECTIVES

Upon completion of this chapter the student will be able to apply the six basic equations of equilibrium, $\Sigma F_x = 0$, $\Sigma F_y = 0$, $\Sigma F_z = 0$, $\Sigma M_x = 0$, $\Sigma M_y = 0$, $\Sigma M_z = 0$, to the three-dimensional systems of:

1. Parallel forces
2. Concurrent forces
3. Nonconcurrent forces

6–1 RESULTANT OF PARALLEL FORCES

A system of parallel forces in three dimensions is shown in Figure 6–1. Gravity is the usual source of parallel forces, but there can be others—for example, water pressure on a flat vertical wall. The location of the resultant of these parallel force systems must be found for applications such as the design of column foundation pads.

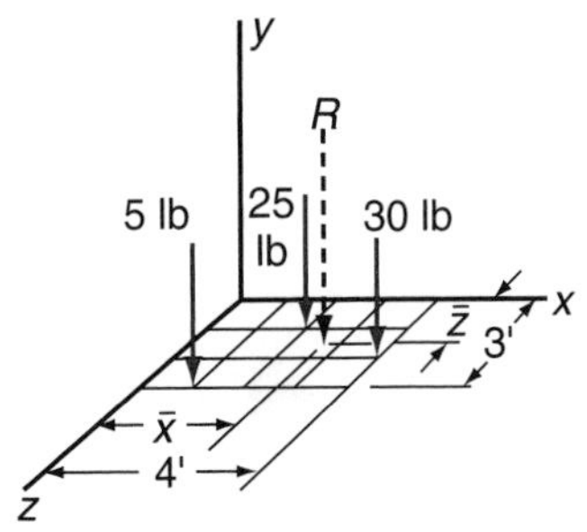

FIGURE 6–1

As in two-dimensional statics, a resultant is a single force that has the same effect as the system of forces that it replaces. In Figure 6–1, the magnitude of the resultant is equal to the algebraic sum of the vertical forces.

$$R = -10 \text{ lb} - 20 \text{ lb} - 30 \text{ lb}$$
$$= -60$$
$$R = 60 \text{ lb} \downarrow$$

One locates the resultant by determining the x and z coordinates and *designating them* $\bar{x}$ and $\bar{z}$. In three-dimensional statics, *moments are always taken about an axis.* This fact may be more obvious in three-dimensional analysis than in two-dimensional analysis, where the axis appears to be a point. The forces shown in Figure 6–1 have moments about both the x- and z-axes, but not about the y-axis.

We are not solving for an equilibrium force, rather a resultant force that replaces all other forces. (It is equal and opposite to the equilibrant force.) It produces the same result as all the other forces so it is, therefore, the sum of them.

Using the sign convention of **clockwise moments as negative and counterclockwise moments as positive**, you can take moments about the x-axis. To visualize the moment direction, imagine yourself looking down the x-axis toward the origin.

The moment of $R\bar{z}$ replaces the moments of all the other forces, so equate $R\bar{z}$ to the sum of the other moments (Figure 6–2).

$$+(R\bar{z}) = +(25 \text{ lb})(1 \text{ ft}) + (30 \text{ lb})(2 \text{ ft}) + (5 \text{ lb})(3 \text{ ft})$$
$$60\bar{z} = 100$$
$$\underline{\bar{z} = 1.67 \text{ ft}}$$

Equate $R\bar{x}$ to the sum of the other moments (Figure 6–3).

$$-(R\bar{x}) = -(5 \text{ lb})(1 \text{ ft}) - (25 \text{ lb})(2 \text{ ft}) - (30 \text{ lb})(4 \text{ ft})$$
$$-60\bar{x} = -5 - 50 - 120$$
$$\underline{\bar{x} = 2.92 \text{ ft}}$$

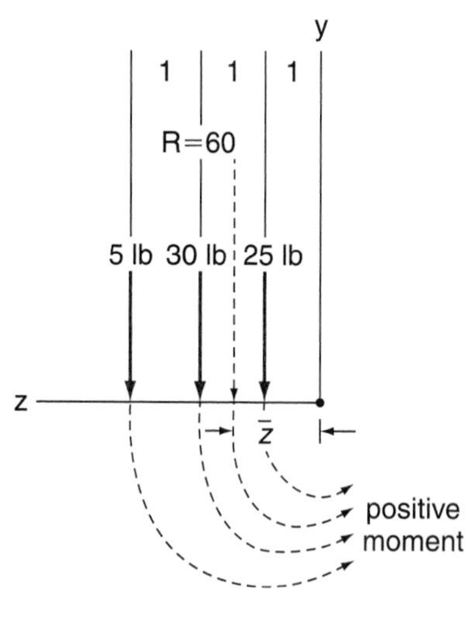

FIGURE 6–2

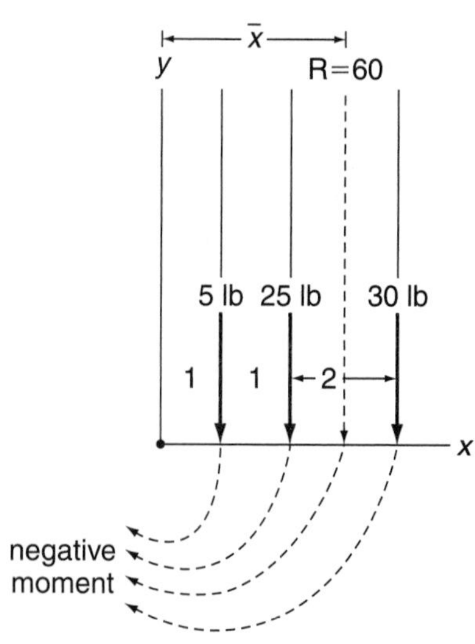

FIGURE 6–3

EXAMPLE 6–1

Determine the magnitude and location of the force system shown in Figure 6–4. The grid dimension is 1 m.

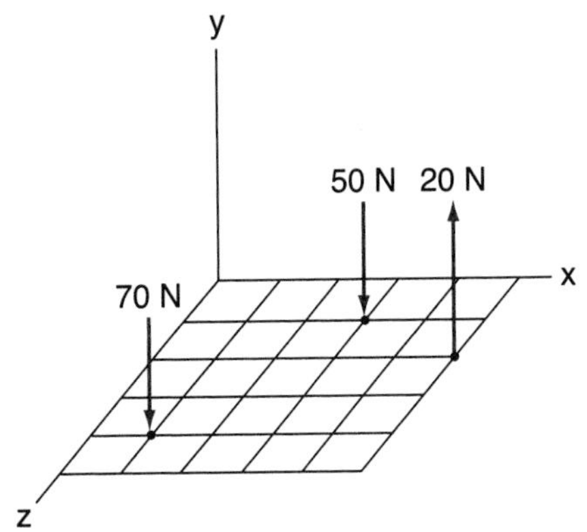

FIGURE 6–4

Step 1. Calculate the resultant

$$R = -70\text{ N} - 50\text{ N} + 20\text{ N}$$
$$= -100$$

or

$$\underline{R = 100\text{ N}\downarrow}$$

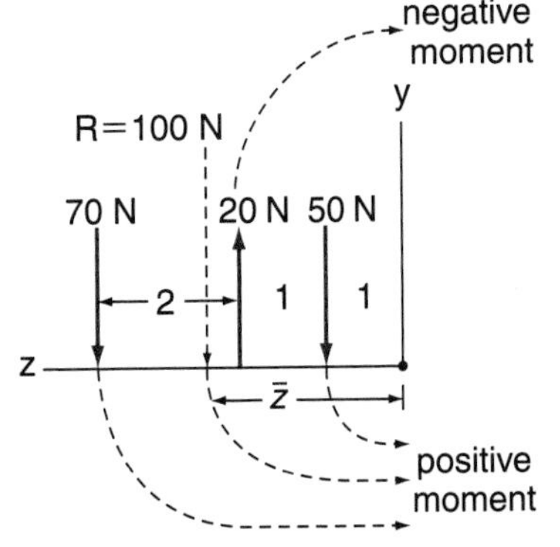

FIGURE 6–5

Step 2. Draw the side view or *z-y* plane (Figure 6–5) (positive and negative moments are shown with broken lines).

Step 3. Equate the moment of *R* to all other moments using the moment sign convention.

$$+(100\text{ N})\bar{z} = +(50\text{ N})(1\text{ m}) + (70\text{ N})(4\text{ m}) - (20\text{ N})(2\text{ m})$$
$$\underline{\bar{z} = 2.5\text{ m}}$$

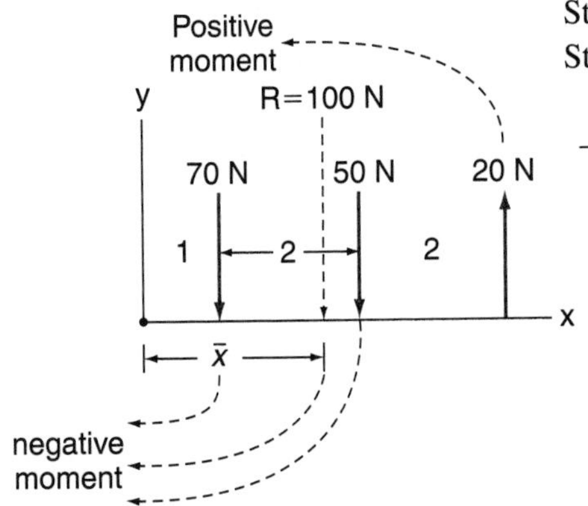

FIGURE 6–6

Step 4. Draw the front view or *x-y* plane (Figure 6–6).

Step 5. Equate $R\bar{x}$ to the other moments.

$$-(100\text{ N})\bar{x} = -(70\text{ N})(1\text{ m}) - (50\text{ N})(3\text{ m}) + (20\text{ N})(5\text{ m})$$
$$\underline{\bar{x} = 1.2\text{ m}}$$

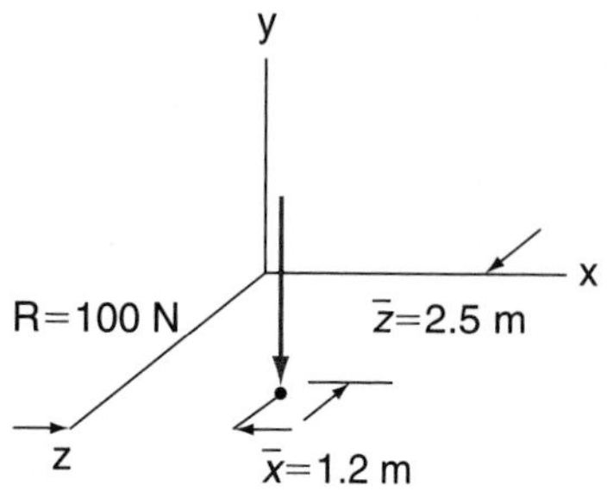

FIGURE 6–7

The magnitude and location of the resultant can be shown (Figure 6–7).

EXAMPLE 6–2

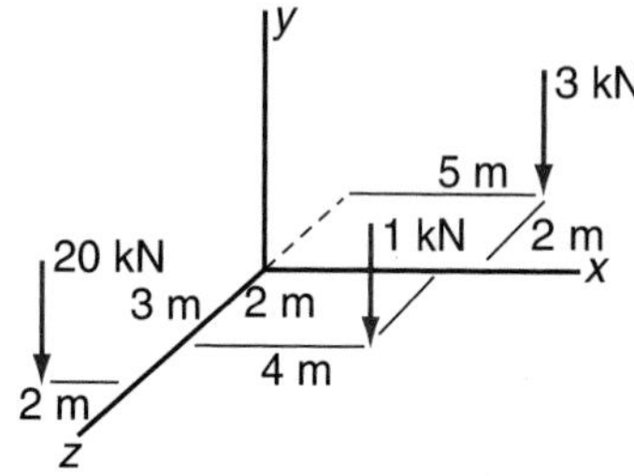

FIGURE 6–8

Determine the magnitude and location of the resultant of the force system shown in Figure 6–8.

$$R = -20 \text{ kN} - 1 \text{ kN} - 3 \text{ kN}$$
$$= -24$$
$$\underline{R = 24 \text{ kN} \downarrow}$$

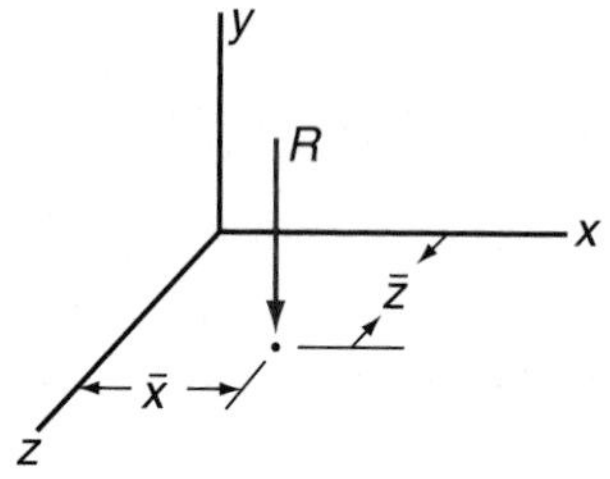

FIGURE 6–9

Assume that R is located as shown in Figure 6–9 and take moments about the x-axis.

$$+R\bar{z} = +(20 \text{ kN})(3 \text{ m}) + (1 \text{ kN})(2 \text{ m}) - (3 \text{ kN})(2 \text{ m})$$
$$24\bar{z} = 60 + 2 - 6$$
$$\underline{\bar{z} = 2.33 \text{ m}}$$

Taking moments about the z-axis, we get

$$-R\bar{x} = +(20 \text{ kN})(2 \text{ m}) - (1 \text{ kN})(4\text{m}) - (3 \text{ kN})(5 \text{ m})$$
$$-24\bar{x} = 40 - 4 - 15$$
$$\underline{\bar{x} = -0.875 \text{ m}}$$

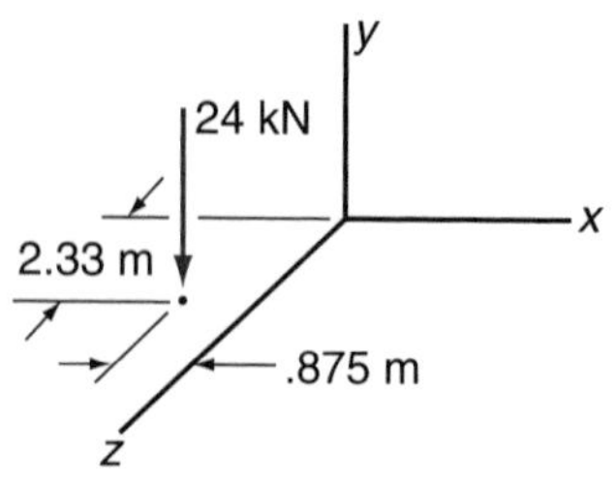

FIGURE 6–10

The negative value of $\bar{x}$ indicates that R is located as shown in Figure 6–10.

6–2 EQUILIBRIUM OF PARALLEL FORCES

The method followed here will allow the solution of only three unknowns. The direction and location of forces are known, so it becomes a matter of writing moment equations about the correct points or axes. The moment equations give simultaneous equations with two unknowns.

EXAMPLE 6–3

A horizontal plate is represented by the grid in Figure 6–11, where each square has sides 1 ft in length. The plate is supported at A, B, and C and has 20 lb applied as shown. Neglect the weight of the plate and determine the reactions at A, B, and C.

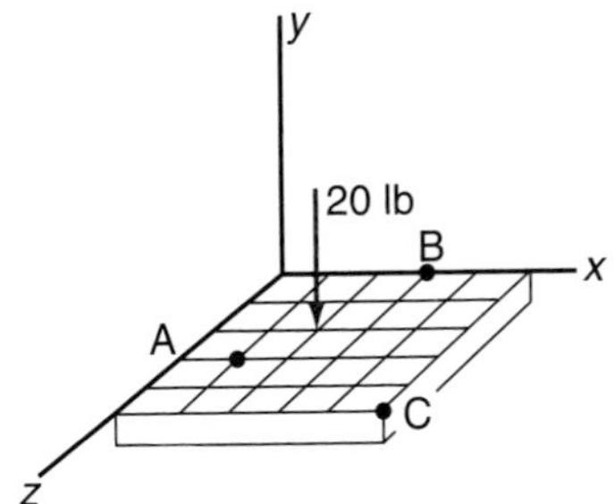

FIGURE 6–11

Side View – *z*-*y* Plane
(viewed down *x*-axis)

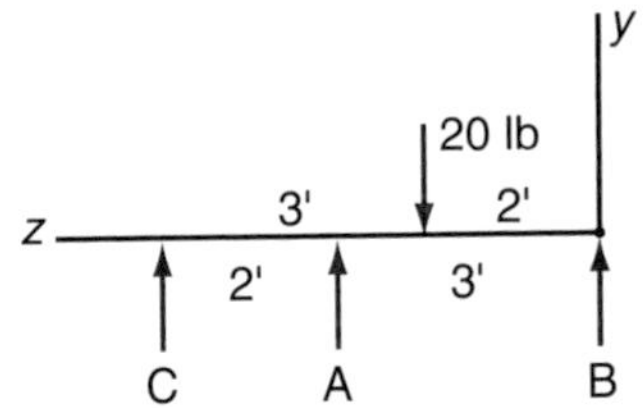

FIGURE 6–12

Front View – *x*-*y* Plane
(viewed down z axis)

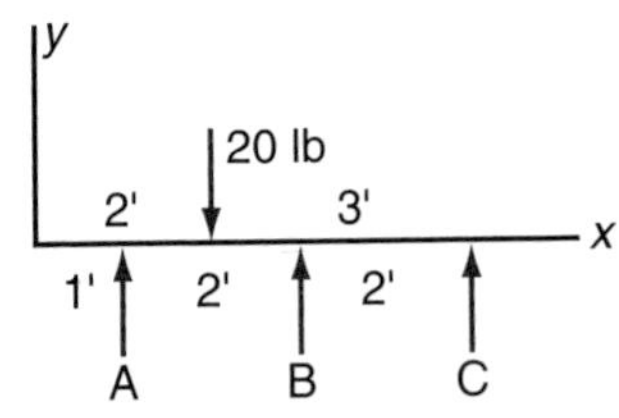

FIGURE 6–13

Similar to Example 6–1, moment equations may be more easily written if we draw a side view and a front view of the plate. A side view (Figure 6–12) consists of looking down the *x*-axis toward the *z*-*y* plane. This is often referred to as "projecting the forces into the *z*-*y* plane." Similarly, a front view (Figure 6–13) consists of looking down the *z*-axis and viewing the forces in the *x*-*y* plane.

Choose one of the three unknown forces—*A*, for example—and take moments about the point through which it passes in each diagram. Simultaneous equations with the unknowns *B* and *C* are then solved.

For the side view (Figure 6–12):

$$\Sigma M_A = 0$$

$$(3\text{ ft})B - (20\text{ lb})(1\text{ ft}) - (2\text{ ft})C = 0$$

$$3B - 2C = 20 \qquad (1)$$

For the front view (Figure 6–13):

$$\Sigma M_A = 0$$

$$(2\text{ ft})B + (4\text{ ft})C - (20\text{ lb})(1\text{ ft}) = 0$$

$$2B + 4C = 20 \qquad (2)$$

Multiplying Equation (1) by 2 and adding Equation (2), we get

$$6B - 4C = 40 \quad (1)\times 2$$

$$\underline{2B + 4C = 20} \quad (2)$$

$$8B + 0 = 60$$

$$\underline{B = 7.5 \text{ lb}\uparrow}$$

Substituting $B = 7.5$ into Equation (1), we get

$$(3 \text{ ft})(7.5 \text{ lb}) - (2 \text{ ft})C = 20$$

$$-2C = -2.5$$

$$\underline{C = 1.25 \text{ lb}\uparrow}$$

The summation of vertical forces will give the value of the third unknown, A.

$\Sigma F_y = 0$

$$A + B + C - 20 = 0$$

$$A + 7.5 \text{ lb} + 1.25 \text{ lb} - 20 \text{ lb} = 0$$

$$\underline{A = 11.25 \text{ lb}\uparrow}$$

6–3 COMPONENTS AND RESULTANTS OF FORCES IN SPACE

When describing the direction of a force in three dimensions, you must use the proper sign convention. The sign convention for each of the x-, y-, and z-axes is shown in Figure 6–14.

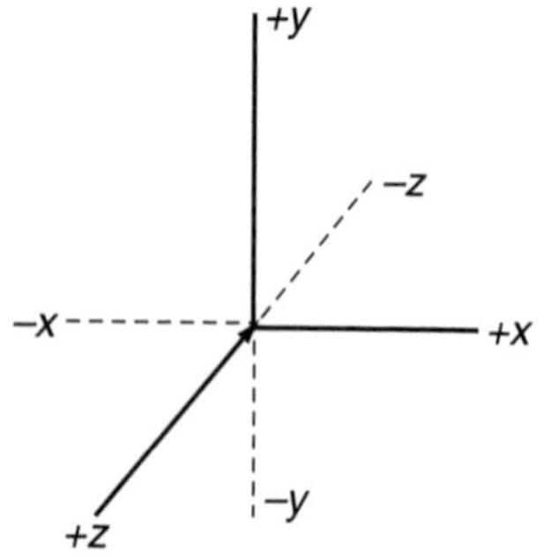

FIGURE 6–14

To solve for the resultant of the three forces shown in Figure 6–15, we could begin by adding 3 N and 4 N as we have been doing in coplanar systems.

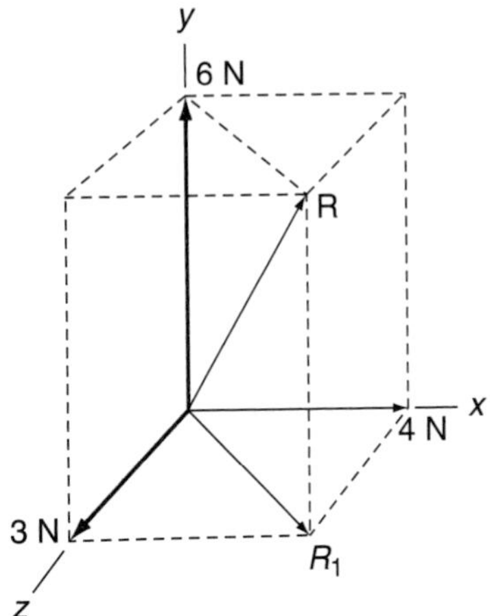

FIGURE 6–15

We would then get

$$R_1 = \sqrt{(3\text{ N})^2 + (4\text{ N})^2}$$
$$R_1 = 5\text{ N}$$

R_1 can now be added to the y component, $R_y = 6$ N, to obtain the final resultant R. Note the shaded plane formed by these two forces (Figure 6–15).

$$R = \sqrt{(5\text{ N})^2 + (6\text{ N})^2}$$
$$R = 7.81\text{ N}$$

R is the diagonal of a rectangular box formed by the three components.

Rather than adding the components in two steps, you could have solved for R in one step.

$$R = \sqrt{(3\text{ N})^2 + (4\text{ N})^2 + (6\text{ N})^2}$$
$$R = 7.81\text{ N}$$

The direction of this resultant is medicated by showing the x, y, z coordinates after the answer as shown.

$$\underline{R = 7.81\text{ N, coordinates } (4, 6, 3)}$$

EXAMPLE 6–4

Determine the resultant of the forces shown in Figure 6–16.

$$R = \sqrt{(8\text{ N})^2 + (20\text{ N})^2 + (10\text{ N})^2}$$
$$= \sqrt{564}$$
$$R = 23.7\text{ N}(-4, 10, 5)$$

The coordinates $(-8, 20, 10)$ can be reduced to $(-4, 10, 5)$.

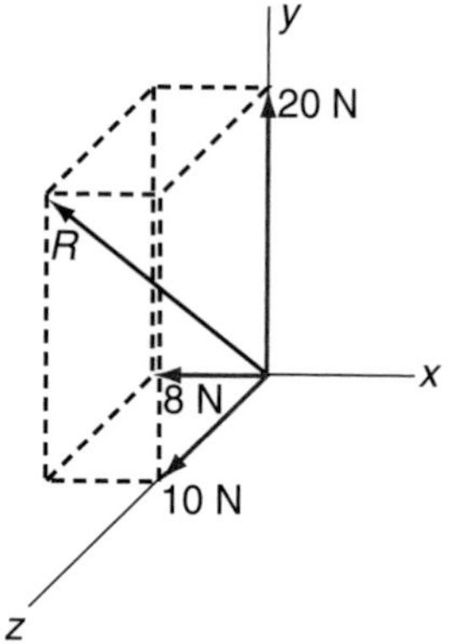

FIGURE 6–16

We now come to the problem of resolving a force in space into components in the x-, y-, and z-directions. Suppose that a 100-lb force in space has coordinates of (8, 4, 2). The length of the diagonal of the box (Figure 6–17) represents a force of 100 lb.

$$\text{diagonal length} = \sqrt{(8)^2 + (4)^2 + (2)^2}$$
$$= \sqrt{64 + 16 + 4}$$
$$= \sqrt{84}$$
$$= 9.16$$

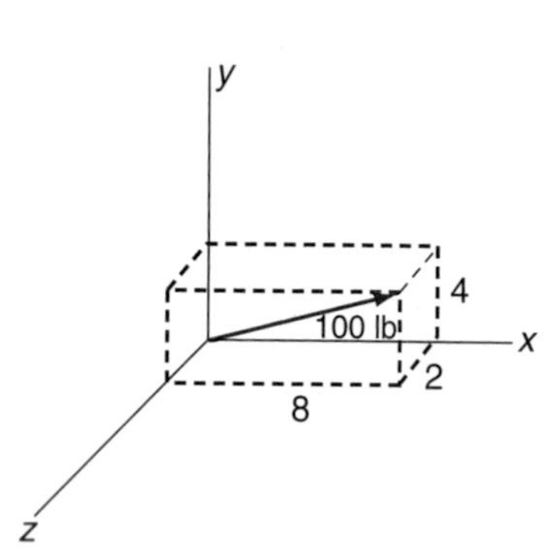

FIGURE 6–17

By proportion, if a diagonal length of 9.16 represents 100 lb, the x dimension of the box (8) represents $8/9.16 \times 100 = 87.4$ lb. Therefore

$$R_x = 87.4\text{ lb} \rightarrow$$

Similarly,

$$R_y = \frac{4}{9.16}(100 \text{ lb})$$

$$R_y = 43.7 \text{ lb} \uparrow$$

$$R_z = \frac{2}{9.16}(100 \text{ lb})$$

$$R_z = 21.8 \text{ lb} \swarrow$$

This is the same principle that we applied in two-dimensional coplanar force systems. A force R with a 3:4 slope has a horizontal component of

$$R_x = \frac{4}{5}R$$

We are now applying the same principle to the third component, R_z. The following equation is a more concise method of showing the relationship among the force components.

$$\frac{R_x}{x} = \frac{R_y}{y} = \frac{R_z}{z} = \frac{R}{\sqrt{x^2 + y^2 + z^2}} \quad (6\text{–}1)$$

or

$$\frac{R_x}{8} = \frac{R_y}{4} = \frac{R_z}{2} = \frac{100}{9.16}$$

$$\underline{R_x = 87.4 \text{ lb} \rightarrow,\ R_y = 43.7 \text{ lb} \uparrow, \text{ and } R_z = 21.8 \text{ lb} \swarrow}$$

EXAMPLE 6–5

Determine the *x, y,* and *z* components of an 800-N force in space that has coordinates (5, −12, 16) (Figure 6–18).

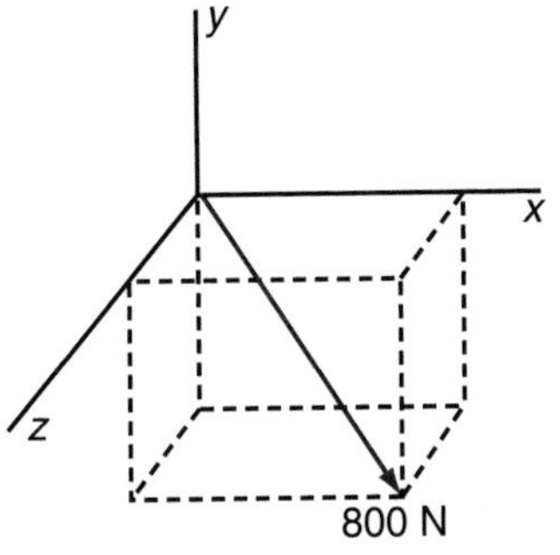

FIGURE 6–18

$$\text{diagonal length} = \sqrt{(5)^2 + (12)^2 + (16)^2}$$

$$= \sqrt{25 + 144 + 256}$$

$$= \sqrt{425}$$

$$= 20.6$$

$$R_x = \frac{5}{20.6}(800 \text{ N})$$

$$\underline{R_x = 194 \text{ N} \rightarrow}$$

$$R_y = \frac{12}{20.6}(800 \text{ N})$$

$$\underline{R_y = 466 \text{ N} \downarrow}$$

$$R_z = \frac{16}{20.6}(800 \text{ N})$$

$$\underline{R_z = 621 \text{ N} \swarrow}$$

6–4 EQUILIBRIUM IN THREE DIMENSIONS

Forces in space are noncoplanar, and may be one of the following:

1. Parallel
2. Nonconcurrent
3. Concurrent

(Parallel forces in space were considered in Sections 6–1 and 6–2.)

The compressive loads in the legs of a camera tripod are an example of *concurrent forces*. These forces are concurrent because they all intersect the camera at a common point. When the forces in three dimensions do not intersect at a common point, they form a noncoplanar, nonconcurrent force system. These are considered in more advanced designs when various member loads are required and where the reactions at supports can be more involved since there may be couples present. Only the simpler reactions with stated assumptions will be considered here. The regular equilibrium equations for forces in three dimensions will be used.

What are the equilibrium equations for concurrent forces in three dimensions? In coplanar force systems, we had three equations, and moments were taken about some point on the free-body diagram.

$$\Sigma F_x = 0$$
$$\Sigma F_y = 0$$
$$\Sigma M_z = 0$$

For forces in three dimensions, we have

$$\Sigma F_x = 0 \qquad \Sigma M_x = 0$$
$$\Sigma F_y = 0 \qquad \Sigma M_y = 0$$
$$\Sigma F_z = 0 \qquad \Sigma M_z = 0$$

Moments are taken about the x-, y-, and z-axes or about any other axis that may be convenient in the problem solution. There is a difference in moment equations: In a two-dimensional coplanar force system, moments can be considered as being taken about a point; in a three-dimensional noncoplanar force system, moments are taken about an axis.

There is more than one method of problem solution. The shortest one for some of the simpler problems consists of drawing a three-dimensional free-body diagram—somewhat like an isometric view—and applying the equilibrium equations. All forces are resolved into their three components and, if possible, are expressed as fractions of the total force.

If you become confused with all the forces shown on one diagram, the following three steps may simplify the problem solution:

1. Project all forces into two or more planes; that is, take front, side, or top views. Any force that has a line of action in the direction that you are viewing is not shown in that view. Another way of expressing this rule is: A force has no component in a plane that is perpendicular to its line of action.
2. If possible, show all these projected forces or components at any point as fractions of the total force at that point.
3. Treat each view as a coplanar force system and apply the three equilibrium equations as before. Moment equations are also often used.

6–5 NONCONCURRENT, THREE-DIMENSIONAL SYSTEMS

EXAMPLE 6–6

The crank in Figure 6–19 has a smooth bearing at B and a ball and socket at D. Calculate all reaction components at B and D.

Note that a ball and socket can support forces in three directions, that is, D_x, D_y, and D_z, but no moments.

The three-dimensional free-body diagram in Figure 6–20 is used here. The rectangular components at A can be found from the diagonal length of the space coordinates.

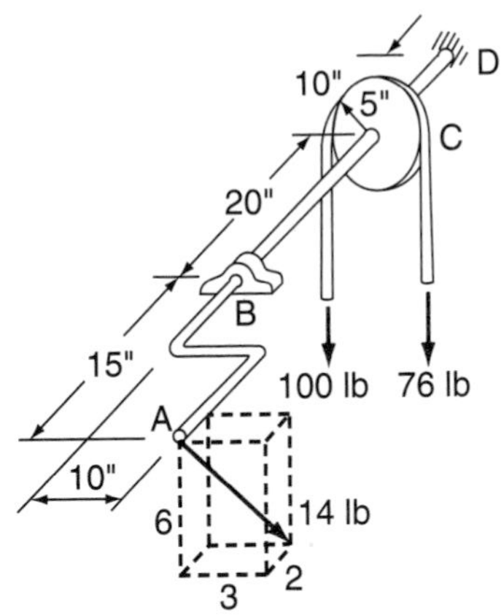

FIGURE 6–19

$$\text{diagonal length} = \sqrt{(3)^2 + (6)^2 + (2)^2}$$
$$= 7$$

$$A_x = \frac{3}{7}(14 \text{ lb}) = 6 \text{ lb}$$

$$A_y = \frac{6}{7}(14 \text{ lb}) = 12 \text{ lb}$$

$$A_z = \frac{2}{7}(14 \text{ lb}) = 4 \text{ lb}$$

As stated before, when using a free-body diagram in three dimensions, one must take moments about an axis, not about a point. Moments about the y-axis at D would give us B_x. (All vertical forces have zero moment about the y-axis.)

Free-Body Diagram of Crank

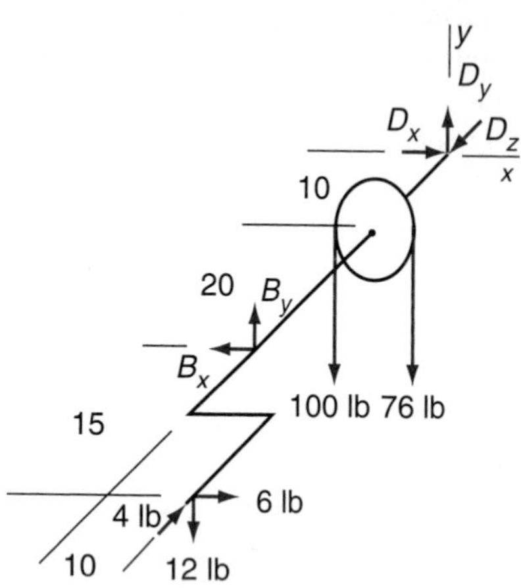

FIGURE 6–20

$\Sigma M_y = 0$

$$-(30\text{ in})B_x + (6\text{ lb})(45\text{ in}) + (4\text{ lb})(10\text{ in}) = 0$$

$$30B_x = 270 + 40$$

$$\underline{B_x = 10.3\text{ lb} \leftarrow}$$

Take moments about the x-axis at D to obtain B_y.

$\Sigma M_x = 0$

$$-(30\text{ in.})B_y + (12\text{ lb})(45\text{ in.}) + (176\text{ lb})(10\text{ in.}) = 0$$

$$30B_y = 540 + 1760$$

$$\underline{B_y = 76.6\text{ lb}\uparrow}$$

The components at D can be found by summation of forces in each of the three directions—x, y, and z.

$\Sigma F_x = 0$

$$D_x + 6 - B_x = 0$$

$$D_x = 10.3\text{ lb} - 6\text{ lb}$$

$$\underline{D_x = 4.3\text{ lb} \rightarrow}$$

$\Sigma F_y = 0$

$$D_y + B_y - 176\text{ lb} - 12\text{ lb} = 0$$

$$D_y = 176 + 12 - 76.6$$

$$\underline{D_y = 111.4\text{ lb}\uparrow}$$

$\Sigma F_z = 0$

$$\underline{D_z = 4\text{ lb} \swarrow}$$

EXAMPLE 6–7

The horizontal platform shown in Figure 6–21 has supporting legs with pinned connections at B and C. Cables AG, GD, and GE are on the same horizontal plane as the platform. Assume that support B carries all of the force in the x direction. Neglecting the weight of the platform, determine all reaction components at B and C.

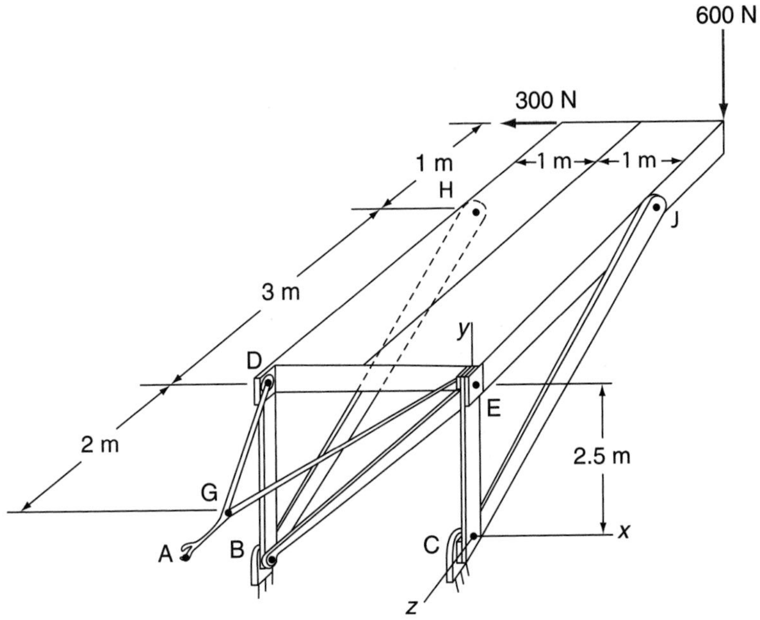

FIGURE 6–21

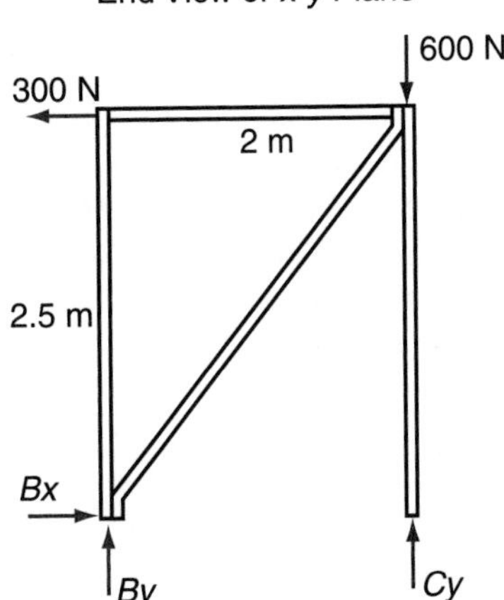

FIGURE 6–22

Begin with an end view and project the forces into the x-y plane (Figure 6–22).

$\Sigma M_c = 0$ (Figure 6–22)

$$-(2\text{ m})(B_y) + (300\text{ N})(2.5\text{ m}) = 0$$

$$\underline{B_y = 375\text{ N}\uparrow}$$

$\Sigma F_y = 0$

$$C_y + 375\text{ N} - 600\text{ N} = 0$$

$$\underline{C_y = 225\text{ N}\uparrow}$$

$\Sigma F_x = 0$

$$-300\text{ N} + B_x = 0$$

$$\underline{B_x = 300\text{ N} \rightarrow}$$

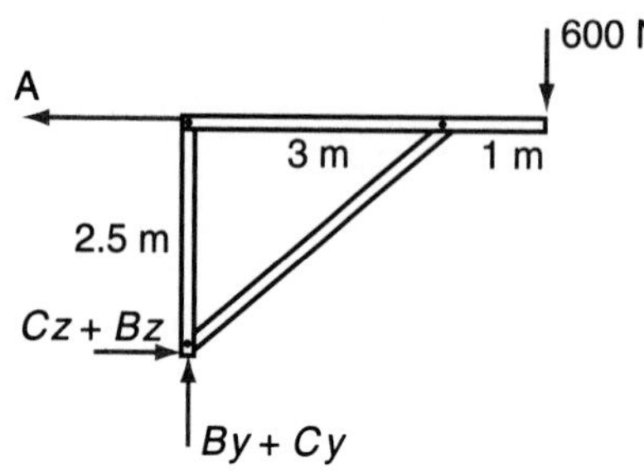

FIGURE 6–23

$$\Sigma M_{BC} = 0 \quad \text{(Figure 6–23)}$$

$$(2.5\text{ m})A - (600\text{ N})(4\text{ m}) = 0$$

$$A = 960\text{ N} \leftarrow$$

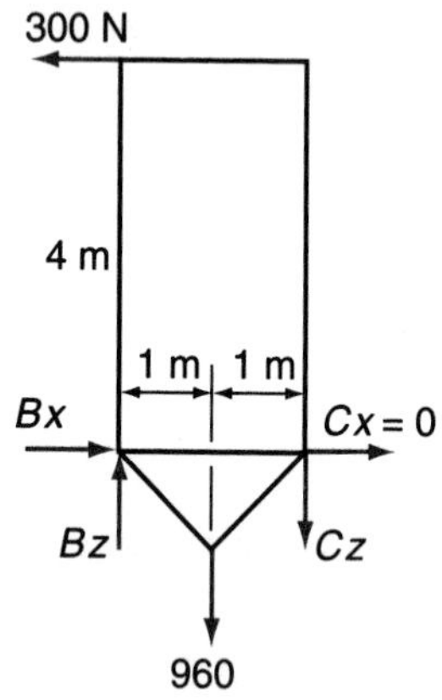

FIGURE 6–24

$$\Sigma M_B = 0 \quad \text{(Figure 6–24)}$$

$$(300\text{ N})(4\text{ m}) - (2\text{ m})C_z - (960\text{ N})(1\text{ m}) = 0$$

$$\underline{C_z = 120\text{ N} \swarrow}$$

$$\Sigma F_z = 0$$

$$-B_z + 960\text{ N} + 120\text{ N} = 0$$

$$\underline{B_z = 1080\text{ N} \nearrow}$$

EXAMPLE 6–8

The hinged platform shown in Figure 6–25 carries a load of 600 N and is held in a horizontal position by cable AB. Neglect the weight of the platform. Determine the load in cable AB and the components of the reactions at points C and D.

$$\text{cable length} = \sqrt{(0.8)^2 + (0.5)^2 + (1.2)^2}$$
$$= 1.53\text{ m}$$

The components at B can be written as

$$B_x = \frac{0.8}{1.53}AB = 0.524AB$$

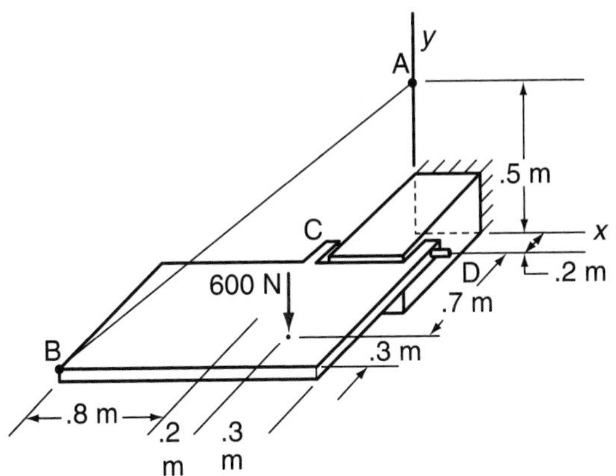

FIGURE 6–25

$$B_y = \frac{0.5}{1.53}AB = 0.328AB$$

$$B_z = \frac{1.2}{1.53}AB = 0.786AB$$

Draw side, front, and top views (Figures 6–26 to 6–28) showing these components of AB. Take moments about CD in the side view (Figure 6–26).

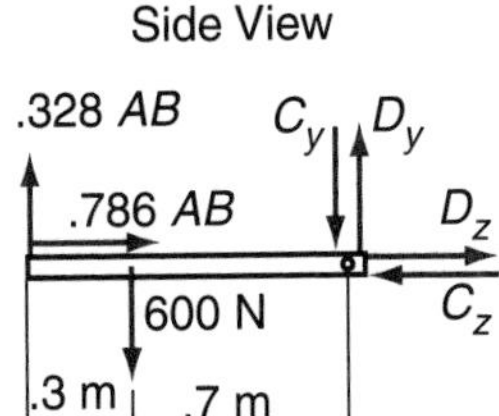

FIGURE 6–26

$\Sigma M_{CD} = 0$

$$-0.328AB(1 \text{ m}) + (600 \text{ N})(0.7 \text{ m}) = 0$$

$$\underline{AB = 1280 \text{ N}T}$$

Now consider the front view (Figure 6–27).

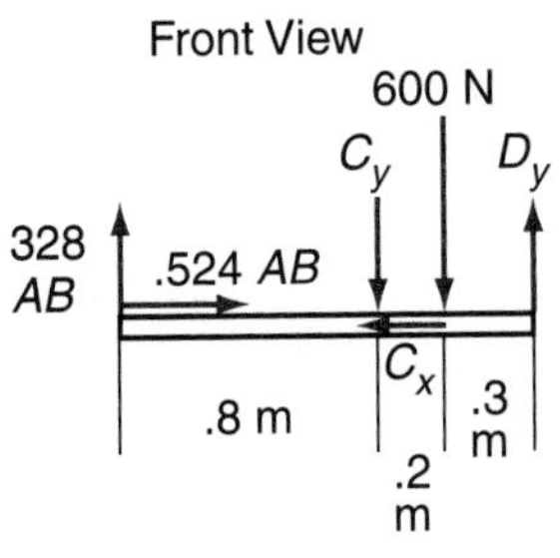

FIGURE 6–27

$\Sigma F_x = 0$

$$C_x = 0.524(1280 \text{ N})$$

$$\underline{C_x = 671 \text{ N} \leftarrow}$$

$\Sigma M_D = 0$

$$(600 \text{ N})(0.3 \text{ m}) + (0.5 \text{ m})C_y - 0.328(1280 \text{ N})(1.3 \text{ m}) = 0$$

$$\underline{C_y = 732 \text{ N} \downarrow}$$

$\Sigma F_y = 0$

$$0.328(1280 \text{ N}) + D_y = 732 \text{ N} + 600 \text{ N}$$

$$\underline{D_y = 912 \text{ N} \uparrow}$$

The top view (Figure 6–28) will give us the remaining unknown components.

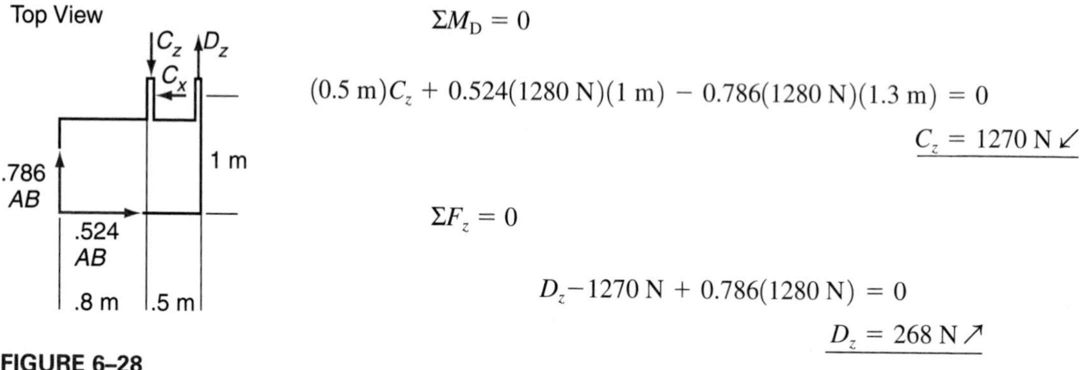

FIGURE 6–28

$$\Sigma M_D = 0$$

$$(0.5\text{ m})C_z + 0.524(1280\text{ N})(1\text{ m}) - 0.786(1280\text{ N})(1.3\text{ m}) = 0$$

$$\underline{C_z = 1270\text{ N} \swarrow}$$

$$\Sigma F_z = 0$$

$$D_z - 1270\text{ N} + 0.786(1280\text{ N}) = 0$$

$$\underline{D_z = 268\text{ N} \nearrow}$$

6–6 CONCURRENT, THREE-DIMENSIONAL SYSTEMS

EXAMPLE 6–9

Determine the tensile load in cable DB and the compressive loads in members AB and BC (Figure 6–29). Neglect the weights of the members.

We will solve this problem by the view method of projecting the forces on a plane. Find the length of diagonal DB.

$$\text{DB} = \sqrt{(6\text{ m})^2 + (6\text{ m})^2 + (3\text{ m})^2}$$
$$= 9\text{ m}$$

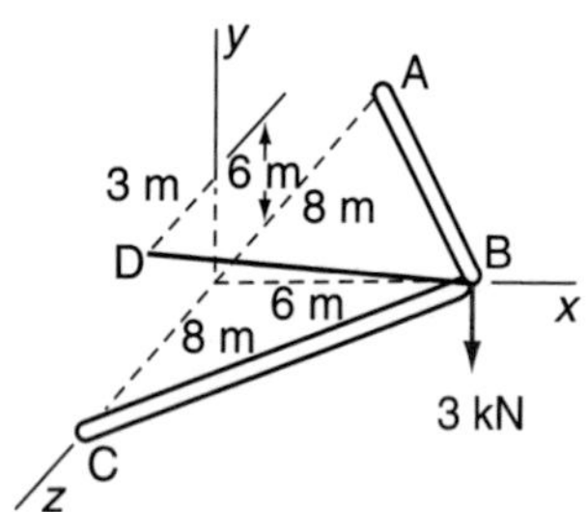

FIGURE 6–29

A side view and a top view can be drawn. The side view consists of looking along the z-axis toward the origin and projecting all the forces onto the x-y plane (Figure 6–30). The top view is obtained by looking down the y-axis and projecting all the forces onto the x-z plane (Figure 6–31).

Note that in the top view (Figure 6–31) the 3-kN force is not shown. In the side view (Figure 6–30), there are no vertical components at points A and C since both members lie in a horizontal plane; therefore, they cannot have vertical components. The weight of the members was also neglected. Since the side view shows the given force of 3 kN, we can start here by taking moments about C.

Side View or x-y Plane

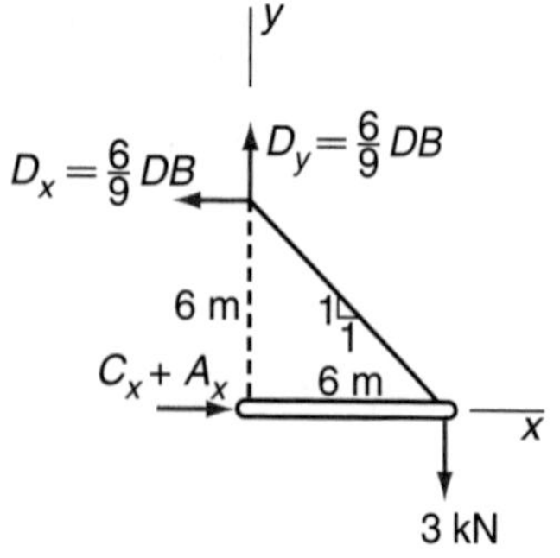

FIGURE 6–30

$\Sigma M_C = 0$ (Figure 6–30)

$$\frac{6}{9}(DB)(6\text{ m}) - (3\text{ kN})(6\text{ m}) = 0$$

$$\underline{DB = 4.5\text{ kN } T}$$

Use $DB = 4.5$ kN in the top view (Figure 6–31) and take moments about point A and then about point C.

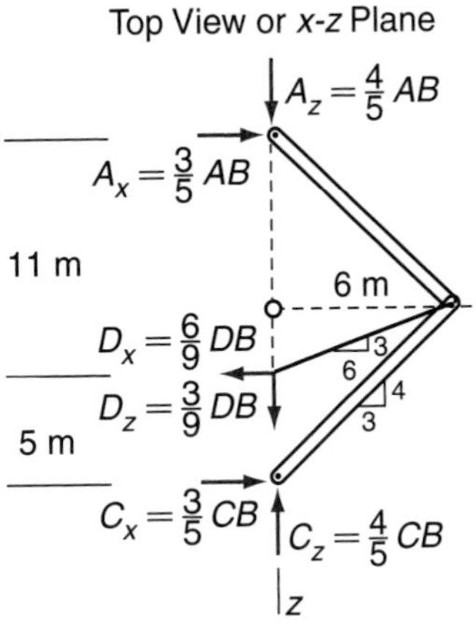

$\Sigma M_A = 0$

$$-\frac{6}{9}DB(11\text{ m}) + \frac{3}{5}CB(16\text{ m}) = 0$$

$$\frac{6}{9}(4.5\text{ kN})(11\text{ m}) = \frac{3}{5}CB(16\text{ m})$$

$$\underline{CB = 3.44\text{ kN } C}$$

$\Sigma M_C = 0$

$$\frac{6}{9}(4.5\text{ kN})(5\text{ m}) - \frac{3}{5}(AB)(16\text{ m}) = 0$$

$$\underline{AB = 1.56\text{ kN } C}$$

FIGURE 6–31

EXAMPLE 6–10

Determine the loads in all members shown in Figure 6–32.

In Example 6–9, we showed each component as a fraction of the total load in the member. In this example, we will simply label each component as being in the x-, y-, or z-direction. You may use either of the two methods.

First find the length of diagonal BC.

$$\text{length of BC} = \sqrt{(6\text{ m})^2 + (8\text{ m})^2 + (4\text{ m})^2}$$
$$= \sqrt{116}$$
$$= 10.77\text{ m}$$

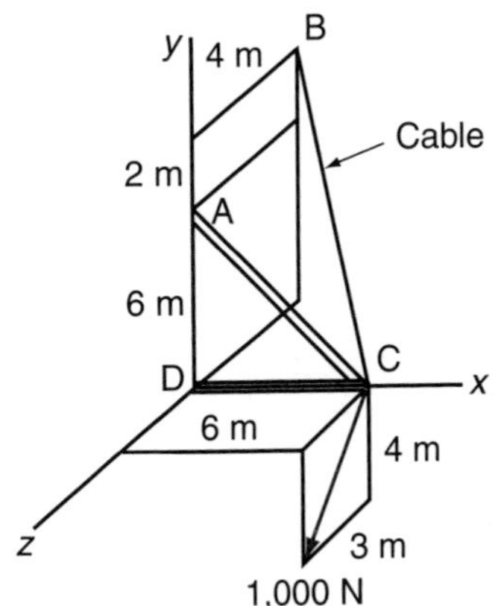

FIGURE 6–32

Top View or x-z Plane

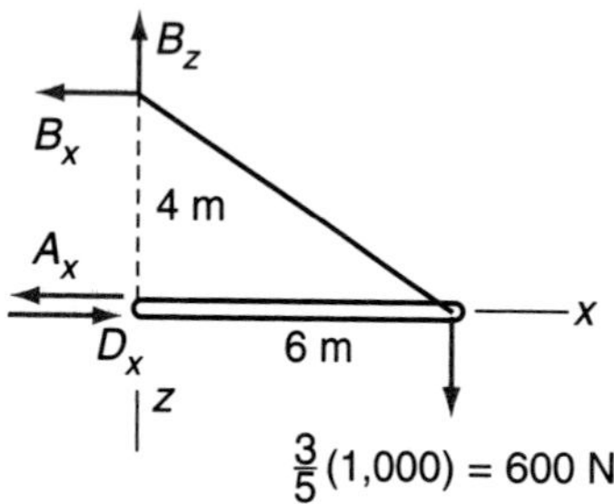

FIGURE 6–33

The top view (Figure 6–33) has the fewest unknowns; taking moments about A will give us B_x:

$\Sigma M_A = 0$

$$(4\text{ m})B_x - (600\text{ N})(6\text{ m}) = 0$$

$$B_x = 900\text{ N}$$

In the x-direction, cable BC has a length of 6 m and a force of 900 N. Since it has a total length of 10.77 m, the total force in BC is found as follows:

$$B_x = \frac{6}{10.77}(BC) \text{ or } BC = \frac{10.77}{6}B_x$$

$$BC = \frac{10.77}{6}(900)$$

$$\underline{BC = 1610\text{ N } T}$$

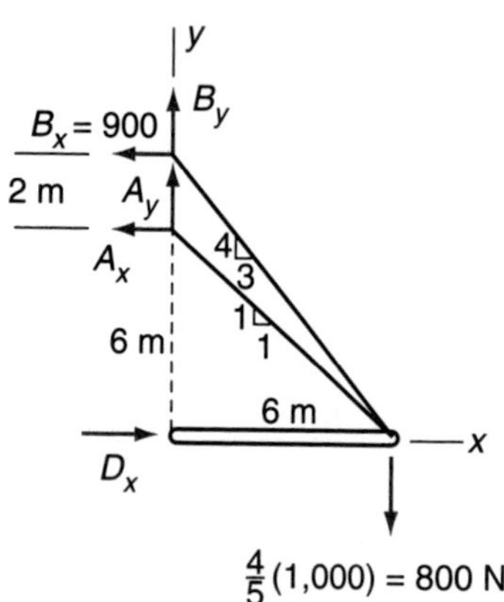

FIGURE 6–34

Go to the side view (Figure 6–34), where $B_x = 900$ N.

$\Sigma M_D = 0$

$$(900\text{ N})(8\text{ m}) + (6\text{ m})A_x - (800\text{ N})(6\text{ m}) = 0$$

$$A_x = -400\text{ N} \leftarrow \; = \; +400\text{ N} \rightarrow$$

Therefore, AC is in compression.

$$\text{length of } AC = \sqrt{(6\text{ m})^2 + (6\text{ m})^2} = 8.48\text{ m}$$

$$\text{force in } AC = \frac{8.48}{6}(400\text{ N})$$

$$\underline{AC = 565\text{ N } C}$$

$\Sigma F_x = 0$

$$D_x - B_x - A_x = 0$$

$$D_x - 900\text{ N} - (-400\text{ N}) = 0$$

$$D_x = 500\text{ N}$$

Since D_y and $D_z = 0$, the total load $\underline{DC = 500\text{ N } C}$.

EXAMPLE 6–11

Solve for the compressive loads in members AD, BD, and CD in Figure 6–35. Points A, B, and C are ball and socket joints.

Calculate the diagonal length of each member.

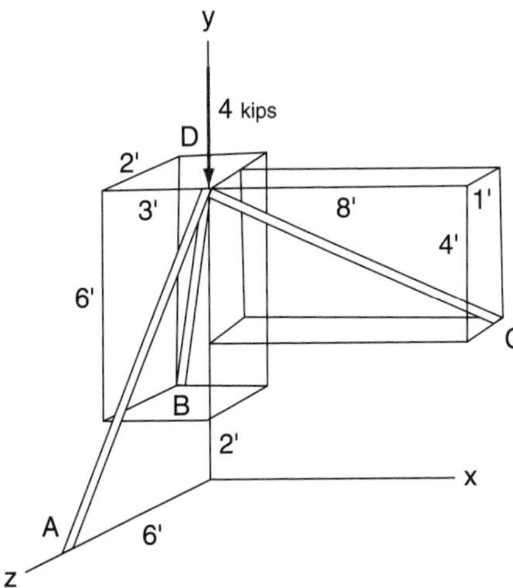

FIGURE 6–35

$$AD = \sqrt{(8)^2 + (6)^2}$$

$$AD = 10 \text{ ft}$$

$$CD = \sqrt{(8)^2 + (4)^2 + (1)^2}$$

$$CD = 9 \text{ ft}$$

$$BD = \sqrt{(3)^2 + (6)^2 + (2)^2}$$

$$BD = 7 \text{ ft}$$

Draw the free-body diagram of the top view (Figure 6–36) and the front view (Figure 6–37) and label each component as a fraction of the total compressive force.

Taking moments about the same point in each diagram will give us simultaneous equations with two unknowns. Take moments about C in the top view (Figure 6–36).

$$\Sigma M_c = 0$$

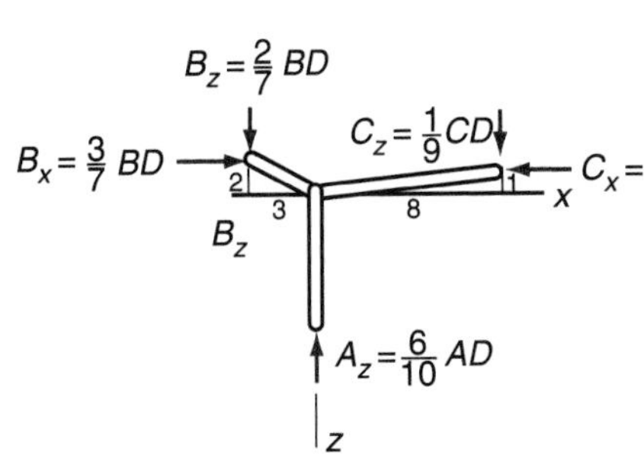

FIGURE 6–36

$$-\frac{6}{10}AD(8 \text{ ft}) - \frac{3}{7}BD(1 \text{ ft}) + \frac{2}{7}BD(11 \text{ ft}) = 0$$

$$4.8AD = 2.71BD$$

$$AD = 0.565BD \quad (1)$$

Take moments about C in the front view (Figure 6–37).

$$\Sigma M_c = 0$$

$$-\frac{8}{10}AD(8 \text{ ft}) - \frac{6}{7}BD(11 \text{ ft})$$

$$+ (4 \text{ kips})(8 \text{ ft}) + \frac{3}{7}BD(2 \text{ ft}) = 0$$

$$-6.4AD - 8.57BD = -32 \qquad (2)$$

Substitute $AD = 0.565BD$: (Equation [1])

$$-3.62BD - 8.57BD = -32$$

$$\underline{BD = 2.63 \text{ kips } C}$$

Front View or x-y Plane

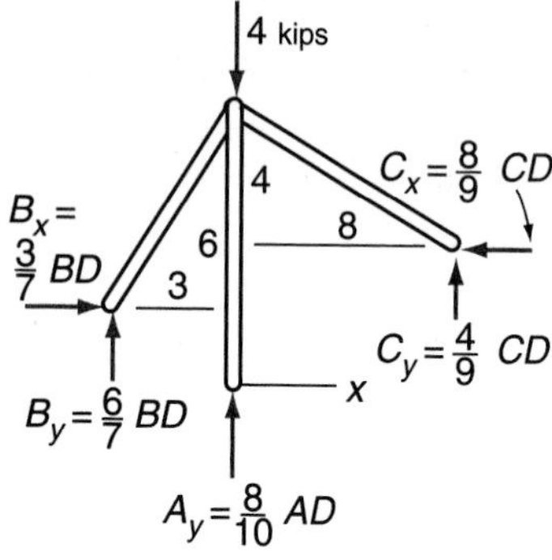

FIGURE 6–37

From Equation (1):

$$AD = 0.565\,(2.63\text{ kips})$$
$$\underline{AD = 1.48\text{ kips }C}$$

Take the sum of the forces in the x-direction in the top view.

$\Sigma F_x = 0$

$$\frac{3}{7}BD = \frac{8}{9}CD$$
$$CD = \frac{3(9)}{7(8)}(2.63\text{ kips})$$
$$\underline{CD = 1.27\text{ kips }C}$$

A check can be made by taking the sum of the forces in the y-direction in the front view.

$\Sigma F_y = 0$

$$\frac{6}{7}(2.63\text{ kips}) + \frac{8}{10}(1.48\text{ kips})$$
$$+\frac{4}{9}(1.27\text{ kips}) - 4\text{ kips} = 0$$
$$2.25 + 1.18 + 0.564 = 4$$
$$\underline{4 = 4\ \textit{check}}$$

HINTS FOR PROBLEM SOLVING

1. Assume the location of the resultant of a parallel force system to have positive coordinates such as $\bar{x}$ and $\bar{z}$. Keep in mind that you are solving for a resultant and not an equilibrium force ($\Sigma M \neq 0$).
2. In two-dimensional problems we found the resultant $R = \sqrt{x^2 + y^2}$. Similarly, in three-dimensional work $R = \sqrt{x^2 + y^2 + z^2}$.
3. For clarity you may prefer to draw top, front, and side views of three-dimensional structures. Label each view as the x-z, x-y, or y-z plane. As a check of your FBD, the subscript of each component must agree with the letters labeling the view or plane.

4. The components at each point in an FBD can be shown
   (a) As fractions of each other, or
   (b) As fractions of the total load of the member at that point.
5. For ease of labeling components of a diagonal force, always calculate the diagonal length in the sequence of x, y, z, that is, $\sqrt{x^2 + y^2 + z^2}$. Consistency in this sequence helps eliminate error in component labeling.

PROBLEMS

APPLIED PROBLEMS FOR SECTION 6–1

6–1—6–5. Determine the magnitude and location of the resultant of the force systems shown in Figures P6–1 to P6–5.

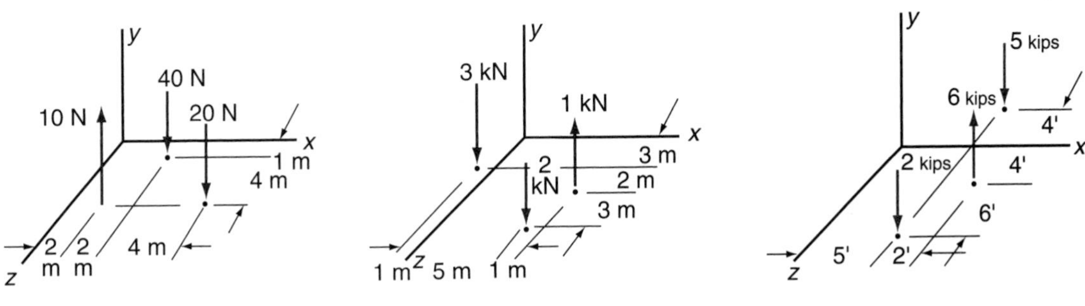

FIGURE P6–1 **FIGURE P6–2** **FIGURE P6–3**

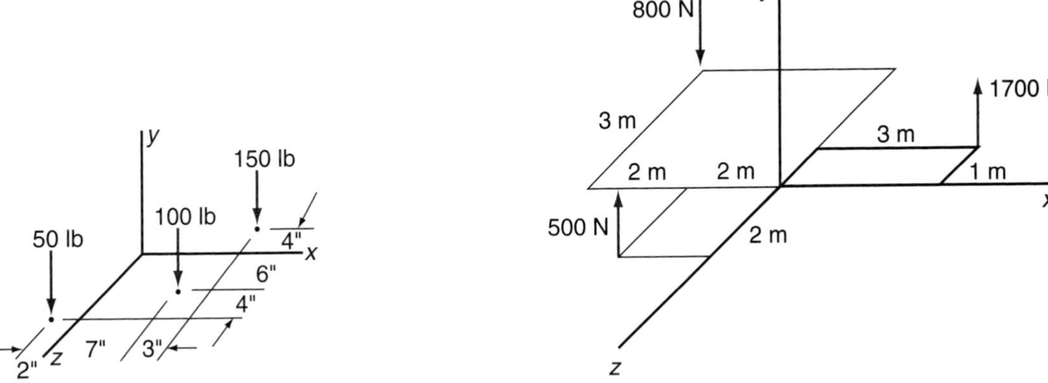

FIGURE P6–4 **FIGURE P6–5**

APPLIED PROBLEMS FOR SECTION 6–2

6–6. Determine the supporting loads at A, B, and C of the 50-kg platform loaded as shown in Figure P6–6.

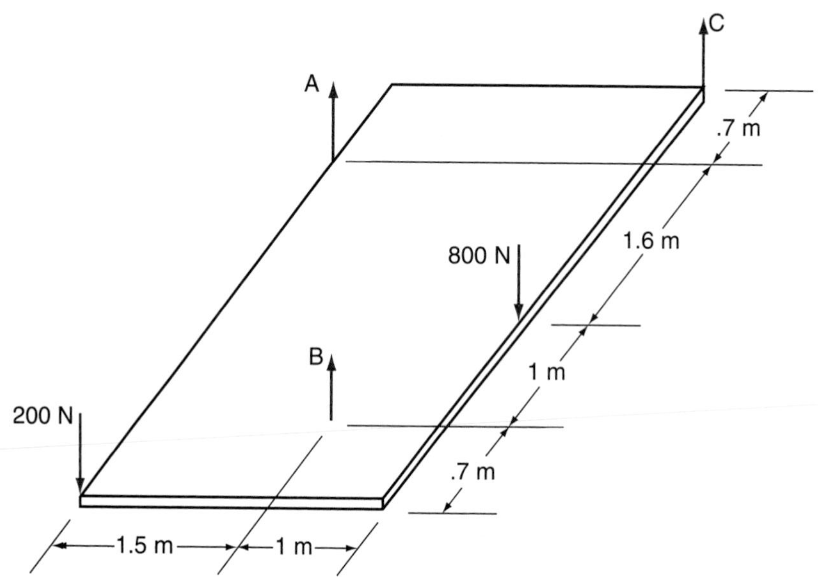

FIGURE P6–6

6–7. The heating unit shown in Figure P6–7 is being hoisted to a roof top when cable B disconnects. Determine the resulting tensions in cables A, C, and D.

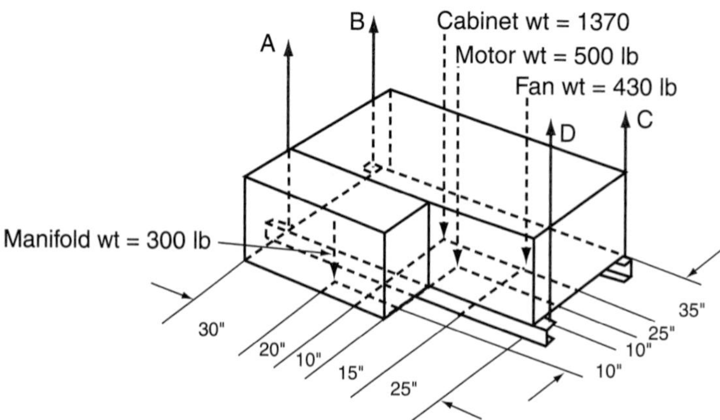

FIGURE P6–7

6–8. The platform, loaded as shown in Figure P6–8, is supported at A, B, and C. Determine the reactions at A, B, and C.

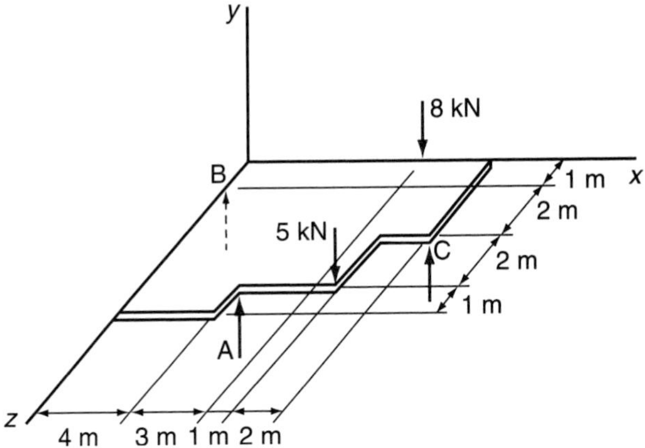

FIGURE P6–8

6–9. A platform supported by three ropes (shown in Figure P6–9) has a mass of 204 kg and carries a load of 4 kN. Calculate the tension in each rope.

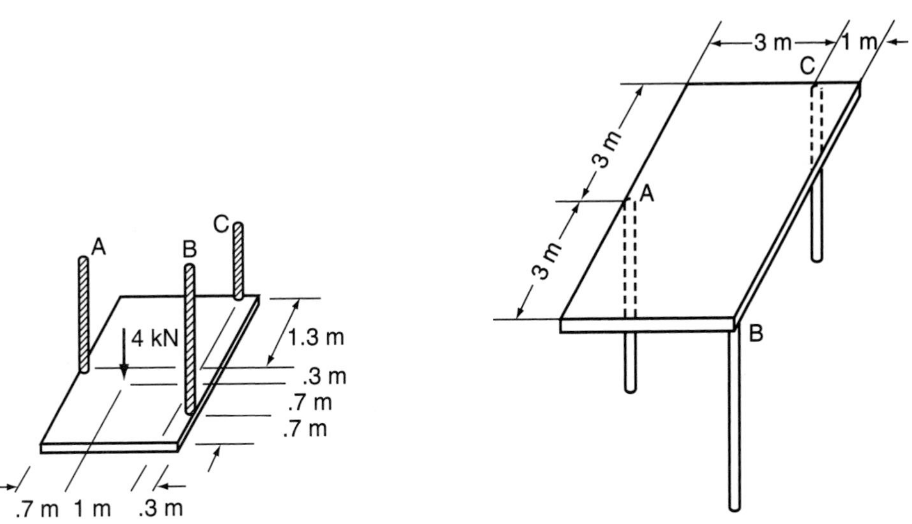

FIGURE P6–9

FIGURE P6–10

6–10. A carport roof is supported as shown in Figure P6–10. If the roof has a mass of 200 kg, determine the load on each support.

6–11. The three-wheeled cart in Figure P6–11 weighs 120 lb. If it is loaded as shown with a crate weighing 200 lb, find the load on each wheel.

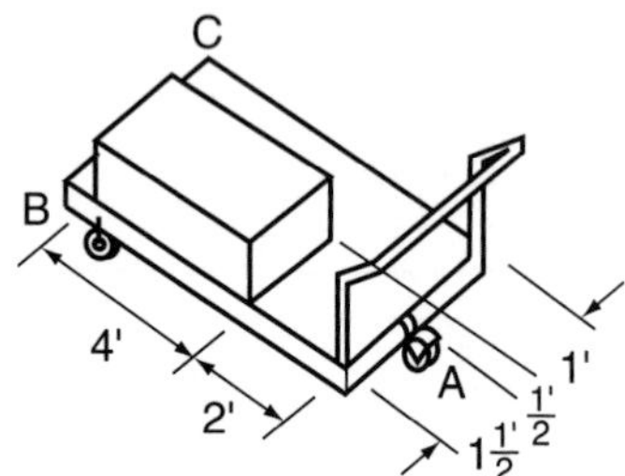

FIGURE P6–11

6–12. The two-compartment tank shown in Figure P6–12 contains material with a specific weight of 50 lb/ft^3. The weight of the empty tank is 800 lb. Determine the supporting forces at A, B, and C.

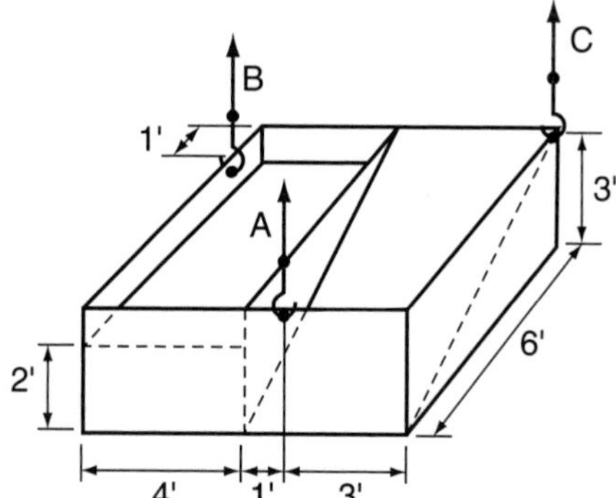

FIGURE P6–12

APPLIED PROBLEMS FOR SECTION 6–3

6–13. Determine the resultant of the forces shown in Figure P6–13. Find the coordinates.

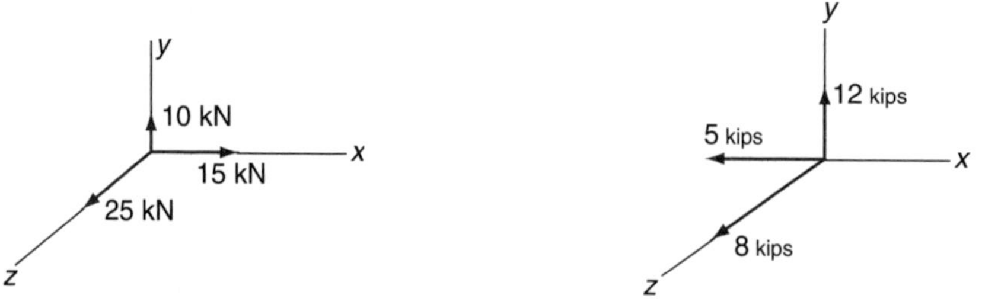

FIGURE P6–13

FIGURE P6–14

6–14. Determine the resultant of the forces shown in Figure P6–14.

6–15. An anchor block is acted upon by the forces shown in Figure P6–15. Determine the resultant force.

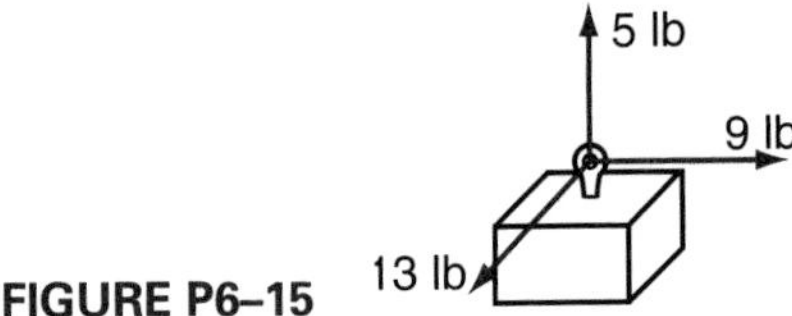

FIGURE P6–15

6–16. Determine the resultant of the forces shown in Figure P6–16.

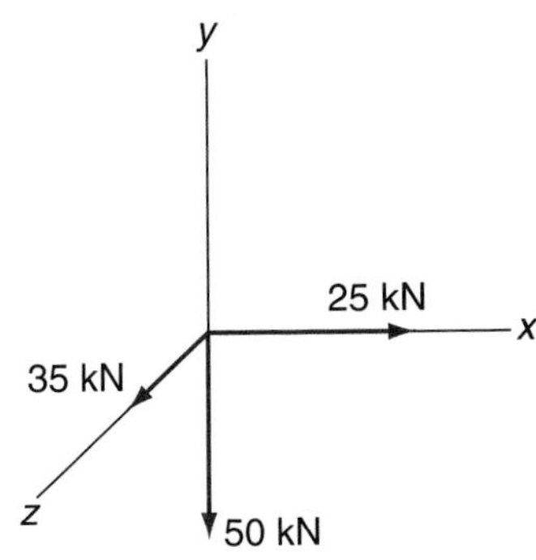

FIGURE P6–16

6–17. Determine the *x*, *y*, and *z* components of the force shown in Figure P6–17.

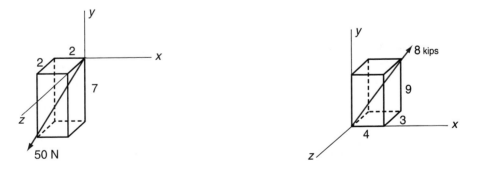

FIGURE P6–17

FIGURE P6–18

6–18. Determine the *x*, *y*, and *z* components of the force shown in Figure P6–18.

6–19. Determine the *x, y,* and *z* components of the force shown in Figure P6–19.

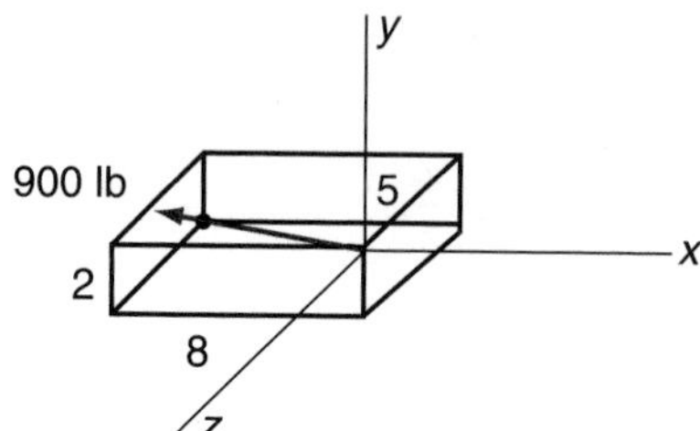

FIGURE P6–19

6–20. Determine the *x, y,* and *z* components of the force shown in Figure P6–20.

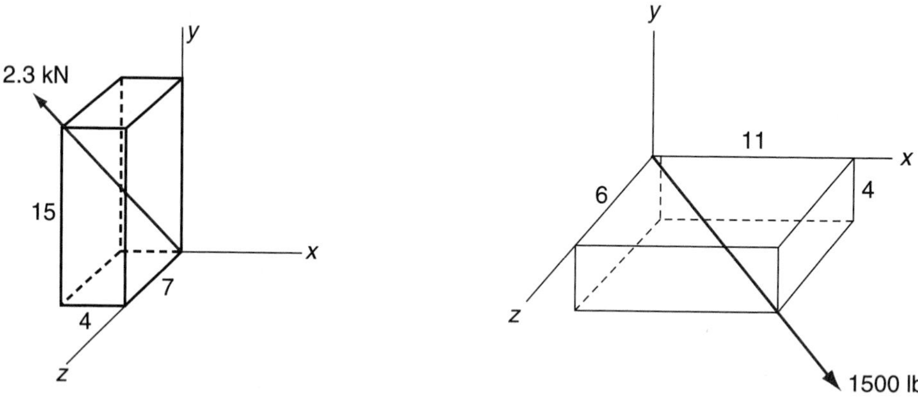

FIGURE P6–20

FIGURE P6–21

6–21. Determine the *x, y,* and *z* components of the force shown in Figure P6–21.

6–22. Leg AB of the derrick shown in Figure P6–22 is subject to a compressive load of 700 lb. Determine the *x, y,* and *z* component forces that it exerts on the ground at A.

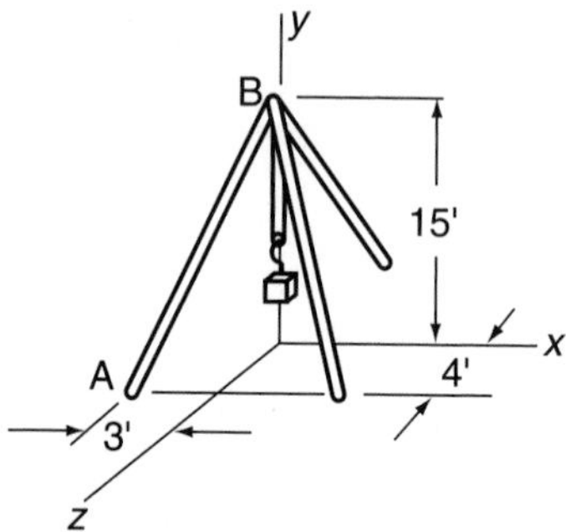

FIGURE P6–22

6–23. Determine the resultant of the forces shown in Figure P6–23.

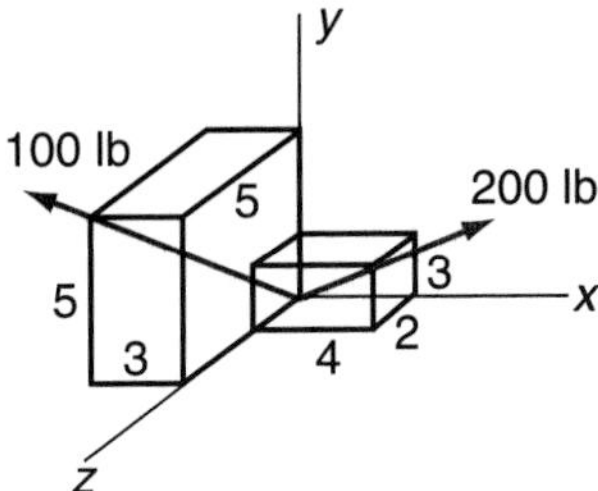

FIGURE P6–23

6–24. Determine the resultant of the forces shown in Figure P6–24.

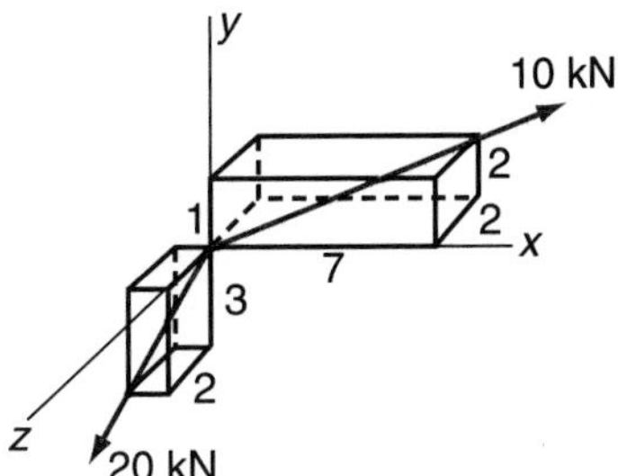

FIGURE P6–24

6–25. Determine the resultant of the forces shown in Figure P6–25.

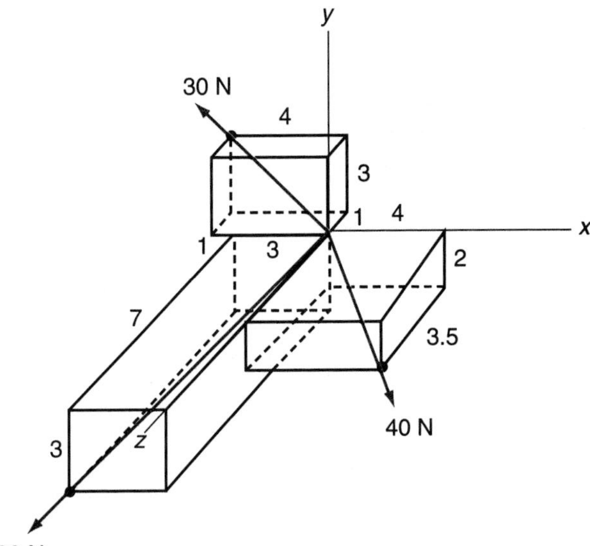

FIGURE P6–25

APPLIED PROBLEMS FOR SECTIONS 6–4 AND 6–5

6–26. The shaft in Figure P6–26 receives an input torque due to the belt tensions shown and it transmits this torque to a machine at D. Determine the y and z components of the bearing reactions at B and C.

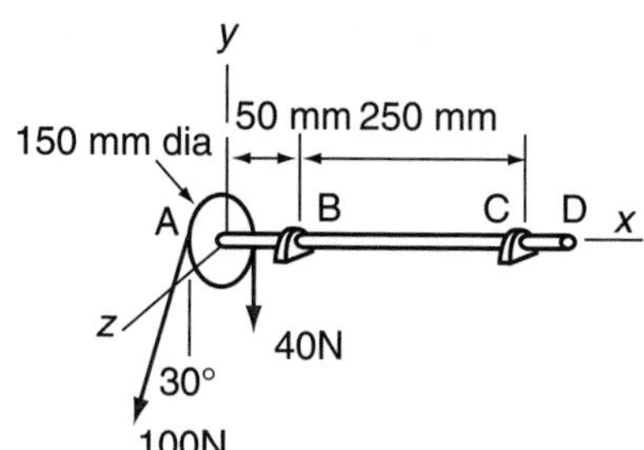

FIGURE P6–26

6–27. A counterweighted mechanism for tightening a belt at pulley E is shown in Figure P6–27. The counterweight weighs 20 lb. At an instant during which the belt is being loosened, the force at G is 5 lb, and the tensile force in AB is 14.5 lb. Determine the x and y components of the reactions at C and D.

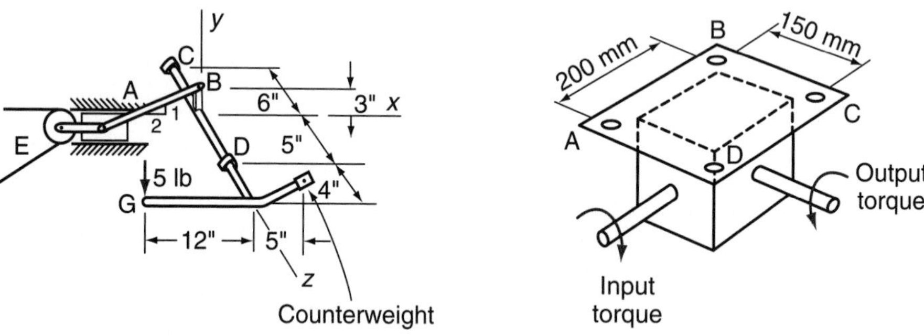

FIGURE P6–27

FIGURE P6–28

6–28. A gear box (Figure P6–28) receives an input torque of 12 N·m. Assume an output torque of the same amount. The gear box is mounted by means of bolts and spacers, which may exert a compressive or tensile force upon the gear box. Calculate the vertical force present at mounting points A, B, C, and D.

6–29. Assume equal z components at A and B (Figure P6–29). A and B are ball- and- socket constructions, capable of supporting in three directions. Determine the x, y, and z components of the forces at A, B, and C.

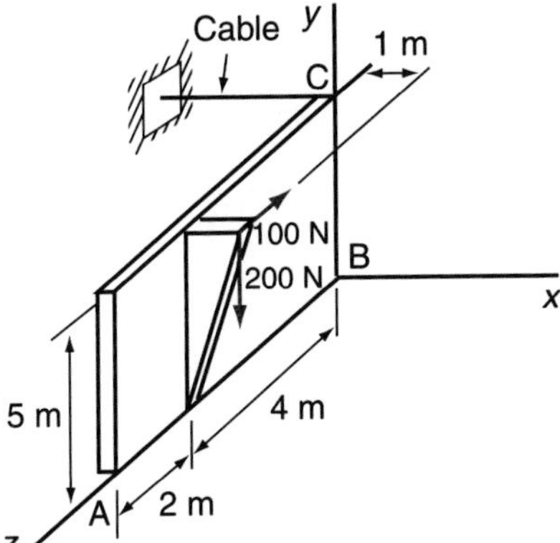

FIGURE P6–29

6–30. The structure in Figure P6–30 is supported at A and B by ball- and- socket connections and at C by a cable. Assuming that $A_z = B_z$, determine the load in cable CD and the reaction components at A and B.

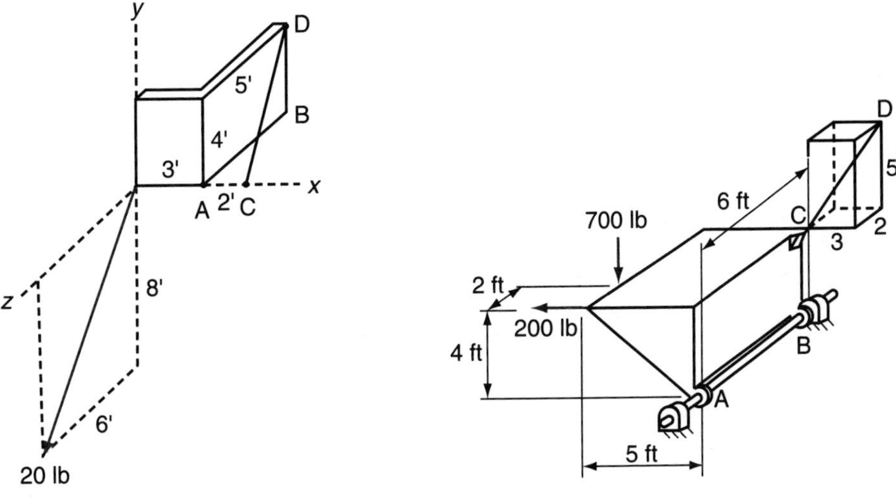

FIGURE P6–30

FIGURE P6–31

6–31. The platform shown in Figure P6–31 hinges on the rod passing through A and B. Determine (a) the tension in cable CD and (b) the reaction components at A and B.

6–32. Member CD has ball and socket connections at each end (Figure P6–32). Determine (a) the reaction components at A and B and (b) the load in member CD.

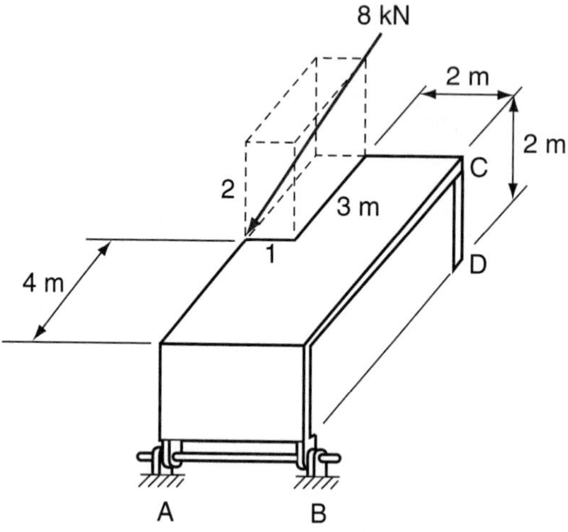

FIGURE P6–32

6–33. The horizontal platform carries 500 lb as shown (Figure P6–33). Determine the reactions at A and B.

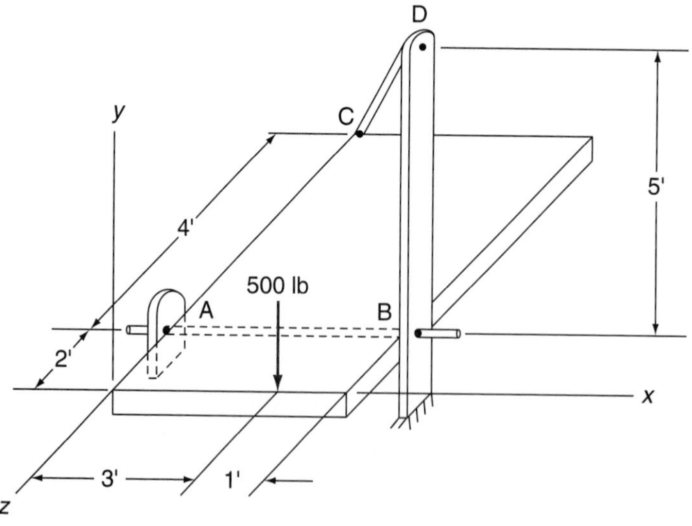

FIGURE P6–33

6–34. Points A, B, and C (Figure P6–34) are in the same vertical plane. Points A, B, D, and E are in a horizontal plane. Assume that supports A and B have equal components in the z direction. Determine the x, y, and z components of the forces at A, B, and C.

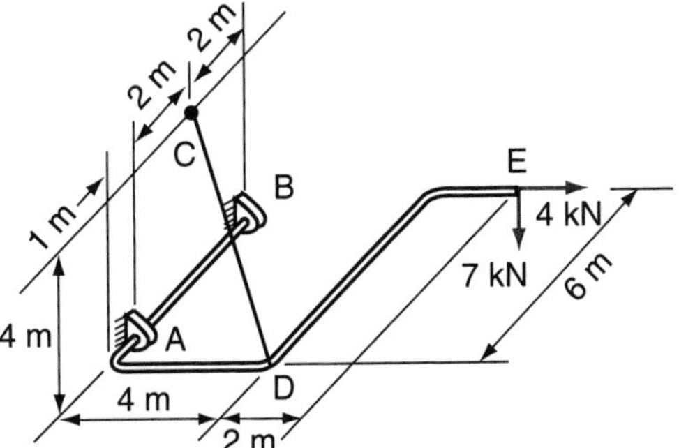

FIGURE P6–34

6–35. The platform shown in Figure P6–35 is supported by bearings at A and B, which equally share the axial load in the z axis, and by pole CD. Determine the reactions at A and B.

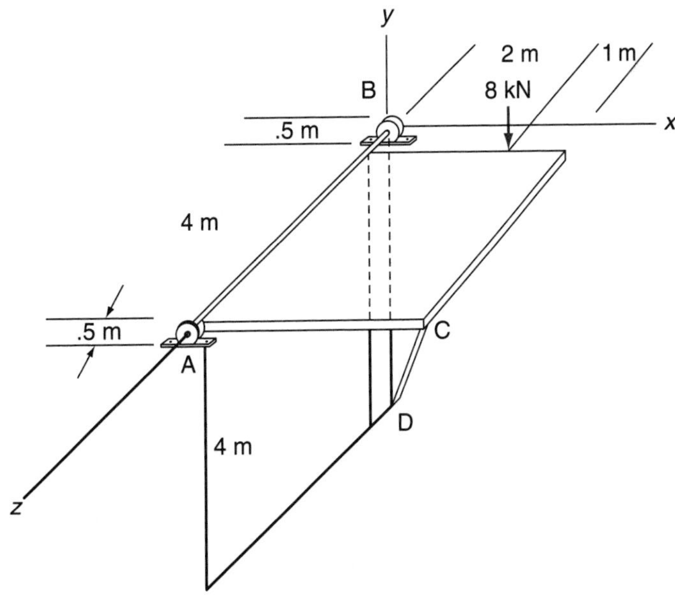

FIGURE P6–35

APPLIED PROBLEMS FOR SECTION 6–6

6–36. Determine the load in each member of the frame shown in Figure P6–36.

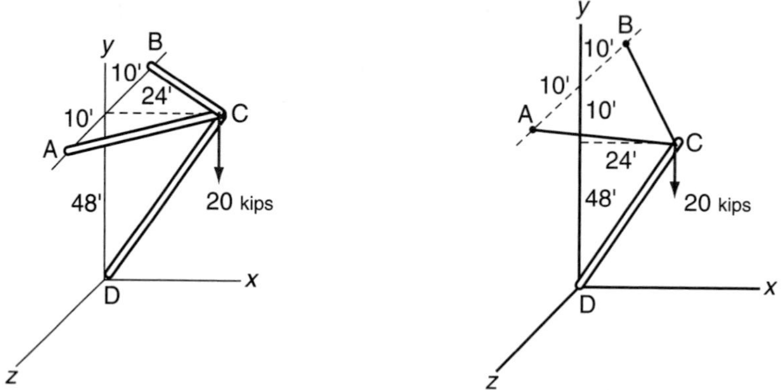

FIGURE P6–36

FIGURE P6–37

6–37. Determine the load in each member of the frame shown in Figure P6–37.

6–38. Determine the load in each member of the frame shown in Figure P6–38.

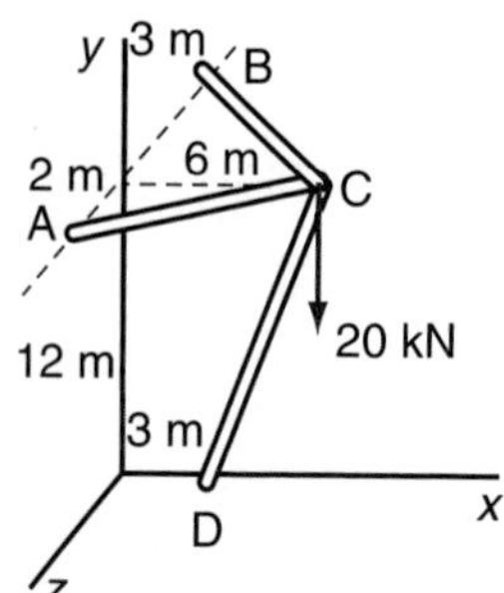

FIGURE P6–38

6–39. Determine the load in each of the cables and pole shown in Figure P6–39. Points A, D, and B lie in a horizontal plane.

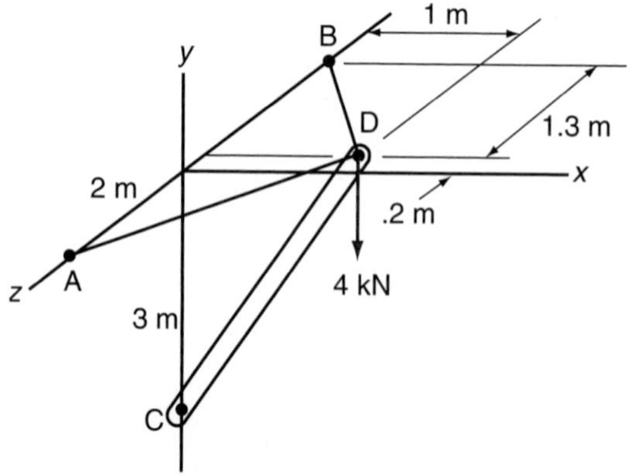

FIGURE P6–39

6–40. Determine the load in members AB, BC, and BD in Figure P6–40.

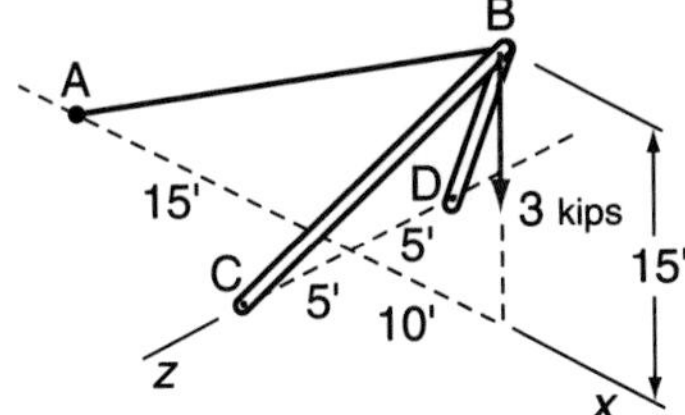

FIGURE P6–40

6–41. Determine load in each of the cables shown in Figure P6–41.

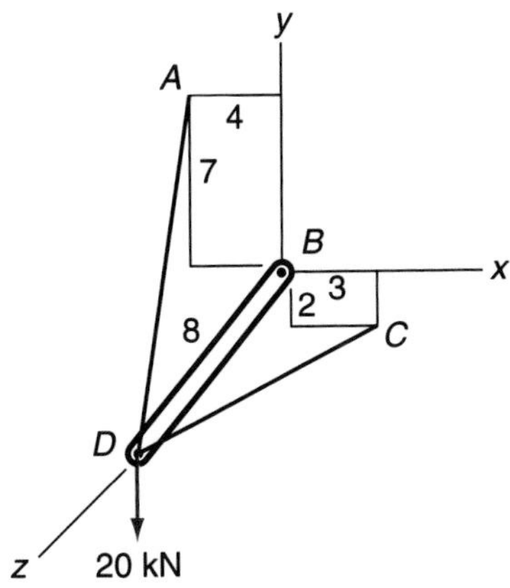

FIGURE P6–41

6–42. Determine the load in each member of the system shown in Figure P6–42.

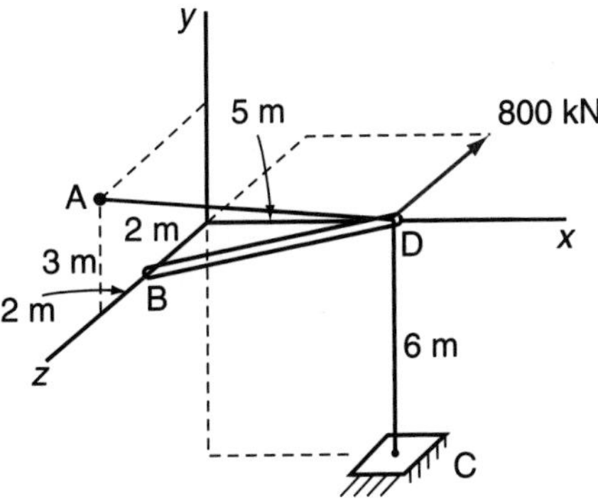

FIGURE P6–42

6–43. A jib crane is supported by a collar D to which three, 5-ft-long supporting members are pinned (Figure P6–43). Points A, B, and C are equidistantly spaced on a 3-ft radius about the center pole. Determine the load in member AD for the position shown.

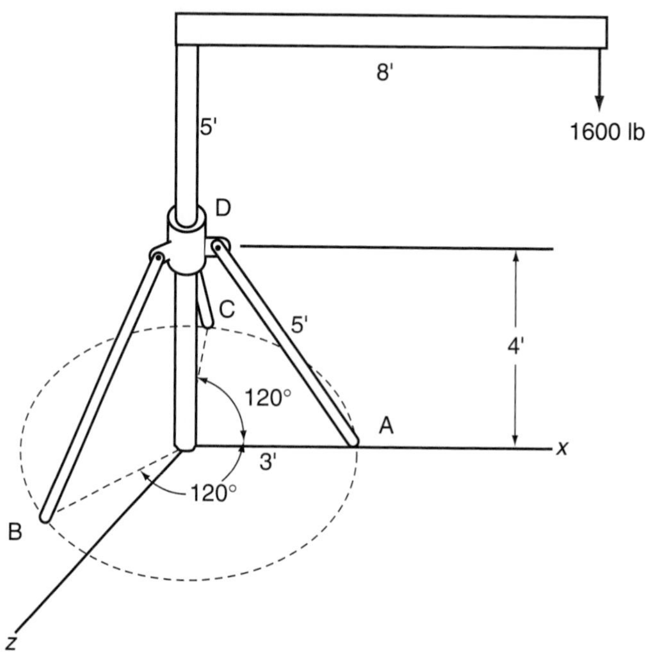

FIGURE P6–43

6–44. The jib crane, loaded as shown in Figure P6–43, is rotated counterclockwise as viewed from above. At some point in this rotation, the load in leg DC will switch from tension to compression. Determine the angle of rotation between the jib arm and the *x*-axis when the load in leg DC is zero.

6–45. Determine the load in each of the three legs in Figure P6–45.

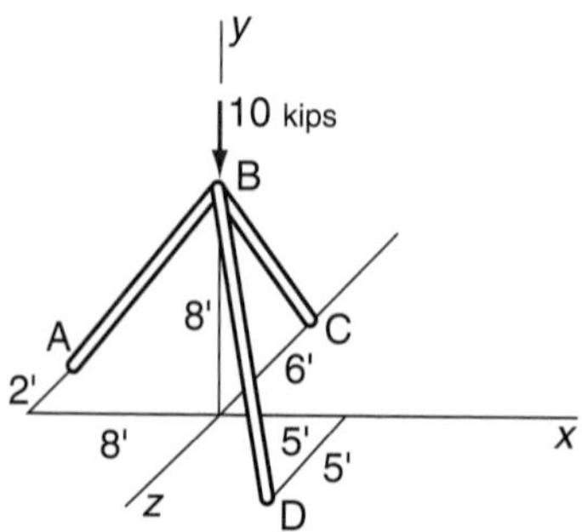

FIGURE P6–45

6–46. Solve for the load in each member of the structure shown in Figure P6–46.

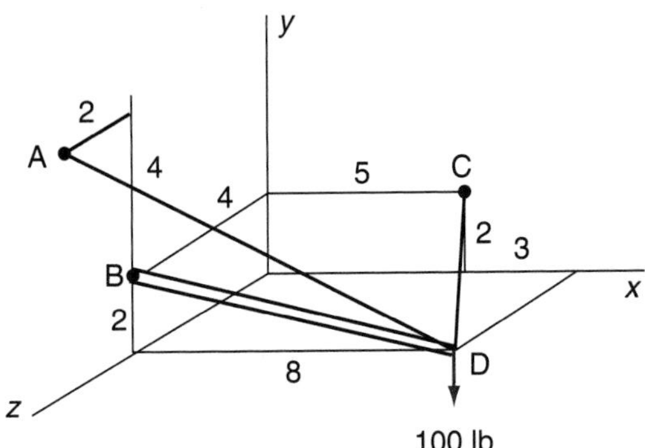

FIGURE P6–46

6–47. Antennas have the supporting cables shown in Figure P6–47. There are no other forces acting on the antennas other than the bottom socket connection. If cable AB is tightened to a tension of 300 lb, what is the tension in cables BC and BD?

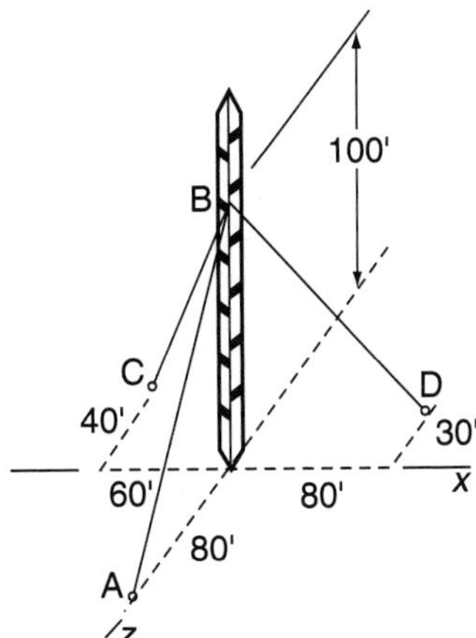

FIGURE P6–47

6–48. Determine the cable tensions if the tower has initial cable tensions of problem 6–47 but subsequently experiences a wind force equivalent to 800 lb at B acting parallel to the z-axis and to the right. Assume no stretching of cable AB.

6–49. Determine the loads in members AD, BD, and CD of the structure loaded as shown in Figure P6–49.

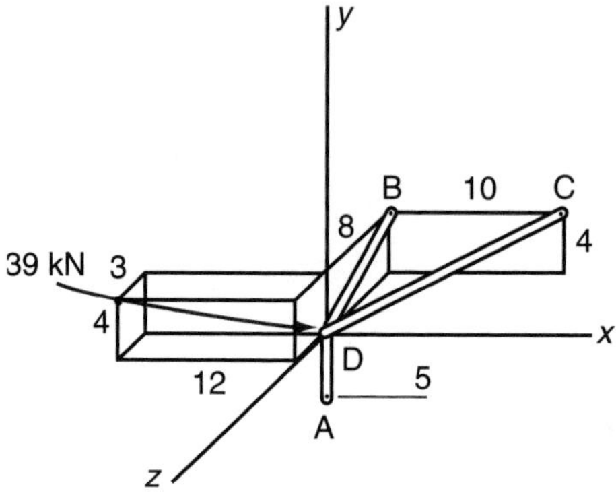

FIGURE P6–49

6–50. Beam BE weighs 200 lb, has a ball- and- socket connection at E, and is supported by cables AB and CD as shown in Figure P6–50. Determine the tension in cables AB and CD.

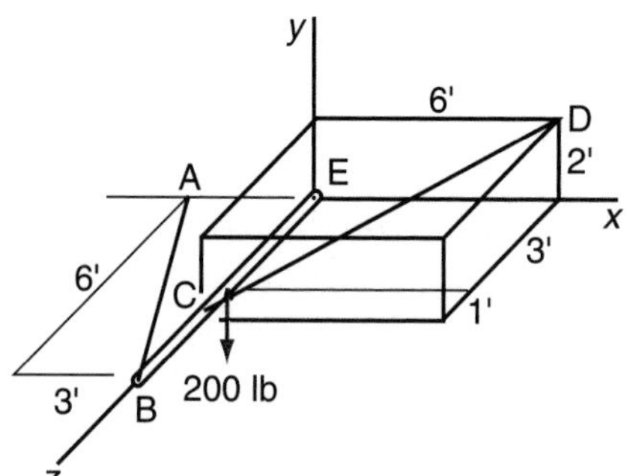

FIGURE P6–50

6–51. A tube handling bulk material in Figure P6–51 pivots at B and is controlled by adjustment of cables AD and DC. The combined weight of the tube and material is 1500 lb, and it acts through point D. Determine the cable tensions AD and CD.

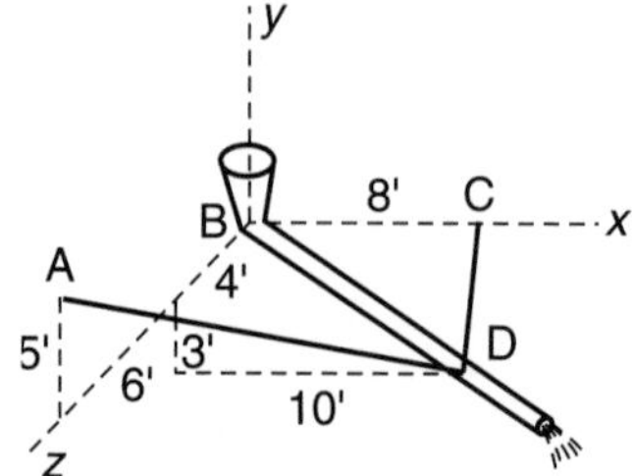

FIGURE P6–51

6–52. Determine the load in each member of the structure shown in Figure P6–52.

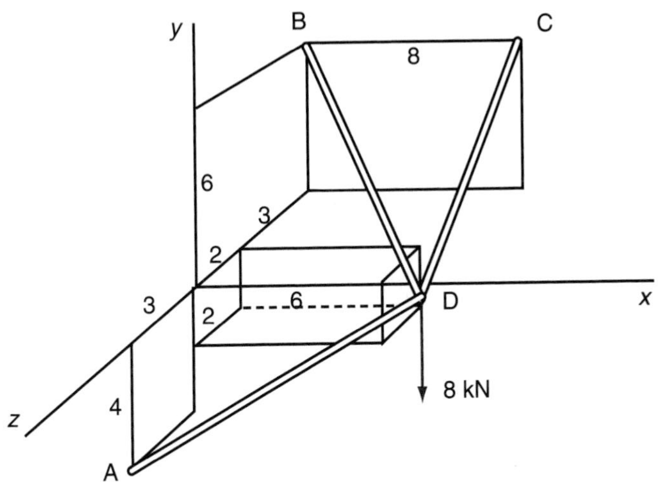

FIGURE P6–52

6–53. Determine the load in each member of the structure shown in Figure P6–53.

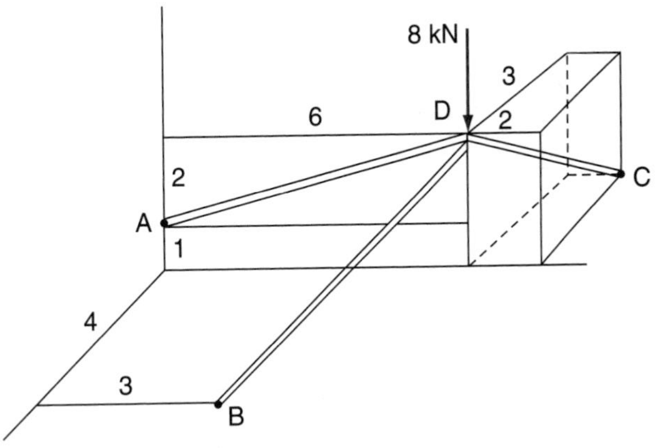

FIGURE P6–53

REVIEW PROBLEMS

R6–1. Determine the magnitude and location of the resultant of the force systems shown in Figure RP6–1.

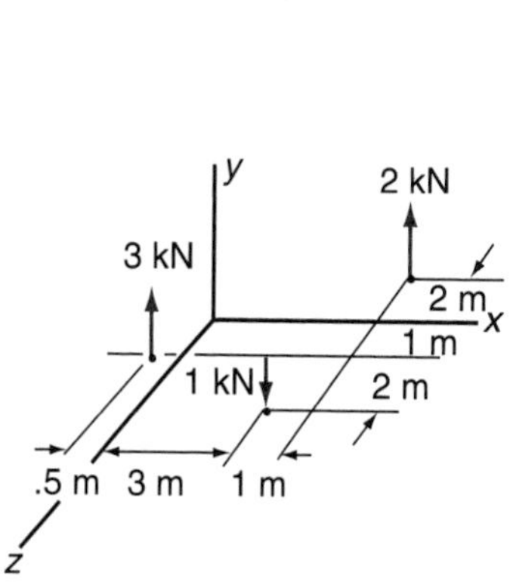

FIGURE RP6–1

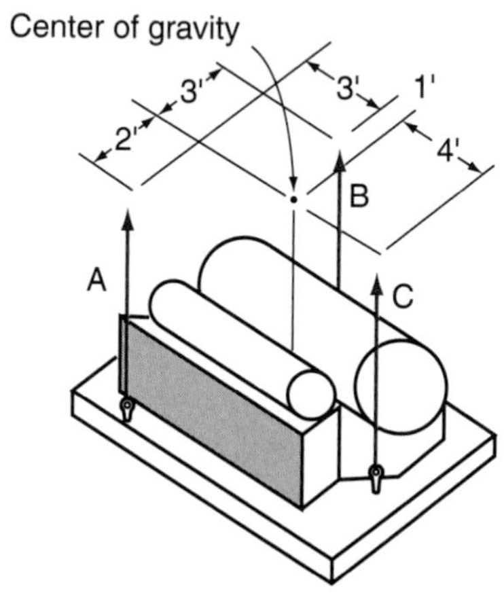

FIGURE RP6–2

R6–2. A piece of machinery weighing 5000 lb is lifted to its rooftop installation site by means of a crane. The vertical crane cables are attached at points A, B, and C. Determine the tension in each cable (Figure RP6–2).

R6–3. Determine the resultant of the forces shown in Figure RP6–3.

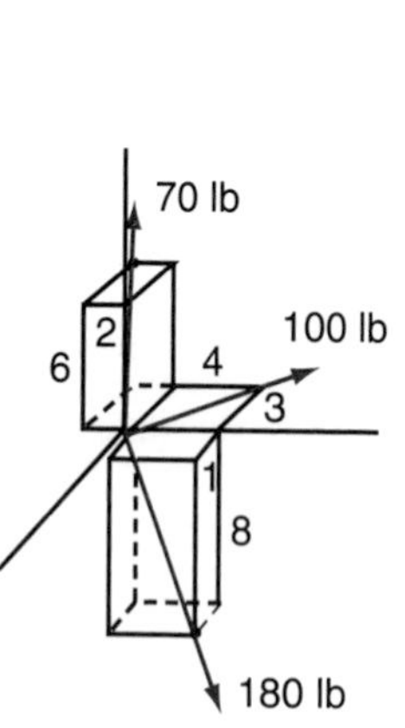

FIGURE RP6–3

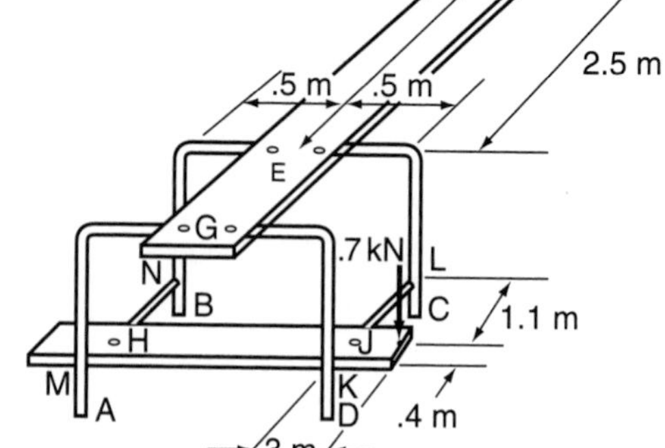

FIGURE RP6–4

R6–4. The diving board shown in Figure RP6–4 is supported by vertical forces at A, B, C, and D. The impact load of a diver at the end of the board is 1.2 kN and a person waiting at the side exerts 0.7 kN. Find the loads at A, B, C, and D.

R6–5. Determine the load in each member of the structure shown in Figure RP6–5.

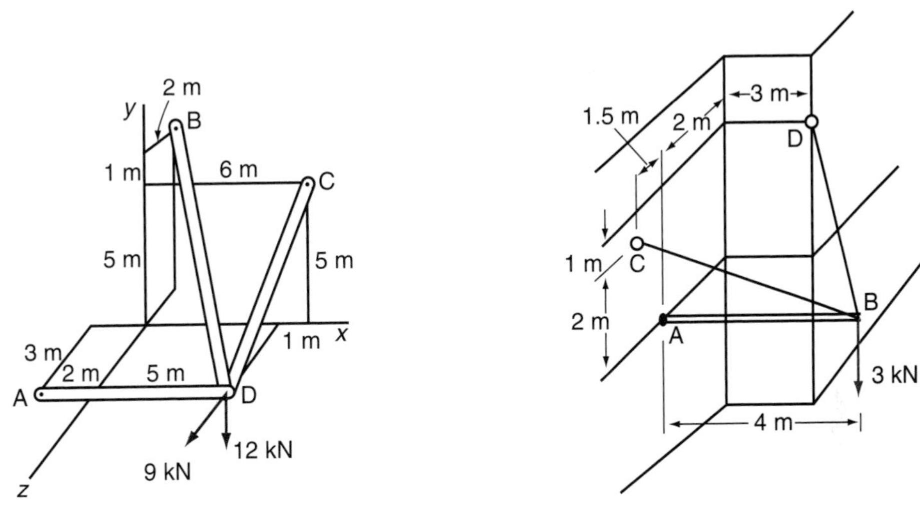

FIGURE RP6–5

FIGURE RP6–6

R6–6. Pole AB is anchored to the side of a building as shown in Figure RP6–6. Determine the load in members AB, CB, and DB.

R6–7. A sign with a mass of 120 kg has a wind force of 400 N on its face as shown (Figure RP6–7). Members BE and FG are very short cable connectors, and point D is a ball-and-socket connection. Determine (a) cable tensions AB, BC, and HK and (b) reaction components at point D.

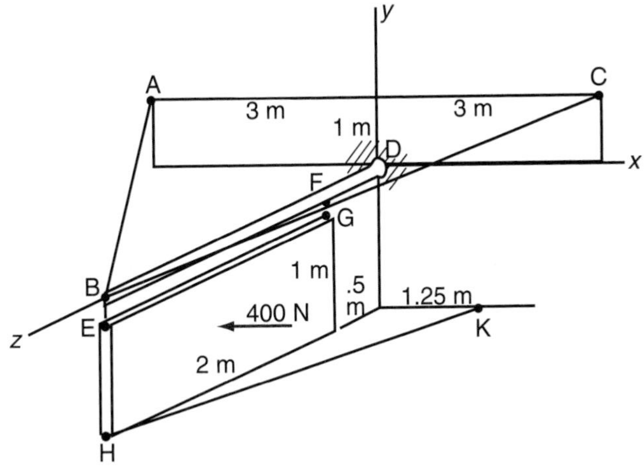

FIGURE RP6–7

CHAPTER 7

Friction

OBJECTIVES

Upon completion of this chapter the student will be able to:

1. Apply the friction laws for dry surfaces to both flat surfaces and flat belts.
2. Determine if motion is impending.
3. Determine whether tipping or sliding will occur.

7–1 INTRODUCTION

Much time and energy has been spent reducing unwanted friction in machines and engines. The internal combustion engine is a prime example of this attempt. Perhaps an equal amount of time has been spent trying to utilize friction. The design of tires and various braking systems is indicative of this effort. The understanding and use of friction are important in so many of our everyday activities—not to mention in equipment design—that the basic friction laws for dry surfaces will now be considered.

7–2 FRICTION LAWS FOR DRY SURFACES

The following discussion will concern only nonlubricated surfaces.

Motion or impending motion of two surfaces in contact causes a reaction force known as a *friction force, F.* This friction force is:

1. Parallel to a flat surface or tangent to a curved surface
2. Opposite in direction to the motion or impending motion
3. Dependent on the force pressing the surfaces together
4. Generally independent of the area of surface of contact
5. Independent of velocity, except for extreme cases not to be considered here
6. Dependent on the nature of the contacting surfaces

Impending motion means that the object being considered is on the verge of moving; a small additional force would cause motion. An object said to be in impending motion is not moving, and therefore is in static equilibrium. A free-body diagram showing all external forces can be drawn.

Friction forces can also exert their effects on an object in motion. This is a case of dynamic equilibrium where velocity and acceleration may be of concern. The dynamics portion of this book covers such a situation.

7–3 COEFFICIENTS OF FRICTION

Consider the case of a block on a horizontal surface being pushed by a force P (Figure 7–1) so that the block has impending motion to the right.

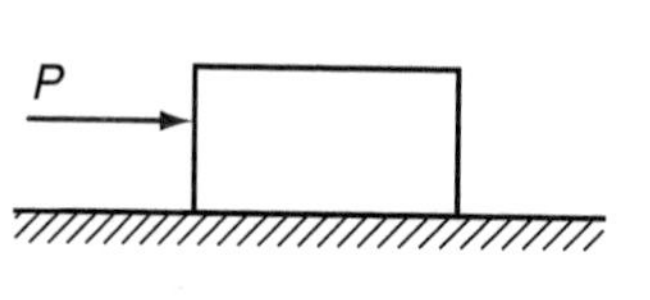

FIGURE 7–1

Free-Body Diagram of Block

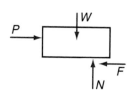

W
P
F
N

FIGURE 7–2

A free-body diagram of the block (Figure 7–2) would have the following forces:

P: as shown

W: weight of the block

F: force of friction

N: normal force

Normal means *at right angles to,* so the normal force here is at right angles to the surface and is equal to the force pressing the two surfaces together. In this case, it is equal to the weight of the block. There can be many cases in which the normal force is not equal to the weight, such as on an inclined surface, when the applied force P is not horizontal, and when other external forces are acting on the object. The use of a free-body diagram and equilibrium equations is necessary for these cases.

Suppose that in Figure 7–2, force P begins as a small value of 10 lb and gradually increases until at $P = 40$ lb, the block is on the verge of sliding. By equating forces in the x direction, when $P = 10$ lb, then $F = 10$ lb; when $P = 20$ lb, then $F = 20$ lb; when $P = 30$ lb, then $F = 30$ lb. Friction force F, as a reacting force, simply matches the applied force P until it reaches its maximum friction force. This maximum force is dependent upon the nature of the surfaces and the amount of the normal force N.

When the friction force reaches its maximum value, we still have static equilibrium but the block is on the verge of moving and therefore has *impending motion.*

In order to describe the friction between two surfaces, we use the relationship between the friction force and the normal force. The friction force depends on the normal force and is always a fraction of the normal force. The friction force is therefore expressed as a fraction or portion of the normal force. This relationship is called the *coefficient of friction* and is represented by the Greek lowercase letter mu (μ).

$$\mu = \frac{F_{max}}{N} \tag{7–1}$$

where F_{max} is the friction force at impending motion, N the normal force, and μ the coefficient of friction (a numerical value with no units).

There can be *static friction forces* (where there is impending motion) or *kinetic friction forces* (where the surfaces are moving with respect to each other). Since both static and kinetic friction forces exist, there are also static and kinetic coefficients of friction, μ_s and μ_k. Static friction will be our prime concern in this chapter, and we will simply use μ without the subscript "s."

The coefficient of friction can be determined for any two materials in contact but varies within a range of values. Because of this variation in values, the coefficient of friction of various surfaces is not given in a tabular form; rather, average values will be given for specific problems. A later example (Example 7–2) shows how the coefficient of friction can be determined experimentally.

7–4 ANGLE OF FRICTION

A second way in which friction is described is by the *angle of friction.* In Figure 7–3 the friction force F_{max} and the normal force N are combined in the resultant R. The friction angle ϕ (phi) is the angle between N and R. Considering the right-angle triangle in which ϕ is located, we can write

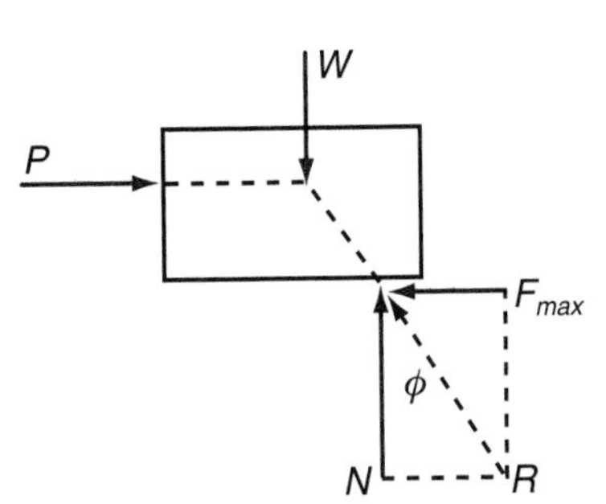

FIGURE 7–3

$$\tan \phi = \frac{F_{max}}{N}$$

but the coefficient of friction $\mu = \dfrac{F_{max}}{N}$; therefore,

$$\mu = \tan \phi \tag{7–2}$$

Friction can be expressed as a coefficient or as the tangent of the friction angle ϕ.

When drawing free-body diagrams in friction problems, you have the choice of showing F_{max} and N separately or in combination as R at some friction angle. The following examples will show that the choice of method is determined by which method offers the easiest solution.

EXAMPLE 7–1

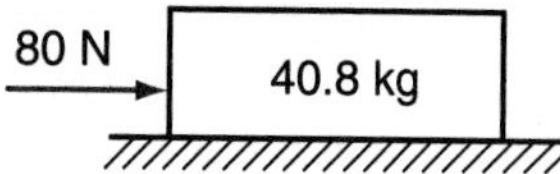

FIGURE 7–4

Free-Body Diagram of Block

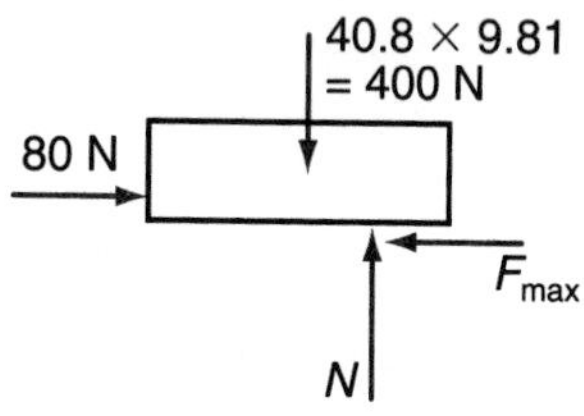

FIGURE 7–5

The 80-N force shown in Figure 7–4 causes impending motion to the right. Determine the static coefficient of friction μ.

Since there are only two vertical forces in the free-body diagram of the block (Figure 7–5):

$$N = 400 \text{ N} \uparrow$$

Similarly, in the horizontal direction,

$$F_{max} = 80 \text{ N} \leftarrow$$

$$\mu = \frac{F_{max}}{N}$$

$$= \frac{80 \text{ N}}{400 \text{ N}}$$

$$\underline{\mu = 0.2}$$

EXAMPLE 7–2

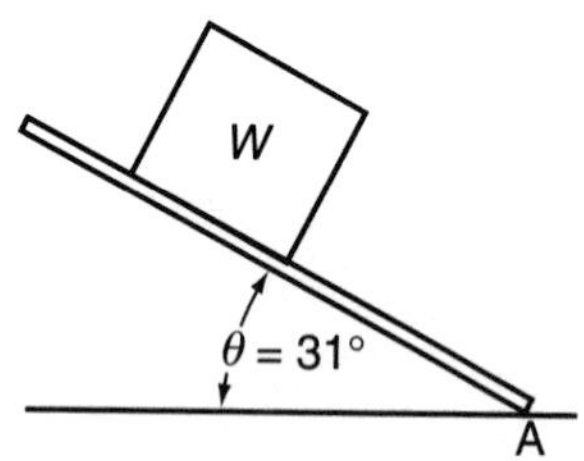

FIGURE 7–6

Free-Body Diagram of Block

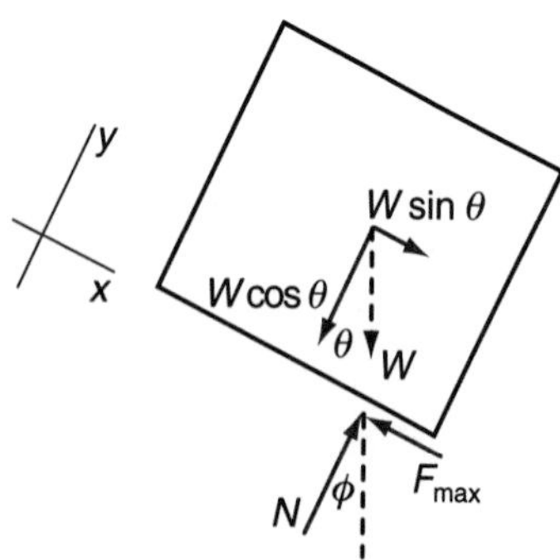

FIGURE 7–7

A block of weight W is placed on a plane that is pivoted at A (Figure 7–6). The plane is tilted until, at $\theta = 31°$, the block begins to slide. What is the coefficient of static friction μ between the block and the plane?

(This is the *angle of repose* method of determining the coefficient of static friction.)

The weight is broken into components that are normal and parallel to the surface (Figure 7–7).

As the plane is tilted upward to 31°, the normal force equals $W \cos \theta$ and the friction force equals $W \sin \theta$. When impending motion at $\theta = 31°$ is reached, then the friction force F is at its maximum; therefore, the friction angle ϕ occurs as shown in Figure 7–7.

$\Sigma F_x = 0$

$$F_{max} = W \sin \theta$$

$\Sigma F_y = 0$

$$N = W \cos \theta$$

but

$$\mu = \frac{F_{max}}{N} = \frac{W \sin \theta}{W \cos \theta}$$

By trigonometric definition

$$\tan \theta = \frac{\sin \theta}{\cos \theta}$$

The weight W cancels, and we have

$$\mu = \tan\theta$$
$$= \tan 31^\circ$$
$$\underline{\mu = 0.6}$$

This is an experimental method of determining the static coefficient of friction by measuring the sloping plane angle that causes impending motion.

EXAMPLE 7–3

The block shown (Figure 7–8) has a force of gravity of 80 N and a coefficient of static friction of 0.5 between it and the sloped surface. Determine the force of friction acting on the block.

There is no mention of impending motion. We will find that the actual friction force cannot be found using $F_{max} = \mu N$. The force of gravity is resolved into components (Figure 7–9).

FIGURE 7–8

$\Sigma F_y = 0$

$$N = 75.2 \text{ N}$$

The maximum available friction force would be

$$F_{max} = \mu N$$
$$= 0.5(75.2 \text{ N})$$
$$F_{max} = 37.6 \text{ N}$$

Free-Body Diagram of Block

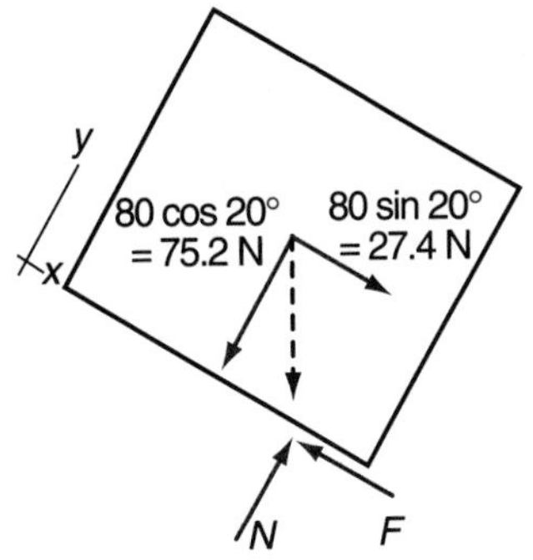

FIGURE 7–9

But from Figure 7–9, the actual friction force is found by

$\Sigma F_x = 0$

$$\underline{F = 27.4 \text{ N}}$$

We have an actual friction force and a maximum friction force since motion is not impending. **At any time up to the point of impending motion, the actual friction force will only be large enough to keep the object in static equilibrium.** At the point of impending motion, it will have reached its maximum value as given by $F_{max} = \mu N$. The friction force in this case is 27.4 N.

You could have also confirmed that motion was not impending by using Equation 7–2:

$$\mu = \tan\theta$$
$$0.5 = \tan\theta$$
$$\theta = 26.6^\circ$$

In this case the sloped surface would have to be raised from a 20° slope to a 26.6° slope before sliding would occur.

EXAMPLE 7–4

By means of a torque on member AB, the block that has a mass of 20.4 kg is in a state of impending motion up the plane (Figure 7–10). Determine the compressive load in member BC if the coefficient of static friction between the block and the plane is 0.2.

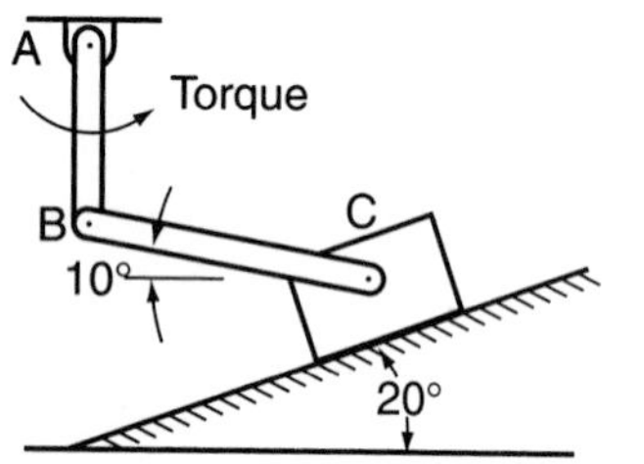

FIGURE 7–10

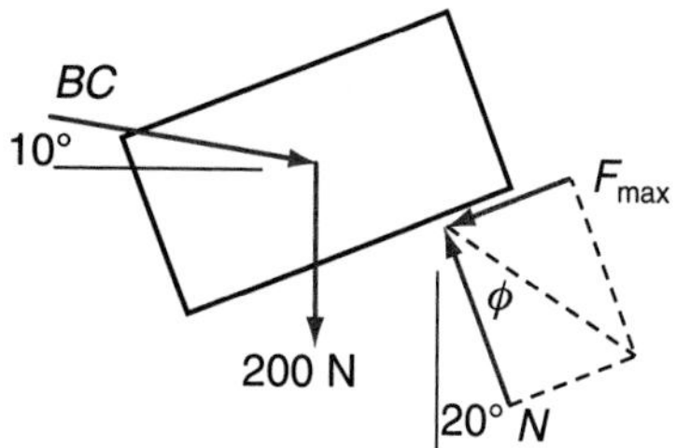

FIGURE 7–11

The block has a force of gravity of 9.81(20.4) = 200 N. As shown in Figure 7–11, the force of friction F_{max} opposes the motion and is acting down the slope. If we attempt to use this diagram, we will have to resolve some forces into components, and simultaneous equations will result.

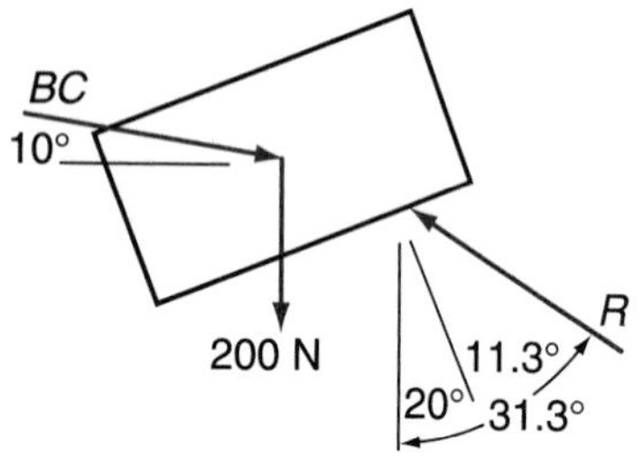

FIGURE 7–12

If F_{max} and N are combined into the resultant R, there are only three forces (Figure 7–12), and a vector triangle solution can be used. The friction angle ϕ is found from the equation

$$\mu = \tan\phi$$

$$0.2 = \tan\phi$$

$$\phi = 11.3°$$

Construct the vector triangle (Figure 7–13) by first drawing the 200-N vector, then vector BC, and finally vector R, which closes the triangle. Applying the sine law, we get

$$\frac{BC}{\sin 31.3°} = \frac{200\text{ N}}{\sin 48.7}$$

$$BC = 200\left(\frac{0.52}{0.751}\right)$$

$$\underline{BC = 138\text{ N } C}$$

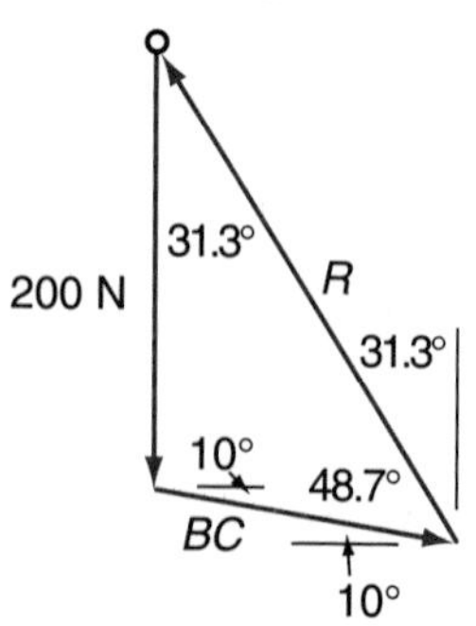

FIGURE 7–13

EXAMPLE 7–5

Determine the force P necessary to produce impending motion to the left (see Figure 7–14). The coefficient of static friction is 0.123.

The friction angle can be calculated using

$$\mu = \tan\phi$$
$$0.123 = \tan\phi$$
$$\phi = 7^\circ$$

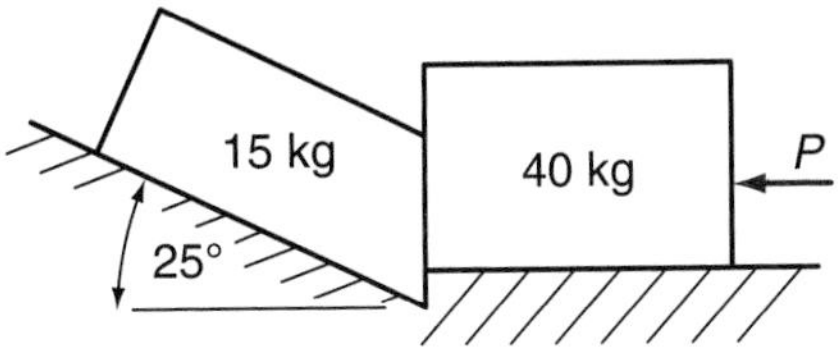

FIGURE 7–14

The friction force on the 15-kg block is downward because it is trying to move upward with respect to the 40-kg block.

We can now draw a free-body diagram of the 15-kg block (see Figure 7–15).

Drawing the vector triangle (Figure 7–16) allows us to solve for R_2.

Free-Body Diagram of 15-kg block

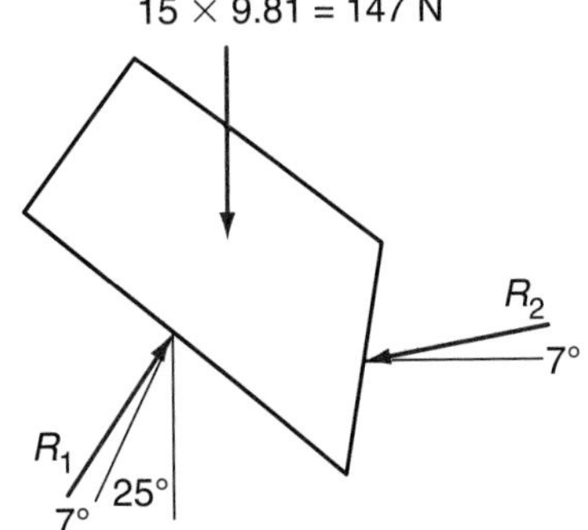

FIGURE 7–15

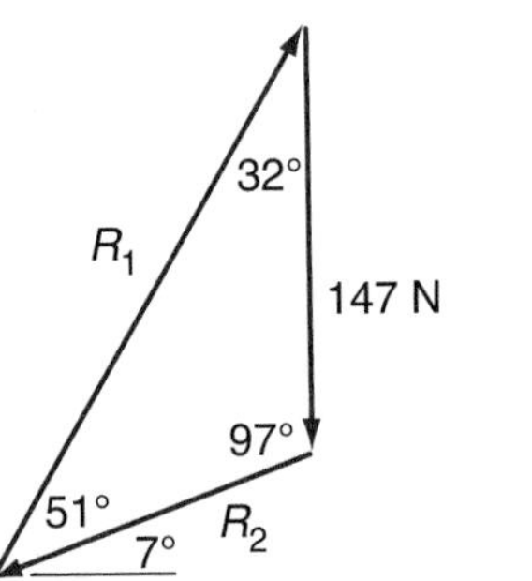

FIGURE 7–16

$$\frac{R_2}{\sin 32} = \frac{147\text{ N}}{\sin 51}$$

$$R_2 = 147\left(\frac{0.53}{0.777}\right)$$

$$R_2 = 100.3\text{ N}$$

Free-Body Diagram of 40-kg block

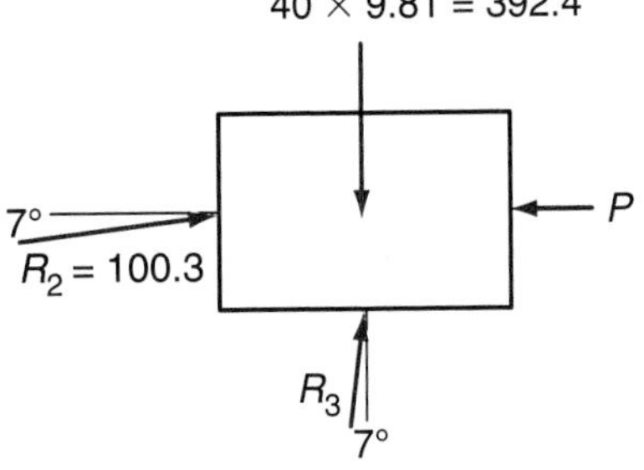

FIGURE 7–17

$\Sigma F_y = 0$ (Figure 7–17)

$$R_3\cos 7^\circ + 100.3\sin 7^\circ - 392.4\text{ N} = 0$$
$$R_3\cos 7^\circ = 392.4 - 12.2$$
$$R_3 = 383\text{ N}$$

$\Sigma F_y = 0$

$$100.3\cos 7^\circ + 383\sin 7^\circ - P = 0$$
$$\underline{P = 146\text{ N} \leftarrow}$$

EXAMPLE 7–6

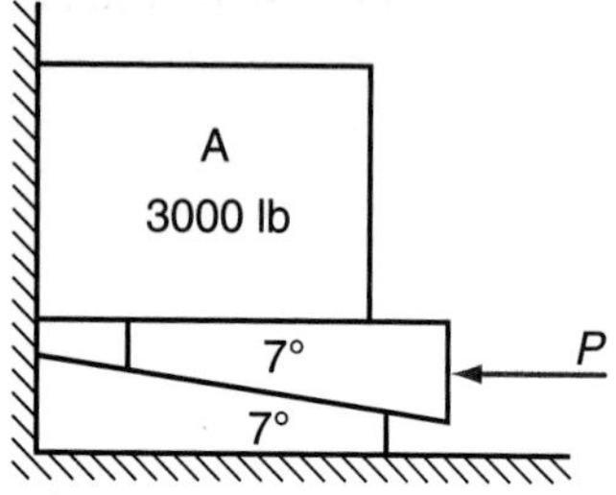

FIGURE 7–18

A 3000-lb weight is raised by two 7° wedges as shown in Figure 7–18. The coefficient of static friction is 0.23 for all surfaces. Determine the minimum force *P*.

Since the friction force and normal force will be combined into a resultant in all free-body diagrams, the friction angle is calculated first.

$$\mu = \tan \phi$$
$$0.23 = \tan \phi$$
$$\phi = 13^\circ$$

Since the only known force is the 3000-lb weight, we start with a free-body diagram of weight A (Figure 7–19). The corresponding vector triangle is shown in Figure 7–20.

Only the value of R_1 is required since this will allow us to draw a free-body diagram of the top wedge and to solve for *P*.

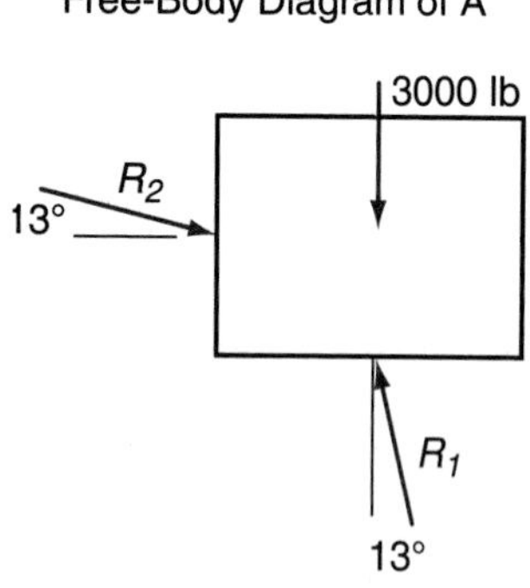

FIGURE 7–19

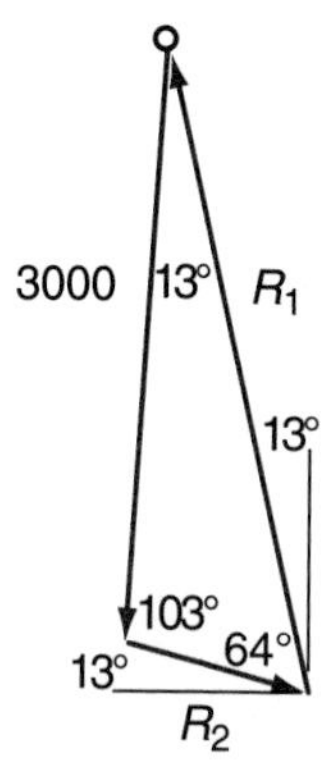

FIGURE 7–20

$$\frac{R_1}{\sin 103^\circ} = \frac{3000 \text{ lb}}{\sin 64^\circ}$$

$$R_1 = 3000\left(\frac{\sin 103^\circ}{\sin 64^\circ}\right)$$

$$= 3000\left(\frac{0.975}{0.899}\right)$$

$$R_1 = 3260 \text{ lb}$$

Now that the value of R_1 acting on block A (Figure 7–19) has been found, it is shown as an equal but opposite in direction force acting on the top wedge (Figure 7–21). From the vector triangle (Figure 7–22),

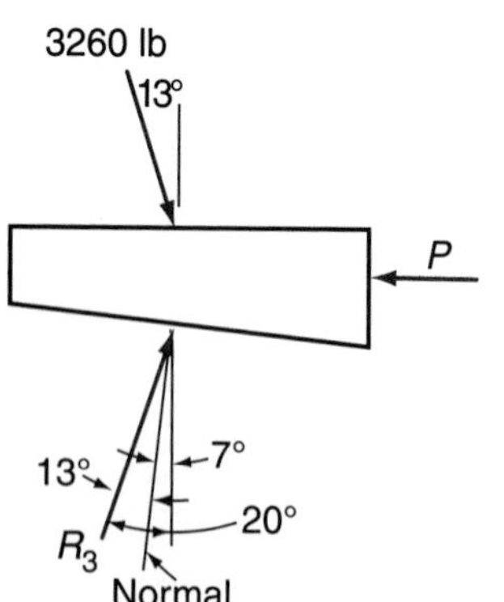

FIGURE 7–21

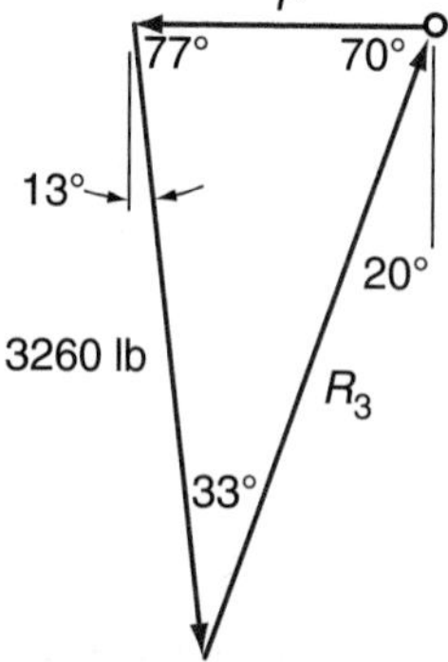

FIGURE 7–22

$$\frac{P}{\sin 33^\circ} = \frac{3260 \text{ lb}}{\sin 70^\circ}$$

$$P = 3260\left(\frac{0.545}{0.94}\right)$$

$$\underline{P = 1890 \text{ lb} \leftarrow}$$

EXAMPLE 7–7

Calculate the force of friction acting upon block A of the system in Figure 7–23. The coefficient of static friction for all surfaces is 0.3.

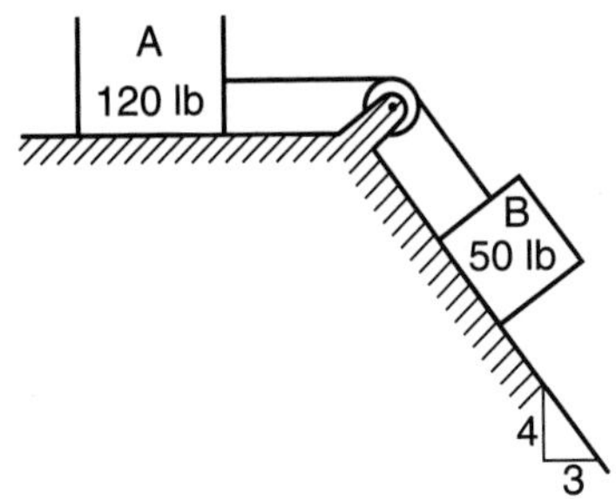

FIGURE 7–23

Note that there is no mention of impending motion. First check block B for impending motion (Figure 7–24).

Free-Body Diagram of B

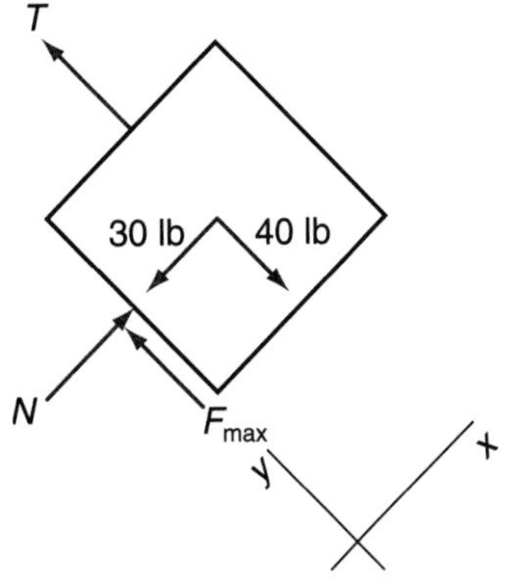

FIGURE 7–24

$\Sigma F_x = 0$

$$N = 30 \text{ lb}$$
$$F_{\text{max}} = \mu N$$
$$= 0.3(30 \text{ lb})$$
$$= 9 \text{ lb}$$

$\Sigma F_y = 0$

$$T + 9 \text{ lb} - 40 \text{ lb} = 0$$
$$T = 31 \text{ lb}$$

When $T = 31$ lb or less, block B will slide down the slope.

Now determine whether $T = 31$ lb acting on block A will cause impending motion (Figure 7–25).

Free-Body Diagram of A

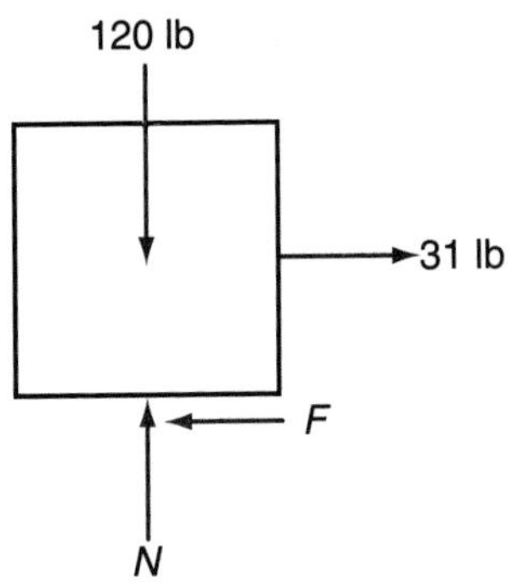

FIGURE 7–25

$\Sigma F_y = 0$

$$N = 120 \text{ lb}$$
$$F_{\text{max}} = \mu N \text{ for impending motion}$$
$$= 0.3(120 \text{ lb})$$
$$= 36 \text{ lb}$$

$\Sigma F_x = 0$

$$\underline{\text{actual } F = 31 \text{ lb} < 36}$$

Therefore, motion is not impending and the actual friction force on A is 31 lb.

EXAMPLE 7–8

Block C rests on rollers and supports bar AB as shown (Figure 7–26). The coefficient of friction for all surfaces is 0.4. Determine the force P and the friction force at B when slipping first occurs at either A or B (neglect the weight of AB).

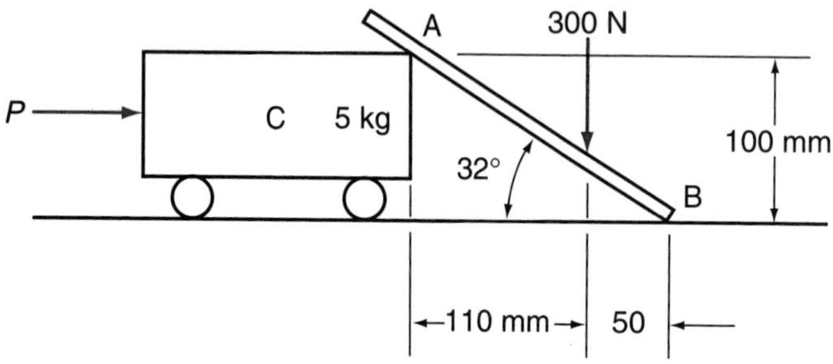

FIGURE 7–26

Free-Body Diagram of AB

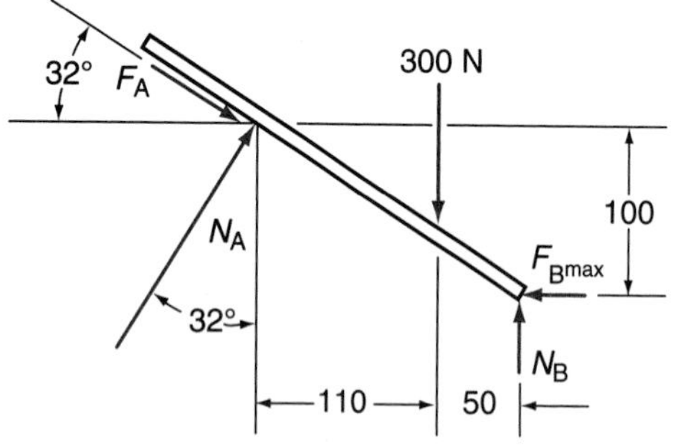

FIGURE 7–27

Referring to the free-body diagram of AB (Figure 7–27) we will have to check for slipping, first at B and then at A. A comparison of values will tell us where slipping actually occurs first.

Step 1. Assume slipping at B,

$\Sigma M_A = 0$

$$(300\text{ N})(110\text{ mm}) + F_B\max(100\text{ mm}) = (160\text{ mm})N_B$$

but for impending motion, $F_B\max = 0.4N_B$.

$$33{,}000 + 0.4N_B(100) = 160N_B$$

$$N_B = 275\text{ N}$$

Therefore,

$$F_B\max = 110\text{ N}$$

Step 2. Assume slipping at A ($F_A\max = 0.4N_A$):

$\Sigma M_B = 0$

$$N_A \cos 32°(160\text{ mm}) + N_A \sin 32°(100\text{ mm}) + 0.4N_A \cos 32°(100\text{ mm}) = 0.4N_A \sin 32°(160) + 300\text{ N}(50\text{mm})$$

$$N_A = 79.5\text{ N}$$

Therefore,

$$F_A\max = 31.8\text{ N}$$

Comparing the sum of the horizontal components of N_A and F_A will indicate where slipping occurs first.

$$\text{horizontal forces at A} = (79.5\text{ N})\sin 32° + (31.8\text{ N})\cos 32°$$
$$= 69\text{ N}$$

which is less than the 110 N at B.

Therefore, slipping will occur at A, and F_B will not be larger than 69 N since slipping at B is not impending.

Free-Body Diagram of C

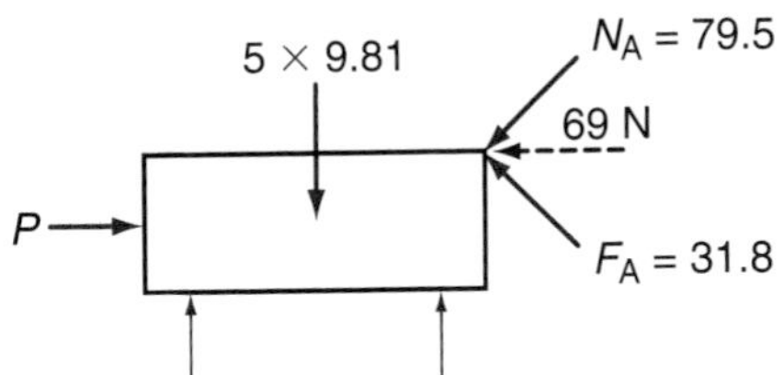

FIGURE 7–28

$$\underline{F_B = 69\text{ N} \leftarrow}$$

From the free-body diagram of C (Figure 7–28) the same horizontal sum of N_A and F_A (69 N) would give $\underline{P = 69\text{ N} \rightarrow}$.

7–5 BELT FRICTION

The two main assumptions in this section are:

1. A rope, cable, or flat belt is used. (Using a notched belt or a V-belt would require further analysis.)
2. Motion is impending—in the manner of a rope wound around a fixed cylinder so that the rope is starting to slip.

Consider the simplified case of a rope passed over a fixed cylindrical beam and a force P causing impending motion of the weight upward (Figure 7–29). Due to friction, force P will be larger than weight W. The rope has two different tensions, a large tension (T_L) and a small tension (T_S). These are shown in a free-body diagram of the portion of rope passing over the cylinder (Figure 7–30). Notice that the friction force is acting in a direction opposite to that of the impending motion of the rope with respect to the cylinder. A free-body diagram of the rope shows the force of friction on the rope due to the cylinder. There is impending motion of the cylinder with respect to the rope, so we show the force of the cylinder on the rope.

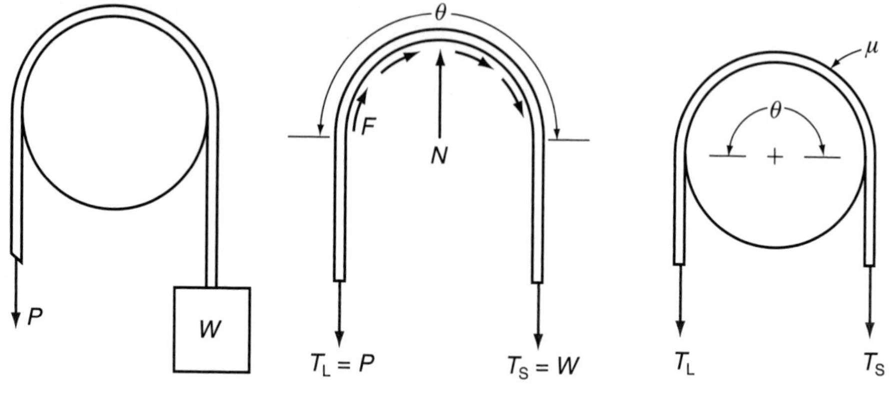

FIGURE 7–29 FIGURE 7–30 FIGURE 7–31

For impending slipping (Figure 7–31), the difference between T_L and T_S depends on:

1. Coefficient of friction μ
2. Angle of contact θ

This is expressed in the equation

$$\log_{10} \frac{T_L}{T_S} = 0.434\,\mu\theta \tag{7–3}$$

where T_L is the large tension, T_S the small tension, μ the static coefficient of friction, and θ the angle of contact in radians.

An alternative equation that uses natural logs is

$$\ln\left(\frac{T_L}{T_S}\right) = \mu\theta \tag{7–4}$$

Recall that if $\ln_e x = y$, then $x = e^y$. Therefore,

$$\frac{T_L}{T_S} = e^{\mu\theta} \tag{7–5}$$

where $e = 2.718$, the Naperian constant, and θ is the angle of contact in radians.

Your choice of which belt friction equation to use will depend on which level of mathematics you are most familiar with. Equation 7–4 is generally preferred. Choose one of the equations and use it consistently to avoid confusion.

EXAMPLE 7–9

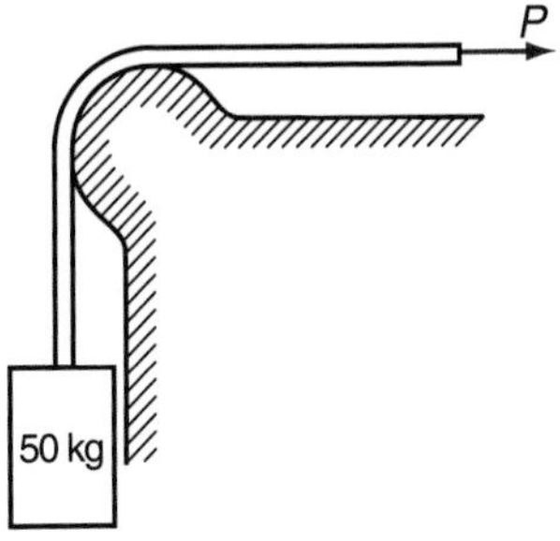

FIGURE 7–32

The 50-kg mass in Figure 7–32 has impending motion upward when $P = 750$ N. Determine the coefficient of friction between the rope and the cylinder.

Since P must overcome friction and lift the 50 kg:

$$T_L = P = 750 \text{ N}$$

and

$$T_S = 9.81(50) = 490 \text{ N}$$

$$\theta = 90° = \frac{\pi}{2} \text{rad}$$

$$\log_{10}\left(\frac{T_L}{T_S}\right) = 0.434\mu\theta \qquad \ln\left(\frac{T_L}{T_S}\right) = \mu\theta$$

$$\log_{10}\left(\frac{750 \text{ N}}{490 \text{ N}}\right) = 0.434\mu\left(\frac{\pi}{2}\right) \qquad \ln\left(\frac{750 \text{ N}}{490 \text{ N}}\right) = \mu\left(\frac{\pi}{2}\right)$$

$$\log_{10} 1.53 = 0.682\mu \qquad \ln 1.53 = 1.57\mu$$

$$0.185 = 0.682\mu \qquad 0.4253 = 1.57\mu$$

$$\underline{\mu = 0.27} \qquad \underline{\mu = 0.27}$$

EXAMPLE 7–10

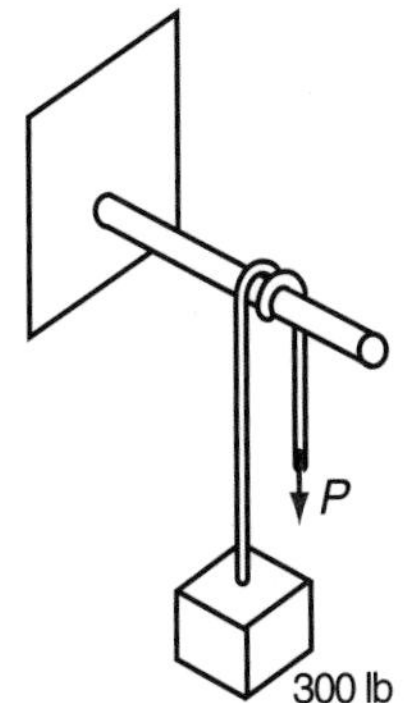

FIGURE 7–33

Determine the minimum force P required to hold the weight of 300 lb in Figure 7–33. There are $1\frac{1}{2}$ turns of rope about the horizontal rod, and the coefficient of static friction is 0.2.

Since motion of the weight is impending downward, friction aids P and

$$T_L = 300 \text{ lb}$$

$$T_S = P$$

$$\mu = 0.2$$

$$\theta = 1.5(2\pi)$$

Using the formula with logs to the base 10, we get

$$\log_{10}\left(\frac{T_L}{T_S}\right) = 0.434\mu\theta$$

$$\log_{10}\left(\frac{300 \text{ lb}}{P}\right) = 0.434(0.2)(1.5)(2\pi)$$

$$\log_{10}\left(\frac{300}{P}\right) = 0.818$$

$$\frac{300}{P} = 6.58$$

$$\underline{P = 45.6 \text{ lb} \downarrow}$$

The alternative method using natural logs is as follows:

$$\ln\left(\frac{T_L}{T_S}\right) = \mu\theta$$

where $\theta = 1.5(2\pi)$ radians.

$$\ln\left(\frac{300 \text{ lb}}{P}\right) = 0.2(1.5)(2\pi)$$

$$= 1.88$$

$$\frac{300}{P} = 6.58$$

$$\underline{P = 45.6 \text{ lb} \downarrow}$$

EXAMPLE 7–11

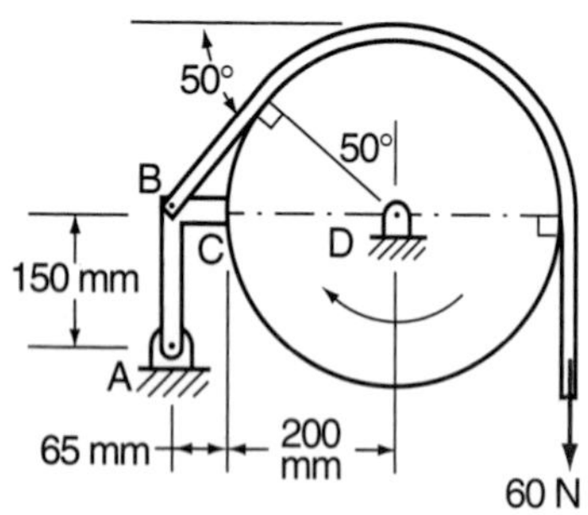

FIGURE 7–34

Lever AB and a belt work together to brake a wheel that is turning clockwise (Figure 7–34). The coefficient of kinetic friction (μ_k) for all surfaces is 0.3. What is the normal force between lever AC and the wheel if the belt is pulled with a force of 60 N?

The friction of the wheel on the belt aids the 60-N force; the small tension $T_S = 60$ N.

$$\ln\left(\frac{T_L}{T_S}\right) = \mu\theta$$

where $\theta = \dfrac{140}{360}\left(\dfrac{2\pi \text{ rad}}{1 \text{ rev}}\right) = 2.44$ rad

$$\ln\left(\frac{T_L}{60 \text{ N}}\right) = 0.3(2.44)$$

$$= 0.732$$

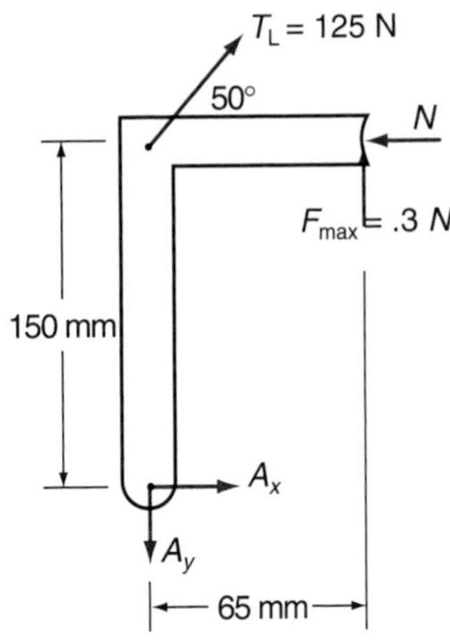

FIGURE 7–35

$$\left(\frac{T_L}{60\text{ N}}\right) = 2.08$$

$$T_L = 125\text{ N}$$

Use a free-body diagram of member AC (Figure 7–35), where

$$\mu_k = \frac{F_{max}}{\text{N}}$$

$$F_{max} = 0.3\text{ N}$$

$$\Sigma M_A = 0$$

$$150\text{ N} + 0.3\text{ N}(65) - (125\text{ N})\cos 50°(150) = 0$$

$$169.5\text{ N} = 12{,}050$$

$$\underline{\text{N} = 71.1\text{ N}}$$

EXAMPLE 7–12

Determine the minimum force P necessary for wheel A to have impending motion clockwise due to an applied torque of 24 N·m clockwise. The coefficient of static friction between the belt and wheel is 0.3 (Figure 7–36).

An initial check shows that the belt tensions T_L and T_S are the connecting link between the wheel and the lever. A free-body diagram of the wheel or the lever would result in two unknowns, T_L and T_S. Expect simultaneous equations. Beginning with a free-body diagram of the wheel (Figure 7–37),

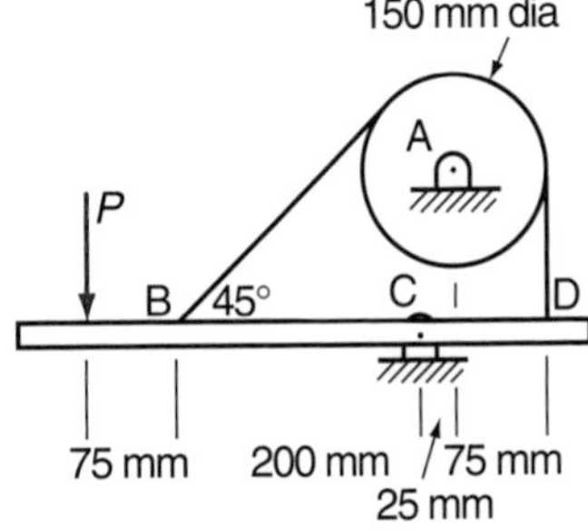

FIGURE 7–36

Free-Body Diagram of Wheel A

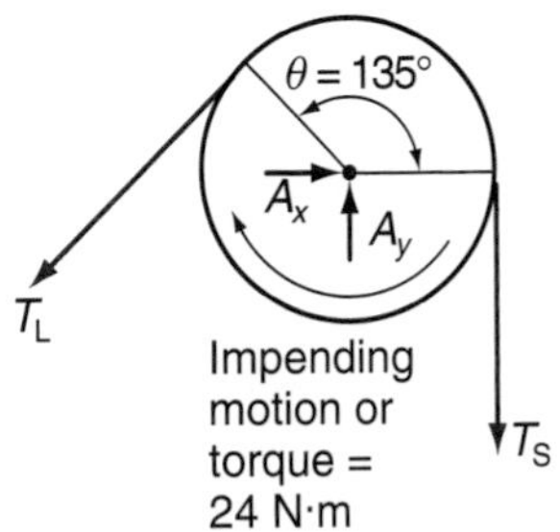

FIGURE 7–37

$$\ln\left(\frac{T_L}{T_S}\right) = \mu\theta$$

$$= 0.3\left(\frac{135}{180}\right)(\pi)$$

$$= 0.707$$

$$\frac{T_L}{T_S} = 2.03$$

$$T_L = 2.03T_S \qquad (1)$$

$$\Sigma M_A = 0$$

$$T_L(r) - T_S(r) - 24 \text{ N} \cdot \text{m} = 0$$
$$T_L(.075 \text{ m}) - T_S(.075 \text{ m}) = 24 \text{ N} \cdot \text{m}$$
$$T_L - T_S = 320 \qquad (2)$$

Substitute (1) into (2):

$$2.03T_S - T_S = 320$$
$$T_S = 310 \text{ N}$$
$$\text{Therefore, } T_L = 631 \text{ N}$$

Free-Body Diagram of the Lever

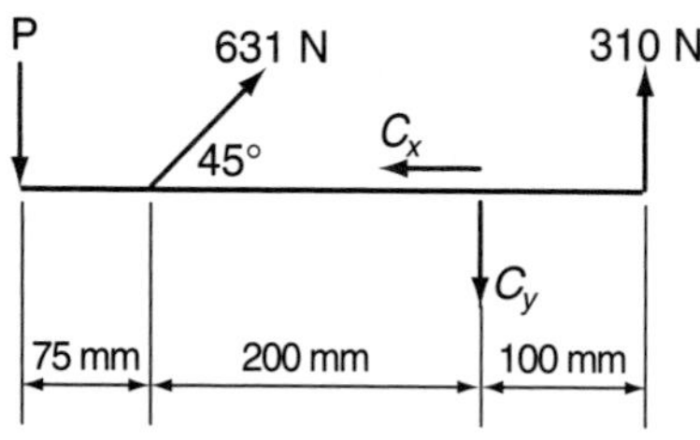

FIGURE 7–38

$$\Sigma M_C = 0 \quad \text{(Figure 7–38)}$$

$$-(631 \text{ N}) \sin 45°(200 \text{ mm}) + (310 \text{ N})(100 \text{ mm}) + P(275 \text{ mm}) = 0$$
$$275P = 89{,}200 - 31{,}000$$
$$\underline{P = 212 \text{ N} \downarrow}$$

HINTS FOR PROBLEM SOLVING

1. The direction of the friction force is always opposite to the impending motion of the object of which you have drawn a free-body diagram. For flat surfaces it is parallel to the surface, and for curved surfaces it is normal to the radius at the point of contact.
2. The friction angle ϕ is between the resultant and the normal force.
3. If motion is not impending, then $\mu \neq F_{max}/N$ and F and N must be treated as any other two unknowns.
4. Don't draw a vector triangle without first drawing a free-body diagram showing the same forces.
5. When solving for T_L or T_S in the equation $\ln(T_L/T_S) = \mu\theta$, keep in mind that (T_L/T_S) must be treated as a single term until both sides of the equation are antilogged. That is, $\ln T_L \neq T_S(\mu\theta)$. Also remember that θ is in radians.

PROBLEMS

APPLIED PROBLEMS FOR SECTIONS 7–1 TO 7–4

7–1. A horizontal force of 80 N is required to start a 30-kg block sliding along a horizontal surface. What is the coefficient of static friction?

7–2. Determine the mass of a block if a force of 15 N is required to start it sliding on a horizontal surface ($\mu = 0.40$).

7–3. Determine the value of P for impending motion (a) down the slope and (b) up the slope in Figure P7–3.

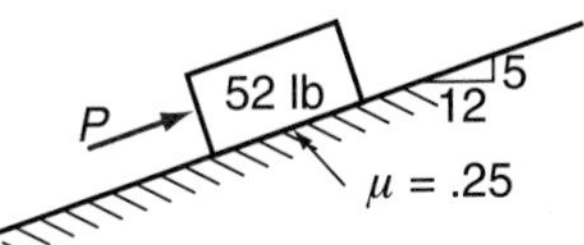

FIGURE P7–3

7–4. Determine the force P for impending motion up the plane shown in Figure P7–4.

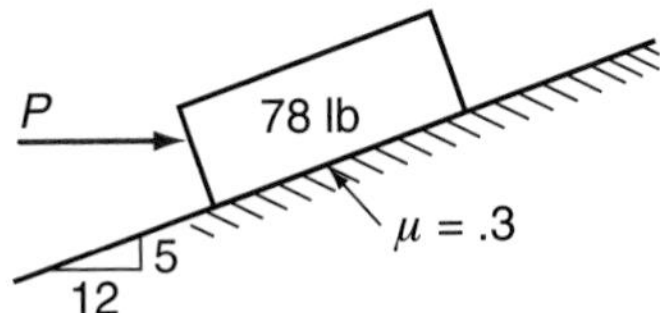

FIGURE P7–4

7–5. The block shown in Figure P7–5 has a mass of 34.7 kg. Determine the horizontal force P for impending motion down the plane.

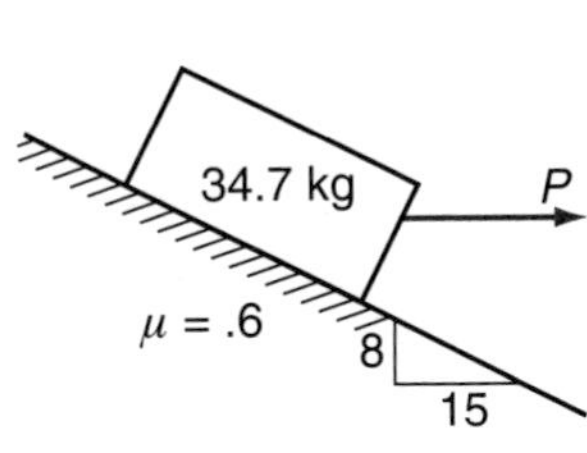

FIGURE P7–5

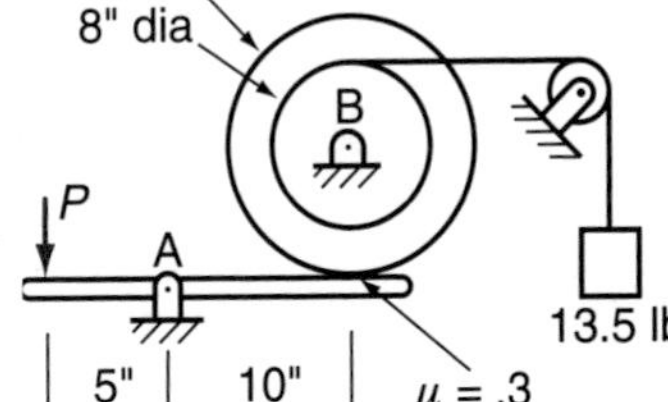

FIGURE P7–7

7–6. A 30-lb block has impending motion down a slope inclined 25° from the horizontal. What is the coefficient of static friction?

7–7. Determine the minimum force P that can hold the 13.5-lb weight shown in Figure P7–7.

7–8. A canned-goods dispenser has a vertical column of 10 cans, each having a mass of 1 kg (Figure P7–8). The cans fit loosely in the vertical slot, and the coefficient of friction is 0.2 for all surfaces. What force P is required to pull the bottom can out?

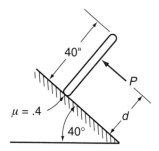

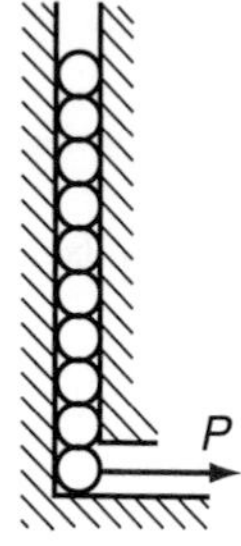

FIGURE P7–8

FIGURE P7–9

7–9. Determine the minimum force P and its location d so that the 100-lb bar shown in Figure P7–9 is in equilibrium with impending motion downward at the lower end.

7–10. For the system shown in Figure P7–9, determine the *maximum* force P and its location d for equilibrium and impending motion at the lower end of the bar.

7–11. Determine (a) the friction force acting on block A of the system shown in Figure P7–11 (note that we do not necessarily have impending motion) and (b) the maximum and minimum weight of A for impending motion.

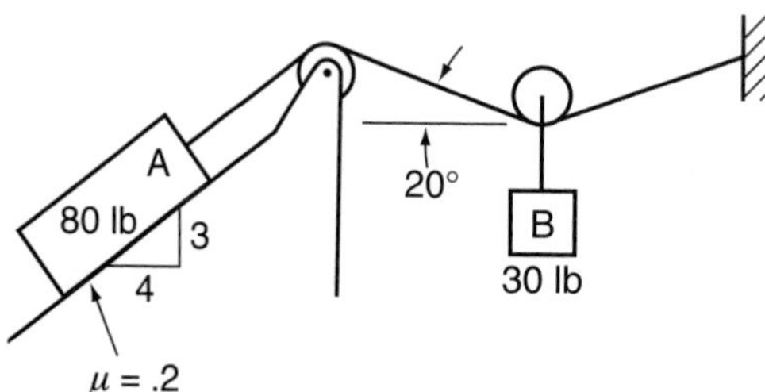

FIGURE P7–11

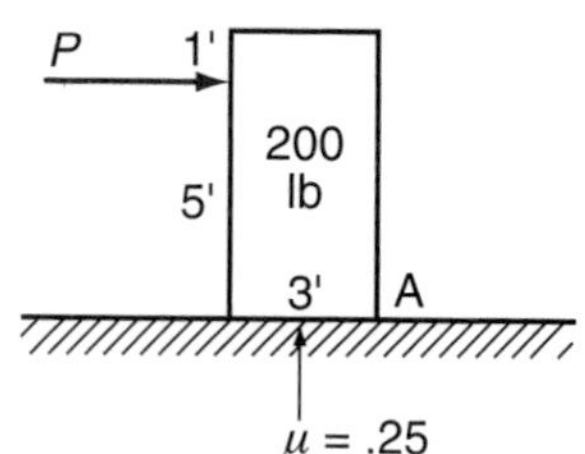

FIGURE P7–12

7–12. Determine the minimum force P that will cause impending motion of the block shown in Figure P7–12. Will the block tip or slide? (If tipping occurs about point A, both the normal force N and the friction force F will be acting through point A.)

7–13. If force P in Figure P7–12 is now applied to the middle of the left side of the block and is acting down to the right at 30° to the horizontal, find the values of P for both tipping and sliding.

7–14. Determine the minimum force P that will cause either tipping or sliding of the block shown in Figure P7–14. The coefficient of static friction is 0.7.

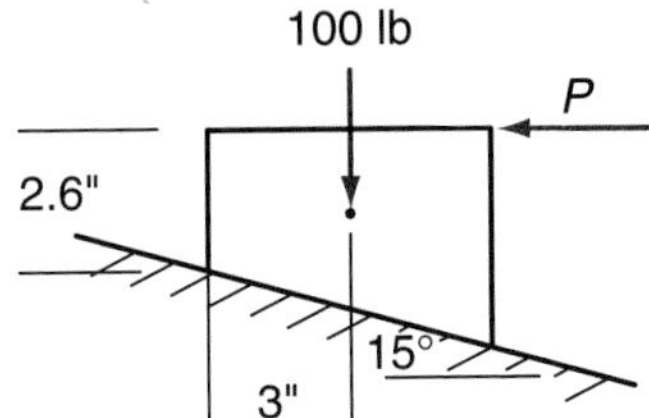

FIGURE P7–14

7–15. Will the block shown in Figure P7–15 tip or slide as angle θ is increased. Find the value of θ when this occurs.

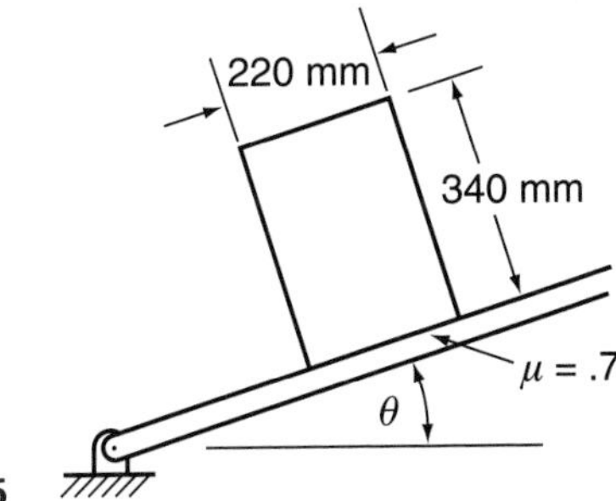

FIGURE P7–15

7–16. Each crate on the sloping chute in Figure P7–16 weighs 39 lb. The coefficient of static friction is 0.3. A spring-loaded roller at B applies a constant force of 54 lb. Determine the maximum number of crates that can be held on the chute before the bottom crate is pushed by the top crates onto the rollers at A.

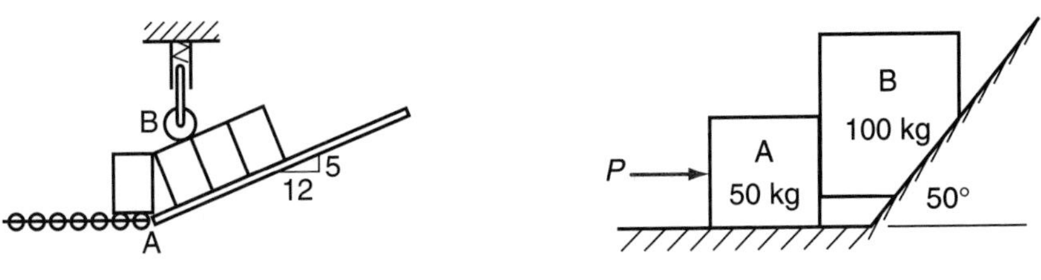

FIGURE P7–16

FIGURE P7–17

7–17. Determine force P for impending motion to the left (Figure P7–17) if the coefficient of friction for all surfaces is 0.15.

7–18. Determine the minimum force P that will pull wedge A to the left in Figure P7–18. Neglect the weight of the wedge.

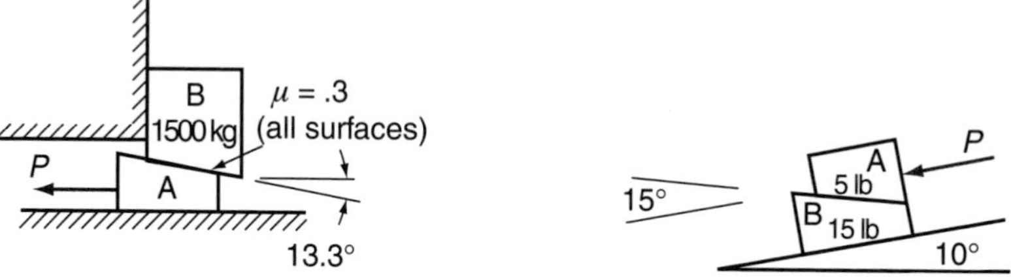

FIGURE P7–18

FIGURE P7–19

7–19. Determine the force P that will cause impending motion in Figure P7–19. The coefficient of static friction for all surfaces is 0.25.

7–20. Determine the force P necessary to cause impending motion of the 40-lb block up the slope in Figure P7–20. The coefficient of static friction of all surfaces is 0.1.

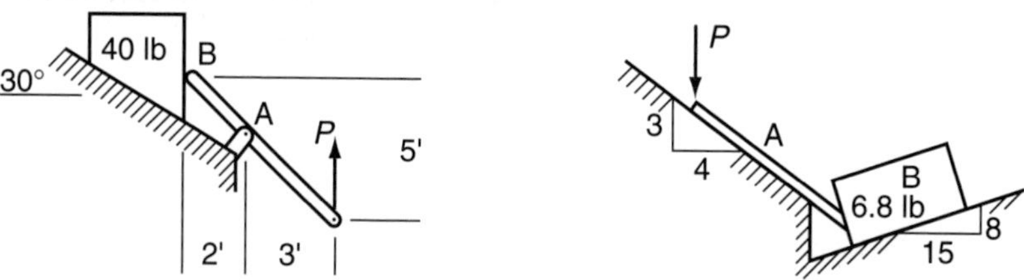

FIGURE P7–20

FIGURE P7–21

7–21. Neglect the weight of A and determine the vertical force P necessary to cause impending motion of block B to the right in Figure P7–21. The coefficient of friction for all surfaces is 0.1.

7–22. Slider A is moved to the right against a compressive spring force of 100 N by means of the eccentric lever B and force P (Figure P7–22). If the coefficient of static friction is 0.2, determine force P. (Neglect the weight of A.)

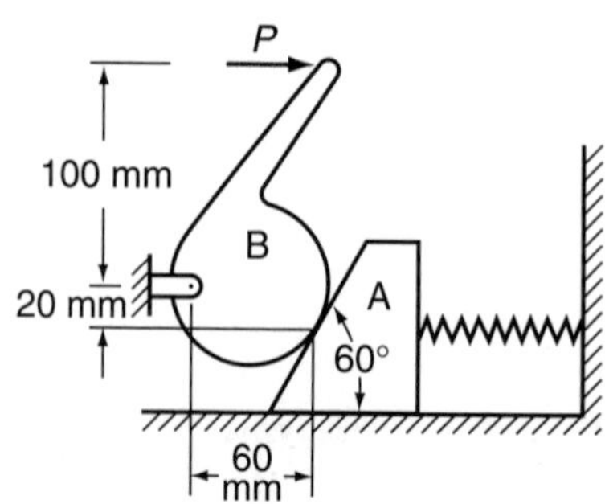

FIGURE P7–22

7–23. Determine the tension T required to produce impending motion of block A (Figure P7–23).

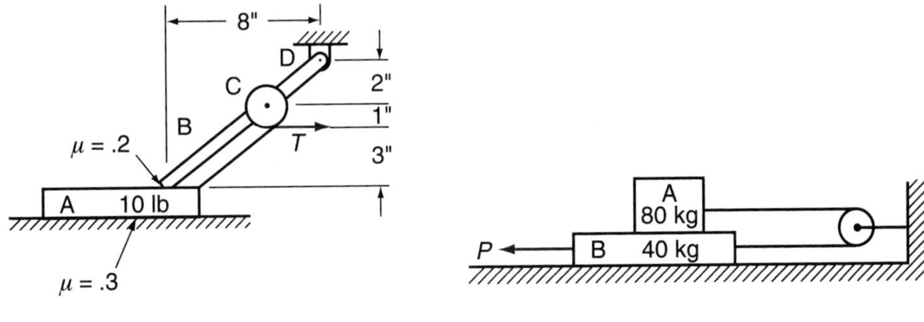

FIGURE P7–23

FIGURE P7–24

7–24. Knowing that $\mu = 0.2$ at all surfaces of contact, determine the magnitude of the force P required to move the 40-kg plate B to the left (Figure P7–24). (Neglect pulley friction.)

7–25. Blocks A and B have masses of 60 kg and 50 kg, respectively (Figure P7–25). Calculate the coefficient of friction between block B and the sloped surface if there is impending motion of B down the slope.

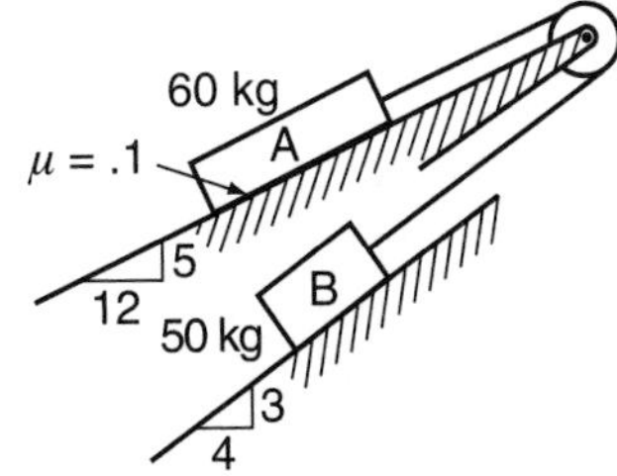

FIGURE P7–25

7–26. A 20-lb uniform ladder rests against a smooth wall at an angle θ between the ladder and the wall. The coefficient of friction between the ladder and the floor is 0.4. If a 50-lb person is at the top of the ladder, determine the largest angle θ for no slipping to occur.

7–27. Determine the normal forces and friction forces acting on member AC (Figure P7–27) if it has impending motion when $P = 700$ N. The coefficient of friction at B is 0.3. Determine μ at A.

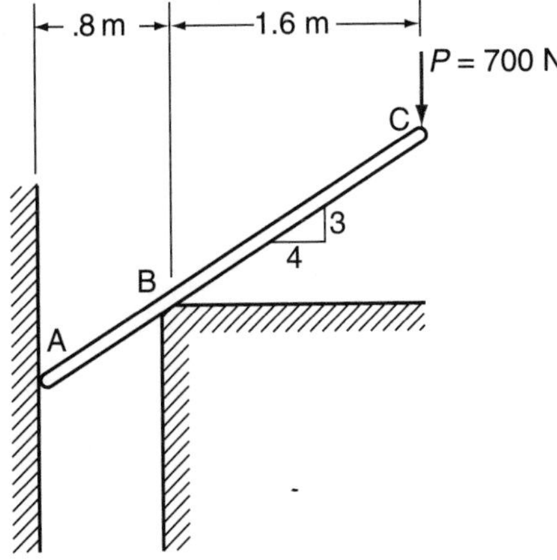

FIGURE P7–27

7–28. A small tractor with 40-in. 1 diameter drive wheels is driven up an incline. If the weight on each wheel is 300 lb and the coefficient of friction is 0.3, determine the maximum slope before slippage occurs. (Prove your answer with a complete free-body diagram.)

7–29. A 200-lb load is applied to the beam system in Figure P7–29, causing a tension of 160 lb in spring DC. Will slipping occur at A if the coefficient of static friction at this point is 0.2? What is the actual value of the friction force at A? What is the minimum allowable coefficient of static friction that will prevent motion at point A?

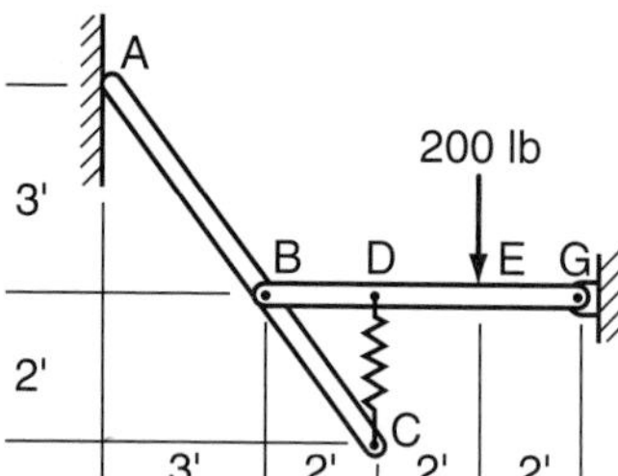

FIGURE P7–29

7–30. A slip clutch with friction plates shown in Figure P7–30 is to transmit a torque of 300 N · m. Assuming that the friction force acts at the average diameter, determine the spring force required to push the plates together.

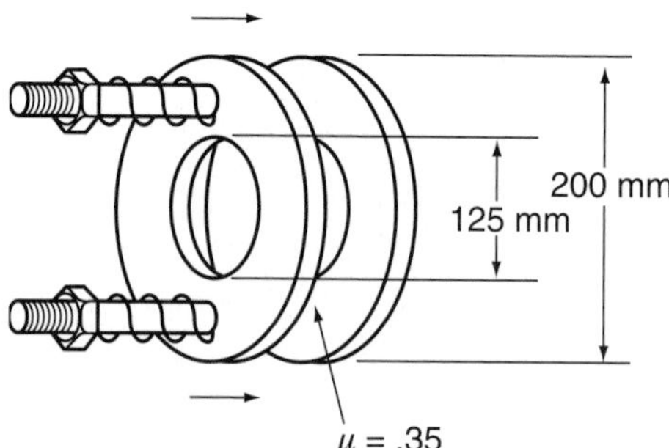

FIGURE P7–30

7–31. The 160-kg block of granite shown in Figure P7–31 has impending motion of tipping to the right due to force *P*. If the coefficient of friction $\mu = 0.35$, determine force *P*.

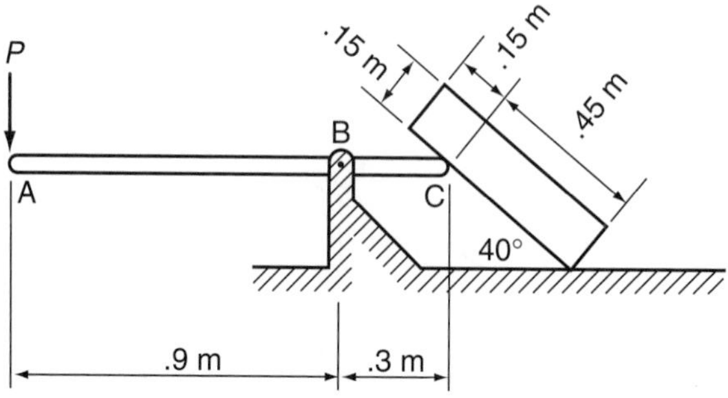

FIGURE P7–31

7–32. The roller weighs 6 lb and member AB, pinned to the roller, weighs 30 lb (Figure P7–32). The coefficient of friction at B is $\mu = 0.5$. Determine the minimum coefficient of friction between the roller and the surface such that the roller will roll clockwise with no slipping at B.

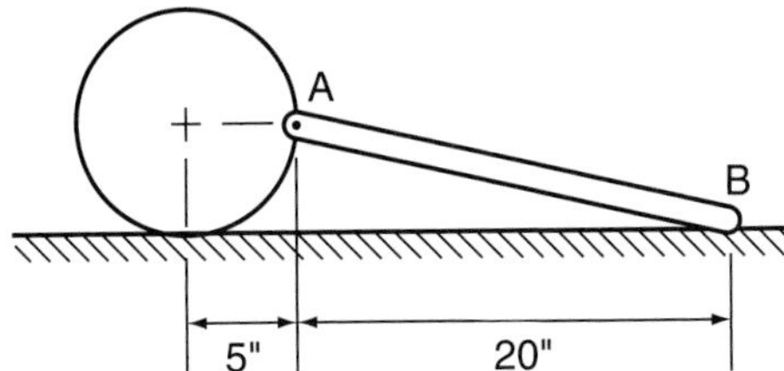

FIGURE P7–32

7–33. The bar shown in Figure P7–33 has a mass of 10 kg. The coefficient of friction for all surfaces is 0.4. Determine the force P required to produce impending motion.

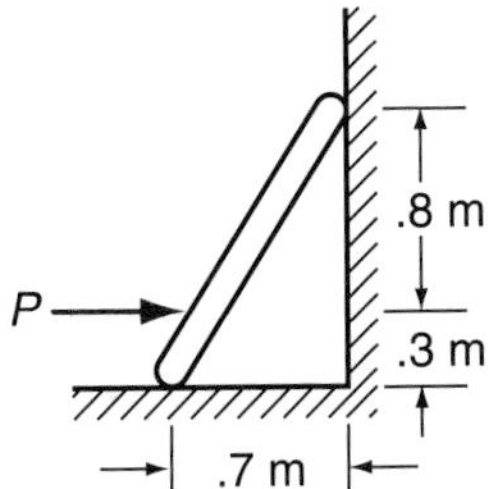

FIGURE P7–33

7–34. Determine force P for impending motion of the system shown in Figure P7–34. The block weighs 20 lb, and the weight of bar AB can be neglected.

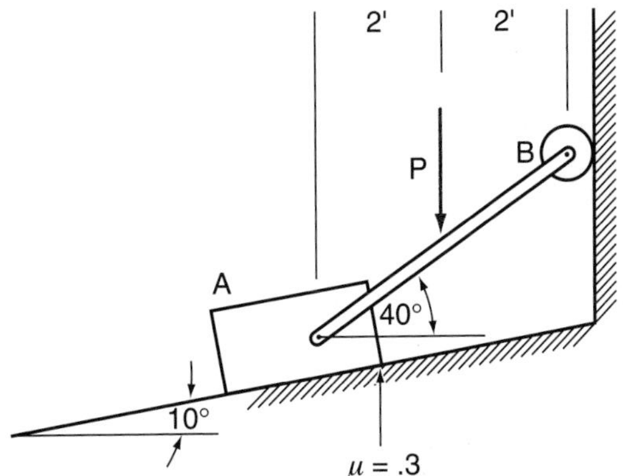

FIGURE P7–34

7–35. Determine the tension T required to cause impending sliding at position A and impending tipping at position B (Figure P7–35). What is the distance *d* at which tipping occurs?

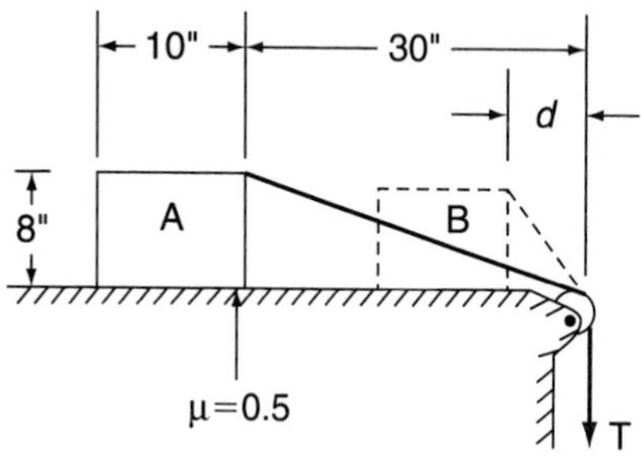

FIGURE P7–35

FIGURE P7–36

7–36. Will the 40-kg cylinder in Figure P7–36 slip at either A or B? If it does slip, determine the actual friction forces at each of these points. The coefficient of static friction is 0.25 at A and 0.18 at B.

7–37. Member AB with a mass of 7 kg and member BC with a mass of 20 kg have impending motion at the position shown in Figure P7–37. Member BC is 1 m long and rests on a fixed shaft at D. Determine the coefficient of static fraction and the pin reactions at B.

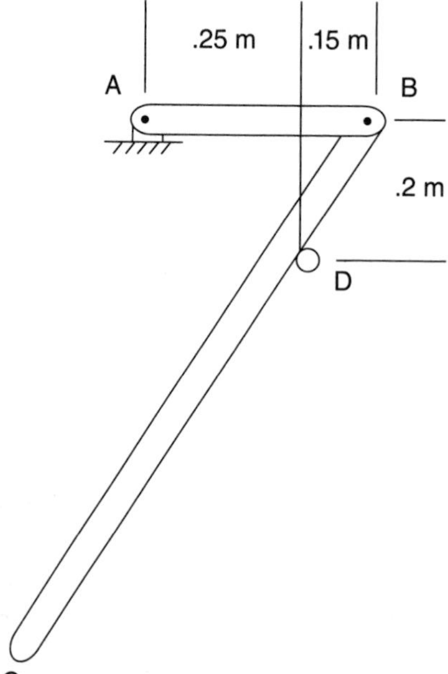

FIGURE P7–37 C

APPLIED PROBLEMS FOR SECTION 7–5

7–38. Determine the force *T* that will produce impending motion upward of the 500-lb weight in Figure P7–38. (Assume a coefficient of friction of 0.23.) Also find the minimum force *T* required to hold the 500-lb weight.

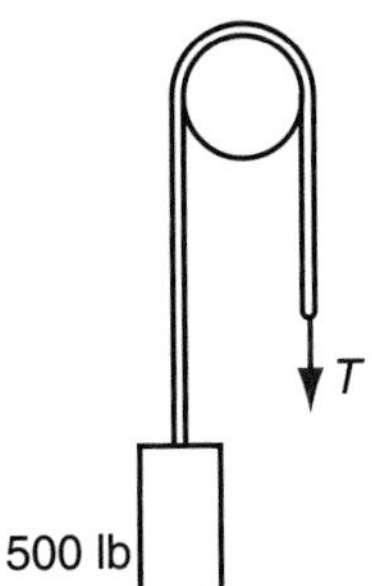

FIGURE P7–38

7–39. Determine the mass that can be lifted by a rope with three-quarters of a turn around a fixed horizontal shaft ($\mu = 0.3$) if a force of 8 kN is applied to its other end.

7–40. When mass B is 340 kg, it produces impending motion of block A (Figure P7–40). What is the coefficient of static friction between the rope and the fixed shaft?

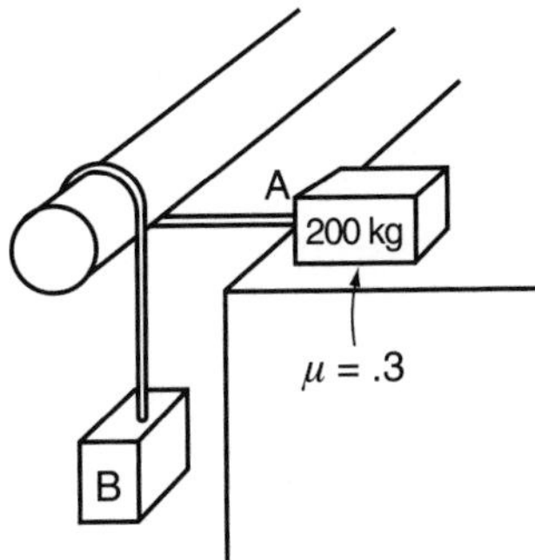

FIGURE P7–40

7–41. How many turns of rope around a horizontal fixed shaft are required if a force of 80 lb on the end of the rope is to hold a weight of 720 lb? Assume the coefficient of static friction to be 0.35.

7–42. Determine the force *P* required to produce impending motion upward of the 90-kg block shown in Figure P7–42. Pulley A is frictionless and cylinder B is fixed.

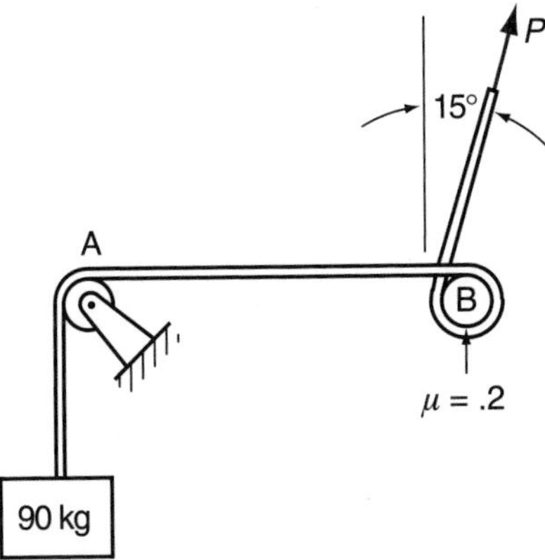

FIGURE P7–42

7–43. The coefficient of static friction for all surfaces in Figure P7–43 is 0.16. Determine weight B that will cause impending motion of block A to the right.

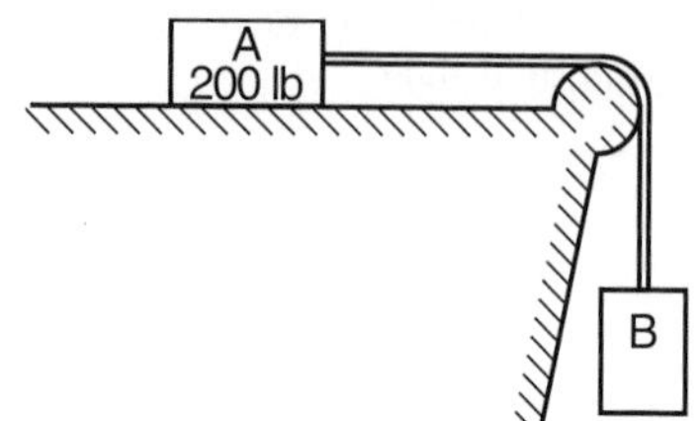

FIGURE P7–43

7–44. Block A in Figure P7–44 has a mass of 500 kg and is in a state of impending motion down the slope. Determine the mass of block B.

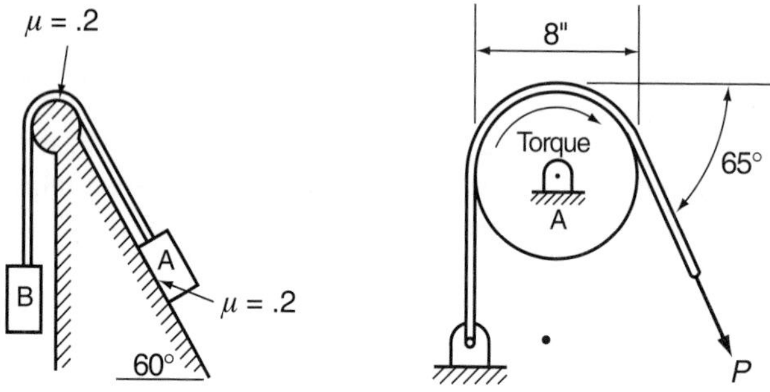

FIGURE P7–44

FIGURE P7–45

7–45. Shaft A has impending motion when a torque of 100 lb-ft is applied (Figure P7–45). If the coefficient of friction is 0.2, determine force *P*.

7–46. Pulley A drives pulley B in Figure P7–46. Assume the idler pulley tightens the belt sufficiently to produce 180° of belt contact on pulley A. If the belt has tensions of 160 lb and 50 lb, determine the torque being transmitted at A and the minimum coefficient of belt fraction.

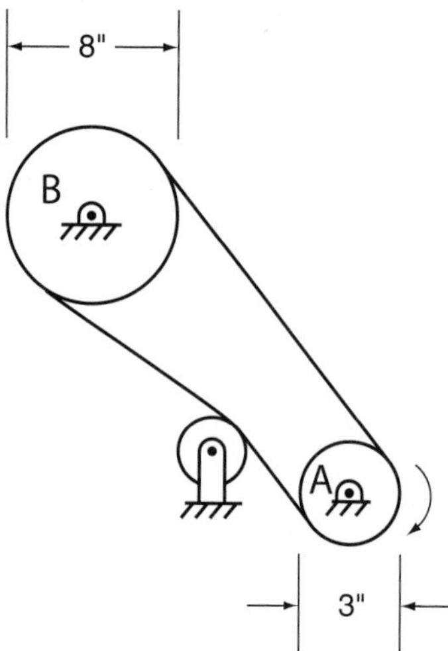

FIGURE P7–46

7–47. The coefficient of static friction is 0.4 between all surfaces in Figure P7–47. Determine the angle θ_1 at which a force of 375 N will cause impending motion of the 40-kg block.

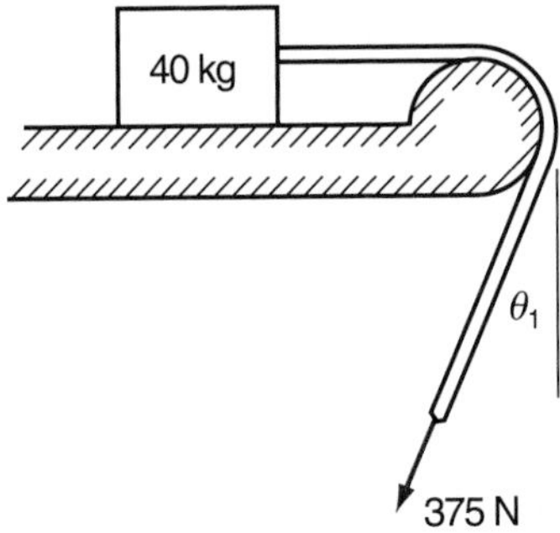

FIGURE P7–47

7–48. Block B of Figure P7–48 has impending motion upward due to cable force $P = 140$ lb. Cylinder A is fixed, the spring force is 30 lb, and μ for all surfaces is 0.2. Determine the angle ϕ.

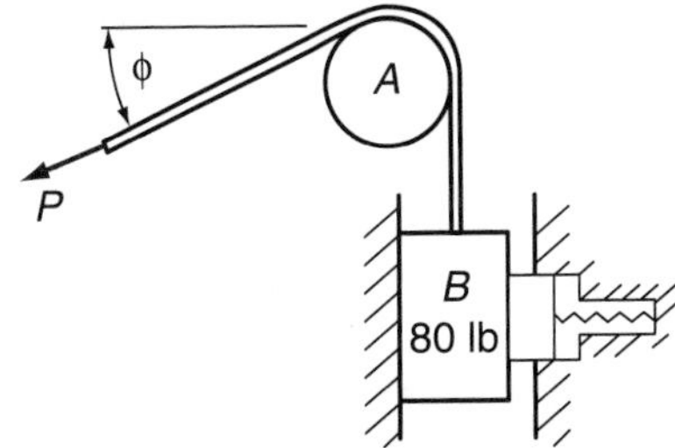

FIGURE P7–48

7–49. The tension in the top portion of the conveyor belt in Figure P7–49 is 1600 N. The coefficient of static friction between the belt and drum A is 0.3. What is the maximum torque that drum A may deliver before slipping occurs?

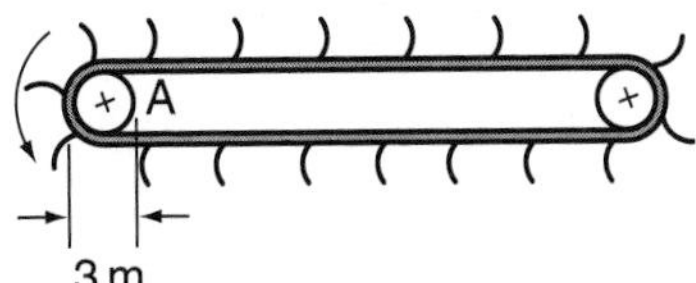

FIGURE P7–49

7–50. A rotating wheel is braked by the rope shown in Figure P7–50. (The coefficient of kinetic friction is 0.12.) Determine the decelerating torque applied to the wheel if $T = 15$ lb.

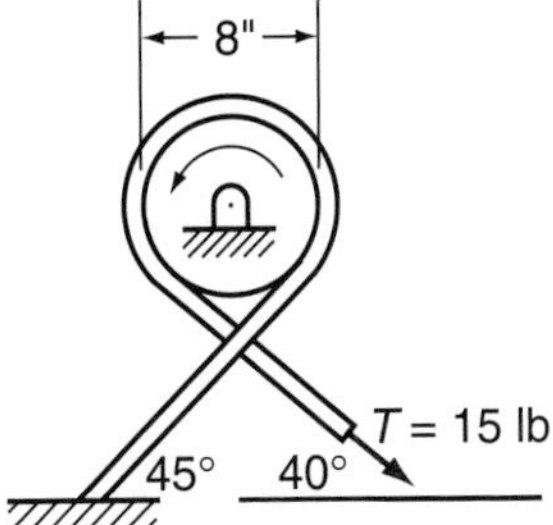

FIGURE P7–50

7–51. A rope tow for skiers has a rotating capstan at A (Figure P7–51) with one turn of rope on it and a coefficient of friction of 0.3.

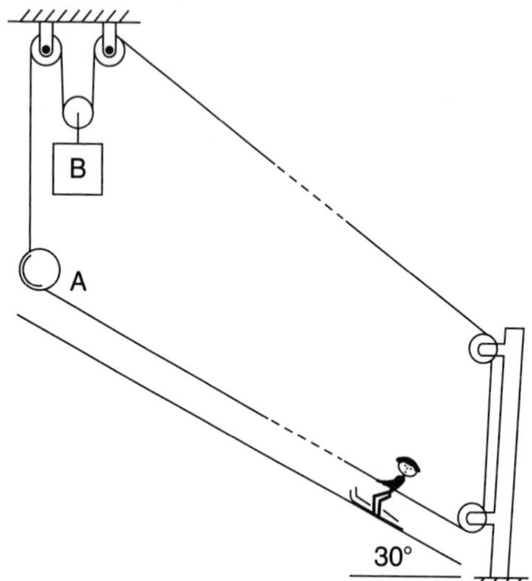

FIGURE P7–51

There are 40 skiers with an average weight of 150 lb being towed up the hill. The coefficient of friction between their skis and snow is 0.05. Determine the minimum weight of counterbalance B.

7–52. Shaft A is driven counterclockwise and has two turns of rope wound on it (Figure P7–52). The rope slips on the shaft if no tension T is applied.
Determine the minimum T required to cause the rotating shaft to pull block B up the slope.

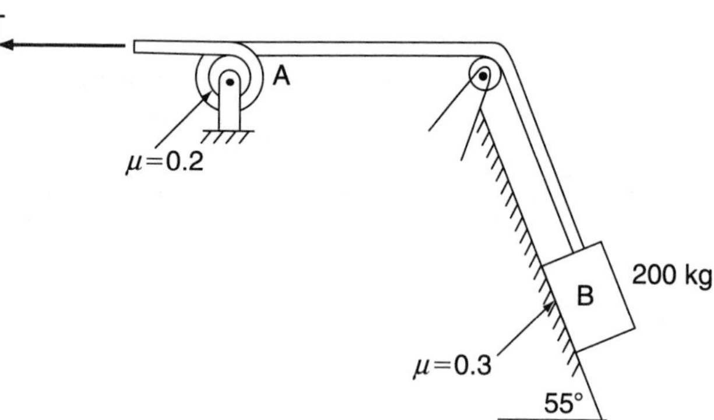

FIGURE P7–52

7–53. Cylinders A and B in Figure P7–53 are fixed and the coefficient of friction is $\mu = 0.3$ for all surfaces. There is less than one turn of cable about cylinder A. Determine the maximum and minimum values of W for equilibrium to exist.

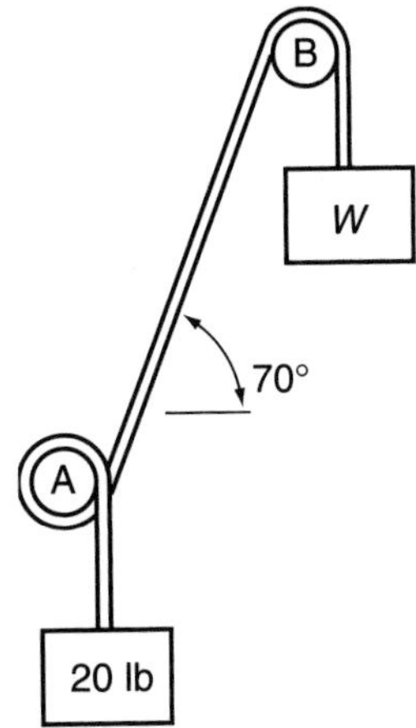

FIGURE P7–53

7–54. The structural shape in Figure P7–54 pivots at point A as it is lowered by slipping on the rope. The structural shape weighs 300 lb, and this weight may be assumed at the center of gravity as shown. If the coefficient of static friction is 0.38, determine the tension T if there is to be impending motion of the structural shape downward.

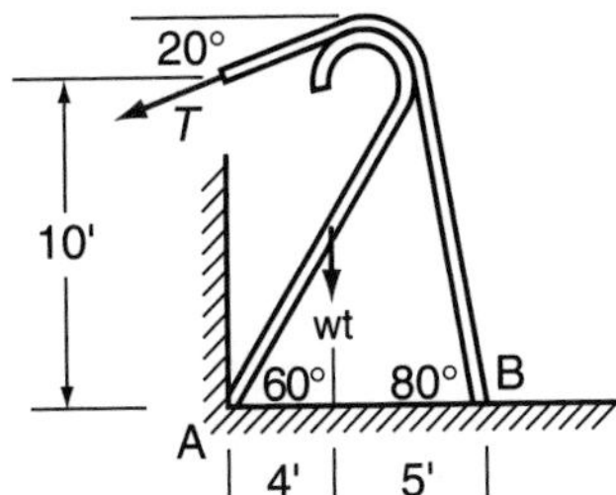

FIGURE P7–54

7–55. The system shown in Figure P7–55 has impending motion of C to the left. Lever AB weighs 30 lb. The coefficient of static friction is 0.3 for all surfaces. Determine force P. (The weight of AB acts through its center of gravity.)

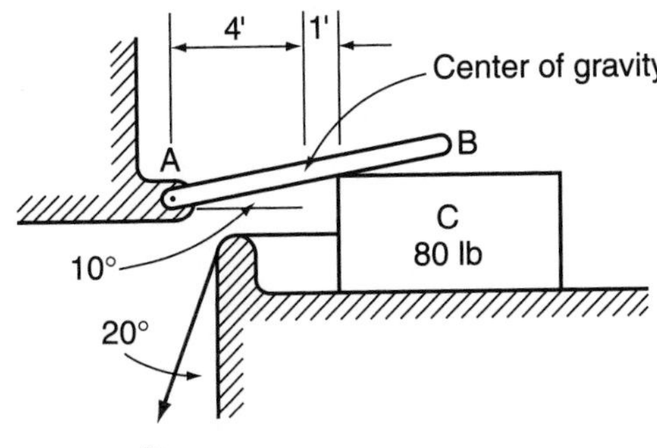

FIGURE P7–55

7–56. For the mechanism shown in Figure P7–56, where does slippage occur first and at what value of *T*? The coefficient of friction is 0.2 for all surfaces. (The 80-lb weight has impending motion upward.)

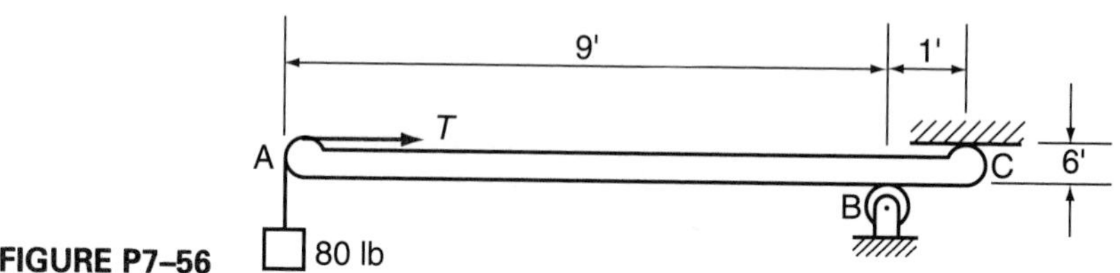

FIGURE P7–56

7–57. Bar AB has a mass of 10 kg (Figure P7–57). Determine if slipping occurs at A or B, and if so, what friction forces are present.

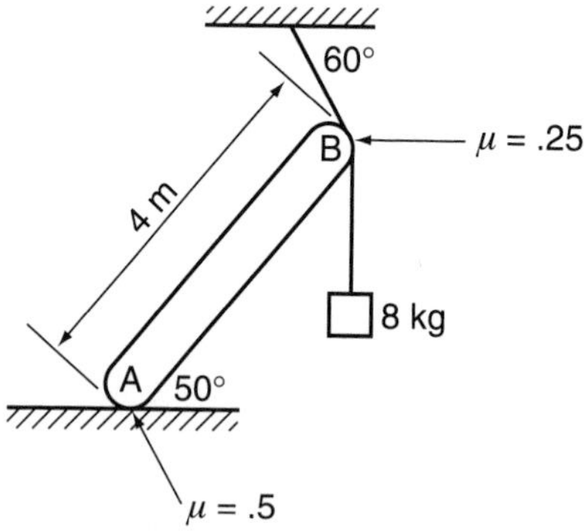

FIGURE P7–57

7–58. Cylinder A with a mass of 25 kg has a fixed hub about which there is one turn of rope as shown in Figure P7–58. (The coefficient of friction is 0.3 for all surfaces.) Determine the minimum force *P* that will prevent slipping of the rope on the hub. Will slipping occur at point A?

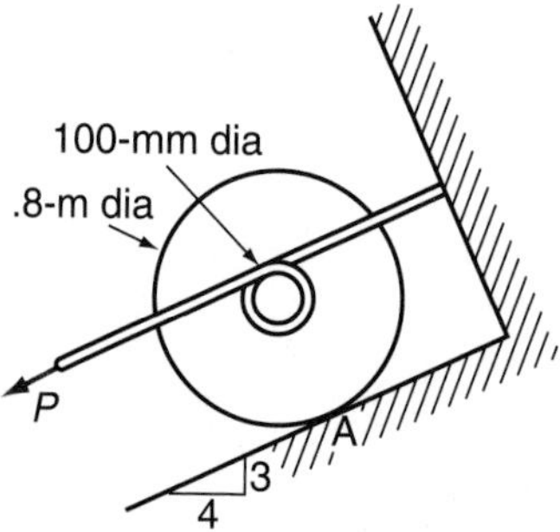

FIGURE P7–58

REVIEW PROBLEMS

R7–1. Will a mass of 10 kg slide—due to its own weight—down a slope inclined 30° to the horizontal if the coefficient of static friction is 0.6?

R7–2. Determine the force *P* that will cause the impending motion of block A (Figure RP7–2). What is the friction force acting on the bottom of block B?

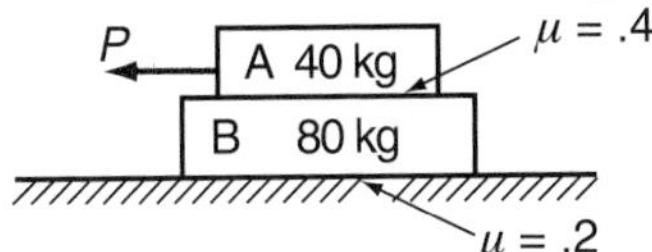

FIGURE RP7–2

R7–3. Block A in Figure RP7–3 has a mass of 4 kg and is on the verge of tipping as it begins to slide due to force *P*. Determine the coefficient of static friction between the block and the horizontal surface.

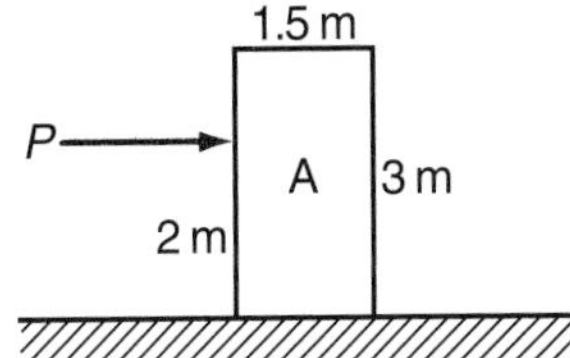

FIGURE RP7–3

R7–4. Neglecting the weight of wedge A, determine the force *P* necessary for impending motion of B to the left (Figure RP7–4). The coefficient of static friction for all surfaces is 0.4.

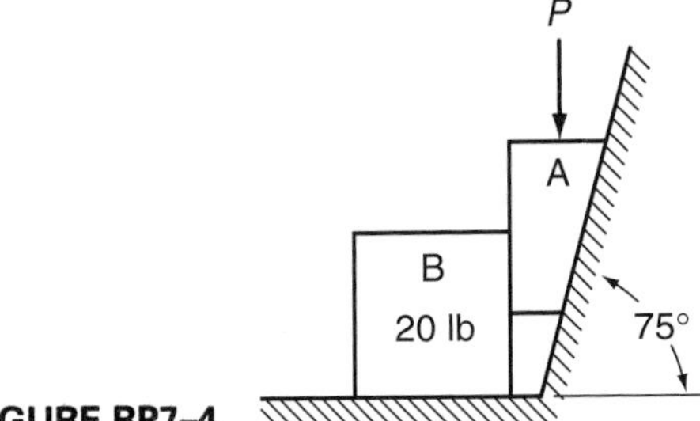

FIGURE RP7–4

R7–5. The coefficient of static friction is 0.45 for all surfaces in Figure RP7–5. Determine the mass of cylinder B that will cause impending motion of both A and B down the plane.

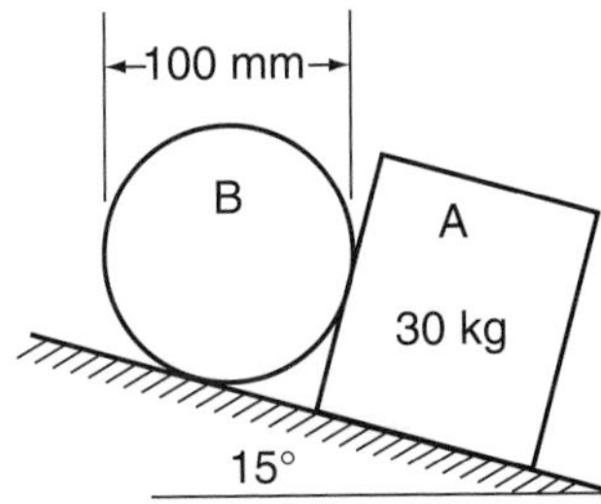

FIGURE RP7–5

R7–6. Block A weighs 80 lb and block B weighs 60 lb. Determine if the system shown in Figure RP7–6 will move when released in the position shown and also what friction force is present at B.

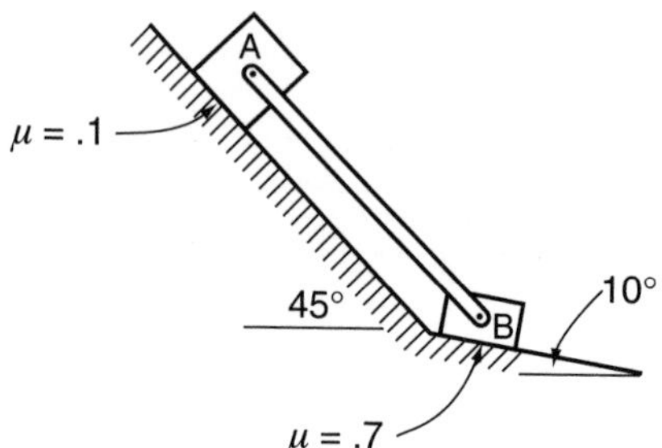

FIGURE RP7–6

R7–7. The drum shown in Figure RP7–7 rotates at 200 rpm and requires a braking torque of 1400 N·m.

This occurs when force $P = 0$, allowing the spring force to be applied. The coefficient of kinetic friction is 0.6. Determine the spring force required. Which pad wears the most?

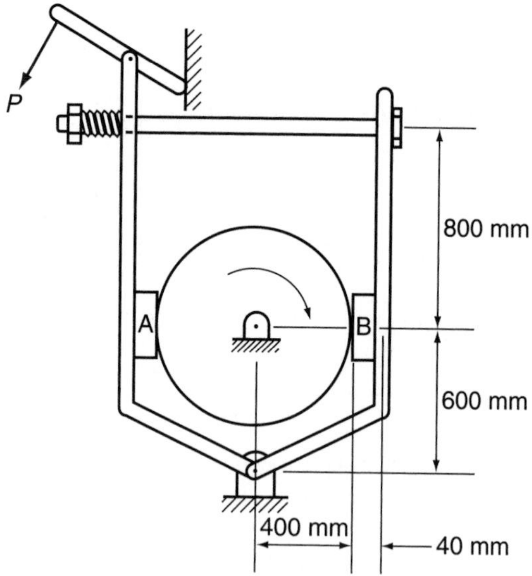

FIGURE RP7–7

R7–8. The bar shown in Figure RP7–8 rests on a wedge, has a mass of 153 kg, and has impending motion downward. Neglecting the weight of the wedge, determine whether slipping occurs at surface B or C. What is the friction force at surface B or C where slipping does not occur?

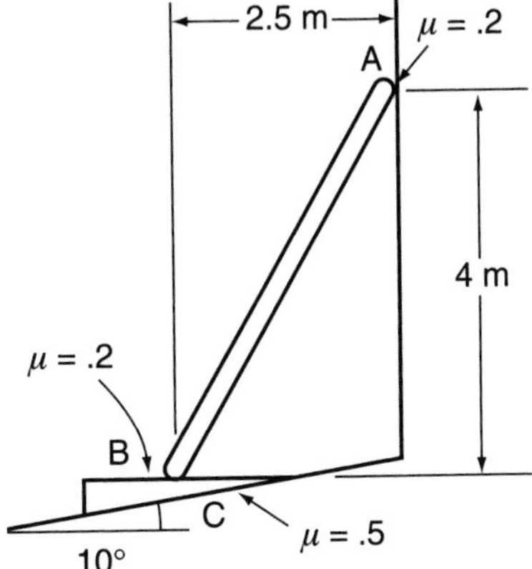

FIGURE RP7–8

R7–9. Block A of the system shown in Figure RP7–9 has impending motion up the slope. Determine (a) the minimum coefficient of belt friction and (b) the torque being applied to drum C.

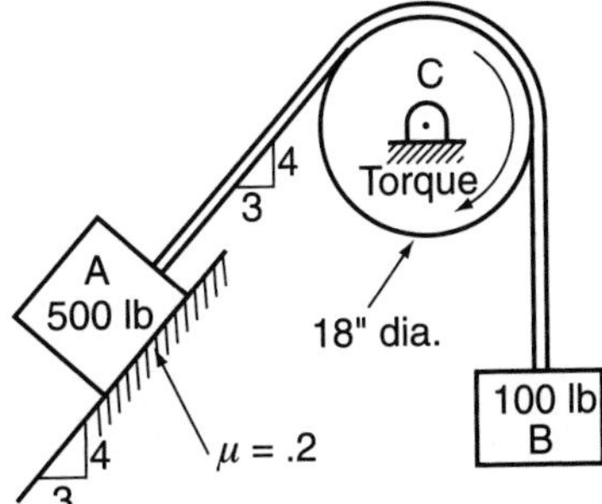

FIGURE RP7–9

R7–10. The drum shown in Figure RP7–10 must be braked by the belt system with a torque of 250 N·m. The coefficient of kinetic friction between the belt and the drum is 0.3. The drum direction of rotation is such that force A is minimum. Determine (a) the force A, (b) the pin reactions at C, and (c) the value of μ if the brake is to become self-locking.

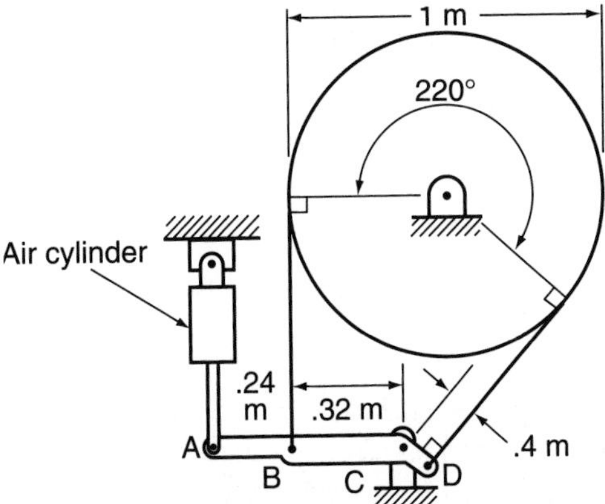

FIGURE RP7–10

R7–11. The rope shown in Figure RP7–11 is wound around fixed cylinder A for less than 1 revolution. Find force P for impending motion of B (tipping or sliding).

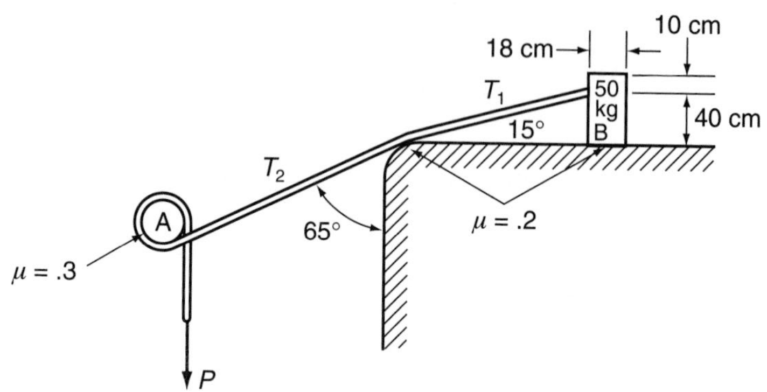

FIGURE RP7–11

R7–12. Determine force P for impending motion of the 15-kg block shown in Figure RP7–12. The coefficient of friction is 0.2 for all surfaces.

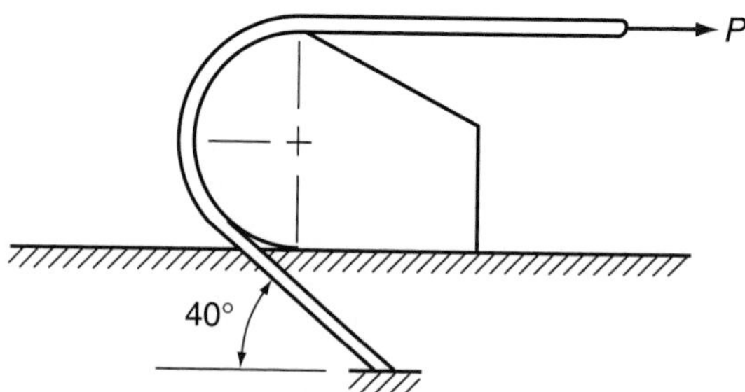

FIGURE RP7–12

R7–13. A shaft, 4 in. in diameter and weighing 100 lb, has impending motion counterclockwise due to an applied torque of 450 lb-in. (The coefficient of static friction for all surfaces is 0.25.) Determine the reactions at A and B on the shaft (Figure RP7–13).

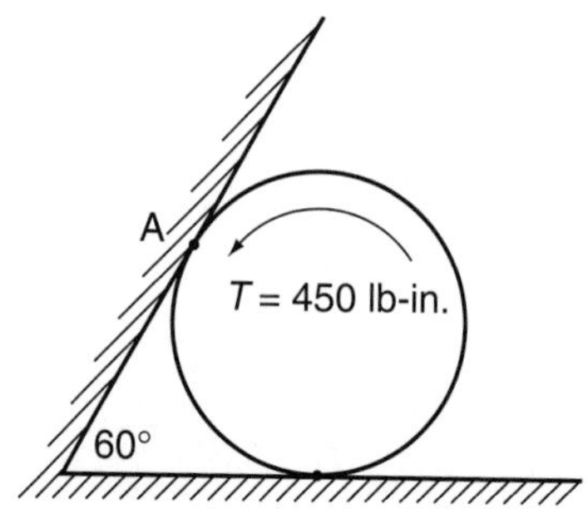

FIGURE RP7–13

CHAPTER 8

Centroids and Center of Gravity

OBJECTIVES

Upon completion of this chapter the student will be able to:

1. Locate centroids of simple areas.
2. Locate the centroid of a composite area by breaking it down into several simple areas.
3. Locate the centroid of lines using a method similar to the one used for determining the centroid of a composite area.

8–1 INTRODUCTION

In Section 4–7, we covered nonuniform loading of beams. When calculating the beam reactions, we were able to show the entire nonuniform load as acting at its center of gravity. The problem became one of determining the amount of the nonuniform load and locating the center of gravity. *Center of gravity* (C of G) or *center of mass* refers to masses or weights and can be thought of as the single point at which the weight could be held and be in balance in all directions.

If the weight or object were homogeneous, the center of gravity and centroid would coincide. In the case of a hammer with a wooden handle, its center of gravity would be close to the heavy metal end. The *centroid,* which is found by ignoring weight and considering only volume, would be closer to the middle of the handle. Due to varying densities, the center of gravity and centroid do not always coincide. Centroid usually refers to lines, areas, and volumes. It is also the central or balance point of the line, area, or volume.

The two main areas in which you will use centroids are in fluid mechanics and stress analysis. An example of centroids used in stress analysis is a beam that is caused to bend due to a load. The beam has a cross-sectional area, and locating the centroid of this area is one of the first steps in solving for the stress or deflection of the beam. In fluid mechanics, a vertical wall, such as a dam, is subjected to water pressure that varies from zero at the top to a maximum at the bottom. Determining the centroid of this pressure distribution is necessary.

8–2 CENTROIDS OF SIMPLE AREAS

To see how centroids of areas are located, consider the area shown in Figure 8–1. Although the answer may be obvious to you as $\bar{x} = 3$ ft and $\bar{y} = 2$ ft, this example is used here for illustration purposes only. An odd-shaped area would result in more involved calculations.

The first step of either the calculus method or the method described below is division of the area into strips of equal width. Divide the area into horizontal strips, arbitrarily selecting 1 ft as the width (Figure 8–2). A vector will represent the area of each strip. The vector equals the strip area and acts in a direction perpendicular to the strip at the center, or centroid, of the strip. Each vector or strip area has a moment about the x-axis. The sum of the moments of all strip areas is equal to the total area multiplied by the centroid distance $\bar{y}$.

$$A_1y_1 + A_2y_2 + A_3y_3 + A_4y_4 = A\bar{y}$$

$$(6\text{ ft}^2)(0.5\text{ ft}) + (6\text{ ft}^2)(1.5\text{ ft}) + (6\text{ ft}^2)(2.5\text{ ft}) + (6\text{ ft}^2)(3.5\text{ ft}) = (24\text{ ft}^2)\bar{y}$$

$$48 = 24\bar{y}$$

$$\underline{\bar{y} = 2\text{ ft}}$$

The centroid distance $\bar{x}$ is found by dividing the area into vertical strips and taking moments about the y-axis (Figure 8–3). Multiply each area by its centroid distance.

$$A_1x_1 + A_2x_2 + A_3x_3 + A_4x_4 + A_5x_5 + A_6x_6 = A\bar{x}$$

$$(4\text{ ft}^2)(0.5\text{ ft}) + (4\text{ ft}^2)(1.5\text{ ft}) + (4\text{ ft}^2)(2.5\text{ ft})$$

$$+ (4\text{ ft}^2)(3.5\text{ ft}) + (4\text{ ft}^2)(4.5\text{ ft}) + (4\text{ ft}^2)(5.5\text{ ft}) = (24\text{ ft}^2)\bar{x}$$

$$\underline{\bar{x} = 3\text{ ft}}$$

To restate the foregoing principle: *the moment of an area equals the algebraic sum of the moments of its component areas.* We will use this principle in locating the centroid of composite areas (Section 8–3).

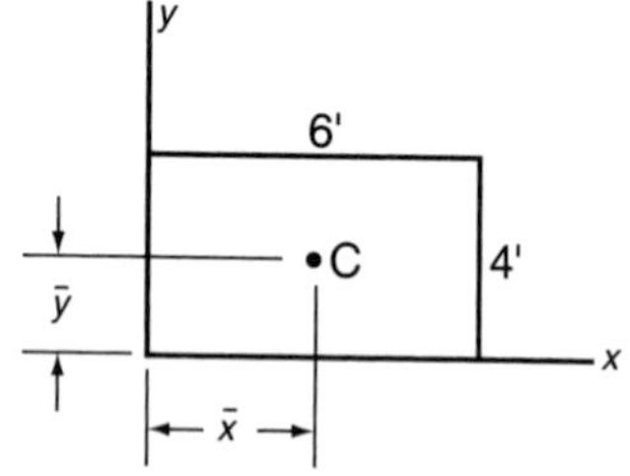

FIGURE 8–1

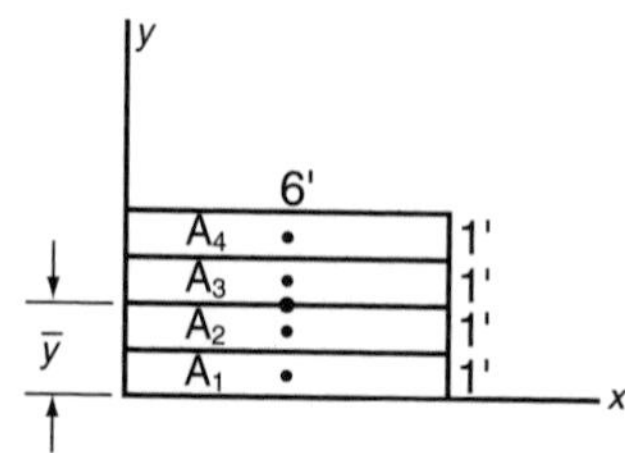

FIGURE 8–2

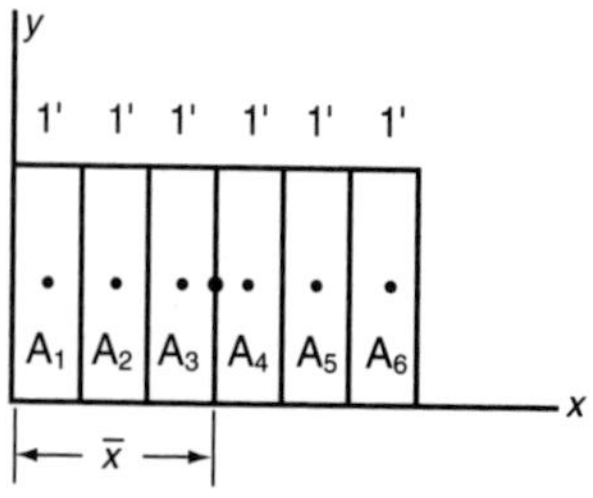

FIGURE 8–3

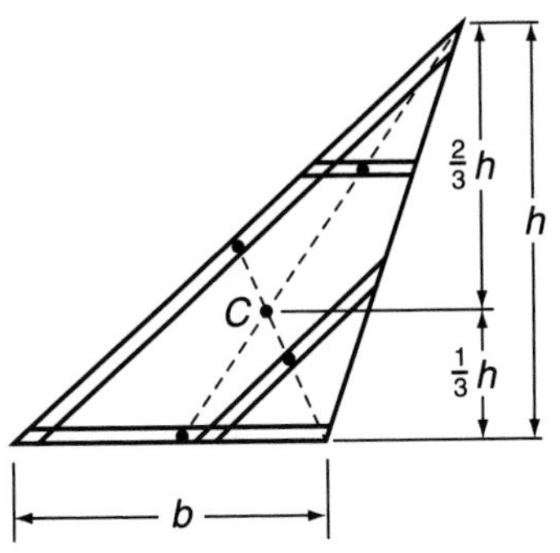

FIGURE 8–4

For the simpler geometric areas, there is a faster method of locating the centroid, *C,* than by taking moments of incremental areas. First observe that a line drawn through the centroids of the horizontal strips (Figure 8–2) and a line drawn through the centroids of the vertical strips (Figure 8–3) intersect at point C, the centroid of the area. Apply the same reasoning to the triangle in Figure 8–4 by taking elemental strips parallel to the base of the triangle and parallel to the left side. The loci of the midpoints of the strips intersect at C, the centroid of the triangle. The indicated proportions of the height h are found to exist. The height is always measured perpendicular to the base. The centroids of some of the more common areas are shown in Table 8–1. For symmetrical areas, the centroid lies on the axis of symmetry.

TABLE 8–1
Centroids Of Areas

Shape	$\bar{x}$	$\bar{y}$	Area
1. Triangle	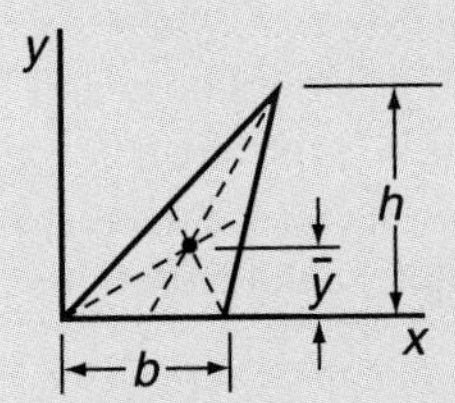	$\frac{h}{3}$	$\frac{1}{2}bh$
2. Semicircle	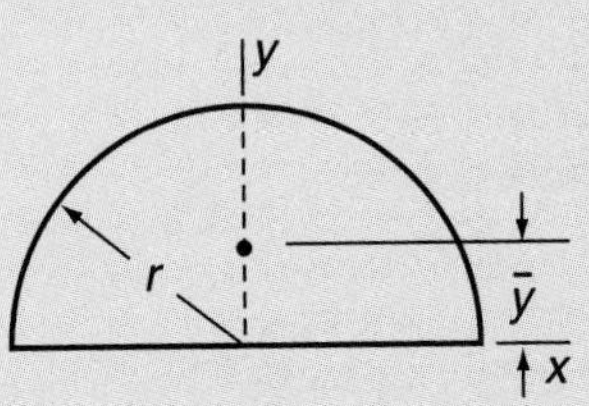 0	$\frac{4r}{3\pi}$	$\frac{\pi r^2}{2}$
3. Quarter circle	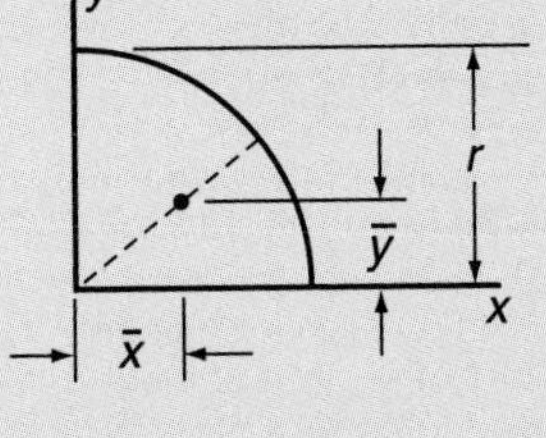 $\frac{4r}{3\pi}$	$\frac{4r}{3\pi}$	$\frac{\pi r^2}{4}$
4. Rectangle	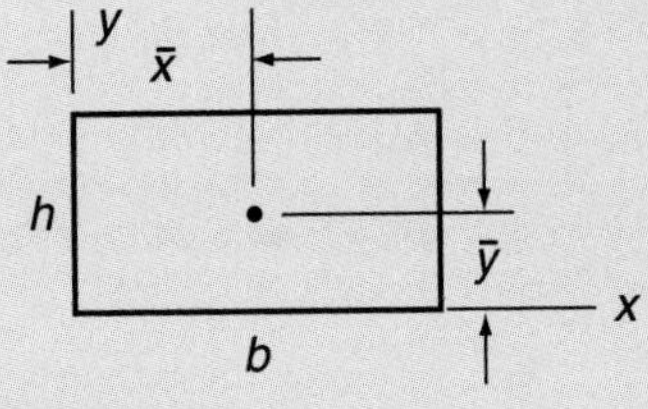 $\frac{b}{2}$	$\frac{h}{2}$	bh

8–3 CENTROIDS OF COMPOSITE AREAS

We have seen how the centroid of an area is located by dividing the area into elemental strips and adding the moments about an axis. Finding the centroid of a more complex area, that is a *composite area*, is done in a similar manner; the difference is that we divide the total area into simple geometric areas that have known centroids. Choosing a convenient axis, we sum the moments of these areas. A cutout area has a negative moment.

EXAMPLE 8–1

Determine the centroid of the composite area shown in Figure 8–5.

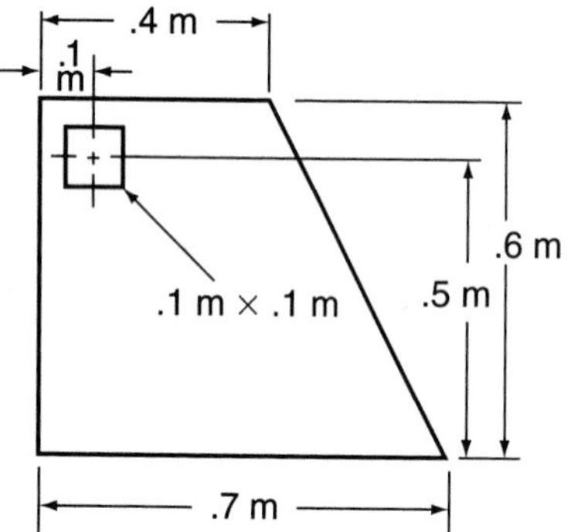

FIGURE 8–5

The first step is to choose and label all component areas as in Figure 8–6. The centroid distances for each area are shown. Any cutout area—such as the square—is considered to have a negative moment. Calculate each area.

$$A_1 = bh \qquad A_2 = -0.01\text{ m}^2 \qquad A_3 = \frac{1}{2}bh$$

$$= 0.4\text{ m}(0.6\text{ m})$$

$$= 0.24\text{ m}^2 \qquad = \frac{1}{2}(0.3\text{ m})(0.6\text{ m})$$

$$= 0.09\text{ m}^2$$

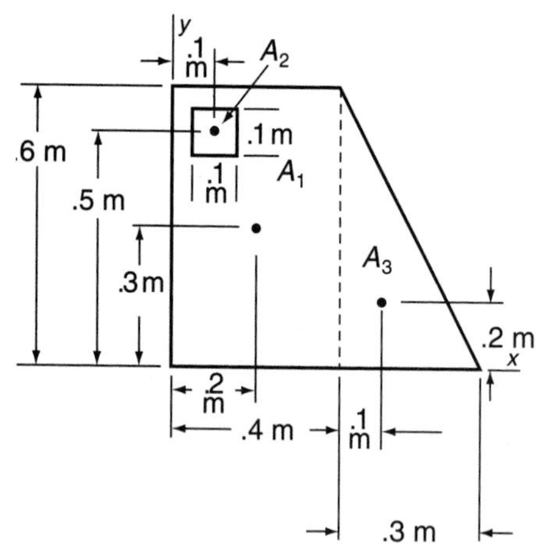

FIGURE 8–6

$$A = A_1 + A_2 + A_3$$
$$= 0.24\text{ m}^2 - 0.01\text{m}^2 + 0.09\text{ m}^2$$
$$= 0.32\text{ m}^2$$

Take moments about the x-axis to find $\bar{y}$ (Figure 8–6).

$$A_1y_1 + A_2y_2 + A_3y_3 = A\bar{y}$$

$$(0.24\text{ m}^2)(0.3\text{ m}) + (-0.01\text{ m}^2)(0.5\text{ m})$$
$$+ (0.09\text{ m}^2)(0.2\text{ m}) = (0.32\text{ m}^2)\bar{y}$$
$$0.072 - 0.005 + 0.018 = 0.32\bar{y}$$
$$\underline{\bar{y} = 0.266\text{ m}}$$

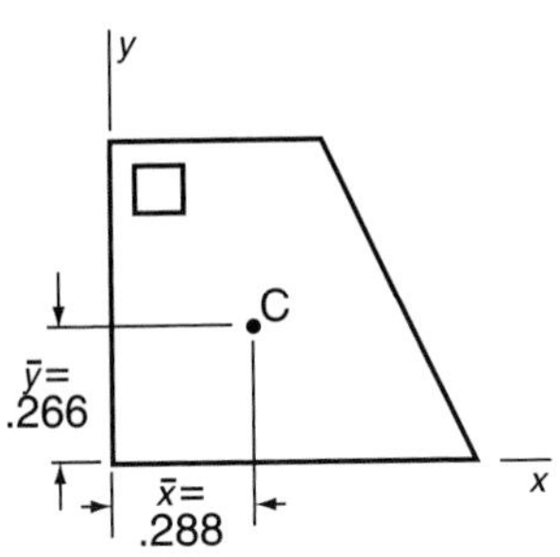

FIGURE 8–7

Moments about the y-axis will give $\bar{x}$ (Figure 8–6).

$$A_1x_1 + A_2x_2 + A_3x_3 = A\bar{x}$$

$$(0.24 \text{ m}^2)(0.2 \text{ m}) + (-0.01 \text{ m}^2)(0.1 \text{ m}) + (0.09 \text{ m}^2)(0.5 \text{ m}) = (0.32 \text{ m}^2)\bar{x}$$

$$\underline{\bar{x} = 0.288 \text{ m}}$$

If this composite area was that of a homogeneous plate of some thickness, the center of gravity would be found by the same method.

EXAMPLE 8–2

Determine the centroid of the composite area of an I-beam and a channel welded together as shown in Figure 8–8.

The centroid distances shown on the left of the diagram and areas are available from tables in various handbooks or stress texts.

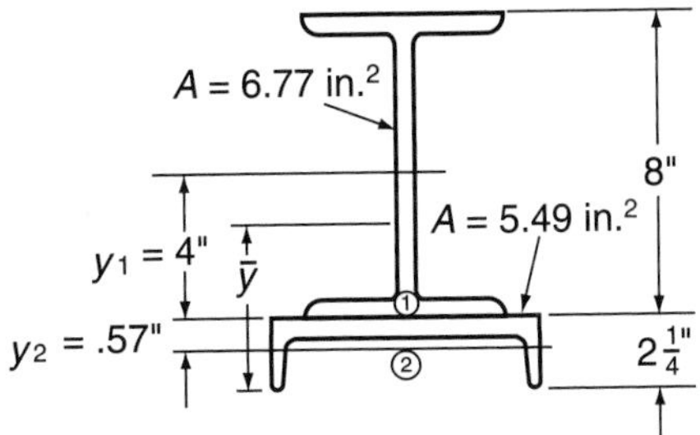

FIGURE 8–8

$$A\bar{y} = A_1y_1 + A_2y_2$$

$$(6.77 \text{ in.}^2 + 5.49 \text{ in.}^2)\bar{y} = 6.77 \text{ in.}^2(4 \text{ in.} + 2.25 \text{ in.}) + 5.49 \text{ in.}^2(2.25 \text{ in.} - 0.57 \text{ in.})$$

$$\bar{y} = \frac{42.31 + 9.22}{12.26}$$

$$\underline{\bar{y} = 4.2 \text{ in.}}$$

The vertical centroidal distance of the composite area is 4.2 in. above the bottom of the area.

8–4 CENTROIDS OF LINES

Centroids are not restricted to areas. A thin material may be in the form of a *complex profile* or *cross section.* The centroid of such a cross section is the *centroid of lines.* The situation is parallel to that encountered when we deal with areas. The total line length multiplied by its centroid distance from an axis is equal to the sum of each segment line length multipled by its centroidal distance. Using $\bar{x}$ and $\bar{y}$ as centroidal distances and L for length, we have

$$L\bar{x} = L_1x_1 + L_2x_2 + L_3x_3 + \cdots$$

$$L\bar{y} = L_1y_1 + L_2y_2 + L_3y_3 + \cdots$$

Some common variations of line centroids are given in Table 8–2.

TABLE 8–2
Centroids of Lines

	Line	$\bar{y}$
1.		$\frac{y}{2}$
2.		$\frac{y}{2}$
3.		$\frac{2r}{\pi}$
4.		$\frac{2r}{\pi}$

EXAMPLE 8–3

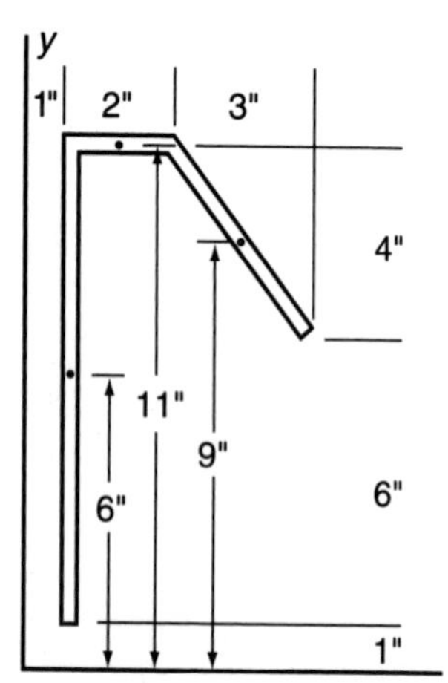

FIGURE 8–9

Locate the centroid of the line shown in Figure 8–9.

total line length = 10 in. + 2 in. + 5 in. = 17 in.

There are three simple lengths or shapes. With the x-axis as our base, we solve for $\bar{y}$.

$$(17 \text{ in.})\bar{y} = (10 \text{ in.})(6 \text{ in.}) + (2 \text{ in.})(11 \text{ in.}) + (5 \text{ in.})(9 \text{ in.})$$

$$\bar{y} = \frac{60 + 22 + 45}{17}$$

$$\underline{\bar{y} = 7.47 \text{ in.}}$$

$$(17 \text{ in.})\bar{x} = (10 \text{ in.})(1 \text{ in.}) + (2 \text{ in.})(2 \text{ in.}) + (5 \text{ in.})(4.5 \text{ in.})$$

$$\bar{x} = \frac{10 + 4 + 22.5}{17}$$

$$\underline{\bar{x} = 2.15 \text{ in.}}$$

EXAMPLE 8–4

Locate the centroid of the line shown in Figure 8–10.

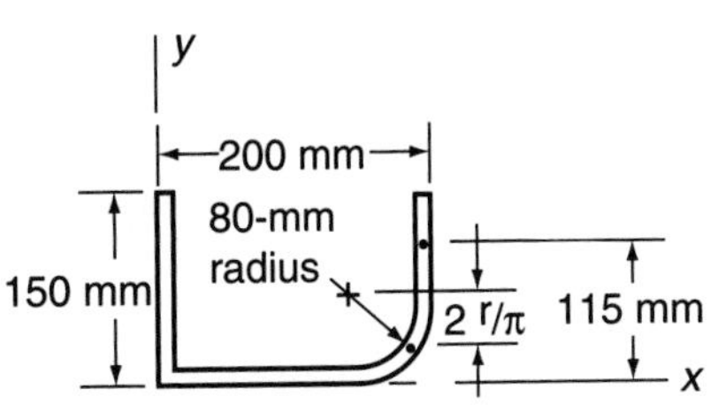

FIGURE 8–10

$$\text{total line length} = 150\text{ mm} + 120\text{ mm} + \frac{2\pi(80\text{ mm})}{4} + 70\text{ mm}$$

$$= 150 + 120 + 126 + 70$$

$$= 466\text{ mm}$$

Since there are four lengths, our equation for $\bar{y}$ will be

$$L\bar{y} = L_1y_1 + L_2y_2 + L_3y_3 + L_4y_4$$

$$(466\text{ mm})\bar{y} = (150\text{ mm})(75\text{ mm}) + (120\text{ mm})(0) + (126\text{ mm})\left(80\text{ mm} - \frac{(2)(80\text{mm})}{\pi}\right) + (70\text{ mm})(115\text{ mm})$$

$$\bar{y} = \frac{11{,}250 + 0 + 3663 + 8050}{466}$$

$$\underline{\bar{y} = 49.3\text{ mm}}$$

$$L\bar{x} = L_1x_1 + L_2x_2 + L_3x_3 + L_4x_4$$

$$(466\text{ mm})\bar{x} = 150\text{ mm}(0) + (120\text{ mm})(60\text{ mm}) + (126\text{ mm})\left(120\text{ mm} + \frac{(2)(80\text{ mm})}{\pi}\right) + (70\text{ mm})(200\text{ mm})$$

$$\bar{x} = \frac{0 + 7200 + 21{,}537 + 14{,}000}{466}$$

$$\underline{\bar{x} = 91.7\text{ mm}}$$

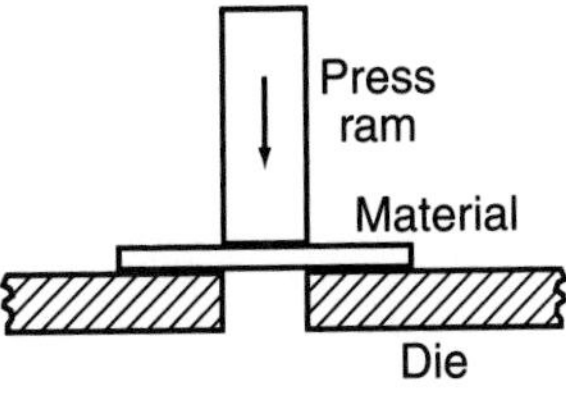

FIGURE 8–11

Punching or shearing an area from a sheet by using a punch press and die is an operation that may require determination of the centroid of lines (Figure 8–11). The cutting forces around the outside of the shape should be symmetrically distributed with respect to the press ram. This is known as the *center of pressure* of the die and is simply the centroid of the outside lines of the shape.

EXAMPLE 8–5

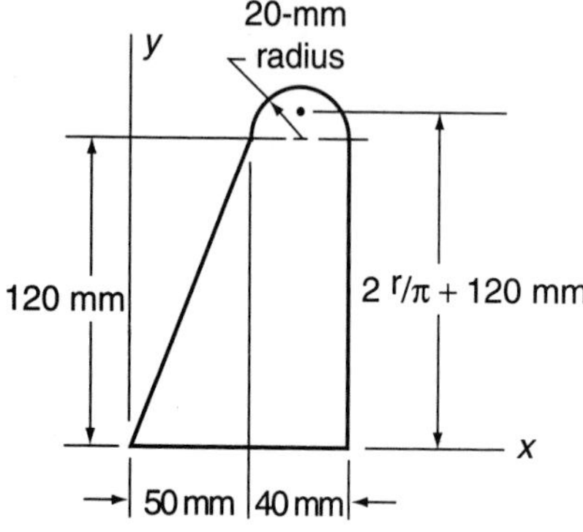

FIGURE 8–12

Determine the center of pressure of the blank in Figure 8–12.

We are to find the centroid of the lines that form the shape or blank shown. There are four simple lines giving us a total line length of

$$90\text{ mm} + 130\text{ mm} + \frac{2\pi(20\text{ mm})}{2} + 120\text{ mm} = 403\text{ mm}$$

The distance from the diameter to the centroid of the semicircle is $2r/\pi$.

$$(403\text{ mm})\bar{y} = (90\text{ mm})(0) + (130\text{ mm})(60\text{ mm}) + \left[\left(\frac{2 \times 20\text{ mm}}{\pi} + 120\text{ mm}\right)62.8\text{ mm}\right] + (120\text{ mm})(60\text{ mm})$$

$$\bar{y} = \frac{0 + 7800 + 8336 + 7200}{403}$$

$$\underline{\bar{y} = 57.9\text{ mm}}$$

$$(403\text{ mm})\bar{x} = (90\text{ mm})(45\text{ mm}) + (130\text{ mm})(25\text{ mm}) + (62.8\text{ mm})(70\text{ mm}) + (120\text{ mm})(90\text{ mm})$$

$$\underline{\bar{x} = 55.8\text{ mm}}$$

HINTS FOR PROBLEM SOLVING

1. Make certain that you have shown a cutout area as a negative value in an equation such as $A\bar{x} = A_1x_1 + A_2x_2 - A_3x_3$.
2. Tabulate the areas and their corresponding centroidal distances before writing your equations. This ensures that when you use the area values in more than one equation, you will have an orderly central source of data.
3. Since "number crunching" increases the chance of simple mathematical errors, slow down and take care with the mathematics of this chapter. When laying out a problem, watch out for the following:
   (a) Don't mix units.
   (b) Be careful of the decimal place.
   (c) Don't omit any areas or line lengths.
   (d) Check each centroidal distance.
   (e) With some calculators you will get better accuracy using, for example, 98 mm^2 rather than 0.000098 m^2.
   (f) Divide a complex area into a minimum number of simple areas.

PROBLEMS

APPLIED PROBLEMS FOR SECTIONS 8–1 TO 8–3

8–1. Determine $\bar{y}$ for the area shown in Figure P8–1.

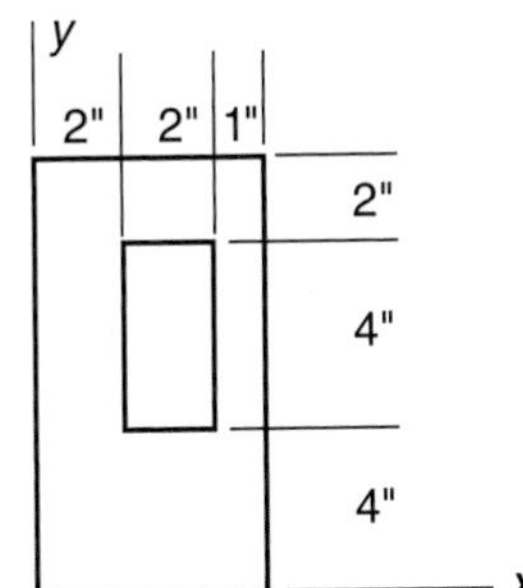

FIGURE P8–1

8–2. Determine $\bar{x}$ for the area shown in Figure P8–2.

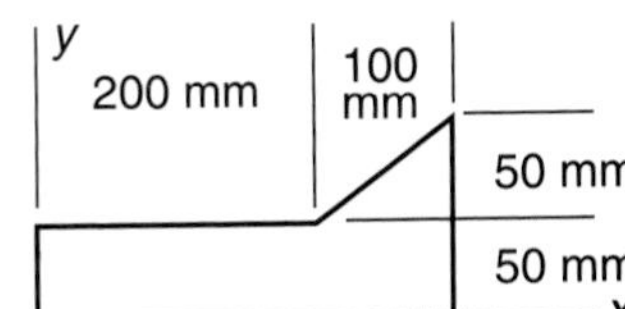

FIGURE P8–2

8–3. Locate the centroid of the area shown in Figure P8–3.

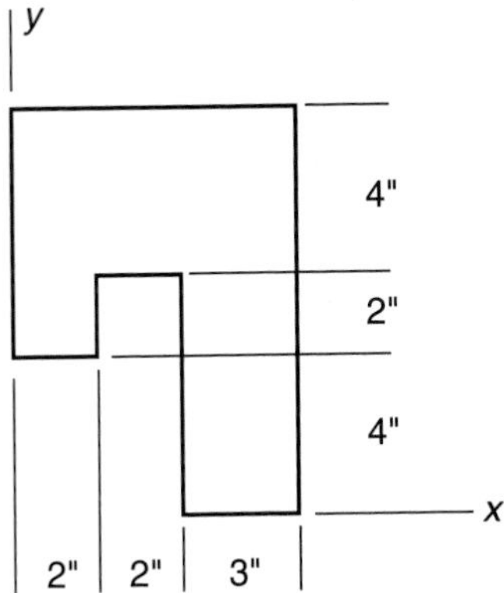

FIGURE P8–3

8–4. Locate the centroid of the area shown in Figure P8–4.

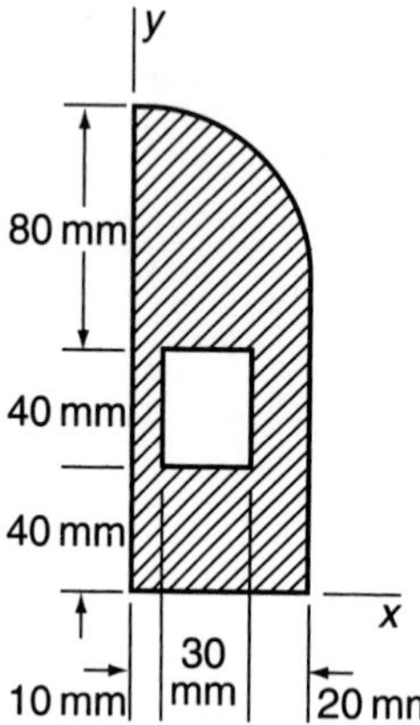

FIGURE P8–4

8–5. Locate the centroid of the area shown in Figure P8–5.

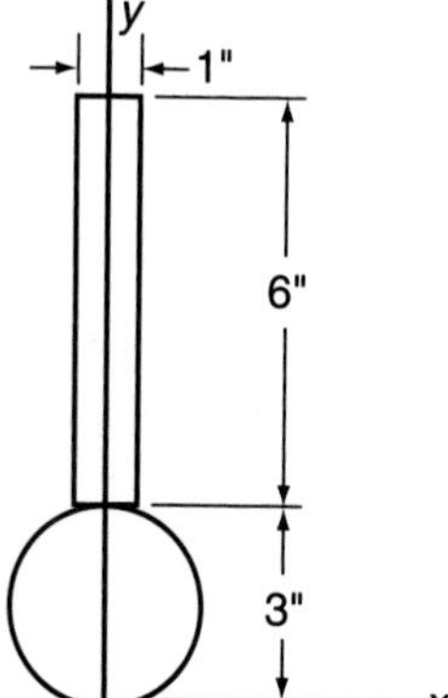

FIGURE P8–5

8–6. Locate the centroid of the channel cross section shown in Figure P8–6.

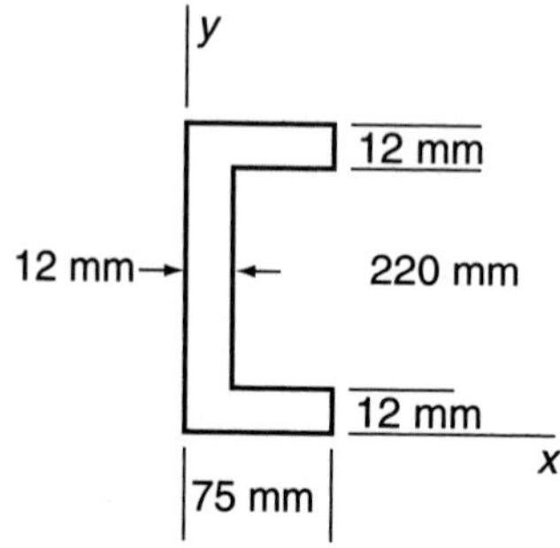

FIGURE P8–6

8–7. Locate the centroid of the cross-sectional area shown in Figure P8–7.

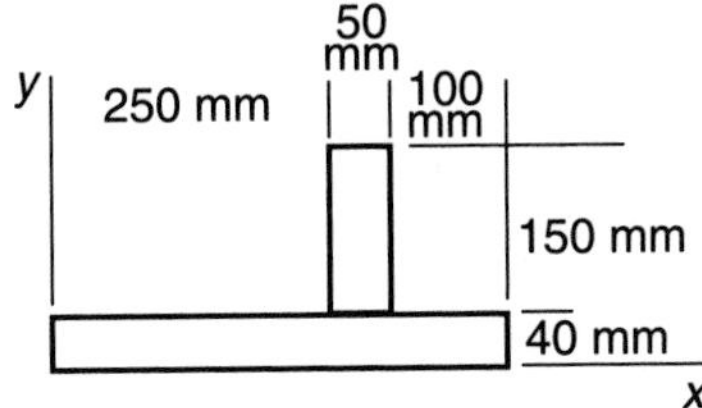

FIGURE P8–7

8–8. Locate the centroid of the cross-sectional area of the fabricated I-beam shown in Figure P8–8.

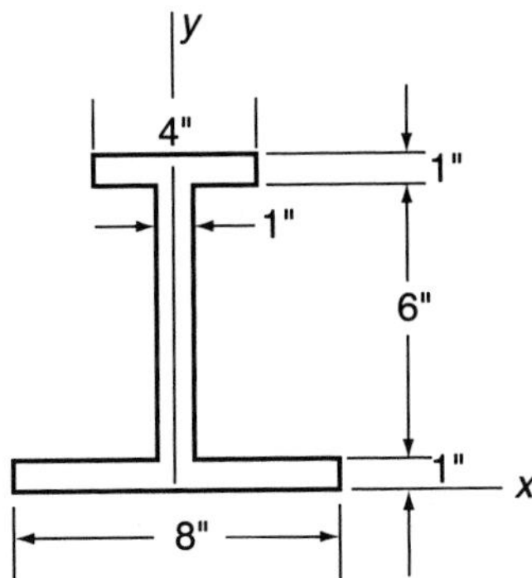

FIGURE P8–8

8–9. Locate the centroid ($\bar{x}$ and $\bar{y}$) for the area shown in Figure P8–9.

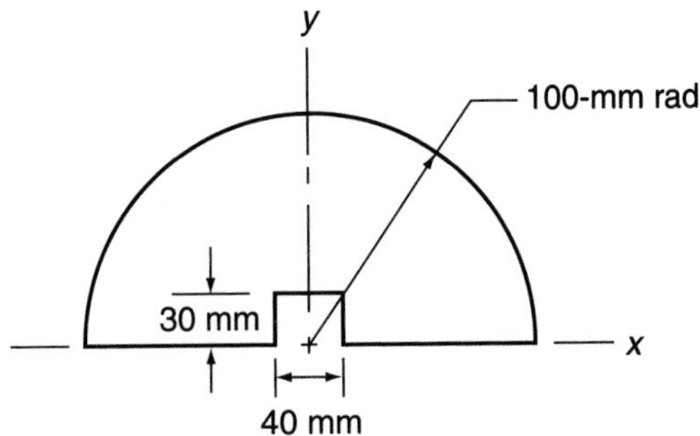

FIGURE P8–9

8–10. Locate the centroid ($\bar{x}$ and $\bar{y}$) for the area shown in Figure P8–10.

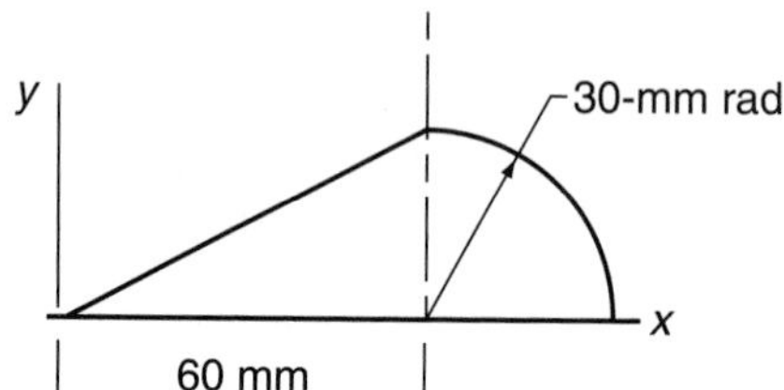

FIGURE P8–10

8–11. Locate the centroid of the area shown in Figure P8–11.

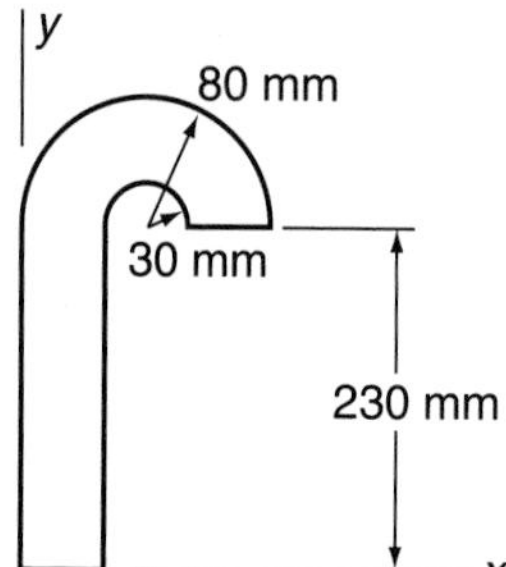

FIGURE P8–11

8–12. Locate the centroid of the area shown in Figure P8–12.

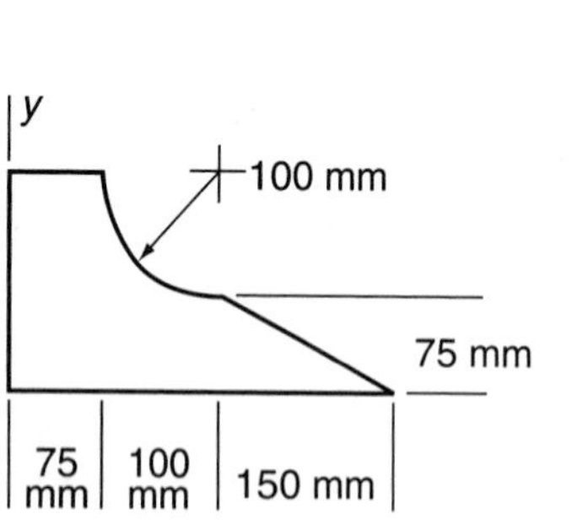

FIGURE P8–12

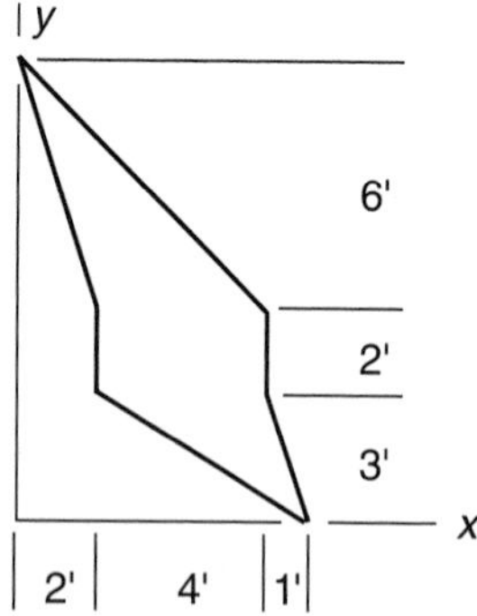

FIGURE P8–13

8–13. Determine $\bar{y}$ for the area shown in Figure P8–13.

8–14. Determine $\bar{y}$ for the combined area of a channel and angle joined as shown in Figure P8–14. The individual areas and centroidal distances are given.

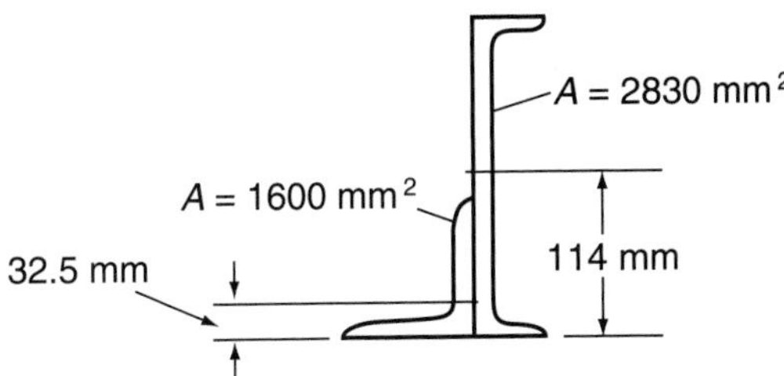

FIGURE P8–14

8–15. Determine the location of the centroidal horizontal axis of the combined structural shapes shown in Figure P8–15.

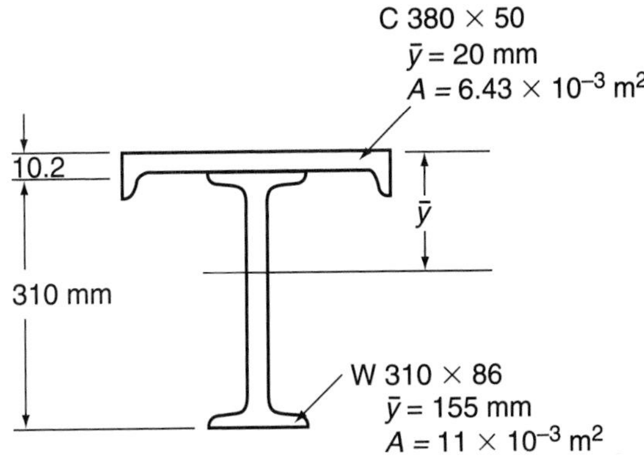

FIGURE P8–15

8–16. Determine the location of the centroidal horizontal axis from the bottom of the combined structural shapes shown in Figure P8–16.

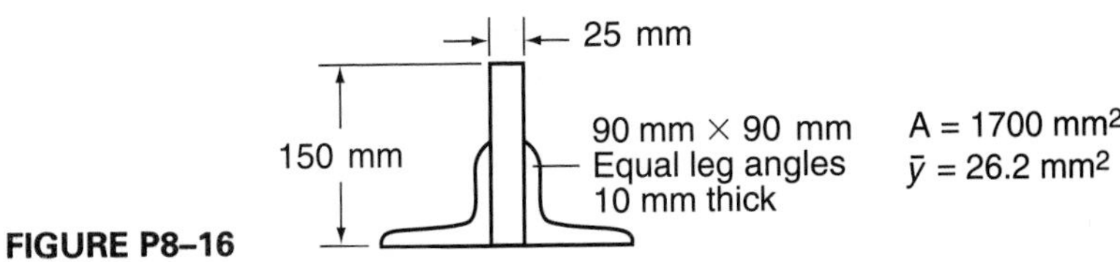

FIGURE P8–16

8–17. Locate the centroidal x-axis for the beam cross sections shown in Figure P8–17.

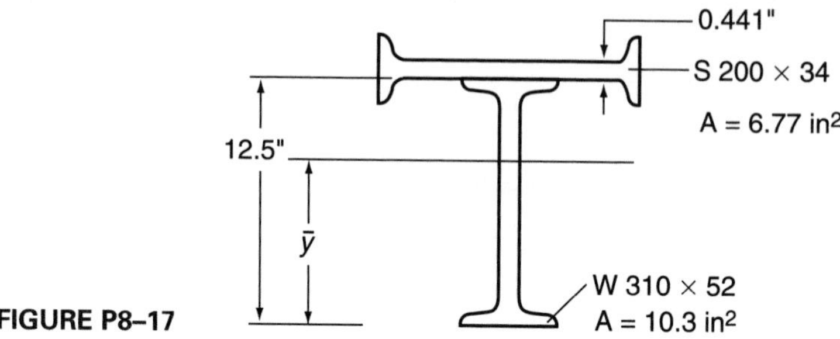

FIGURE P8–17

8–18. Locate the centroid of the extruded shape shown in Figure P8–18.

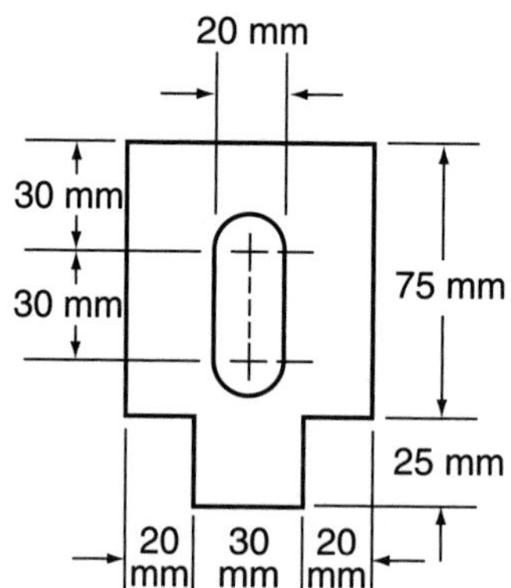

FIGURE P8–18

8–19. Determine the centroid of the area shown in Figure P8–19.

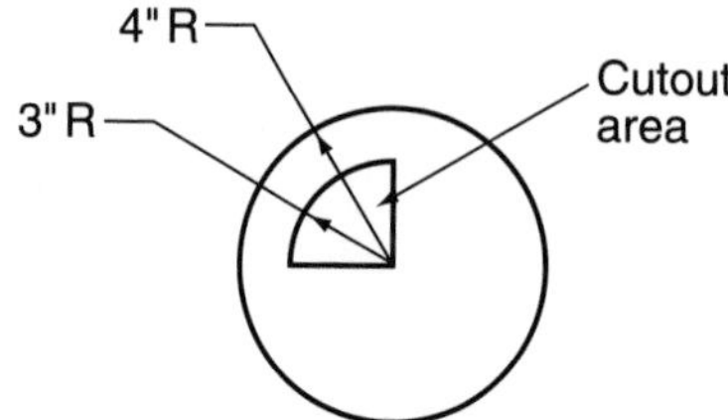

FIGURE P8–19

APPLIED PROBLEMS FOR SECTION 8–4

8–20. Locate the centroid of the line shown in Figure P8–20.

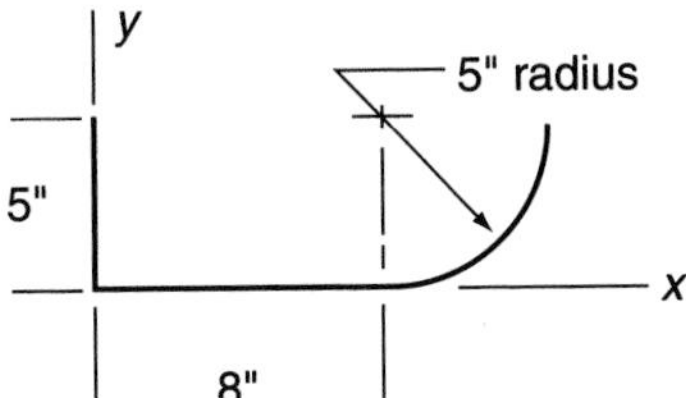

FIGURE P8–20

8–21. Sheet material is formed and welded into the cross section shown in Figure P8–21. Locate the centroid.

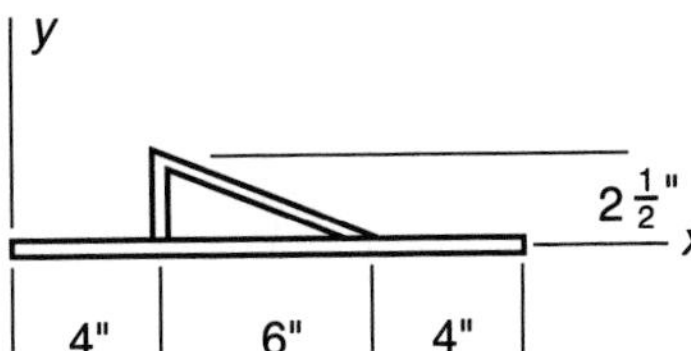

FIGURE P8–21

8–22. Locate the centroid of a rod bent to the shape shown in Figure P8–22.

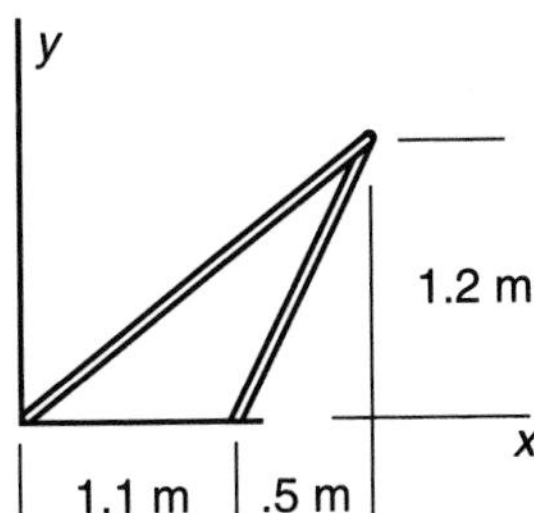

FIGURE P8–22

8–23. The line profile of the cross section formed by fabricated thin sheet metal is shown in Figure P8–23. Locate the centroid.

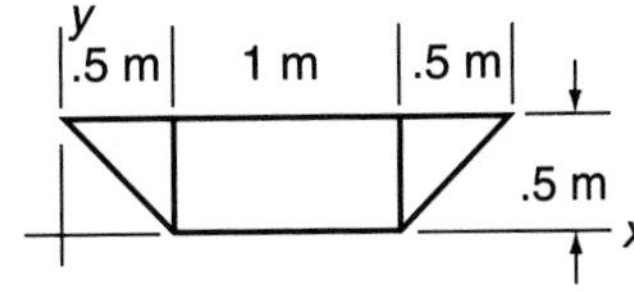

FIGURE P8–23

8–24. The cross-sectional area shown in Figure P8–24 is to be punched from sheet metal with a punch press. Determine the center of pressure.

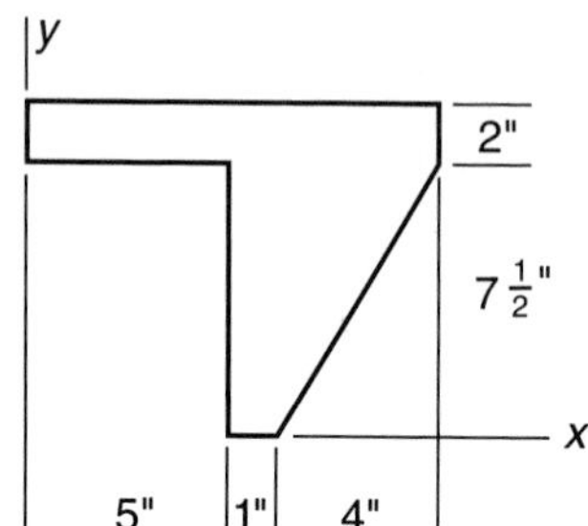

FIGURE P8–24

8–25. Determine the center of pressure for the die stamping pattern shown in Figure P8–25.

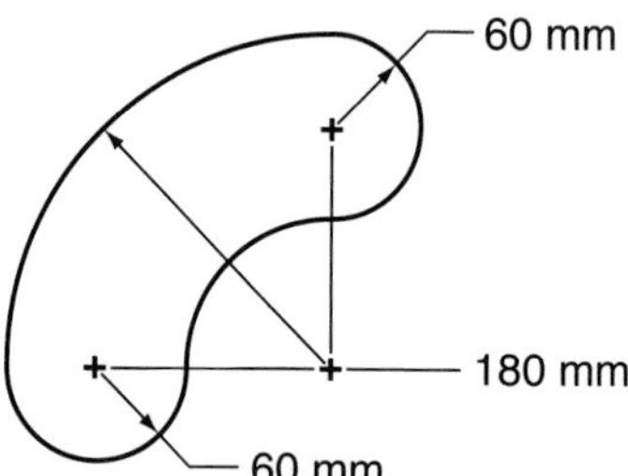

FIGURE P8–25

8–26. Determine the center of pressure for the die stamping pattern shown in Figure P8–26.

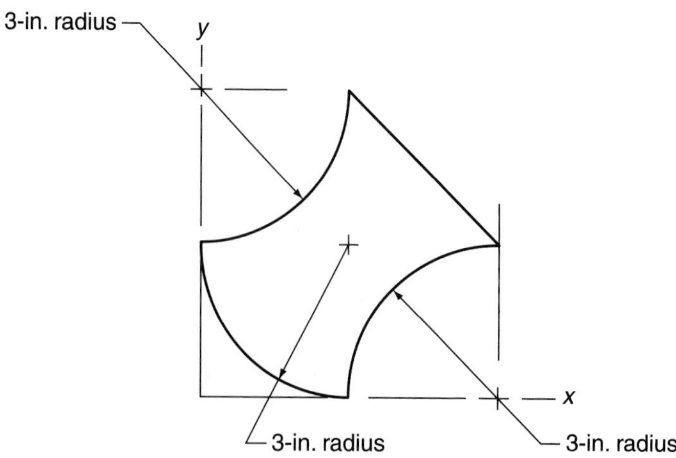

FIGURE P8–26

8–27. A thin wire is bent to form the shape shown in Figure P8–27. Locate the centroid.

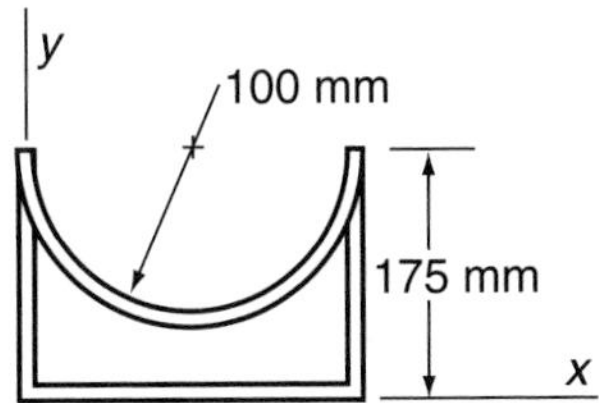

FIGURE P8–27

REVIEW PROBLEMS

R8–1. Locate the centroid of the area shown in Figure RP8–1.

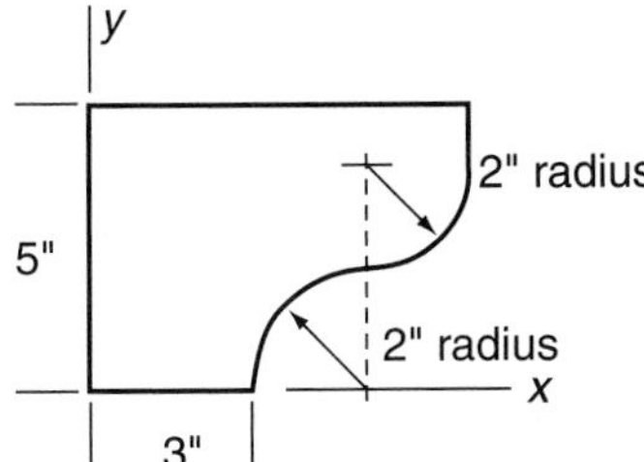

FIGURE RP8–1

R8–2. Locate the centroid of the cross-sectional area of the beam shown in Figure RP8–2.

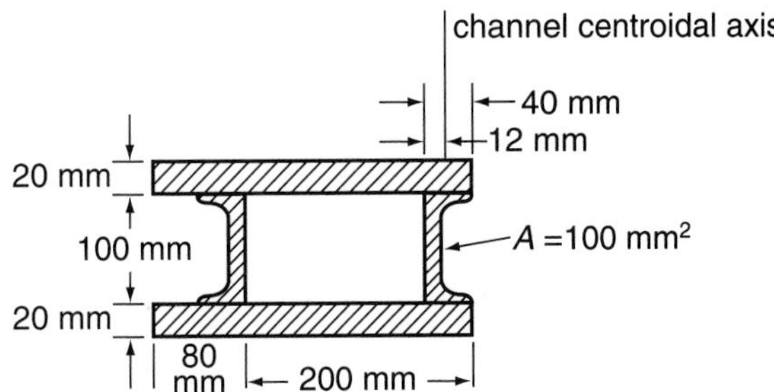

FIGURE RP8–2

R8–3. Locate the centroid of the line that is the perimeter of the area shown in Figure RP8–3.

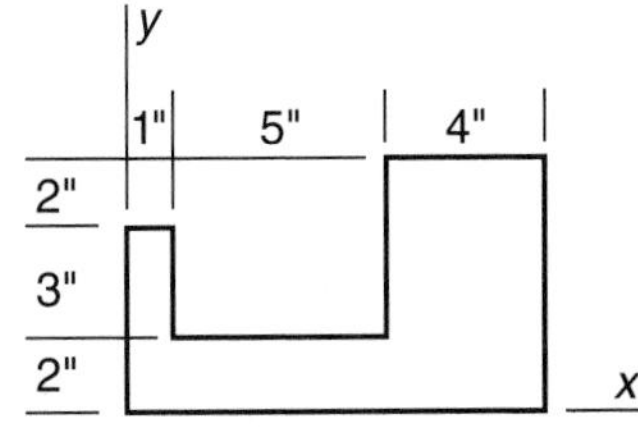

FIGURE RP8–3

CHAPTER 9

Moment of Inertia

OBJECTIVES

Upon completion of this chapter the student will be able to:

1. Determine the moment of inertia of a simple area.
2. Transfer the centroidal moment of inertia to a parallel axis.
3. Determine the moment of inertia of a composite area.
4. Define radius of gyration of an area and use it to calculate the second moment of an area.
5. Determine the mass moment of inertia of simple shapes.
6. Determine the mass moment of inertia of composite shapes, using the parallel axis equation.
7. Calculate the mass moment of inertia using the radius of gyration.

9–1 MOMENT OF INERTIA OF AN AREA

The quantity to be discussed in this section is called *second moment of area* or *moment of inertia of an area.* It is used in the calculation of stresses in beams and columns and is often referred to simply as *moment of inertia.* It measures the effect of the cross-sectional shape of a beam on the beam's resistance to a bending moment. The usual units are in.^4 or mm^4.

Referring to Figure 9–1, we determine the moment of inertia of an area as follows:

1. Divide the total area into incremental strips (ΔA) parallel to an axis, the x-axis in this case (Figure 9–1).
2. Multiply each area by the square of the distance to the x-axis.

$$I_x = y^2 \Delta A$$

3. Sum these terms for the total area.

$$I_x = \Sigma y^2 \Delta A \qquad (9\text{–}1)$$

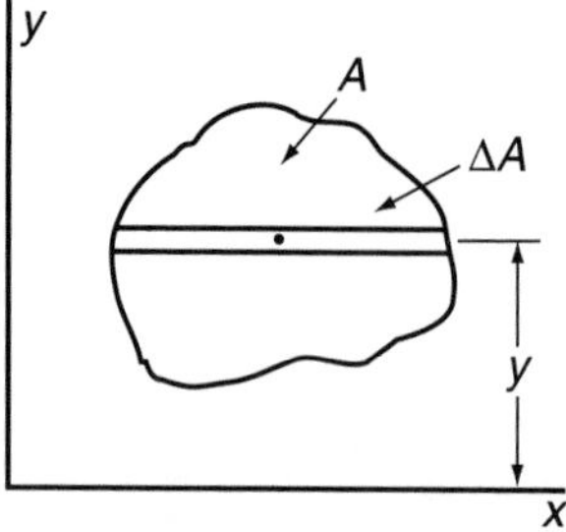

FIGURE 9–1

To determine the moment of inertia about the y-axis, use vertical strips in the same manner; the total area is expressed as

$$I_y = \Sigma x^2 \Delta A \tag{9–2}$$

For an area such as the one in Figure 9–1, the narrower the strip, the greater the accuracy of the moment of inertia. For this reason, the use of calculus generally yields the most accurate value. Most areas considered will be simple geometric areas or composite areas of geometric shapes. Calculus has been used to determine the moment of inertia formulae for these simple geometric areas (Table 9–1).

From Equations (9–1) and (9–2), where we have distance squared times area, it can be seen that the units for inertia of an area will be a unit of length to the fourth power, such as in.^{4} or mm^4. The millimeter is a small unit of length and tends to give large numerical values. The use of meters is not necessarily that much better. For example, a value of $I =$ 625 in.^{4} is equal to 0.00026 m^4 or 260,000,000 mm^4. Neither one of these terms is well suited for convenient calculation.

The centimeter may be more convenient to use for inertia calculations. The final answer is converted from cm^4 to mm^4 or m^4.

$$\begin{aligned} 1 \text{ cm} &= 10 \text{ mm} \\ 1 \text{ cm}^4 &= (10 \text{ mm})^4 \\ &= 10^4 \text{ mm}^4 \\ 1 \text{ cm} &= 1 \times 10^{-2} \text{ m} \\ 1 \text{ cm}^4 &= 1 \times 10^{-8} \text{ m}^4 \end{aligned}$$

9–2 PARALLEL AXIS THEOREM

As you may have noticed in Table 9–1, the moment of inertia is given about both the centroidal axis and the x-axis. We will now consider how the moment of inertia about the centroidal axis can be used to calculate the moment of inertia about any parallel axis, such as the x-axis.

TABLE 9–1
Moments of Inertia and Radii of Gyration of Simple Areas

Area	Moment of Inertia	Radius of Gyration
1. Rectangle 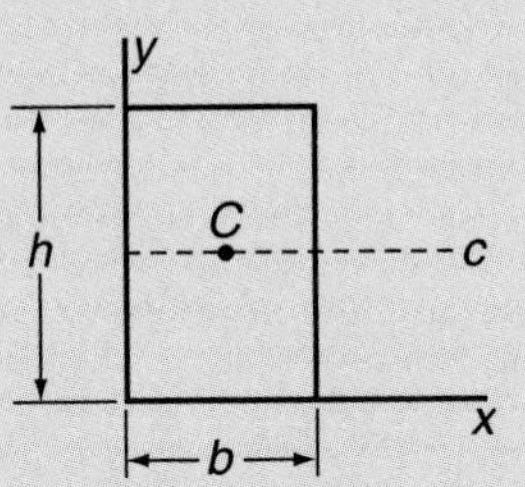	$I_c = \dfrac{bh^3}{12}$ $I_x = \dfrac{bh^3}{3}$	$k_c = \dfrac{h}{\sqrt{12}}$ $k_x = \dfrac{h}{\sqrt{3}}$
2. Triangle 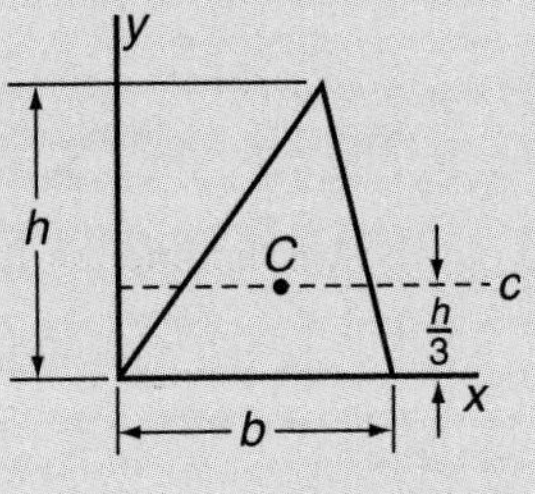	$I_c = \dfrac{bh^3}{36}$ $I_x = \dfrac{bh^3}{12}$	$k_c = \dfrac{h}{\sqrt{18}}$ $k_x = \dfrac{h}{\sqrt{6}}$
3. Circle 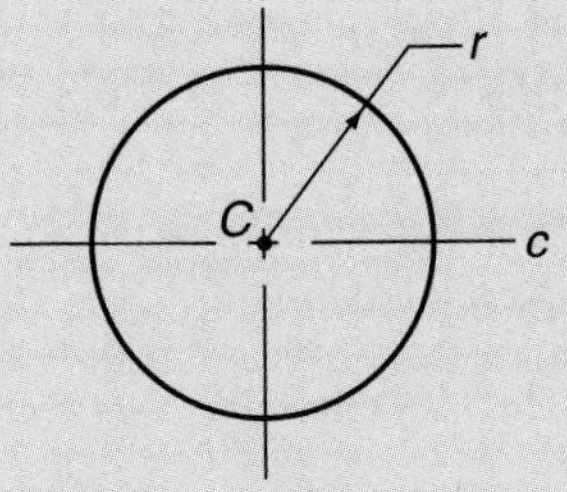	$I_c = \dfrac{\pi r^4}{4}$	$k_c = \dfrac{r}{2}$
4. Semicircle 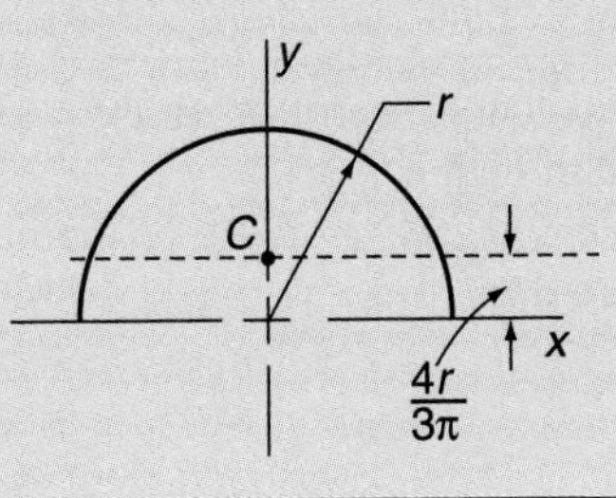	$I_c = 0.11r^4$ $I_x = \dfrac{\pi r^4}{8}$	$k_c = 0.264r$ $k_x = \dfrac{r}{2}$

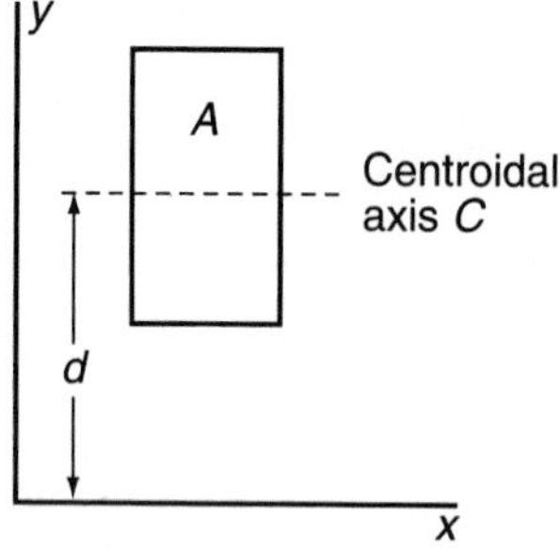

FIGURE 9–2

Suppose that in Figure 9–2 we wish to know the moment of inertia about the x-axis some distance d from the centroidal axis. Let

I_c = moment of inertia about the centroidal axis; mm^4 or $in.^4$

I_x = required moment of inertia about the x-axis; mm^4 or $in.^4$

d = perpendicular distance between parallel axes; mm or in.

A = total area; mm^2 or $in.^2$

The *parallel axis theorem* states these terms as

$$I_x = I_c + Ad^2 \tag{9–3}$$

You may also see this equation referred to as the *transfer formula,* because it transfers the moment of inertia from the centroidal axis to a parallel axis.

EXAMPLE 9–1

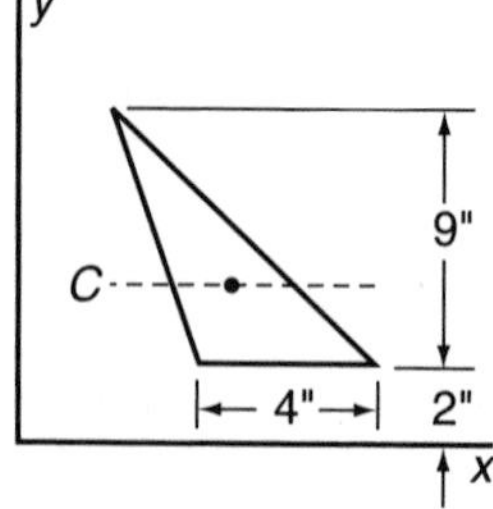

FIGURE 9–3

Determine the moment of inertia about the x-axis for the triangular area shown in Figure 9–3.

From Table 9–1, the moment of inertia about an axis through the centroid is

$$I_c = \frac{bh^3}{36}$$

$$= \frac{(4\text{ in.})(9\text{ in.})^3}{36}$$

$$I_c = 81\text{ in.}^4$$

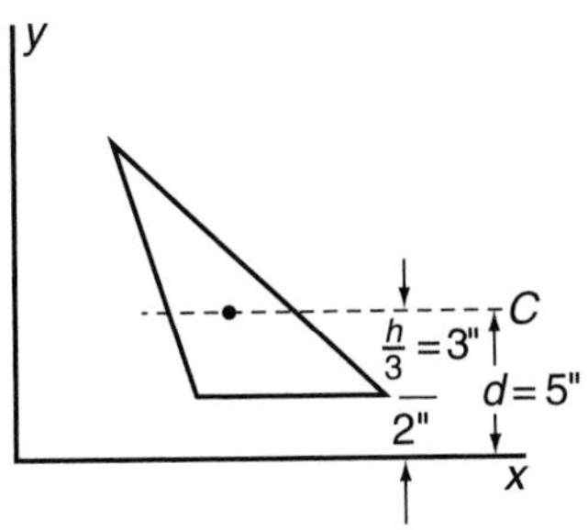

FIGURE 9–4

The distance from the base of the triangle to its centroid is

$$\frac{h}{3} = \frac{9 \text{ in.}}{3} = 3 \text{ in.} \qquad \text{(Figure 9–4)}$$

The distance d between parallel axes is 5 in. Next, we find the area of the triangle.

$$A = \frac{1}{2}bh$$
$$= \frac{1}{2}(4 \text{ in.})(9 \text{ in.})$$
$$A = 18 \text{ in.}^2$$

Now we apply the parallel axis equation (Equation 9–3).

$$I_x = I_c + Ad^2$$
$$= 81 \text{ in.}^4 + (18 \text{ in.}^2)(5 \text{ in.})^2$$
$$= 81 + 450$$
$$\underline{I_x = 531 \text{ in.}^4}$$

EXAMPLE 9–2

Determine the moment of inertia about the x-axis for the semicircular area shown in Figure 9–5.

From Table 9–1, the centroidal moment of inertia is

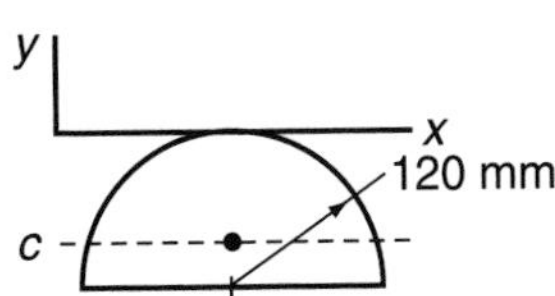

FIGURE 9–5

$$I_c = 0.11r^4$$

Using units of 12 cm rather than 120 mm (Figure 9–6), we get

$$I_c = 0.11(12 \text{ cm})^4$$
$$I_c = 2281 \text{ cm}^4$$

The distance from the base to the centroid is

$$\frac{4r}{3\pi} = \frac{4(12 \text{ cm})}{3(\pi)} = 5.09 \text{ cm}$$

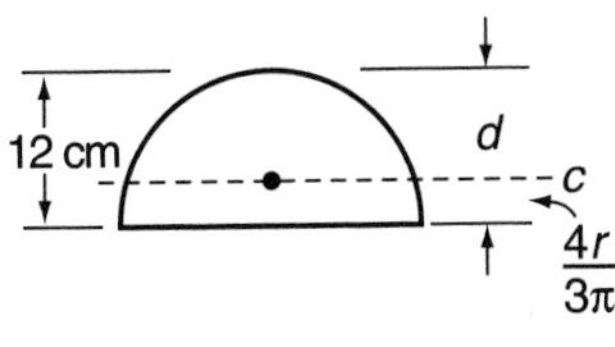

FIGURE 9–6

The centroidal moment of inertia will be transferred a distance of

$$d = 12 \text{ cm} - 5.09 \text{ cm}$$
$$= 6.91 \text{ cm}$$

Applying the parallel axis equation, we get

$$I_x = I_c + Ad^2$$

$$= 2281 \text{ cm}^4 + \left(\frac{\pi(12 \text{ cm})^2}{2}\right)(6.91 \text{ cm})^2$$

$$I_x = 13{,}100 \text{ cm}^4$$

or, since $1 \text{ cm}^4 = 10^4 \text{ mm}^4$,

$$\underline{I_x = 131 \times 10^6 \text{ mm}^4}$$

To find the moment of inertia about any new axis, always use the centroidal moment of inertia as the starting point. **Moment of inertia cannot be transferred from any axis but the centroidal axis.** The following example illustrates this point.

EXAMPLE 9–3

As shown in Figure 9–7, area A is 200 cm² and has a moment of inertia of 175×10^6 mm⁴ about axis x_1. Find its moment of inertia about axis x_2.

We cannot go directly from x_1 to x_2 but must first find I_c and then, by means of the parallel axis equation, find I_{x2}. Convert $I_{x1} = 175 \times 10^6 \text{ mm}^4 = 17{,}500 \text{ cm}^4$.

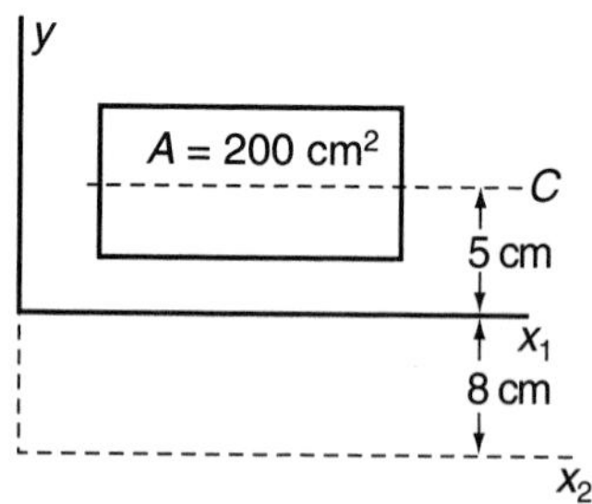

FIGURE 9–7

$$I_{x1} = I_c + Ad^2$$

$$17{,}500 \text{ cm}^4 = I_c + (200 \text{ cm}^2)(5 \text{ cm})^2$$

$$I_c = 12{,}500 \text{ cm}^4$$

$$I_{x2} = I_c + Ad^2$$

$$= 12{,}500 \text{ cm}^4 + (200 \text{ cm}^2)(13 \text{ cm})^2$$

$$= 46{,}300 \text{ cm}^4$$

$$\underline{I_{x2} = 463 \times 10^6 \text{ mm}^4}$$

9–3 MOMENT OF INERTIA OF COMPOSITE AREAS

The method of finding the moment of inertia of composite areas is very similar to the one used for centroids of composite areas. The usual sequence of steps is as follows:

1. Divide the composite area into simple areas.
2. For each simple area, find the area and the moment of inertia about its centroidal axis.

3. Transfer each centroidal moment of inertia to a parallel reference axis.
4. The sum of the moments of inertia about the reference axis is the moment of inertia of the composite area. Any cutout area has a negative moment; all other areas are positive.

EXAMPLE 9–4

Determine the moment of inertia about the x-axis for the area shown in Figure 9–8.

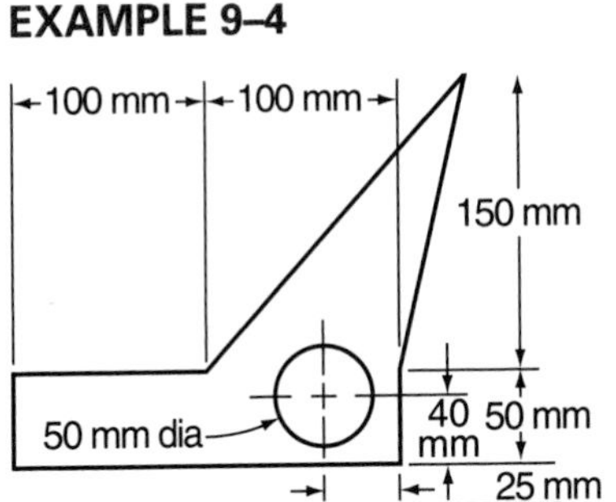

FIGURE 9–8

Divide the total area into simple areas (Figures 9–9, 9–10, and 9–11). Find I_c and the area of each of the simple areas.

Rectangle (Figure 9–9):

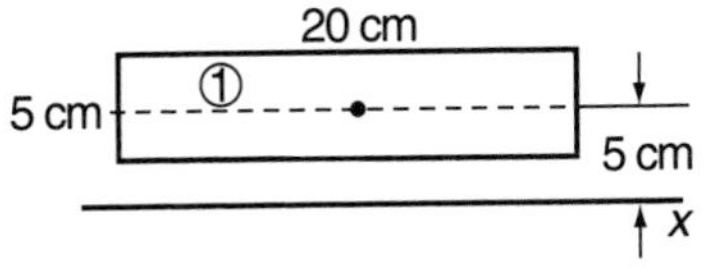

FIGURE 9–9

$$I_c = \frac{bh^3}{12}$$

$$= \frac{(20\text{ cm})(5\text{ cm})^3}{12}$$

$$I_c = 208.3\text{ cm}^4$$

$$A = 100\text{ cm}^2$$

Circle (Figure 9–10):

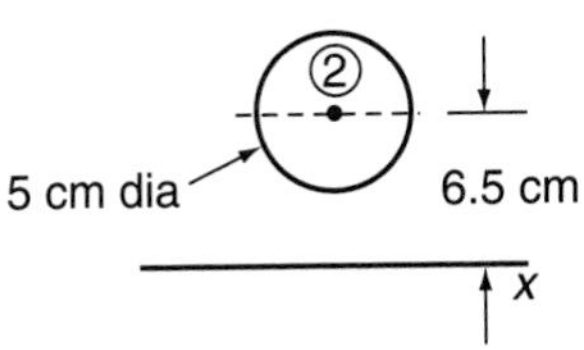

FIGURE 9–10

$$I_c = \frac{\pi r^4}{4}$$

$$= \frac{\pi(2.5\text{ cm})^4}{4}$$

$$I_c = 30.7\text{ cm}^4$$

$$A = \frac{\pi d^2}{4} = \frac{\pi(5\text{ cm})^2}{4}$$

$$A = 19.6\text{ cm}^2$$

Triangle (Figure 9–11):

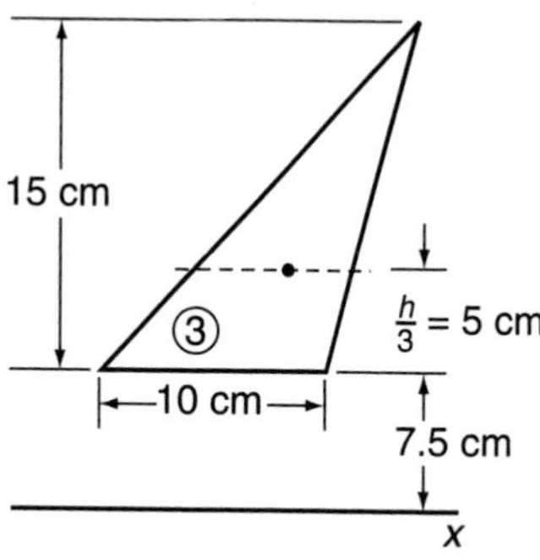

FIGURE 9–11

$$I_c = \frac{bh^3}{36}$$

$$= \frac{(10\text{ cm})(15\text{ cm})^3}{36}$$

$$I_c = 937\text{ cm}^4$$

$$A = \frac{1}{2}bh$$

$$= \frac{1}{2}(10\text{ cm})(15\text{ cm})$$

$$A = 75\text{ cm}^2$$

Apply the parallel axis equation (Equation 9–3) to each area and add the moments about the x-axis. Note that I_x of the circle is negative since it is a cutout area.

$$I_x = I_{x1} - I_{x2} + I_{x3}$$

$$= (I_c + Ad^2)_1 - (I_c + Ad^2)_2 + (I_c + Ad^2)_3$$

$$= [(208.3\text{ cm}^4) + (100\text{ cm}^2)(5\text{ cm})^2]$$

$$- [30.7\text{ cm}^4 + (19.6\text{ cm}^2)(6.5\text{ cm})^2]$$

$$+ [937\text{ cm}^4 + (75\text{ cm}^2)(12.5\text{ cm})^2]$$

$$= 2708 - 859 + 12{,}650$$

$$= 14{,}500\text{ cm}^4 = 14{,}500 \times 10^8\text{ mm}^4$$

$$\underline{I_x = 145 \times 10^6\text{ mm}^4}$$

EXAMPLE 9–5

Determine the moment of inertia about the horizontal centroidal axis of the composite area in Figure 9–12.

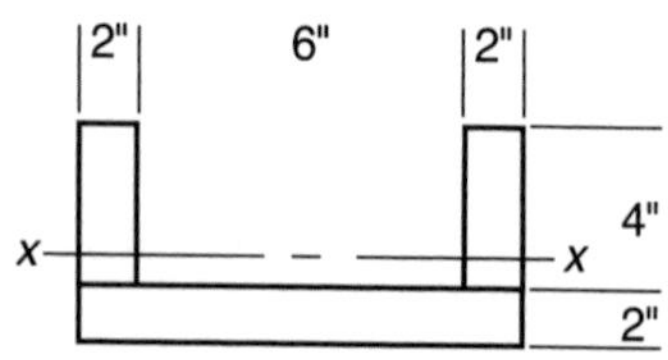

FIGURE 9–12

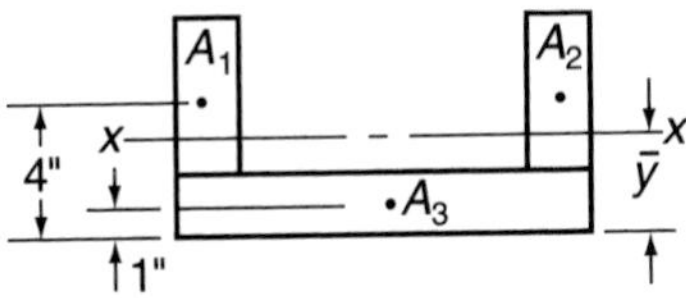

FIGURE 9–13

The solution is similar to that of Example 9–4; here, however, we must first find the location of the centroidal axis x-x (Figure 9–13).

$$A = A_1 + A_2 + A_3$$
$$= 8 \text{ in.}^2 + 8 \text{ in.}^2 + 20 \text{ in.}^2$$
$$A = 36 \text{ in.}^2$$
$$A\bar{y} = A_1y_1 + A_2y_2 + A_3y_3$$
$$(36 \text{ in.}^2)\bar{y} = (8 \text{ in.}^2)(4 \text{ in.}) + (8 \text{ in.}^2)(4 \text{ in.}) + (20 \text{ in.}^2)(1 \text{ in.})$$
$$\bar{y} = \frac{84}{36} = 2.33 \text{ in.}$$

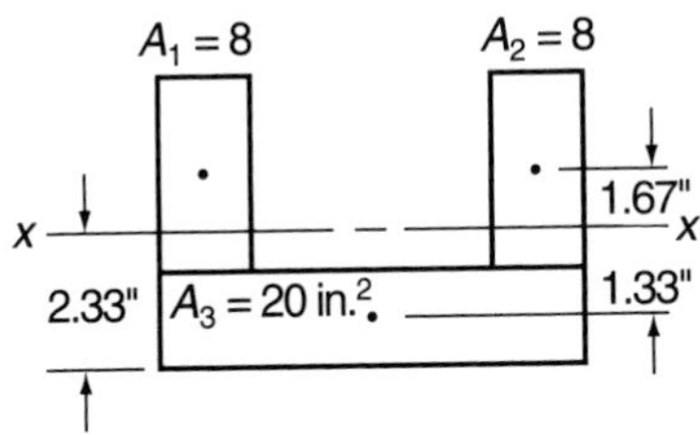

FIGURE 9–14

Determine the centroidal inertia for each area and transfer it to the x-axis (Figure 9–14). This can be done with the one equation.

$$I_x = (I + Ad^2)_1 + (I + Ad^2)_2 + (I + Ad^2)_3$$
$$= \frac{(2 \text{ in.})(4 \text{ in.})^3}{12} + (8 \text{ in.}^2)(1.67 \text{ in.})^2 + \frac{(2 \text{ in.})(4 \text{ in.})^3}{12} + (8 \text{ in.}^2)(1.67 \text{ in.})^2$$
$$+ \frac{(10 \text{ in.})(2 \text{ in.})^3}{12} + (20 \text{ in.}^2)(1.33 \text{ in.})^2$$
$$= 10.67 + 22.31 + 10.67 + 22.31 + 6.67 + 35.4$$
$$\underline{I_x = 108 \text{ in.}^4}$$

9–4 RADIUS OF GYRATION

The *radius of gyration, k*—the distance from the centroidal axis of an area at which the entire area could be concentrated and still have the same moment of inertia—is defined by the following sequence of steps:

1. Calculate I_c for an area A.
2. Assume that area A is concentrated at some distance k from the centroidal axis in such a manner that we still have a moment of inertia equal to I_c.
3. Recall that area moment of inertia was defined by the equation, moment of inertia = (area)(distance)^2. Therefore, we have

$$I_c = Ad^2$$

Since the distance d equals k:

$$k = \sqrt{\frac{I_c}{A}} \tag{9–4}$$

where

I_c = moment of inertia; in.^{4}, mm^4, or cm^4
A = total area; in.^{2}, mm^2, or cm^2
k = radius of gyration; in., mm, or cm

The radius of gyration is a mathematical expression used in conjunction with the slenderness ratio in column design. In column design, the beam will fail by bending about the centroidal axis of the minimum k.

EXAMPLE 9–6

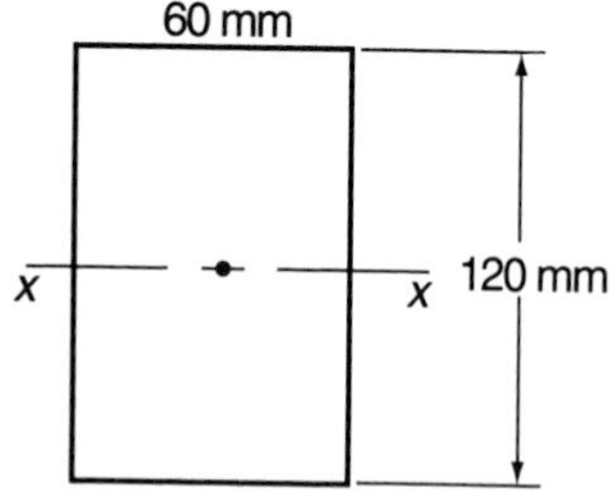

FIGURE 9–15

Determine the radius of gyration about the centroidal x-axis of the area shown in Figure 9–15.

From Table 9–1, the equation for centroidal moment of inertia is

$$I_c = \frac{bh^3}{12}$$

Using cm rather than mm, we get

$$I_c = \frac{(6\text{ cm})(12\text{ cm})^3}{12} = 864\text{ cm}^4$$

$$A = (6\text{ cm})(12\text{ cm}) = 72\text{ cm}^2$$

The radius of gyration is the distance from the x-axis at which the area of 72 cm^2 can be concentrated and still have a moment of inertia of 864 cm^4 about the x-axis.

$$k = \sqrt{\frac{I_c}{A}}$$

$$= \sqrt{\frac{864\text{ cm}^4}{72\text{ cm}^2}}$$

$$= 3.46\text{ cm}$$

$$\underline{k = 34.6\text{ mm}}$$

EXAMPLE 9–7

A tube is used as a column; it has the cross-sectional area shown in Figure 9–16. Determine the radius of gyration about the x-axis.

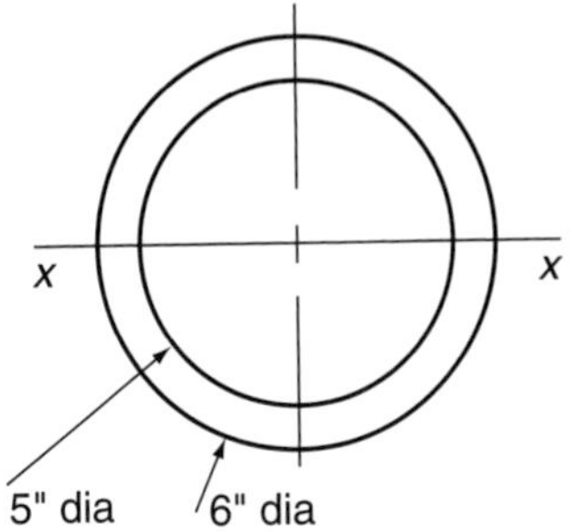

FIGURE 9–16

Find I_x of the composite area by considering the area due to the 5-in. diameter to be negative. Using the inertia formula from Table 9–1, we have

$$I_x = \frac{\pi(r_1)^4}{4} - \frac{\pi(r_2)^4}{4}$$

$$= \frac{\pi}{4}[(3\text{ in.})^4 - (2.5\text{ in.})^4]$$

$$I_x = 33\text{ in.}^4$$

Calculate the cross-sectional area.

$$A = \frac{\pi(d_1)^2}{4} - \frac{\pi(d_2)^2}{4}$$

$$= \frac{\pi}{4}[(6\text{ in.})^2 - (5\text{ in.})^2]$$

$$A = 8.63\text{ in.}^2$$

$$k = \sqrt{\frac{I}{A}}$$

$$= \sqrt{\frac{33\text{ in.}^4}{8.63\text{ in.}^2}}$$

$$\underline{k = 1.95\text{ in.}}$$

9–5 MASS MOMENT OF INERTIA

Mass moment of inertia is a measure of resistance to rotational acceleration and will be used later in dynamics of rotational motion. Referring to Figure 9–17, we define mass moment of inertia by the equation

$$\Delta I = r^2 \Delta m$$

$$I = \Sigma r^2 \Delta m \tag{9–5}$$

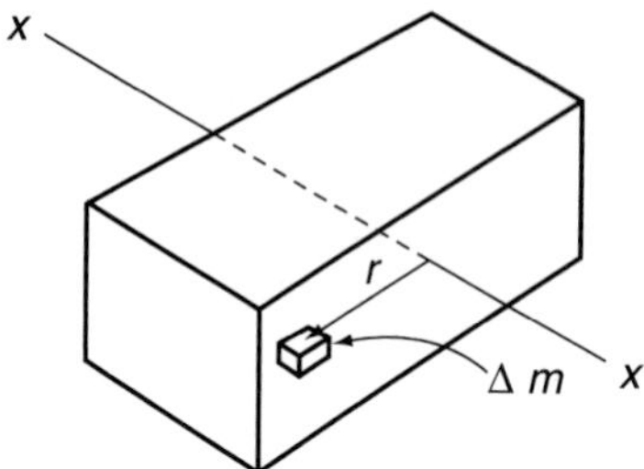

FIGURE 9–17

where in the English system

m = mass; slugs

$$= \frac{W}{g} = \frac{\text{weight}}{\text{gravity acceleration}} = \frac{\text{lb}}{\text{ft/s}^2} = \frac{\text{lb-s}^2}{\text{ft}}$$

r = perpendicular distance from the axis to Δm; ft

I_x = mass moment of inertia; ft-lb-s^2 or slug-ft^2

$$\left[(\text{mass})(\text{distance})^2 = \left(\frac{\text{lb-s}^2}{\text{ft}}\right)(\text{ft})^2 = \text{ft-lb-s}^2\right]$$

All dimensions used in connection with mass moment of inertia in the English system must be in feet since mass has units of slugs. In the SI system, the units for Equation (9–5) are

m = mass; kg

r = distance; m

I_x = mass moment of inertia; kg · m^2

Recall that

$$\text{mass (kg)} = \frac{\text{force of gravity (N or kg}\cdot\text{m/s}^2)}{\text{acceleration of gravity of 9.81 m/s}^2}$$

While all final answers should be shown in units of kg·m^2, to avoid cumbersome numbers, all intermediate calculations can use cm or mm with the conversion to meters being made for the final answer. Mass moment of inertia values can be converted from English to SI units by using the conversion factor 1 ft-1b-s^2 = 1.356 kg·m^2.

As with the other types of inertia of areas that we have discussed, mass moment of inertia is accurately computed by means of calculus. Examples of calculated mass inertia of simple shapes are listed in Table 9–2. We will be considering composite bodies that can be broken into these shapes.

TABLE 9–2
Mass Moment of Inertia

Shape	Mass Moment of Inertia
1. Circular cylinder 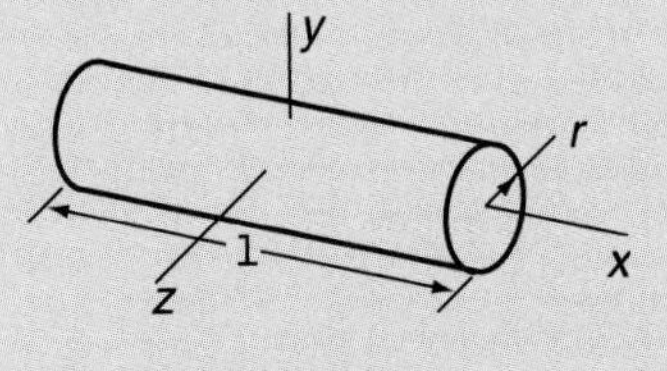	$I_x = \frac{1}{2}mr^2$ $I_y = \frac{1}{12}m(3r^2 + l^2)$ $I_z = \frac{1}{12}m(3r^2 + l^2)$
2. Slender rod 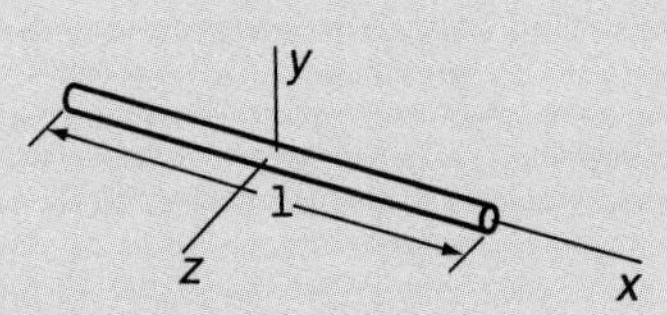	$I_x = 0$ $I_y = \frac{1}{12}ml^2$ $I_z = \frac{1}{12}ml^2$
3. Thin disc 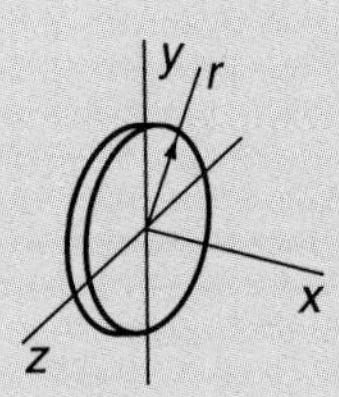	$I_x = \frac{1}{2}mr^2$ $I_y = \frac{1}{4}mr^2$ $I_z = \frac{1}{4}mr^2$
4. Right circular cone 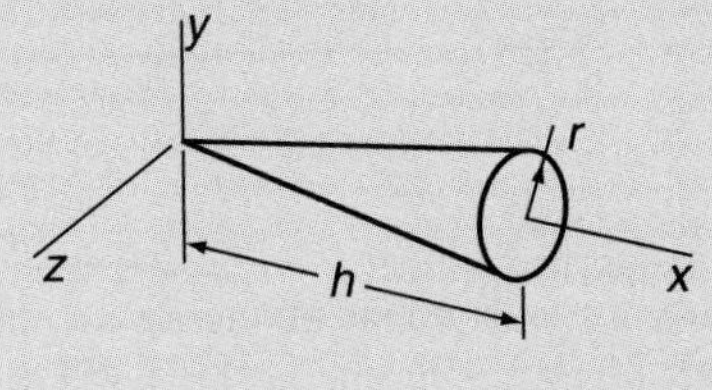	$I_x = \frac{3}{10}mr^2$ $I_y = \frac{3}{5}m\left(\frac{1}{4}r^2 + h^2\right)$ $I_z = \frac{3}{5}m\left(\frac{1}{4}r^2 + h^2\right)$
5. Sphere 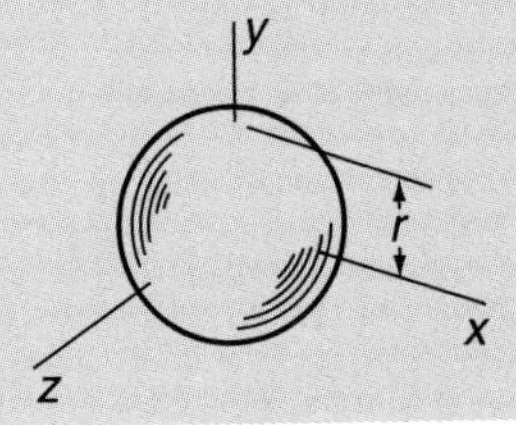	$I_x = \frac{2}{5}mr^2$ $I_y = \frac{2}{5}mr^2$ $I_z = \frac{2}{5}mr^2$

TABLE 9–2
(continued)

Shape	Mass Moment of Inertia
6. Hemisphere	$I_x = \frac{2}{5}mr^2$ $I_y = \frac{2}{5}mr^2$ $I_z = \frac{2}{5}mr^2$
7. Rectangular prism	$I_x = \frac{1}{12}m(a^2 + b^2)$ $I_y = \frac{1}{12}m(a^2 + l^2)$ $I_z = \frac{1}{12}m(b^2 + l^2)$
8. Hollow cylinder	$I_x = \frac{1}{2}m(r_1^2 + r_2^2)$

EXAMPLE 9–8

Determine the mass moment of inertia about the longitudinal axis of a shaft 80 mm in diameter that has a mass of 25 kg.

From Table 9–2,

$$I = \frac{1}{2}mr^2$$

where, using meters, we get

$$m = 25 \text{ kg}$$
$$r = 40 \text{ mm} = 0.04 \text{ m}$$

$$I = \frac{1}{2}(25\text{ kg})(4 \times 10^{-2}\text{ m})^2$$
$$= \frac{25}{2}(16)(10^{-4})$$
$$= 200(10^{-4})$$
$$\underline{I = 0.02\text{ kg}\cdot\text{m}^2}$$

or using mm, we get

$$m = 25\text{ kg}$$
$$r = 40\text{ mm}$$
$$I = \frac{1}{2}(25\text{ kg})(40\text{ mm})^2$$
$$I = 20{,}000\text{ kg}\cdot\text{mm}^2$$

but $1\text{ m}^2 = 10^6\text{ mm}^2$. Therefore,

$$\underline{I = 0.02\text{ kg}\cdot\text{m}^2}$$

9–6 MASS MOMENT OF INERTIA OF COMPOSITE BODIES

Like moment of inertia of area, mass moment of inertia can be transferred from the centroidal axis to any other axis of rotation by means of the equation

$$I = I_c + md^2 \tag{9–6}$$

where

I = mass moment of inertia about some new axis; ft-lb-s^2 or kg·m^2
I_c = mass moment of inertia about the centroidal axis; ft-lb-s^2 or kg·m^2
m = mass; slugs or kg
d = perpendicular distance between axes; ft or m

EXAMPLE 9–9

In the composite body shown in Figure 9–18, A weighs 25 lb and B weighs 10 lb. Calculate the mass moment of inertia about the z-axis.

Divide the composite body into bodies A and B and calculate moment of inertia of each about the z-axis, that is, I_{zA} and I_{zB}.

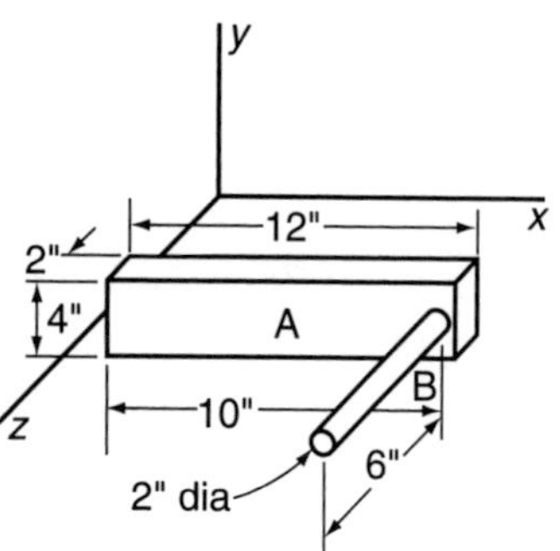

FIGURE 9–18

Applying the appropriate equation from Table 9–2 to our condition of body A (Figure 9–19), we obtain

$$I_c = \frac{1}{12}m(b^2 + l^2)$$

$$= \frac{1}{12}\left(\frac{W}{g}\right)(b^2 + l^2)$$

$$= \frac{1}{12}\left(\frac{25 \text{ lb}}{32.2 \text{ ft/s}^2}\right)[(0.33 \text{ ft})^2 + (1 \text{ ft})^2]$$

$$I_c = 0.072 \text{ ft-lb-s}^2$$

$$I_{zA} = I_c + md^2$$

$$= 0.072 \text{ ft-lb-s}^2 + \left(\frac{25 \text{ lb}}{32.2 \text{ ft/s}^2}\right)(0.5 \text{ ft})^2$$

$$I_{zA} = 0.266 \text{ ft-lb-s}^2$$

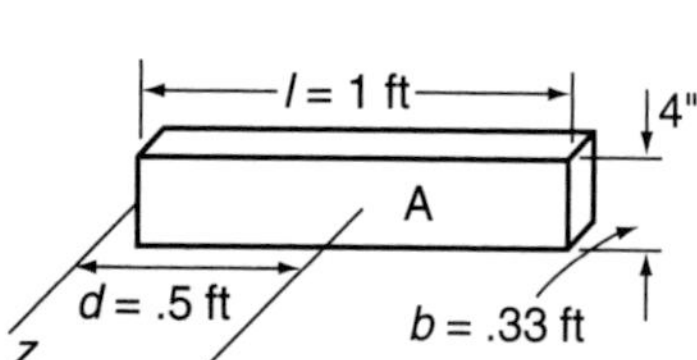

FIGURE 9–19

Applying the same procedure to body B (Figure 9–20), we obtain

$$I_c = \frac{1}{2}mr^2$$

$$= \frac{1}{2}\left(\frac{10 \text{ lb}}{32.2 \text{ ft-lb-s}^2}\right)\left(\frac{1}{12}\text{ft}\right)^2$$

$$I_c = 0.0011 \text{ ft-lb-s}^2$$

$$I_{zB} = I_c + md^2$$

$$= 0.0011 + \left(\frac{10 \text{ lb}}{32.2 \text{ ft/s}^2}\right)\left(\frac{10}{12}\text{ft}\right)^2$$

$$I_{zB} = 0.216 \text{ ft-lb-s}^2$$

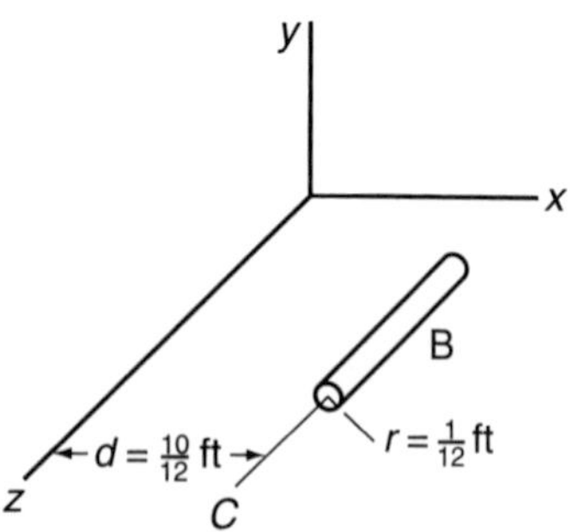

FIGURE 9–20

The total moment of inertia with respect to the z-axis is

$$I_z = I_{zA} + I_{zB}$$
$$= 0.266 \text{ ft-lb-s}^2 + 0.216 \text{ ft-lb-s}^2$$
$$\underline{I_z = 0.482 \text{ ft-lb-s}^2}$$

EXAMPLE 9–10

Calculate the mass moment of inertia about the x-axis for the thin rectangular plate shown in Figure 9–21. The plate had a mass of 5 kg before the hole material (1.07 kg) was cut out.

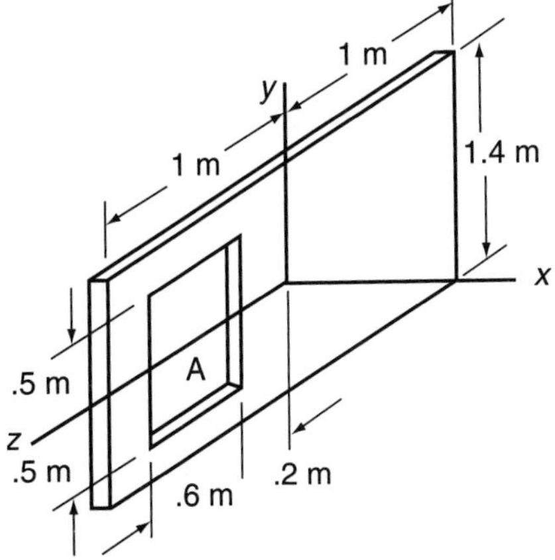

FIGURE 9–21

The method required here is treatment of the hole as a negative mass moment of inertia. Figure 9–22 illustrates the hole treated as an equivalent piece of plate.

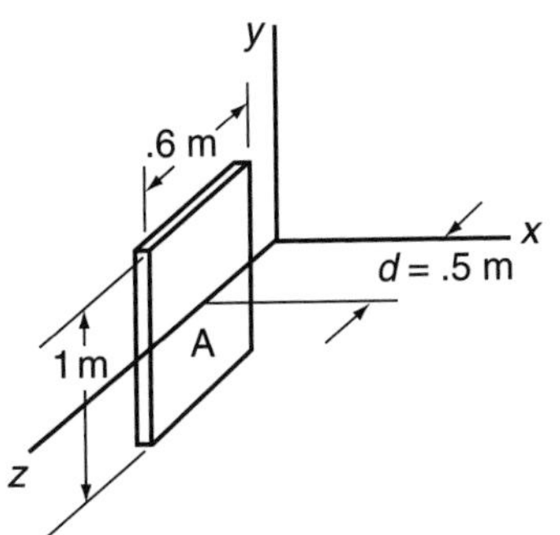

FIGURE 9–22

$$I_c = \frac{1}{12} m(a^2 + b^2)$$
$$= \frac{1}{12}(1.07 \text{ kg})[(0.6 \text{ m})^2 + (1 \text{ m})^2]$$
$$I_c = 0.121 \text{ kg} \cdot \text{m}^2$$
$$I_{xA} = I_c + md^2$$
$$= 0.121 \text{ kg} \cdot \text{m}^2 + 1.07 \text{ kg}(0.5 \text{ m})^2$$
$$I_{xA} = 0.388 \text{ kg} \cdot \text{m}^2$$

Find I_x for the plate without the hole.

$$I_x = I_c = \frac{1}{12} m(a^2 + b^2)$$
$$= \frac{1}{12}(5 \text{ kg})[(2 \text{ m})^2 + (1.4 \text{ m})^2]$$
$$I_x = 2.48 \text{ kg} \cdot \text{m}^2$$

The moment of inertia of the plate with the hole is

$$I_x = 2.48\ \text{kg}\cdot\text{m}^2 - 0.388\ \text{kg}\cdot\text{m}^2$$

$$\underline{I_x = 2.1\ \text{kg}\cdot\text{m}^2}$$

9–7 RADIUS OF GYRATION OF BODIES

Once again—similar to the radius of gyration of areas—the radius of gyration of a mass is defined as:

$$k = \sqrt{\frac{I}{m}} \tag{9–7}$$

where

m = mass; slugs or kg
k = radius of gyration; ft or m
I = mass moment of inertia; ft-lb-s^2 or kg·m^2

For the radius of gyration of a simple body, I_c is used to find the moment of inertia. To obtain the radius of gyration of a composite body, substitute the final I value into Equation (9–7). You cannot add the individual radii of gyration of the simple bodies to get the radius of gyration of a composite body.

Another often-used form of Equation (9–7) is

$$I = k^2 m \tag{9–8}$$

HINTS FOR PROBLEM SOLVING

1. The attention to detail and method of problem layout as suggested in Chapter 8 hints also applies to this chapter.
2. Use the transfer equation $I_x = I_c + Ad^2$ to transfer a moment of inertia only from the *centroidal* axis to another parallel axis.

PROBLEMS

APPLIED PROBLEMS FOR SECTIONS 9–1 AND 9–2

9–1—9–6. Determine the moment of inertia about the x-axis of each of the areas shown in Figures P9–1 to P9–6.

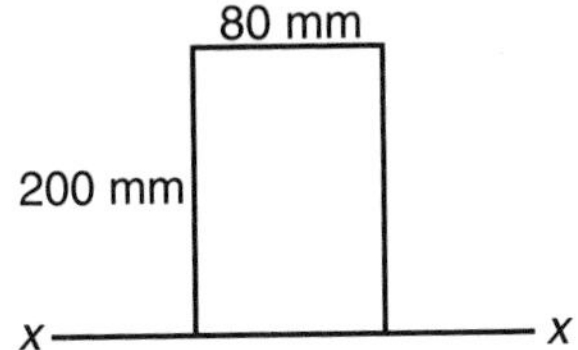

FIGURE P9–1

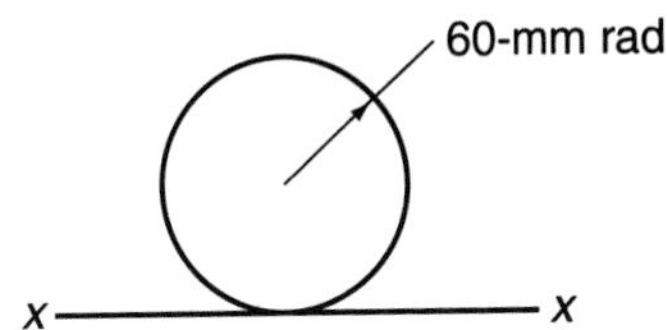

FIGURE P9–2

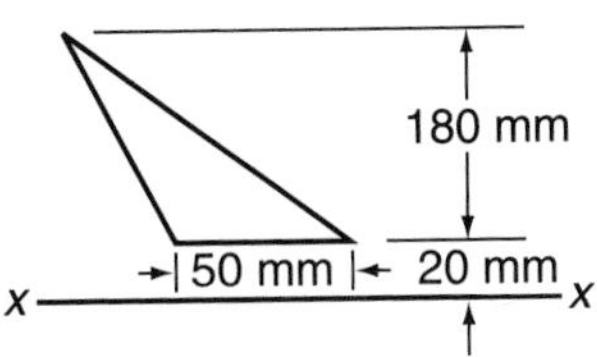

FIGURE P9–3

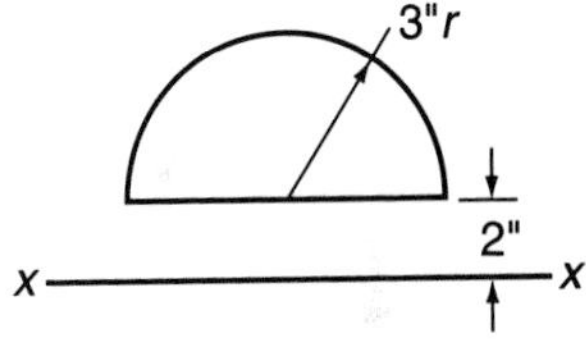

FIGURE P9–4

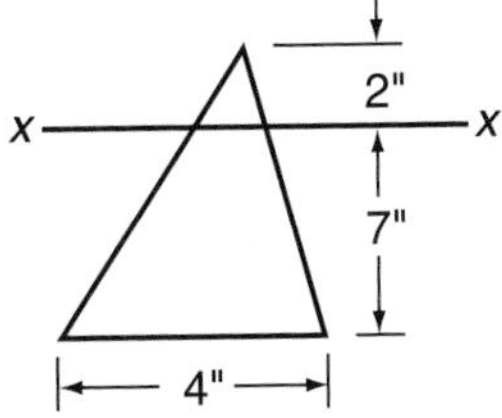

FIGURE P9–5

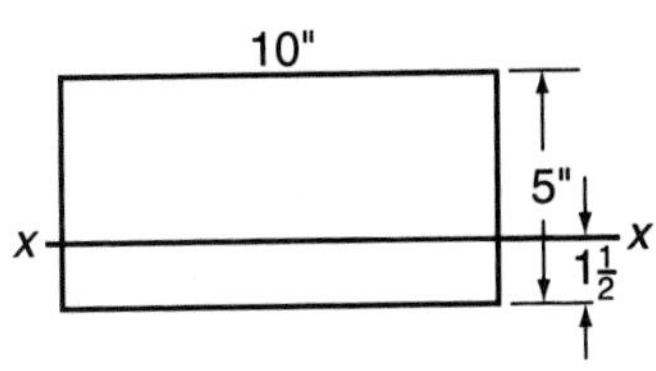

FIGURE P9–6

9–7. The moment of inertia $I_{x1} = 18.67$ in.^{4} for the area shown in Figure P9–7. Find I_{x2}.

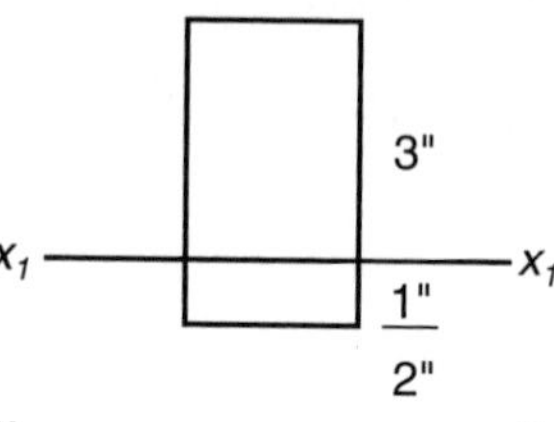

FIGURE P9–7

9–8. The cross section of the beam shown in Figure P9–8 has a thin center web of negligible moment of inertia. Determine dimension b if the total horizontal centroidal moment of inertia is 1344 in.^{4} If the center web dimension is increased from 4 in. to 6 in., determine the new horizontal centroidal moment of inertia.

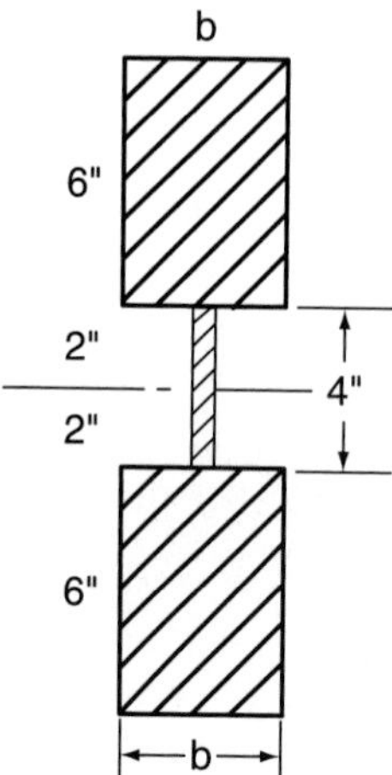

FIGURE P9–8

APPLIED PROBLEMS FOR SECTIONS 9–3 AND 9–4

9–9. Determine the moment of inertia I_x for the area shown in Figure P9–9.

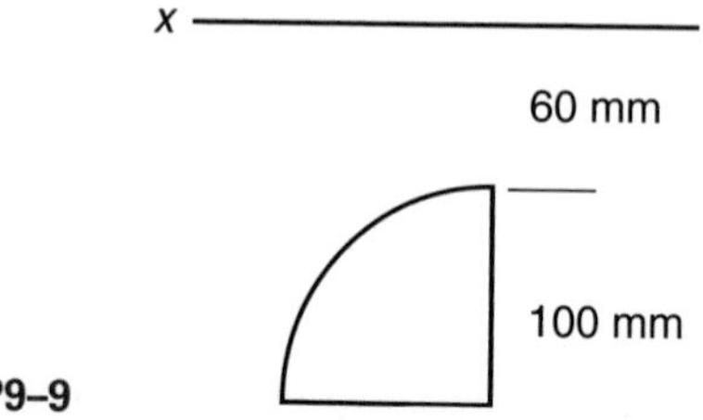

FIGURE P9–9

9–10—9–13. Determine the moments of inertia about the x-axis for the areas shown in Figures P9–10 to P9–13.

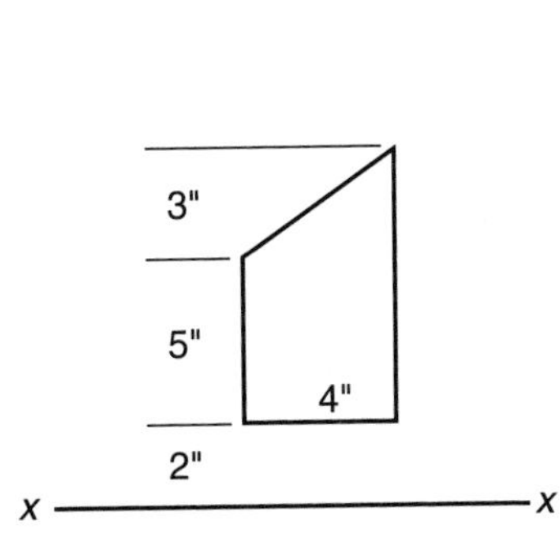

FIGURE P9–10

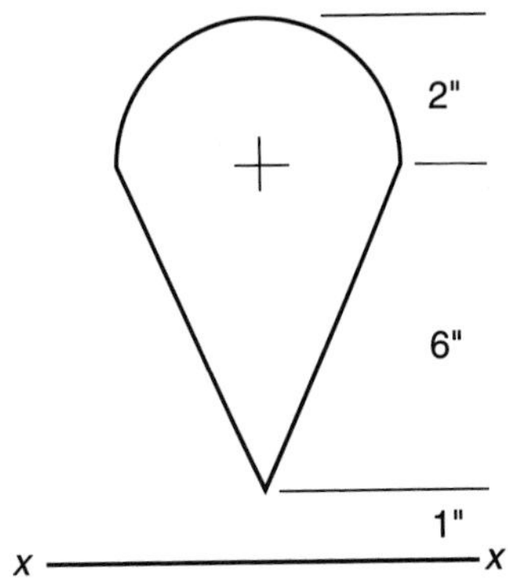

FIGURE P9–11

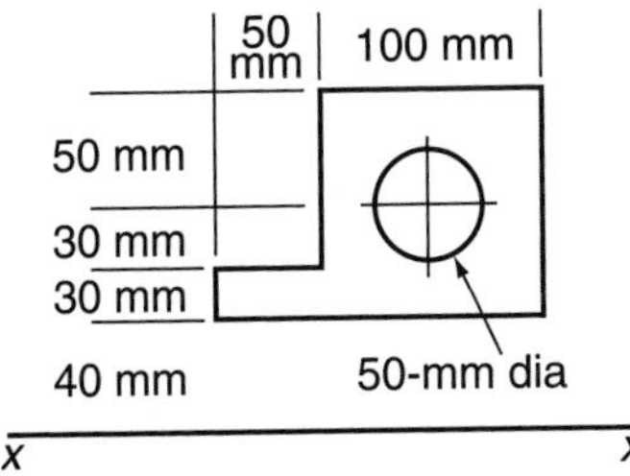

FIGURE P9–12

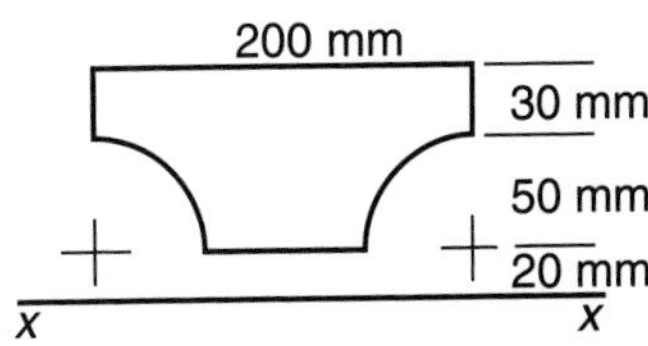

FIGURE P9–13

9–14. Determine the moment of inertia of the area shown in Figure P9–14 about its centroidal x-axis.

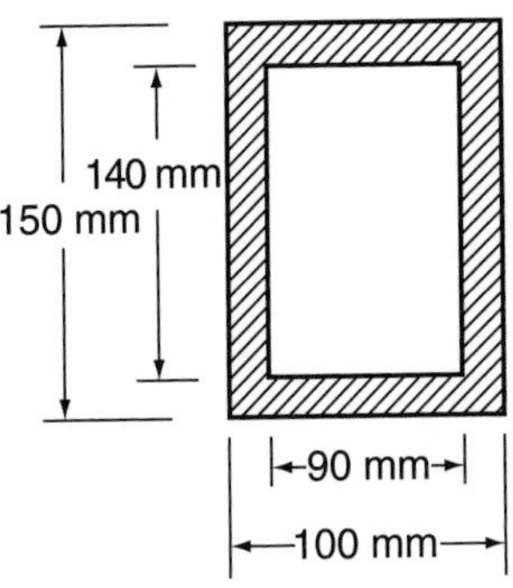

FIGURE P9–14

9–15. Determine the radius of gyration for problem 9–14.

9–16—9–23. Determine the moment of inertia about the centroidal x-axis of the areas shown in Figures P9–16 to P9–23.

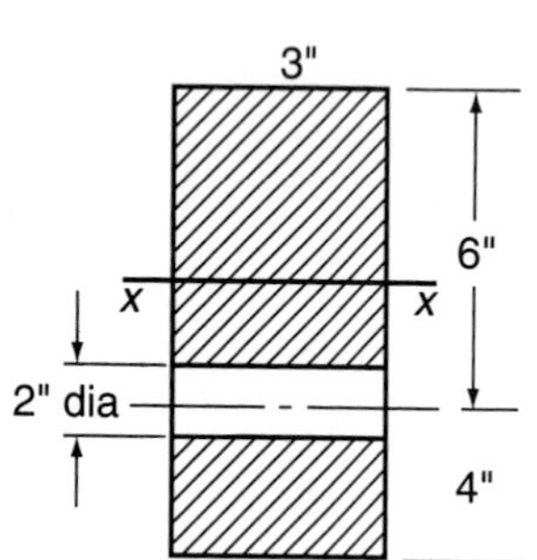

FIGURE P9–16

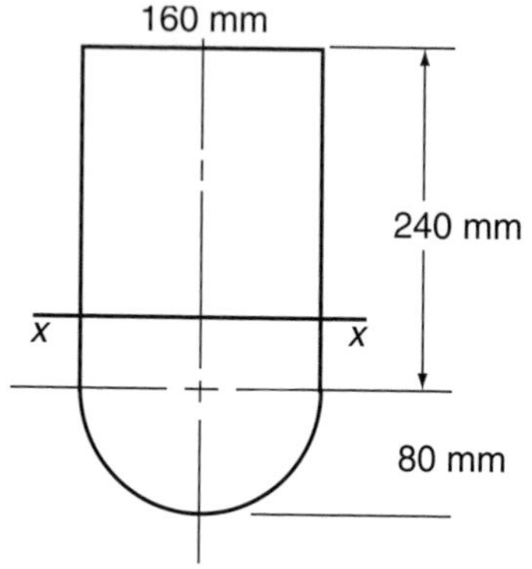

FIGURE P9–17

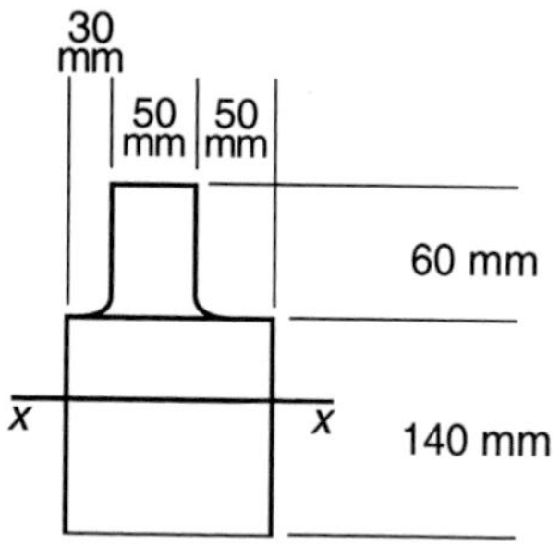

FIGURE P9–18

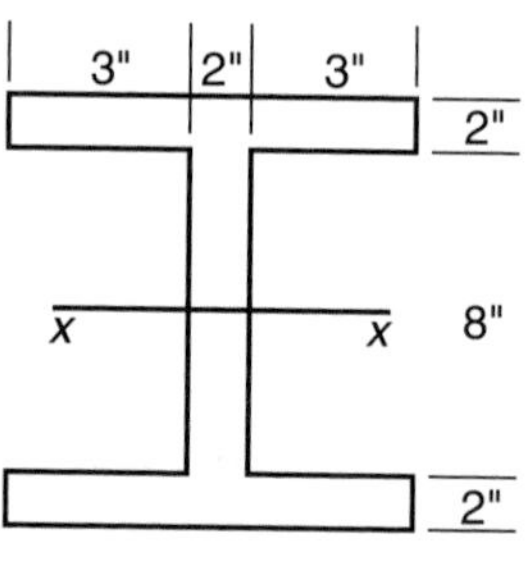

FIGURE P9–19

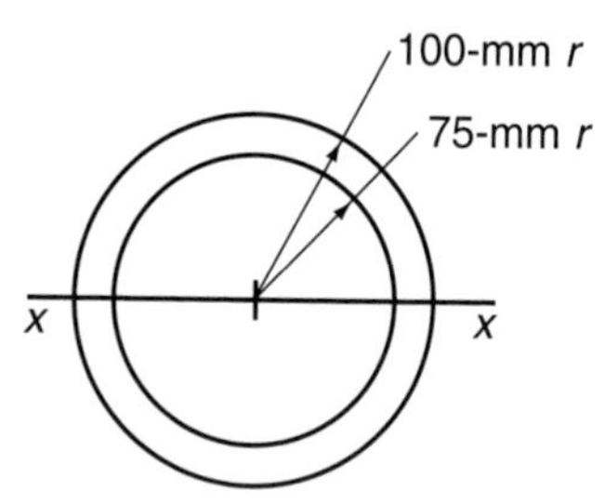

FIGURE P9–20

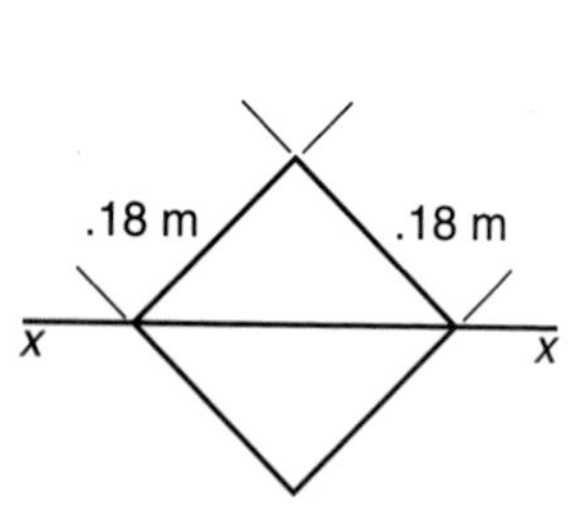

FIGURE P9–21

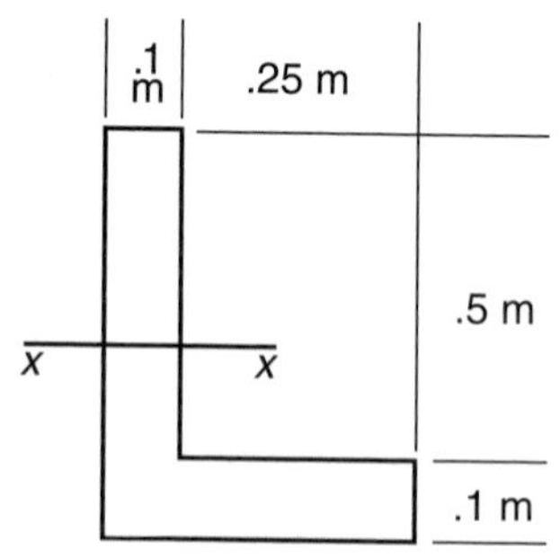

FIGURE P9–22

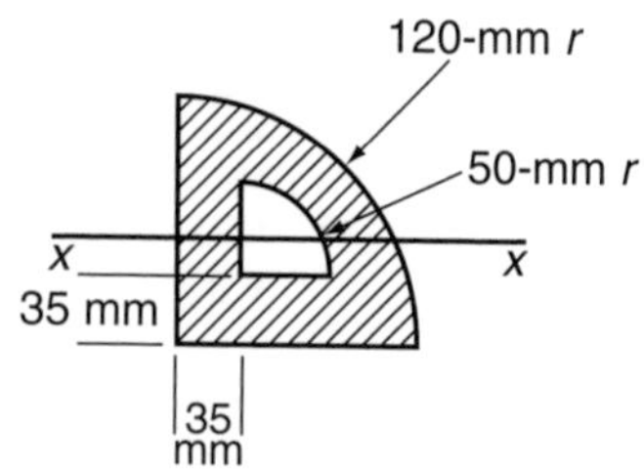

FIGURE P9–23

9–24. The cast frame of a press is subjected to a bending stress and has a cross-sectional area as shown in Figure P9–24. As a first step in the calculation of bending stress, determine the moment of inertia about the centroidal y-axis.

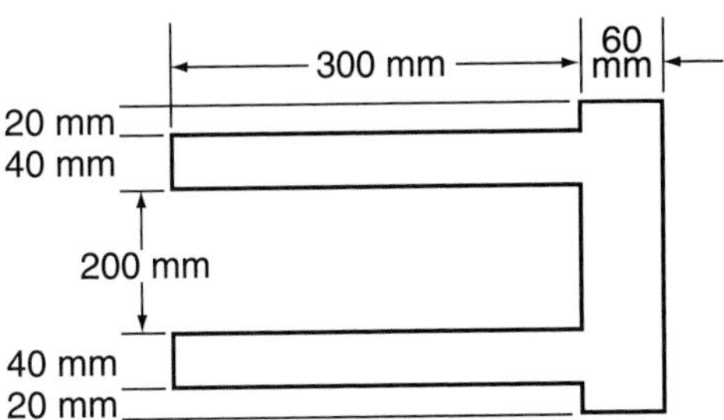

FIGURE P9–24

9–25. Determine the moment of inertia about the centroidal x-axis of the composite area shown in Figure P9–25.

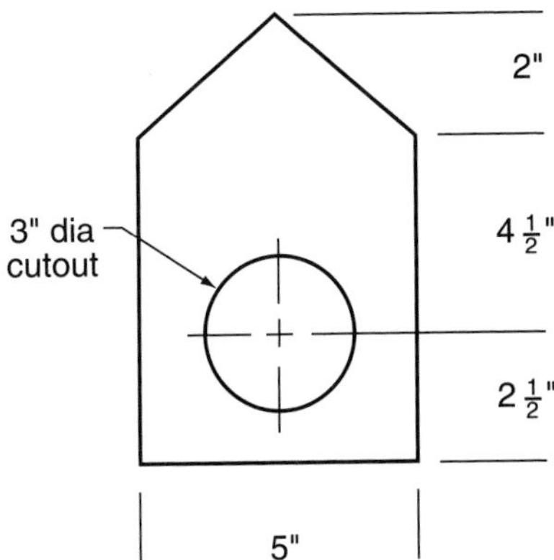

FIGURE P9–25

9–26. Determine the moment of inertia about the centroidal x-axis of the composite area shown in Figure P9–26.

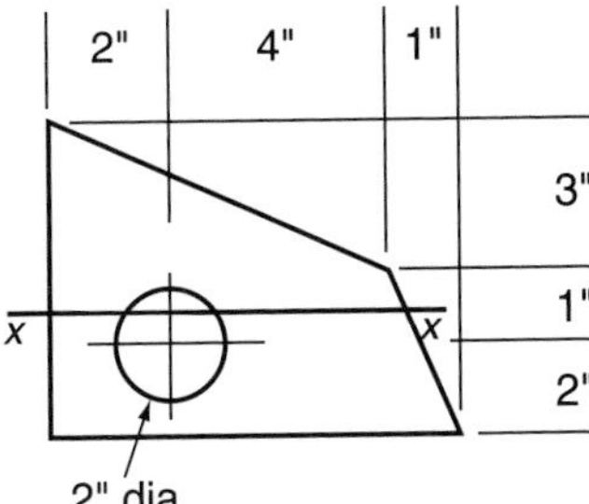

FIGURE P9–26

9–27. The plate shown in Figure P9–27 has a slot cut in it. Determine the area moment of inertia about its centroidal x-axis.

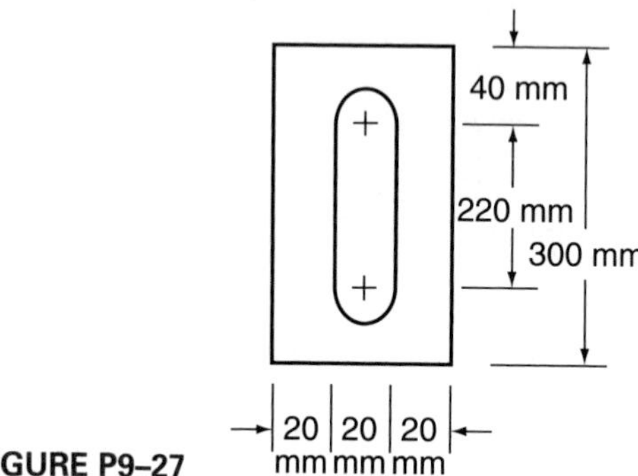

FIGURE P9–27

9–28. Determine the radius of gyration for problem 9–27.

9–29. Determine the moment of inertia of the composite area shown about the horizontal centroidal axis (Figure P9–29).

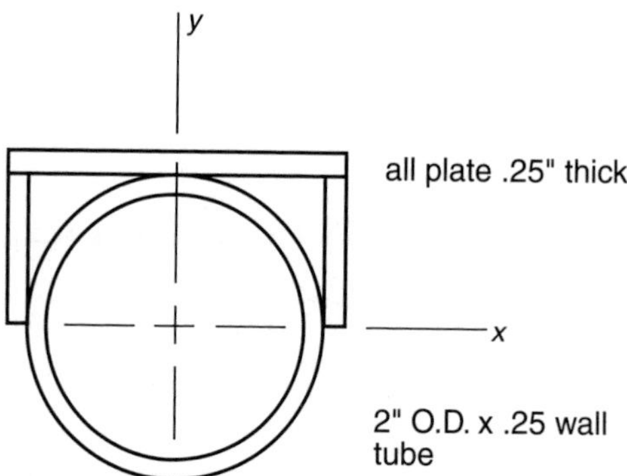

FIGURE P9–29

9–30. Two channels (C 150 × 12) are welded together as shown in Figure P9–30. Determine the moment of inertia about both the x and y centroidal axes. For each channel $I_x = 13.1$ in.^{4} and $I_y = 0.7$ in.^{4}

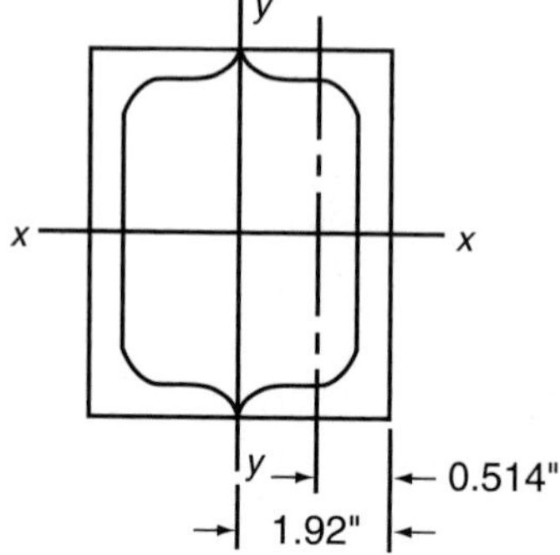

FIGURE P9–30

9–31. Determine the moment of inertia about the centroidal x-axis of the fabricated area shown in Figure P9–31.

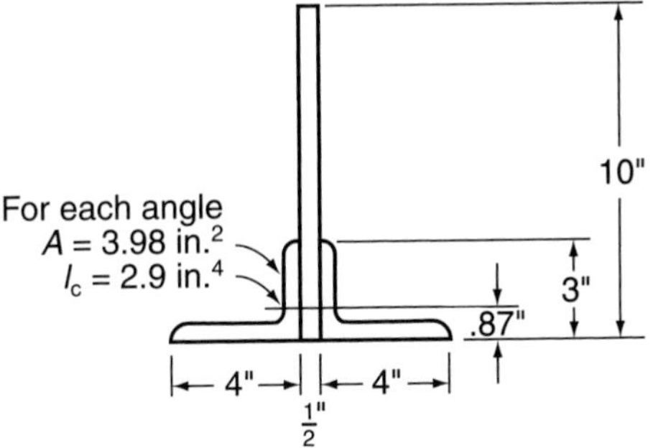

FIGURE P9–31

9–32—9–34. Determine the moment of inertia about the centroidal x-axis for the shapes shown in Figures P8–15 to P8–17.

9–35. Determine the moment of inertia of the composite area shown about the horizontal centroidal axis (Figure P9–35).

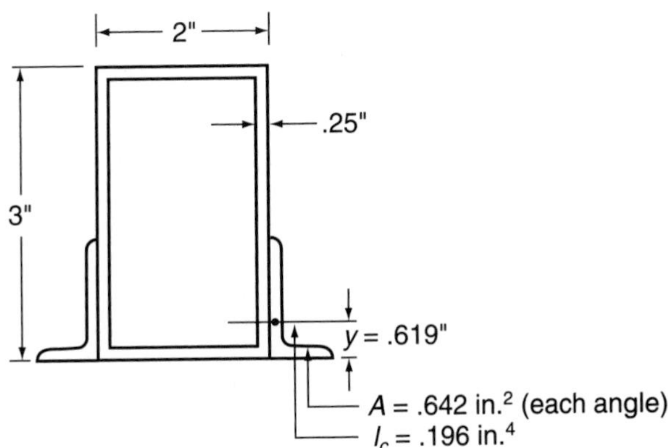

FIGURE P9–35

9–36. Determine the moment of inertia about the centroidal x-axis (Figure P9–36).

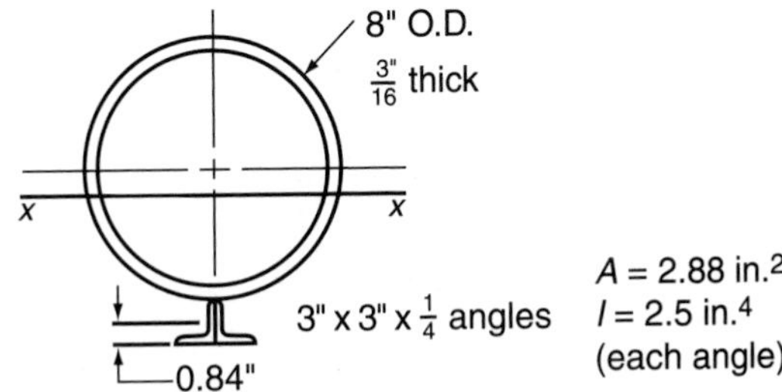

FIGURE P9–36

9–37. Determine the moment of inertia about the centroidal x-axis of the cross-sectional area of the fabricated beam shown in Figure P9–37.

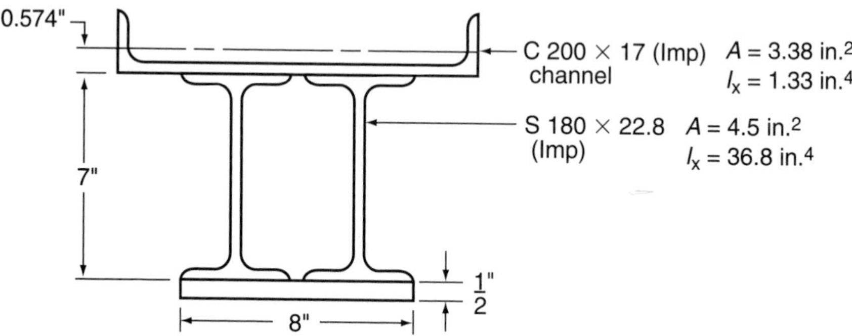

FIGURE P9–37

9–38. Determine the radius of gyration of the area shown in Figure P9–17 with respect to the x-axis.

9–39. Determine the radius of gyration of the area shown in Figure P9–18 with respect to the x-axis.

9–40. Determine the radius of gyration of the area shown in Figure P9–19 with respect to the x-axis.

9–41. Determine the radius of gyration of the area shown in Figure P9–22 with respect to the x-axis.

APPLIED PROBLEMS FOR SECTIONS 9–5 TO 9–7

9–42. Determine the mass moment of inertia and the radius of gyration about the centroidal axis of a sphere 2 ft in diameter and weighing 64.4 lb.

9–43. Determine the mass moment of inertia and the radius of gyration about the centroidal longitudinal axis of a shaft that has a mass of 100 kg and a diameter of 120 mm.

9–44. Determine the mass moment of inertia of a solid cylinder, 4 ft in diameter, about an axis on the surface of the cylinder and parallel to its centroidal axial axis. The cylinder weighs 96.6 lb.

9–45. A slender rod 0.6 m long rotates about an axis perpendicular to its length and 0.14 m from its center of gravity. If the rod has a mass of 8 kg, determine its mass moment of inertia about this axis.

9–46. The right circular cone in Figure P9–46 has a mass of 90 kg. Determine the mass moment of inertia about the x-axis.

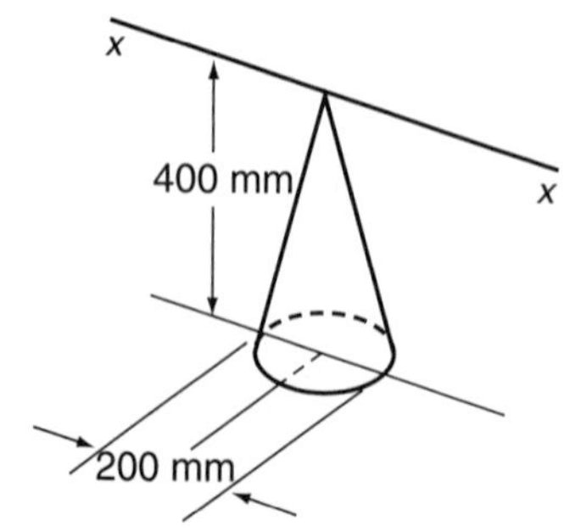

FIGURE P9–46

9–47. A flywheel can be considered as being composed of a thin disc and a rim. The rim weighs 322 lb and has diameters of 24 in. and 30 in. The disc weighs 64.4 lb. Determine the mass moment of inertia about the centroidal axis about which the flywheel rotates.

9–48. The plate in Figure P9–48 has a mass of 3000 kg/m^3. Determine the mass moment of inertia about the x-axis.

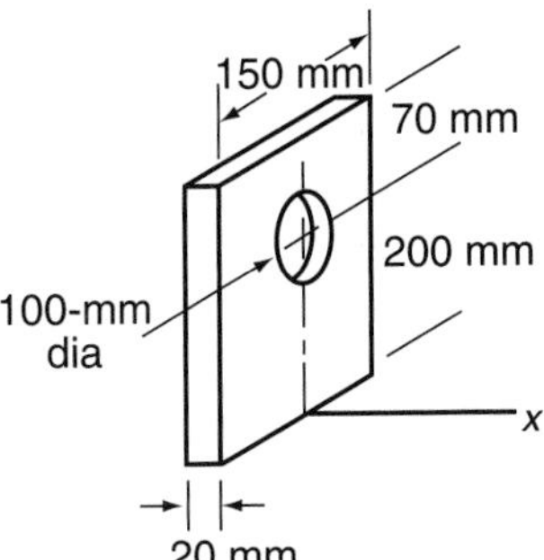

FIGURE P9–48

9–49. The shape in Figure P9–49 weighs 0.2 lb/in.^{3} Determine the radius of gyration about the y-axis. The volume of a right circular cone is $\frac{1}{3}\pi r^2 h$.

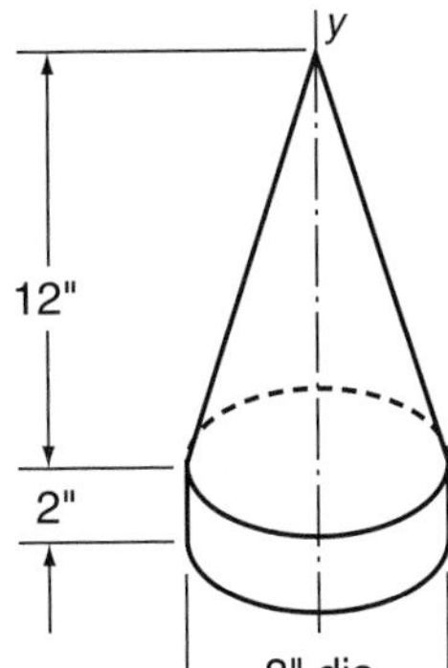

FIGURE P9–49

9–50. The shaft of a shredder has cutter blades welded to the shaft as shown in Figure P9–50. The blades are 10 mm thick. The shaft and blades have a mass of 8000 kg/m^3. Determine the mass moment of inertia about the longitudinal centroidal axis.

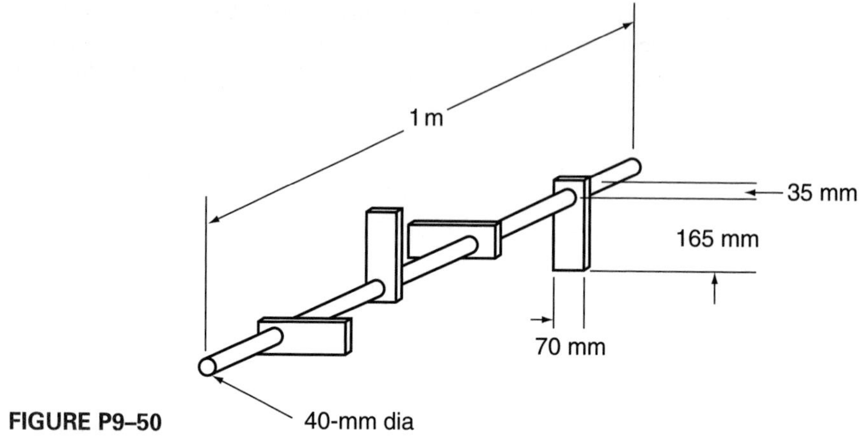

FIGURE P9–50

REVIEW PROBLEMS

R9–1. Determine the moment of inertia about the *y*-axis of the area shown in Figure RP9–1.

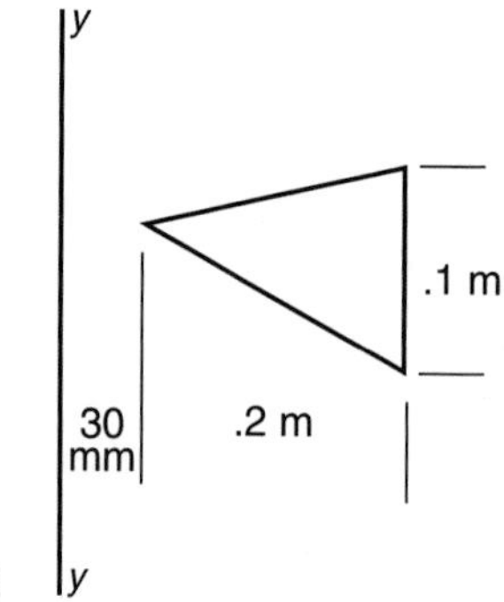

FIGURE RP9–1

R9–2. The moment of inertia is $I_{x1} = 1203 \text{ in.}^4$ for the area shown in Figure RP9–2. Find I_{x2}.

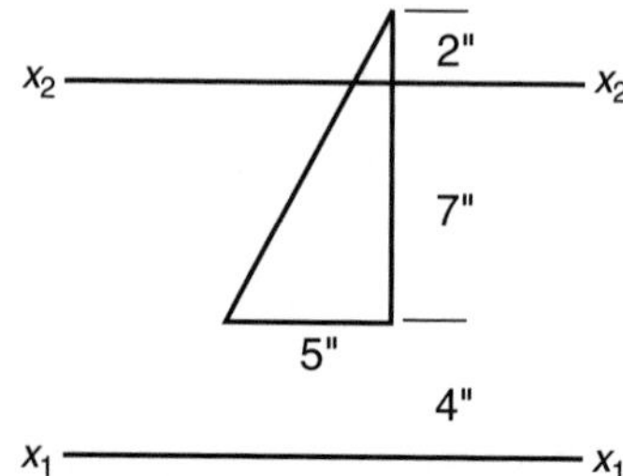

FIGURE RP9–2

R9–3. A composite wood beam has a 1/2"-thick panel of web glued into slots to form the cross section shown in Figure RP9–3. Determine the moment of inertia about the centroidal x-axis.
If the web dimension of 6 in. is increased 33% to 8 in., what is the percent increase in the moment of inertia about the centroidal x-axis?

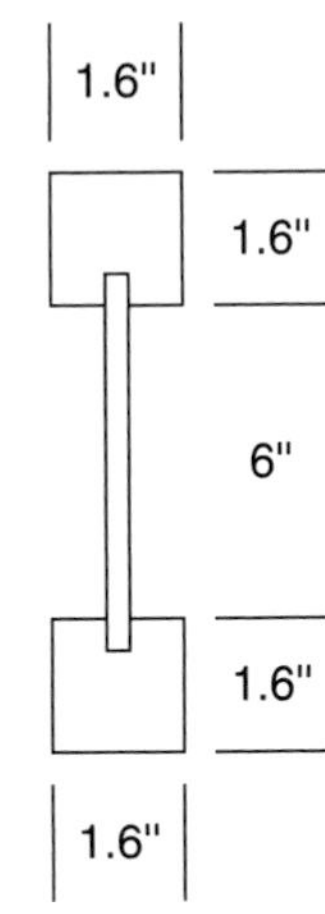

FIGURE RP9–3

R9–4. Determine the moment of inertia of the area shown in Figure RP9–4 about its centroidal x-axis.

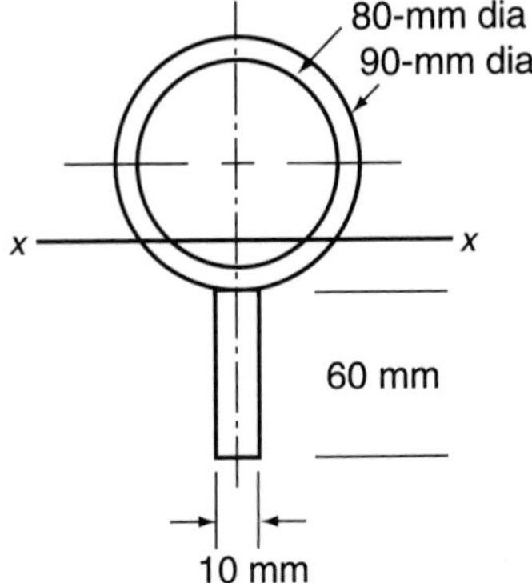

FIGURE RP9–4

R9–5. Determine the radius of gyration for Figure RP9–4.

R9–6. Four equally spaced masses are positioned on a light frame as shown in Figure RP9–6. Find the moment of inertia of the system about the *x*-axis. Neglect the mass of the frame.

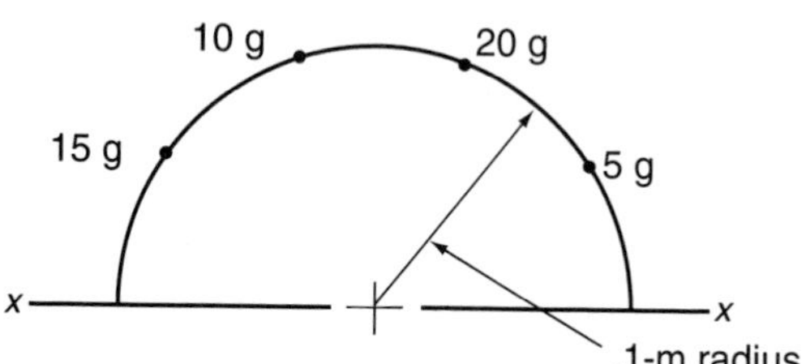

FIGURE RP9–6

R9–7. In the system shown in Figure RP9–7, the sphere weighs 10 lb, the rod weighs 8 lb, and the plate weighs 4 lb. Determine the mass moment of inertia about the *y*-axis.

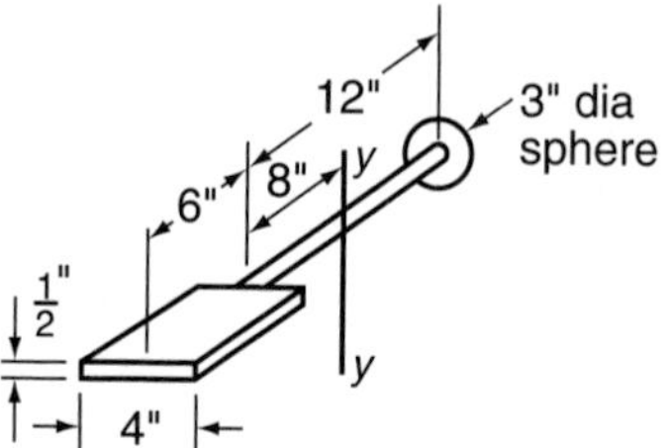

FIGURE RP9–7

R9–8. A 6-in.-diameter cylinder originally weighing 30 lb has a 2-in.-diameter hole drilled in it longitudinally as shown in Figure RP9–8. Determine the moment of inertia of the drilled cylinder about the new centroidal, longitudinal axis.

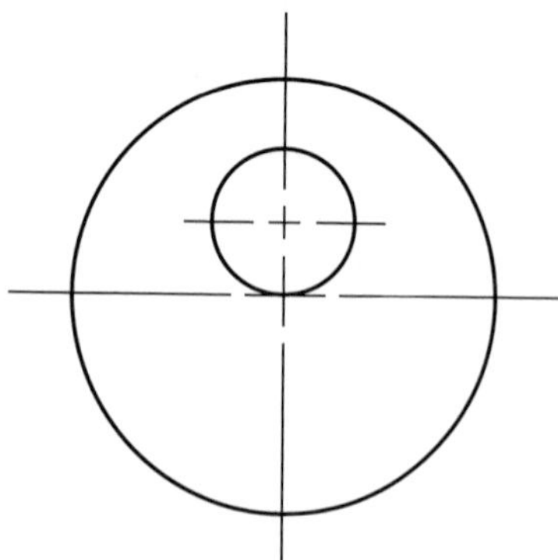

FIGURE RP9–8

CHAPTER 10

Kinematics: Rectilinear Motion

OBJECTIVES

Upon completion of this chapter the student will be able to:

1. Distinguish between distance and displacement, speed and velocity.
2. Calculate average acceleration due to a change in direction of velocity.
3. Solve for displacement, velocity, and acceleration using the three equations of constant acceleration rectilinear kinematics for objects in motion, including projectile motion.

10–1 INTRODUCTION

Previously, we were concerned with *statics*—bodies at rest or with uniform velocity. We now come to *dynamics*—the study of bodies in motion. Dynamics consists of kinematics and kinetics. *Kinematics* is the analysis of the geometry of motion without concern for the forces causing the motion; it involves quantities such as displacement, velocity, acceleration, and time. *Kinetics* is the study of motion and the forces associated with motion; it involves the determination of the motion resulting from given forces.

In our study of kinematics, we will consider motion in one plane only; such motion will be one of three types.

1. *Rectilinear* or *translational motion.* The particle or body moves in a straight line and does not rotate about its center of mass. The piston, pin C in Figure 10–1, has rectilinear or straight-line motion.
2. *Circular motion.* A particle follows the path of a perfect circle. In Figure 10–1, pin B on the end of arm AB has circular motion.
3. *General plane motion.* A particle may follow a path that is neither straight nor circular (point D in Figure 10–1). This also applies to a body that may have both rotating and rectilinear motion simultaneously.

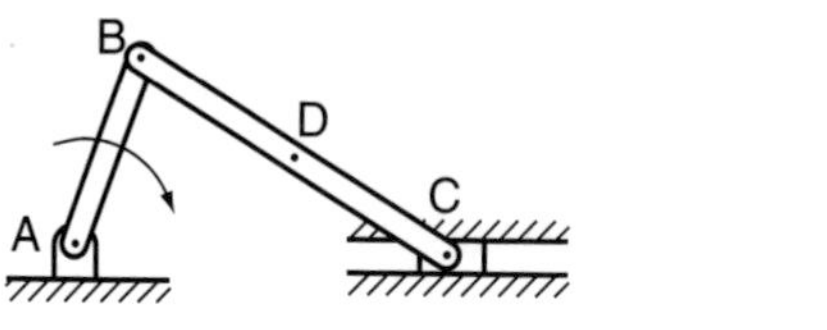

FIGURE 10–1

FIGURE 10–2

Link BC has both rectilinear and circular motion since it partially rotates as it moves downward to the right. General plane motion may be even more random and undefined, as shown by the particle moving from A to B in Figure 10–2. Only particles will be considered in the kinematics of this chapter. A *particle* here refers either to a small concentrated object or to a large object whose center of mass has motion identical to that of all other parts of the object.

A half-loaded tanker truck could not be considered to possess particle motion since the center of mass changes as the tank moves on its springs and the load shifts inside the tank. When an object has rotation about its center of mass, it too cannot possess particle motion since not all portions of the object have identical motion. This type of motion is considered in Chapter 12. The term *particle* does not, therefore, necessarily mean a very small object; it could also refer to a very large object if all portions of that object have the same motion as its center of mass.

10–2 DISPLACEMENT

There is a distinct difference between *distance* and *displacement.* Distance is a scalar quantity, and displacement is a vector quantity. To travel from point A to B (Figure 10–3), any one of three paths may be used. Each path involves a different distance, but they all have the same displacement, 10 m to the right. Displacement is merely the difference between original position and some later position. Equations that involve velocity, acceleration, and time will also have displacement values.

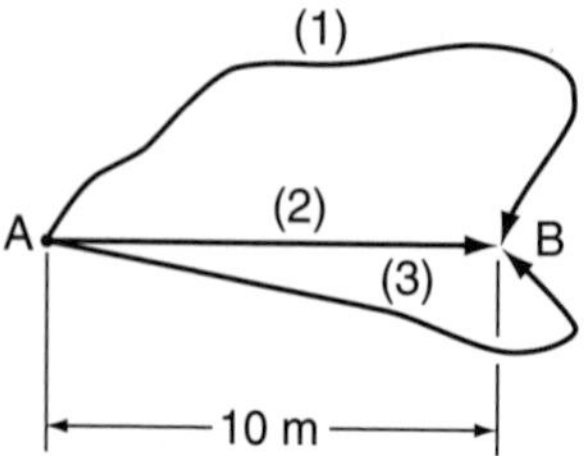

FIGURE 10–3

EXAMPLE 10–1

A car is driven 8 km north, 9 km east, and then another 4 km north. Calculate the displacement of the car and the distance traveled.

The total distance is, naturally, $8 + 9 + 4 = 21$ km. The path traveled can be shown as in Figure 10–4. By the geometry of similar triangles, the lengths of the remaining sides of the two triangles can be found. The displacement of the car is 15 km $\measuredangle^{4}_{3}$.

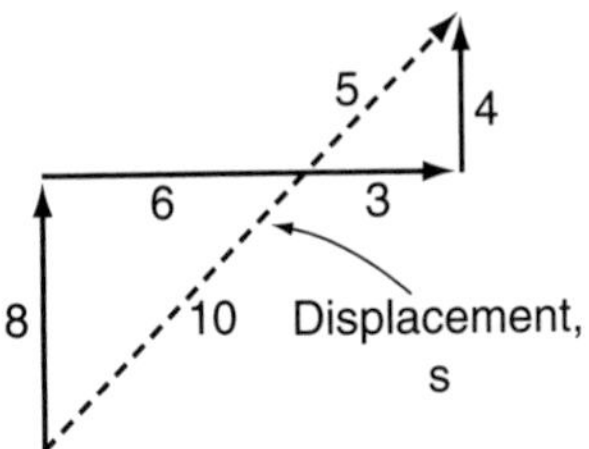

FIGURE 10–4

10–3 VELOCITY

There is also a basic distinction between *speed* and *velocity.* Speed is a scalar quantity; velocity is a vector quantity. Speed is the change of distance per unit of time, such as meters per second (m/s), kilometers per hour (km/h), feet per minute (ft/min), or feet per second (ft/s). None of these units of speed indicates any direction.

Velocity, however, indicates both speed and direction. Velocity is the rate of change of displacement with respect to time and has the same units as speed, that is, m/s, km/h, ft/s, and ft/min. In equation form, we have

$$v = \frac{s}{t} = \frac{\Delta s}{\Delta t} \qquad \text{10–1}$$

where

$$v = \text{velocity (average)}$$
$$s = \text{displacement}$$
$$t = \text{time}$$
$$\Delta s = s_2 - s_1$$
$$\Delta t = t_2 - t_1$$

The velocity obtained by this equation is an average velocity, since there are no values given for intermediate values of displacement and time. A driver of a car that travels 60 km from point A to B in 1 hour may have traveled at a constant speed of 60 km/h or may have stopped for a half-hour and then traveled at 120 km/h. The velocity equation tells us that his end result is an *average velocity* of 60 km/h.

EXAMPLE 10–2

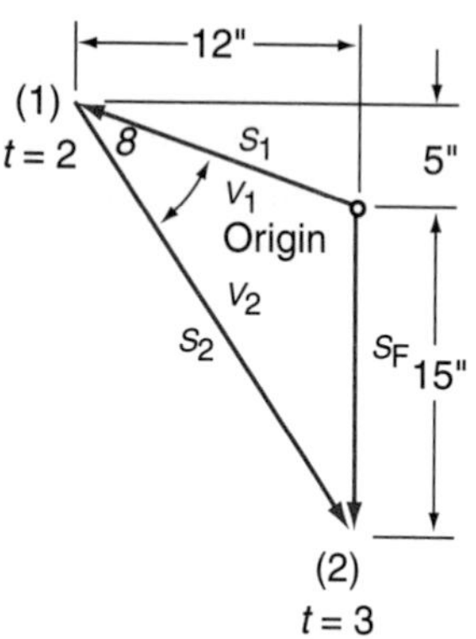

FIGURE 10–5

The path followed by a pin in a printing press mechanism is shown in Figure 10–5. Starting from the origin, it reaches point (1) in 2 seconds and then point (2) after a total elapsed time of 3 seconds. Determine

(a) the velocity from the origin to (1)
(b) the velocity from (1) to (2)
(c) the *average* velocity from the origin to (2)

$$v_1 = \frac{s_1}{t} = \frac{13 \text{ in.}}{2 \text{ s}}$$

$$\underline{v_1 = 6.5 \text{ in./s} \quad (5, 12 \text{ slope})}$$

Displacement s_2 occurs from $t = 2$ to $t = 3$ seconds; therefore

$$s_2 = \sqrt{(12 \text{ in.})^2 + (20 \text{ in.})^2}$$

$$= 23.3 \text{ in.} \quad (5, 3 \text{ slope})$$

and

$$v_2 = \frac{s_2}{t} = \frac{23.3 \text{ in.}}{1 \text{ s}}$$

$$\underline{v_2 = 23.3 \text{ in./s} \quad (5, 3 \text{ slope})}$$

Displacement s_f (final displacement) occurs from $t = 0$ to $t = 3$ seconds.

$$s_F = 15 \text{ in.} \downarrow$$

and $$v_F = \frac{s_F}{t} = \frac{15 \text{ in.}}{3 \text{ s}}$$

$$\underline{v_F = 5 \text{ in./s} \downarrow} \text{ (average velocity only)}$$

10–4 ACCELERATION

Acceleration is the rate of change of velocity with respect to time. As was discussed previously, velocity has both direction and magnitude. A change of velocity either in direction or in magnitude constitutes acceleration. The velocity change with which we are most familiar is that of magnitude. A car whose speed increases from 30 to 60 km/h is said to accelerate. If the car now changes direction while maintaining a speed of 60 km/h, it again experiences acceleration; this time, acceleration is due to the change in direction. Both types of acceleration can be calculated by means of the formula

$$a = \frac{\Delta v}{\Delta t} \tag{10–2}$$

where

$$\begin{aligned} a &= \text{acceleration (average)} \\ v &= \text{velocity change} \\ &= v_2 - v_1 \\ \Delta t &= \text{time for velocity change} \\ &= t_2 - t_1 \end{aligned}$$

Substituting the units of velocity and time into this equation, we see that acceleration is expressed in terms of displacement/time/time. The common units are mm/s^2, m/s^2, in./s^2, and ft/s^2. Acceleration is a vector quantity that has the same direction as the velocity change. Equation (10–2) indicates an average acceleration for a given period.

EXAMPLE 10–3

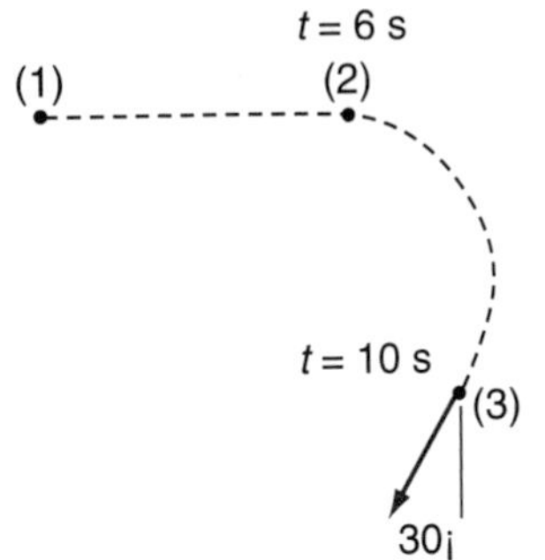

FIGURE 10–6

The car in Figure 10–6 starts from rest at point (1) and uniformly accelerates due east for 6 seconds, reaching a speed of 40 km/h at point (2). Maintaining the speed of 40 km/h, it reaches point (3) at $t = 10$ seconds, traveling a direction of south 30° west. Determine the acceleration from point (1) to point (2) and from point (2) to point (3).

The first important step is to obtain uniform units: convert km/h to m/s.

$$40 \text{ km/h} = \frac{(40 \text{ km/h})(1000 \text{ m/km})}{(60 \text{ min/hr})(60 \text{ sec/min})} = 11.1 \text{ m/s}$$

Applying the acceleration equation (Equation 10–2) between points (1) and (2), we get

$$\begin{aligned} a &= \frac{\Delta v}{\Delta t} = \frac{v_2 - v_1}{t_2 - t_1} \\ &= \frac{11.1 \text{ m/s} - 0}{6\text{s} - 0} \\ \underline{a} &\underline{= 1.85 \text{ m/s}^2 \rightarrow} \end{aligned}$$

Between points (2) and (3),

$$a = \frac{\Delta v}{\Delta t} = \frac{v_3 - v_2}{t_3 - t_2}$$

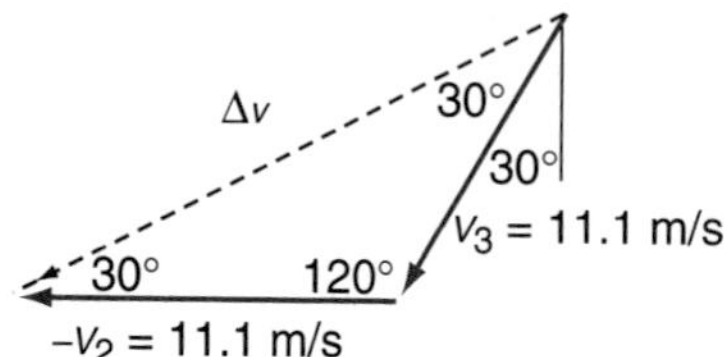

FIGURE 10–7

To find $v_3 - v_2$ vectorially, we add a negative v_2 to v_3; that is, $v_3 + (-v_2)$ (Figure 10–7).

By using the sine law, we get

$$\frac{\Delta v}{\sin 120°} = \frac{11.1 \text{ m/s}}{\sin 30°}$$

$$\Delta v = \frac{11.1(0.866)}{0.5}$$

$$\Delta v = 19.2 \text{ m/s } \overline{30°\!/}$$

Substituting this value into the acceleration equation, we get

$$a = \frac{v_3 - v_2}{t_3 - t_2}$$

$$= \frac{19.2 \text{ m/s}}{10\text{s} - 6\text{s}}$$

$$\underline{a = 4.8 \text{ m/s}^2 \ \overline{30°\!/}}$$

Note here that the direction of the acceleration vector must be the same as the direction of the velocity change, Δv. Remember that these are average velocities and average accelerations. The instantaneous values of acceleration between points (2) and (3) are constantly changing direction.

10–5 RECTILINEAR MOTION WITH UNIFORM ACCELERATION

Motion in a straight line with uniform or constant acceleration is relatively common and easy to analyze. A free-falling body is one example of this type of motion since the acceleration due to gravity (g) is assumed constant at $g = 9.81 \text{ m/s}^2$ (32.2 ft/sec^2) at sea level.

The three equations that relate the variables of time, displacement, velocity, and acceleration to each other are

$$s = v_0 t + \frac{1}{2}at^2 \tag{10–3}$$

$$v = v_0 + at \tag{10–4}$$

$$v^2 = v_0^2 + 2as \tag{10–5}$$

where, in common units,

s = displacement; m or ft
v_0 = initial velocity; m/s or ft/s
v = final velocity; m/s or ft/s
a = constant acceleration; m/s^2 or ft/s^2
t = time; seconds

Keep the following points in mind when you use these equations:

1. Acceleration, although it may be in any direction, must be constant. Constant velocity is a special case in which acceleration is constant at zero.
2. A free-falling body has an acceleration of $a = g = 9.81$ m/s^2 (32.2 ft/s^2).
3. Designate the direction that is to be positive. The direction of the initial velocity or displacement is often used as the positive direction.
4. An object that decelerates or slows down in the positive direction is treated as having a negative acceleration.

These same three Equations (10–3, 10–4, and 10–5) could also be derived using a calculus approach as follows:

$$v = \frac{\Delta s}{\Delta t}$$

$$\text{or } v = \lim_{t \to 0} \frac{\Delta s}{\Delta t}$$

$$v = \frac{ds}{dt} \tag{10–6}$$

Similarly

$$a = \lim_{t \to 0} \frac{\Delta v}{\Delta t}$$

$$a = \frac{dv}{dt} \tag{10–7}$$

Multiplying by $\frac{ds}{ds}$

$$a = \frac{dv\,ds}{dt\,ds} \quad \text{where } \frac{ds}{dt} = v$$

Therefore

$$a = v\frac{dv}{ds} \qquad (10\text{–}8)$$

Integration can be performed on each of these equations as follows: Rearranging Equation (10–7)

$$dv = a\,dt$$

$$\text{or} \int_{v_0}^{v} dv = a\int_0^t dt$$

$$v - v_0 = at$$

$$v = v_0 + at \qquad \text{(previously Equation 10–4)}$$

Using Equation (10–6)

$$ds = v\,dt \quad \text{where} \quad v = v_0 + at$$

$$ds = (v_0 + at)dt$$

$$\int_0^s ds = \int_0^t (v_0 + at)dt$$

$$s = v_0t + \frac{1}{2}at^2 \qquad \text{(previously Equation 10–3)}$$

Using Equation (10–8)

$$v\,dv = a\,ds$$

$$\int_{v_0}^{v} v\,dv = a\int_0^s ds$$

$$\frac{v^2}{2} - \frac{v_0^2}{2} = as$$

$$v^2 = v_0^2 + 2as \qquad \text{(previously Equation 10–5)}$$

EXAMPLE 10–4

An object dropped from the top of a building strikes the ground 7 seconds later. What was the height of the building, and with what velocity did the object strike the ground?

A good rule to follow with this type of problem is to tabulate all given information as follows so that the equation selection becomes more obvious.

Using Equation (10–3), we obtain

$t = 7$ s

$a = 9.81 \text{ m/s}^2$

$s = ?$

$v = ?$

$v_0 = 0$

$$s = v_0 t + \frac{1}{2}at^2$$

$$= 0 + \frac{1}{2}(9.81 \text{ m/s}^2)(7\text{s})^2$$

$$\underline{s = 240 \text{ m}}$$

Using Equation (10–4), we obtain

$$v = v_0 + at$$

$$= 0 + (9.81 \text{ m/s}^2)(7\text{s})$$

$$\underline{v = 68.7 \text{ m/s} \downarrow}$$

EXAMPLE 10–5

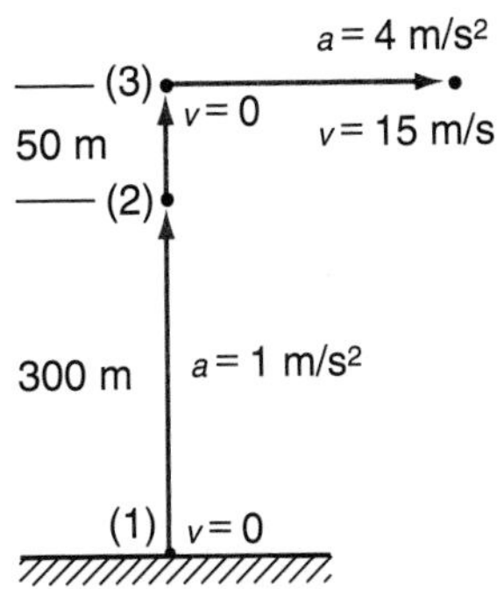

FIGURE 10–8

A helicopter accelerates uniformly upward at 1 m/s^2 to a height of 300 m. By the time it reaches 350 m, it has decelerated to zero vertical velocity. It then accelerates horizontally at 4 m/s^2 to a velocity of 15 m/s. Determine the total time required for this sequence.

A suggestion for the solution here is to draw a sketch of the several stages of flight, as in Figure 10–8 and to label the given information. Now calculate the time for each stage.

Stage (1) to (2):

$s = 300$ m

$v_0 = 0$

$a = 1 \text{ m/s}^2$

$v = v_2$

$$s = v_0 t + \frac{1}{2}at^2$$

$$300 \text{ m} = 0 + \frac{1}{2}(1 \text{ m/s}^2)(t^2)$$

$$t = 24.5 \text{ s}$$

Let the velocity at (2) be v_2.

$$v = v_0 + at$$
$$v_2 = 0 + 1\ \text{m/s}^2(24.5\ \text{s})$$
$$\underline{v_2 = 24.5\ \text{m/s} \uparrow}$$

Stage (2) to (3):

$v_0 = 24.5\ \text{m/s} \uparrow$ $\quad v_2^2 = v_0^2 + 2as$

$s = 50\ \text{m}$ $\quad 0 = (24.5\ \text{m/s}^2)^2 + 2(a)(50\ \text{m})$

$v = 0$ $\quad a = -6\ \text{m/s}^2 \uparrow$

(The minus sign indicates a deceleration in the direction of the arrow.)

$$v = v_0 + at$$
$$0 = (24.5\ \text{m/s}) - (6\ \text{m/s}^2)$$
$$\underline{t = 4.08\ \text{s}}$$

Stage (3) to (4):

$v_0 = 0$ $\quad v = v_0 + at$

$v = 15\ \text{m/s} \rightarrow$ $\quad 15\ \text{m/s} = 0 + (4\ \text{m/s}^2)t$

$a = 4\ \text{m/s}^2 \rightarrow$ $\quad \underline{t = 3.75\ \text{s}}$

$$\text{total time elapsed} = 24.5 + 4.08 + 3.75$$
$$\text{total time elapsed} = 32.3\ \text{s}$$

EXAMPLE 10–6

A particle starting from rest and traveling to the right in a straight line accelerates uniformly to a velocity of 30 ft/s in 3 seconds. It then uniformly decelerates so that, from its initial place of rest to its final position, the displacement is 15 ft to the right and the total distance traveled is 155 ft. Determine the total time interval and the final velocity.

Once again, a labeled sketch of points defining the motion of this particle will aid in the problem's solution (Figure 10–9). We cannot be sure whether the final position, point (4), is to the left or to the right of point (2), so we will solve for the distance

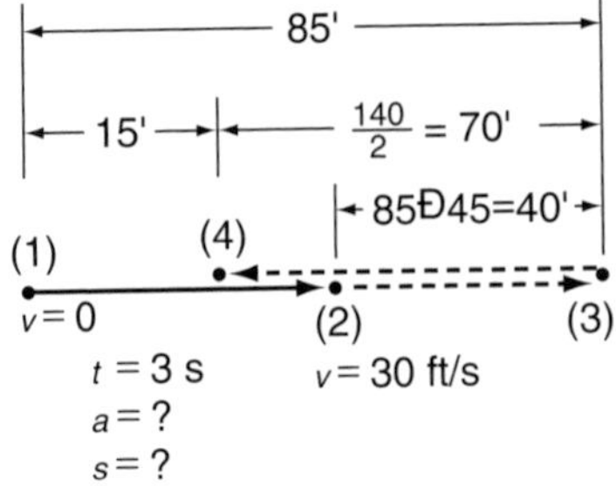

FIGURE 10–9

between points (1) and (2). Initial movement is to the right, so the positive direction throughout the solution will be to the right. Between (1) and (2):

$$v_0 = 0 \qquad v = v_0 + at$$

$$v = 30\text{ ft/s} \qquad 30\text{ ft/s} = 0 + a(3\text{ s})$$

$$t = 3\text{ s} \qquad a = 10\text{ ft/s}^2 \rightarrow$$

$$a = ? \qquad s = v_0t + \frac{1}{2}at^2$$

$$s = ? \qquad = 0 + \frac{1}{2}(10\text{ ft/s}^2)(3\text{ s})^2$$

$$s = 45\text{ ft}$$

To get from point (2) to the final position, point (4), the particle must reverse direction; that is, it must decelerate. Since at point (2) it is traveling 30 ft/s to the right, it must decelerate to zero velocity at some point—point (3)—and, at this same deceleration, return to the left, arriving at point (4) with some final velocity.

If the total distance traveled from points (1) to (4) is 155 ft, then the distance between (3) and (4) is (155 − 15)/2 = 70 ft (Figure 10–9). The distance between (2) and (3) is then (15 + 70) − 45 = 40 ft. Solve for the deceleration rate between points (2) and (3).

$$v_0 = 30\text{ ft/s} \qquad v^2 = v_0^2 + 2as$$

$$v = 0 \qquad 0 = (30\text{ ft/s})^2 + 2(a)(40\text{ ft})$$

$$a = ? \qquad a = -11.25\text{ ft/s}^2$$

$$s = 40\text{ ft}$$

You could now solve for individual velocities and times between points (2) and (3) and points (3) and (4), but a shorter method would be to calculate directly the velocity between points (2) and (4).

$$s = -30\text{ ft} \qquad v^2 = v_0^2 + 2as$$

$$v_0 = 30\text{ ft/s} \qquad v^2 = (30\text{ ft/s})^2 + (2)(-11.25\text{ ft/s}^2) \times (-30\text{ ft})$$

$$v = ?$$

$$a = -11.25\text{ ft/s}^2 \qquad \underline{v = 39.7\text{ ft/s} \leftarrow}$$

Since v was squared, we cannot know the sign of this answer, but by observation we know that the direction of the velocity is to the left. Once again, between points (2) and (4):

$$v = v_0 + at$$

$$-39.7 \text{ ft/s} = 30 \text{ ft/s} - (11.25 \text{ ft/s}^2)t$$

$$t = 6.2 \text{ s}$$

$$\text{total time} = 3 + 6.2$$

$$\underline{\text{total time} = 9.2 \text{ s}}$$

10–6 PROJECTILES

A golf ball driven down the fairway, a baseball hit for a home run, and a rocket fired from a launching pad are all examples of a projectile tracing a path (*trajectory*) during its travel. This *projectile motion* consists of two rectilinear motions occurring simultaneously; projectile motion is comprised of both vertical and horizontal motion. Each of these motions can be represented by displacement, velocity, and acceleration vectors; each motion can therefore be treated separately, or they can be added together vectorially.

The main assumption here will be that of zero air resistance. The only factors affecting projectile motion will thus be initial velocity, the projectile's direction, and the pull of gravity. Acceleration is constant in both directions—zero horizontally and approximately 9.81 m/s^2 (32.2 ft/s^2) vertically. The direction of displacements, velocities, and accelerations will be shown by the previously used sign convention; that is, the direction of the initial velocity is positive.

EXAMPLE 10–7

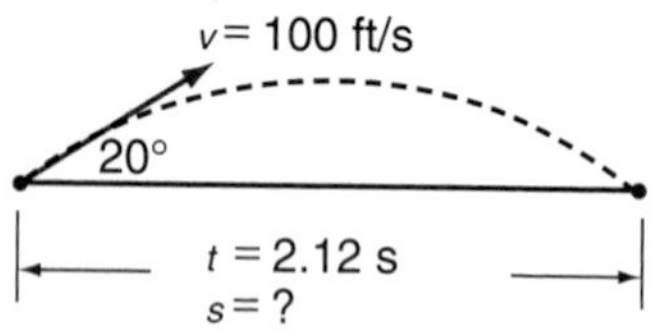

FIGURE 10–10

A projectile is fired at an angle of 20° from the horizontal, with a velocity of 100 ft/s. If it lands 2.12 seconds later at the same elevation, how far did it travel horizontally?

From Figure 10–10, the horizontal component of the velocity is

$$v_x = v \cos \theta$$

$$= (100 \text{ ft/s})\cos 20°$$

$$v_x = 94 \text{ ft/s}$$

$$s = v_0 t + \frac{1}{2}at^2$$

$$= (94 \text{ ft/s})(2.12 \text{ s}) + 0$$

$$\underline{s = 199 \text{ ft}}$$

EXAMPLE 10–8

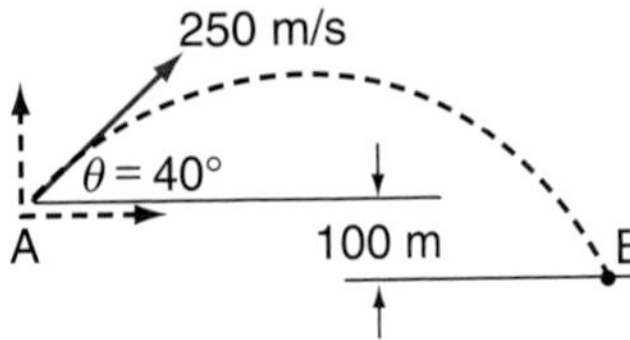

FIGURE 10–11

A projectile that is fired with an initial velocity of 250 m/s is inclined upward at an angle of 40°. It lands at a point 100 m lower than the initial point. Determine (a) the time of flight, (b) the horizontal displacement, and (c) the final velocity at the point of landing. Figure 10–11 shows the given information in sketch form.

(a) Between points A and B in the vertical direction, we have

$$v_0 = v\sin\theta = 250 \sin 40°$$

$$v_0 = 161 \text{ m/s} \uparrow \qquad s = v_0 t + \frac{1}{2}at^2$$

$$s = -100 \text{ m}$$

$$a = -9.81 \text{ m/s}^2 \qquad -100 \text{ m} = (161 \text{ m/s})t + \left(\frac{1}{2}\right)(-9.81 \text{ m/s}^2)t^2$$

$$t = ? \qquad 4.9t^2 - 161t - 100 = 0$$

Using the quadratic equation, we have

$$t = \frac{-b \pm \sqrt{b^2 - 4ac}}{2a}$$

$$= \frac{161 \pm \sqrt{(161)^2 - 4(4.9)(-100)}}{2(4.9)}$$

$$= \frac{161 \pm 167}{9.8}$$

$$t = 33.4 \text{ (or } -0.61) \text{ s}$$

The projectile is in flight for 33.4 s.

(b) Between points A and B in the horizontal direction, we have

$$v_0 = 250 \cos 40°$$

$$v_0 = 192 \text{ m/s} \rightarrow \qquad s = v_0 t + \frac{1}{2}at^2$$

$$t = 33.4 \text{ s} \qquad = (192 \text{ m/s})(33.4 \text{ s}) + 0$$

$$a = 0 \qquad s = 6400 \text{ m}$$

$$s = ?$$

The horizontal distance between A and B is 6400 m.

(c) The final velocity at B is the resultant of the vertical and horizontal velocities. The horizontal velocity remains constant $v_x = 192$ m/s. Between A and B in the vertical direction, we get

$$v_0 = 161 \text{ m/s}\uparrow \qquad v = v_0 + at$$
$$v = v_y \qquad v_y = 161 \text{ m/s} - (9.81 \text{ m/s}^2)(33.4 \text{ s})$$
$$a = -9.81 \text{ m/s}^2 \qquad = -167 \text{ m/s}\uparrow$$
$$t = 33.4 \text{ s} \qquad v_y = 167 \text{ m/s}\downarrow$$

Adding v_x and v_y vectorially, we get

$$\underline{v = 254 \text{ m/s} \searrow 41^\circ}$$

EXAMPLE 10–8 (ALTERNATIVE SOLUTION)

To avoid using the quadratic equation, you could deal with a situation such as the one in Example 10–8 by analyzing the trajectory in small segments AC and CB shown in Figure 10–12. Point C is the top of the trajectory, the point at which velocity in the vertical direction is zero.

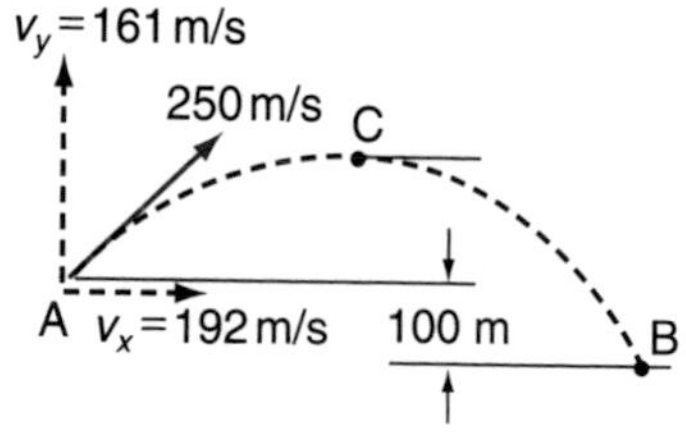

FIGURE 10–12

Step 1. A to C in the vertical direction:

$$v_0 = 161 \text{ m/s}\uparrow \qquad v = v_0 + at$$
$$v = 0 \qquad 0 = 161 \text{ m/s} - (9.81 \text{ m/s}^2)t$$
$$a = -9.81 \text{ m/s}^2 \qquad \underline{t = 16.4 \text{ s}}$$
$$t = ?$$

$$s = v_0 t + \frac{1}{2}at^2$$

$$= (161 \text{ m/s})(16.4 \text{ s}) + \frac{1}{2}(-9.81 \text{ m/s}^2)(16.4 \text{ s})^2$$

$$s = 1321 \text{ m}$$

Step 2. C to B in the vertical direction:

$$s = 1321 + 100$$

$$= 1421 \text{ m} \qquad s = v_0 t + \frac{1}{2}at^2$$

$$a = -9.81 \text{ m/s}^2$$

$$v_0 = 0 \qquad -1421 \text{ m} = 0 + \frac{1}{2}(-9.81 \text{ m/s}^2)t^2$$

$$t = ? \qquad t = 17 \text{ s}$$

$$\underline{\text{total time} = 16.4 + 17 = 33.4 \text{ s}}$$

The horizontal distance and the final velocity would be determined just as they were in the first solution.

EXAMPLE 10–9

A projectile just clears the far side of a building at point B as shown in Figure 10–13. Determine (a) horizontal distance d and (b) the maximum width of the building so that the projectile does not strike it at point C.

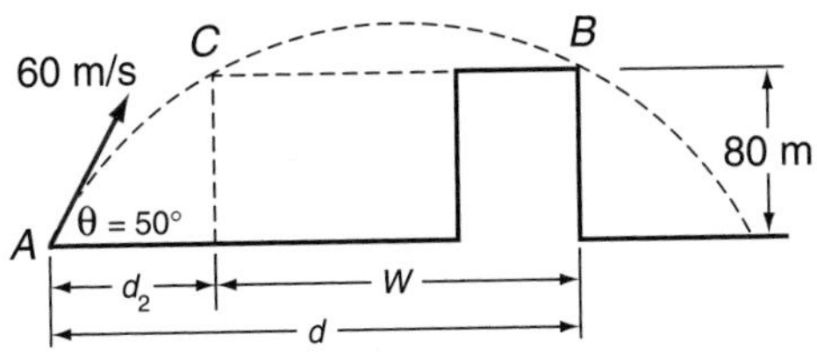

FIGURE 10–13

(a) Between points A and B in the vertical direction, we have

$$v_0 = 60 \sin 50° = 45.96 \text{ m/s}$$
$$a = -9.81 \text{ m/s}^2$$
$$s = 80 \text{ m}$$
$$t = ?$$

Using Equation (10–3), we have

$$s = v_0 t + \frac{1}{2}at^2$$

$$80 \text{ m} = (45.96 \text{ m/s}) + \left(\frac{1}{2}\right)(-9.81 \text{ m/s}^2)t^2$$

$$t^2 - 9.37t + 16.31 = 0$$

Using the quadratic equation

$$t = \frac{-b \pm \sqrt{b^2 - 4ac}}{2a}$$

$$= \frac{9.37 \pm \sqrt{(9.37)^2 - 4(1)(16.31)}}{2}$$

$$= \frac{9.37 \pm 4.754}{2}$$

$= 7.06$ s or 2.31 s (point B or point C)

(b) Between A and B in the horizontal direction, we have

$$v_0 = v = 60 \cos 50° = 38.56 \text{ m/s}$$

$$a = 0$$

$$t = 7.06 \text{ s}$$

$$s = d$$

Using Equation (10–3)

$$s = v_0 t + \frac{1}{2}at^2$$

$$d = (38.56 \text{ m/s})(7.06 \text{ s})$$

$$d = 272 \text{ m}$$

(c) Between A and C in the horizontal direction, we have

$$v_0 = v = 38.56 \text{ m/s}$$

$$s = d_2$$

$$t = 2.31 \text{ s}$$

$$s = v_0 t + \frac{1}{2}at^2$$

$$d_2 = (38.56 \text{ m/s})(2.31 \text{ s})$$

$$= 89 \text{ m}$$

Therefore maximum width $W = 272 \text{ m} - 89 \text{ m} = 183 \text{ m}.$

HINTS FOR PROBLEM SOLVING

1. Since it is important to know not only the answer to a problem, but how it was obtained, the problem solution should be in an orderly sequence as follows:
   (a) Tabulate given information. Watch your sign convention. The direction of the initial velocity is the positive direction.
   (b) For a projectile problem draw a sketch, name the two points for which you are tabulating data, and specify the direction (vertical or horizontal). (Be careful not to mix horizontal and vertical data.)
   (c) Write out the *entire* equation to be used.
   (d) Substitute data into the equation and solve.
   (e) A negative acceleration value indicates deceleration. Continue to use as a negative value in later calculations.
2. For projectile problems that appear to have too many unknowns or to result in simultaneous equations, look for a common factor between points such as time or displacement, which can then be equated.

PROBLEMS

APPLIED PROBLEMS FOR SECTIONS 10–1 TO 10–4

10–1. A 7-ft-long auger is rotated 90° to lie along the side of a grain cart while the cart simultaneously moves 20 ft forward (Figure P10–1). Determine the displacement of point A at the outer end of the auger.

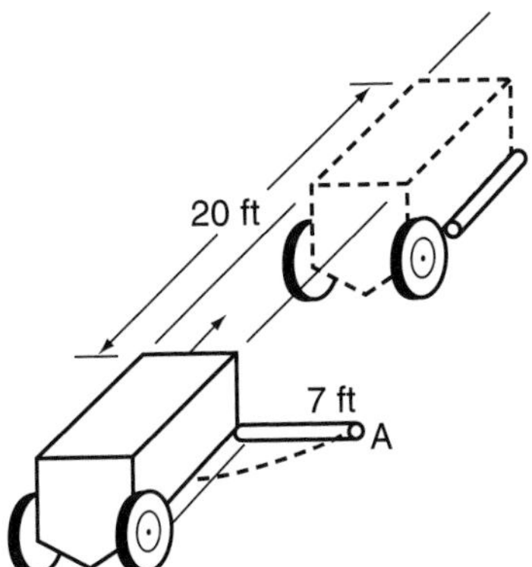

FIGURE P10–1

10–2. A forklift truck lifts a pallet 2 m off the floor, moves 7 m ahead, and sets the pallet on a stack 1.5 m high. Determine the displacement of the pallet and the distance it has traveled.

10–3. A conveyor 60 m long is inclined at an angle of 18° to the horizontal and deposits material 8 m below its top end. Determine the distance and displacement of material carried by this conveyor.

10–4. Calculate the distance and displacement of a plane that flies 100 km east and then 200 km southeast.

10–5. An object travels from A to B on the portable conveyor shown in Figure P10–5. The conveyor belt speed is 1 m/s. The complete portable conveyor is traveling at 9 km/h to the left. Determine the total displacement of the object as it travels from A to B.

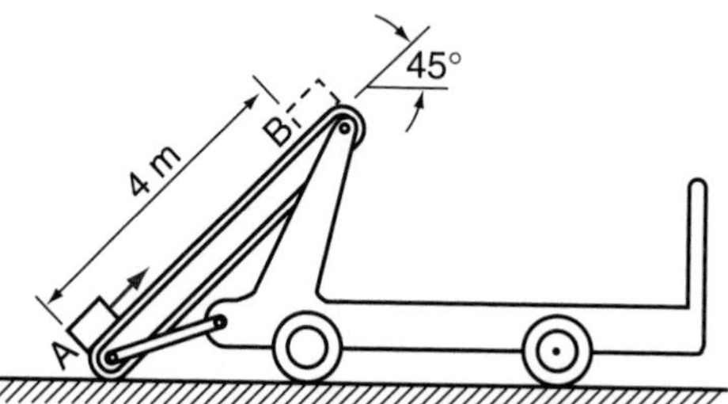

FIGURE P10–5

10–6. A reach forklift is initially at the position shown in Figure P10–6. Determine the displacement of the forks when the boom is extended 16 ft and lifted from 12° to 35°.

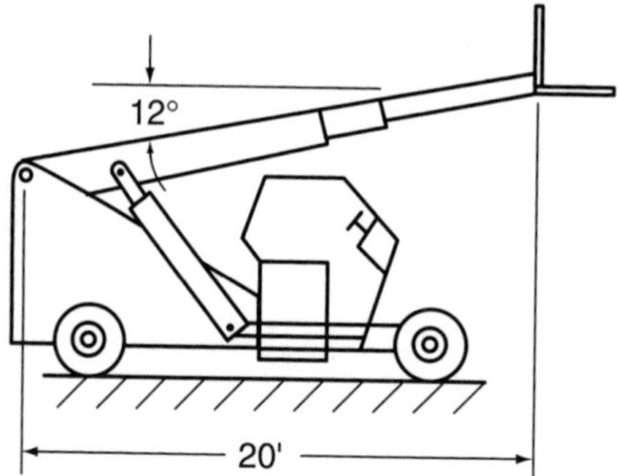

FIGURE P10–6

10–7. A stunt driver travels a total distance of 1172 ft moving at a constant speed of 40 mph while weaving through an obstacle course. If the total displacement of the car is 600 ft from the start of the course to the finish, determine the car's average velocity.

10–8. An object travels from the origin to point (1) in 5 seconds and then to point (2) in a total time of 8 seconds for a final displacement of 5 ft as shown in Figure P10–8. Determine the velocity from point (1) to (2).

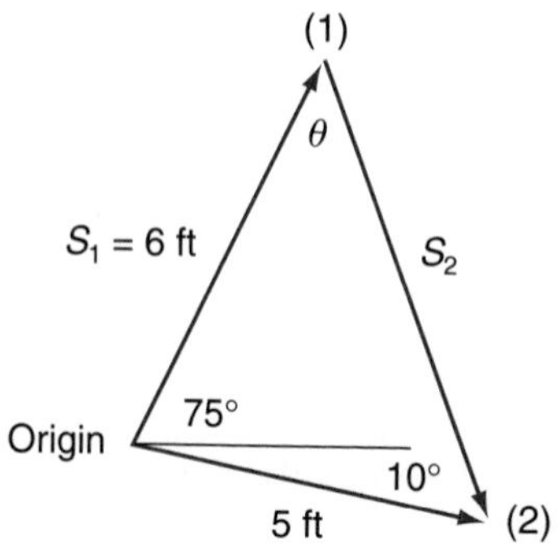

FIGURE P10–8

10–9. A roller coaster car traveling at a constant velocity of 8 m/s goes around a 45° horizontal curve in 5 seconds. Determine the average acceleration.

10–10. A car traveling at a constant speed of 90 km/h takes 8 seconds to travel through a constant radius curve as shown in Figure P10–10. Determine the average acceleration of the car.

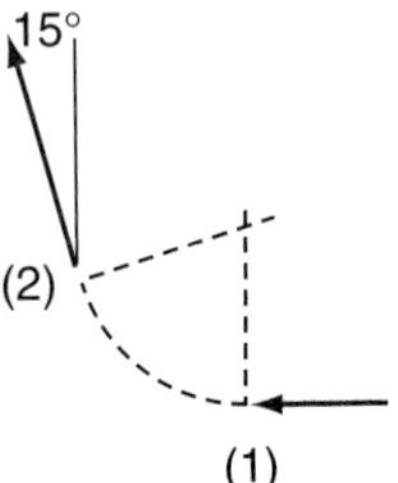

FIGURE P10–10

10–11. The object shown in Figure P10–11 moves from point (1) to (2) in a circular path at a constant speed of 30 m/s in a time of 4 seconds. Determine the average acceleration.

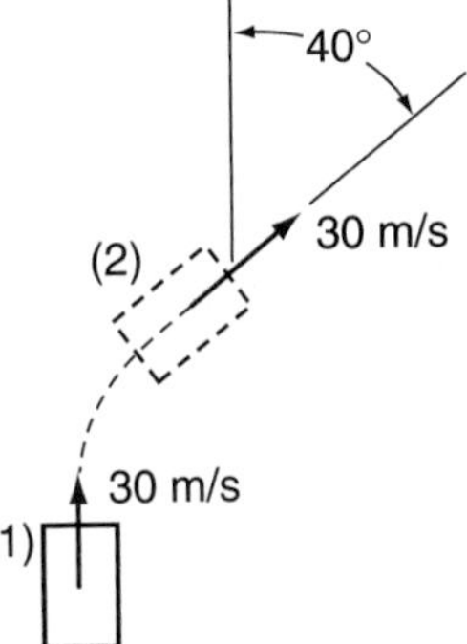

FIGURE P10–11

10–12. A speed skater, when viewed from above, has an initial velocity of 25 ft/s ↑ and 4 seconds later a final velocity of 25 ft/s 40°↗. Determine the skater's average acceleration.

10–13. The velocity of a particle at $t = 2$ s is 50 m/s 40°↘; at $t = 6$ s, it is 50 m/s ∠15°. Calculate the average acceleration during the time interval.

10–14. A pipeline is filled with oil traveling at a constant velocity of 5 ft/s. How long will it take for oil to reach a point 10 miles away?

10–15. A conveyor carries 0.25 m^3 of gravel per meter of length. This conveyor dumps the gravel into a truck and can fill a 6-m^3 truck in 10 seconds. What is the speed of the conveyor?

10–16. A car traveling 40 mph is 140 ft from the near side of an intersection when the light turns yellow. The intersection is 80 ft wide, and the light remains yellow for 4 seconds. If the car maintains its speed, will it be clear of the intersection when the light turns red? If so, with what distance to spare?

10–17. A rocket initially at rest attains a velocity of 40 m/s in 5 seconds. What is its average acceleration?

APPLIED PROBLEMS FOR SECTION 10–5

10–18. The elevator system shown (Figure P10–18) has a car weight of 800 lb and carries passengers weighing 2500 lb. The counterbalance weight is 2000 lb. There are 3 cables connecting the car to the counterweights. The system is designed to reach a maximum speed of 450 ft/min. with equal acceleration and deceleration rates of 2.5 ft/s^2. The distance between floors is 12 ft but maximum speed is not reached when moving only one floor, as the car accelerates in the first 6 ft and decelerates in the final 6 ft. (a) Determine the time to travel between floors and the maximum speed reached. (b) What is the total time to go from the third floor to the seventh floor if there are no stops at other floors?

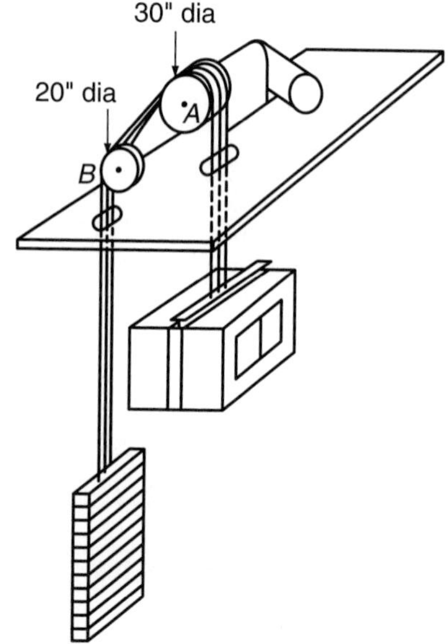

FIGURE P10–18

10–19. The ball of a pinball machine is accelerated from rest to 11 in./s in a distance of 4 in. Determine the acceleration of the ball.

10–20. A motorcycle accelerates at 1.6 m/s^2 to reach a velocity of 20 m/s in a distance of 50 m. Determine its initial velocity and time elapsed.

10–21. Car A, traveling at 50 mph and located 700 ft ahead of car B, decelerates at a constant rate of 5 ft/s^2. Car B has a speed of 40 mph in the same direction as car A and is accelerating at 8 ft/s^2. How far does car B travel in order to pass car A?

10–22. A truck accelerates uniformly in one direction from an initial velocity of 2 m/s (point A) to a final velocity of 8 m/s at point B. This velocity is maintained for 15 seconds to reach point C. It is then braked uniformly at 0.7 m/s^2 to come to rest at point D. If the total displacement from A to D is 240 m, determine (a) the acceleration rate from A to B and (b) the total time from A to D.

10–23. A train with a maximum speed of 105 km/h has an acceleration rate of 0.25 m/s^2 and a deceleration rate of 0.7 m/s^2. Determine the minimum running time between stations 7 km apart, if it stops at all stations.

10–24. At $t = 0$, a car with an initial velocity of 80 ft/s begins to coast to a stop as it proceeds up a hill. It comes to a stop in a distance of 500 ft and then starts to accelerate down the slope. Determine the *displacement* of the car at $t = 16$ s. The magnitude of deceleration up the hill is equal to the magnitude of acceleration down the hill.

10–25. An object with negligible air resistance is dropped from a height of 60 m. Determine how long it takes to land and the velocity at which it lands.

10–26. A person decides to measure the approximate height of a bridge above the water by dropping a stone and measuring the time it takes to hit the water. If the measured time is 3 seconds, what is the height of the bridge above the water?

10–27. A boy throws a stone vertically and it lands 5 seconds later. Assume that the stone landed at the same level as the boy's hand that released it. How high did he throw it ($g = 9.81$ m/s^2)?

10–28. A heavy object with negligible buoyancy is dropped from a height of 100 m above the water level of a lake. Assuming a constant deceleration rate of 25 m/s^2 after hitting the water, how deep will the object be in the water when its velocity has decreased to 5 m/s?

10–29. An astronaut jumps a distance of 3 ft vertically onto the surface of the moon ($g = 5.31$ ft/s^2). At what velocity does he hit the ground? What would be his velocity if he were subject to the earth's gravity ($g = 32.2$ ft/s^2)?

10–30. An object with an initial velocity of 25 m/s upward lands 80 m below its starting point. Find its maximum height, its total time in the air, and the velocity at the time of its landing. Compare the time and final velocity if the initial velocity is downward.

10–31. A skydiver jumps from a balloon and falls freely for 6 seconds. Her parachute then opens, decelerating her at a constant rate for 3 seconds to a velocity of 18 ft/s, which she maintains until she lands. (Neglect air resistance during the free fall.) If the skydiver jumped from an elevation of 6000 ft, determine (a) her maximum velocity and (b) the total time elapsed during her travel from the balloon to the ground.

APPLIED PROBLEMS FOR SECTION 10–6

10–32. A golfer hits a ball, giving it a maximum height of 20 m. If the ball lands 130 m away at the level from which he hit it, what was its initial velocity? (Neglect air resistance.)

10–33. The golfer wants to hit the ball the same horizontal distance of 130 m and 20 m above his tee-off point as in Problem 10–32, but in this case, the tee-off point is 10 m above the landing spot of the ball. What is the ball's initial velocity?

10–34. A projectile is fired from the top of a building as shown in Figure P10–34. Determine (a) the horizontal distance d and (b) its velocity just prior to striking point B.

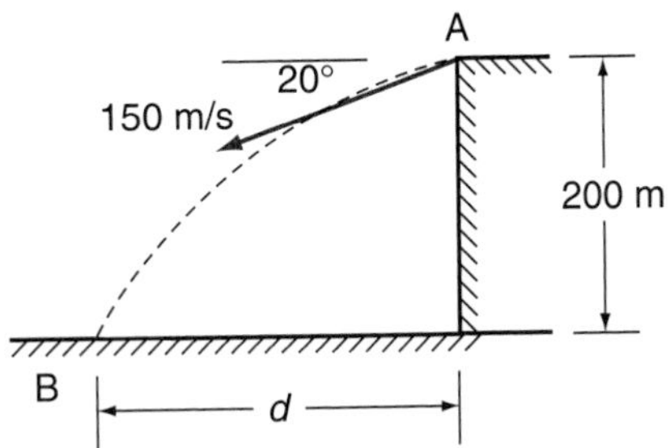

FIGURE P10–34

10–35. A projectile leaves the top of a tower with a horizontal velocity of 13 m/s. Determine its horizontal displacement and vertical displacement if it lands at 70° from horizontal.

10–36. Determine the maximum horizontal distance that a projectile will have at a height of 60 m above its origin if it is fired at an angle at 70° from horizontal with a velocity of 50 m/s.

10–37. A ball is thrown with an initial velocity of 100 ft/s and at an angle of 40° with the horizontal. Determine the minimum distance from a 50-ft-high building from which the ball can be thrown and not hit the side of the building.

10–38. The projectile shown in Figure P10–38 reaches its maximum elevation at point A. Determine the initial velocity and angle θ.

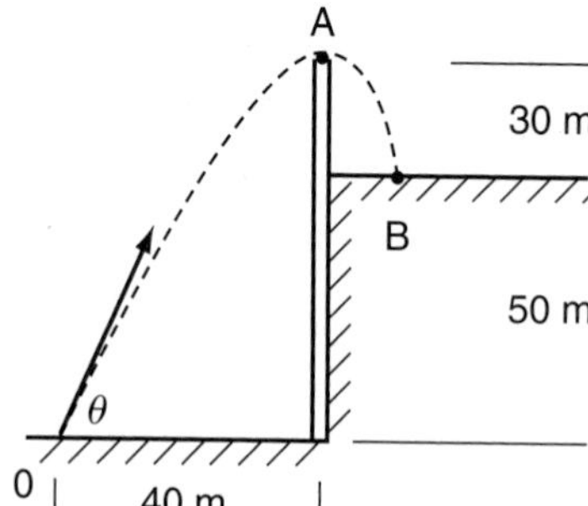

FIGURE P10–38

10–39. Determine the maximum distance d (Figure P10–39) such that the projectile will just clear point C.

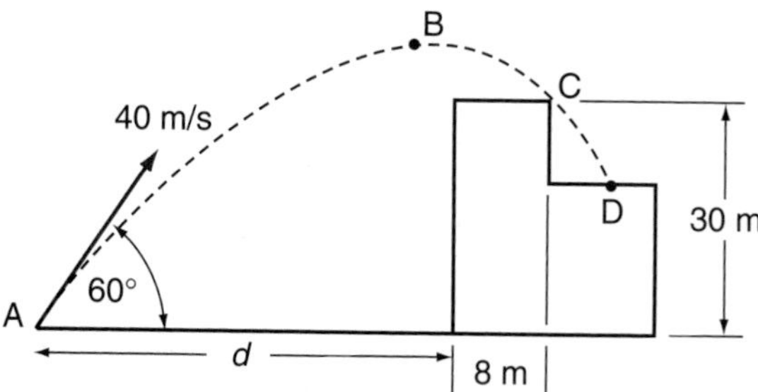

FIGURE P10–39

10–40. Water from a fire hose follows the path shown in Figure P10–40. Determine the velocity of the water as it leaves the hose.

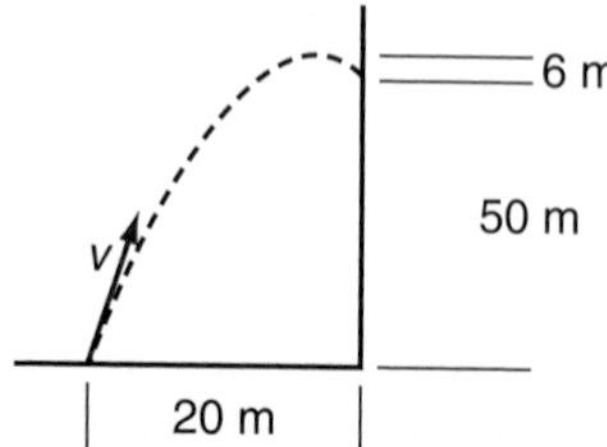

FIGURE P10–40

10–41. The projectile shown in Figure P10–41 lands at point C with a velocity of 132 m/s $\diagdown 13.8°$. Determine the initial velocity v_1 and the angle θ.

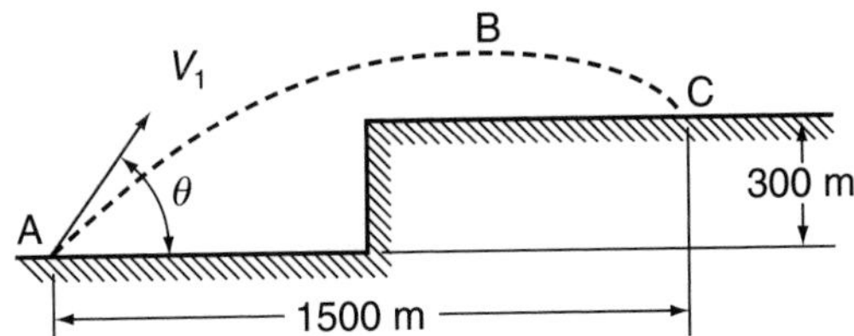

FIGURE P10–41

10–42. A basketball player throws a basketball from a height of 5 ft above the floor, with a velocity of 23 ft/s, at an angle of 65° above horizontal. If the ceiling height if 11 ft, at what angle will the ball strike the ceiling?

10–43. Grain pours from the end of a chute and lands in a container below (Figure P10–43). Determine the minimum velocity that it can have as it leaves the chute and still land in the container.

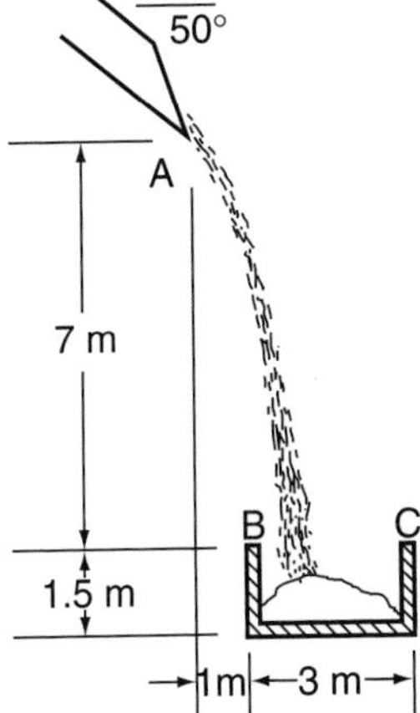

FIGURE P10–43

10–44. The projectile shown in Figure P10–44 just clears the 20-ft wall. Determine the minimum horizontal distance x and the velocity of the projectile just prior to landing at C.

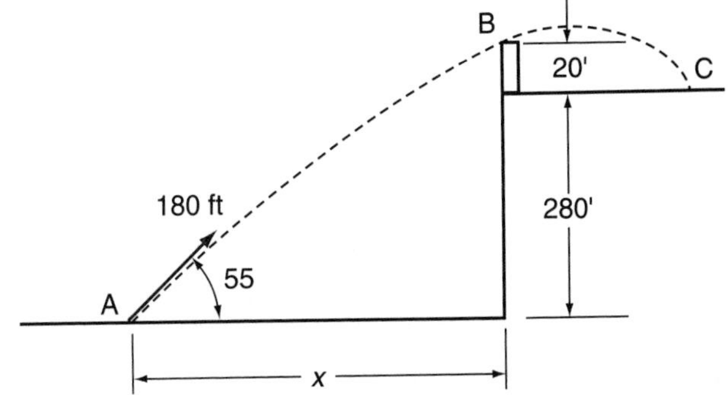

FIGURE P10–44

10–45. Projectile A is fired with a velocity of 20 m/s at an angle of 30° below horizontal as shown in Figure P10–45 and lands 2.31 seconds later. At what velocity must a second projectile B be fired at an angle of 50° above horizontal if it is to land at the same location as A?

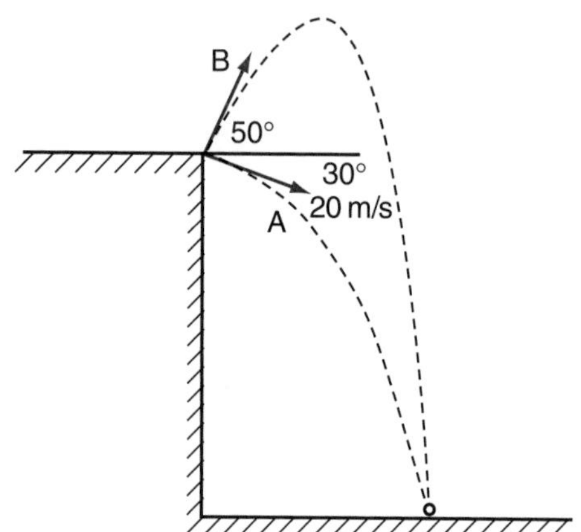

FIGURE P10–45

10–46. Determine the minimum and maximum distances d such that the projectile shown (Figure P10–46) will strike point C.

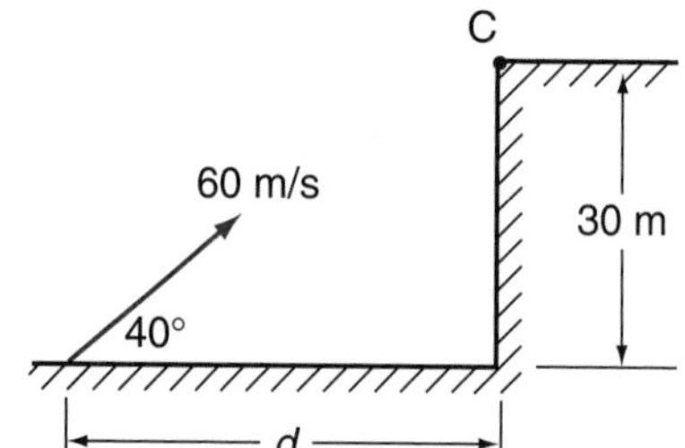

FIGURE P10–46

10–47. A bale-thrower attachment of a farm haybaler throws a completed bale into a wagon as shown in Figure P10–47. If $\theta = 48°$ and the velocity can be adjusted depending on the weight of the bale, what is the maximum velocity allowable before the bale will be thrown over the back of the trailer?

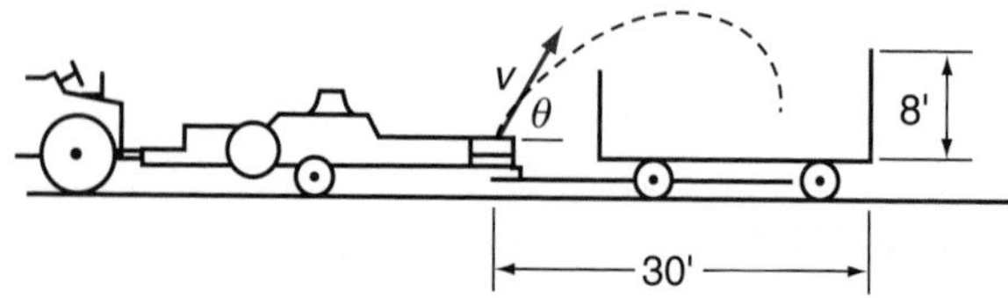

FIGURE P10–47

10–48. A projectile is fired at an angle of 30° to the horizontal up a slope that is at 10° to the horizontal. The firing velocity is 450 m/s. How far along the slope and with what velocity will the projectile strike the ground?

10–49. The projectile is fired as shown in Figure P10–49. Determine the horizontal distance from the origin to the point that the projectile lands.

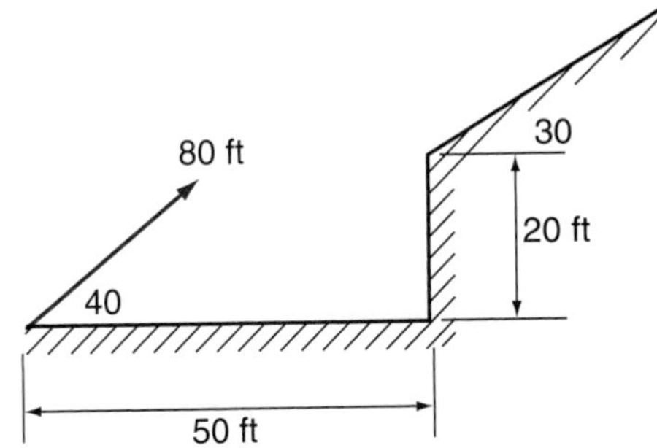

FIGURE P10–49

10–50. Determine at which landing point B the projectile shown in Figure P10–50 will land. Calculate the projectile velocity just prior to landing.

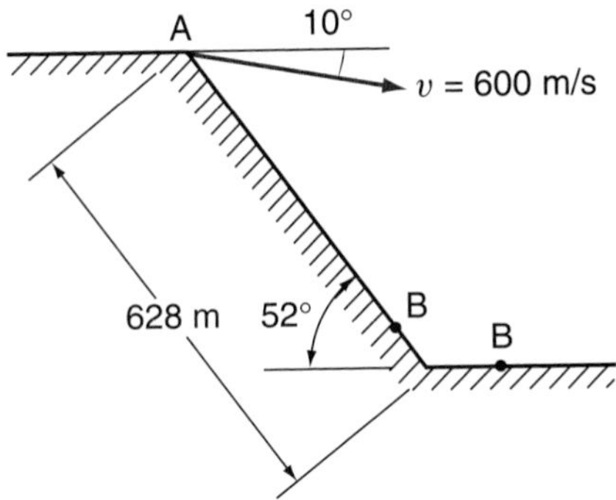

FIGURE P10–50

REVIEW PROBLEMS

R10–1. A dump truck hoists a box to an angle of 25° from the horizontal and pulls ahead 10 ft. Calculate the displacement of point A (Figure RP10–1).

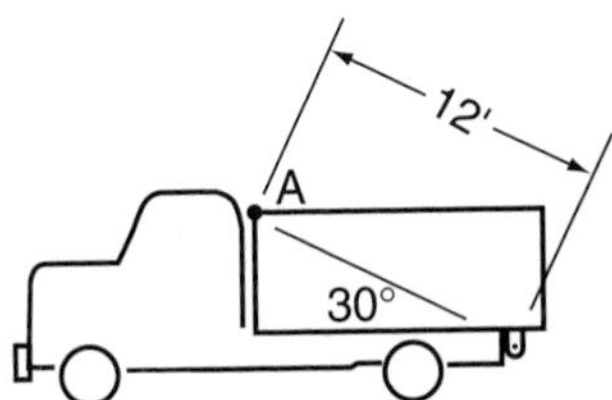

FIGURE RP10–1

R10–2. An object moves from the origin to point (1) in 3 seconds and then from point (1) to point (2) in another 4 seconds (Figure RP10–2). Determine (a) the displacement from origin to (2) and (b) average velocity from origin to (2).

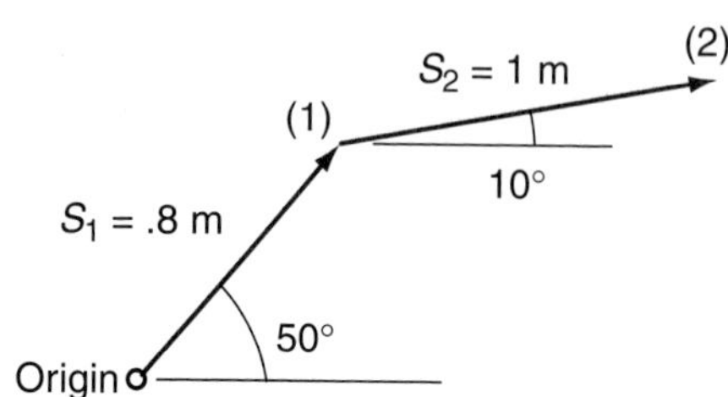

FIGURE RP10–2

R10–3. A plane traveling 250 mph, while approaching the airport, banks into a 90° turn while maintaining its altitude and speed. Determine the average acceleration of the plane if it takes 50 seconds to complete the turn.

R10–4. A car traveling at 120 km/h passes by a parked police car. If it takes 5 seconds to start the police car, which then accelerates at 3 m/s^2 to a maximum speed of 150 km/h, how far does the police car travel in overtaking the speeding car, which maintains a speed of 120 km/h?

R10–5. Car A is 900 ft ahead of car B and both cars are stopped at traffic lights. The light at A turns green, and 10 seconds later the light at B is green. Car A accelerates uniformly at a rate of 3 ft/s^2 to a constant velocity of 40 mph. Car B accelerates uniformly at a rate of 4 ft/s^2 to a constant velocity of 60 mph. How long after car A starts out will it be overtaken by car B?

R10–6. Body A is projected upward at 100 ft/s from the top of a 600-ft-high building. Body B is projected downward at 160 ft/s, 7 seconds later. Determine the point at which the two bodies are abreast of each other and their respective velocities at this instant.

R10–7. Determine whether the projectile shown in Figure RP10–7 follows path A or B, and having decided this, calculate the corresponding displacement d_1 or d_2.

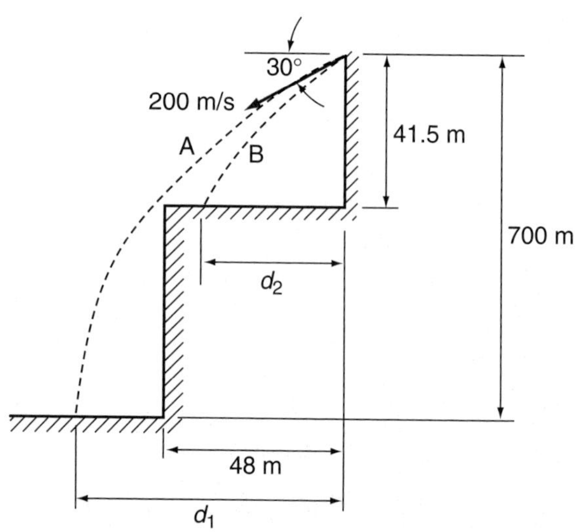

FIGURE RP10–7

R10–8. Will the projectile shown in Figure RP10–8 land at B or C? Calculate the horizontal distance from the origin to the point of landing.

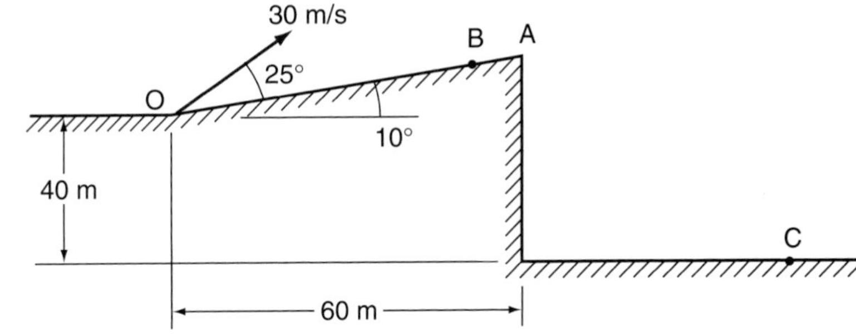

FIGURE RP10–8

R10–9. Determine the distance *d* for the projectile to land as shown in Figure RP10–9.

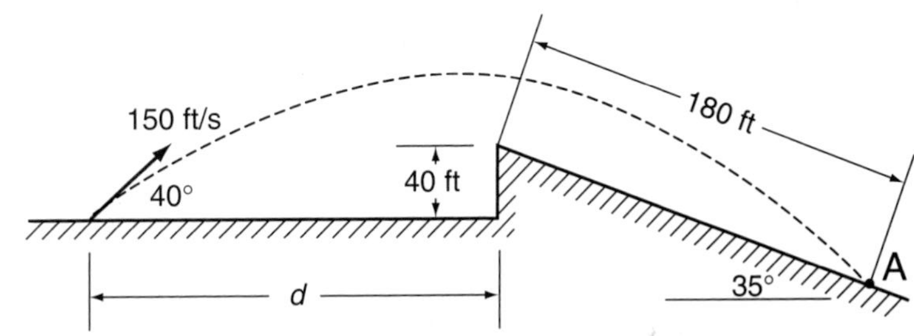

FIGURE RP10–9

CHAPTER 11

Kinematics: Angular Motion

OBJECTIVES

Upon completion of this chapter the student will be able to:

1. Solve for values of angular displacement, velocity, and acceleration using the three equations of angular motion with uniform acceleration.
2. Distinguish between and calculate values of tangential acceleration, normal acceleration, and total acceleration.

11–1 INTRODUCTION

Any object like a rotor or a lever can rotate about an axis or pin connection. As it rotates, it turns through some angle that can be expressed in degrees, radians, or revolutions. For this reason, *rotational motion* of a body is also called *angular motion.*

Similar to rectilinear motion, angular motion is described in terms of distance, displacement, velocity, and acceleration.

11–2 ANGULAR DISPLACEMENT

The movement of lever AB is measured in terms of the angle through which it turns. It may turn a few degrees or several revolutions. A more useful unit used to measure angular displacement is the radian (rad). In Figure 11–1, AB turns through an angle of 1 *radian* when point B moves a distance on the circumference equal to radius AB. Since circumference is equal to $2\pi r$, there are 2π radians in one revolution. The radian is a dimensionless unit; it will be used later in Section 11–6 for conversion between rectilinear and angular motion. The angular displacement units are:

$$\text{one revolution} = 360 \text{ degrees} = 2\pi \text{ radians}$$

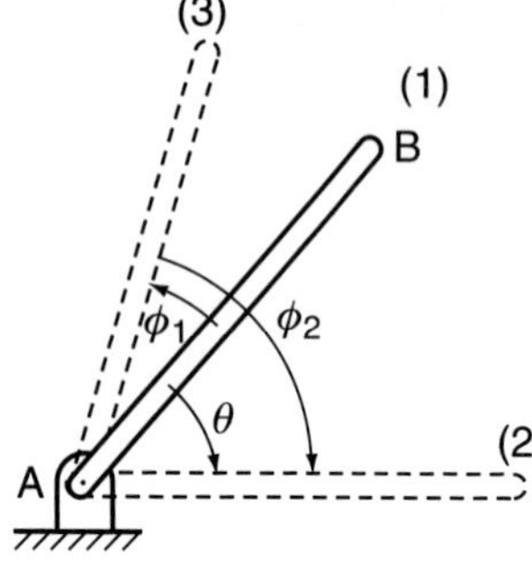

FIGURE 11–1

The symbol applied to angular displacement is the Greek lowercase letter theta (θ).

In Figure 11–1, suppose that lever AB moves from position (1) to position (2). The angular displacement is θ radians. If, instead, the lever moved from position (1) to (3) and then to position (2), the angular displacement would still be θ radians, but the angular distance would be $(\phi_1 + \phi_2)$ radians. This distinction between displacement and distance is similar to the distinction that we made with rectilinear motion in Section 10–5.

EXAMPLE 11–1

An ammeter needle, starting from a zero reading, deflects 40° clockwise and then returns 15° counterclockwise to indicate a final reading. What is its angular displacement and angular distance in terms of radians?

$$\text{angular displacement} = 40° - 15°$$
$$= 25° \text{ clockwise}$$

since $360° = 2\pi$ rad

$$25° = \frac{25}{360 \text{ deg/rev}}(2\pi \text{ rad/rev})$$
$$\underline{\theta = 0.437 \text{ rad} \circlearrowright}$$
$$\text{angular distance} = 40° + 15°$$
$$\text{converting to radians: } 55° = \frac{55}{360 \text{ deg/rev}}(2\pi \text{ rad/rev})$$
$$\underline{= 0.96 \text{ rad}}$$

11–3 ANGULAR VELOCITY

Angular velocity is the rate of change of angular displacement and is represented by the Greek lowercase letter omega (ω). Like average rectilinear velocity, average angular velocity can be defined as

$$\omega = \frac{\Delta\theta}{\Delta t} \tag{11–1}$$

where ω is the average angular velocity, θ the angular displacement, and t the time. The most widely used units of ω are rad/s and rev/min (rpm).

Angular velocity, expressed in revolutions per minute (rpm), is not consistent with the SI metric system because the unit "minute" does not conform to the SI system and is classified as a "permitted" unit. Strictly speaking, the time units in the SI system are in seconds and multiples or submultiples, such as kiloseconds and milliseconds. Since the terms *minute* and *hour* are so universally used, the SI system is modified to permit their continued use.

EXAMPLE 11–2

The crankshaft of an engine turns 800 revolutions clockwise in 4 minutes at constant velocity. What is its angular velocity in rad/s?

$$\omega = \frac{\Delta\theta}{\Delta t} = \frac{800 \text{ rev}}{4 \text{ min}}$$

$$= 200 \frac{\text{rev}}{\text{min}}\left(\frac{2\pi \text{ rad/rev}}{60 \text{ s/min}}\right)$$

$$\underline{\omega = 20.9 \text{ rad/s} \circlearrowright}$$

11–4 ANGULAR ACCELERATION

Angular acceleration is the rate of change of angular velocity and is represented by the Greek lowercase letter alpha (α). Like rectilinear motion, angular acceleration can be defined as

$$\alpha = \frac{\Delta\omega}{\Delta t} \tag{11–2}$$

where α is the average angular acceleration. The usual units of α are rad/s^2.

EXAMPLE 11–3

A flywheel accelerates uniformly from rest to a speed of 20 rpm counterclockwise in 5 seconds. What is the average angular acceleration in rad/s^2?

$$\Delta\omega = \left(200 \frac{\text{rev}}{\text{min}}\right)\left(\frac{2\pi \text{ rad/rev}}{60 \text{ s/min}}\right)$$

$$= 2.09 \text{ rad/s}$$

$$\alpha = \frac{\Delta\omega}{\Delta t} = \frac{2.09 \text{ rad/s}}{5 \text{ s}}$$

$$\underline{\alpha = 0.42 \text{ rad/s}^2 \circlearrowleft}$$

11–5 ANGULAR MOTION WITH UNIFORM ACCELERATION

In Section 10–5 we had three equations for rectilinear motion with uniform acceleration. These three equations are repeated in Table 11–1 with their analogous angular motion equations. The angular terms θ, ω, and α are simply used in the original rectilinear equations. Keep in mind that all of these equations are based on uniform acceleration.

TABLE 11–1

Rectilinear	Angular	
$s = v_0 t + \frac{1}{2}at^2$	$\theta = \omega_0 t + \frac{1}{2}\alpha t^2$	(11–3)
$v = v_0 + at$	$\omega = \omega_0 + \alpha t$	(11–4)
$v^2 = v_0^2 + 2as$	$\omega^2 = \omega_0^2 + 2\alpha\theta$	(11–5)

In common units:

$$\theta = \text{displacement; rad}$$
$$\omega_0 = \text{initial velocity; rad/s}$$
$$\omega = \text{final velocity; rad/s}$$
$$\alpha = \text{uniform acceleration; rad/s}^2$$
$$t = \text{time; seconds}$$

EXAMPLE 11–4

When a small plane touches down on a landing strip, its wheels accelerate from rest to a speed of 1100 rpm in 3 seconds. Calculate (a) the average angular acceleration, (b) the number of revolutions of each wheel in this time period.

Listing the information given:

$$\omega_0 = 0$$
$$\omega = 1100\frac{\text{rev}}{\text{min}}\left(\frac{2\pi\ \text{rad/rev}}{60\,\text{s/min}}\right)$$
$$= 115\ \text{rad/s}$$
$$t = 3\ \text{s}$$
$$\alpha = ?$$

$$\omega = \omega_0 + \alpha t$$
$$115\ \text{rad/s} = 0 + \alpha(3\ \text{s})$$
$$\underline{\alpha = 38.3\ \text{rad/s}^2}$$

$$\theta = \omega_0 t + \frac{1}{2}\alpha t^2$$
$$= 0 + \frac{1}{2}(38.3 \text{ rad/s}^2)(3 \text{ s})^2$$
$$= 172 \text{ rad}$$
$$= \frac{172 \text{ rad}}{2\pi \text{ rad/rev}}$$
$$\underline{\theta = 27.4 \text{ rev}}$$

EXAMPLE 11–5

A propeller fan 2 m in diameter used in a cooling tower comes to rest with uniform deceleration from a clockwise speed of 600 rpm. If it turns through 15 revolutions while stopping, calculate the time that it requires to stop.

$$\omega_0 = 600 \text{ rpm} = \frac{600 \text{ rev/min}(2\pi \text{ rad/rev})}{60 \text{ s/min}} = 62.8 \text{ rad/s}$$
$$\omega = 0$$
$$\theta = (15 \text{ rev})(2\pi \text{ rad/rev}) = 94.2 \text{ rad}$$

The deceleration α must be found first.

$$\omega^2 = \omega_0^2 + 2\alpha\theta$$
$$0 = (62.8 \text{ rad/s})^2 + 2\alpha\,(94.2 \text{ rad})$$
$$-3940 = 188.4\alpha$$
$$\alpha = -21 \text{ rad/s}^2 \text{ clockwise}$$

or $$\alpha = 21 \text{ rad/s}^2 \text{ counterclockwise}$$

The minus sign indicates that there is deceleration in the direction of the rotation. Using Equation 11–4 to calculate time, we get

$$\omega = \omega_0 + \alpha t$$
$$0 = 62.8 \text{ rad/s} - (21 \text{ rad/s}^2)t$$
$$\underline{t = 3s}$$

EXAMPLE 11–6

A flywheel rotating clockwise at 60 rpm has a torque applied to it that decelerates it to a stop, then accelerates it to 40 rpm counterclockwise in 20 seconds. If the deceleration and acceleration rates are equal and constant, determine (a) the angular displacement and (b) the revolutions in each direction.

If the initial angular velocity is positive, then the flywheel will have negative acceleration until it reaches a final negative angular velocity.

$$\text{(a)} \quad \omega_0 = \frac{(60 \text{ rev/min})(2\pi \text{ rad/rev})}{(60 \text{ s/m})} = 6.28 \text{ rad/s}$$

$$\omega = \frac{-(40 \text{ rev/min})(2\pi \text{ rad/rev})}{(60 \text{ s/min})} = -4.18 \text{ rad/s}$$

$$t = 20 \text{ s}$$

Using $\omega = \omega_0 + \alpha t$

$$-4.18 \text{ rad/s} = 6.28 \text{ rad/s} + \alpha(20 \text{ s})$$
$$\alpha = -0.523 \text{ rad/s}^2$$

Solve for displacement using

$$\theta = \omega_0 t + \frac{1}{2}\alpha t^2$$

$$\theta = (6.28 \text{ rad/s})(20 \text{ s}) + \left(\frac{1}{2}\right)(-0.523 \text{ rad/s}^2)(20 \text{ s})^2$$
$$= 125.6 - 104.6$$
$$\underline{\theta = 21 \text{ rad clockwise}}$$

(b) Consider each stage of rotation separately, starting with deceleration to zero velocity.

$$\omega_0 = +6.28 \text{ rad/s}$$
$$\omega = 0$$
$$\alpha = -0.523 \text{ rad/s}^2$$

Using

$$\omega^2 = \omega_0^2 + 2\alpha\theta$$
$$0 = (6.28 \text{ rad/s})^2 + (2)(-0.523 \text{ rad/s}^2)\theta$$
$$\underline{\theta = 37.7 \text{ rad clockwise}}$$

The flywheel now accelerates from rest to 40 rpm or 4.18 rad/s counterclockwise

$$\omega_0 = 0$$
$$\omega = -4.18 \text{ rad/s}$$
$$\alpha = -0.523 \text{ rad/s}^2$$

$$\omega^2 = \omega_0^2 + 2\alpha\theta$$
$$(-4.18 \text{ rad/s})^2 = 0 + (2)(-0.523 \text{ rad/s}^2)\theta$$
$$\theta = -16.7 \text{ rad}$$

or

$$\underline{\theta = 16.7 \text{ rad counterclockwise}}$$

The calculations check because it has turned 37.7 rad CW, then 16.7 rad CCW, resulting in a *displacement* of $\underline{37.7 - 16.7 = 21 \text{ rad CW.}}$

11–6 RELATIONSHIP BETWEEN RECTILINEAR AND ANGULAR MOTION

There are many situations in which rectilinear and angular or rotational motion are combined. Belts on pulleys and revolving car wheels are two such examples. We must be able to convert between rectilinear values (s, v, a) and angular values (θ, ω, α).

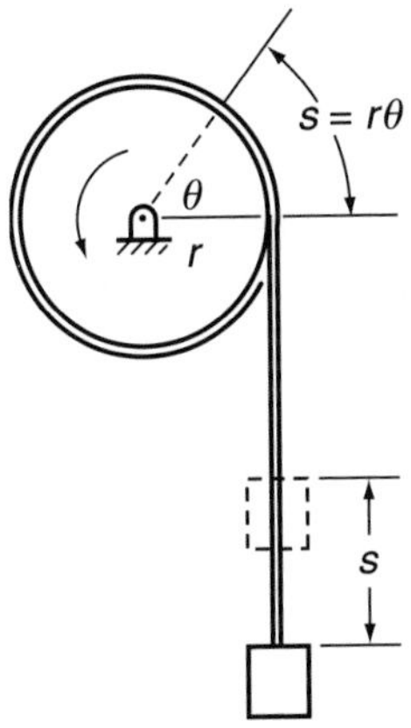

FIGURE 11–2

A hoist drum with a cable wound around it can rotate and lift a weight (Figure 11–2). If the drum turns counterclockwise through an angle of 1 rad, the amount of cable wound onto the drum is equal to the radius; the weight is also lifted a distance s, equal to the radius. For 2 rad of drum rotation, the cable wound onto the drum equals $2r$ and distance $s = 2r$. Distance s depends on the radius and amount of rotation measured in radians or, in equation form,

$$s = r\theta \tag{11–6}$$

Where s and r must have the same units, such as meters or feet; θ is expressed in radians. Dividing both sides by t, we have

$$\frac{s}{t} = r\frac{\theta}{t}$$

$$v = r\omega \tag{11–7}$$

Dividing both sides by t again, we have

$$\frac{v}{t} = r\frac{\omega}{t}$$

$$a = r\alpha \tag{11–8}$$

A feature common to all three equations is that a rectilinear value equals the radius multiplied by the angular value.

EXAMPLE 11–6

Wire for a transmission line is unrolled from a reel at a rate of 750 ft/min. The radius of the reel is 2.5 ft for the instant considered. What is the angular speed of the reel in rpm?

$$\begin{aligned} v &= r\omega \\ 750 \text{ ft/min} &= (2.5 \text{ ft})\omega \\ \omega &= 300 \text{ rad/min} \\ &= \frac{300 \text{ rad/min}}{2\pi \text{ rad/rev}} \\ \omega &= 47.8 \text{ rpm} \end{aligned}$$

EXAMPLE 11–7

As shown in Figure 11–3, weight D is suspended by a rope wrapped around pulley C. Pulleys B and C are fastened together and pulley A is belt-driven by pulley B. Starting from rest,

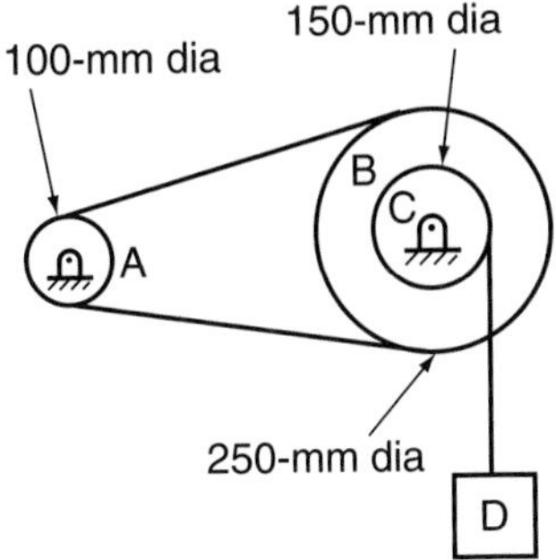

FIGURE 11–3

weight D drops 18 m in 3 seconds. For each pulley, determine (a) the number of revolutions, (b) the angular velocity, and (c) the angular acceleration at $t = 3$ seconds.

Calculate the rectilinear values before converting to angular values.

$$s = 18 \text{ m}$$
$$t = 3 \text{ s}$$
$$v_0 = 0$$
$$v = ?$$
$$a = ?$$

$$s = v_0 t + \frac{1}{2}at^2$$
$$18 \text{ m} = 0 + \frac{1}{2}(a)(3 \text{ s})^2$$
$$a = 4 \text{ m/s}^2 \downarrow$$
$$v = v_0 + at$$
$$= 0 + (4)(3)$$
$$v = 12 \text{ m/s} \downarrow$$

Considering pulley C first (Figure 11–4), we have

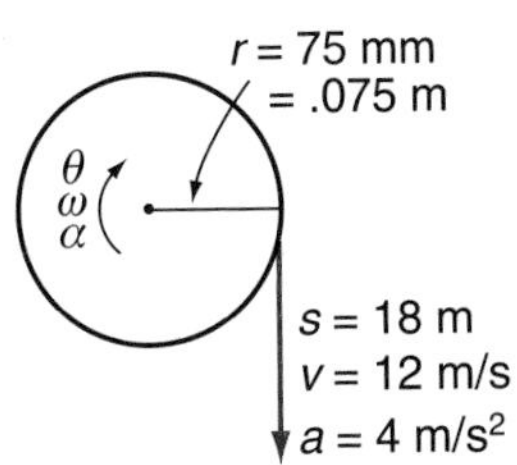

FIGURE 11–4

$$s = r\theta$$
$$18 \text{ m} = (0.075 \text{ m})\theta$$
$$\theta = 240 \text{ rad}/2\pi \frac{\text{rad}}{\text{rev}}$$
$$\underline{\theta = 38.2 \text{ rev} \circlearrowright}$$
$$v = r\omega$$
$$12 \text{ m/s} = (0.075 \text{ m})\omega$$
$$\underline{\omega = 160 \text{ r/s} \circlearrowright}$$
$$a = r\alpha$$
$$4 \text{ m/s}^2 = (0.075 \text{ m})\alpha$$
$$\underline{\alpha = 53.3 \text{ rad/s}^2 \circlearrowright}$$

Since pulley B is fastened to pulley C, it must have the same angular values as pulley C; that is,

$$\underline{\theta = 38.2 \text{ rev} \circlearrowright}$$
$$\underline{\omega = 160 \text{ rad/s} \circlearrowright}$$
$$\underline{\alpha = 53.3 \text{ rad/s}^2 \circlearrowright}$$

The belt has the same speed throughout its length; therefore, the tangential velocity of A equals the tangential velocity of B.

$$v_A = v_B$$

$$r_A \omega_A = r_B \omega_B$$

$$\omega_A = \frac{r_B}{r_A}(\omega_B)$$

$$= \frac{0.125 \text{ m}}{0.050 \text{ m}}(160 \text{ rad/s})$$

$$\underline{\omega_A = 400 \text{ rad/s} \circlearrowright}$$

Angular displacement and acceleration can be found by means of the same ratio of radii.

$$\theta_A = \frac{r_B}{r_A}\theta_B$$

$$= \frac{0.125 \text{ m}}{0.050 \text{ m}}(38.2 \text{ rev})$$

$$\underline{\theta_A = 95.5 \text{ rev} \circlearrowright}$$

$$\alpha_A = \frac{0.125 \text{ m}}{0.050 \text{ m}}(53.3 \text{ rad/s}^2)$$

$$\underline{\alpha_A = 133 \text{ rad/s}^2 \circlearrowright}$$

11–7 NORMAL AND TANGENTIAL ACCELERATION

As we demonstrated in Section 11–6, a point on the rim of a wheel—with radius r and angular acceleration α—has a rectilinear acceleration of $r\alpha$. The direction of the motion is tangent to the arc of travel and is therefore perpendicular to the radius (Figure 11–5). We will place special emphasis on *tangential acceleration* by designating it a_t, since there is another acceleration present, normal acceleration, which we will designate a_n.

Tangential acceleration a_t is due to the wheel's speed changing, that is, changing *magnitude* of its v and ω values. Another acceleration, *normal acceleration,* is present even with constant wheel speed. Normal acceleration is due to the change in *direction* of velocity, as was discussed in Section 10–4.

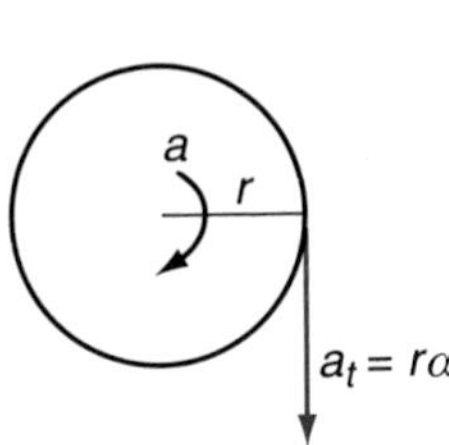

FIGURE 11–5

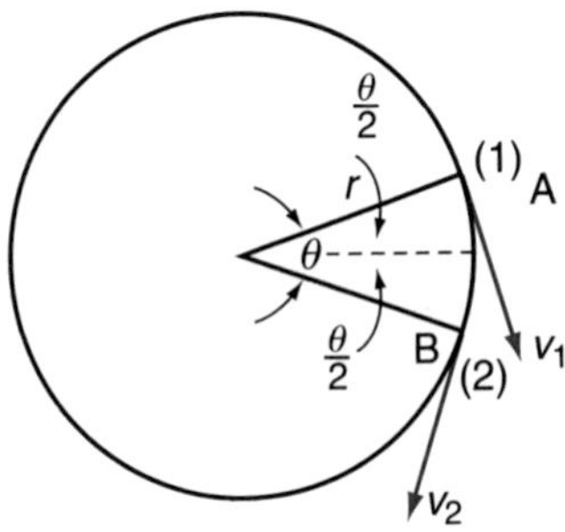

FIGURE 11–6

In Figure 11–6, a wheel turns at constant speed, and a point on the rim travels from point A to point B. The tangential velocities, v_1 and v_2, are equal in magnitude.

$$\text{normal acceleration} = \text{velocity change (in direction)}$$

$$a_n = \frac{\Delta v}{\Delta t}$$

where

$$\begin{aligned} \Delta v &= v_2 - v_1 \\ &= v_2 + (-v_1) \end{aligned}$$

Figure 11–7 shows the vector addition of $v_2 + (-v_1)$ to obtain Δv. From this vector triangle, we have

$$v = v_2 \text{ or } v_1$$

Figure 11–8 is an enlargement of the triangle portion of Figure 11–6 where, for *very* small angles, arc length AB is equal to the chord length AB.

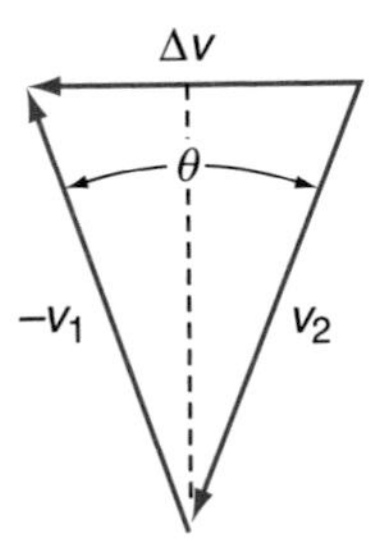

FIGURE 11–7

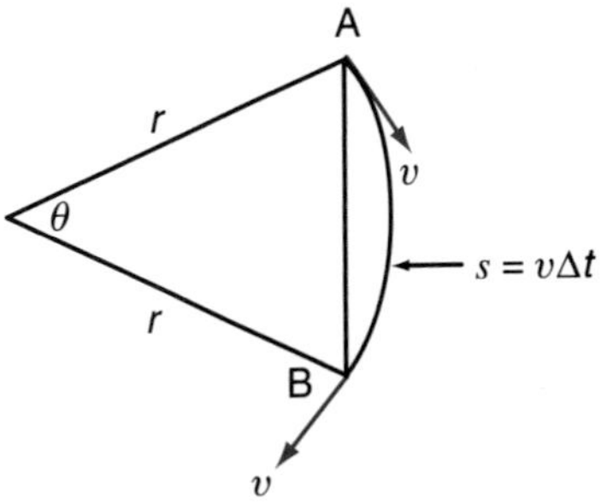

FIGURE 11–8

Since Figures 11–7 and 11–8 are similar triangles, the ratios $\frac{\Delta v}{v}$ and $\frac{AB}{r}$ can be equated and written as

$$\frac{\Delta v}{v} = \frac{s}{r}$$

$$\frac{\Delta v}{v} = \frac{v\Delta t}{r}$$

$$\frac{\Delta v}{\Delta t} = \frac{v^2}{r} \text{ but } a_n = \frac{\Delta v}{\Delta t}$$

or, for normal acceleration,

$$a_n = \frac{v^2}{r} \tag{11–11}$$

The direction of a_n must be the same as that of Δv, which was along the radius toward the center of rotation (horizontally to the left).

Since $v = r\omega$, another version of the equation for normal acceleration is

$$a_n = \frac{r^2\omega^2}{r}$$

or

$$a_n = \omega^2 r \tag{11–12}$$

The reason this characteristic is called normal acceleration is that it acts inwardly along the radius, that is, at right angles to—or normal to—the circular path at that instant. Since there may be both normal and tangential acceleration simultaneously, the total acceleration would have to be the vector sum of a_n and a_t. The following examples will illustrate this.

EXAMPLE 11–8

Determine the normal acceleration of a car traveling around a circle with a radius of 200 ft at a speed of 30 mph. (60 mph = 88 ft/s).

$$v = 30 \text{ mph} = \frac{30}{60}(88 \text{ ft/s}) = 44 \text{ ft/s}$$

$$a_n = \frac{v^2}{r}$$

$$= \frac{(44 \text{ ft/s})^2}{200 \text{ ft}}$$

$$\underline{a_n = 9.68 \text{ ft/s}^2}$$

EXAMPLE 11–9

A pulley with a radius of 80 mm is turning at a velocity of 150 rpm clockwise and is accelerating at 100 rad/s². Determine the total acceleration of a point on the rim at the instant during which it is at the top of the pulley (Figure 11–9).

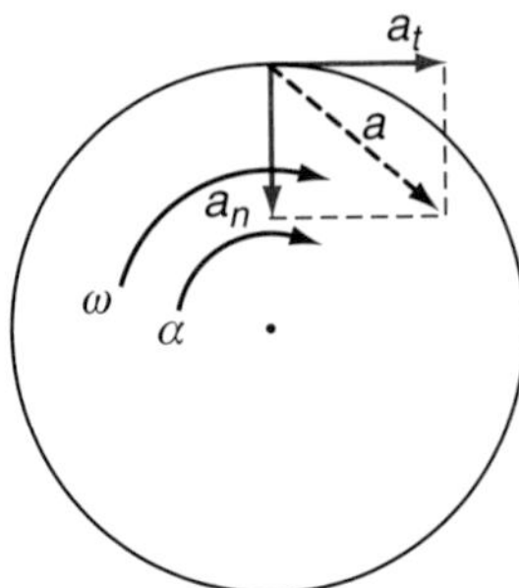

FIGURE 11–9

$$\omega = 150 \text{ rpm} = \frac{(150 \text{ rev/min})(2\pi \text{ rad/rev})}{60 \text{ s/min}} = 15.7 \text{ rad/s}$$

$$\begin{aligned} a_n &= \omega^2 r \\ &= (15.7 \text{ rad/s})^2(0.08 \text{ m}) \\ a_n &= 19.7 \text{ m/s}^2 \downarrow \end{aligned} \qquad \begin{aligned} a_t &= r\alpha \\ &= (0.08 \text{ m})(100 \text{ rad/s}^2) \\ a_t &= 8 \text{ m/s}^2 \rightarrow \end{aligned}$$

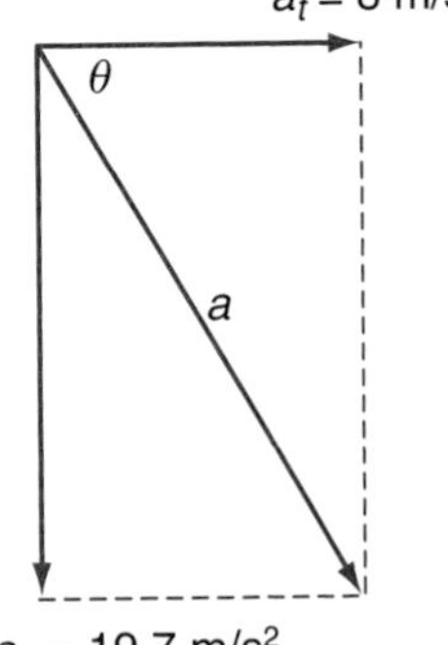

FIGURE 11–10

The total acceleration a is the sum of a_n and a_t.

$$a = \sqrt{(8 \text{ m/s}^2)^2 + (19.7 \text{ m/s}^2)^2} \quad \text{(Figure 11–10)}$$

$$\underline{a = 21.3 \text{ m/s}^2 \quad \diagdown 67.9^\circ}$$

$$\tan\theta = \frac{19.7}{8}$$

$$\theta = 67.9^\circ$$

EXAMPLE 11–10

A 1.2-m wheel, mounted in a vertical plane, accelerates uniformly from rest at 3 rad/s² for 5 seconds and then maintains uniform velocity in a clockwise direction. Determine the normal and tangential acceleration of a point at the top of the wheel for $t = 0$ and $t = 6$.

For Figure 11–11:

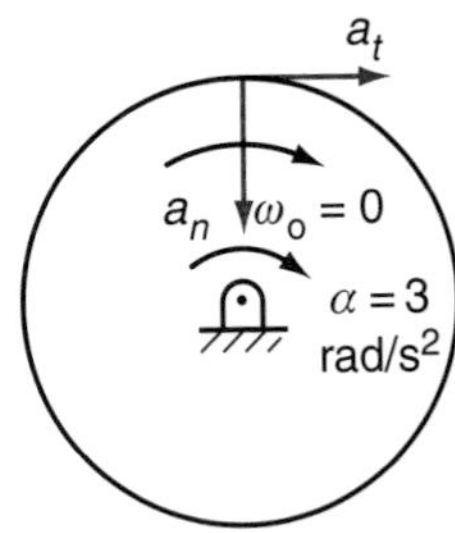

FIGURE 11–11

at $t = 0$, $\omega_0 = 0$ $\qquad a_n = \omega^2 r$

therefore $\underline{a_n = 0}$

$$\begin{aligned} a_t &= r\alpha \\ &= \left(\frac{1.2 \text{ m}}{2}\right)(3 \text{ rad/s}^2) \\ a_t &= 1.8 \text{ m/s}^2 \rightarrow \end{aligned}$$

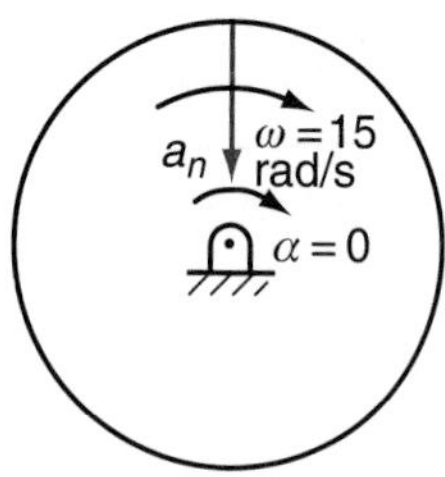

FIGURE 11–12

Now determine the velocity at $t = 5$ seconds:

$$\omega_0 = 0 \qquad \omega = \omega_0 + \alpha t$$
$$\omega = ? \qquad \omega = 0 + (3\ \text{rad/s}^2)(5\text{s})$$
$$\alpha = 3 \qquad \omega = 15\ \text{rad/s} \circlearrowright$$
$$t = 5$$

For Figure 11–12, at $t = 6$ seconds:

$$\alpha = 0 \quad a_t = r\alpha$$
$$\text{therefore } \underline{a_t = 0}$$

$$a_n = \omega^2 r$$
$$= (15\ \text{rad/s})^2(0.6\ \text{m})$$
$$\underline{a_n = 135\ \text{m/s}^2 \downarrow}$$

HINTS FOR PROBLEM SOLVING

1. As before, pay careful attention to sign convention of both given data and calculated values. If initial velocity or displacement is positive, then acceleration is positive and deceleration is negative.
2. Mating pulleys and gears have a common tangential velocity.
3. For a rotating object:
   (a) If angular velocity is zero, $a_n = 0$ but a_t may be present.
   (b) If angular acceleration is zero, $a_t = 0$ but a_n may be present.

PROBLEMS

APPLIED PROBLEMS FOR SECTIONS 11–1 TO 11–5

11–1. The minute hand of a clock travels from the 12 to the 9, whereupon it is discovered that the clock is fast. The minute hand is turned back and now points at the 6. What is the angular distance and displacement of the hand in radians?

11–2. Convert 15 revolutions to radians.

11–3. Convert 1480 degrees to (a) revolutions and (b) radians.

11–4. Convert 90 rpm to radians per second.

11–5. A gear rotates 280° clockwise in 1.5 seconds. Determine its angular velocity in rad/s.

11–6. The handle of a winch is turned 70 revolutions in 1.5 minutes. Determine the angular speed in rad/s.

11–7. It is calculated that a fan may be rotated at 220 rad/s before stress limits are exceeded. What is this speed in rpm?

11–8. Rod A in Figure P11–8 reciprocates up and down 50 times in 40 seconds due to the action of the rotating cam B. Determine the angular speed of cam B in rad/min.

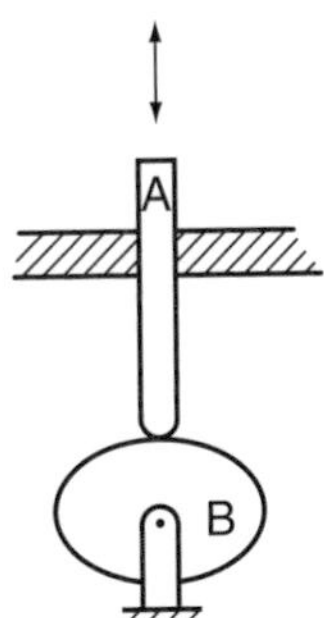

FIGURE P11–8

11–9. Starting from rest, a shaft accelerates uniformly to 800 rpm in 10 seconds. Determine the angular acceleration in rad/s^2.

11–10. A miter saw blade is braked from 4800 rpm to a stop in 1.5 seconds. Calculate the deceleration rate and the number of revolutions turned during the braking.

11–11. The wheel on a boat trailer has an angular speed of 1400 rpm when the trailer is towed at 90 km/h. Calculate the angular deceleration if the trailer decelerates uniformly to 40 km/h in 1.2 minutes. (*Hint:* Angular speed is proportional to linear speed.)

11–12. A rotating drive shaft decelerates uniformly from 900 rpm to 650 rpm in 6 seconds. Determine the angular deceleration and the total number of revolutions in the 6-second interval.

11–13. A wheel turns through 500 revolutions while accelerating from 80 rad/s to 110 rad/s. Determine the angular acceleration and the time required.

11–14. A revolving vane anemometer turns 40 revolutions in 30 seconds at constant velocity. Determine angular velocity in (a) rpm and (b) rad/s.

11–15. A pulley rotating at 80 rpm clockwise changes its rotation to 120 rpm counterclockwise in 3 seconds. Determine (a) the angular deceleration and (b) the total number of revolutions in the 3-second period.

11–16. A bicycle wheel held up off the ground is spun at 350 rpm and is then allowed to coast to a stop while turning through 400 revolutions. How long does it take to stop? (Assume constant deceleration.)

11–17. A flywheel accelerates for 8 seconds at 1.3 rad/s^2 from a speed of 40 rpm. Determine (a) the total number of revolutions and (b) the final angular speed.

11–18. A direct-connected pump and motor accelerate uniformly from rest to 1750 rpm in 0.3 second. Determine the angular acceleration.

11–19. A wheel rotating at 10 rpm is accelerated at 5 rad/s^2 for 6 seconds. It is then braked to a stop in 2 seconds. Determine (a) the maximum rpm or rad/s reached, (b) the total number of revolutions from $t = 0$ to $t = 8$ seconds, and (c) the deceleration rate.

11–20. A hand-started motor is accelerated from rest to 150 rpm in 1 second when the rope is pulled. It then starts and, under load, accelerates uniformly to 3600 rpm in another 7 seconds. Determine (a) the angular acceleration in each case and (b) the total number of revolutions of the motor.

11–21. A large saw blade in a sawmill accelerates from rest to 800 rpm while turning 160 revolutions. Determine the time and the angular acceleration.

11–22. A motorcycle wheel accelerates to 600 rpm in 5 seconds and then is braked to a stop in 2 seconds. Calculate the total number of revolutions.

11–23. Starting from rest, a water turbine turns 130 revolutions in accelerating uniformly to its operating speed of 200 rpm. Determine (a) the angular acceleration, (b) the total time required, and (c) the turbine speed after the first 40 seconds.

11–24. A drive pulley decelerates at 5 rad/s^2 from 300 rpm to 180 rpm. It then accelerates to 260 rpm in 2 seconds. Determine (a) the total time required, (b) the acceleration rate, and (c) the total number of revolutions.

11–25. A pulley turns initially at 30 rpm clockwise, then reverses its direction of rotation to 40 rpm counterclockwise in 5 seconds, at a constant deceleration and acceleration rate. Determine (a) the deceleration rate, (b) the total number of revolutions of the pulley during the 5-second interval, and (c) the angular displacement of the pulley at $t = 5$ seconds.

11–26. Starting from rest, a wheel accelerates uniformly to 300 rpm in 5 seconds. After rotating at 300 rpm for some time, it is braked to a stop in 90 seconds. If the total number of revolutions of the wheel is 800, calculate the total time of rotation.

APPLIED PROBLEMS FOR SECTION 11–6

11–27. For the system shown in Figure P11–27, determine the angular acceleration of the wheel and the tangential velocity of point A.

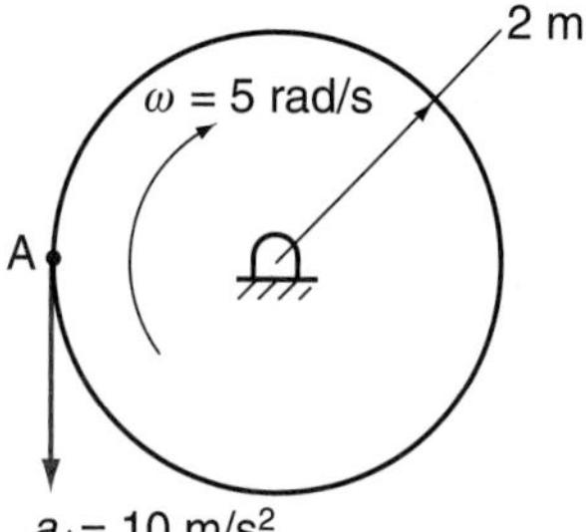

FIGURE P11–27

11–28. Determine the rpm of an 18-m-diameter Ferris wheel if the speed of the seats is 3.8 m/s.

11–29. Determine the length of an arc that has a radius of 50 m and an included angle of 2.5 rad.

11–30. A 600-m length of curved highway has a 35° change in direction. Determine the radius of the curve.

11–31. An automatic pipewelder can weld 40 in./min. How long does it take to complete one pass around a pipe 4 ft in diameter?

11–32. By means of a stroboscope, the speed of a pulley 200 mm in diameter is found to be 1600 rpm. Determine the speed of a belt passing over this pulley.

11–33. If a spotlight rotating in a horizontal plane at 2 rpm is located 130 m from you, at what speed would the light beam flash across you?

11–34. The braking mechanism of a winch fails while lowering a heavy object. If the drum diameter of the winch is 8 in. and the object accelerates at 25 ft/s^2, determine the angular acceleration of the winch drum.

11–35. In the manufacture of sheet steel, the sheet, which has a velocity of 12 m/s, is drawn between two rollers. If the rollers are 180 mm in diameter, determine their speed of rotation in rpm.

11–36. A pulley 10 in. in diameter is belt-driven by a pulley 6 in. in diameter. A pulley 4 in. in diameter is used for the belt tightener. If the 10-in.-diameter pulley is turning at 120 rpm, determine the belt speed and the rpm of the other two pulleys.

11–37. A hand-sling psychrometer is composed of two thermometers with a handle at one end and a wet wick at the other. By whirling the thermometers about the handle, one obtains a relative humidity reading. If the distance from the handle to the wick is 400 mm and the wick must have a velocity of 5 m/s for a reliable reading, at what rpm must the thermometers be whirled?

11–38. The flywheel in Figure P11–38 is decelerating at 3 rad/s^2 while turning at 12 rad/s clockwise. Determine the tangential acceleration and velocity of point P on the rim.

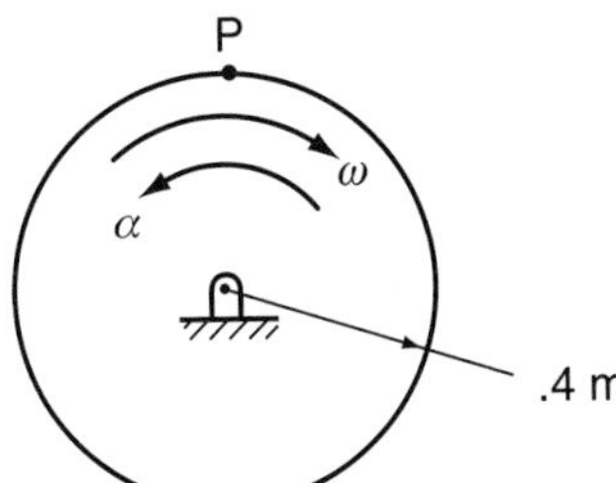

FIGURE P11–38

11–39. A lathe chuck 8 in. in diameter accelerates uniformly from rest to 400 rpm in 1 second. What is the tangential acceleration of a point on the rim?

11–40. Pulleys C and D in Figure P11–40 are fastened together. Weights A and B are supported by ropes wound around the pulleys as shown. If weight A is accelerating downward at 2 m/s^2, determine (a) the angular acceleration of pulleys C and D and (b) the linear acceleration of weight B.

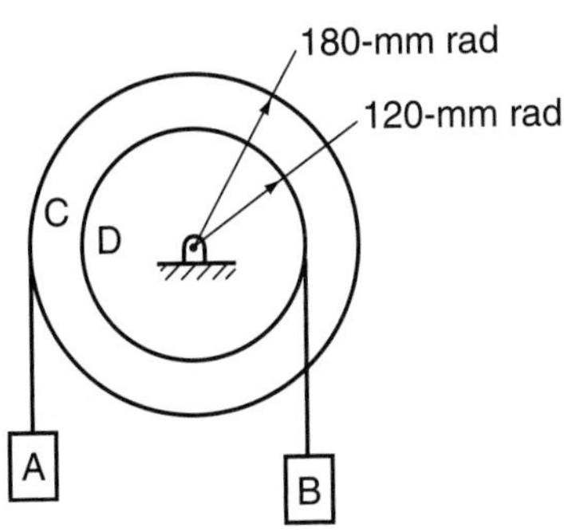

FIGURE P11–40

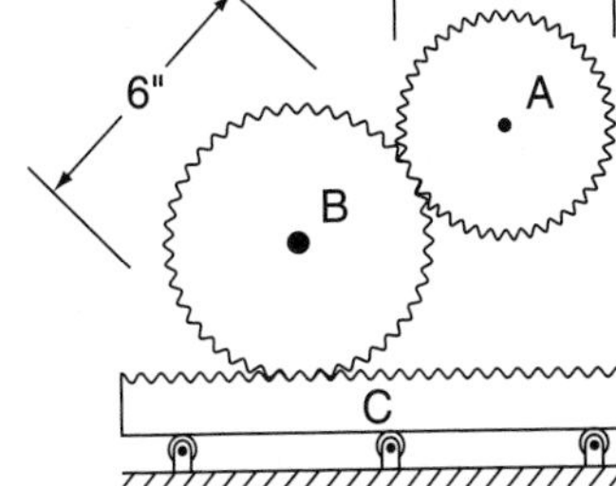

FIGURE P11–41

11–41. Gear A accelerates clockwise from rest to 100 rpm in 2 seconds. Determine the angular acceleration of gear B and the linear acceleration of rack C.

11–42. The combined pulley shown in Figure P11–42 has two cables wound around it at different diameters and fastened to point C and block E, respectively. If the velocity of block E is 240 mm/s downward, determine (a) the angular velocity of D, (b) the velocity of point C, (c) the angular velocity of lever AC, and (d) the velocity of point A, at the instant shown.

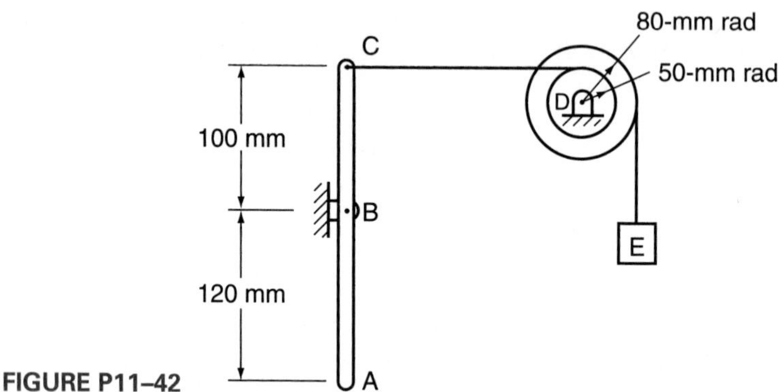

FIGURE P11–42

11–43. Weight D in Figure P11–43 is lifted from rest at constant acceleration to a height of 50 ft in 20 seconds by means of a cable wound around drum A. Assume that the diameter of A remains constant and determine (a) the angular distance traveled by gears B and C, (b) the angular acceleration of gears B and C, and (c) the tangential acceleration of the point of contact between gears B and C. (Number of gear teeth shown is not representative.)

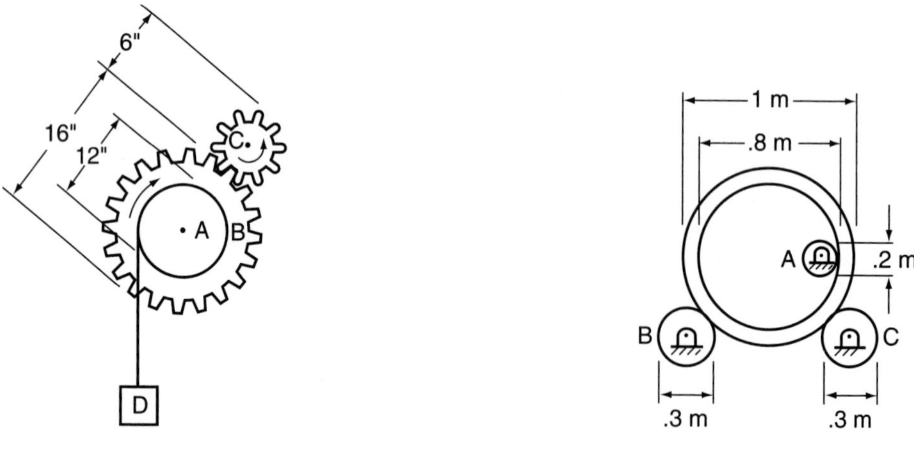

FIGURE P11–43

FIGURE P11–44

11–44. A large pipe rests on rollers B and C (Figure P11–44) and is rotated clockwise by roller A. If roller A rotates at 150 rpm, determine the angular velocities of the pipe and roller C.

11–45. Gear A rotates at 300 rpm clockwise (Figure P11–45a).

(a) Determine the angular velocity of gear C.

(b) Determine the tangential velocity of the teeth of gear C.

(c) Will the above values change if the diameter of gear B changes?

When gears A and B are moved inside gear C, a planetary gear system is created (Figure P11–45b), where the gears are called the sun (A), planets (B), and planet carrier or ring gear (C).

(d) Determine the angular velocity of gear C, the ring gear, when the sun gear rotates at 300 rpm clockwise.

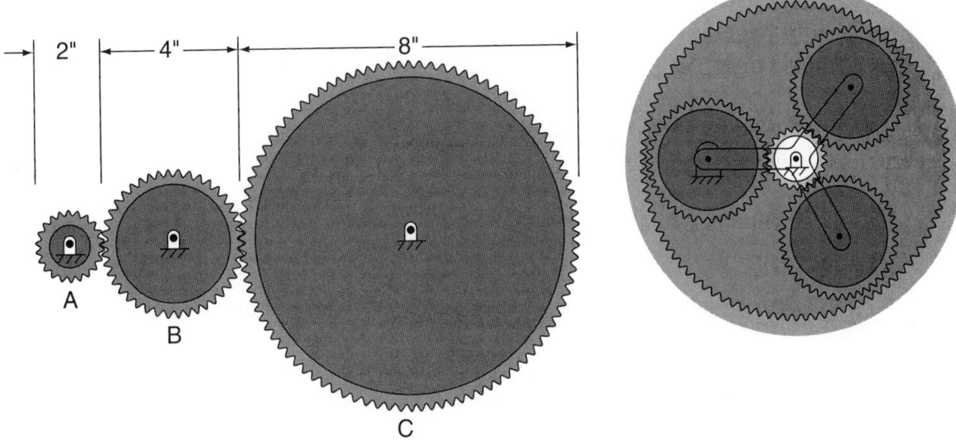

FIGURE P11–45a

FIGURE P11–45b

11–46. Block A (Figure P11–46) is dropping at a velocity of 600 mm/s and an acceleration of 100 mm/s^2. Determine the linear velocity and acceleration of block B.

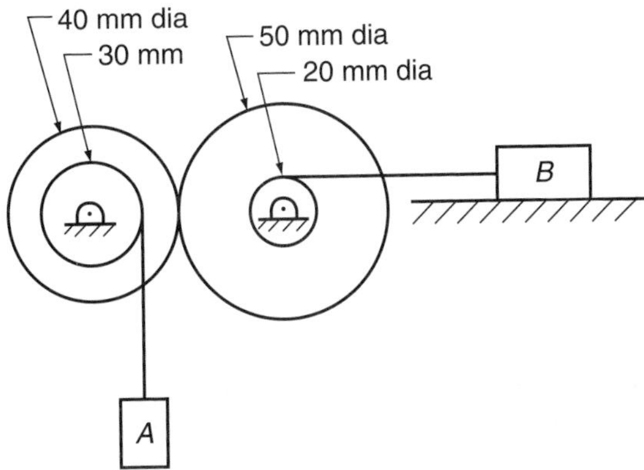

FIGURE P11–46

11–47. Wheel A accelerates from rest to 25 rpm clockwise in 3 seconds. Determine the displacement and velocity of block D at $t = 3$ seconds (Figure P11–47).

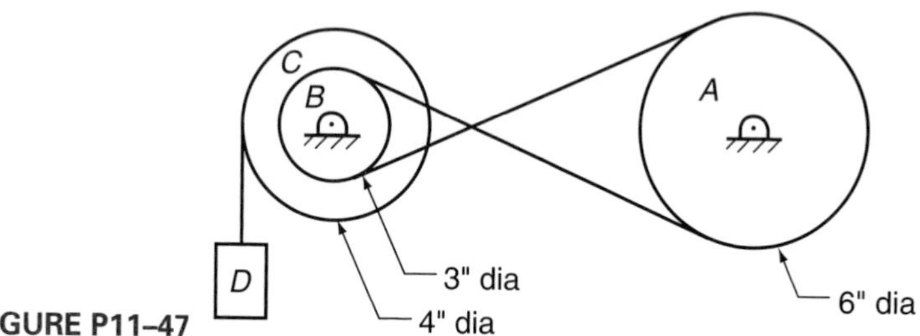

FIGURE P11–47

11–48. A cylinder 2 ft in diameter is rolled along a flat surface. If it turns four revolutions, how far has it moved along the flat surface?

11–49. The rear wheels of a tractor are 1.8 m in diameter, and the front wheels are 0.8 m. If the rear wheels are turning at 40 rpm, determine (a) the velocity of the tractor and (b) the angular speed of the front wheels. Assume no slipping.

11–50. A tractor has worn tires of 48 in. diameter, replaced by new tires of 52 in. diameter. What will be the percent increase in ground speed when this tractor is operated in the same gear and at the same engine rpm as before?

11–51. Determine the maximum rpm and angular acceleration of pulleys A and B for the elevator shown in Figure P10–18.

APPLIED PROBLEMS FOR SECTION 11–7

11–52. What is the normal acceleration of a point 9 in. from the center of a rotor that is turning at 4500 rpm?

11–53. A point 150 mm from the center of rotation has a normal acceleration of 4000 m/s^2. Determine the speed of rotation.

11–54. A weight is whirled on the end of a 2.2-m rope. The rope has an angular speed of 600 rad/min. Determine the normal acceleration of the weight.

11–55. The blade tips of a turbine are designed to withstand a normal acceleration of 250,000 ft/s^2. If the radius of the turbine wheel is 19 in., determine the maximum rpm allowed.

11–56. The blades on a fan will start to deform at a speed of 2000 rpm due to the normal acceleration. If you want a factor of safety of 2, or one-half of the normal acceleration, what should be the maximum operating rpm?

11–57. The material in the tubes of a centrifuge rotates at 450 rpm and is at a radius of 8 in. Determine the normal acceleration acting on this material.

11–58. Determine the normal and tangential accelerations for point P in Figure P11–58.

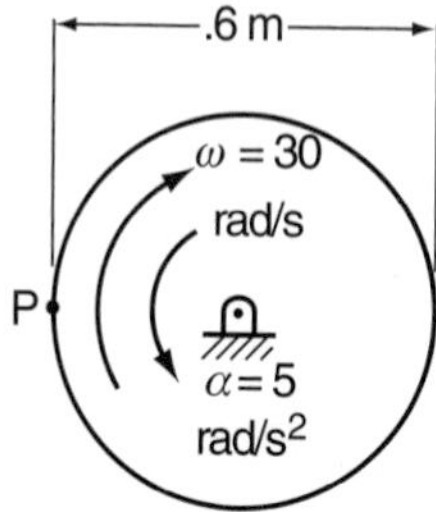

FIGURE P11–58

11–59. A 3-ft-long member, pinned at one end, accelerates from 2 rpm to 10 rpm in 6 seconds. Determine the total acceleration of the outer end when $t = 6$ seconds.

11–60. A car accelerates on a 500-m-radius curve at 1.2 m/s^2. What is the speed of the car when it has a total acceleration of 2 m/s^2?

11–61. Point P has a total acceleration of 8 m/s^2 at the instant shown in Figure P11–61. Determine the angular velocity and angular acceleration of the wheel.

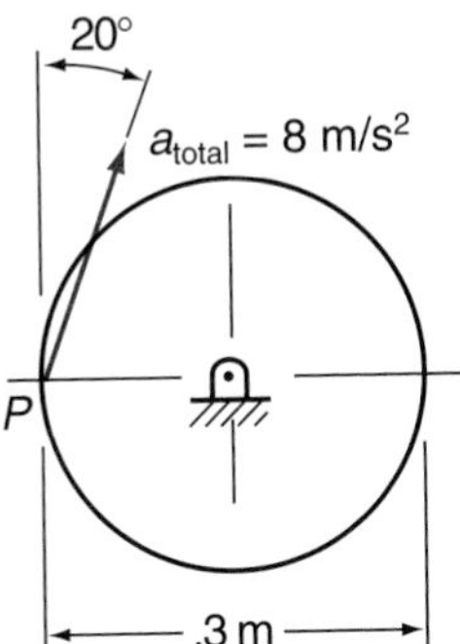

FIGURE P11–61

11–62. The object shown in Figure P11–62 rotates about A and accelerates from an initial speed of 10 rpm to 40 rpm in 5 seconds. At $t = 5$ seconds, determine the total acceleration of point B for the position shown.

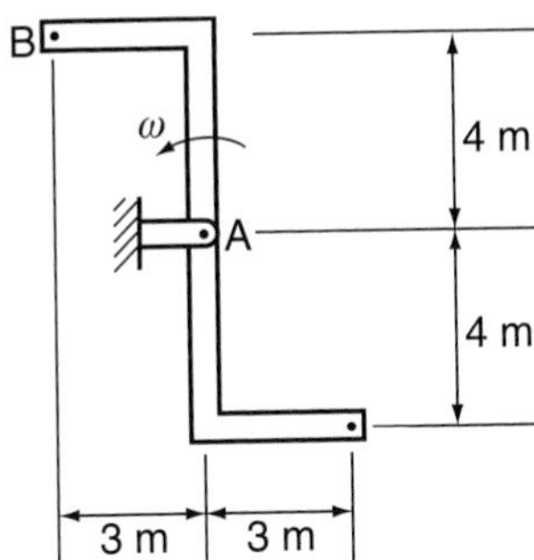

FIGURE P11–62

11–63. A car accelerates uniformly from rest to 50 mph over a distance of $\frac{1}{8}$ mile along a curve of 700-ft radius. Determine the total acceleration at the instant that the speed of 50 mph is reached.

11–64. A horizontal disc can rotate about a vertical axis through its center. A 16-lb block rests on this disc and is 10 in. from the vertical center of rotation. The disc is accelerated from rest to 80 rpm in 3 seconds. The block slides off the disc at $t = 2$ seconds. Determine the coefficient of static friction between the block and the disc.

11–65. Point A has a total acceleration of 240 in./s^2 $\overline{35°\!\!/}$ (Figure P11–65). For member ABC, rotating clockwise, determine (a) angular velocity ω and (b) angular acceleration α.

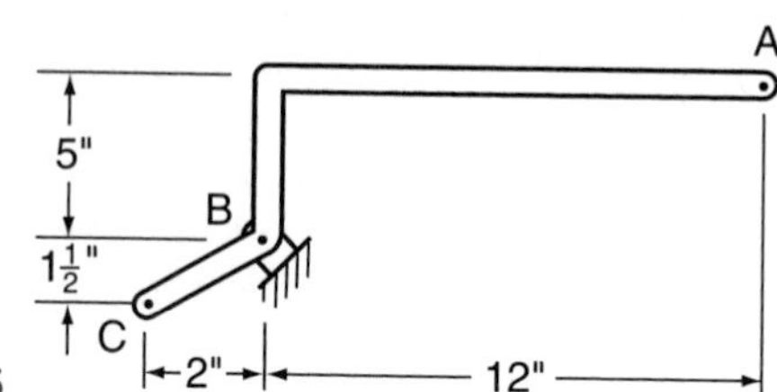

FIGURE P11–65

11–66. The total acceleration of point A is 4 m/s^2 $\angle 45°$ (Figure P11–66). For member ABCD determine (a) angular velocity ω and (b) angular acceleration α.

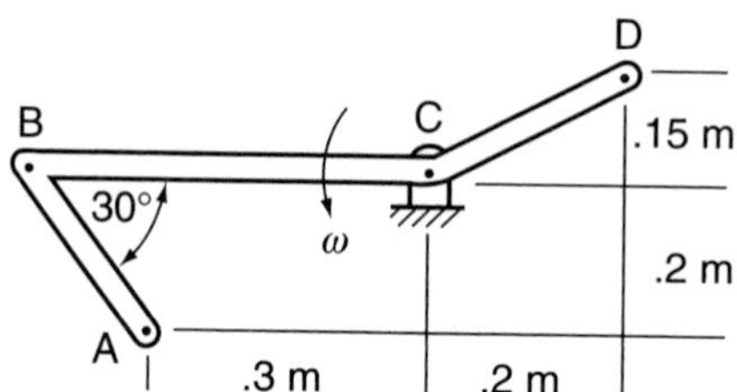

FIGURE P11–66

REVIEW PROBLEMS

R11–1. When a power lawn mower is stalled, its blade is brought to rest from a speed of 2000 rpm, turning 250 revolutions in the process. Determine (a) the angular deceleration and (b) the time required for the lawn mower to stop.

R11–2. A large flywheel used on a punch press is slowed from 70 rpm to 65 rpm in 2 seconds. Determine its rate of deceleration.

R11–3. A wheel that is rotating initially at 20 rpm clockwise is decelerated at a uniform rate for 6 seconds, at which time it is rotating at 10 rpm counterclockwise. Determine the angular deceleration and the total number of revolutions of the wheel during the 6-second interval.

R11–4. Rocker arms AB and CD in Figure RP11–4 rock the horizontal deck of a combine back and forth. Assume that each arm accelerates uniformly to the midpoint of its arc and then decelerates uniformly to the end of the arc. A complete cycle takes 2 seconds. Determine (a) the maximum angular velocity and (b) the angular acceleration.

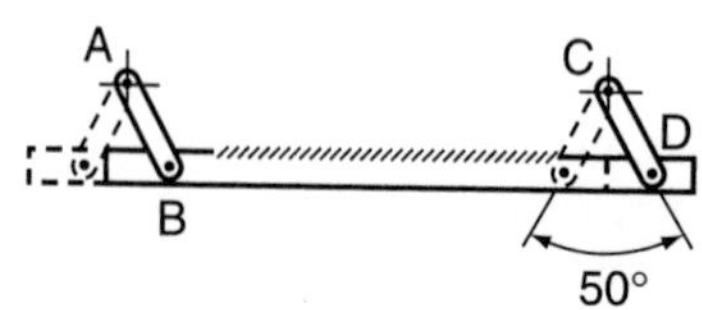

FIGURE RP11–4

R11–5. A 1.2-m-diameter barrel rolls down a slope with an initial velocity of 0.8 m/s. Ten seconds later it has a velocity of 3.4 m/s. Determine (a) the angular acceleration of the barrel and (b) the angular velocity at $t = 10$ seconds.

R11–6. The trailer shown in Figure RP11–6 has a pulley C fastened to wheel B. Drum A is belt driven by pulley C. If the trailer is pulled with a velocity of 9 km/h, determine the angular velocity of drum A.

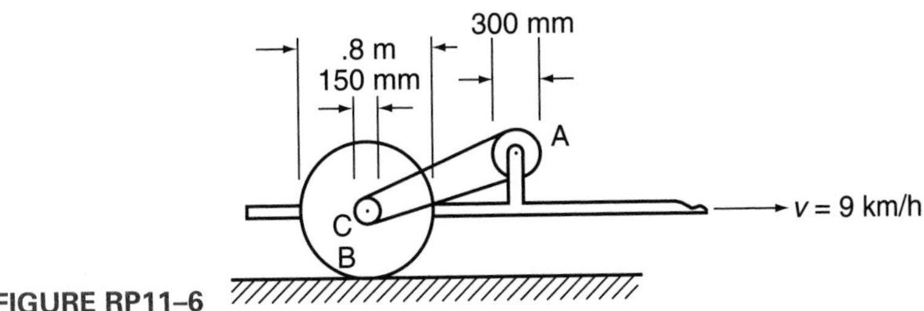

FIGURE RP11–6

R11–7. Weight A in Figure RP11–7 drops 30 in. from rest, at a uniform acceleration of 200 in./s^2. Determine (a) the linear acceleration of B, (b) the angular acceleration of E, (c) the angular displacement of E, and (d) the angular velocity of C.

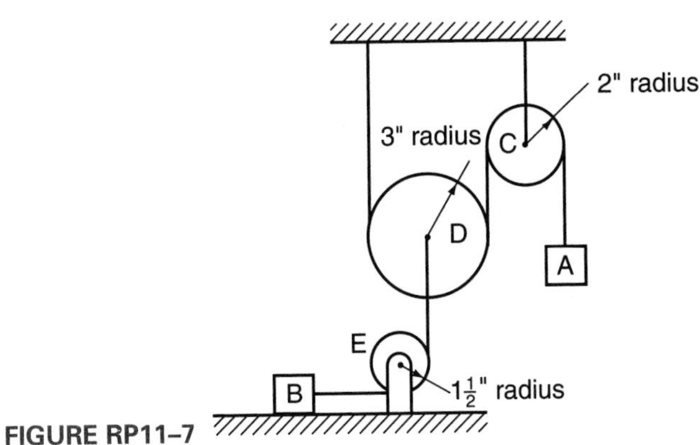

FIGURE RP11–7

R11–8. A highway curve with a radius of 1200 ft is designed in such a way that it is unsafe for a car on this curve to have a normal acceleration greater than 5 ft/s^2. What is the maximum velocity that a car may attain safely on this curve?

R11–9. In an amusement ride, people sit in a bullet-shaped compartment, such as A in Figure RP11–9, and are rotated in a vertical circle. Determine the normal and tangential accelerations due to the rotation for the position shown at $t = 10$ seconds. (The ride accelerates from rest to 3 rad/s in 10 seconds.)

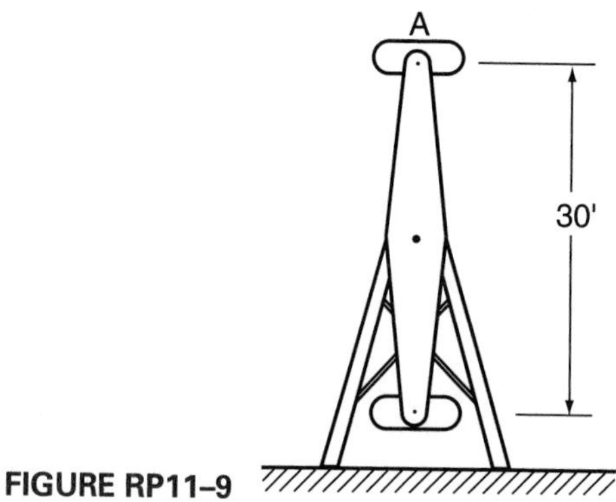

FIGURE RP11–9

R11–10. A car accelerates uniformly from 20 mph to 80 mph in 6 seconds while traveling on a curve with a radius of 500 ft. Determine its total acceleration at $t = 4$ seconds.

CHAPTER 12

Plane Motion

OBJECTIVES

Upon completion of this chapter the student will be able to:

1. Solve for linear values of displacement, velocity, or acceleration in either absolute or relative terms.
2. Define and locate an instantaneous center of an object or mechanism.
3. Determine both linear and angular velocities of various mechanisms by means of instantaneous centers.

12–1 RELATIVE MOTION

When a body is in motion, it may have displacement, velocity, and acceleration. The following discussion of relative motion applies to all three of these variables; for easier visualization, we will begin with velocity.

A common assumption often made about velocity is that velocity is stated with respect to earth. The reason for this common reference is that the earth appears stationary to us. For most engineering calculations, this is a safe assumption, and we will use it here. Since the earth is considered stationary, a velocity measured with respect to earth is an *absolute velocity.*

Relative velocity comes into play when the velocity of one object is related to that of another reference object that is also moving. For example, if car A were traveling at 40 km/h (absolute) and were passed by car B traveling at 60 km/h (absolute), car B would have a velocity of 20 km/h relative to car A. Car B, therefore, would have an absolute velocity of 60 km/h and a relative velocity of 20 km/h with respect to car A. As you can appreciate, a relative velocity has no meaning unless the reference or point to which the velocity is relative is stated.

Since we are concerned with the velocities of cars A and B and the velocity of B with respect to A, the following notation will be used for absolute and relative velocity:

$$v_A = \text{velocity of A}$$

$$v_B = \text{velocity of B}$$

$$v_{B/A} = \text{velocity of B with respect to A}$$

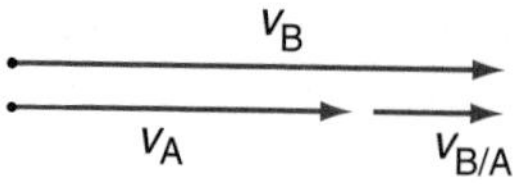

v_B

v_A $v_{B/A}$

FIGURE 12–1

For the situation in our example (Figure 12–1):

$$v_B = v_A + v_{B/A}$$
$$60 = 40 + 20$$
$$60 = 60$$

Keep in mind that we could have also written

$$s_B = s_A + s_{B/A}$$
$$a_B = a_A + a_{B/A}$$

When the velocities are not in the same direction, but on some angle to one another, we must employ vector addition—possibly using the sine law.

Throughout this chapter we will consider various moving objects or linkages at a very specific "split second" position as though we had photographed a still picture of them. This means we are only considering *this* instant and no earlier or later times.

EXAMPLE 12–1

Starting from the same point, car A travels north at 40 km/h and car B travels east at 60 km/h. Determine the velocity of B with respect to A.

To visualize the velocity of B with respect to that of A, you would have to be in car A looking at car B (Figure 12–2). Our previous equation applies, but directions must be shown as well.

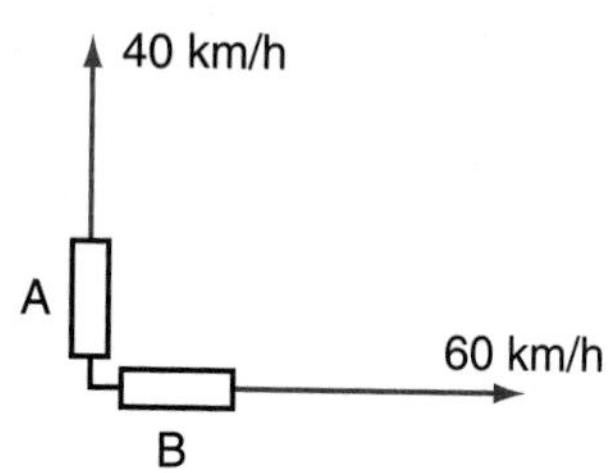

FIGURE 12–2

$$v_B = v_A + v_{B/A}$$
$$\overset{\rightarrow}{60\text{ km/h}} = \overset{\uparrow}{40\text{ km/h}} + v_{B/A}$$

(Immediately beneath the equation show the known direction and magnitude of each vector quantity.) We construct the vector triangle (Figure 12–3) as indicated by the equation; that is, v_A and $v_{B/A}$ are tip to tail.

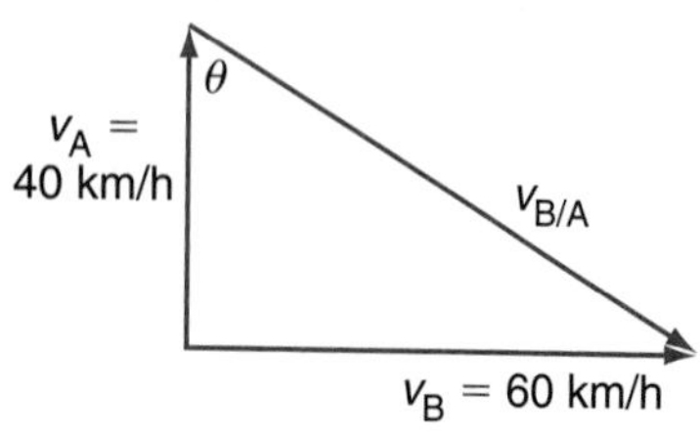

FIGURE 12–3

$$v_{B/A} = \sqrt{(40\text{ km/h})^2 + (60\text{ km/h})^2}$$

$$v_{B/A} = 72\text{ km/h} \measuredangle 56.3^\circ$$

$$\tan\theta = \frac{60}{40}$$
$$\theta = 56.3^\circ$$

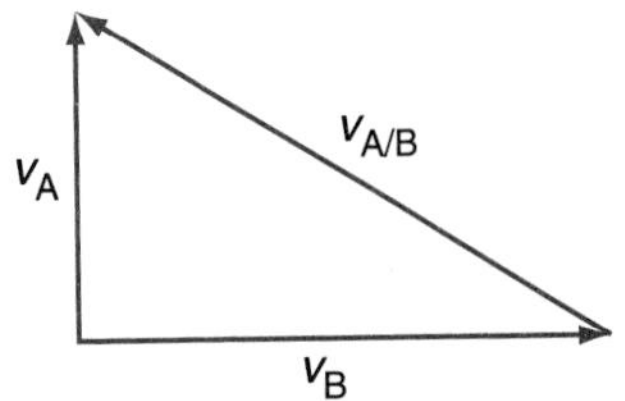

FIGURE 12–4

The velocity of A with respect to B ($v_{A/B}$) is equal and opposite to $v_{B/A}$ and could have been solved for by means of the vector triangle shown in Figure 12–4 and the following equation.

$$v_A = v_B + v_{A/B}$$

$$\overset{\uparrow}{40 \text{ km/h}} = \overset{\rightarrow}{60 \text{ km/h}} + v_{A/B}$$

Note that a way of ensuring that you have the correct equation is to check that the subscripts on the right side of the equation cancel out to equal the subscript on the left.

$$v_A = v_{\not{B}} + v_{A/\not{B}}$$
$$v_A = v_A$$

Now that we have looked at relative velocity, or motion between two separate objects, let us investigate the relative motion between two points on the same object. This will occur when an object moves with general plane motion consisting of simultaneous translation and rotation.

An example of plane motion is a bar leaning against a wall, as in Figure 12–5; the bottom of the bar is slipping to the right. The *translational motion* consists of A moving downward and B moving to the right. *Rotational motion* is also evident as the bar rotates in a counterclockwise direction about its center. The complete plane motion can be analyzed and divided into its individual components: translational motion and then rotational (Figure 12–6), or rotational motion and then translational (Figure 12–7).

In Figure 12–6, the displacement of A with respect to B, $s_{A/B}$, is the arc of a circle or

$$s_{A/B} = r\theta$$
$$= (AB)\theta$$

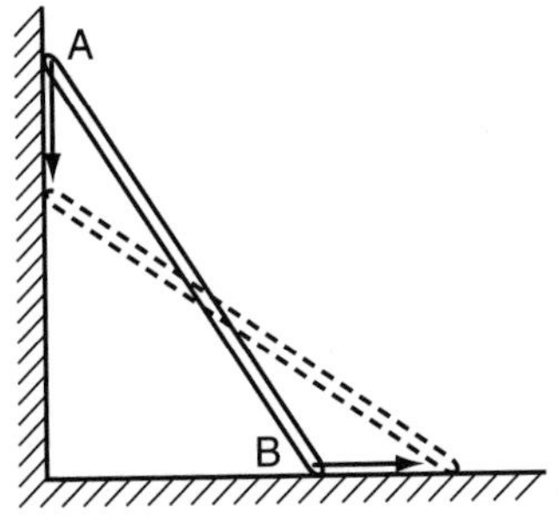

FIGURE 12–5

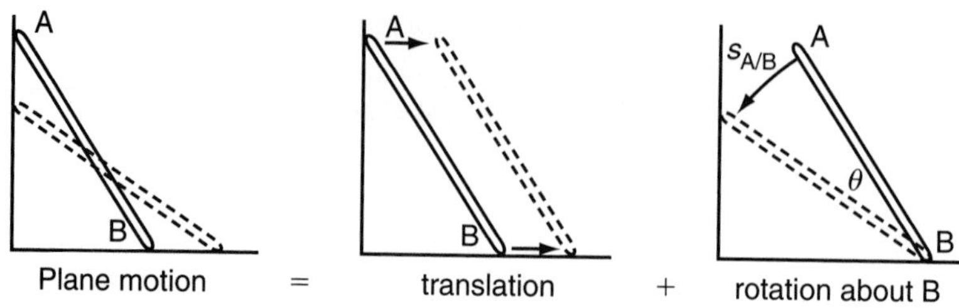

FIGURE 12–6

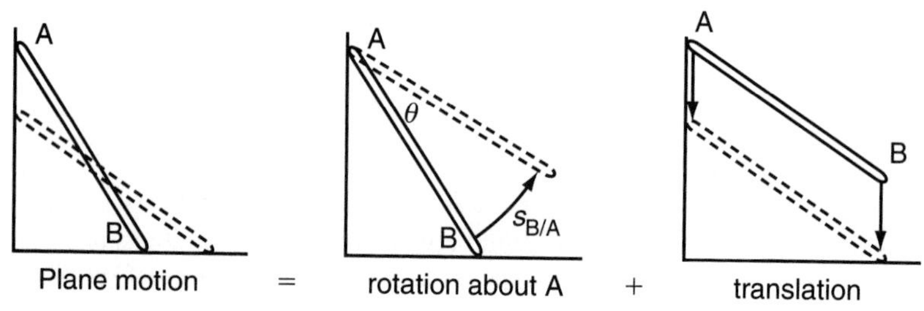

FIGURE 12–7

Similarly, in Figure 12–7,

$$s_{B/A} = (AB)\theta$$

The equation describing the plane motion at the instant shown would be

$$s_A = s_B + s_{A/B}$$

$$\downarrow = \rightarrow + \swarrow$$

or

$$s_B = s_A + s_{B/A}$$

$$\rightarrow = \downarrow + \nearrow$$

EXAMPLE 12–2

A football receiver runs straight downfield (north), turns 40° to his right, and maintaining his speed of 20 ft/s, catches a football that is traveling at 50 ft/s due north (Figure 12–8). Determine the velocity of the ball with respect to the receiver.

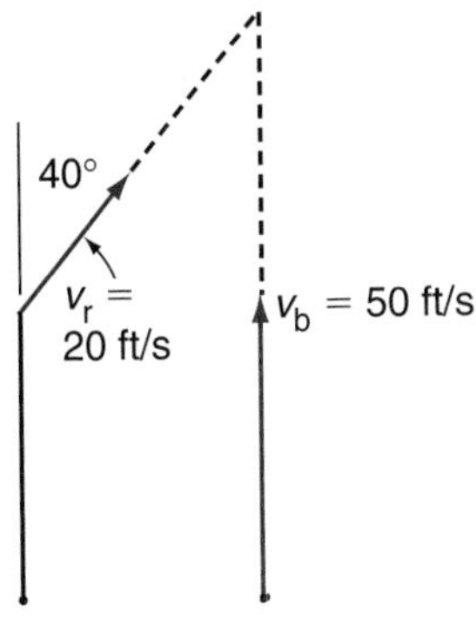

FIGURE 12–8

Use the following notation.

$$v_r = \text{velocity of the receiver}$$

$$v_b = \text{velocity of the ball}$$

$$v_{b/r} = \text{velocity of the ball with respect to the receiver}$$

$$v_b = v_r + v_{b/r}$$

$$50\uparrow = \underline{|40°}\nearrow + 20$$

Showing the known values and directions of each velocity under the relative velocity equation can often be helpful when drawing the vector triangle.

Constructing the vector triangle (Figure 12–9) and applying the cosine law, we get

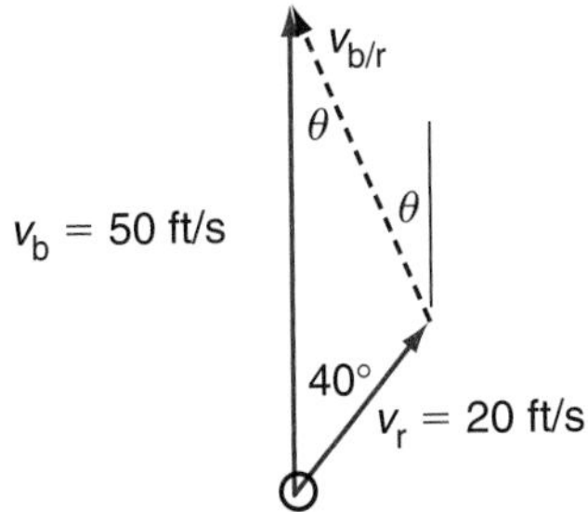

FIGURE 12–9

$$(v_{b/r})^2 = (20\text{ ft/s})^2 + (50\text{ ft/s})^2 - 2(20\text{ ft/s})(50\text{ ft/s})(\cos 40°)$$

$$v_{b/r} = 37\text{ ft/sec} \quad \angle 20.3°$$

$$\frac{20\text{ ft/s}}{\sin\theta} = \frac{37\text{ ft/s}}{\sin 40°}$$

$$\theta = 20.3°$$

EXAMPLE 12–3

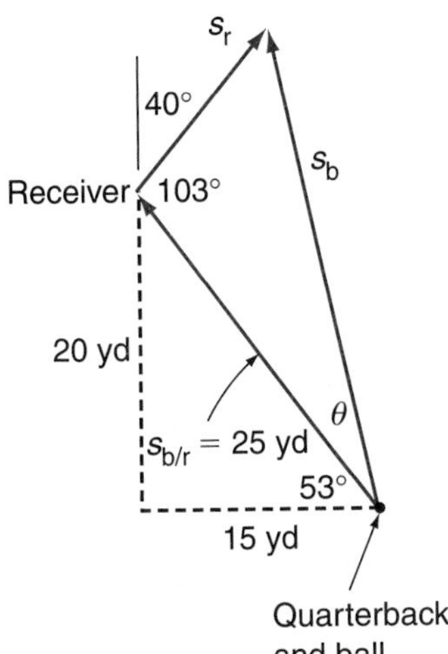

FIGURE 12–10

Suppose now that the quarterback and receiver are located as in Figure 12–10. As before, the receiver and the ball have velocities of 20 ft/s and 50 ft/s, respectively, and the ball is released the instant the receiver makes his 40° turn. Determine the distance traveled by the ball before it is caught and the angle θ at which the quarterback must lead his receiver.

Distances s_r and s_b are covered in the same time, so

$$t = \frac{s_r}{v_r} = \frac{s_b}{v_b}$$

$$\frac{s_r}{20\text{ ft/s}} = \frac{s_b}{50\text{ ft/s}}$$

$$s_r = 0.4s_b$$

By the relative velocity formula and the vector triangle, we have

$$s_b = s_r + s_{b/r}$$

Using the cosine law, we obtain

$$(s_b)^2 = (s_r)^2 + (s_{b/r})^2 - 2s_r s_{b/r} \cos 103°$$

$$(s_b)^2 = (0.4s_b)^2 + (25 \text{ yd})^2 - (2)(0.4s_b)(25 \text{ yd})(-0.225)$$

$$s_b^2 - 5.36s_b - 745 = 0$$

Solving by means of the quadratic equation, we obtain

$$\underline{s_b = 30 \text{ yd}}$$

$$s_r = 0.4s_b$$

$$= 0.4(30 \text{ yd})$$

$$s_r = 12 \text{ yd}$$

Using the sine law, we obtain

$$\frac{s_b}{\sin 103°} = \frac{s_r}{\sin \theta}$$

$$\sin \theta = \frac{s_r}{s_b} \sin 103°$$

$$= 0.4(0.975)$$

$$\theta = 23°$$

The ball travels 30 yd, and the quarterback leads the receiver by 23°.

Examples 12–2 and 12–3 are based on quite independent values of displacement and velocity for the receiver and the ball, since they were not physically connected in any way. Consider now the relationship between two points that are connected, such as points A and B, which represent the opposite ends of a bar as shown in Figure 12–11.

EXAMPLE 12–4

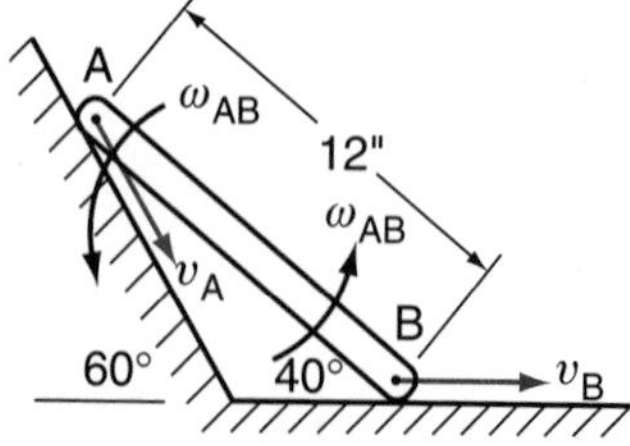

FIGURE 12–11

Determine (a) the linear velocity of end A of the bar shown in Figure 12–11 if the velocity of B is 16 in./s to the right and (b) the angular velocity of AB.

For bar AB we can write the equation

$$v_A = v_B + v_{A/B}$$

To understand how we can show the velocity of A with respect to B($v_{A/B}$), imagine yourself to be riding on point B and observing point A. Point A is not getting any closer to or farther

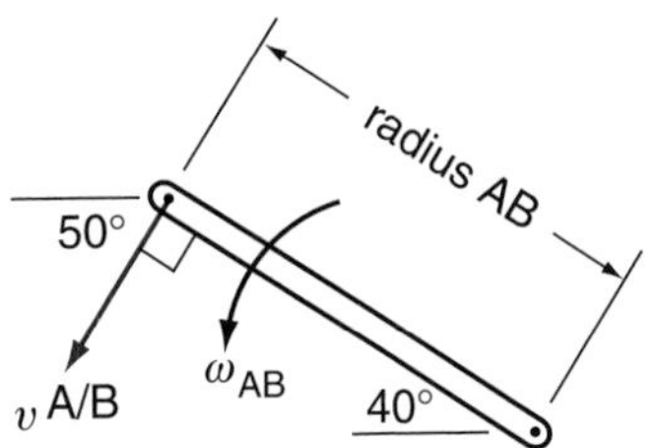

FIGURE 12–12

from you, but appears to be rotating downward as a tangential velocity at the end of bar AB.

As shown in Figure 12–12, the equation relating linear velocity to angular velocity is:

$$\text{linear velocity} = (\text{radius})(\text{angular velocity})$$

or

$$v_{A/B} = (\text{length AB})(\omega_{AB})$$

The direction of the tangential velocity $v_{A/B}$ is normal to the radius.

The relative velocity equation can be rewritten and all possible magnitudes and directions of each velocity can be shown immediately below each velocity.

$$v_A = v_B + v_{A/B}$$

$$\overline{\diagdown 60^\circ} = \overrightarrow{16\ \text{in./s}} + \overline{50^\circ \diagup}$$

Construct a vector triangle (Figure 12–13). Solve for v_A using the sine law.

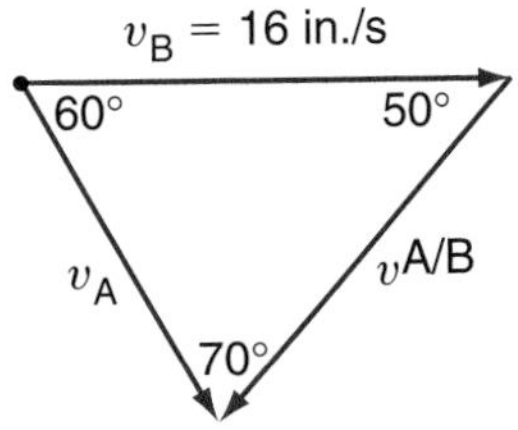

FIGURE 12–13

$$\frac{v_A}{\sin 50^\circ} = \frac{16\ \text{in./s}}{\sin 70^\circ}$$

$$v_A = \frac{16}{\sin 70^\circ}(\sin 50^\circ)$$

$$\underline{v_A = 13\ \text{in./s}\ \overline{\diagdown 60^\circ}}$$

Similarly solving for $v_{A/B}$

$$\frac{v_{A/B}}{\sin 60^\circ} = \frac{16\ \text{in./s}}{\sin 70^\circ}$$

$$v_{A/B} = 14.7\ \text{in./s}$$

Solving for the angular velocity now

$$\omega_{AB} = \frac{v_{A/B}}{AB}$$

$$= \frac{14.7\ \text{in. s}}{12\ \text{in.}}$$

$$\underline{\omega_{AB} = 1.23\ \text{rad/s} \circlearrowright}$$

EXAMPLE 12–5

Bar AB of the linkage shown in Figure 12–14 rotates at 20 rad/s clockwise. For the position shown, determine the velocity of pin C and the angular velocity of link CD.

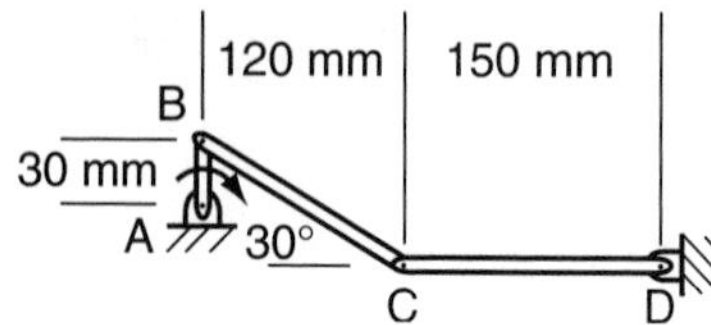

FIGURE 12–14

Considering bar AB we have

$$v_B = r\omega$$
$$= (30\text{ mm})(20\text{ rad/s})$$
$$v_B = 600\text{ mm/s} \rightarrow$$

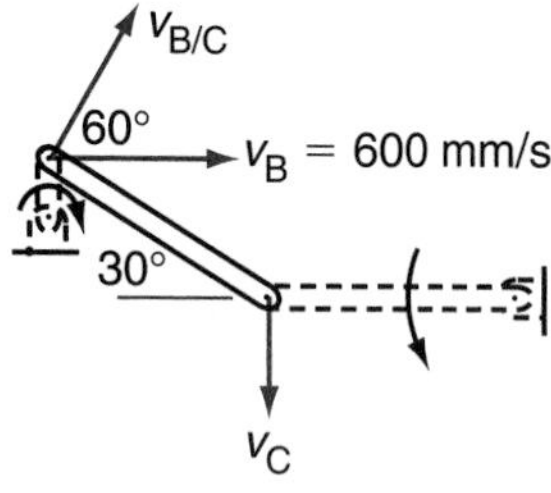

FIGURE 12–15

For bar BC (Figure 12–15), we can write the equation

$$v_B = v_C + v_{B/C}$$

Note that $v_{B/C}$ must be perpendicular to bar BC (Figure 12–16). Imagine yourself to be at C observing B. Point B will appear to rotate upward about C; therefore, $v_{B/C}$ will appear as a tangential velocity at right angles to the radius BC.

Show all known magnitudes and directions below the relative velocity equation.

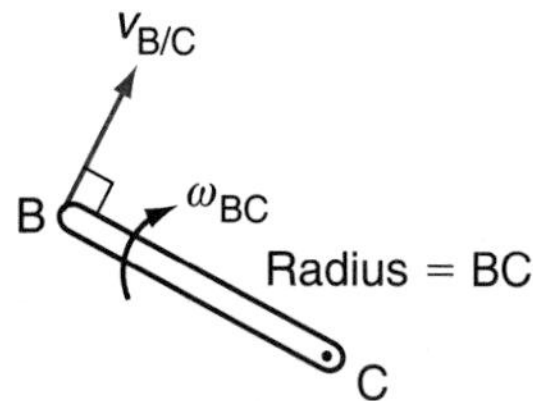

FIGURE 12–16

$$v_B = v_C + v_{B/C}$$
$$\underset{\rightarrow}{600} = \downarrow + \angle 60^\circ$$

The vector triangle is drawn according to the equation; that is, v_C and $v_{B/C}$ are tip-to-tail as in Figure 12–17.

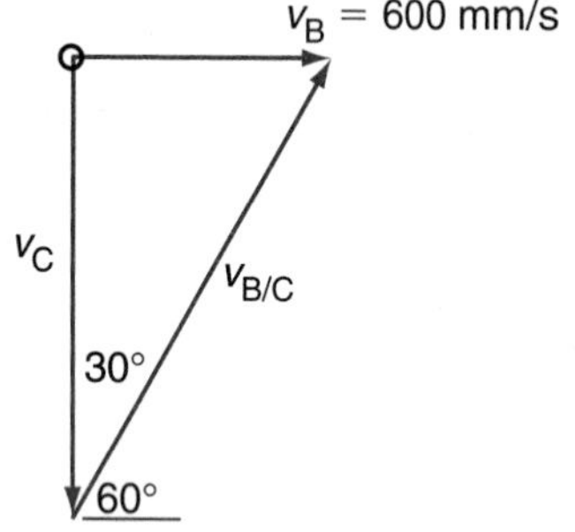

FIGURE 12–17

$$\tan 30^\circ = \frac{v_B}{v_C}$$
$$0.577 = \frac{600\text{ mm/s}}{v_c}$$
$$v_c = 1040\text{ mm/s} \downarrow$$
$$\underline{v_c = 1.04\text{ m/s} \downarrow}$$

From Figure 12–18, we have

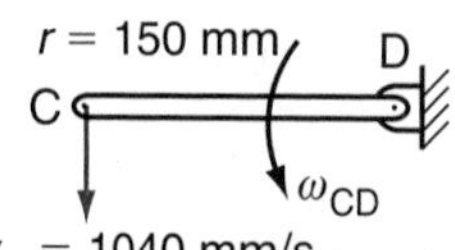

FIGURE 12–18 v_C = 1040 mm/s

$$v_C = r\omega_{CD}$$
$$\omega_{CD} = \frac{1040\text{ mm/s}}{150\text{ mm}}$$
$$\underline{\omega_{CD} = 6.93\text{ rad/s} \circlearrowleft}$$

EXAMPLE 12–6

Bar BC in Figure 12–19 has an angular velocity of 3.73 rad/s clockwise. Determine the angular velocity of AB and the velocity of the piston at A.

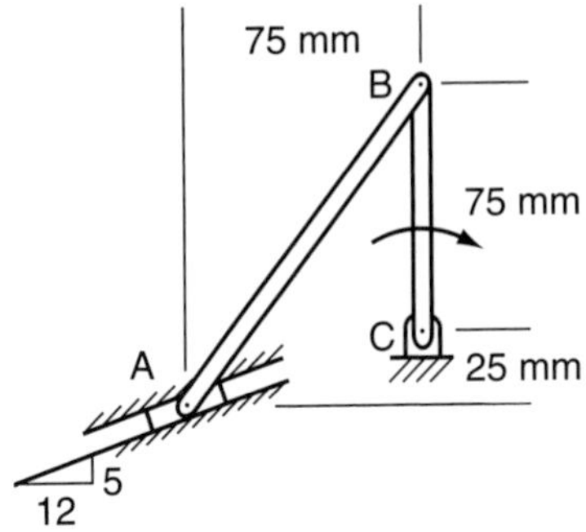

FIGURE 12–19

$$v_B = r\omega \qquad \text{(Figure 12–20)}$$

$$= (75\text{ mm})(3.73\text{ rad/s})$$

$$v_B = 280\text{ mm/s} \rightarrow$$

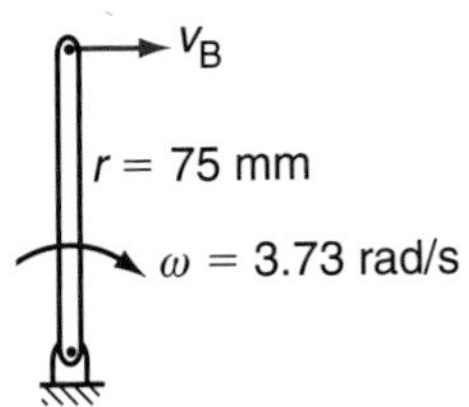

FIGURE 12–20

Considering member AB in its position in Figure 12–21, we can see that although it has transitional motion, it rotates clockwise. The relative velocity between points A and B, which causes this rotation, can be expressed either as $v_{A/B}$ or as $v_{B/A}$. Both velocities must be at right angles to bar AB since they are tangential velocities at the end of radius arm AB. Either velocity may be used in our equation; let us arbitrarily use $v_{A/B}$.

$$v_A = v_{A/B} + v_B$$

$$\overset{5}{\underset{12}{\nearrow}} = 3\,\underset{4}{\nwarrow} + \overset{\rightarrow}{280}\text{ mm/s}$$

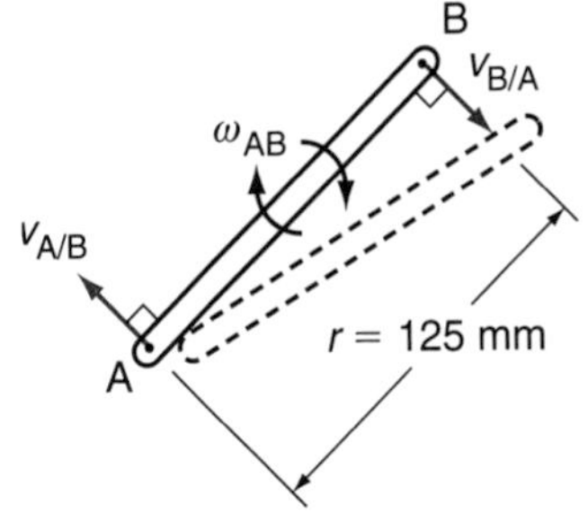

FIGURE 12–21

After constructing the vector triangle (Figure 12–22), we can calculate the internal angles of the triangle and use the sine law; alternatively, we can use the method described below.

The common component of two of the vectors, vertical in this case, can have a common denominator.

The slope numbers of each vector are increased as follows:

v_A: 13, 5, 12 becomes v_A: 39, 15, 36, and $v_{A/B}$: 3, 5, 4 becomes $v_{A/B}$: 15, 25, 20

Combining these two triangles that have larger slope numbers gives us Figure 12–23.

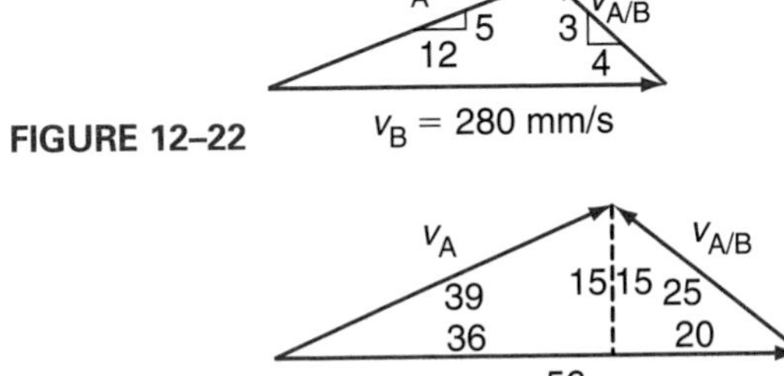

FIGURE 12–22

FIGURE 12–23

$$v_A = \frac{39}{56}(280\text{ mm/s})$$

$$v_A = 195\text{ mm/s} \ \overset{5}{\underset{12}{\nearrow}}$$

$$v_{A/B} = \frac{25}{56}(280\text{ mm/s})$$

$$= 125\text{ mm/s} \ 3\,\underset{4}{\nwarrow}$$

$$v_{A/B} = (AB)\omega_{AB}$$

$$125\text{ mm/s} = (125\text{ mm/s})\omega_{AB}$$

$$\underline{\omega_{AB} = 1\text{ rad/s} \curvearrowright}$$

EXAMPLE 12–7

The velocity of point A is 10 m/s to the right, for the system shown (Figure 12–24). Using the relative velocity method, determine the angular velocity of AC and the linear velocity of point C.

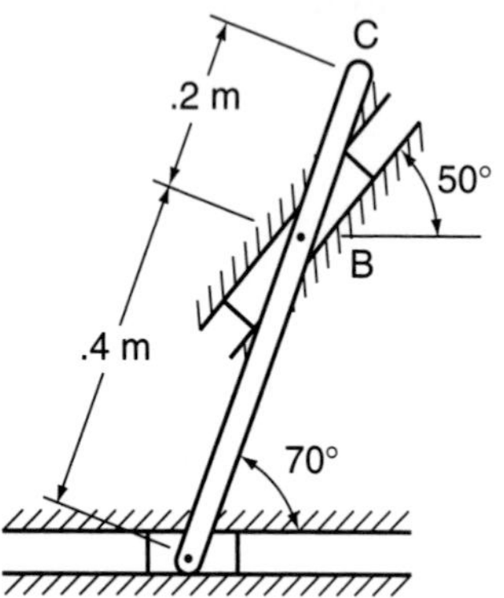

FIGURE 12–24

Referring to Figure 12–25, we have

$$v_B = v_A + v_{B/A}$$

$$\underline{\angle 50^\circ} = \overset{\rightarrow}{10}\text{ m/s} + \underline{20^\circ\diagdown}$$

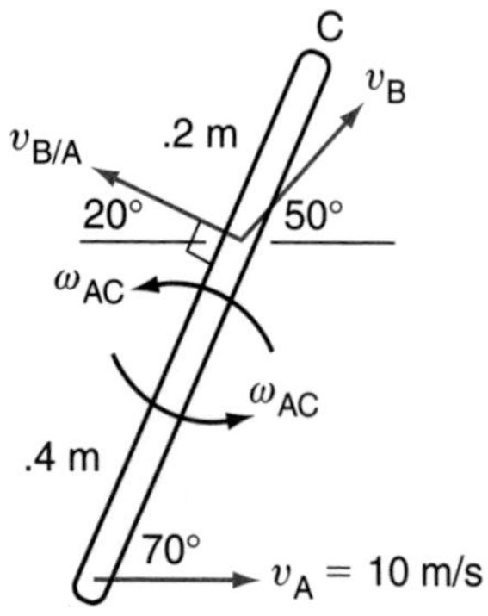

FIGURE 12–25

Constructing the vector triangle (Figure 12–26) yields

$$\frac{v_{B/A}}{\sin 50^\circ} = \frac{10 \text{ m/s}}{\sin 110^\circ}$$

$$v_{B/A} = 8.15 \text{ m/s}$$

$$\omega_{AC} = \omega_{AB} = \frac{v_{B/A}}{BA} = \frac{8.15 \text{ m/s}}{0.4 \text{ m}}$$

$$\underline{\omega_{AC} = 20.4 \text{ rad/s} \circlearrowleft}$$

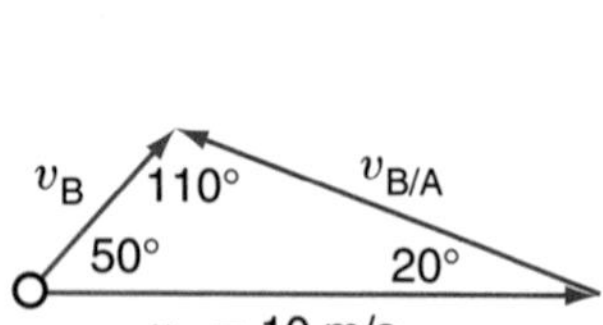

FIGURE 12–26

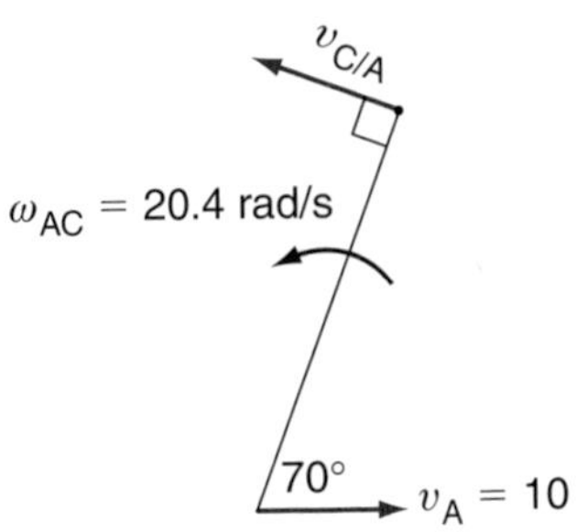

FIGURE 12–27

Referring to Figure 12–27 gives us

$$v_{C/A} = AC(\omega_{AC})$$
$$= (0.6 \text{ m})(20.4 \text{ rad/s})$$
$$= 12.2 \text{ m/s } \underline{20°\diagdown}$$

Using the relative velocity equations now, we have

$$v_C = v_A + v_{C/A}$$
$$\overrightarrow{10} \text{ m/s} + \underline{20°\diagdown}\ 12.2 \text{ m/s}$$

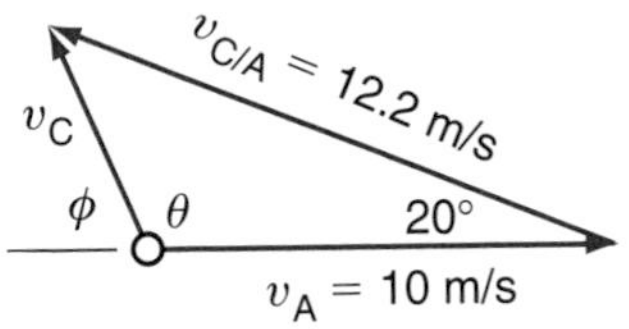

FIGURE 12–28

From the vector triangle (Figure 12–28),

$$v_C^2 = (12.2 \text{ m/s})^2 + (10 \text{ m/s})^2$$
$$- 2(12.2 \text{ m/s})(10 \text{ m/s})(\cos 20°)$$
$$= 19.55$$
$$v_C = 4.42 \text{ m/s } \underline{70.7°\diagdown}$$

$$\frac{12.2 \text{ m/s}}{\sin \theta} = \frac{4.42 \text{ m/s}}{\sin 20°}$$
$$\theta = 109.3° \text{ or } \phi = 180 - 109.3 = 70.7°$$

Relative acceleration can be handled by using the type of equation that was used for distance and velocity; for example,

$$a_B = a_{B/A} + a_A$$

When points A and B are on the same object, such as a link of a mechanism, each acceleration term in the equation above will consist of two components, a *tangential component* and a *normal component*. We are now dealing with six acceleration values rather than just three, as the equation may indicate. As only limited coverage will be given to this topic, the following example illustrates the added complexity of solving for acceleration values.

EXAMPLE 12–8

Bar AB rotates at 5 rad/s clockwise and accelerates at 2 rad/s^2 clockwise. Determine the linear acceleration of C at its position in Figure 12–29.

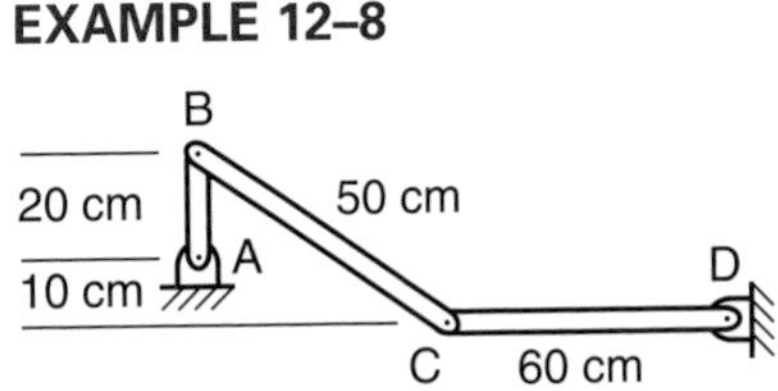

FIGURE 12–29

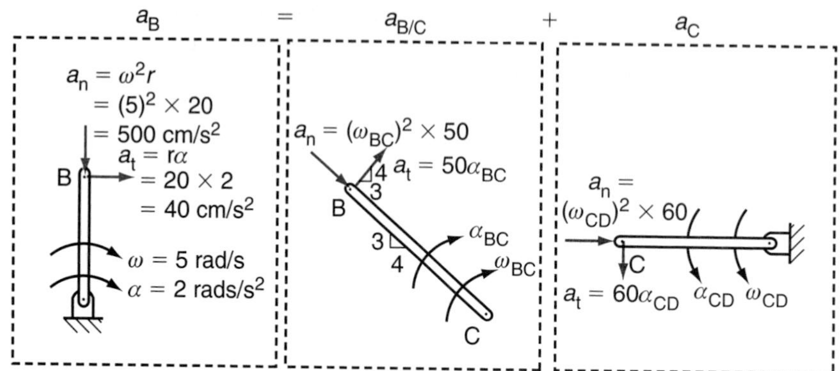

FIGURE 12–30

Figure 12–30 shows the equation to be used as well as vector sketches of normal and tangential acceleration in all cases.

From Figure 12–30, we can see that the unknown velocities, ω_{BC} and ω_{CD}, must be found before any normal acceleration values can be determined. Figure 12–31 shows the relative velocity equation used in solving for these velocities.

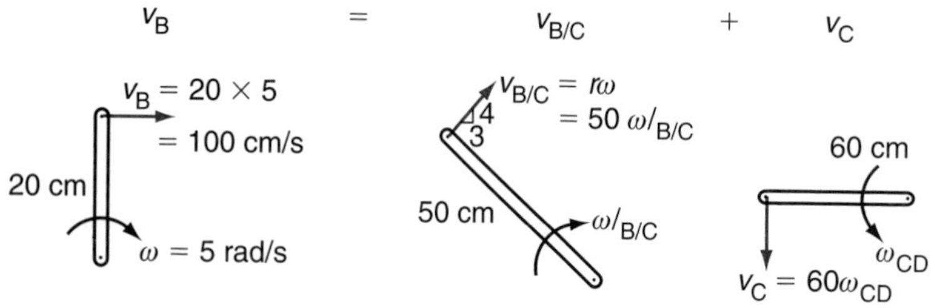

vB = vB/C + vC
vB = 20 × 5
= 100 cm/s
20 cm
ω = 5 rad/s
vB/C = rω
= 50 ω/B/C
4
3
50 cm
ω/B/C
60 cm
ωCD
vC = 60ωCD

FIGURE 12–31

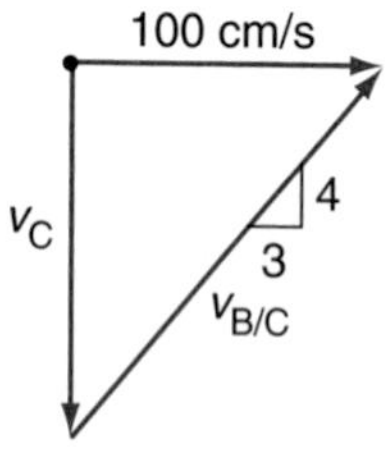

FIGURE 12–32

From the vector triangle of Figure 12–32, we have:

$$v_C = \frac{4}{3}(100\ \text{cm/s}) = 133\ \text{cm/s}$$

$$v_{B/C} = \frac{5}{3}(100\ \text{cm/s}) = 167\ \text{cm/s}$$

$$\omega_{CD} = \frac{v_C}{r_{CD}} = \frac{133\ \text{cm/s}}{60\ \text{cm/s}} = 2.22\ \text{rad/s}$$

$$\omega_{BC} = \frac{v_{B/C}}{r_{BC}} = \frac{167\ \text{cm/s}}{50\ \text{cm/s}} = 3.34\ \text{rad/s}$$

Return now to the acceleration equation in Figure 12–30 with the angular velocity values calculated above. Note that the equation shows two vectors (a_n and a_t) of B equal to the sum of four other vectors. Now write an equation showing the horizontal components of the two vectors equal to the horizontal components of the four vectors.

$$0 + 40\ \text{cm/s}^2 = \left[\frac{4}{5}(\omega_{BC})^2(50\ \text{cm})\right] + \left[\frac{3}{5}(50\ \text{cm})(\alpha_{BC})\right]$$

$$+ [(\omega_{CD})^2(60\ \text{cm})] + [0]$$

$$40 = \left[\frac{4}{5}(3.34\ \text{rad/s})^2(50\ \text{cm})\right] + \left[\frac{3}{5}(50\ \text{cm})(\alpha_{BC})\right]$$

$$+ [(2.22\ \text{rad/s})^2(60\ \text{cm})]$$

$$40 = 446 + 30\alpha_{BC} + 296$$

$$\alpha_{BC} = -23.4\ \text{rad/s}^2$$

The minus sign indicates that the direction of α_{BC} that we assumed was incorrect, but we can substitute this value into the next equation as a minus value.

Now consider all vertical components.

$$-(500\ \text{cm/s}^2) + 0 = \left[-\frac{3}{5}(\omega_{BC})^2(50\ \text{cm})\right] + \left[\frac{4}{5}(50\ \text{cm})(\alpha_{BC})\right]$$

$$+ [0] - [(60\ \text{cm})(\alpha_{CD})]$$

$$-500 = \left[-\frac{3}{5}(3.34\ \text{rad/s})^2(50\ \text{cm})\right]$$

$$+ \left[\frac{4}{5} \times (50\ \text{cm})(-23.4\ \text{rad/s}^2)\right]$$

$$- [(60\ \text{cm})(\alpha_{CD})]$$

$$-500 = -335 - 936 - 60\alpha_{CD}$$

$$\alpha_{CD} = -12.8\ \text{rad/s}^2 \circlearrowright$$

Therefore,

$$\alpha_{CD} = 12.8\ \text{rad/s}^2 \circlearrowleft$$

We can now find the linear acceleration of C by considering link CD alone (Figure 12–33).

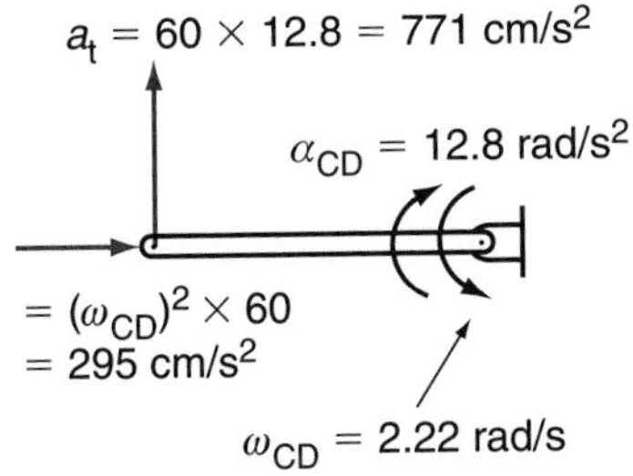

FIGURE 12–33

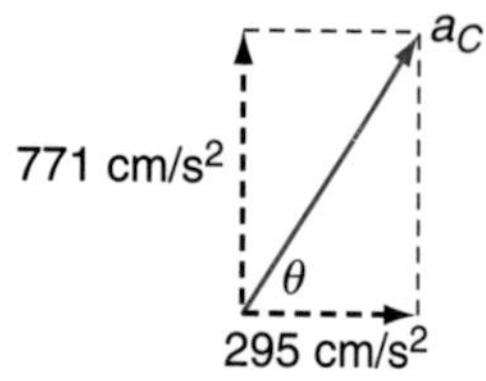

FIGURE 12–34

The total acceleration of C (Figure 12–34) is

$$\alpha_{CD} = 12.8 \text{ rad/s}^2 \circlearrowright$$

$$a_C = \sqrt{(295 \text{ cm/s}^2)^2 + (771 \text{ cm/s}^2)^2}$$

$$= 825 \text{ cm/s}^2$$

$$\tan\theta = \frac{771}{295}$$

$$\theta = 69°$$

$$\underline{a_C = 8.25 \text{ m/s}^2 \angle 69°}$$

12–2 THE ROLLING WHEEL

An everyday example of plane motion is a rolling wheel. If a wheel is rolled from position (1) to position (2) in Figure 12–35, it has not only rotational motion but also translational motion. If it had been pivoted at its center and were not rolling, then its motion would have been only rotational. Since there is translational motion to the right, every point on the wheel must have some velocity to the right.

To visualize the speed at which the wheel would move to the right, consider a wheel with a radius of 0.5 m held slightly off the ground and rotated at 8 rad/s clockwise (Figure 12–36). The velocity of A is the velocity of A with respect to C, or

$$v_{A/C} = r\omega$$

$$= 0.5(8)$$

$$v_{A/C} = 4 \text{ m/s} \leftarrow$$

The wheel is now lowered to the ground while rotating. (We shall assume that there is no slippage.) Point A is stationary at this instant, but there is still a relative velocity between A and C, $v_{C/A}$.

$$v_{C/A} = -v_{A/C}$$

$$v_{C/A} = 4 \text{ m/s} \rightarrow$$

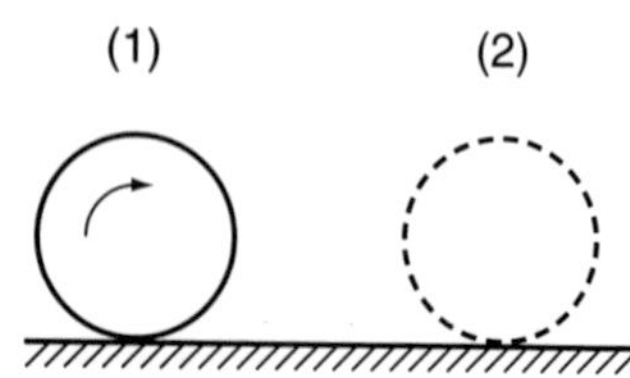

FIGURE 12–35

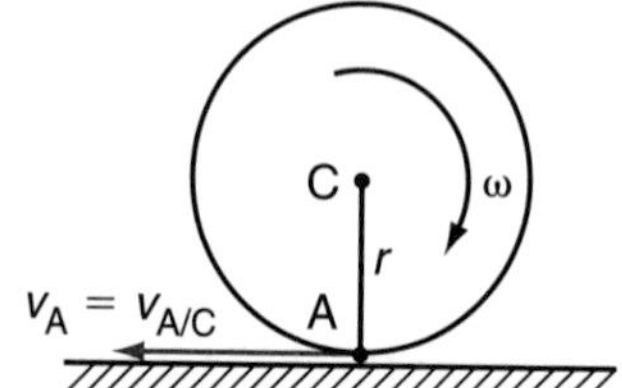

FIGURE 12–36

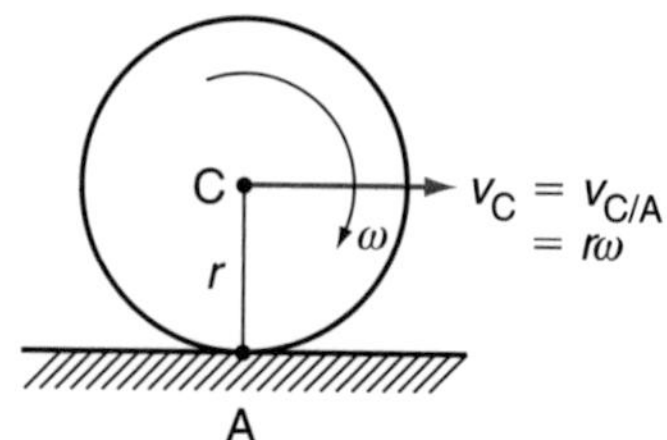

FIGURE 12–37

The translational speed of the center of a rolling wheel is therefore equal to $r\omega$ (Figure 12–37).

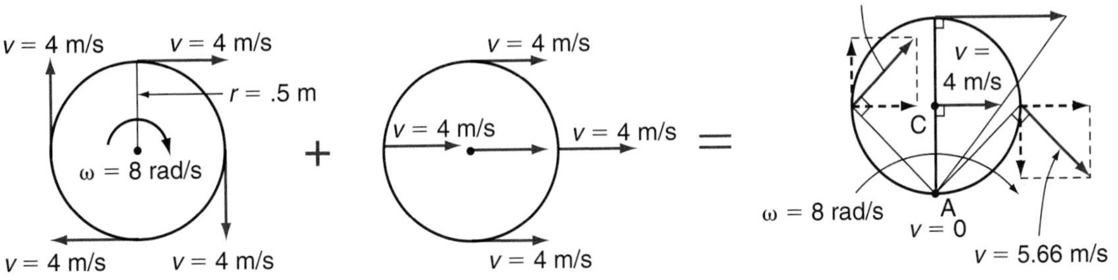

FIGURE 12–38 **FIGURE 12–39** **FIGURE 12–40**

As before, plane motion can be broken into its rotational (Figure 12–38) and translational (Figure 12–39) components and shown for given points on the wheel. Figure 12–40 shows these two motions superimposed; the result is *total plane motion.* Note that each velocity vector is at a right angle to the line from its origin to point A. The velocity of any point on a rolling wheel can be found by multiplying the radius distance to point A by the angular speed ω. Point A is the point of contact between the wheel and the ground, and each distance measured from A is essentially a radius with a tangential velocity at a right angle to it.

EXAMPLE 12–9

The 0.8-m diameter wheel in Figure 12–41 rolls to the right with an angular speed of 5 rad/s. Determine the velocities of points A, B, C, and D (Figure 12–42).

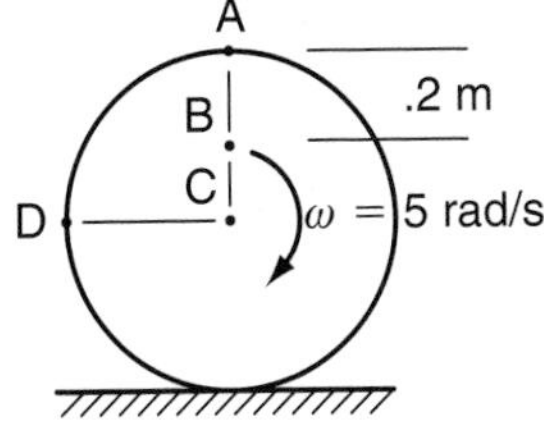

FIGURE 12–41

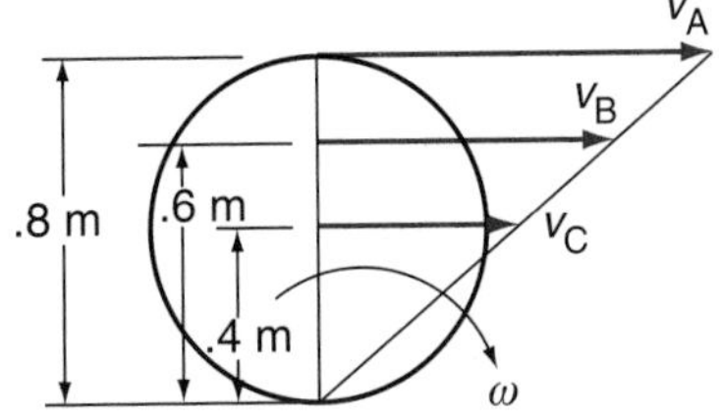

FIGURE 12–42

$$v_C = r\omega$$

$$= (0.4 \text{ m})(5 \text{ rad/s})$$

$$\underline{v_C = 2 \text{ m/s} \rightarrow}$$

$$v_A = (0.8 \text{ m})(5 \text{ rad/s})$$

$$\underline{v_A = 4 \text{ m/s} \rightarrow}$$

$$v_B = (0.6 \text{ m})(5 \text{ rad/s})$$

$$\underline{v_B = 3 \text{ m/s} \rightarrow}$$

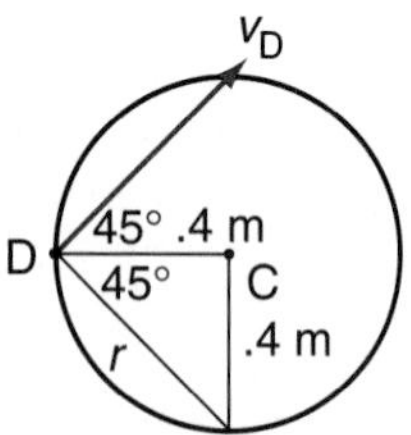

FIGURE 12–43

For point D (Figure 12–43):

$$r = \sqrt{(0.4\text{ m})^2 + (0.4\text{ m})^2}$$
$$r = 0.566\text{ m}$$
$$v_D = (0.566\text{ m})(5\text{ rad/s})$$
$$\underline{v_D = 2.83\text{ m/s} \angle 45°}$$

An alternative approach would be:

$$v_D = v_C + v_{D/C}$$
$$\nearrow = \underset{\rightarrow}{2\text{ m/s}} + \uparrow 2\text{ m/s}$$
$$v_D = \sqrt{2^2 + 2^2} = 2.83\text{ m/s} \ \underline{\angle 45°}$$

EXAMPLE 12–10

A wheel 6 in. in diameter fits through a slot and rolls on its hub, which has a diameter of 2 in. (Figure 12–44). The wheel has an angular speed of 10 rad/s clockwise. Determine the velocities of points A, B, and C.
For point C:

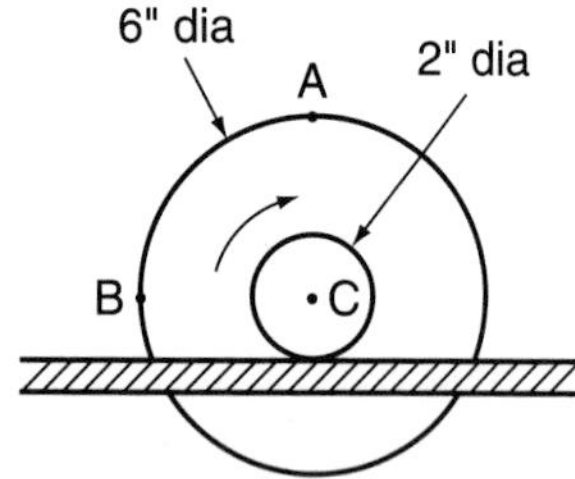

FIGURE 12–44

$$v_C = r\omega$$
$$= (1\text{ in.})(10\text{ rad/s})$$
$$\underline{v_C = 10\text{ in./s} \rightarrow}$$

For point A, which has a radius = 3 in. + 1 in. = 4 in.:

$$v_A = (4\text{ in.})(10\text{ rad/s})$$
$$\underline{v_A = 40\text{ in./s} \rightarrow}$$

For point B (Figure 12–45):

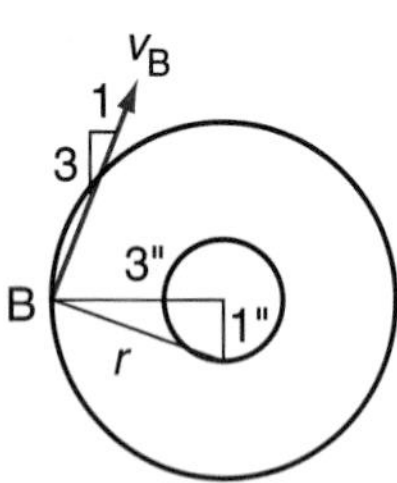

FIGURE 12–45

$$r = \sqrt{(1\text{ in.})^2 + (3\text{ in.})^2}$$
$$r = 3.16\text{ in.}$$
$$v_B = (3.16\text{ in.})(10\text{ rad/s})$$
$$\underline{v_B = 31.6\text{ in./s} \ \angle^{3}_{1}}$$

12–3 INSTANTANEOUS CENTER OF ROTATION

Consider a rigid body with plane motion consisting of movement downward and counterclockwise rotation. Points A, B, and C have absolute velocities as shown in Figure 12–46. This body will appear to have pure rotation if it is viewed from a point at which all the velocities are tangential velocities. Construct a radius arm perpendicular to each velocity vector. From point O and for the instant shown, the body would appear to have pure rotation with zero translational motion. The point O, about which all velocities appear as tangential velocities, is called the *instantaneous center of rotation* and has zero velocity.

To illustrate this with a more concrete example, let us reconsider the bar in Figure 12–5 (reproduced here in Figure 12–47). For the bar to remain in contact with the wall and floor, A and B must have the velocities shown in the figure. For v_A to appear as a tangential velocity, the instantaneous center O must be somewhere on a horizontal line that passes through point A.

Similarly, the radius for tangential velocity v_B must be a vertical line at a right angle to v_B. The intersection of these two lines gives the instantaneous center at point O. For any point on the bar (including point C),

$$v = r\omega$$

or

$$\omega = \frac{v}{r}$$

$$\omega = \frac{v_A}{AO} = \frac{v_B}{OB} = \frac{v_C}{OC}$$

This ω value is also equal to the angular velocity of AB, that is, ω_{AB}. Consider triangle AOB (Figure 12–47): All lines and points within the triangle rotate about instantaneous center O and have an angular velocity ω. Therefore,

$$\omega_{OA} = \omega_{OB} = \omega_{AB}$$

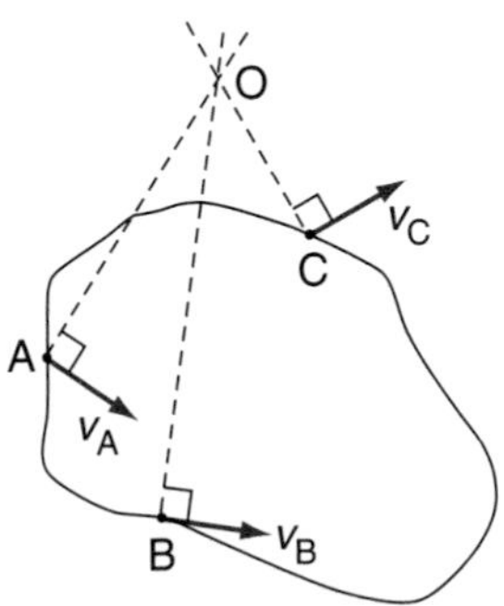

FIGURE 12–46

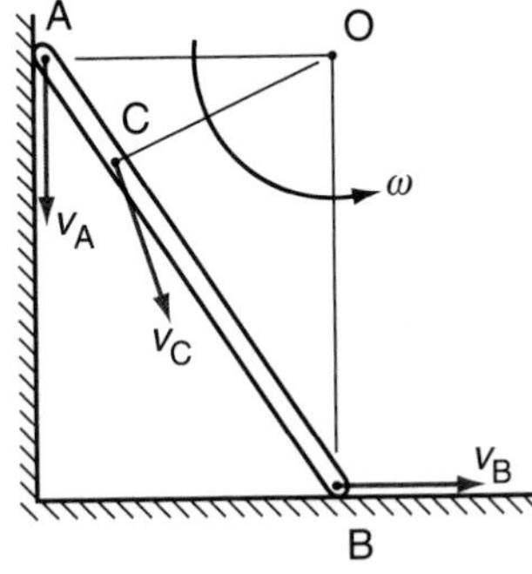

FIGURE 12–47

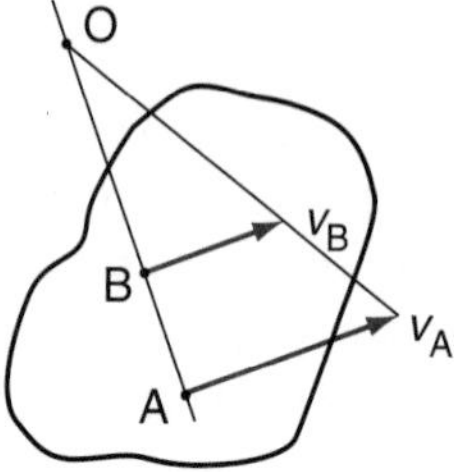

FIGURE 12–48

Previously, we would have solved for ω_{AB} by using

$$v_A = v_B + v_{A/B}$$

Then

$$\omega_{AB} = \frac{v_{A/B}}{AB}$$

Note that we used the instantaneous center of rotation earlier with the rolling wheel (Section 12–2). You can see that the instantaneous center of rotation is the point at which the wheel contacts the flat surface. All velocities are tangent to radii from this instantaneous center. Note that only absolute velocities are involved and that the instantaneous center is only applicable at—as the name suggests—the instant shown.

Another word of warning: This theory does not apply to acceleration; it applies only to velocity since it is an instantaneous center of zero velocity (not necessarily zero acceleration) due to the possibility of normal acceleration. Do not apply it to acceleration vectors.

If the velocities of points A and B on a body are parallel, the radii lie upon each other, so there will be no point of intersection, and the result will be that similar triangles exist, such as those in Figure 12–48.

Whether you are using the relative velocity method or the instantaneous center method, you will draw triangles. The relative velocity method uses vector triangles that are composed of velocity vectors. The instantaneous center method uses triangles that are composed of radii that are drawn at right angles to velocity vectors. Geometry and trigonometry are required in each case.

EXAMPLE 12–11

For the system shown in Figure 12–49 determine (a) the angular velocity of AC, (b) the velocity of point A, and (c) the angular velocity of the roller.

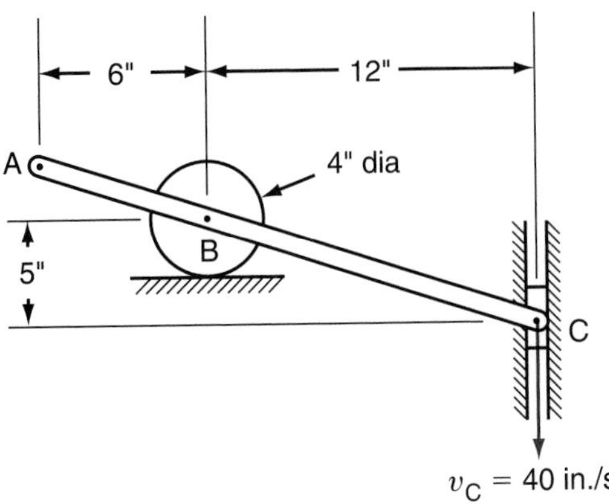

FIGURE 12–49

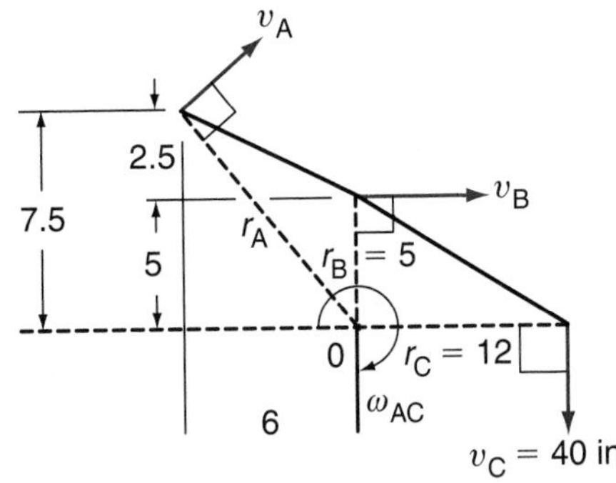

FIGURE 12–50

Considering Figure 12–50, if $v_C = 40$ in./s downward, then v_B will be horizontally to the right, parallel to the horizontal surface. Projecting radius lines at right angles to each of these velocities will give us intersection O, which is the instantaneous center of rotation for member ABC.

$$\omega_{AC} = \frac{v_C}{r_C} = \frac{v_B}{r_B} = \frac{v_A}{r_A}$$

Using $v_C = 40$ in./s

$$\omega_{AC} = \frac{40 \text{ in./s}}{12 \text{ in.}}$$

$$\underline{\omega_{AC} = 3.33 \text{ rad/s} \circlearrowright}$$

Solving for the radius r_A

$$r_A = \sqrt{(6 \text{ in.})^2 + (7.5 \text{ in.})^2}$$
$$= 9.6 \text{ in.}$$

Therefore

$$v_A = r_A \omega_{AC}$$
$$= (9.6 \text{ in.})(3.33 \text{ rad/s})$$
$$\underline{v_A = 32 \text{ in./s } \angle 38.7^\circ}$$

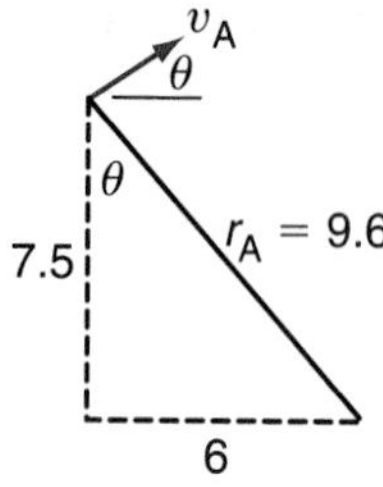

FIGURE 12–51

The angle is determined as in Figure 12–51, where

$$\tan \theta = \frac{6 \text{ in.}}{7.5 \text{ in.}}$$
$$\theta = 38.7°$$

Since all velocities are tangential about the instantaneous center

$$v_B = r_B \omega_{AC}$$
$$= (5 \text{ in.})(3.33 \text{ rad/s})$$
$$v_B = 16.7 \text{ in./s} \rightarrow$$

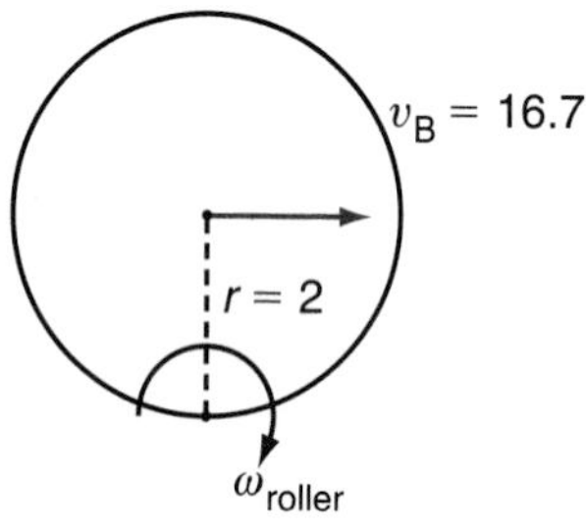

FIGURE 12–52

As shown in Figure 12–52

$$v_B = r\omega_{roller}$$
$$16.7 \text{ in./s} = (2 \text{ in.})\omega_{roller}$$
$$\underline{\omega_{roller} = 8.35 \text{ rad/s}}$$

EXAMPLE 12–12

Pin C of the linkage in Figure 12–53 has a velocity of 5 m/s downward. Use the method of instantaneous centers to determine the velocity of B and the angular velocity of AB.

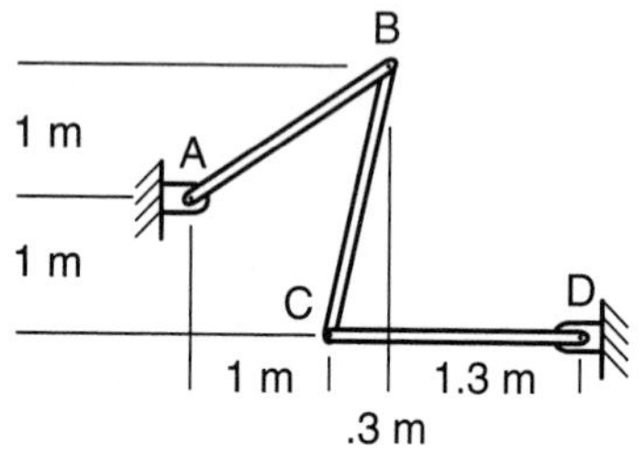

FIGURE 12–53

Redraw the linkages as a line diagram (Figure 12–54).

Velocity v_C is perpendicular to radius CD. For v_C to appear as a tangential velocity, it must be perpendicular to radius CD or an extension of CD. Line CD can therefore be extended both left and right.

Since all points on member BC have the same instantaneous center, consider point B now and show its velocity perpendicular to AB. Extend line AB to get an intersection at O, the instantaneous center for BC.

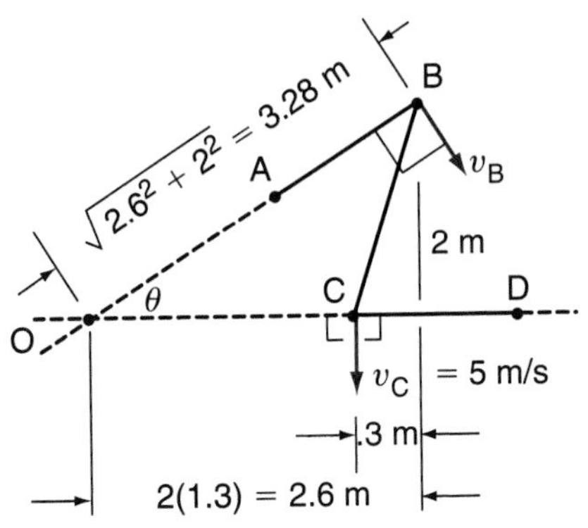

FIGURE 12–54

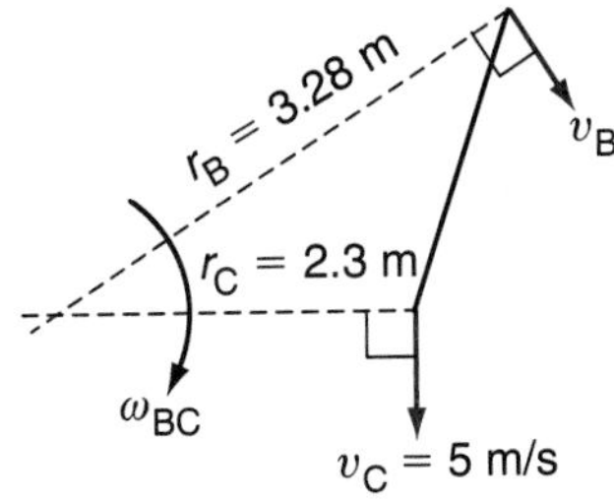

FIGURE 12–55

From the dimensions shown in Figure 12–54 the lengths OB and OC can be calculated and shown as r_B and r_C in Figure 12–55.

$$\omega_{BC} = \frac{v_B}{r_B} = \frac{v_C}{r_C}$$

or

$$\frac{v_B}{3.28 \text{ m}} = \frac{5 \text{ m/s}}{2.3 \text{ m}}$$

$$\underline{v_B = 7.13 \text{ m/s} \diagdown 52.4°}$$

$$\tan\theta = \frac{2 \text{ m}}{2.6 \text{ m}} \quad \text{(Figure 12–54)}$$

$$\theta = 37.6 \text{ or } 90 - 37.6 = 52.4°$$

Member AB pivots at A and has a length of 1.64 m.

Therefore

$$\omega_{AB} = \frac{v_B}{r_{AB}} = \frac{7.13 \text{ m/s}}{1.64 \text{ m}}$$

$$\underline{\omega_{AB} = 4.35 \text{ rad/s} \circlearrowright}$$

EXAMPLE 12–13

Link BC in Figure 12–56 is pinned to a cylinder rolling at 12 in./s to the right. Calculate the velocity of pin C for the instant shown.

Point A is the instantaneous center of rotation for the cylinder (Figure 12–57).

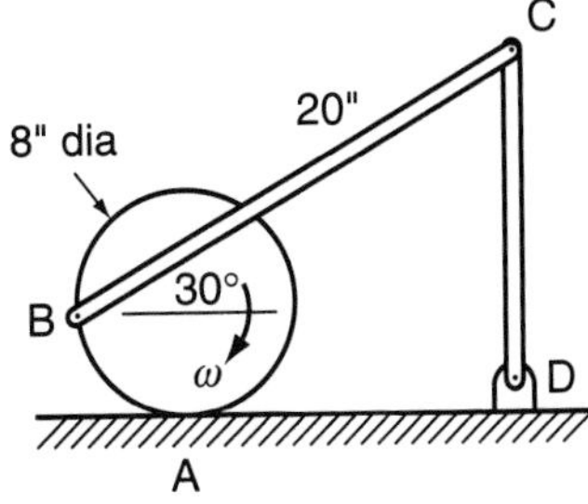

FIGURE 12–56

$$\text{radius AB} = \sqrt{(2 \text{ in.})^2 + (2 \text{ in.})^2}$$
$$= 5.66 \text{ in.}$$

Therefore,

$$\omega \text{ about point A} = \frac{v}{r} = \frac{v_B}{AB}$$

$$\frac{12 \text{ in./s}}{4 \text{ in.}} = \frac{v_B}{5.66 \text{ in.}}$$

$$v_B = 17 \text{ in./s}$$

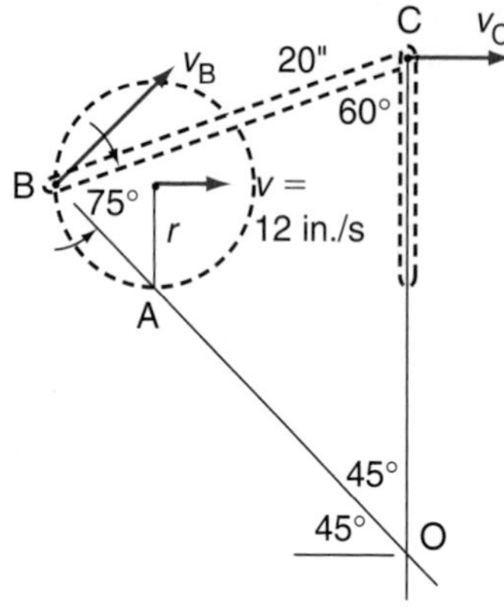

FIGURE 12–57

We now locate point O, the instantaneous center for v_B and v_C, by drawing lines perpendicular to v_B and v_C.

$$\frac{\text{OB}}{\sin 60°} = \frac{20\text{ in.}}{\sin 45°} = \frac{\text{OC}}{\sin 75°}$$

$$\text{OB} = 24.5\text{ in.}$$

$$\text{OC} = 27.4\text{ in.}$$

$$\omega\text{ about O} = \frac{v_B}{\text{OB}} = \frac{v_C}{\text{OC}}$$

$$\frac{17\text{ in./s}}{24.5\text{ in.}} = \frac{v_C}{27.4\text{ in.}}$$

$$\underline{v_C = 19\text{ in./s} \rightarrow}$$

EXAMPLE 12–14

Member AC rotates at 2 rad/s counterclockwise (Figure 12–58). Use the method of instantaneous centers to determine the velocity of point D, angular velocity of DC, and velocity of point G.

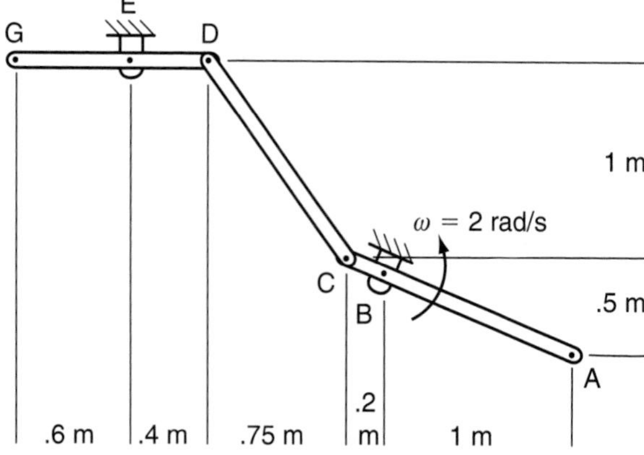

FIGURE 12–58

Two of the required unknowns involve member DC, so solving for v_C will be the first step.

By considering member CA (Figure 12–59), we get

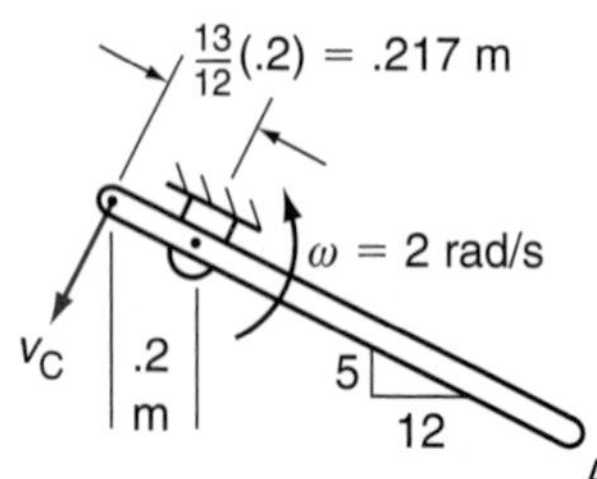

FIGURE 12–59

$$v_C = r\omega$$

$$= (0.217\text{ m})\,(2\text{ rad/s})$$

$$v_C = 0.433\text{ m/s} \;\measuredangle^{12}_{5}$$

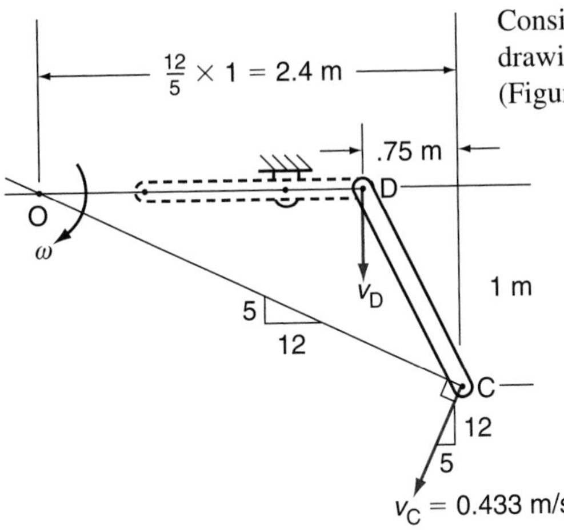

FIGURE 12–60

Consider member DC and locate the instantaneous center by drawing radii perpendicular to the velocities v_c and v_D (Figure 12–60).

$$\text{radius OC} = \frac{13}{5}(1\text{ m}) = 2.6\text{ m}$$

$$\text{radius OD} = 2.4\text{ m} - 0.75\text{ m} = 1.65\text{ m}$$

$$\omega = \frac{v_D}{OD} = \frac{v_C}{OC}$$

$$\frac{v_D}{1.65\text{ m}} = \frac{0.433\text{ m/s}}{2.6\text{ m}}$$

$$v_D = 0.275\text{ m/s}\downarrow$$

All lines in triangle ODC have the same angular velocity:

$$\omega = \frac{v_C}{OC} = \omega_{DC}$$

Therefore,

$$\omega_{DC} = \frac{0.433\text{ m/s}}{2.6\text{ m}}$$

$$\underline{\omega_{DC} = 0.167\text{ rad/s} \circlearrowright}$$

For Figure 12–61, we have

$$\omega_{GD} = \frac{v_D}{ED} = \frac{v_G}{EG}$$

or

$$\frac{0.275\text{ m/s}}{0.4\text{ m}} = \frac{v_G}{0.6\text{ m}}$$

$$\underline{v_G = 0.412\text{ m/s}\uparrow}$$

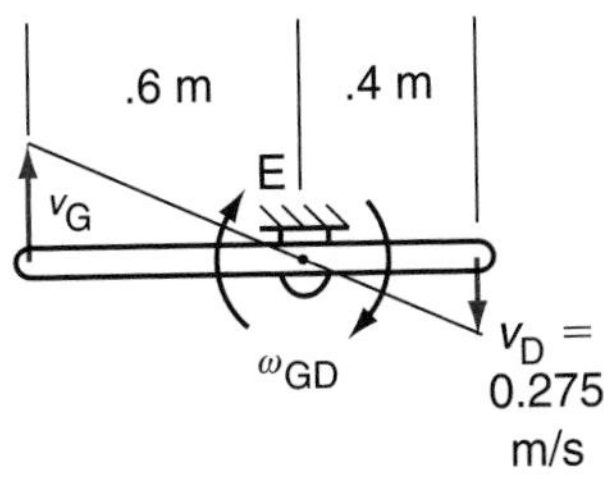

FIGURE 12–61

EXAMPLE 12–15

Link AB of the mechanism shown in Figure 12–62 rotates at 5 rad/s counterclockwise. Point C, a point on link BD, is assumed to be directly over pin E for the purposes of this example. Determine the velocity of points B, C, and D, and the angular velocity of link BD.

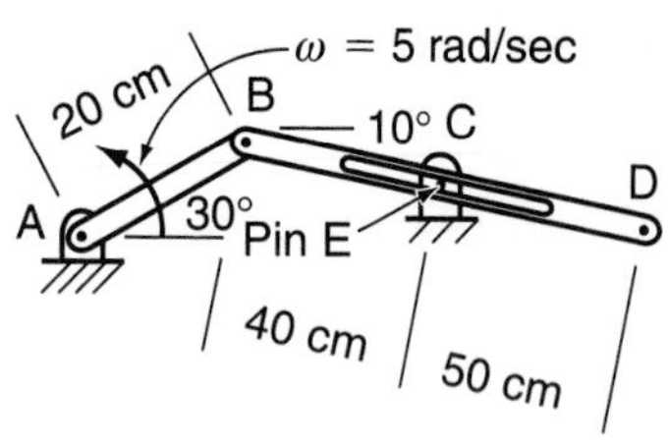

FIGURE 12–62

Considering link AB (Figure 12–63), we have

$$v_B = r\omega$$

$$= (20\text{ cm})(5\text{ rad/s})$$

$$v_B = 100\text{ cm/s } 60°$$

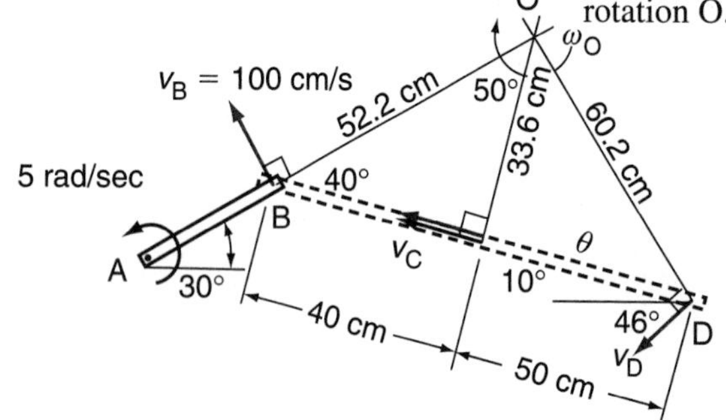

FIGURE 12–63

Drawing perpendicular lines from the known velocity directions of B and C, we can locate the instantaneous center of rotation O. In triangle OBC:

$$\tan 40° = \frac{OC}{40\text{ cm}}$$

$$OC = 33.6\text{ cm}$$

$$\cos 40° = \frac{40\text{ cm}}{OB}$$

$$OB = 52.2\text{ cm}$$

In triangle OCD:

$$OD = \sqrt{(33.6\text{ cm})^2 + (50\text{ cm})^2}$$

$$OD = 60.2\text{ cm}$$

$$\sin\theta = \frac{33.6\text{ cm}}{50\text{ cm}}$$

$$\theta = 34°$$

All velocities have the same angular velocity ω about O; therefore,

$$\omega_O = \frac{v_B}{OB} = \frac{v_C}{OC} = \frac{v_D}{OD}$$

$$\frac{100\text{ cm/s}}{52.2\text{ cm}} = \frac{v_C}{33.6\text{ cm}}$$

$$v_C = 64.4\text{ cm/s } 10°$$

and

$$\frac{100\text{ cm/s}}{52.2\text{ cm}} = \frac{v_D}{60.2\text{ cm}}$$

$$v_D = 115\text{ cm/s } 46°$$

For triangle BOD (Figure 12–63),

$$\omega_{BD} = \omega_O = \frac{v_B}{OB}$$

Therefore,

$$\omega_{BD} = \frac{100 \text{ cm/s}}{52.2 \text{ cm}}$$

$$\underline{\omega_{BD} = 1.92 \text{ rad/s} \curvearrowright}$$

HINTS FOR PROBLEM SOLVING

1. A correct relative velocity equation will have subscripts that cancel out.
2. A relative velocity such as $v_{A/B}$ is the velocity of A with respect to B. It is the velocity that A appears to have when viewed by someone stationed on B.
3. $v_{A/B}$ is equal and opposite to $v_{B/A}$.
4. After writing a relative velocity equation, show all the possible information of vector direction and magnitude under each term of the equation. Drawing a vector triangle is the next step.
5. The velocity direction of a point and its radius to the instantaneous center are always at right angles to one another.
6. The instantaneous center of rotation can only be used on points that are on the same object or member. Remember this when dealing with linkages and mechanisms.
7. The instantaneous center of rotation does not apply to acceleration vectors.
8. Calculations for all problems are based upon what is happening at *this* instant, not in the *next* instant.

PROBLEMS

APPLIED PROBLEMS FOR SECTION 12–1

12–1. A high-speed escalator travels at 180 ft/min. What is the absolute velocity of a person running at 700 ft/min (a) in the same direction as the escalator's motion and (b) in a direction opposite to that of the escalator's motion?

12–2. A river flows from north to south at 10 mph. A boat is to cross this river from west to east at a speed of 25 mph (speed of the boat with respect to the water). At what angle must the boat be pointed upstream such that it will proceed directly across the river? Draw a vector triangle to prove your answer.

12–3. If weight A moves 4 m downward (Figure P12–3), determine the distance that A moves with respect to B.

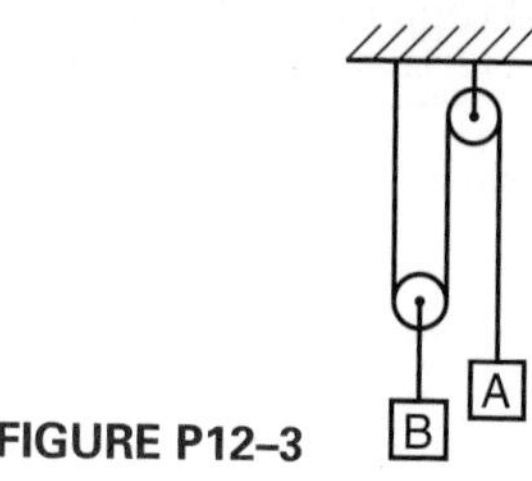

FIGURE P12–3

12–4. The acceleration of B in Figure P12–4 is 12 ft/s^2 downward. Determine the acceleration of A with respect to B.

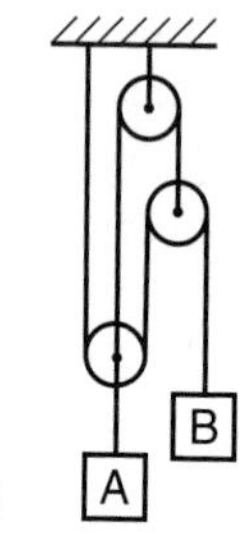

FIGURE P12–4

12–5. A car traveling at 25 m/s eastward passes under a railway overpass that intersects the highway at right angles. If a train is traveling at 18 m/s southward on the overpass, determine the velocity of the train with respect to the car.

12–6. Two roller coaster cars on adjacent tracks appear on the same line of sight, as in Figure P12–6. If the velocity of car A is 4 m/s and the velocity of car B is 15 m/s, determine the velocity of car A with respect to car B.

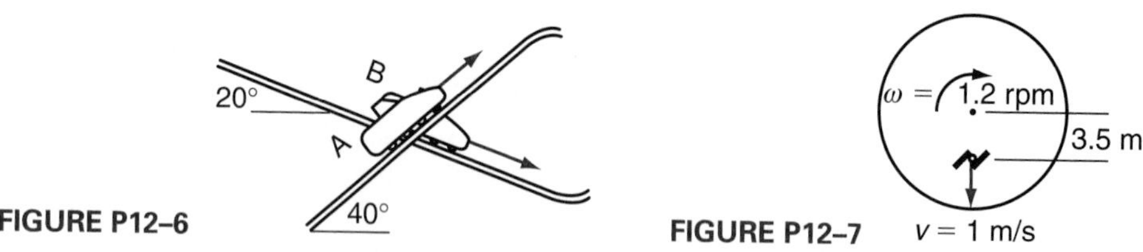

FIGURE P12–6

FIGURE P12–7

12–7. An entertainer walks forward with a velocity of 1 m/s while 3.5 m from the center of a stage that is rotating at 1.2 rpm (Figure P12–7). Determine his absolute velocity.

12–8. An automated feeding system has a conveyor that moves at 3 km/h parallel to the feed trough and has a belt speed of 1.8 m/s (Figure P12–8). Determine the absolute velocity of the material on the conveyor.

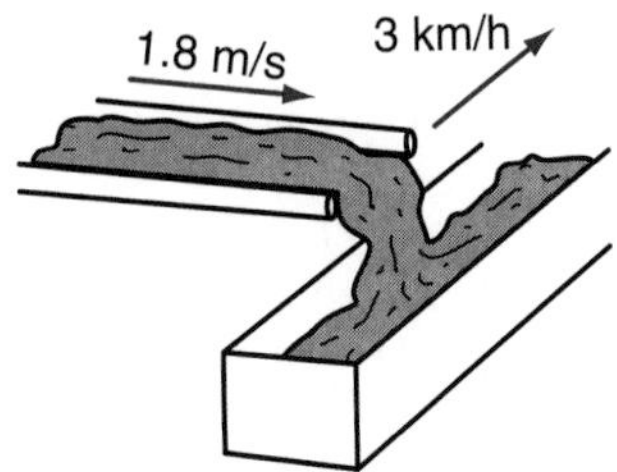

FIGURE P12–8

12–9. Boats A and B will collide in 5 seconds if the velocities shown in Figure P12–9 are maintained. What is the distance between the boats at this instant? What is the velocity of B with respect to A?

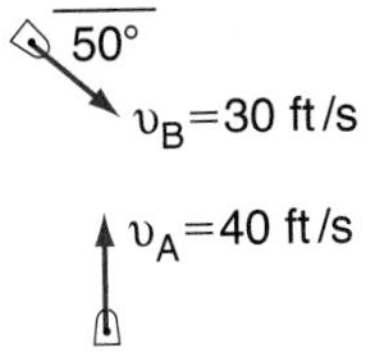

FIGURE P12–9

12–10. The angular velocity of AB in Figure P12–10 is 400 rpm clockwise. Determine the angular velocity of BC and the velocity of C when (a) $\theta = 0°$ and (b) $\theta = 90°$.

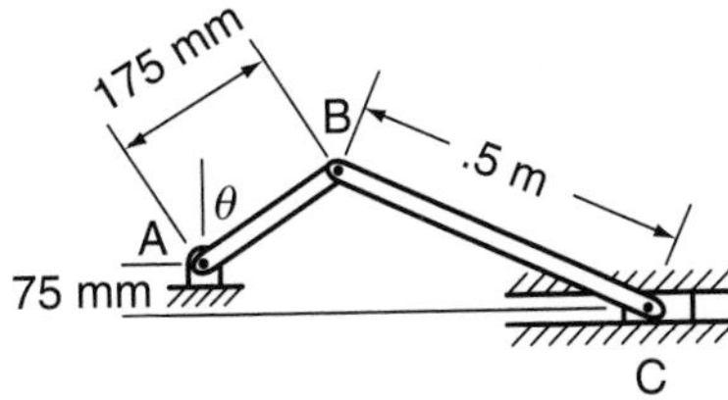

FIGURE P12–10

12–11. The velocity of slider C is 26 in./s downward (Figure P12–11). Determine the linear velocity of B and the angular velocity of AB.

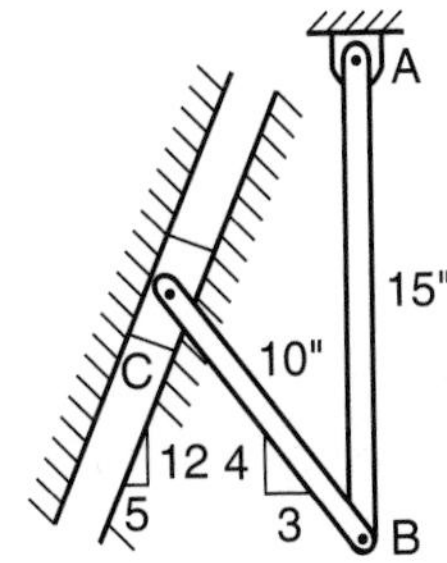

FIGURE P12–11

12–12. If slider C in Figure P12–12 moves downward at 0.7 m/s, determine (a) the angular velocity of AB and (b) the velocity of D.

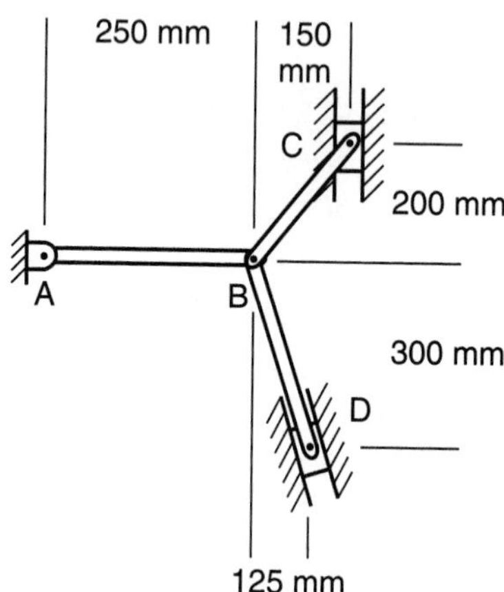

FIGURE P12–12

12–13. The angular velocity of DE is 4 rad/s clockwise (Figure P12–13). Determine the angular velocity of DB at the instant shown.

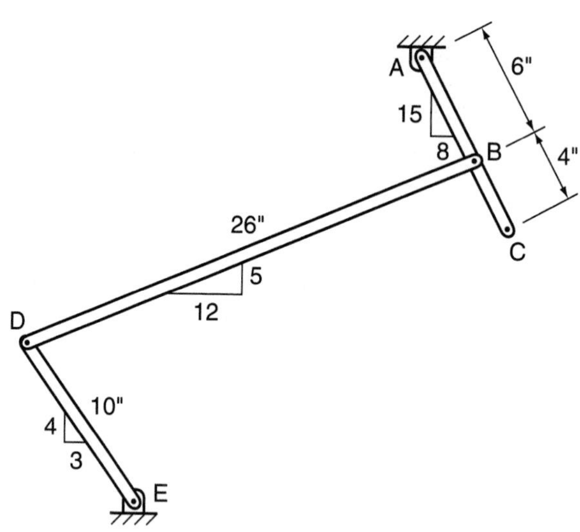

FIGURE P12–13

12–14. The velocity of point B is 260 mm/s downward in the slot (Figure P12–14). Using the relative velocity equation method, determine the angular velocity of AB.

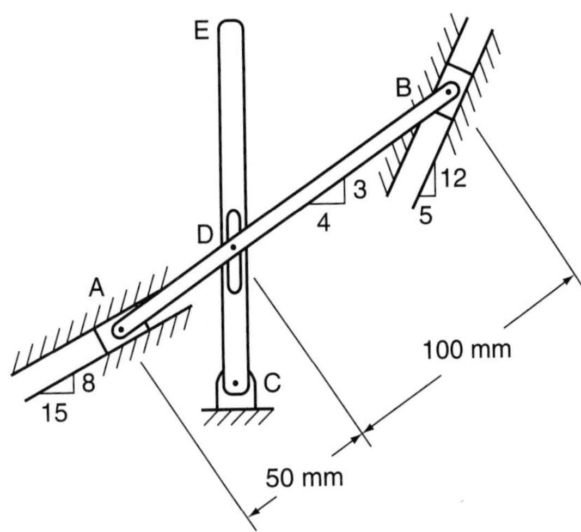

FIGURE P12–14

12–15. Pin B of the linkage shown in Figure P12–15 has a velocity of 10 m/s to the right at the instant shown. Determine (a) the velocity of pin C (b) the angular velocity of AC, and (c) the angular velocity of BD.

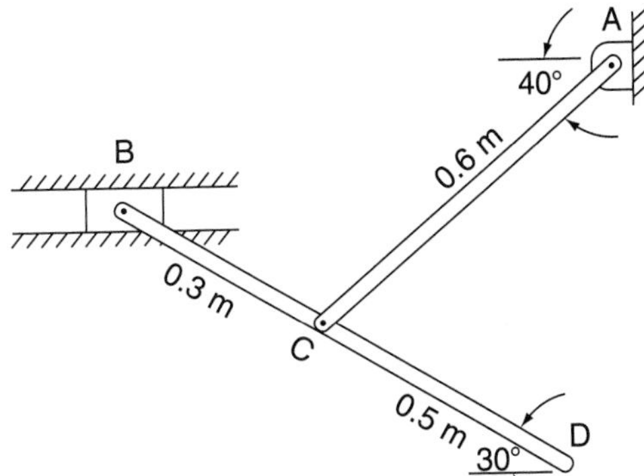

FIGURE P12–15

12–16. For the toggle linkage shown in Figure P12–16, OA = 0.625 in., AB = 2.5 in., BC = 1.75 in., and BD = 2.25 in. If the angular velocity of OA is 5 rad/s counterclockwise, determine the linear velocity of points B and D.

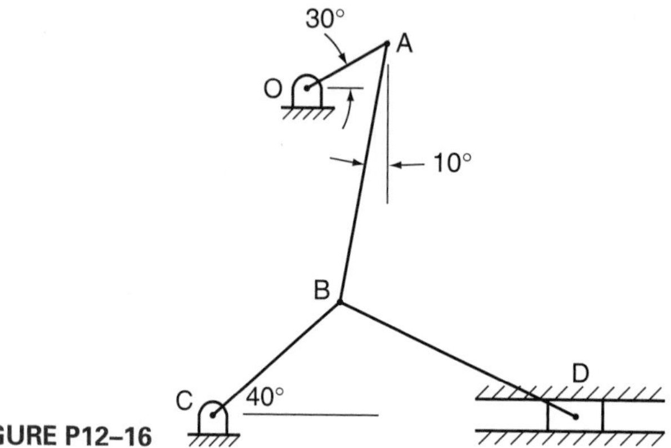

FIGURE P12–16

12–17. If the bicycle in Figure P12–17 is being pedaled so that it is accelerating and has a velocity of 20 mph for the instant shown, what is the absolute velocity of pedal A?

FIGURE P12–17

12–18. The gear drive of a food mixer is shown in Figure P12–18. The outer gear is fixed, and gear A, which has a diameter of 40 mm, rotates on the end of arm AB. Member DE is welded to gear A. If ω_{AB} = 8 rad/s, determine the absolute velocity of D and E at the instant shown.

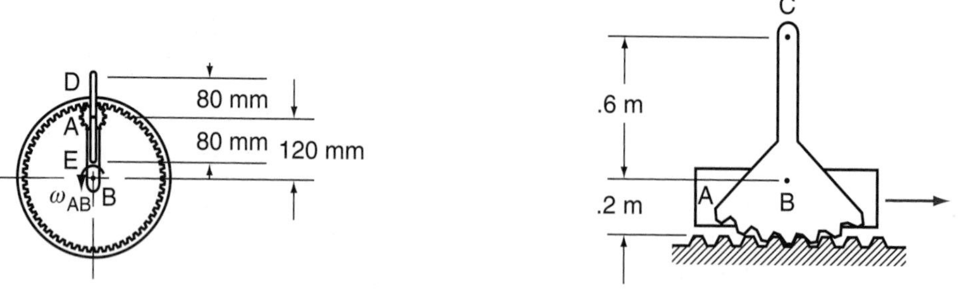

FIGURE P12–18

FIGURE P12–19

12–19. The gear in Figure P12–19 pivots at B on block A. If block A is moving to the right with a velocity of 12 mm/s, determine the velocity of C with respect to B. What is the absolute velocity of C?

12–20. Member AC (Figure P12–20) has an angular velocity of 8 rad/s counterclockwise about point B. Member BD has an angular velocity of 2 rad/s counterclockwise about point D. Determine the absolute linear velocity of point C and the absolute angular velocity of AC.

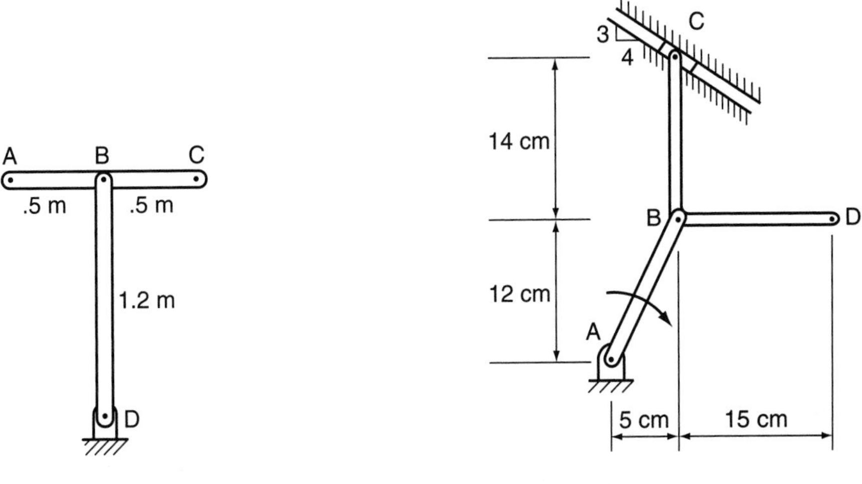

FIGURE P12–20

FIGURE P12–21

12–21. The angular velocity of AB = 2 rad/s clockwise in Figure P12–21. Using the relative velocity equation, determine (a) the velocity of point C, (b) the angular velocity of CBD, and (c) the velocity of point D.

12–22. The velocity of point C is 26 in./sec downward (Figure P12–22). Using the relative velocity equation, determine the angular velocity of BC and the linear velocity of point A.

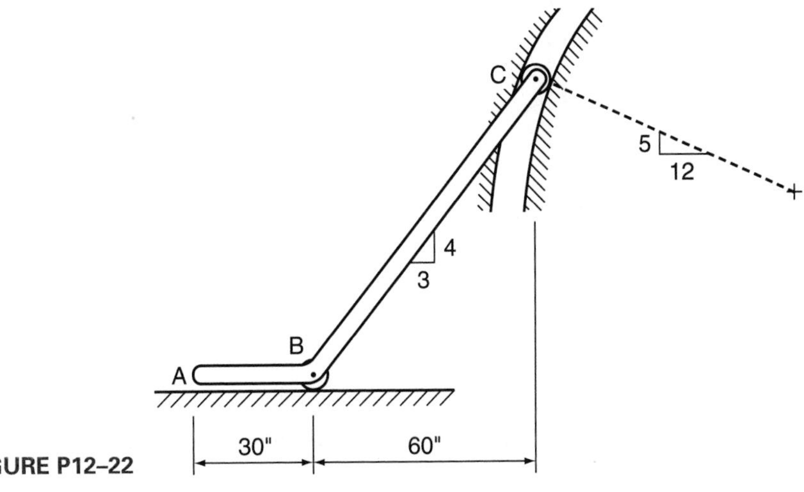

FIGURE P12–22

12–23. Determine the velocity of point D (Figure P12–23) using the relative velocity equation method.

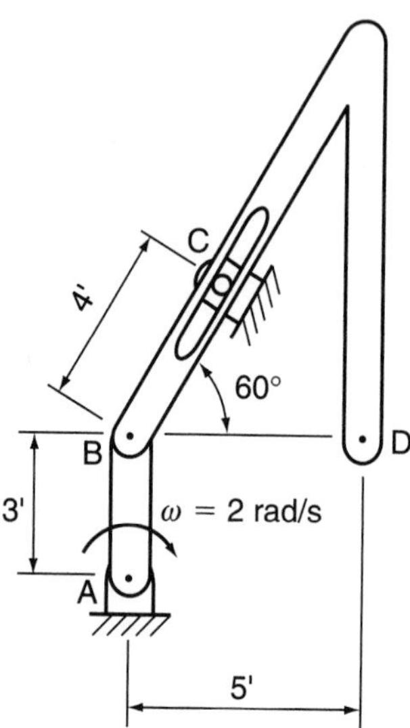

FIGURE P12–23

12–24. Determine the velocity of point D of the linkage shown in Figure P12–15.

12–25. The angular velocity of AB is 3 rad/s counterclockwise for the mechanism shown in Figure P12–25. Determine (a) the linear velocity of C, (b) the angular velocity of DBC, and (c) the linear velocity of D.

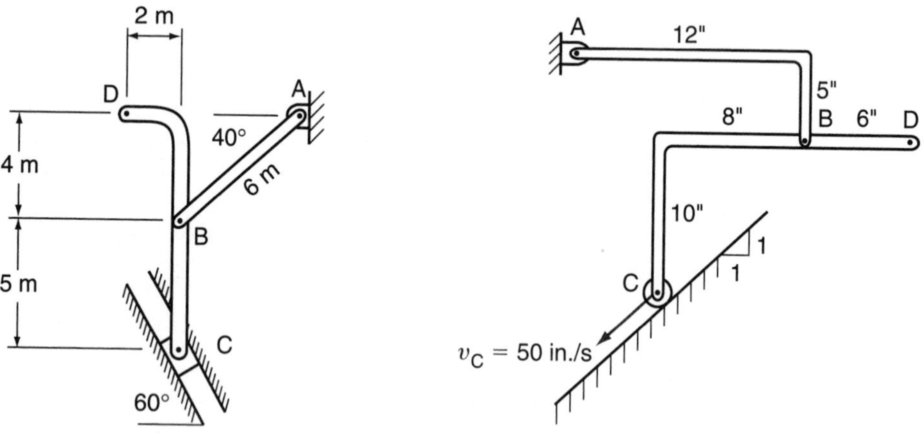

FIGURE P12–25

FIGURE P12–26

12–26. Determine the angular velocity of CBD and the linear velocity of D for the mechanism shown in Figure P12–26.

12–27. Using the relative velocity equation, determine the velocity of point D and the angular velocity of BCD for the system shown in Figure P12–27.

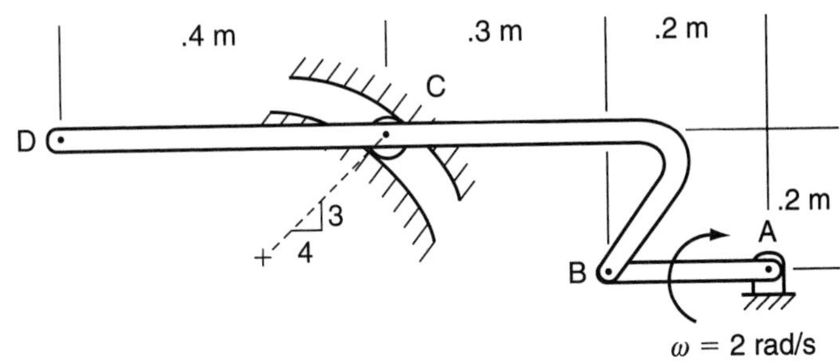

FIGURE P12–27

12–28. For the mechanism in Figure P12–28, AB rotates at 4 rad/s clockwise and is accelerating at 10 rad/s^2. Determine the linear acceleration of C.

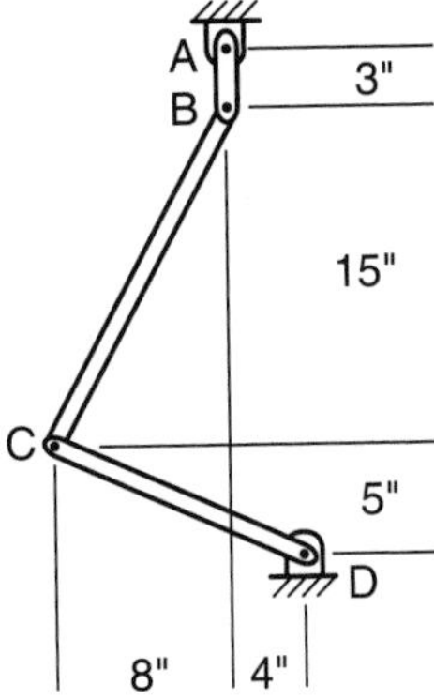

FIGURE P12–28

12–29. At the instant shown in Figure P12–29, AB has an angular acceleration of 8 rad/s^2 clockwise. Determine the acceleration of C and the angular acceleration of BC if AB has an angular velocity of 3 rad/s clockwise.

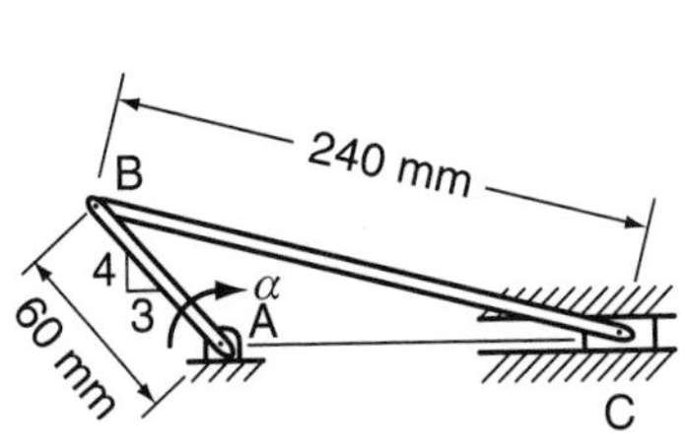

FIGURE P12–29

FIGURE P12–30

12–30. At the position shown in Figure P12–30, AB has an angular velocity of 2 rad/s counterclockwise and an angular acceleration of 6 rad/s^2 counterclockwise. Determine the linear acceleration of C.

APPLIED PROBLEMS FOR SECTION 12–2

12–31. A car wheel 26 in. in diameter turns without slipping at a speed of 900 rpm. What is the speed of the car?

12–32. The cylinder shown in Figure P12–32 rolls to the right with a velocity of 6 m/s. For the instant shown, determine (a) the angular velocity of the cylinder and (b) the linear velocity of point B.

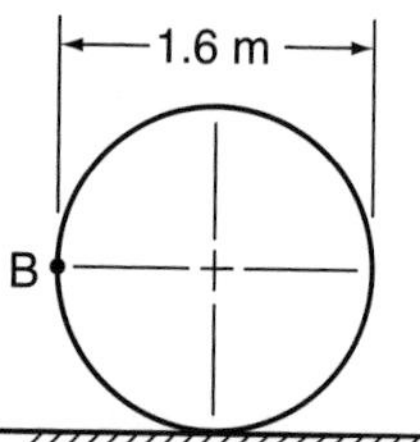

FIGURE P12–32

12–33. The cylinder shown in Figure P12–33 rolls to the right at 8 m/s. Determine the velocity of point B at the position shown.

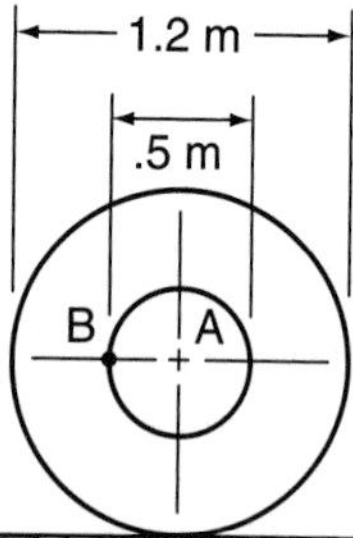

FIGURE P12–33

12–34. A cord is wound in the slot of cylinder A in Figure P12–34. Mass B moves downward with a velocity of 6 m/s. Assume no slipping of the cylinder and determine the velocities of points D, E, and C on cylinder A. If B drops 4 m, how far does cylinder A move to the right?

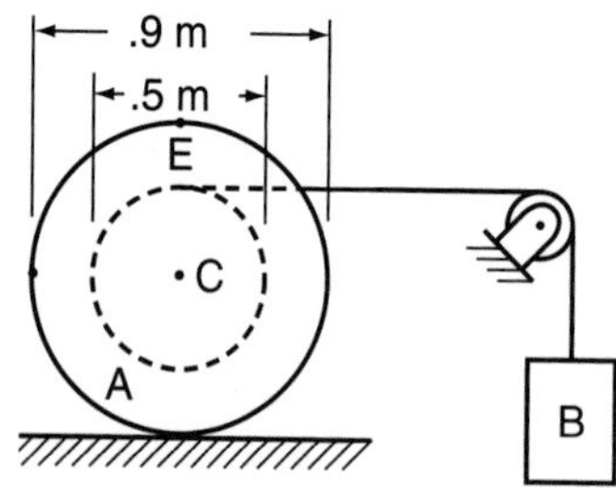

FIGURE P12–34

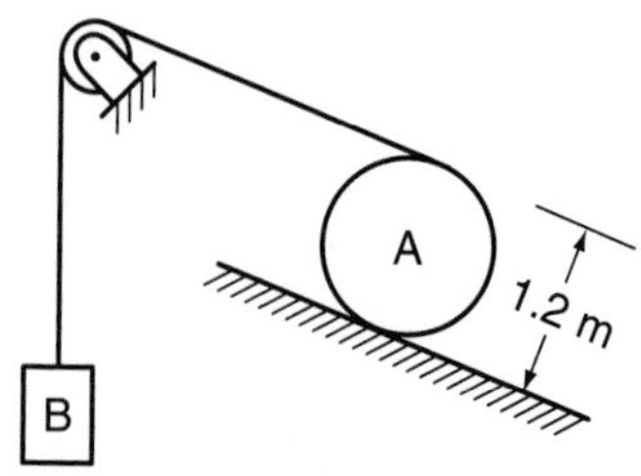

FIGURE P12–35

12–35. Cylinder A in Figure P12–35 rolls 5 m down the slope. What distance is mass B lifted?

12–36. The cable of the system shown in Figure P12–36 is wound around the hub of A and connected to the center of cylinder B. Starting from rest, cylinder A rolls 2 m down the slope in 3 seconds. Determine the angular velocity and angular acceleration of B at $t = 3$ s.

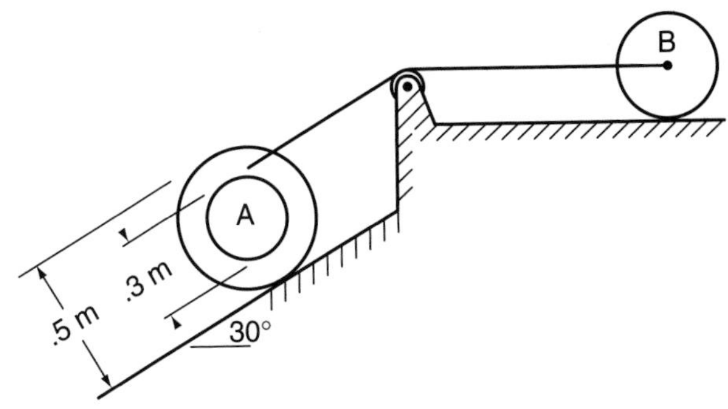

FIGURE P12–36

12–37. Determine the velocity of point B on the cylinder shown in Figure P12–37 if weight A is dropping at a velocity of 2 m/s.

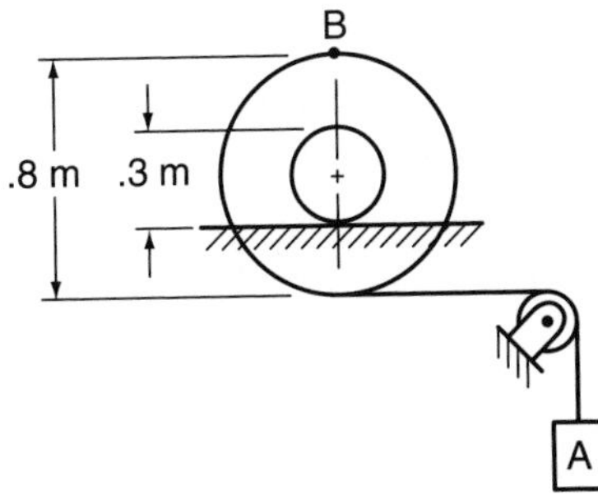

FIGURE P12–37

12–38. A barrel 30 in. in diameter rolls down a slope with an acceleration of 5 ft/s^2. Determine the barrel's angular acceleration.

12–39. The wheel in Figure P12–39 turns at 8 rad/s clockwise. Determine the velocity of B.

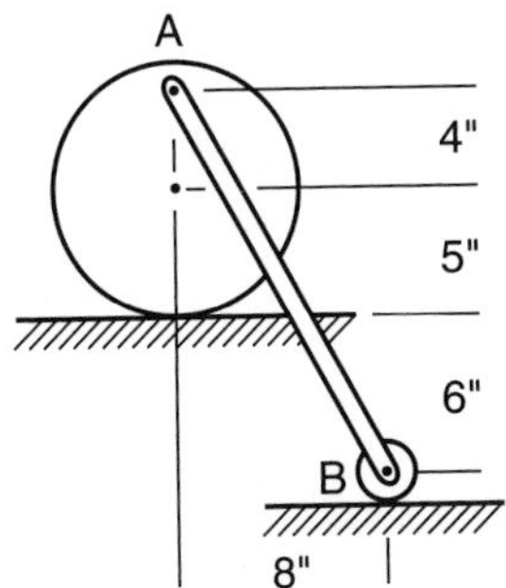

FIGURE P12–39

12–40. Member BC of the mechanism shown in Figure P12–40 is 0.8 m long. Using the relative velocity equation method, determine the velocity of point C.

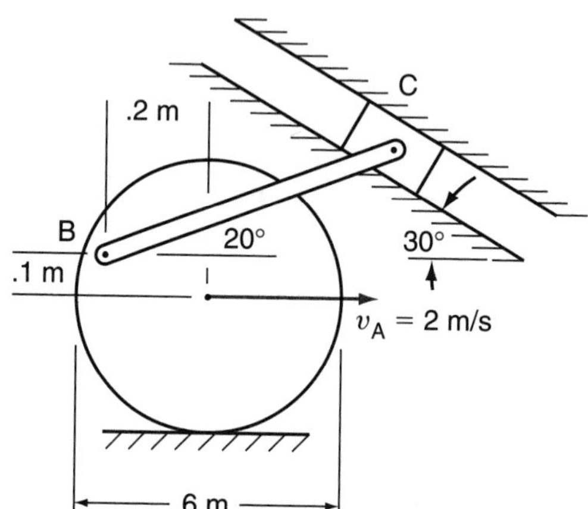

FIGURE P12–40

12–41. For the system shown in Figure P12–41, determine (a) the velocity of D and (b) the angular velocity of CD.

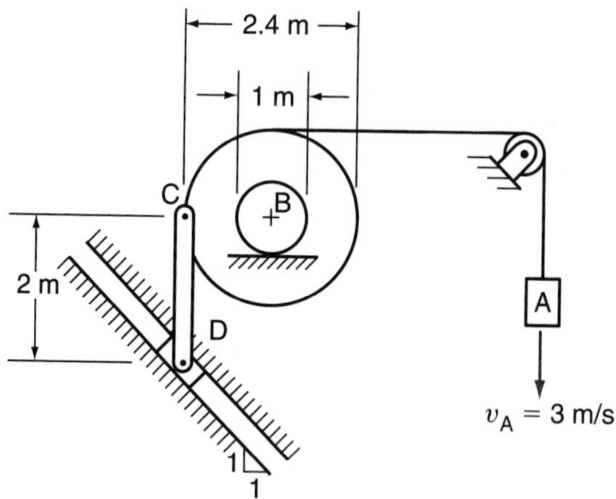

FIGURE P12–41

12–42. The cylinder shown in Figure P12–42 rolls downward to the left at 2 m/s. Determine the angular velocity of AC and the linear velocity of D.

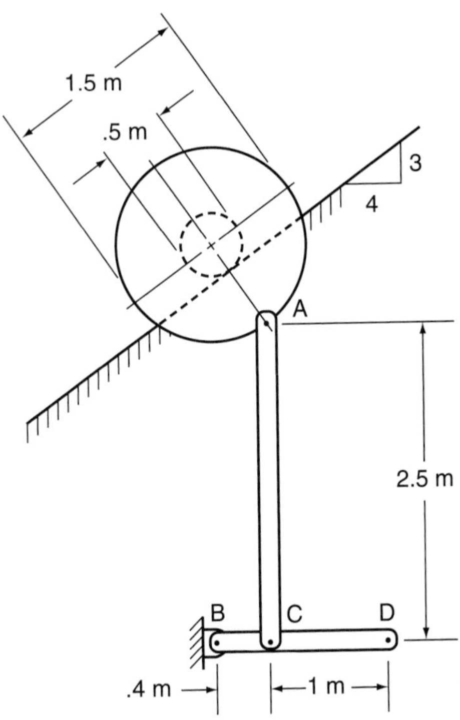

FIGURE P12–42

12–43. Cylinder A (Figure P12–43) rolls down the slope with an angular velocity of 2 rad/s. Determine the velocity of C and the angular velocity of BC.

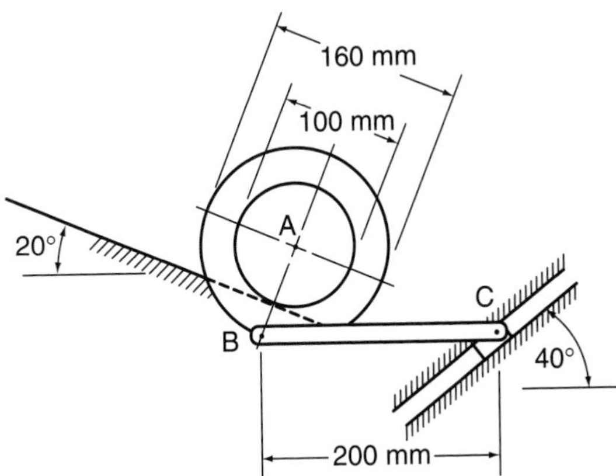

FIGURE P12–43

12–44. The cylinder shown in Figure P12–44 rolls to the right at 10 in./s. Determine the angular velocity of BC and CD.

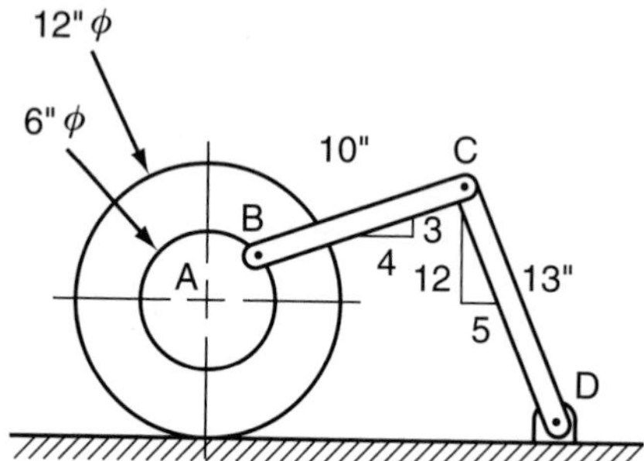

FIGURE P12–44

12–45. Cylinder A rolls to the right at 0.2 m/s. Using the relative velocity method, determine the velocity of point E (Figure P12–45).

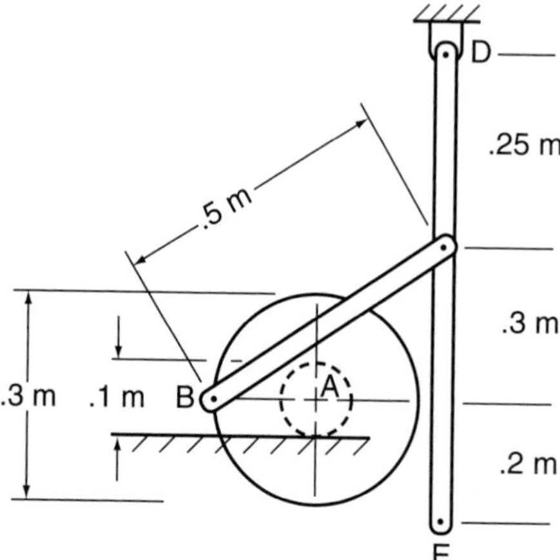

FIGURE P12–45

APPLIED PROBLEMS FOR SECTION 12–3

12–46. Draw the instantaneous centers for member BC and DE (Figure P12–46). Label and show the direction of velocities v_B, v_C, ω_{BC}, v_E, v_D, and ω_{DE}.

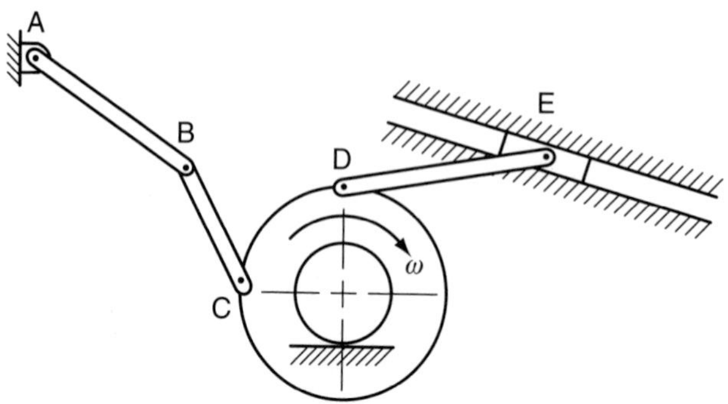

FIGURE P12–46

12–47. Locate and label all necessary instantaneous centers and radii and list the sequence of steps you would take in solving for the velocity of point D in Figure P12–26.

12–48. Point D is the center of rotation for roller A of Figure P12–48. If point C has a velocity of 40 in./s to the left, determine (a) the linear velocity of A, (b) the angular velocity of AC, and (c) the linear velocity of B.

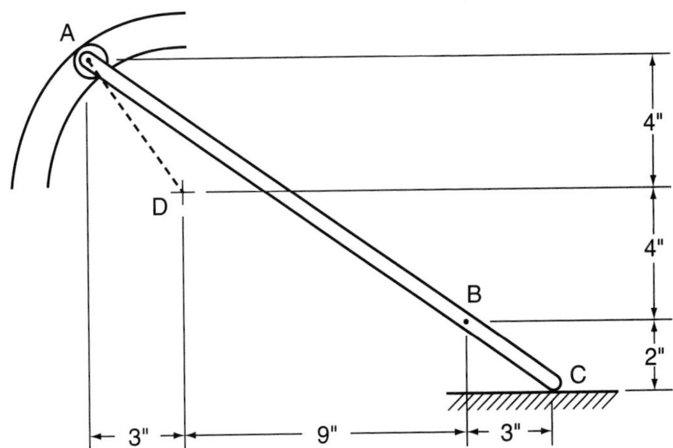

FIGURE P12–48

12–49. Using instantaneous centers, determine the velocity of D (Figure P12–49) and the angular velocity of BD. Lengths are as follows: AC = 4 in., DE = 4 in., BD = 5 in., CD = 2 in.

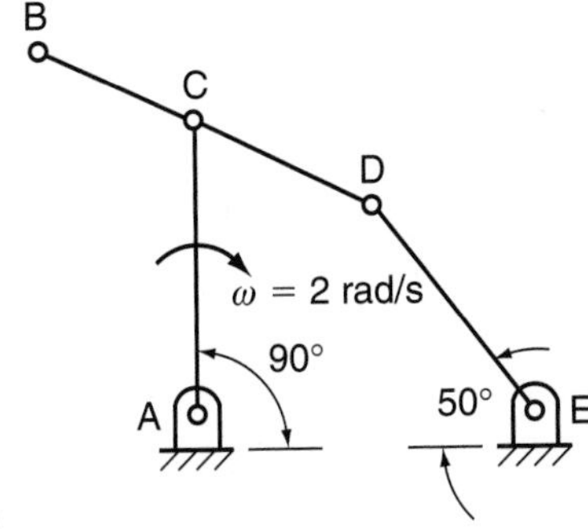

FIGURE P12–49

12–50. For the system shown in Figure P12–50, determine (a) the velocity of pin C, (b) the angular velocity of BCD, and (c) the velocity of point D.

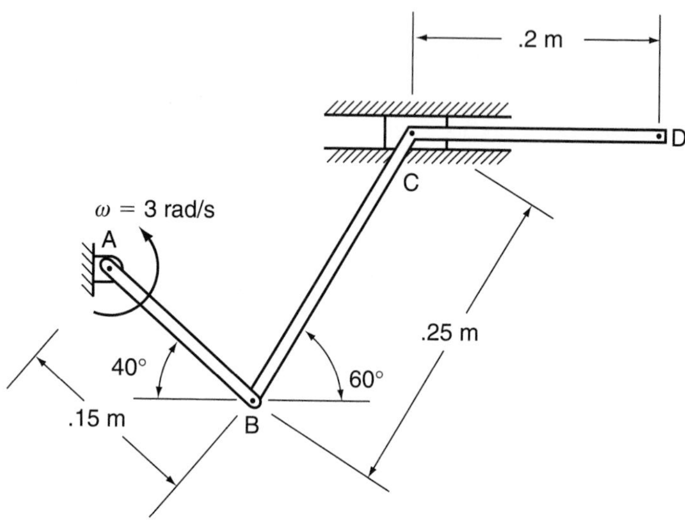

FIGURE P12–50

12–51. For the system shown in Figure P12–51, the velocity of point D is 51 in./s. Using the method of instantaneous centers, determine (a) the velocity of point A, (b) the angular velocity of BC, and (c) the angular velocity of AD.

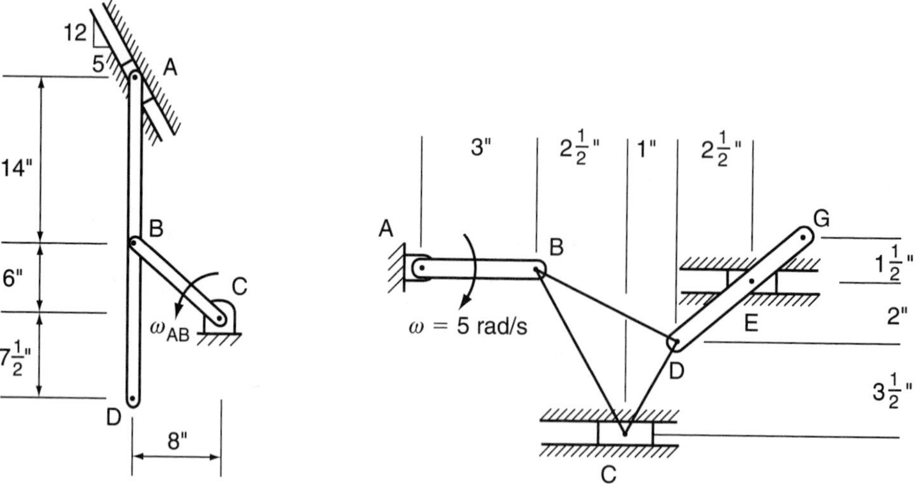

FIGURE P12–51

FIGURE P12–52

12–52. Using the method of instantaneous centers, determine (a) the angular velocity of DG and (b) the velocity of point C (Figure P12–52).

12–53. Block B in Figure P12–53 has a velocity of 10 ft/s to the left. Determine the velocities of A and C.

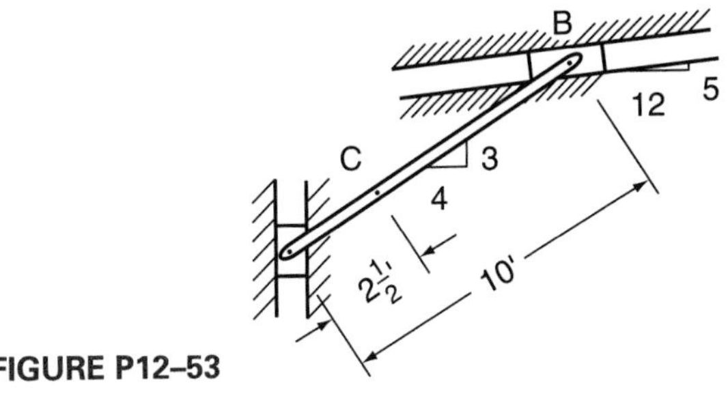

FIGURE P12–53

12–54. Roller A in Figure P12–54 has a velocity of 2 m/s to the right. Use the method of instantaneous centers to determine the velocity of point D.

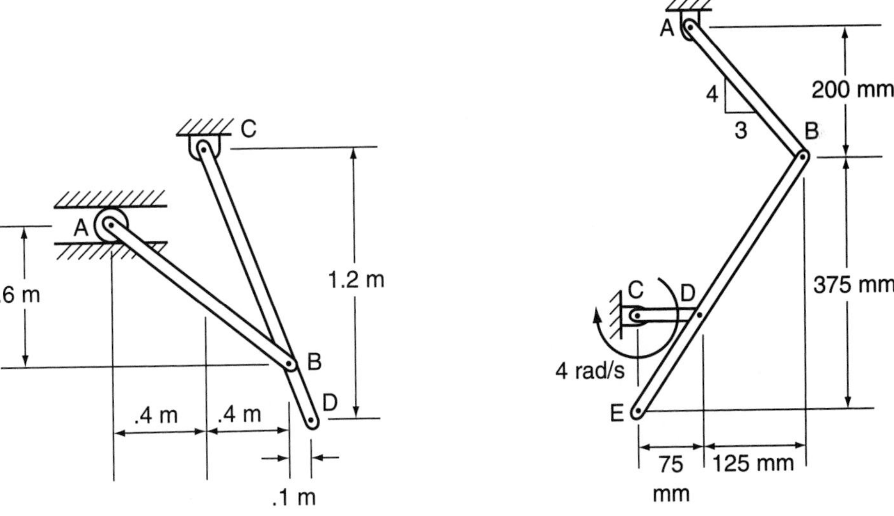

FIGURE P12–54

FIGURE P12–55

12–55. Using the instantaneous center method (Figure P12–55), determine the velocity of point E and the angular velocities of EB and AB.

12–56. Using the method of instantaneous centers for the system shown in Figure P12–56, determine (a) velocity of pin C, (b) velocity of point D, and (c) angular velocity of EC.

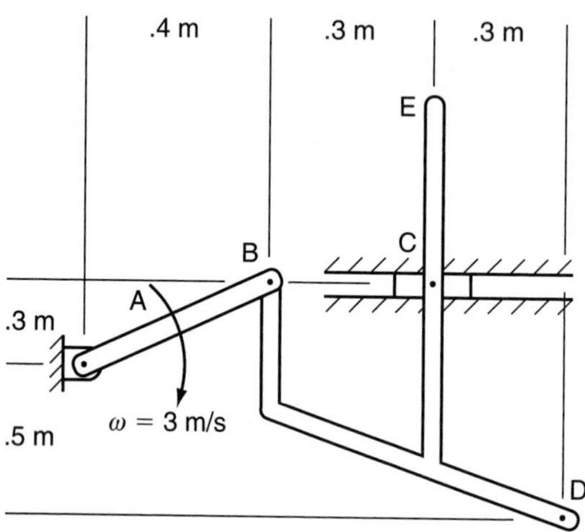

FIGURE P12–56

12–57. Roller A of Figure P12–57 moves downward to the left at 45 in./s. Using the method of instantaneous centers, determine the linear velocity of point E.

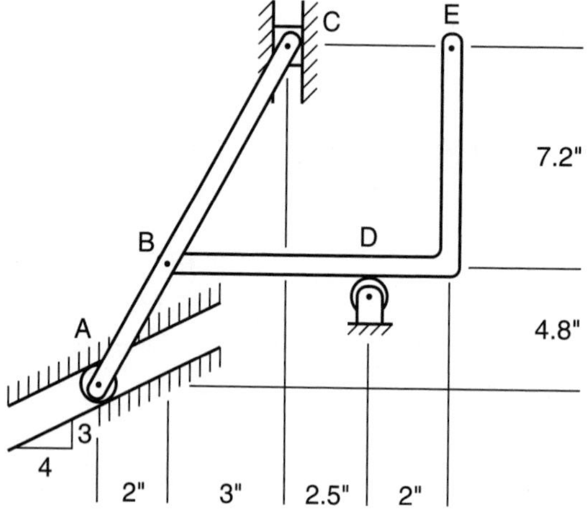

FIGURE P12–57

12–58. The angular velocity of AB in Figure P12–58 is 4 rad/s clockwise. Use instantaneous centers to determine the velocity of points C and E.

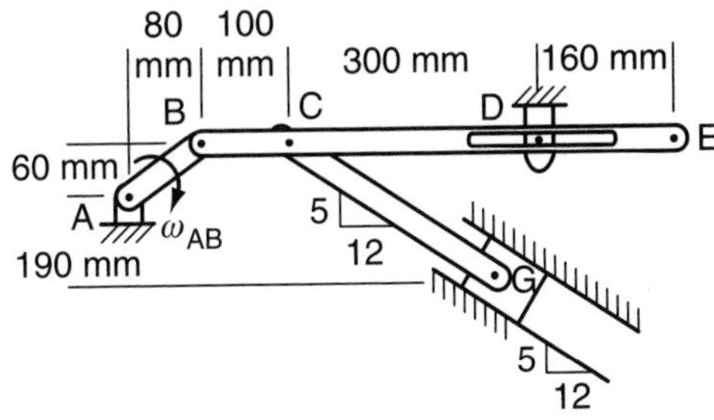

FIGURE P12–58

12–59. Use instantaneous centers to determine the velocity of point G (Figure P12–58).

12–60. Point C of the mechanism shown in Figure P12–60 has a velocity of 14 in./s to the right. Using the method of instantaneous centers, determine (a) the angular velocity of CD, (b) the linear velocity of B, and (c) the linear velocity of A.

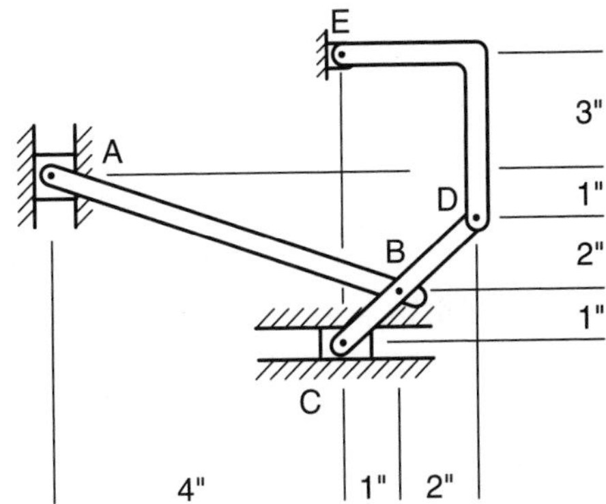

FIGURE P12–60

12–61. Determine (a) the angle θ (Figure P12–61) to produce horizontal motion of point D at the instant shown, and (b) the velocity of D.

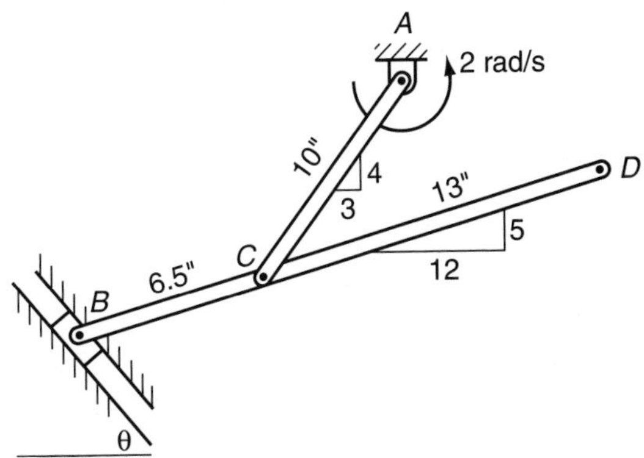

FIGURE P12–61

12–62. Using the method of instantaneous centers, determine the velocity of point C of the mechanism shown in Figure P12–62. Show where you would locate pivot E if point C were to have vertical velocity at this instant.

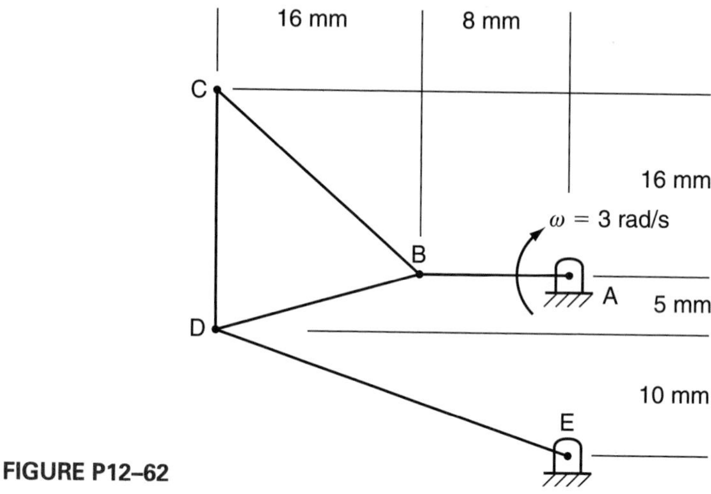

FIGURE P12–62

REVIEW PROBLEMS

R12–1. Starting from rest, car A accelerates at 2.5 m/s^2 to the east. Starting at the same point at the same time, car B accelerates at 2 m/s^2 to the north. At t = 10 seconds, determine the distance, velocity, and acceleration of B with respect to A.

R12–2. A turnover device is shown in Figure RP12–2. Member AD is 20 in. long and has an angular velocity of 4 rad/s clockwise. Member CB is 20.5 in. long. Without using instantaneous centers, determine (a) the velocity of point C and (b) the angular velocity of CD.

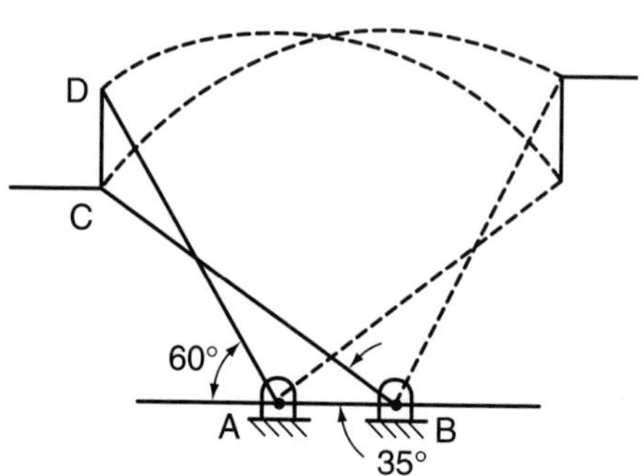

FIGURE RP12–2

R12–3. For the system shown in Figure RP12–3, determine (a) the absolute velocity of C, (b) the absolute velocity of E, and (c) the linear acceleration of E with respect to D.

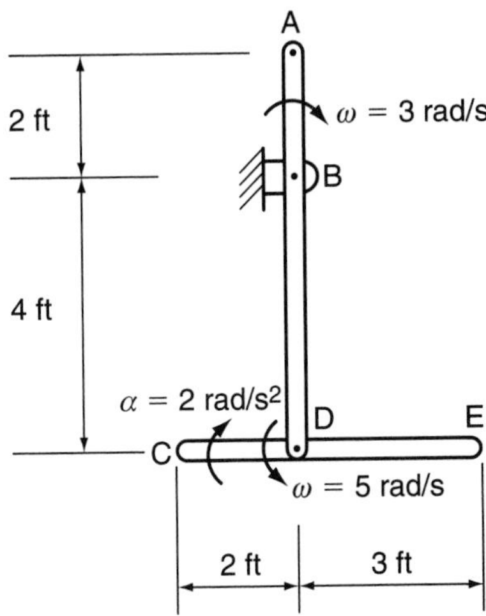

FIGURE RP12–3

R12–4. Member AC of Figure RP12–4 is 375 mm long and is bolted to the wheel shown at A and B. If the angular velocity of the wheel is 10 rad/s clockwise, determine (a) the linear velocity of E, (b) the angular velocity of CE, and (c) the angular velocity of AC.

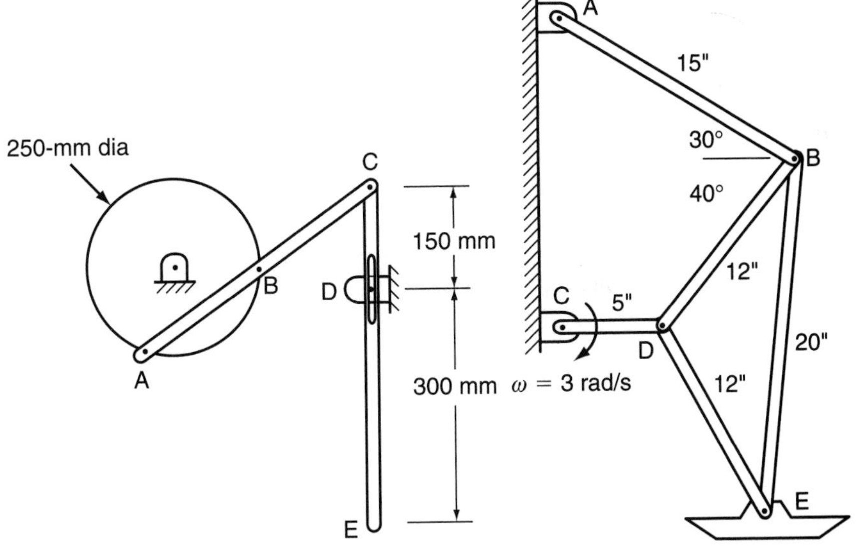

FIGURE RP12–4

FIGURE RP12–5

R12–5. The angular velocity of CD is 3 rad/s for the walking link device shown in Figure RP12–5. Using the relative velocity equation, determine (a) the velocity of point B, (b) the angular velocity of BD, and (c) the velocity of point E.

R12–6. Wheel A in Figure RP12–6 rolls without slipping; weight B has a velocity of 20 ft/s downward. Determine the velocities of points D, E, and C on cylinder A. If B drops 10 ft, how far does cylinder A move to the right?

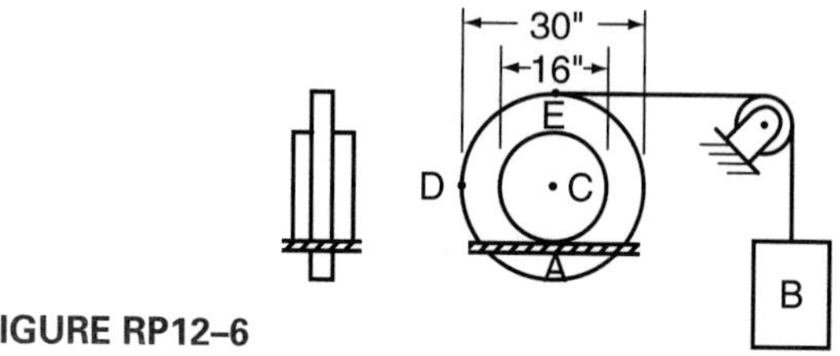

FIGURE RP12–6

R12–7. Cylinder D (Figure RP12–7) rolls to the right at 250 mm/s. Determine the angular velocity of BC.

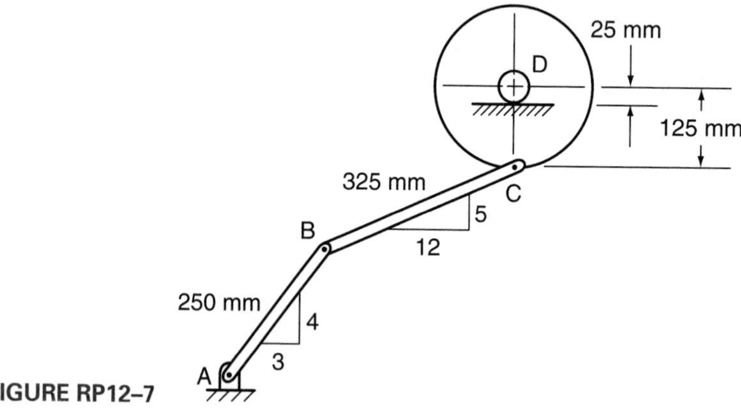

FIGURE RP12–7

R12–8. The velocity of point B is 3 m/s at the instant shown in Figure RP12–8. Using the relative velocity equation method, determine (a) the angular velocity of member ABD and (b) the linear velocity of the center of mass of cylinder E.

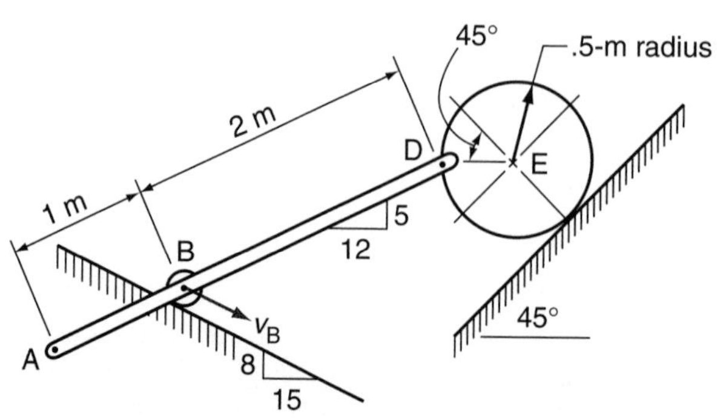

FIGURE RP12–8

R12–9. For the system shown in Figure RP12–9, use the method of instantaneous centers to determine the velocity of weight C (neglect inertia and assume no slipping occurs).

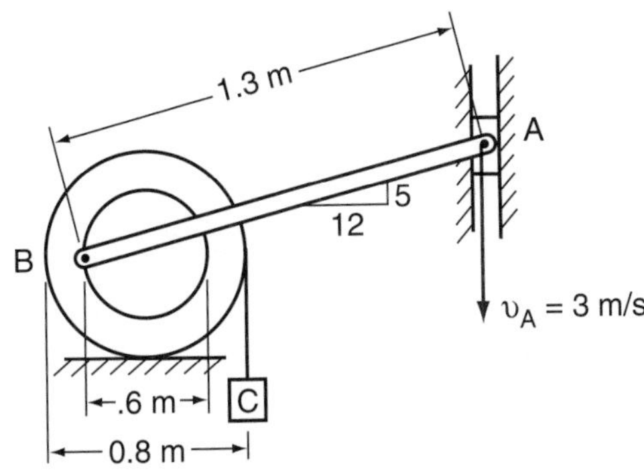

FIGURE RP12–9

R12–10. Using the method of relative velocity, determine the angular velocity of member CD in Figure RP12–10. Using the method of instantaneous centers, determine (a) angular velocity of ABC, (b) linear velocity of A, and (c) angular velocity of cylinder E.

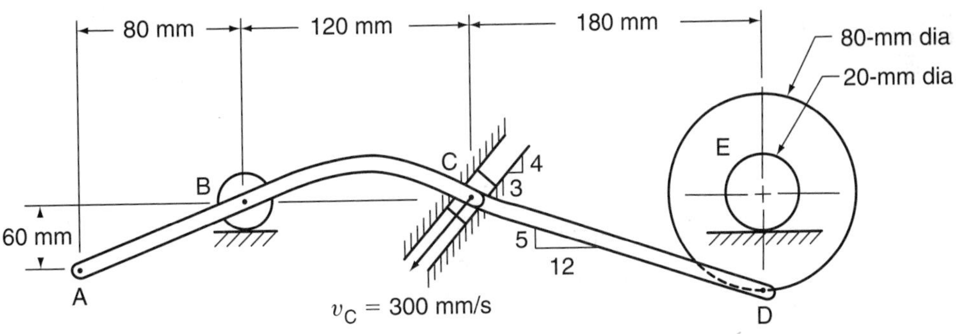

FIGURE RP12–10

R12–11. Member BF of the system shown in Figure RP12–11 has an angular velocity of 15 rad/s counterclockwise about point F. Using instantaneous centers, determine the velocity of points A and C. Using the relative velocity equation, determine the angular velocity of DE and the linear velocity of point D.

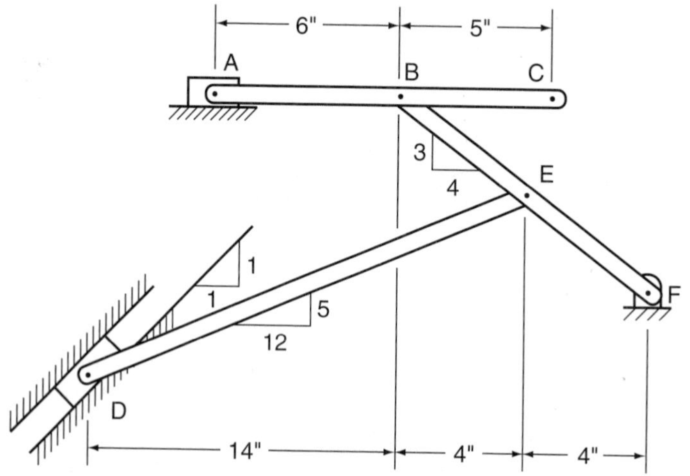

FIGURE RP12–11

R12–12. For the system shown in Figure RP12–12, use the method of instantaneous centers to determine the velocities of C and D. Using the relative velocity equation, determine the angular velocity ω_{DB}.

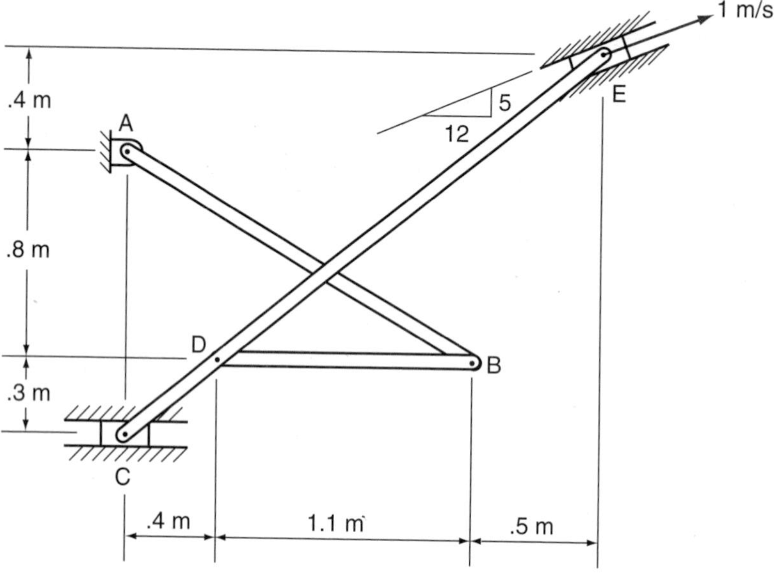

FIGURE RP12–12

CHAPTER 13

Kinetics

OBJECTIVES

Upon completion of this chapter the student will be able to solve for:

1. Forces and acceleration of linear motion.
2. Torque and acceleration of angular motion.
3. Force, torque, linear acceleration, and angular acceleration of plane motion.

13–1 INTRODUCTION

Kinetics concerns not only velocity and acceleration but also the accompanying unbalanced forces that cause the motion. At this point, it would be convenient to summarize Newton's three laws of motion.

1. Every object remains at rest or maintains a constant velocity in a straight line unless an unbalanced force acts upon it.

This is the law that we applied in statics when we balanced force systems. Another case of the same force balance that we have not yet considered is that of an object traveling at constant velocity in a straight line.

2. A body that has a resultant unbalanced force acting upon it behaves as follows:
   (a) The acceleration is proportional to the resultant force.
   (b) The acceleration is in the direction of the resultant force.
   (c) The acceleration is inversely proportional to the mass of the body.
3. For every action there is an equal and opposite reaction.

Kinetics, the study of unbalanced forces causing motion, can be analyzed by three methods (as the following chapters will show):

1. Inertia force or torque (dynamic equilibrium)
2. Work and energy
3. Impulse and momentum

This chapter will cover the first method as applied to both *linear* and *angular motion.* (*Linear motion* is another term for rectilinear or translational motion.)

13–2 LINEAR INERTIA FORCE

Newton's second law can be examined more closely and placed in equation form. Let

$$\begin{aligned} a &= \text{acceleration} \\ F &= \text{resultant force} \\ m &= \text{mass of body} \\ W &= \text{force of gravity (weight in the U.S. Customary system)} \end{aligned}$$

1. $a \propto F$ or $a = (F)(\text{constant})$
2. a and F are in the same direction
3. $a \propto \dfrac{1}{m}$ or $a \dfrac{\text{constant}}{m}$

These three statements can be combined and stated in one equation:

$$F = ma \tag{13–1}$$

where in the SI system

$$\begin{aligned} F &= \text{force in newtons, N} \\ m &= \text{mass in kilograms, kg} \\ a &= \text{acceleration, m/s}^2 \end{aligned}$$

This is consistent with the original definition of 1 newton being the force that causes a mass of 1 kg to accelerate at 1 m/s^2.

$$1\text{N} = (1\ \text{kg})(1\ \text{m/s}^2)$$

$$1\text{N} = 1\ \text{kg} \cdot \text{m/s}^2$$

This force can also be the force of gravity on an object (customarily called weight). Taking the acceleration of gravity as 9.81 m/s^2 and using W to represent weight or force of gravity, from

$$\text{force} = (\text{mass})(\text{acceleration})$$

we get

$$\begin{aligned} \text{weight} &= (\text{mass})(\text{acceleration of gravity}) \\ W &= mg \end{aligned}$$

For a 1-kg mass,

$$W = 1\ \text{kg}(9.81\ \text{m/s}^2)$$
$$W = 9.81\ \text{N}$$

Although mass and weight are often confused in the U.S. Customary system, they are distinguished as follows. The *weight* of an object is a measure of the pull of gravity on it. The acceleration due to gravity, g, may vary depending on the location of the object on the earth's surface. Since the weight of an object can vary with gravity, another term, *mass,* a quantity that does not change with changing gravity, is used to describe an object. Mass is constant for an object and is defined by the following equation:

$$\text{mass} = \frac{\text{weight}}{\text{acceleration due to gravity}}$$

$$m = \frac{W}{g} \qquad (13\text{–}2)$$

where

$m = \text{lb}/(\text{ft/s}^2) = \text{slugs}$

$W = \text{lb}$

$g = 32.2\ \text{ft/s}^2$ (an approximate value for our calculations)

From Equation (13–2) you can see that although there may be one-half the weight at a particular location due to one-half the acceleration or pull of gravity, the mass would still be the same.

In summary, we now have a situation in which a force or unbalance of several forces causes a body to move with changing velocity; that is, the body accelerates. The equations describing the body, the force, and the acceleration are

$$F = ma \qquad \text{for a single force}$$

or

$$\Sigma F = ma \qquad \text{for several forces}$$

13–3 LINEAR INERTIA FORCE: DYNAMIC EQUILIBRIUM

There are two methods of dealing with inertia and acceleration for linear motion. The *inertia-force method of dynamic equilibrium* will be used here since it often gives rise to a shorter, easier solution. The other method, which will not be covered, involves summing all forces to obtain the force that causes unbalance and equating that sum with mass times acceleration.

In order to visualize the inertia-force method, recall Newton's third law: For every action there is an equal and opposite reaction. The opposing reaction or force is the inertia force, which is equal to *ma.* Whether an object is accelerating from zero velocity or from some existing velocity, the inertia force will act opposite to the acceleration. If the velocity direction changes, producing normal acceleration, then the inertia force will act opposite to

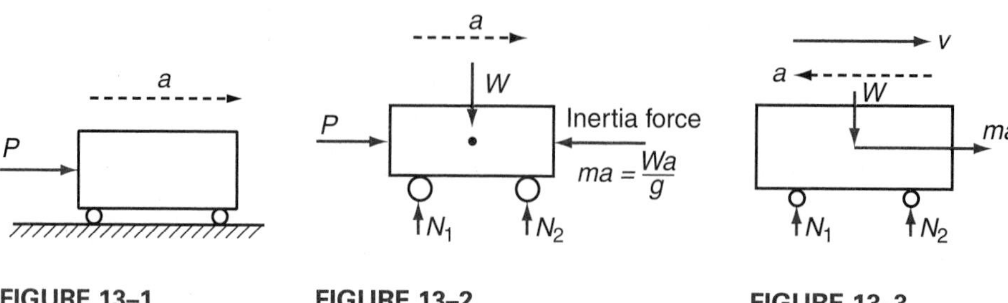

FIGURE 13–1 FIGURE 13–2 FIGURE 13–3

the normal acceleration. Keep in mind that inertia of an object tends to maintain the present state of the object; it opposes change. Suppose that we have an acting force P that causes acceleration of the block to the right in Figure 13–1. The opposing force (or *inertia force*) is shown in the opposite direction to the acceleration vector (Figure 13–2).

Just as the weight of an object is taken as acting through the center of gravity, the inertia force also is thought of as acting through the center of gravity (or *mass center*). The other vectors of weight and normal forces are shown as they would be in statics. The block is now in *dynamic equilibrium* and can be treated exactly as it was in statics. All methods and equations of statics apply.

To illustrate that the inertia force always acts in a direction opposite to acceleration, consider the block moving to the right but decelerating (Figure 13–3). Deceleration always has an opposite direction to velocity, so the acceleration vector now acts to the left. The inertia force is opposite to acceleration and is acting to the right. The inertia force tends to keep the block moving; it is opposite to acceleration and is therefore shown acting to the right. Just remember that all forces must be shown on the free-body diagram just as in statics. By adding an additional force to account for inertia, we produce a situation of dynamic equilibrium, and all the previous methods apply.

EXAMPLE 13–1

A 10-lb block is lifted vertically by a force of 18 lb. Determine its acceleration.

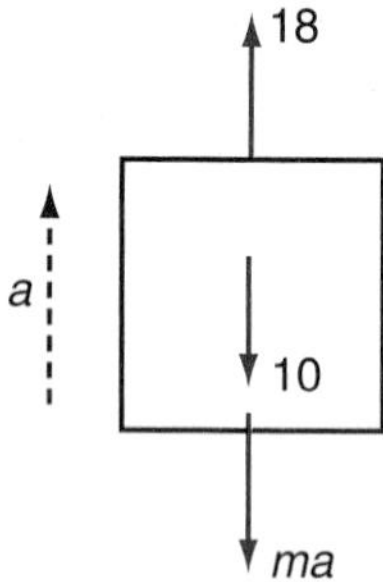

FIGURE 13–4

A free-body diagram of the weight (Figure 13–4) shows an unbalance of applied vertical forces. Since the block will accelerate upward, the inertia force, *ma,* is shown acting downward. We now have dynamic equilibrium and the vertical forces can be written in equation form:

$$\Sigma F_y = 0$$

$$18 - 10 - ma = 0$$

$$18 - 10 - \frac{W}{g}a = 0$$

$$18\text{ lb} - 10\text{ lb} - \frac{10\text{ lb}}{32.2\text{ ft/s}^2}a = 0$$

$$\underline{a = 25.8\text{ ft/s}^2\uparrow}$$

EXAMPLE 13–2

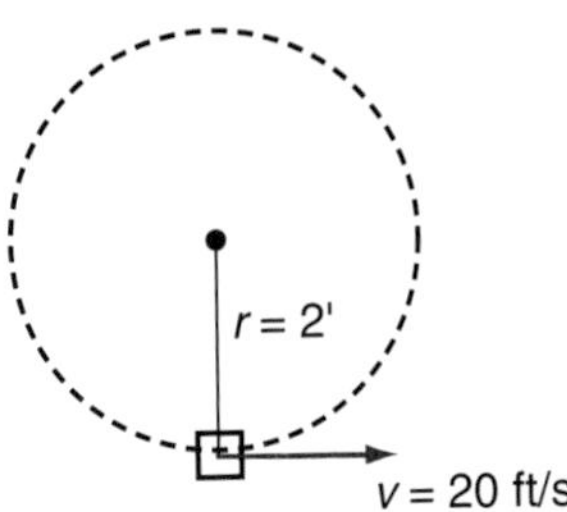

FIGURE 13–5

A 5-lb weight is swung in a vertical circle on the end of a 2-ft rope. If the velocity of the weight at the bottom of the circle is 20 ft/s, determine the tension in the rope at this point (Figure 13–5).

Because the weight moves in a circular fashion, the inward-acting normal acceleration produces an outward-acting inertia force, sometimes referred to as a *centrifugal force.*

The free-body diagram of the weight is drawn just as in statics, but with the addition of the inertia force. This gives us dynamic equilibrium (Figure 13–6).

The acceleration in this case is normal acceleration.

$$a_n = \frac{v^2}{r} = \frac{(20\text{ ft/s})^2}{2} = 200\text{ ft/s}^2$$

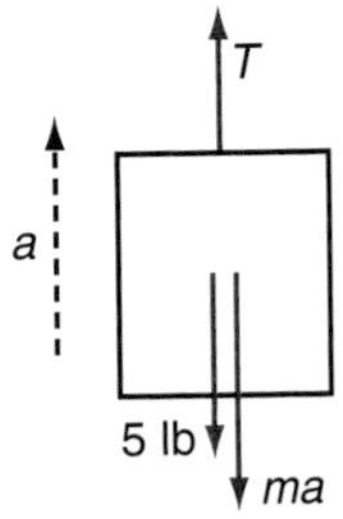

FIGURE 13–6

$$\Sigma F_y = 0$$

$$T - 5\text{ lb} - \left(\frac{5\text{ lb}}{32.2\text{ ft/s}^2}\right)(200\text{ lb}) = 0$$

$$\underline{T = 36\text{ lb}}$$

EXAMPLE 13–3

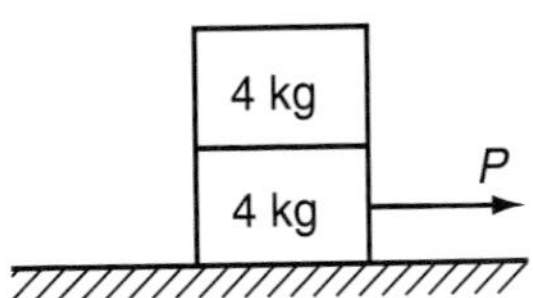

FIGURE 13–7

Two blocks, each with a mass of 4 kg, are placed one on top of the other. The coefficient of static friction for all surfaces is 0.3. Calculate the minimum P required to pull the bottom block out from beneath the top one without moving the top one horizontally (Figure 13–7).

The free-body diagram of the top block (Figure 13–8) shows that the inertia must be sufficient to overcome the friction force of the bottom block on the top one.

$$\Sigma F_y = 0$$

$$\begin{aligned} N_1 &= 39.2\text{ N} \\ F_{\max} &= \mu N_1 \\ &= (0.3)(39.2\text{ N}) \\ &= 11.8\text{ N} \end{aligned}$$

Free-Body Diagram of Top Block

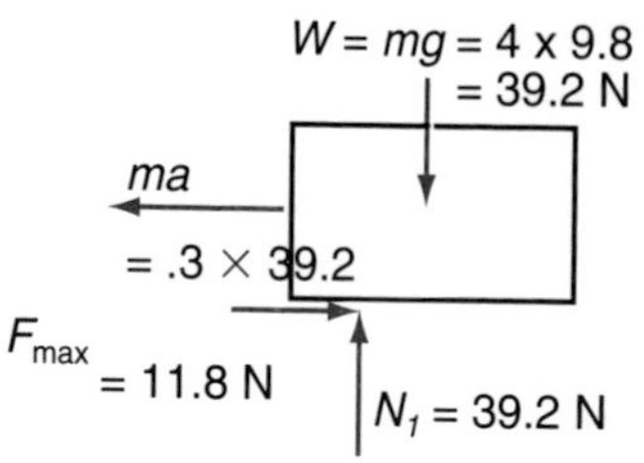

FIGURE 13–8

$\Sigma F_x = 0$

$$11.8\text{ N} - ma = 0$$

$$a = \frac{11.8}{m}$$

$$= \frac{11.8\text{ N}}{4\text{ kg}} = \frac{11.8\text{ kg}\cdot\text{m/s}^2}{4\text{ kg}}$$

$$\underline{a = 2.95\text{ m/s}^2 \rightarrow}$$

The bottom block must have an acceleration of 2.95 m/s^2 or greater. The total normal force and friction force on this bottom block is shown in Figure 13–9.

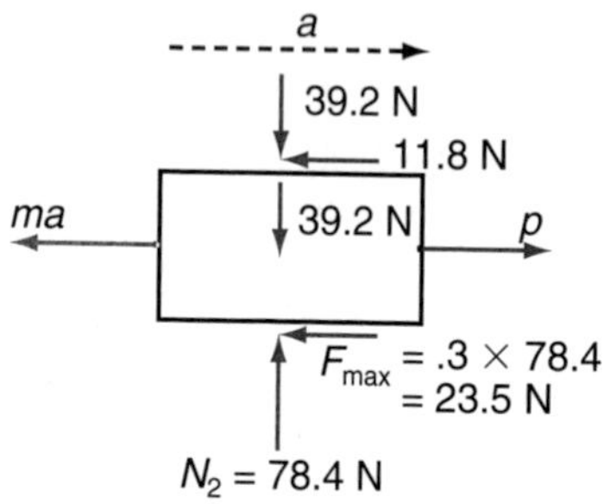

Free-Body Diagram of Bottom Block

a
39.2 N
11.8 N
ma
39.2 N
p
$F_{max} = .3 \times 78.4$
$= 23.5$ N
$N_2 = 78.4$ N

FIGURE 13–9

$\Sigma F_y = 0$

$$N_2 - 39.2\text{ N} - 39.2\text{N} = 0$$

$$N_2 = 78.4\text{ N}$$

$$F_{max} = \mu\text{N}$$

$$= (0.3)(78.4\text{ N})$$

$$= 23.5\text{ N}$$

$\Sigma F_x = 0$ (Figure 13–9)

$$P - 11.8\text{ N} - 23.5\text{ N} - (4\text{ kg})(2.95\text{ m/s}^2) = 0$$

$$\underline{P = 47.1\text{ N} \rightarrow}$$

EXAMPLE 13–4

A car bumper is designed to bring a 4000-lb car to a stop from a speed of 5 mph while deforming 6 in. Assume constant deceleration and determine the average force on the bumper during this stop (60 mph = 88 ft/s).

Referring to the directions shown in Figure 13–10

$$v_0 = \frac{5\text{ mph}}{60\text{ mph}}(88\text{ ft/s}) = 7.33\text{ ft/s}$$

$$v = 0$$

$$s = 0.5\text{ ft}$$

$$a = ?$$

$$v^2 = v_0^2 + 2as$$
$$0 = (7.33\ \text{ft/s})^2 + 2a(0.5\ \text{ft})$$
$$a = -53.8\ \text{ft/sec}^2 \rightarrow$$
$$a = 53.8\ \text{ft/sec}^2 \leftarrow$$

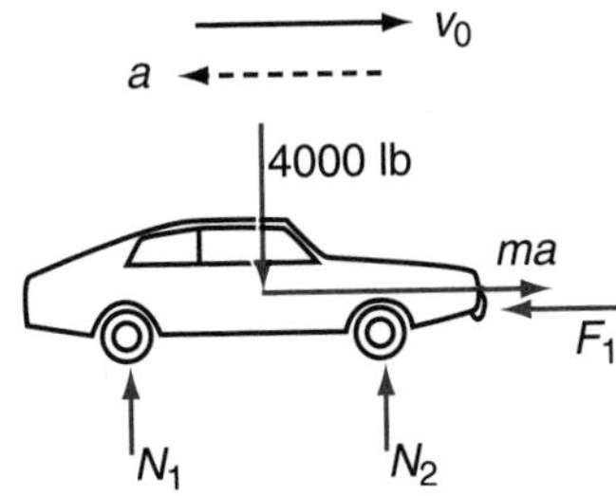

FIGURE 13–10

$\Sigma F_x = 0$ (Figure 13–10)

$$ma - F_1 = 0$$
$$F_1 = \frac{4000\ \text{lb}}{32.2\ \text{ft/s}^2}(53.8\ \text{ft/s}^2)$$
$$\underline{F_1 = 6680\ \text{lb}}$$

The bumper will deform 6 inches when subjected to a force of 6680 lb.

EXAMPLE 13–5

Neglect the inertia of the pulleys and the rolling resistance of mass B and determine the acceleration of masses A and B when the system is released from rest (Figure 13–11).

Notice that if mass A drops 1 m, mass B will move 2 m, B moves twice as far as A does in the same time interval; thus, the acceleration of B is twice that of A.

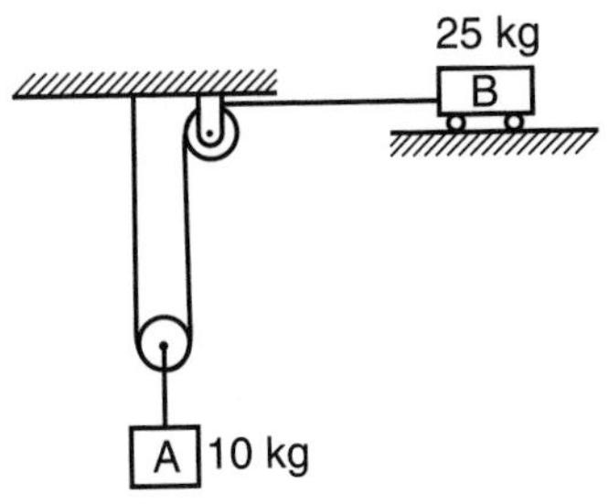

FIGURE 13–11

$$a_B = 2a_A$$

(When A drops 1 m, the two cables above A each lengthen 1 m for a total of 2 m. This pulls B 2 m to the left.)

From Figure 13–12:

$\Sigma F_y = 0$

$$2T + ma_A - 98.1 = 0$$
$$2T + (10\ \text{kg})a_A - 98.1\ \text{N} = 0$$
$$T = 49 - 5a_A \quad (1)$$

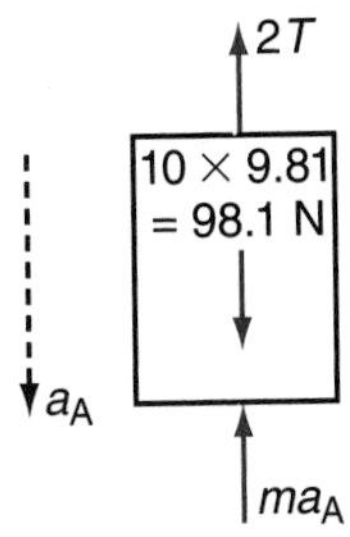

FIGURE 13–12

In Figure 13–13, there is twice the acceleration; therefore, the mass times acceleration term becomes $m(2a_A)$.

$\Sigma F_x = 0$

$$m(2a_A) = T$$
$$(25\ \text{kg})(2a_A) = T$$
$$T = 50a_A \quad (2)$$

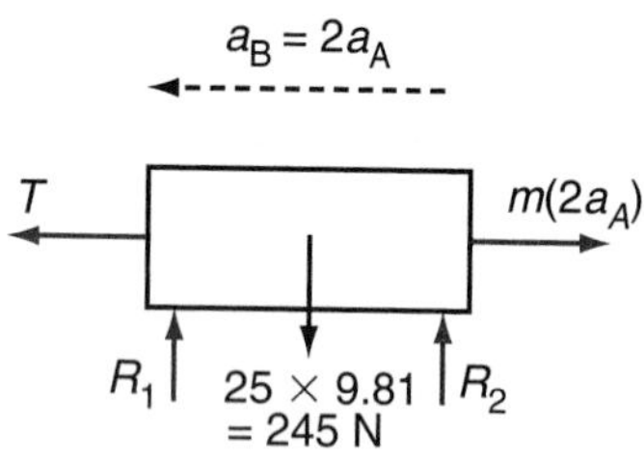

FIGURE 13–13

Equating Equations (1) and (2), we get

$$50a_A = 49 - 5a_A$$

$$a_A = \frac{49}{55}$$

$$\underline{a_A = 0.89\ \text{m/s}^2 \downarrow}$$

Therefore $a_B = 2a_A$

$$= 2(0.89)$$

$$\underline{a_B = 1.78\ \text{m/s}^2 \leftarrow}$$

13–4 ANGULAR INERTIA

The main limitation that we impose on the angular inertia of a rotating body in this section is that the axis of rotation coincide with the center of mass. We therefore have a homogeneous body with an angular motion and inertia about the body's center of mass. This by no means covers all types of rotation. In Section 13–6 we deal with the topic in more depth.

Consider the portion of the body (Δm) shown in Figure 13–14; the larger Δm is and/or the larger the radius, the larger the body's angular inertia will be. To give this wheel a clockwise angular acceleration, a clockwise torque would have to be applied. This torque would accelerate many particles of mass Δm, each at its own radius r. Each Δm has an acceleration tangent to the radius and an inertia force (Δma) in the opposite direction. The torque for each Δm is $(\Delta ma)r$, where

$$a = r\alpha \quad \text{or} \quad \text{torque} = (\Delta m r\alpha)r$$

The total torque that will accelerate all elements of Δm is

$$\text{torque} = \Sigma(\Delta m r\alpha)r$$
$$= \Sigma r^2\alpha\Delta m$$

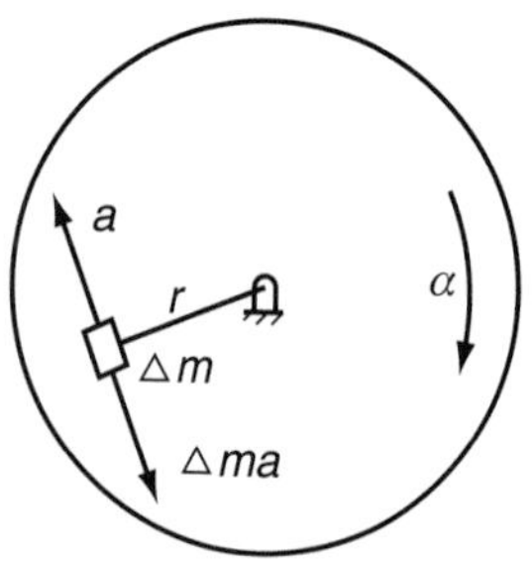

FIGURE 13–14

Since α is constant

$$\text{torque } T = \alpha \Sigma r^2 \Delta m$$

From Section 9–5 we have

$$\text{mass moment of inertia, } I = \Sigma r^2 \Delta m$$

Therefore, for rotation about the center of mass

$$T = I_C \alpha \tag{13–3}$$

where

T = torque; N · m
I_C = mass moment of inertia about the center of mass; kg · m^2
α = angular acceleration; rad/s^2

In the U.S. Customary system,

T = torque; lb-ft
I_C = mass moment of inertia about the center of mass; slug-ft^2 or ft-lb-s^2
α = angular acceleration; rad/s^2

I can be obtained from Table 9–2 or from the radius of gyration equation (Section 9–7).

$$k = \sqrt{\frac{I}{m}} \quad or \quad I = k^2 m \qquad \text{(Equations 9–6 and 9–7)}$$

EXAMPLE 13–6

What torque is required to accelerate a wheel about its center (I_C = 2 ft-lb-s^2) at an angular acceleration of 30 rad/s^2?

$$T = I_C \alpha$$
$$= (2 \text{ ft-lb-s}^2)(30 \text{ rad/s}^2)$$
$$\underline{T = 60 \text{ lb-ft}}$$

EXAMPLE 13–7

A rotor with a mass moment of inertia (I_C) of 6 kg · m^2 about its center of mass has a torque of 90 N · m applied to it. Determine the angular acceleration of the rotor.

$$T = I_C \alpha$$
$$90 \text{ N} \cdot \text{m} = (6 \text{ kg} \cdot \text{m}^2)\alpha$$
$$\underline{\alpha = 15 \text{ rad/s}^2}$$

13–5 ANGULAR DYNAMIC EQUILIBRIUM

With angular dynamic equilibrium, we are again faced with a situation analogous to rectilinear and curvilinear motion (Section 13–3). We have shown an inertia force opposite in direction to acceleration. This gives us dynamic equilibrium and the ability to solve in the same manner as previous statics problems.

The same situation exists here since we now show the angular inertia *torque* opposite in direction to the angular acceleration; thus, dynamic equilibrium results. But there is an important distinction: whereas we had a *force* (ma) for rectilinear and curvilinear motion, we have a *torque* ($I\alpha$) for angular motion. The importance of this distinction will become more evident after you have dealt with the following examples.

EXAMPLE 13–8

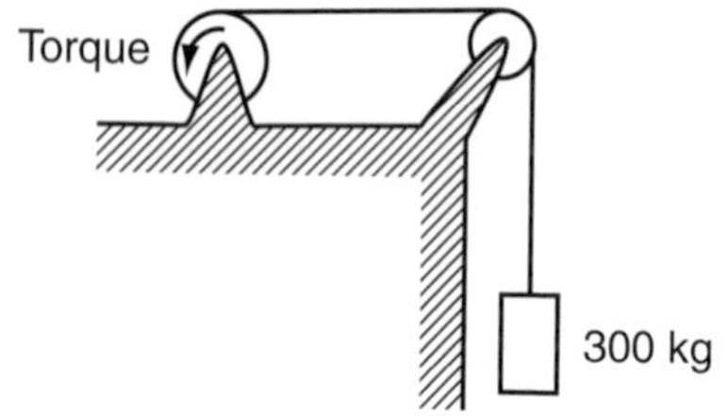

FIGURE 13–15

A power-driven winch is used to raise a mass of 300 kg with an acceleration 2 m/s^2. The winch drum is 0.5 m in diameter and has a mass moment of inertia about its center I_C equal to 8 kg · m^2. What torque must be applied to the winch drum (Figure 13–15)?

The rope tension T must first be found. Use a free-body diagram of the 300-kg mass (Figure 13–16).

Free-Body Diagram of Weight

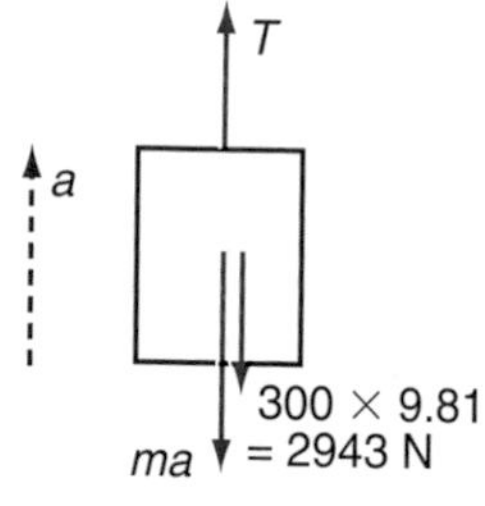

FIGURE 13–16

$\Sigma F_y = 0$

$$T - 2943 - ma = 0$$
$$T - 2943\text{ N} - (300\text{ kg})(2\text{ m/s}^2) = 0$$
$$T = 3543\text{ N}$$

The drum (Figure 13–17) has a tangential acceleration equal to the 2 m/s^2 acceleration of the 300-kg mass.

$$a = r\alpha$$
$$\alpha = \frac{a}{r} = \frac{2\text{ m/s}^2}{0.25\text{ m}} = 8\text{ rad/s}^2 \circlearrowleft$$

Free-Body Diagram of Drum

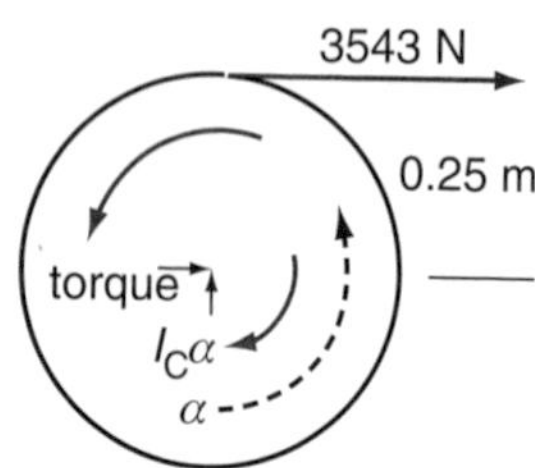

FIGURE 13–17

Taking moments about the center of the drum, we get

$\Sigma M_c = 0$

$$\text{torque} - (3543\text{ N})(0.25\text{ m}) - (I_C\alpha) = 0$$
$$\text{torque} - 886 - (8\text{ kg}\cdot\text{m}^2)(8\text{ rad/s}^2) = 0$$
$$\underline{\text{torque} = 950\text{ N}\cdot\text{m} \circlearrowleft}$$

Note that $I_C\alpha$ is opposite in direction to α. Also note that $I_C\alpha$ is a torque and is not multiplied by a radius as is the force of 3543 N.

EXAMPLE 13–9

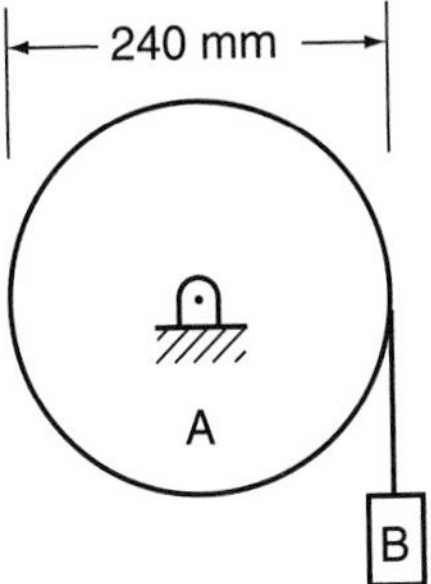

FIGURE 13–18

Wheel A has a mass of 22 kg and a radius of gyration of 180 mm. The mass of B is 10 kg. If the system starts from rest and has no bearing friction, determine the angular acceleration of A and the tension in the rope (Figure 13–18).

Pay special attention here, as your training in statics might lead to two possible pitfalls:

1. The tension in the rope is not $10 \times 9.81 = 98.1$ N; rather, it is somewhat less because B is accelerating downward.
2. The acceleration of B is not 9.81 m/s^2; it is less due to its restraint by the rope, which in turn is accelerating cylinder A.

For mass B (Figure 13–19), we have

$\Sigma F_y = 0$

$$T + (10\text{ kg})a - 98.1\text{ N} = 0$$
$$T + 10a = 98.1$$

Free-Body Diagram of B

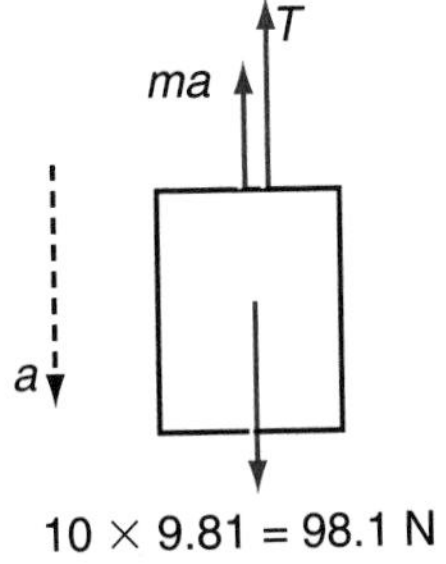

FIGURE 13–19

but wheel A has a radius of 0.12 m and an angular acceleration α.

$$a = r\alpha = (0.12\text{ m})\alpha$$

Substituting 0.12α, we get

$$T + 10(0.12\alpha) = 98.1$$
$$T + 1.2\alpha = 98.1 \qquad (1)$$

Consider the cylinder now (Figure 13–20).

$$I_C = k^2m \qquad \text{(Equation 9–7)}$$
$$= (0.18\text{ m})^2(22\text{ kg})$$
$$I_C = 0.713\text{ kg}\cdot\text{m}^2$$

$\Sigma M_c = 0$

$$-Tr + I_C\alpha = 0$$
$$-T(0.12\text{ m}) + (0.713\text{ kg}\cdot\text{m}^2)\alpha = 0$$
$$T = 5.94\alpha \qquad (2)$$

Substituting Equation (2) into Equation (1), we get

$$5.94\alpha + 1.2\alpha = 98.1$$
$$\underline{\alpha = 13.7\text{ rad/s}^2}$$

and from Equation (2) $T = (5.94)(13.7)$

$$\underline{T = 81.7\text{ N}}$$

Free-Body Diagram of A

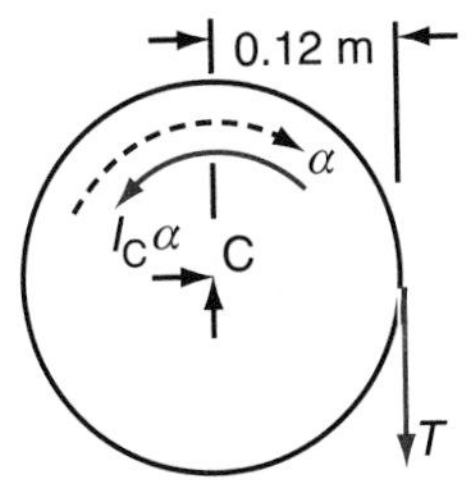

FIGURE 13–20

EXAMPLE 13–10

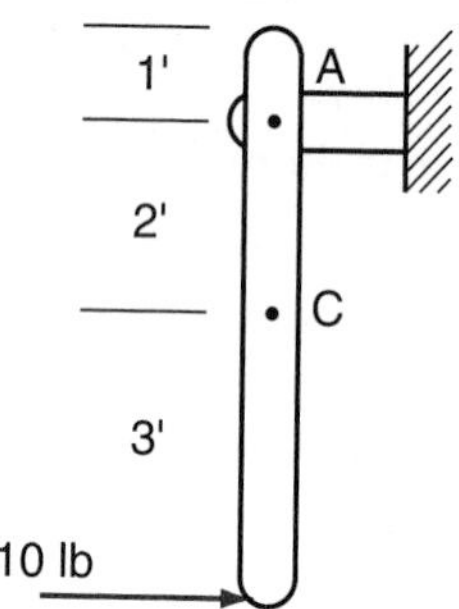

FIGURE 13–21

A 6-ft slender rod weighing 64.4 lb is initially at rest when the force of 10 lb is applied as shown in Figure 13–21. Determine the reactions at A and the angular acceleration of the rod for this instant.

The inertia of the rod will resist motion, so its mass moment of inertia must be found. From Table 9–2, the mass moment of inertia about the center of the rod is

$$I_C = \frac{1}{12}ml^2$$

$$= \frac{1}{12}\left(\frac{64.4\text{ lb}}{32.2\text{ ft/s}^2}\right)(6\text{ ft})^2$$

$$I_C = 6\text{ ft-lb-s}^2$$

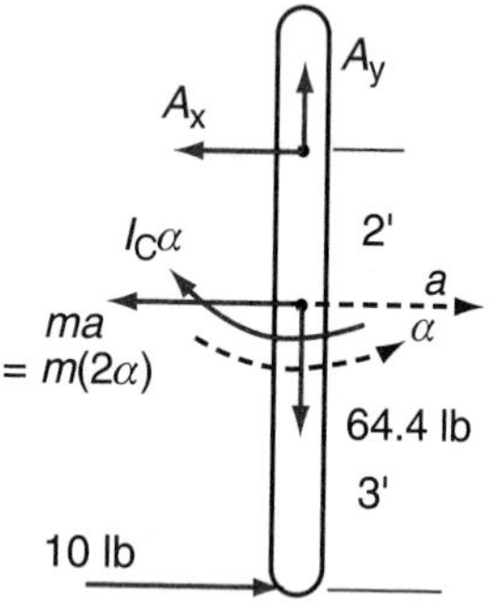

FIGURE 13–22

Show the rod in dynamic equilibrium (Figure 13–22), where the linear acceleration of its mass center is

$$a = r\alpha = (2\text{ ft})\alpha$$

$\Sigma F_y = 0$

$$\underline{A_y = 64.4\text{ lb}\uparrow}$$

$\Sigma M_A = 0$

$$(10)(5) - I_C\alpha - (ma)2 = 0$$

$$(10\text{ lb})(5\text{ ft}) - (6\text{ ft-lb-s}^2)\alpha - \left(\frac{64.4\text{ lb}}{32.2\text{ ft/s}^2}\right)(2\text{ ft})\alpha(2\text{ ft}) = 0$$

$$\underline{\alpha = 3.57\text{ rad/s} \circlearrowleft}$$

$\Sigma F_x = 0$

$$A_x + \frac{64.4\text{ lb}}{32.2\text{ ft/s}^2}(2\text{ ft})(3.57\text{ rad/s}) - 10\text{ lb} = 0$$

$$= -4.29\text{ lb} \leftarrow$$

$$\underline{A_x = 4.29\text{ lb} \rightarrow}$$

13–6 PLANE MOTION

There are many types of plane motion. The rectilinear, curvilinear, and angular motions that we have been considering separately are basically isolated or restricted forms of plane motion. Although we will be unable to cover all types of plane motion, the following list will illustrate the scope of this class of motion and the portion that we are considering.

(A) Constrained plane motion
  1. Translational
    (a) Rectilinear (Section 13–3)
    (b) Curvilinear (Section 13–3)
  2. Centroidal rotation (angular) (Section 13–4). The axis of rotation coincides with the mass center.
  3. Translational and centroidal rotation. A rolling wheel or connecting rod is typical of this motion. There is a definite relationship between the translational acceleration and the angular acceleration.
  4. Translational and noncentroidal rotation

(B) Unconstrained plane motion
  1. Translational and unrelated centroidal rotation. A wheel simultaneously rolling and sliding is typical of this motion. There is no relationship between the translational acceleration and the angular acceleration.

The combination of rectilinear motion and angular motion in one object gives plane motion such as a rolling wheel. This plane motion has the following features:

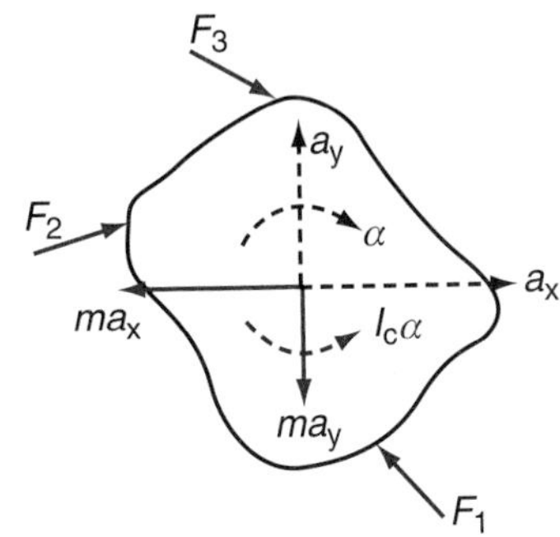

FIGURE 13–23

1. There is a definite relationship between the angular acceleration and the translational acceleration (a_x or a_y).
2. It can be resolved into its component motions, translation and rotation.
3. Using free-body diagrams and the principle of dynamic equilibrium, we can use the three equations: $\Sigma F_x = 0$, $\Sigma F_y = 0$, $\Sigma M = 0$.

Moments can be taken about any point, although the instantaneous center of rotation is often used. Figure 13–23 shows an unbalance of forces, causing three simultaneous accelerations of a body (a_x, a_y, and α).

EXAMPLE 13–11

A solid cylinder 4 ft in diameter weighing 96.6 lb rolls down the slope shown in Figure 13–24 without slipping. Determine the angular acceleration.

From Table 9–2, the mass moment of inertia of a cylinder about its center of mass C is found by the following equation.

$$I_C = \frac{1}{2}mr^2$$

$$= \frac{1}{2}\left(\frac{96.6 \text{ lb}}{32.2 \text{ ft/s}^2}\right)(2 \text{ ft})^2$$

$$I_C = 6 \text{ ft-lb-s}^2$$

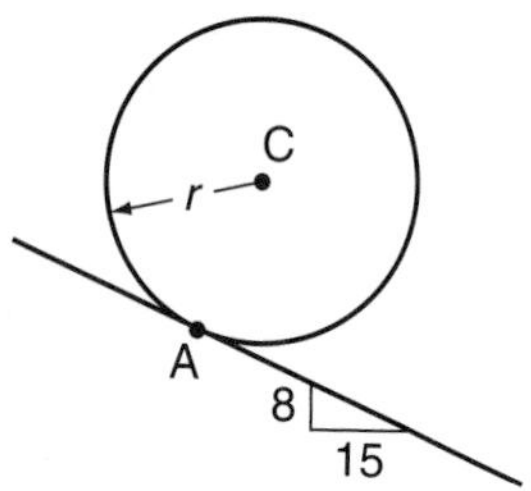

FIGURE 13–24

Dynamic equilibrium is shown in Figure 13–25; there, you can see that both translational inertia (ma_C) and angular inertia ($I_C\alpha$) are involved. Since there is no slippage,

$$a_C = r\alpha = (2 \text{ ft})\alpha = 2\alpha$$

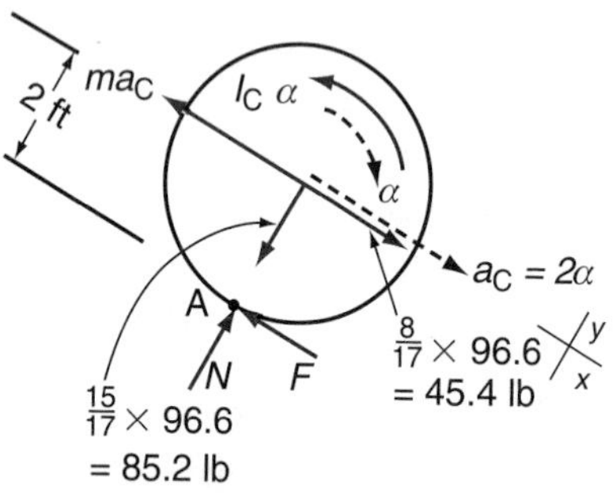

FIGURE 13–25

Note that since the coefficient of friction and the friction force are unknown, we cannot use either

$$\Sigma F_x = 0$$

or

$$\Sigma M_C = 0$$

Moments are taken about A, and the torque $I_C\alpha$ is treated as a couple.

$\Sigma M_A = 0$

$$+I_C\alpha + 2(ma_C) - 45.4(2) = 0$$

$$(6 \text{ ft-lb-s}^2)\alpha + (2 \text{ ft})\left(\frac{96.6 \text{ lb}}{32.2 \text{ ft/s}^2}\right)2\alpha = (45.4 \text{ lb})(2 \text{ ft})$$

$$6\alpha + 12\alpha = 90.8$$

$$\underline{\alpha = 5.05 \text{ rad/s}^2} \circlearrowright$$

EXAMPLE 13–12

A 20-kg cylinder has a cord wound in a 0.8-m-diameter groove at its center (Figure 13–26). Assume that there is no slippage and that a tension of 100 N is applied. Determine the acceleration a_C of the center of mass of the wheel, and the minimum coefficient of static friction.

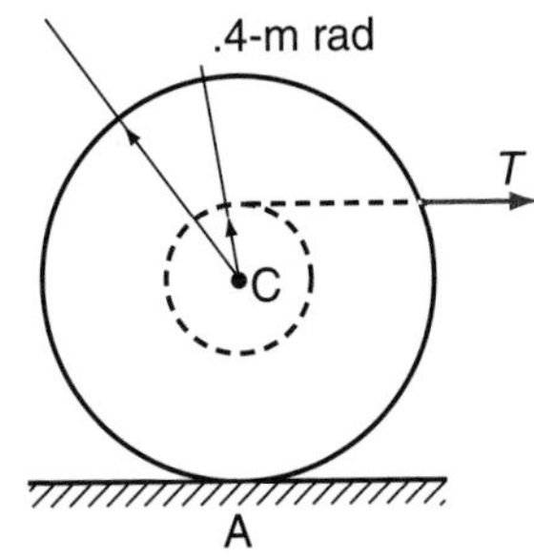

FIGURE 13–26

From Table 9–2, we have

$$I_C = \frac{1}{2}mr^2$$

$$= \frac{1}{2}(20\text{ kg})(1.5\text{ m})^2$$

$$I_C = 22.5\text{ kg}\cdot\text{m}^2$$

As shown in Figure 13–27,

$$a_C = r\alpha$$

$$\alpha = \frac{a_C}{1.5\text{ m}}$$

Taking moments about A, we get

$$\Sigma M_A = 0$$

$$ma_C(1.5) + I_C\alpha - 100(1.9) = 0$$

$$(20\text{ kg})a_C(1.5\text{ m}) + (22.5\text{ kg}\cdot\text{m}^2)\left(\frac{a_c}{1.5\text{ m}}\right) = (100\text{ N})(1.9\text{ m})$$

$$\underline{a_C = 4.22\text{ m/s}^2 \rightarrow}$$

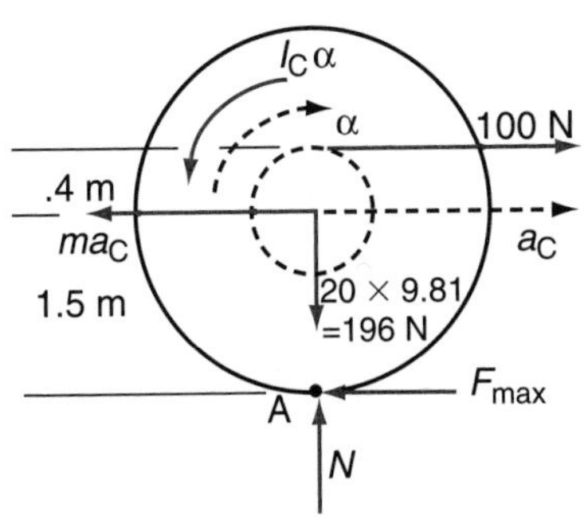

FIGURE 13–27

Now solving for friction, we obtain

$$\Sigma F_x = 0$$

$$100\text{ N} - ma_C - F_{max} = 0$$

$$F_{max} = 100\text{ N} - (20\text{ kg})(4.22\text{ m/s}^2)$$

$$F_{max} = 15.6\text{ N} \leftarrow$$

$$\Sigma F_y = 0$$

$$N = 196\text{ N}$$

$$\mu = \frac{F_{max}}{N} = \frac{15.6\text{ N}}{196\text{ N}}$$

$$\underline{\mu = 0.079}$$

EXAMPLE 13–13

Cylinder B has a mass of 40 kg and is 1.2 m in diameter. Block D has a mass of 30 kg (Figure 13–28). Assume the mass and friction of the cable and pulley to be negligible. If this system is

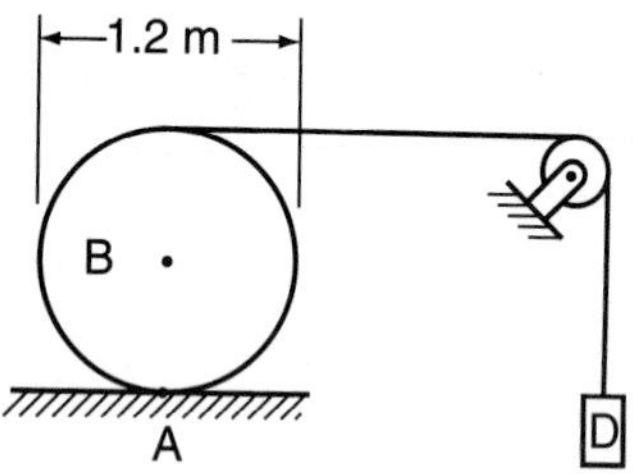

FIGURE 13–28

released from rest, determine the tension in the cable and the acceleration of D. (There is no slippage at A.)

Considering the cylinder first, we have

$$I_C = \frac{1}{2}mr^2 = \frac{1}{2}(40\text{ kg})(0.6\text{ m})^2$$

$$I_C = 7.2\text{ kg}\cdot\text{m}^2$$

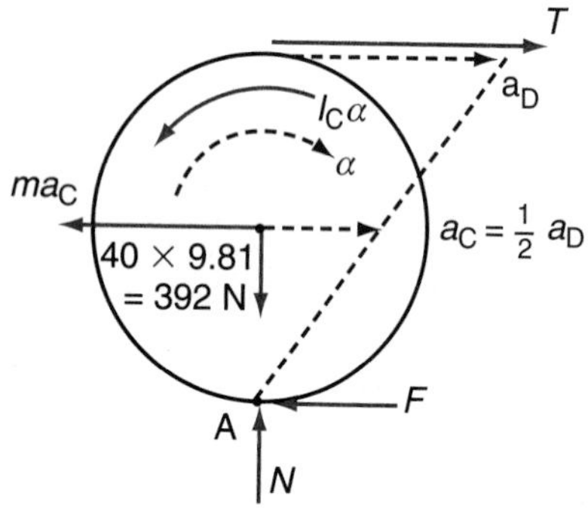

FIGURE 13–29

Show both the angular and the linear accelerations with their corresponding inertia torque and inertia force (Figure 13–29). Note that the point at the top of the cylinder at which T is acting has twice the acceleration of a_C since A is the instantaneous center of rotation. This is the same as the acceleration of D or

$$a_D = 2a_C \quad \text{or} \quad a_C = \frac{1}{2}a_D$$

$$\text{and also } a_C = r\alpha \qquad or \qquad \alpha = \frac{a_C}{r} = \frac{a_D}{2r}$$

Taking moments about A will give an equation with the two required unknowns, a_D and T.

$\Sigma M_A = 0$

$$-T(1.2) + ma_C(0.6) + I_C\alpha = 0$$

$$(1.2\text{ m})T = (40\text{ kg})\left(\frac{1}{2}a_D\right)(0.6\text{ m}) + (7.2\text{ kg}\cdot\text{m}^2)\left(\frac{a_D}{(2)(0.6\text{ m})}\right)$$

$$T = 15a_D \tag{1}$$

Drawing a free-body diagram of D (Figure 13–30), we have

$\Sigma F_y = 0$

$$T + ma_D - 294 = 0$$

$$T + (30\text{ kg})a_D = 294\text{ N}$$

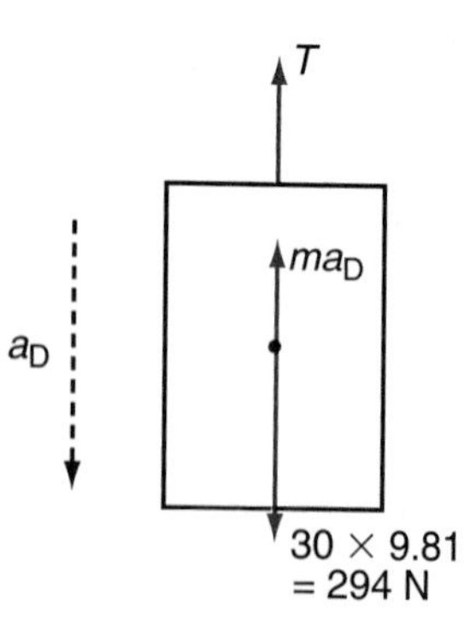

FIGURE 13–30

Substituting Equation (1),

$$15a_D + 30a_D = 294$$

$$\underline{a_D = 6.54\text{ m/s}^2 \downarrow}$$

and $\quad T = 15a_D = 15(6.54)$

$$\underline{T = 98.1\text{ N}}$$

EXAMPLE 13–14

The rod AB in Figure 13–31 is 10 in. long and weighs 16.1 lb. For the instant shown, A is at rest and has an acceleration of 15 ft/s^2 downward. Neglect the weight and friction of the slider blocks at A and B. Determine the force P that is partially restraining the motion, and the reaction forces at A and B.

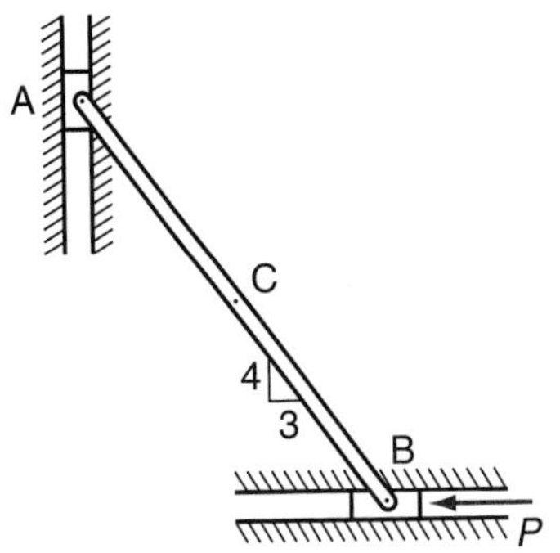

FIGURE 13–31

The first items to solve for will be the vertical and horizontal accelerations of the rod's center of mass, C. We must determine the angular acceleration as well. The relative acceleration can be written (Figure 13–32)

$$\underset{\downarrow}{a_A} = \underset{\rightarrow}{a_B} + \underset{\downarrow}{a_{A/B}}$$

From the vector triangle of Figure 13–33, we have

$$a_B = \frac{4}{3}(a_A) = \frac{4}{3}(15\ \text{ft/s}^2) = 20\ \text{ft/s}^2$$

$$a_{A/B} = \frac{5}{3}(15\ \text{ft/s}^2) = 25\ \text{ft/s}^2$$

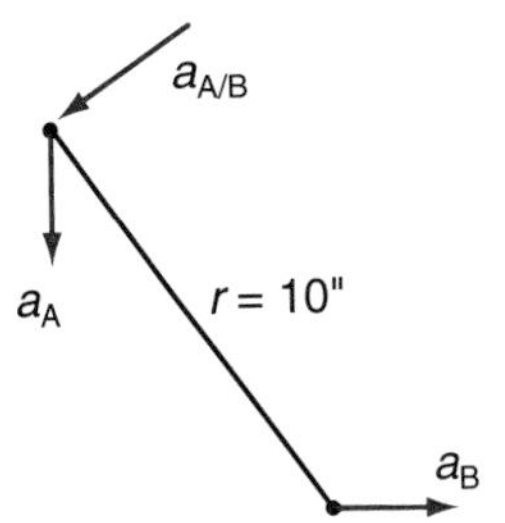

FIGURE 13–32

Since

$$a_{A/B} = r\alpha$$

$$\alpha = \frac{25\ \text{ft/s}^2}{(10/12)\text{ft}} = 30\ \text{rad/s}^2$$

Since point C is at the midpoint of AB, it has one-half the horizontal and vertical acceleration that the ends have; therefore,

$$a_x = 10\ \text{ft/s}^2$$

$$a_y = 7.5\ \text{ft/s}^2$$

From Table 9–2, I_C for the rod is

$$I_C = \frac{1}{12}mL^2$$

$$= \frac{16.1\ \text{lb}}{12(32.2\ \text{ft/s}^2)}\left(\frac{10\ \text{in.}}{12\ \text{in./ft}}\right)^2$$

$$I_C = 0.0289\ \text{ft-lb-s}^2$$

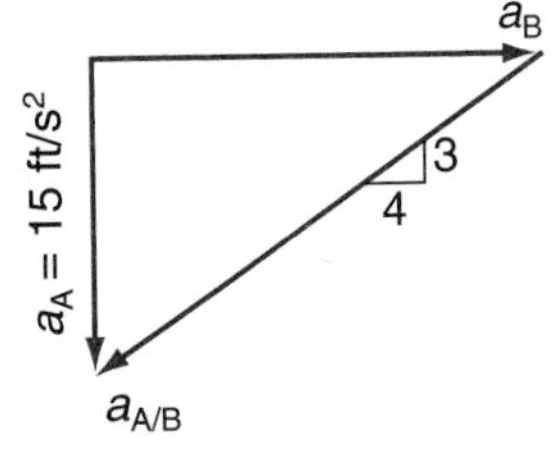

FIGURE 13–33

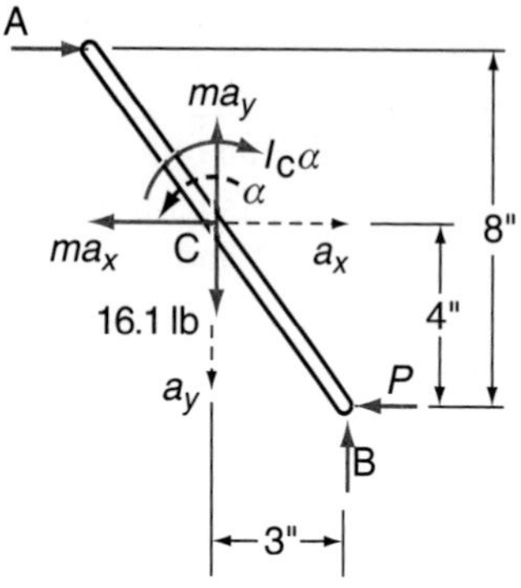

FIGURE 13–34

Now turn to Figure 13–34, which shows the dynamic equilibrium of the rod. We can apply any one of the three main equations:

$$\Sigma F_y = 0, \quad \Sigma F_x = 0, \quad \text{or } \Sigma M = 0$$

$\Sigma F_y = 0$

$$B + ma_y - 16.1 = 0$$

$$B + \left(\frac{16.1 \text{ lb}}{32.2 \text{ ft/s}^2}\right)(7.5 \text{ ft/s}^2) - 16.1 \text{ lb} = 0$$

$$\underline{B = 12.35 \text{ lb} \uparrow}$$

$\Sigma M_B = 0$

$$-(A)\left(\frac{8}{12}\right) - I_C\alpha - ma_y\left(\frac{3}{12}\right) + ma_x\left(\frac{4}{12}\right) + 16.1\left(\frac{3}{12}\right) = 0$$

$$-(A)\left(\frac{8}{12}\right)\text{ft} - (0.0289 \text{ ft-lb-s}^2)(30 \text{ rad/s}^2) - \left(\frac{16.1 \text{ lb}}{32.2 \text{ ft/s}^2}\right)(7.5 \text{ ft/s}^2)\left(\frac{3}{12}\right)\text{ft}$$

$$+\left(\frac{16.1 \text{ lb}}{32.2 \text{ ft/s}^2}\right)(10 \text{ ft/s}^2)\left(\frac{4}{12}\right)\text{ft} + (16.1 \text{ lb})\left(\frac{3}{12}\right)\text{ft} = 0$$

$$\underline{A = 5.84 \text{ lb} \rightarrow}$$

$\Sigma F_x = 0$

$$5.84 \text{ lb} - P - \left(\frac{16.1 \text{ lb}}{32.2 \text{ ft/s}^2}\right)(10 \text{ ft/s}^2) = 0$$

$$\underline{P = 0.84 \text{ lb} \leftarrow}$$

The main restriction placed on the above example was that rod AB was at rest. In Figure 13–32, there was one relative acceleration, $a_{A/B}$, at point A and it was perpendicular to AB. If AB had had an angular velocity, there would have been a normal component of $a_{A/B}$ acting along AB.

HINTS FOR PROBLEM SOLVING

1. A free-body diagram showing dynamic equilibrium is a "static equilibrium" free-body diagram with the addition of either an inertia force and/or an inertia torque.
2. For linear motion the inertia force is
   (a) Equal to *ma*
   (b) Acting through the center of gravity
   (c) Opposite in direction to the acceleration *a*
3. For rotational motion the inertia torque is
   (a) Equal to $I_C\alpha$
   (b) Opposite in direction to the angular acceleration α
4. When using $I_C\alpha$ in a moment equation, remember that it is a moment, not a force, and is like a couple that has the same moment about *any* point.
5. To ensure the correct direction of the inertia force or torque on a free-body diagram, show the direction of the acceleration immediately beside the free-body diagram and then the inertia opposite to it. (Do not include the acceleration vector in your force summation equation.)
6. In static pulley systems, the cable tension is equal to the weight it supports. When motion occurs, the tension will be more or less than the weight, depending on the direction of motion.
7. The acceleration of bodies in pulley systems may be multiples of one another. Determine these acceleration relationships, show them on the free-body diagrams, and then solve for one of them and the cable tension by using simultaneous equations.
8. Normal acceleration acts toward the center of rotation, and the inertia or centrifugal force acts outward from the center.
9. For a plane motion problem such as a rolling cylinder, try to
   (a) Equate or relate linear acceleration to angular acceleration.
   (b) Take moments at the rolling point of contact with the surface.

PROBLEMS

APPLIED PROBLEMS FOR SECTIONS 13–1 TO 13–3

13–1. Determine the acceleration of the 150-lb block in Figure P13–1 if the coefficient of kinetic friction is 0.4.

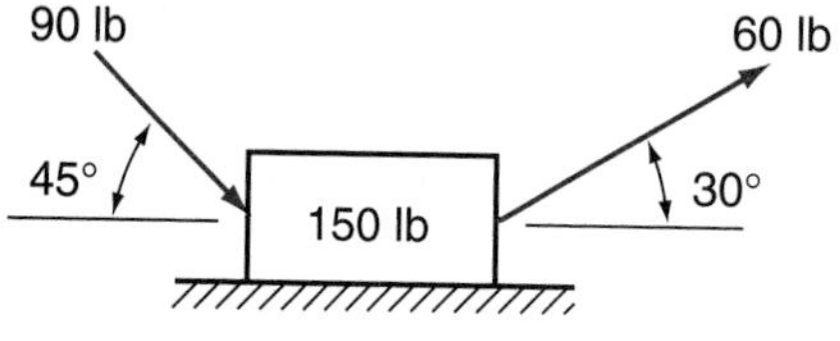

FIGURE P13–1

FIGURE P13–2

13–2. Determine the acceleration of the 30-kg block shown in Figure P13–2.

13–3. A 130-kg cart is accelerated horizontally by a 250-N force pulling at an angle of 20° above horizontal. Neglecting rolling resistance, determine the acceleration of the cart.

13–4. Determine the acceleration of block A down the slope in Figure P13–4.

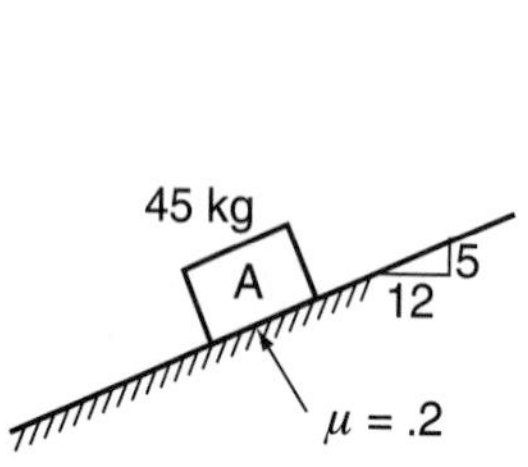

FIGURE P13–4

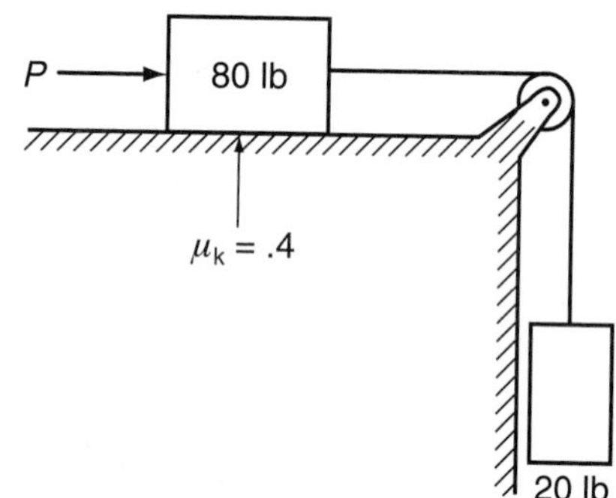

FIGURE P13–10

13–5. A body slides down a plane inclined at 40° to the horizontal, with an acceleration of 2.3 m/s^2. Determine the coefficient of kinetic friction.

13–6. A body takes twice as long to slide down a plane inclined at 30° to the horizontal as compared to the time required if the plane were frictionless. Determine the coefficient of kinetic friction.

13–7. At what maximum acceleration rate can a 500-N test-strength cable lift a 40-kg mass?

13–8. What force does a 180-lb man exert on the floor of an elevator that is moving downward and decelerating at 15 ft/s^2?

13–9. Determine the downward acceleration of an elevator so that a mass of 50 kg exerts only a force of 400 N upon its floor.

13–10. Determine the force P to accelerate the 80-lb block at 6 ft/s^2 to the right (Figure P13–10).

13–11. An elevator weighing 1000 lb is carrying a load of 600 lb and descending at 20 ft/s. The cable supporting the elevator has a design load of 2500 lb. Determine the shortest distance in which the elevator can be stopped.

13–12. The system shown in Figure P13–12 is released from rest. Determine the acceleration of each mass. (Neglect beam inertia.)

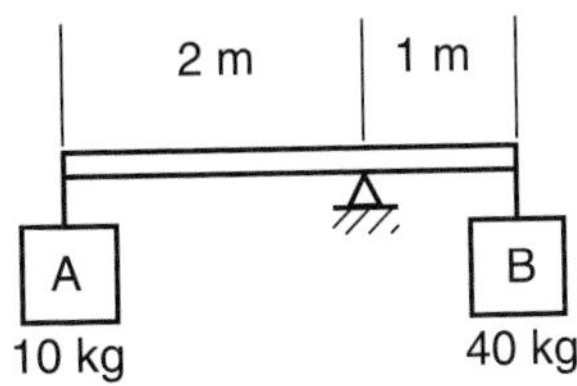

FIGURE P13–12

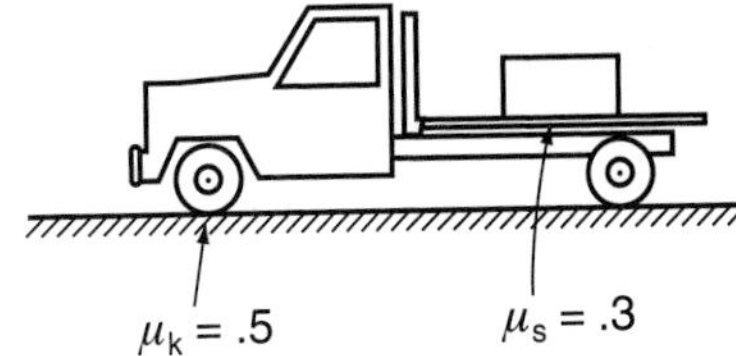

FIGURE P13–13

13–13. The truck in Figure P13–13 weighs 3500 lb, and the crate on it weighs 1500 lb. Assume equal weight distribution on all wheels and determine (a) the maximum deceleration rate of the loaded truck (assume that the crate does not slide) and (b) the maximum deceleration before the crate does slide. Would the declaration rates change if the truck were 500 lb lighter and the crate were 200 lb heavier?

13–14. A crate in the back of a truck with an open tailgate has impending sliding as the truck accelerates from rest to 60 mph in 10 seconds on a horizontal surface. Determine the maximum acceleration up a 10° slope so that the crate does not slide off the truck.

13–15. A 4000-lb car traveling at 60 mph decelerates at a constant rate and comes to a stop over a distance of 200 ft. Determine the minimum coefficient of friction between the pavement and the car's tires.

13–16. A car comes to rest in a controlled skid from a speed of 120 km/h. The skid length measures 220 m. The car mass is 900 kg. Determine the average friction force applied by the road on the tires during the skid.

13–17. A crate falls from the back of a truck traveling at 50 mph up a hill sloped 14° from horizontal. If the coefficient of friction is 0.3, what will be the length of the skid?

13–18. A 6-kg mass is whirled in a vertical circle on the end of a 1.2-m rope. What is the maximum and minimum tension in the rope during one revolution if it has a constant angular velocity of 5 rad/s?

13–19. A 4-kg ball fastened to a cord swings in a horizontal circle with a 2-m diameter with a velocity of 2.5 m/s (Figure P13–19). Determine (a) the tension in the cord, (b) the angle θ, and (c) the length of cord AB.

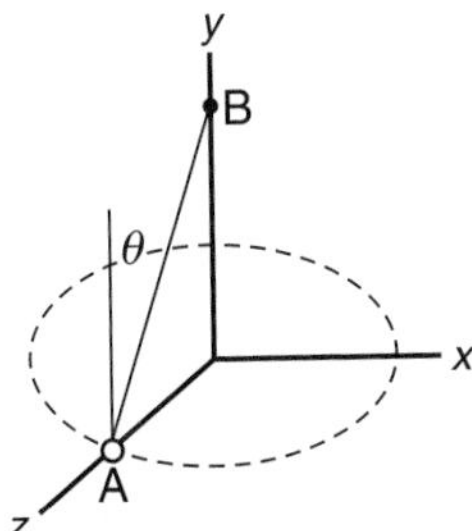

FIGURE P13–19

13–20. The coefficient of kinetic friction for mass B in Figure P13–20 is 0.25. Determine the acceleration of mass A if it has a mass of (a) 30 kg and (b) 50 kg.

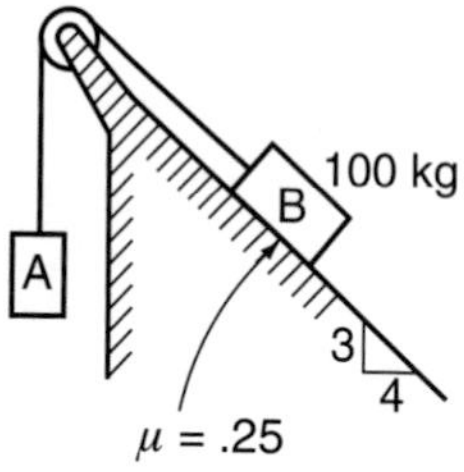

FIGURE P13–20

FIGURE P13–21

13–21. Determine the maximum acceleration of A (Figure P13–21) to the right, during which B will not slip. The coefficient of static friction for all surfaces is 0.35. (Neglect pulley and cable inertia.)

13–22. Block A has a mass of 53.1 kg; block B has a mass of 13.3 kg. Determine the velocity of B at $t = 2$ seconds if the system is initially at rest and is released from the position shown in Figure P13–22.

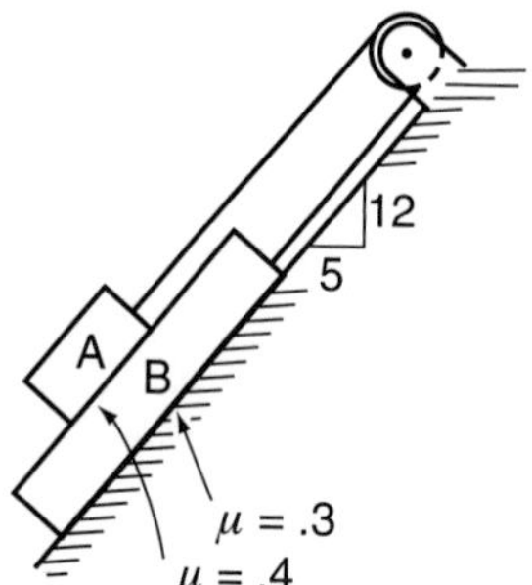

FIGURE P13–22

13–23. A skater travels at 8 m/s around a 30-m-radius curve. Determine how much his body is inclined to the vertical for maximum stability.

13–24. A horizontal disc accelerates from rest at 5 rad/s^2. The coefficient of static friction between it and a 2.5-kg block is 0.20. At what angular speed of the disc will the block begin to slide (Figure P13–24)?

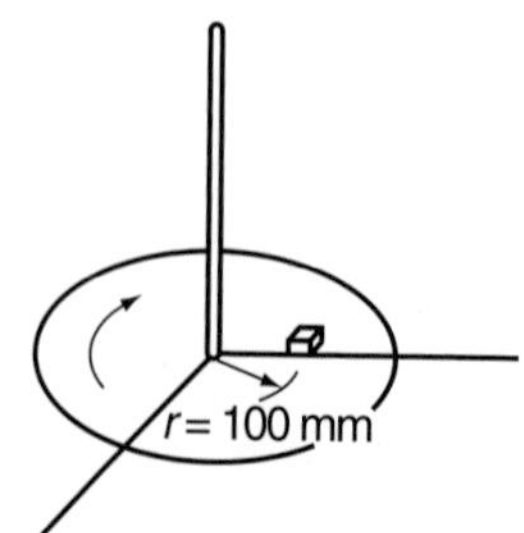

FIGURE P13–24

13–25. A car travels at 65 mph around a highway curve that has a radius of 800 ft and a slope or bank of 15°. The coefficient of friction between the tires and dry pavement is 0.7, but partially around the curve, the car passes over some snow with a coefficient of friction of 0.1. Prove by calculations whether or not the car will begin to skid.

13–26. The 2-kg block shown in Figure P13–26 is fastened at A by a cord and is free to slide outward on the rotating horizontal disc as it speeds up. If the coefficient of kinetic friction is 0.25, determine the rpm of the disc for the position shown.

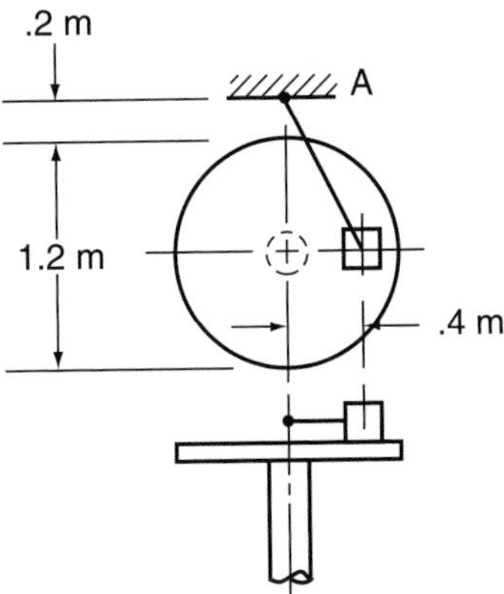

FIGURE P13–26

13–27. Crate A in Figure P13–27 is given an initial velocity of 6 ft/s down the inclined plane. If the coefficient of kinetic friction is 0.5, how far will the crate slide down the plane before coming to a stop?

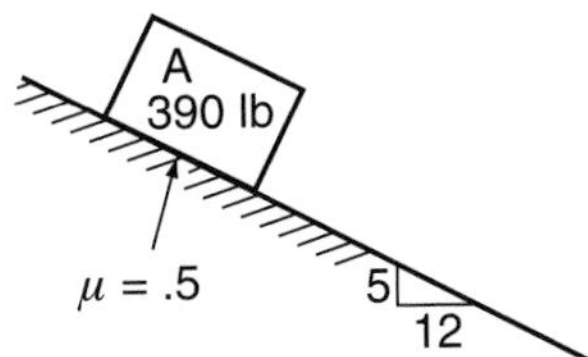

FIGURE P13–27

13–28. A 10-lb parcel has a velocity of 7 ft/s as it leaves the conveyor and begins sliding down a 20-ft long chute that is sloped at 20°. If the velocity of the parcel is 12 ft/s at the bottom of the chute, determine the coefficient of kinetic fraction.

13–29. Block A in Figure P13–29 is given an initial velocity of 6 m/s up the incline. It comes to a stop in distance d and then slides back down the incline with uniform acceleration. Determine (a) the distance d and (b) the velocity of A when it returns to its original point.

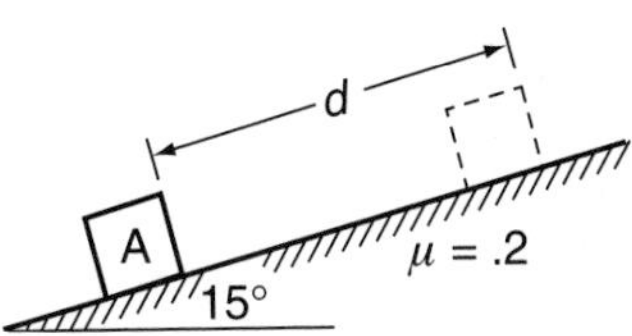

FIGURE P13–29

13–30. Neglect pulley inertia and determine the acceleration of (a) mass A and (b) mass B in Figure P13–30.

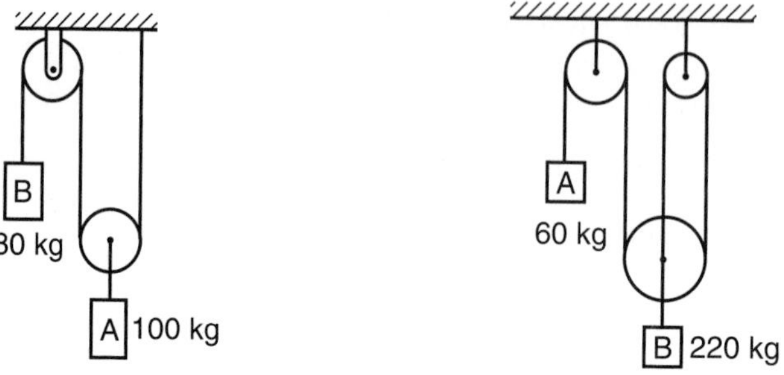

FIGURE P13–30

FIGURE P13–31

13–31. Neglect pulley inertia, and with the system initially at rest, determine the distance that mass B in Figure P13–31 will move in 4 seconds.

13–32. Determine the cable tensions that support (a) the elevator car and (b) the counterweight for the elevator system (shown in Figure P10–18) when the elevator car is accelerating upward.

13–33. Neglect pulley inertia and determine the acceleration of mass B in Figure P13–33.

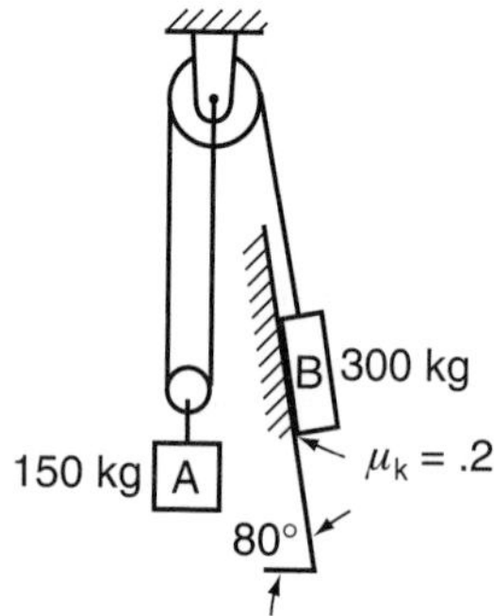

FIGURE P13–33

13–34. The weight of A is 2 lb greater than that required to produce impending motion. If the system accelerates from the position shown, determine the velocity at which weight A strikes the floor (Figure P13–34).

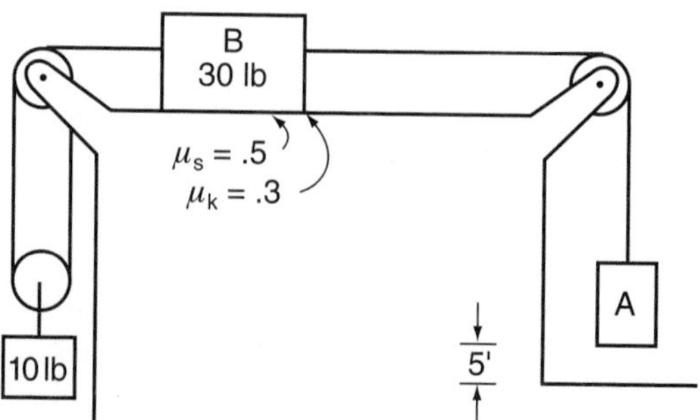

FIGURE P13–34

APPLIED PROBLEMS FOR SECTIONS 13–4 AND 13–5

13–35. A 50-lb sphere with a radius of gyration of 0.4 ft is accelerated about its centroidal axis. Determine the torque required to accelerate it from rest to 800 rpm in 15 seconds.

13–36. A 1000-kW generator has a 3500-lb rotor that is accelerated from rest to 3600 rpm in 10 seconds. Determine the torque required. Assume the rotor to be a solid cylinder 40 in. in diameter.

13–37. A belt-driven machine has a moment of inertia of 240 ft-lb-s^2. The belt tensions differ by 80 lb as the belt passes over the 18-in.-diameter driver pulley. Calculate the speed it reaches if it starts from rest and accelerates uniformly for 40 seconds.

13–38. A 0.8-m-diameter, 50-kg flywheel with a radius of gyration of 0.283 m ($I = K^2m$) must be braked from 60 rpm to rest in 2 seconds. Calculate the tangential force required to accomplish this braking action.

13–39. It takes 10 seconds to accelerate a wheel from rest to 900 rpm. If the wheel has a mass moment of inertia of 2 kg $\cdot$ m^2, determine the constant torque required to produce this acceleration.

13–40. An electric motor has a rotor with a mass moment of inertia of 35 ft-lb-s^2. What torque is required to accelerate it from rest to 1160 rpm in 1.5 seconds?

13–41. A 150-mm-diameter shaft with a mass of 20 kg is rotating at 900 rpm. A pulley mounted on the shaft has a mass moment of inertia of 0.15 kg $\cdot$ m^2. If the shaft and the pulley coast to a stop due to a tangential frictional force of 8 lb at the outer radius of the shaft, determine the time required.

13–42. Cylinder A has a mass of 25 kg and a radius of gyration of $k = 0.5$ m. At the instant shown in Figure P13–42, it is turning clockwise and is being braked by the application of a force, P, of 40 N. Determine the angular deceleration of the wheel.

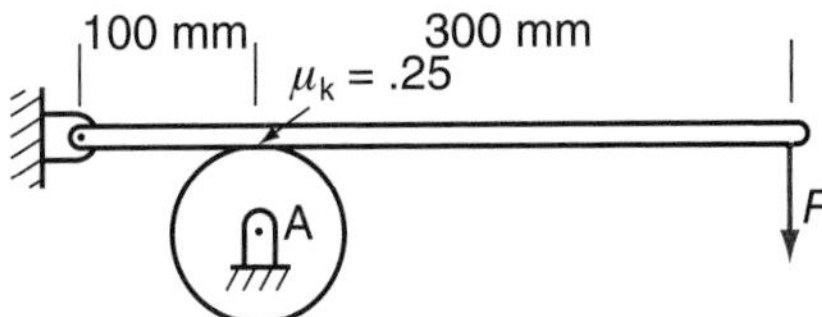

FIGURE P13–42

13–43. A large mixer requires a torque of 10 lb-ft when operating under fully loaded conditions. An additional torque is required during startup due to the mixer having a mass moment of inertia of 1.66 ft-lb-s^2. How long will it take the 8-in.-diameter drive pulley to accelerate from rest to 90 rpm under loaded conditions if the difference in belt tensions applied to it is 40 lb?

13–44. The rotation of winch drum A in Figure P13–44 causes motion of weight B to the right. The winch drum has a mass moment of inertia of 10-ft-lb-s^2. Determine the torque that must be applied to the drum to cause weight B to accelerate at 4 ft/s^2.

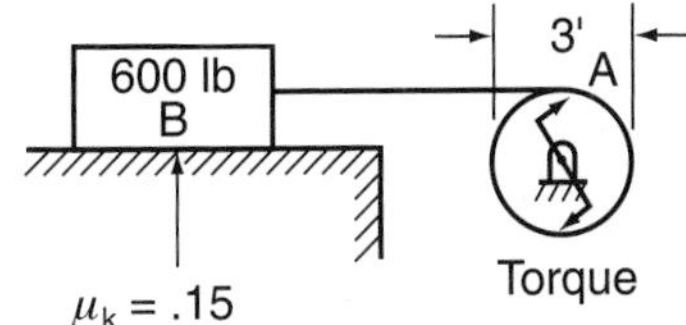

FIGURE P13–44

13–45. A 100-lb wheel, 2 ft in diameter, is accelerated from rest by weight A moving downward (Figure P13–45). Wheel motion is retarded by a bearing friction moment of 20 lb-ft and a braking force due to a spring force of 80 lb. How far will A fall in 6 seconds? (Assume that $I = \frac{1}{2}mr^2$ for B.)

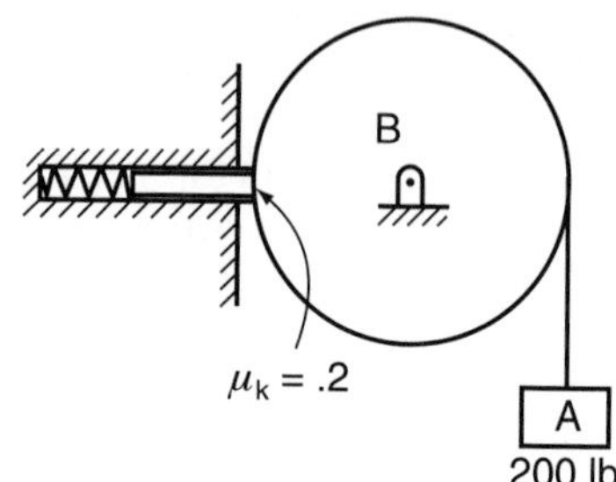

FIGURE P13–45

13–46—13–47. The two pulleys fastened together in Figures P13–46 and P13–47 have diameters of 1.5 and 1 m. Their combined mass moment of inertia is 35 kg · m^2. Determine (a) the angular acceleration of the pulleys and (b) the tension T in the cord supporting the 612-kg mass when it is released from rest.

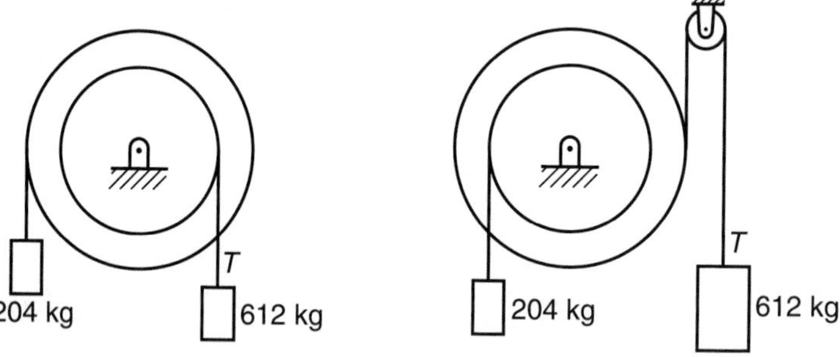

FIGURE P13–46 FIGURE P13–47

13–48. The long, slender rod in Figure P13–48 has a mass of 4.5 kg and, while at rest, is acted upon by a force, $F = 90$ N. Determine the reaction components at A.

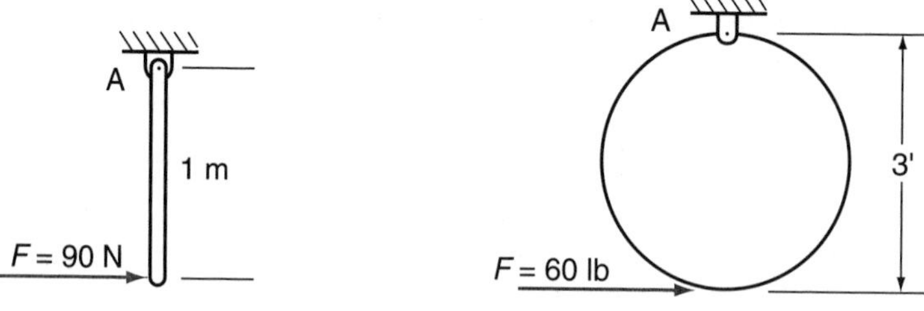

FIGURE P13–48 FIGURE P13–49

13–49. The solid cylinder in Figure P13–49 weighs 96.6 lb and, while at rest, is acted upon by a force, $F = 60$ lb. Determine (a) the reaction components at A and (b) the angular acceleration of the cylinder about A.

13–50. The hollow cylinder in Figure P13–50 has a mass of 14 kg and, while at rest, is acted upon by a force, $F = 350$ N. Determine (a) the reaction components at A and (b) the angular acceleration of the cylinder about A.

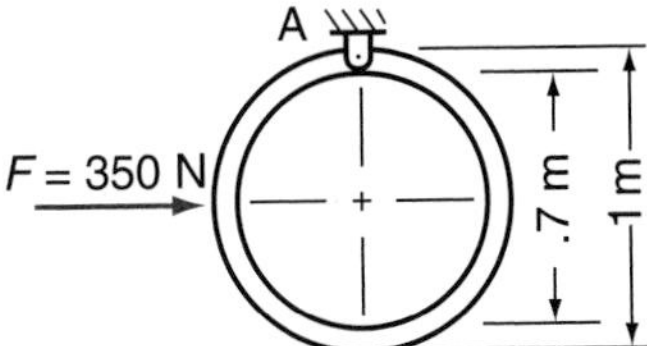

FIGURE P13–50

APPLIED PROBLEMS FOR SECTION 13–6

13–51. What force P is required to accelerate the 4000-lb sewer pipe in Figure P13–51 at 1 ft/s^2 to the right?

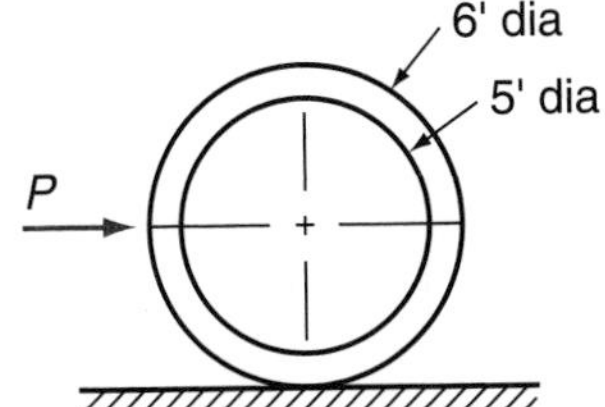

FIGURE P13–51

13–52—13–54. The cylinder and hub in Figures P13–52 to P13–54 have a total mass of 30 kg and a radius of gyration of 0.5 m. Assume no slippage. Determine the acceleration of the mass center of the cylinder and hub when $P = 90$ N.

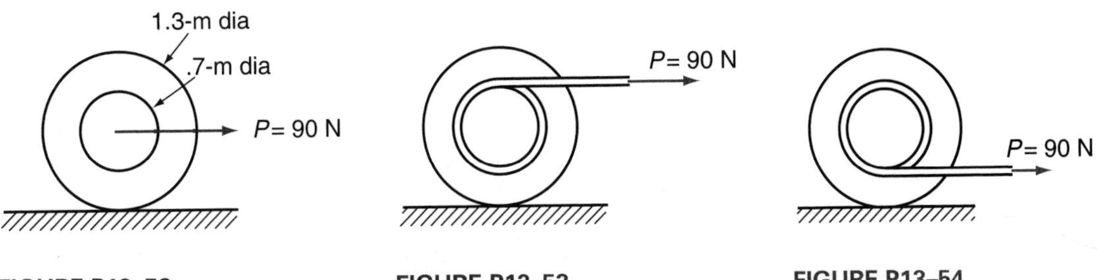

FIGURE P13–52

FIGURE P13–53

FIGURE P13–54

13–55. Cylinder A in Figure P13–55 has a diameter of 0.6 m and a mass of 260 kg. Assume that there is no slippage of A and that mass B is released from rest; determine (a) the tension in the rope and (b) the distance that B will drop in 20 seconds. (Neglect the mass and inertia of the rope and pulley.)

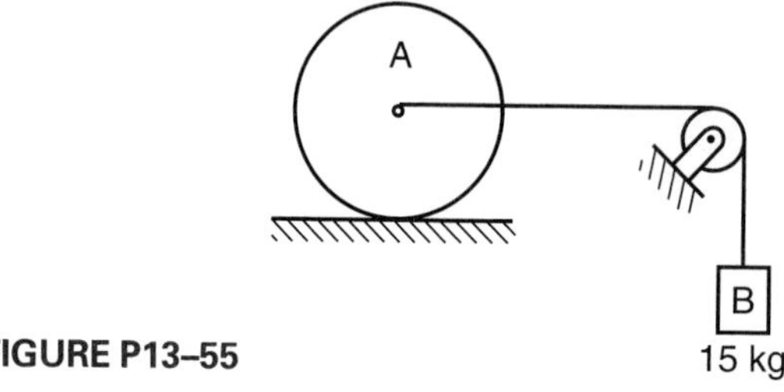

FIGURE P13–55

13–56. The roller-hub system shown in Figure P13–56 has a mass of 3 kg and $I_C = 0.06\ \text{kg}\cdot\text{m}^2$. Assuming no slipping, determine (a) the linear acceleration of the center of the roller and (b) the minimum coefficient of friction.

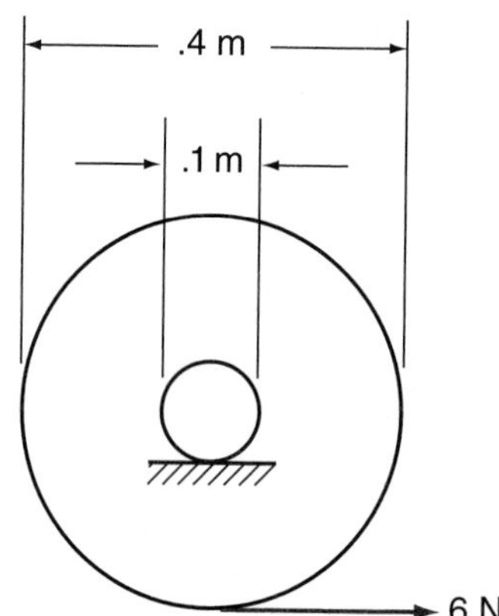

FIGURE P13–56

13–57. Block B weighs 100 lb (Figure P13–57). The mass moment of inertia of rotating part D is 40 ft-lb-s^2. Calculate the required pull P to give part D a counterclockwise angular acceleration of 5 rad/s^2.

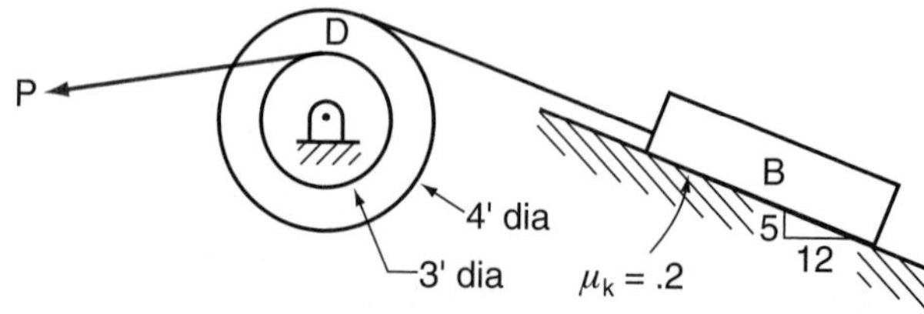

FIGURE P13–57

13–58. Cylinder A weighs 500 lb and is 4 ft in diameter (Figure P13–58). Block B weighs 450 lb. Assuming no slipping and equal linear acceleration for A and B, determine the angular acceleration of cylinder A.

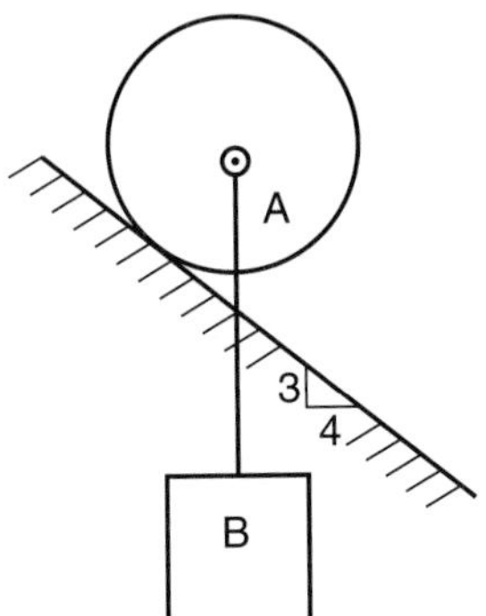

FIGURE P13–58

13–59. The system shown in Figure P13–59 is released from rest. B weighs 500 lb. Cylinder A weighs 900 lb (I_C = 30 ft-lb-s^2). Determine (a) the tension T, (b) the linear acceleration of B, and (c) the angular acceleration of A.

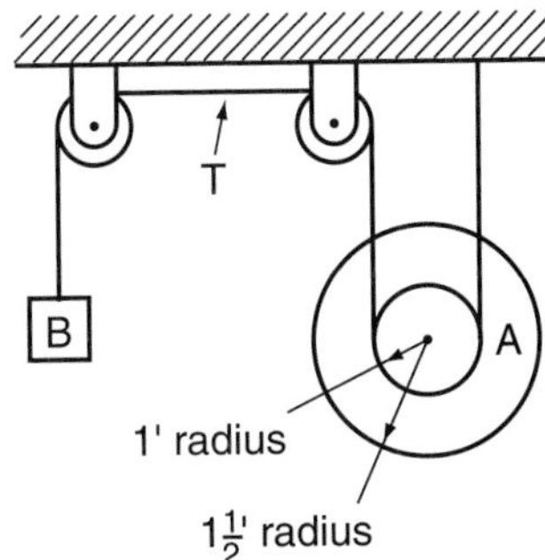

FIGURE P13–59

13–60. The mass of cylinder A is 80 kg (Figure P13–60), and it has a cord wound in a slot as shown. Determine the acceleration of B when the system is released from rest.

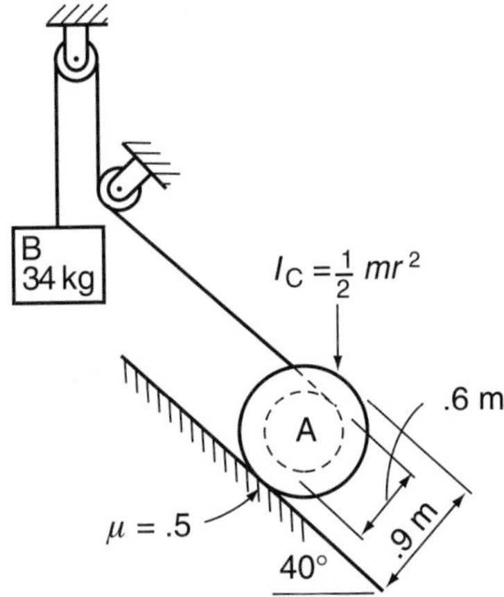

FIGURE P13–60

13–61. The cart shown in Figure P13–61 starts from rest and rolls down the slope. For each wheel, $I_C = 0.8$ ft-lb-s^2, weight = 30 lb, and diameter = 32 in. The body of the cart weighs 60 lb. Assuming no slipping and neglecting friction at the axles, determine the acceleration of the cart down the slope.

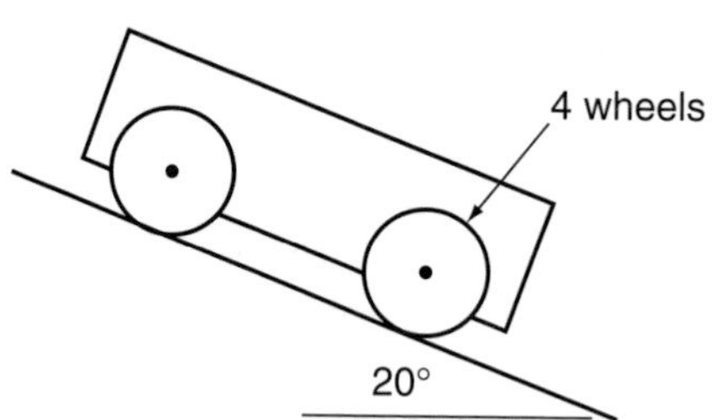

FIGURE P13–61

13–62. The cart shown is accelerated from rest by weight A. The cart body weight is 100 lb. Each of the four wheels weighs 20 lb ($I_C = 0.31$ ft-lb-s^2). Neglect the inertia of the pulleys. Determine the acceleration of the cart (Figure P13–62).

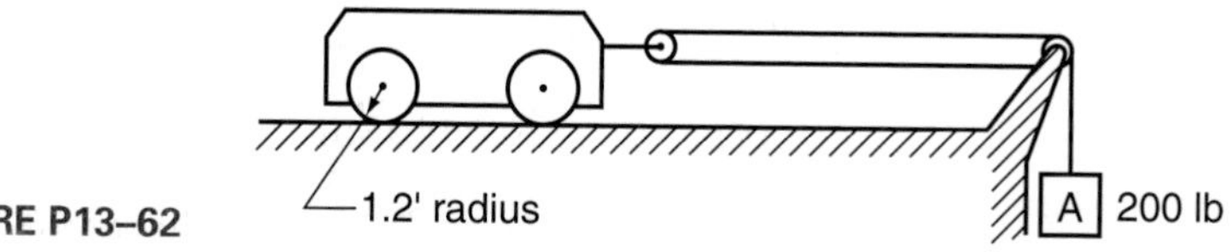

FIGURE P13–62

13–63. The system shown in Figure P13–63 is released from rest. Determine the acceleration of mass A.

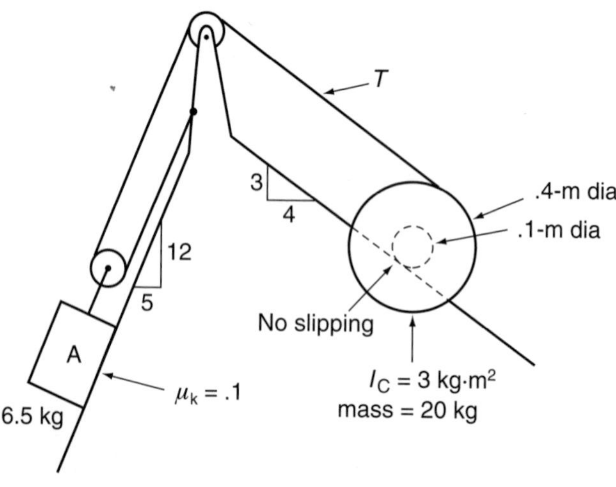

FIGURE P13–63

13–64. Determine the acceleration of block B (Figure P13–64) if the system is released from rest. Assume cylinder A does not slip.

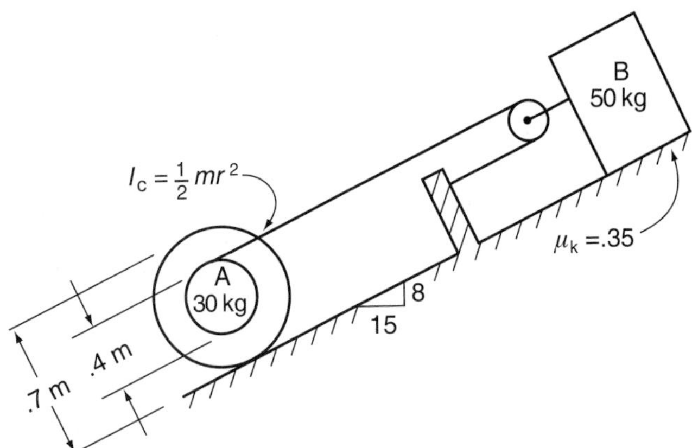

FIGURE P13–64

13–65. A cable is unwound from a reel by anchoring the cable and moving to the left with the reel and its carriage. At the instant shown in Figure P13–65, the reel weighs 3220 lb and has a radius of gyration of 1.8 ft. The reel carriage weighs 1000 lb. Neglect the angular inertia of the carriage wheels and the friction at the axle of the reel. Assume a constant rolling resistance of the carriage of 100 lb. Determine the acceleration of the carriage when a force, $P = 600$ lb, is applied.

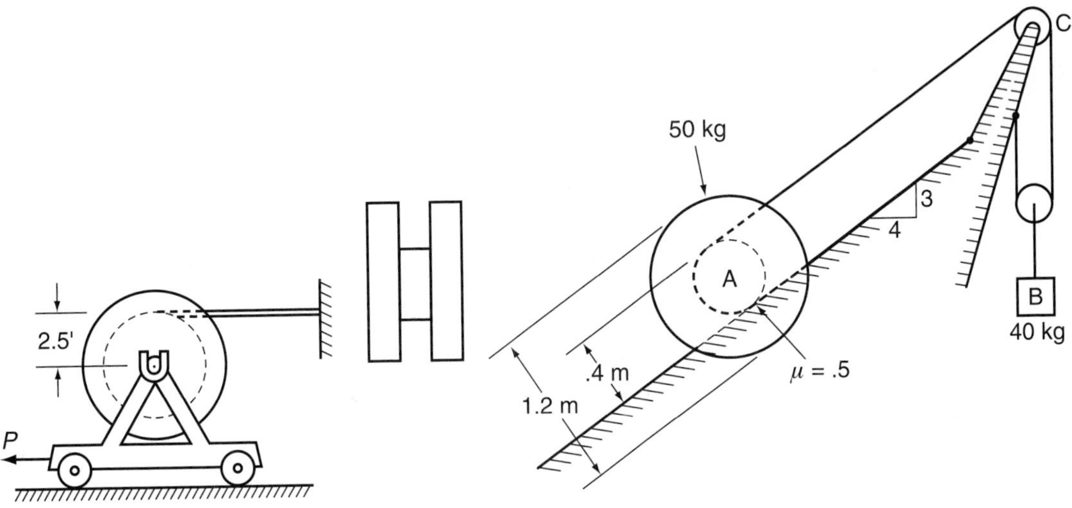

FIGURE P13–65 **FIGURE P13–66**

13–66. The cable supporting mass B (Figure P13–66) is wound around the hub of cylinder A (I_C = 8 kg · m^2). Neglect the inertia of the cable and pulley C. If the system is released from rest, determine the tension in the cable and the acceleration of mass B.

13–67. A cylinder 1.2 m in diameter, having a mass of 40 kg and a rope wrapped around it (Figure P13–67), is released from rest. Determine the velocity of the mass center of the cylinder after it has dropped 2.5 m. What would be the velocity of the mass center of the cylinder if it were dropped without the rope attached?

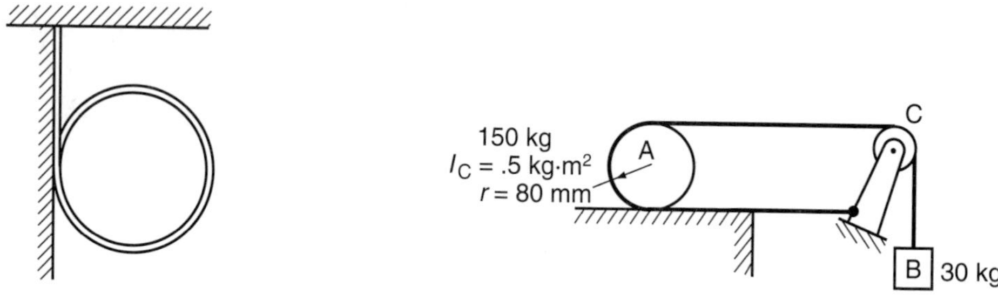

FIGURE P13–67 **FIGURE P13–68**

13–68. Neglect pulley friction and inertia and determine the acceleration of the mass center of cylinder A in Figure P13–68 when the system is released from rest.

REVIEW PROBLEMS

R13–1. The governor shown in Figure RP13–1 regulates the speed of a motor by having the weights shown, rotating on a vertical shaft. Collar C is fixed to the shaft and collar B moves down due to centrifugal action of the weights. The weights at D and E are 0.5 lb each and the weight of the arms can be neglected. For the position shown at 400 rpm, determine the compressive load on the spring.

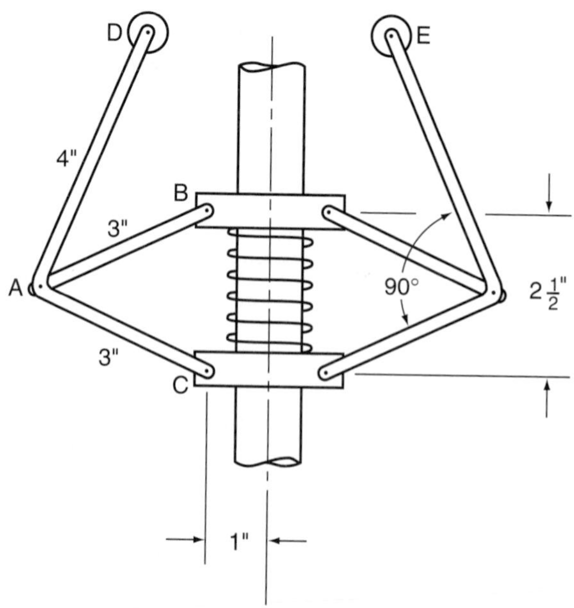

FIGURE RP13–1

R13–2. An amusement ride consists of a large horizontal circular platform turning at 9 rpm. People attempt to sit on this platform without slipping toward the outer edge. The coefficient of static friction is 0.15. Determine the distance from the center of the platform that a 120-lb person may sit before slipping occurs. Does the weight of a person have any effect on this distance?

R13–3. The system shown in Figure RP13–3 has a coefficient of kinetic friction of 0.3 for all surfaces. Neglecting the inertia of the cables and pulleys, determine (a) the linear acceleration of block A and (b) the tension *T*.

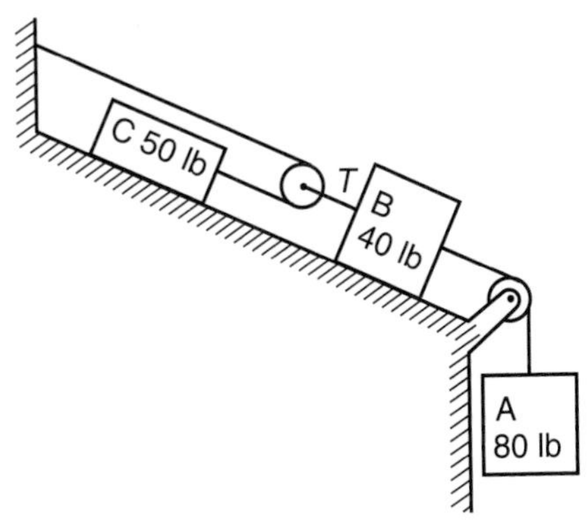

FIGURE RP13–3

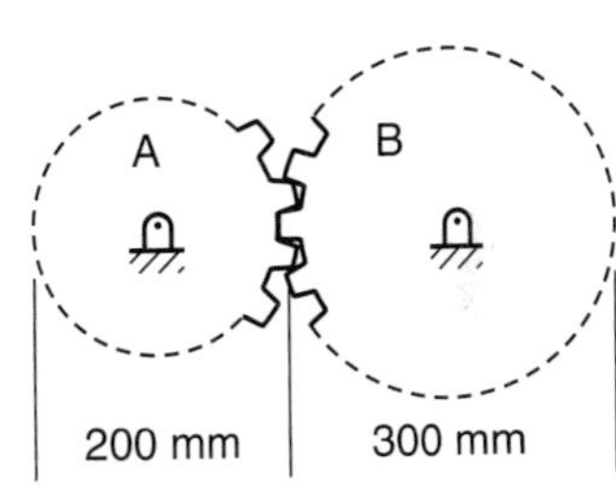

FIGURE RP13–4

R13–4. For gear A, $I = 0.4 \text{ kg} \cdot \text{m}^2$, for gear B, $I = 1.2 \text{ kg} \cdot \text{m}^2$ (Figure RP13–4). If B is to be accelerated at 10 rad/s^2 clockwise, what torque must be applied to gear A?

R13–5. The 20-ft-long horizontal arm of a crane weighs 1500 lb $(I_C = \frac{1}{12}ml^2)$, pivots at one end, and has a center of gravity 10 ft from its end. Calculate the torque required to accelerate this crane arm horizontally at 0.5 rad/s^2.

R13–6. The cylinder-hub system shown in Figure RP13–6 has a mass of 60 kg $(I_C = 1.1 \text{ kg} \cdot \text{m}^2)$. Assume no slipping of the cylinder. Determine (a) the acceleration of the center of the cylinder and (b) the minimum coefficient of static friction for no slipping to occur.

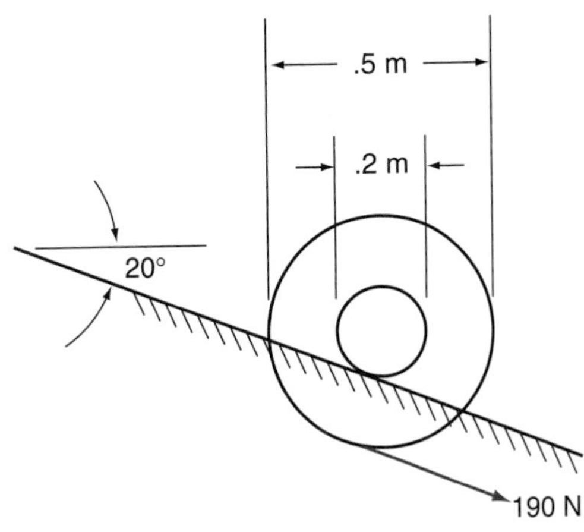

FIGURE RP13–6

R13–7. The cylinder and hub shown in Figure RP13–7 has a total weight of 340 lb ($I_C = 30$ ft-lb-s^2). Assume no slippage. Determine (a) the acceleration of the mass center of the cylinder and hub and (b) the minimum coefficient of friction when $P = 350$ lb.

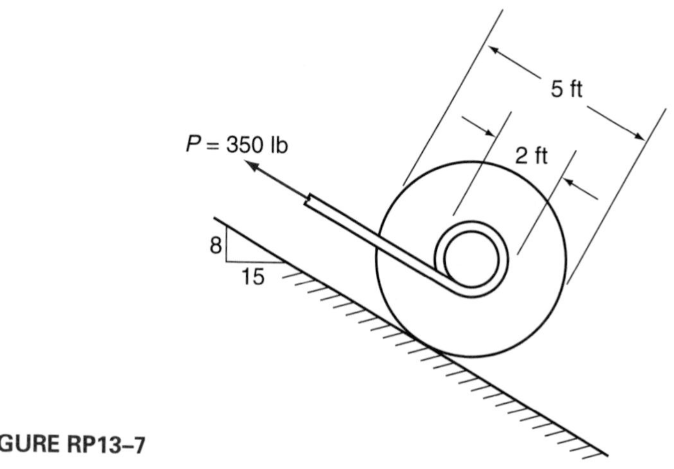

FIGURE RP13–7

R13–8. The 18-kg cylinder in Figure RP13–8, when released from rest, slips on the inclined surface. Determine the angular acceleration of the cylinder ($I_C = 1.84$ kg · m^2).

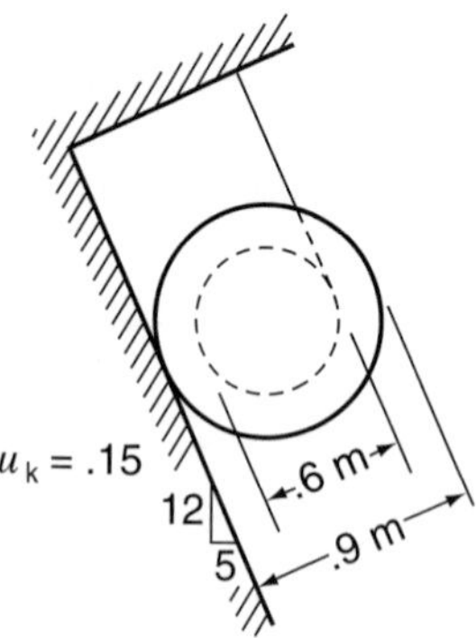

FIGURE RP13–8

R13–9. Determine the linear acceleration of the cylinder shown if the system is released from rest (Figure RP13–9). Assume no slipping of the cylinder and neglect the weights of members AB and AE.

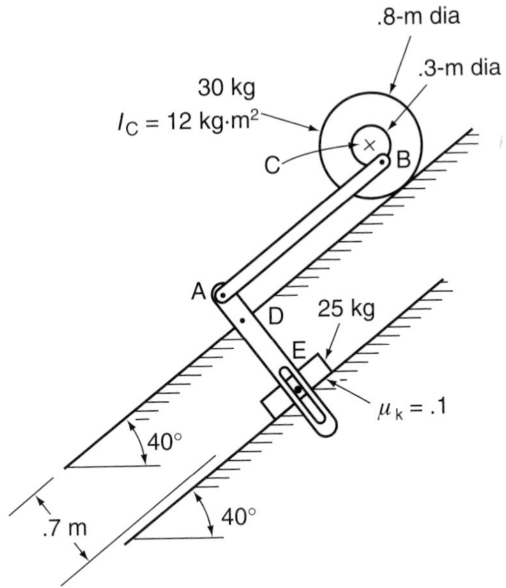

FIGURE RP13–9

R13–10. The 70-kg cylinder ($I_C = 1.3\ \text{kg} \cdot \text{m}^2$) shown in Figure RP13–10 has a 1-m-long bar with a mass of 20 kg pinned at the center of the cylinder. Determine the acceleration of the center of the cylinder if the system is released from rest.

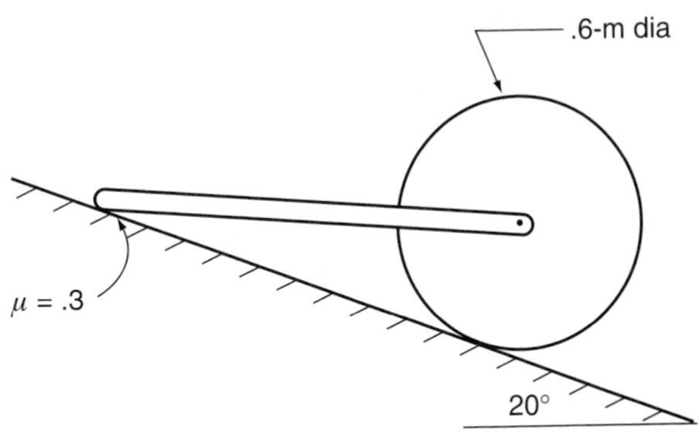

FIGURE RP13–10

CHAPTER 14

Work, Energy, and Power

OBJECTIVES

Upon completion of this chapter the student will be able to:

1. Calculate the work of a constant force.
2. Calculate the work of a variable force such as a spring force.
3. Apply the conservation of energy principles to linear, angular, and plane motion.
4. Calculate power and efficiency.

14–1 INTRODUCTION

In Chapter 13, when we made use of the inertia-force method, we were concerned with force and acceleration. If distance and velocity were further required, we would employ one or two of the three basic kinematic equations (Section 10–5). The work-energy method gives us a direct measure of distance and velocity values. If acceleration is required, the three main kinematic equations are used again.

In Chapter 13 we also looked at bodies or systems of bodies in motion. They were in motion due to an unbalance of forces acting on them. A body in motion possesses *energy*, receiving it from a force F acting on it through a distance s. This quantity of energy we call *work*. We now come to a method that is an accounting process for all quantities of energy. According to the law of the conservation of energy, energy cannot be lost—merely converted from one form to another. Three common types of energy are work, potential energy, and kinetic energy. One or two of these types of energy may initiate motion from one point to another and be converted to another form of energy in the process. Our equations will account for all these energies.

14–2 WORK OF A CONSTANT FORCE

Work (denoted as U) is due to an applied force acting over some distance. The most familiar work would be that shown in Figure 14–1, where a force F moves an object a distance s.

$$\text{work} = (\text{force})(\text{distance})$$

$$U = Fs \tag{14–1}$$

where

U = work; joules (1 J = 1 N·m)
F = force; newtons (N)
s = distance; meters (m)

or in the U.S. Customary system,

U = work; ft-lb
F_1 = force; lb
s = distance; ft

In the SI metric system, with work expressed in units of joules and torque in units of newton-meters (N·m), the units provide more of a differentiation between the terms than is provided by the terms in the U.S. Customary system (ft-lb of work and lb-ft of torque).

If the force is applied as in Figure 14–2, then $U = (F \cos \theta)(s)$ since only the horizontal component causes motion in the direction in which distance s is measured. There is no vertical movement, so $F \sin \theta$ does no work. In these two examples, you will have noted that the force and distance are in the same direction and that the force is constant.

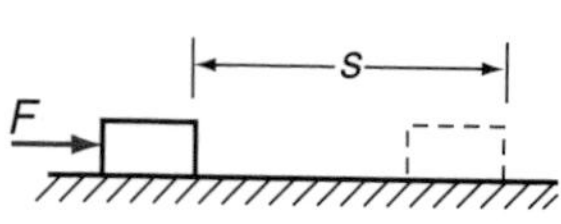

FIGURE 14–1

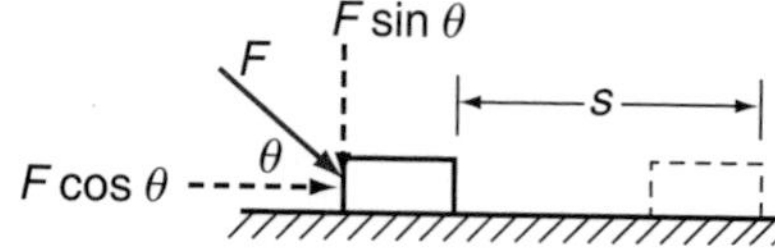

FIGURE 14–2

EXAMPLE 14–1

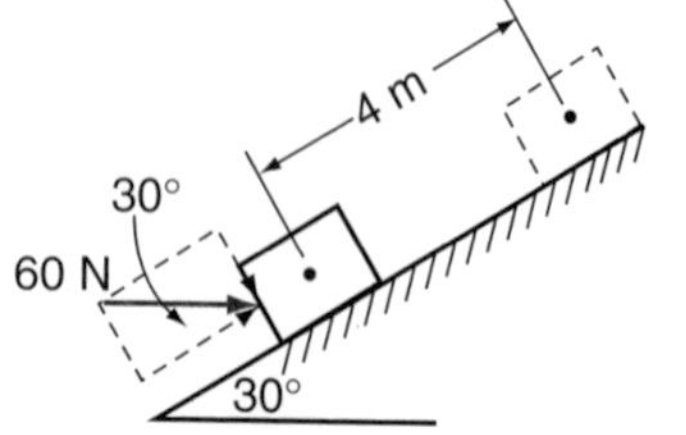

FIGURE 14–3

An 8-kg block is pushed 4 m up a 30° inclined plane by a horizontal force of 60 N. Determine the work done by this force.

Using the forces and the distance shown in Figure 14–3, we get

$$U = (60\text{ N})(\cos 30°)(4\text{ m})$$
$$= 52(4)$$
$$\underline{U = 208\text{ J}}$$

EXAMPLE 14–2

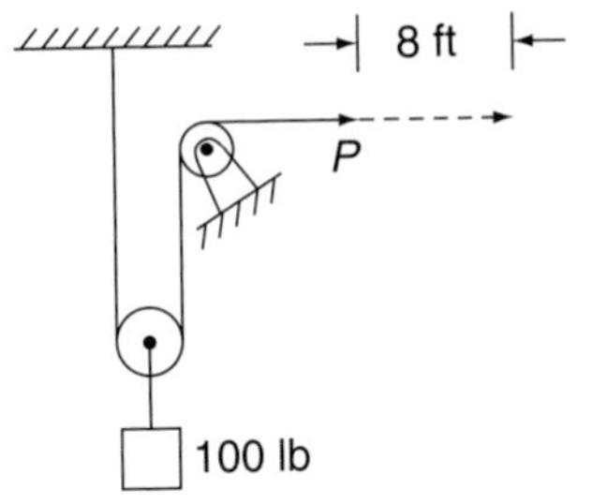

FIGURE 14–4

Force P slowly lifts the 100-lb weight as it moves 8 ft to the right. [Figure 14-4] Determine the work done.

There are two equally valid ways of calculating the work done.

Method 1

By observing the bottom pulley tension

$$P = 100/2 = 50 \text{ lb}$$

Therefore

$$U = Fs$$
$$= (50 \text{ lb})(8 \text{ ft})$$
$$\underline{U = 400 \text{ ft-lb}}$$

Method 2

Observing the bottom pulley once again, because there are two supporting cables the distance moved by the 100 lb is 8 ft/2 = 4 ft.

Therefore

$$U = Fs$$
$$= (100 \text{ lb})(4 \text{ ft})$$
$$\underline{U = 400 \text{ ft-lb}}$$

14–3 WORK OF A VARIABLE FORCE

Work is not always due to a constant force acting through a distance. There are many practical applications of work resulting from a force that varies as it moves through distance s. To describe some of the more complex variations of force requires either a calculus approach or the drawing of nonlinear curves on a force-displacement diagram. We will not discuss these more complex variable forces but will limit ourselves to forces that vary in a linear fashion.

The force required to stretch or compress most springs is a force that varies linearly. For example, if a force of 10 lb compresses a spring 2 in., then a force of 30 lb compresses that spring 6 in. This relationship expressed in equation form is

$$F = kx \tag{14–2}$$

where

F = force exerted on the spring; lb

k = spring constant; lb/in.

x = change in length of the spring; in., as measured from its unloaded or free length

The corresponding SI units are

F = spring force; newtons

k = spring constant; newtons/meter (N/m)

x = change in length; meters (m)

EXAMPLE 14–3

What is the spring constant of a spring that is compressed 50 mm by a force of 25 N? What force is required to compress it a total of 125 mm?

$$F = kx$$

$$25\text{ N} = k(0.05\text{ m}) \qquad \text{(using length in meters)}$$

$$\underline{k = 500\text{ N/m}}$$

$$F = kx$$

$$= (500\text{ N/m})(0.125\text{ m})$$

$$\underline{F = 62.5\text{ N}}$$

Consider now the work required to compress the spring in Figure 14–5. The spring is compressed a distance x. The force acting through this distance is varied from 0 to the final force F. Therefore, the average force over this distance is $F/2$, but $F = kx$ and $F/2 = kx/2$. Substituting into the equation

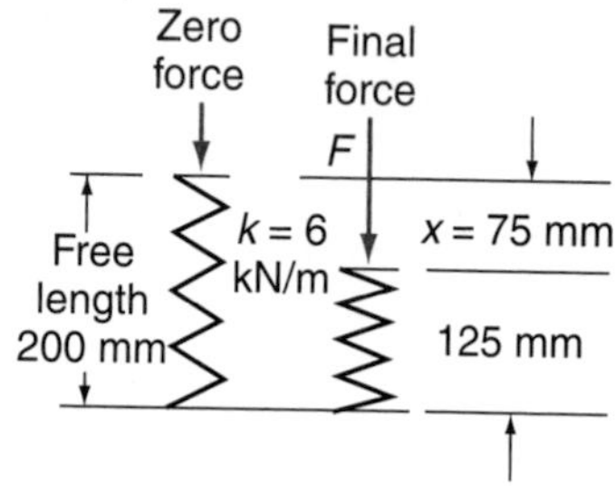

FIGURE 14–5

$$\text{work} = (\text{average force})(\text{distance})$$

$$\text{work} = \left(\frac{F}{2}\right)(x)$$

or

$$\text{spring work} = \frac{kx}{2}(x)$$

$$U = \frac{1}{2}kx^2 \tag{14–3}$$

Using the values in Figure 14–5, we get

$$U = \frac{1}{2}(6000\text{ N/m})(0.075\text{ m})^2$$

$$U = 16.9\text{ J}$$

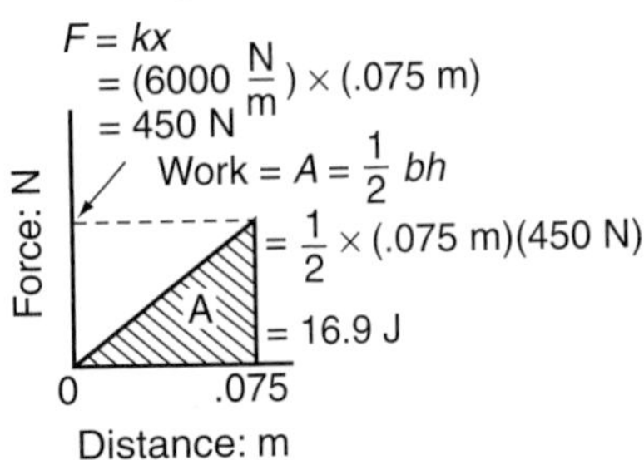

FIGURE 14–6

Another way of expressing this equation and its corresponding calculations is on a work diagram such as Figure 14–6, where the area on the diagram is equal to the work. For larger springs, the spring constant may be kN/m. Although units of meters may seem awkward to use for a spring that has a deflection of only a few mm, meters are consistent with the SI convention and will yield spring work (Equation 14–3) in joules.

EXAMPLE 14–4

A spring has a spring constant of 80 lb/in. and a free length of 10 in. The spring is stretched to point A, where its total length is 13 in. What work would now be required to stretch it to point B, where it would be 15 in. in length?

The spring is stretched 3 in. and then 5 in. beyond its free length. The work required to stretch the spring from 3 in. to 5 in. is the difference between the work to stretch it 5 in. and the work required to stretch it 3 in., or

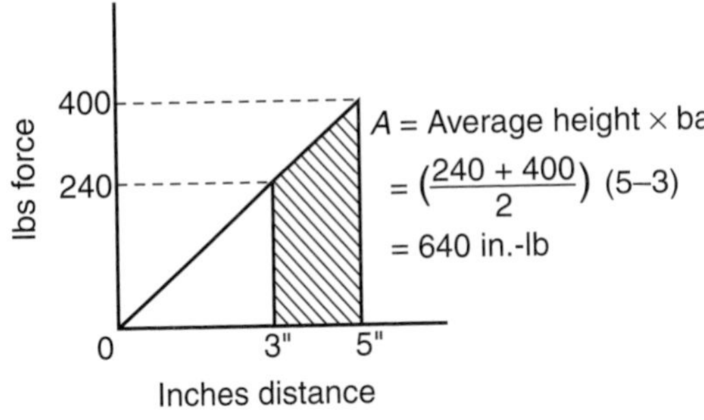

FIGURE 14–7

$$U = \frac{1}{2}kx_2^2 - \frac{1}{2}kx_1^2$$

$$= \frac{1}{2}k(x_2^2 - x_1^2)$$

$$= \frac{1}{2}(80\ \text{lb/in.})[(5\ \text{in.})^2 - (3\ \text{in.})^2]$$

$$\underline{U = 640\ \text{in.-lb}}$$

This work value could also have been shown on the work diagram, Figure 14–7. Note that the units of work are in.-lb and may often have to be converted to ft-lb in further problem calculations.

EXAMPLE 14–5

An elevator starts at the first floor with a load of eight crates at 600 N each. As the elevator goes up, a crate is left at each floor. The first crate is left at the second floor and the final crate at the ninth floor, with the elevator reaching the tenth floor empty. The height between floors is 3 m and the elevator weighs 7 kN. Determine the minimum work done in lifting the elevator and crates.

The elevator makes nine stops while traveling a distance of 9(3) = 27 m.

$$\begin{aligned}
\text{work of lifting empty elevator} &= Fd \\
&= (7000\ \text{N})(27\ \text{m}) \\
&= 184{,}000\ \text{J} \\
\text{maximum crate weight} &= 4800\ \text{N} \\
\text{average crate weight} &= \frac{4800 + 0}{2} = 2400\ \text{N} \\
\text{work of lifting crates} &= (\text{average force})(\text{distance}) \\
&= (2400\ \text{N})(27\ \text{m}) \\
&= 64{,}800\ \text{J} \\
\text{total work} &= 184{,}000\ \text{J} + 64{,}800\ \text{J} \\
\underline{U} &\underline{= 249\ \text{kJ}}
\end{aligned}$$

14–4 POTENTIAL AND KINETIC ENERGY: TRANSLATIONAL

A weight being lowered from one level to another has the ability or potential to do work since it has a force (weight) acting through a distance (height). For this reason, the object is said to have *potential energy* (PE): It has the potential either to do work or to convert its energy into another form, such as heat or motion.

$$\text{PE} = Wh \tag{14–4}$$

where

$$\begin{aligned}
\text{PE} &= \text{potential energy; joules (J)} \\
W &= \text{force of gravity (or weight); N} \\
h &= \text{vertical height; m}
\end{aligned}$$

and in the U.S. Customary system,

$$\begin{aligned}
\text{PE} &= \text{potential energy; ft-lb} \\
W &= \text{weight; lb} \\
h &= \text{vertical height; ft}
\end{aligned}$$

The reference level at which we measure height is arbitrary since we are only concerned with a change in height or potential energy. For this reason, if an object moves from one level to another, we often use either one of these levels as a reference datum. When an object moves up against the weight, the weight's potential energy increases. As the weight falls, it loses potential energy.

EXAMPLE 14–6

A 1000-lb elevator moves upward from the tenth floor to the fourteenth floor, a distance of 44 ft. What is the increase in the potential energy of the elevator?

Using the tenth floor as the reference datum from which we measure height, we have

$$\text{PE} = Wh$$
$$= (1000\text{ lb})(44\text{ ft})$$
$$\underline{\text{PE} = 44{,}000\text{ ft-lb}}$$

The potential energy of the elevator increased by 44,000 ft-lb; therefore, a corresponding amount of work must have been done to raise the elevator to the fourteenth floor.

EXAMPLE 14–7

While driving through a valley, a 1600-kg car has a resulting elevation drop of 400 m. Determine its decrease in potential energy.

The force of gravity or weight

$$W = (1600\text{ kg})(9.81\text{ m/s}^2)$$
$$W = 15.7\text{ kN}$$
$$\text{PE} = Wh$$
$$= (15.7 \times 10^3\text{ N})(400\text{ m})$$
$$= 6{,}280{,}000\text{ J}$$
$$\underline{\text{PE} = 6.28\text{ MJ}}$$

Work or energy is required to start an object moving—and to stop a moving object as well. The energy of a moving object is *kinetic energy.* Consider now the work required to accelerate an object from an initial velocity v_0 to some final velocity v.

$$U = Fs \quad \text{and} \quad F = ma$$

Therefore,

$$U = mas \tag{1}$$

but

$$v^2 = v_0^2 + 2as \quad \text{or} \quad a = \frac{v^2 - v_0^2}{2s} \tag{2}$$

Substituting Equation (2) into Equation (1), we get

$$U = m\left(\frac{v^2 - v_0^2}{2s}\right)s$$

$$U = \frac{1}{2}m(v^2 - v_0^2)$$

The kinetic energy change for a given speed change is

$$\text{KE} = \frac{1}{2}m(v^2 - v_0^2) \tag{14–5}$$

For an initial velocity of zero,

$$\text{KE} = \frac{1}{2}mv^2$$

where, in the SI system,

KE = kinetic energy; J

m = mass; kg

v = velocity; m/s

and in the U.S. Customary system,

KE = kinetic energy; ft-lb

m = mass; slugs or $\text{lb}/(\text{ft/s}^2) = \dfrac{\text{lb} \cdot \text{s}^2}{\text{ft}}$

v = velocity; ft/s

This equation does not apply to rotational motion; it applies only to translational motion. (Translational motion may be either rectilinear or curvilinear.)

EXAMPLE 14–8

A "slap shot" in the game of hockey can cause an increase in the velocity of a 5-oz puck from 10 mph to 100 mph (60 mph = 88 ft/s). What would be the corresponding increase in the kinetic energy of the puck (1 lb = 16 oz)?

$$v_0 = \frac{10}{60}(88 \text{ ft/s}) = 14.67 \text{ ft/s}$$

$$v = \frac{100}{60}(88 \text{ ft/s}) = 146.7 \text{ ft/s}$$

$$\text{KE} = \frac{1}{2}m(v^2 - v_0^2)$$

$$= \frac{1}{2}\left(\frac{(5/16 \text{ lb})}{(32.2 \text{ ft/s}^2)}\right)[(146.7 \text{ ft/s})^2 - (14.67 \text{ ft/s})^2]$$

$$\underline{\text{KE} = 100 \text{ ft-lb}}$$

EXAMPLE 14–9

A 100-g sample in a centrifuge is rotated at a speed of 3600 rpm. If the sample is 200 mm from the center, what is its kinetic energy?

In this case the translational velocity follows the arc of a circle (tangential velocity) rather than a straight line.

$$v = r\omega$$

$$= (0.2 \text{ m})\left[\frac{(2\pi \text{ rad/rev})(3600 \text{ rev/min})}{60 \text{ s/min}}\right]$$

$$v = 75.4 \text{ m/s}$$

$$\text{KE} = \frac{1}{2}mv^2$$

$$= \frac{1}{2}(0.1 \text{ kg})(75.4 \text{ m/s})^2$$

$$\underline{\text{KE} = 284 \text{ J}}$$

14–5 CONSERVATION OF ENERGY: TRANSLATIONAL

The mechanical forms of energy that we have looked at are work, potential energy, and kinetic energy. Another form of energy, perhaps not so obvious, is the work of the force of friction acting through a distance. This energy is, in turn, dissipated as heat.

By *conservation of energy,* we mean that no matter what motion a given system has, the total initial energy equals the total final energy; no matter what form the energy takes or changes to, all quantities of energy must be accounted for. With the accounting system that we will use here, we can handle systems that are decelerating, accelerating, and at constant velocity; we can even deal with the change of position of a system of objects that has both initial and final velocities of zero.

The method consists of considering only initial and final conditions and writing an energy equation—the intermediate values or conditions are of no concern. The first step of our method of attack will be to analyze the motion of the system and to ask ourselves, "What is the change in energy that causes the resulting motion?" Some object or portion of a system causes motion; the energy of this portion is converted into other forms of energy. We therefore equate this activating energy with all the other amounts of energy into which it has been converted in the system.

For example, consider a block that slides down a plane and reaches the bottom with some velocity, v (Figure 14–8). The amount of energy that caused this motion is accounted for by the loss of potential energy of the block since the weight acted vertically over a height, h. This energy was converted to or reappears as kinetic energy at point (2); and the work of the friction force dissipated as heat over distance, s.

$$\Delta PE = \Delta KE + \text{energy lost to friction}$$

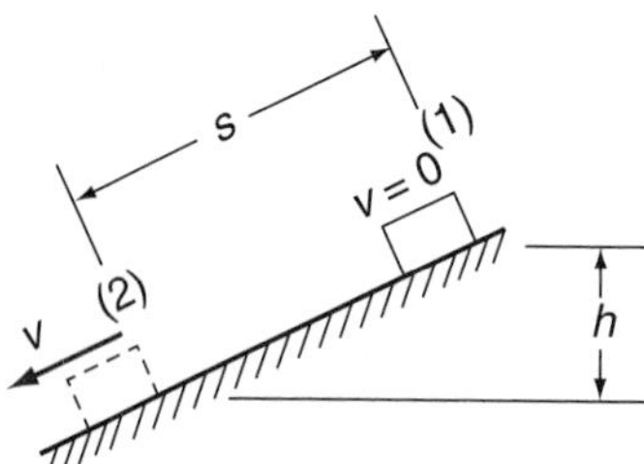

FIGURE 14–8

EXAMPLE 14–10

A 100-lb cart starts up a 30° slope with a velocity of 20 ft/s (Figure 14–9). If it has a constant rolling resistance of 5 lb, determine the distance it will travel on the slope before coasting to a stop. (Neglect the angular kinetic energy of the wheels.)

Rolling resistance is similar to the resistance of a friction force, although it is much less.

The activating energy is the initial kinetic energy of the cart, which is then dispersed to the work of the rolling resistance and the increase in potential energy.

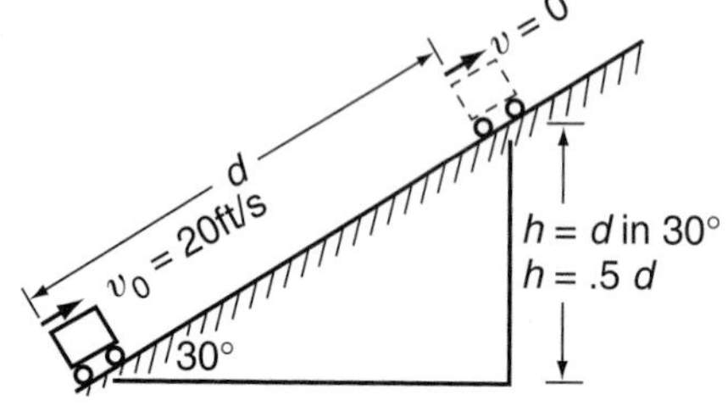

FIGURE 14–9

$$\text{activating energy} = \text{resulting energies}$$

$$\text{KE of cart} = \text{rolling resistance work} + \text{PE gain}$$

$$\frac{1}{2}\left(\frac{W}{g}\right)(v_0^2 - v^2) = (F)(d) + (W)(h) \qquad \text{(Figure 14–9)}$$

$$\frac{1}{2}\left(\frac{100\text{ lb}}{32.2\text{ ft/s}^2}\right)[(20\text{ ft/s})^2 - 0^2] = (5\text{ lb})d + (100\text{ lb})(.5d)$$

$$621 = 55d$$

$$\underline{d = 11.3\text{ ft}}$$

EXAMPLE 14–11

For the system shown in Figure 14–10, B weighs 200 lb and A weighs 50 lb. The spring is at its free length and has a spring constant of 25 lb/ft. If the system is initially at rest and the weight and inertia of the cable and pulley are neglected, determine the velocity of weight B after it has dropped 3 ft.

The activating energy that causes the motion results from the loss of potential energy of B. Equating this energy with all the other forms of energy into which it is dispersed, we have

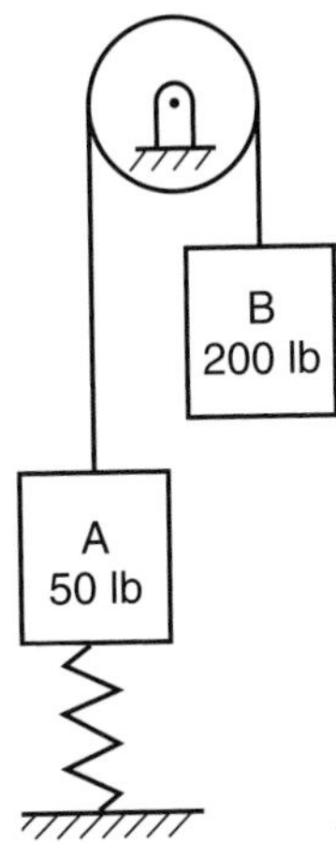

FIGURE 14–10

$$\text{activating energy} = \text{resulting energies}$$

$$\text{PE loss of B} = \text{KE gain of B} + \text{KE gain of A} + \text{PE gain of A} + \text{spring work}$$

$$W_B \times h = \left(\frac{1}{2}m_B v^2\right) + \left(\frac{1}{2}m_A v^2\right) + (W_A)(h) + \left(\frac{1}{2}ks^2\right)$$

$$(200\text{ lb})(3\text{ ft}) = \left[\frac{1}{2}\left(\frac{200\text{ lb}}{32.2\text{ ft/s}^2}\right)v^2\right] + \left[\frac{1}{2}\left(\frac{50\text{ lb}}{32.2\text{ ft/s}^2}\right)v^2\right] + (50\text{ lb})(3\text{ ft}) + \left[\frac{1}{2}(25\text{ lb/ft})(3\text{ ft})^2\right]$$

$$\underline{v = 9.33\text{ ft/s}\downarrow}$$

You may have noted that the answer was expressed in ft and seconds since the units of g are ft/s^2.

EXAMPLE 14–12

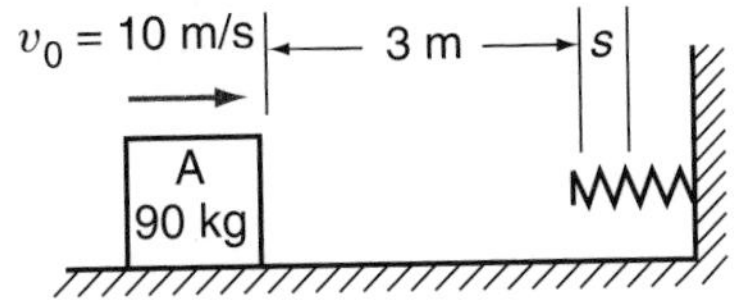

FIGURE 14–11

Block A in Figure 14–11 has a mass of 90 kg and an initial velocity of 10 m/s to the right. The spring constant is 12 kN/m, and the coefficient of kinetic friction is 0.15. How much will the spring deflect in bringing the block to rest?

From Figure 14–12, the force of friction is

$$F = \mu N$$

$$F = 0.15\,(883\text{ N}) = 132\text{ N}$$

Activating energy is equal to resulting energy.

$$\text{KE loss of A} = \text{energy lost to friction} + \text{spring work}$$

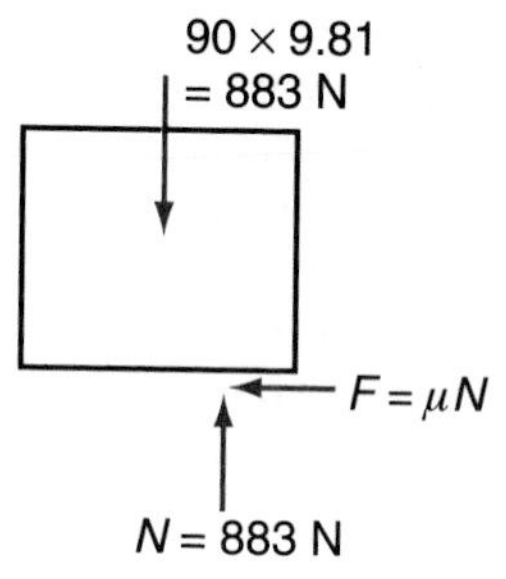

FIGURE 14–12

$$\frac{1}{2}mv^2 = F(3 + s) + \frac{1}{2}ks^2$$

$$\frac{1}{2}(90\text{ kg})(10\text{ m/s})^2 = (132\text{ N})(3\text{ m} + s) + \frac{1}{2}(12{,}000\text{ N/m})s^2$$

$$45.45s^2 + s - 31.1 = 0$$

$$s = \frac{-1 \pm \sqrt{(1)^2 - (4)(45.45)(-31.1)}}{2(45.45)}$$

$$\underline{s = 0.816\text{ m}}$$

EXAMPLE 14–13

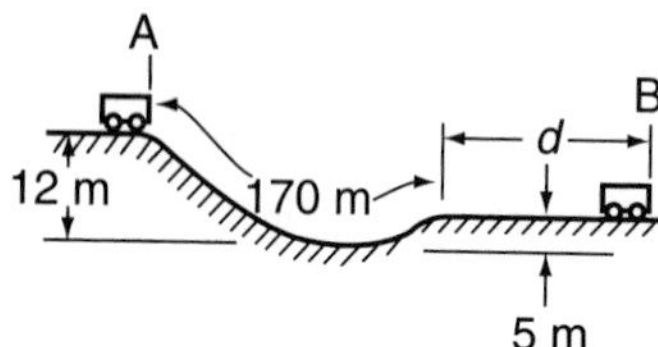

FIGURE 14–13

A 1300-kg car starts from rest at A, coasts 170 m through a dip, and comes to a stop at B after coasting a distance d on level ground (Figure 14–13). If the rolling resistance is a constant force of 220 N, determine the distance d.

The kinetic energy is zero at both initial and final conditions, and we do not have to be concerned with the velocity reached at some intermediate stage. What we do have to realize is that a loss of potential energy was the activating energy; the potential energy was dispersed as work, overcoming the rolling resistance force.

$$\text{PE loss} = \text{work of rolling resistance}$$

$$(1300\text{ kg})(9.81\text{ m/s}^2)(12\text{ m} - 5\text{ m}) = (170\text{ m} + d)220\text{ N}$$

$$\underline{d = 236\text{ m}}$$

14–6 KINETIC ENERGY: ANGULAR

Our equation for rectilinear kinetic energy is $\text{KE} = \frac{1}{2}mv^2$. We will now convert this equation into angular terms by considering Figure 14–14. There, a mass m rotates about point A with an angular velocity ω. Consider mass Δm at a distance r from the center of rotation A, and write the kinetic energy equation as

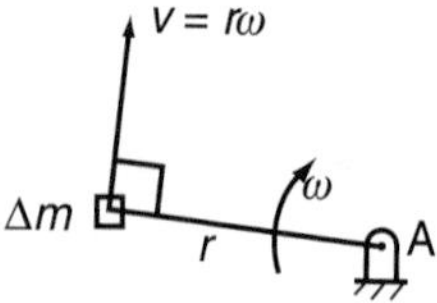

FIGURE 14–14

$$KE = \frac{1}{2}\Delta m(r\omega)^2$$

The kinetic energy of the total mass m is

$$KE = \Sigma\frac{1}{2}mr^2\omega^2$$

$$KE = \frac{\omega^2}{2}\Sigma mr^2$$

But as we saw in Section 9–5, mass moment of inertia is given by

$$I = \Sigma mr^2$$

Therefore, for angular kinetic energy,

$$KE = \frac{1}{2}I\omega^2 \qquad (14\text{–}6)$$

where, in the SI system,

KE = kinetic energy; joules
ω = angular velocity; rad/s
I = mass moment of inertia about the center of rotation; kg·m^2

and in the U.S. Customary system,

KE = kinetic energy; ft-lb
ω = angular velocity; rad/s
I = mass moment of inertia about the center of rotation, slug-ft^2 or ft-lb-s^2

Tables and formulas for mass moment of inertia are usually set up, taking as their reference the center of mass of the object. For cases in which the center of mass and the center of rotation are not coincident, we will consider the total kinetic energy as being equal to the angular KE, using the mass moment of inertia (I_C) about the center of mass, plus the rectilinear KE of the center of mass. In Section 14–8 we cover this in more detail; examples will show what we mean by this.

EXAMPLE 14–14

A 96.6-lb rotor is rotated at 1500 rpm while on a dynamic balancing machine. Assume the rotor to be a cylinder with a radius of 6 in. and determine its kinetic energy at this speed.

From Table 9–2,

$$I_C = \frac{1}{2}mr^2$$

$$= \frac{1}{2}\left(\frac{96.6 \text{ lb}}{32.2 \text{ ft/s}^2}\right)(0.5 \text{ ft})^2$$

$$I_C = 0.375 \text{ ft-lb-s}^2$$

$$\text{KE} = \frac{1}{2}I_C\omega^2$$

$$= \frac{1}{2}(0.375 \text{ ft-lb-s}^2)\left[\frac{(1500 \text{ rev/min})(2\pi \text{ rad/rev})}{60 \text{ s/min}}\right]^2$$

$$\underline{\text{KE} = 4620 \text{ ft-lb}}$$

EXAMPLE 14–15

A 150-mm-diameter shaft is being turned on a lathe at 80 rpm. If the shaft mass is 210 kg, determine its kinetic energy.

From Table 9–2,

$$I_C = \frac{1}{2}mr^2$$

$$= \frac{1}{2}(210 \text{ kg})(0.075 \text{ m})^2$$

$$I_C = 0.591 \text{ kg} \cdot \text{m}^2$$

$$\text{KE} = \frac{1}{2}I_C\omega^2$$

$$= \frac{1}{2}(0.591 \text{ kg} \cdot \text{m}^2)\left[\frac{(80 \text{ rev/min})(2\pi \text{ rad/rev})}{60 \text{ s/min}}\right]^2$$

$$\underline{\text{KE} = 20.7 \text{ J}}$$

Another form of energy often encountered in rotational motion is when a torque is applied during a given number of revolutions. In Figure 14–15, a force of 300 N is applied to a wheel with a radius of 0.5 m as the wheel turns through three revolutions. The work could be calculated in one of two ways:

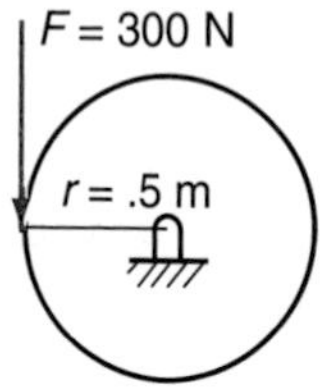

$F = 300$ N

$r = .5$ m

FIGURE 14–15

1. $U = Fs$

$$= F(3 \text{ rev})(\text{circumference})$$

$$= 300(3 \text{ rev})(2\pi \text{ rad/rev})(0.5 \text{ m})$$

$$\underline{U = 2830 \text{ J}}$$

2. $U = Fs$ but $s = r\theta$

$$U = Fr\theta \quad \text{but torque } T = Fr$$

$$U = \text{torque}(\theta) \qquad (14\text{–}7)$$

where

$T =$ torque, N·m (or lb-ft)
$s =$ distance; radians
$U =$ work, J (or ft-lb)

In this case

$$U = [(300\text{ N})(0.5\text{ m})][(3\text{ rev})(2\pi\text{ rad/rev})]$$
$$\underline{U = 2830\text{ J}}$$

EXAMPLE 14–16

A torque of 250 lb-in. is transmitted by a shaft rotating at 800 rpm. What is the work in ft-lb transmitted by the shaft in 1 minute?

$$U = \text{torque}(\theta)$$
$$= \frac{250\text{ lb-in.}}{12\text{ in./ft}}(800\text{ rev/min})(2\pi\text{ rad/rev})$$
$$\underline{U = 104{,}700\text{ ft-lb/min}}$$

14–7 CONSERVATION OF ENERGY: ANGULAR

We must account for all energies in rotational motion—just as we did in translational motion. In this case, the list of possible energies will be:

- spring work
- energy lost due to friction
- potential energy change
- translational kinetic energy $\frac{1}{2}mv^2$

with the addition of:

- angular kinetic energy $\frac{1}{2}I_C\omega^2$

Since both translational and angular motions are occurring, they can be related to one another using the equations $s = r\theta$, $v = r\omega$, and $a_t = r\alpha$.

EXAMPLE 14–17

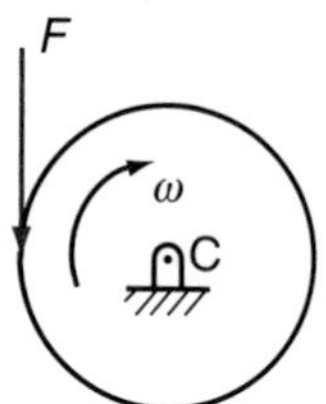

FIGURE 14–16

A wheel with a radius of 15 in. ($I_C = 0.8$ ft-lb-s^2) is braked by a force applied at its outer diameter (Figure 14–16). If the wheel speed is reduced from 6 rad/s to 2 rad/s as it turns three revolutions, determine the braking force.

The activating energy is accounted for by the loss of kinetic energy of the wheel, and this energy is completely absorbed by the braking force.

$$\frac{1}{2}I(\omega_2^2 - \omega_1^2) = \text{torque}(\theta)$$

$$\frac{1}{2}(0.8\text{ ft-lb-s}^2)[(6\text{ rad/s})^2 - (2\text{ rad/s})^2] = \left[F\left(\frac{15}{12}\right)\text{ft}\right][(3\text{ rev})(2\pi\text{ rad/rev})]$$

$$\underline{F = 0.544\text{ lb}}$$

Let us now solve Example 13–9 using the conservation of energy method instead of the angular inertia and torque method used previously.

The given information is repeated here for easy reference.

EXAMPLE 14–18

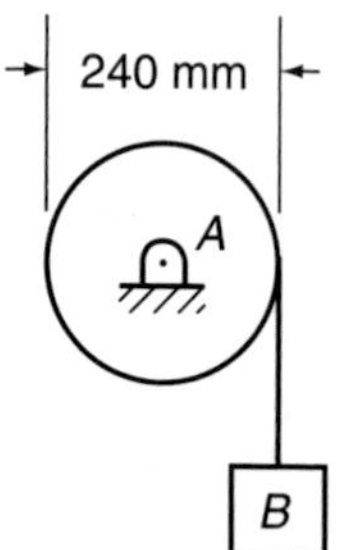

FIGURE 14–17

Wheel A has a mass of 22 kg and a radius of gyration of 180 mm. The mass of B is 10 kg. If the system starts from rest and has no bearing friction, determine the angular acceleration of A and the tension in the rope (Figure 14–17).

Because the conservation of energy method uses velocities, we will solve for velocity and then acceleration. Weight B will drop and we can assume a convenient distance such as 1 meter.

The potential energy loss of B is distributed to linear kinetic energy of B and angular kinetic energy of A.

$$\text{PE}_{\text{loss of B}} = \Delta\text{KE}_\text{B} + \Delta\text{KE}_\text{A}$$

$$Wh = \frac{1}{2}mv^2 + \frac{1}{2}I\omega^2$$

where $\quad v = r\omega = (0.12\text{ m})\omega$

and $\quad \text{I} = k^2m$

$$(10\text{ kg})(9.81\text{ m/s}^2)(1\text{ m}) = \left(\frac{1}{2}\right)(10\text{ kg})(0.12\omega)^2 + \left(\frac{1}{2}\right)(0.18\text{ m})^2(22\text{ kg})\omega^2$$

$$\underline{\omega = 15.1\text{ rad/s} \circlearrowleft}$$

Now list the known values for wheel A.

$$\omega_0 = 0$$
$$\omega = 15.1 \text{ rad/s}$$
$$\theta = \frac{s}{r} = \frac{1 \text{ m}}{0.12 \text{ m}} = 8.33 \text{ rad}$$
$$\omega^2 = \omega_0^2 + 2\alpha\theta$$
$$(15.1 \text{ rad/s})^2 = 0 + (2)(\alpha)(8.33 \text{ rad})$$
$$\underline{\alpha = 13.7 \text{ rad/s}^2 \circlearrowright}$$

To solve for tension *T*, consider block B alone.

$$\text{PE loss of B} = \Delta\text{KE}_\text{B} + \text{rope tension work}$$
$$Wh = \frac{1}{2}mv^2 + T \times h$$

where

$$v = r\omega$$
$$= (.12 \text{ m})(15.1 \text{ rad/s})$$
$$= 1.81 \text{ m/s}$$
$$(10 \text{ kg})(9.81 \text{ m/s}^2)1\text{m} = \left(\frac{1}{2}\right)(10 \text{ kg})(1.81 \text{ m/s})^2 + T(1 \text{ m})$$
$$\underline{T = 81.7 \text{ N}}$$

EXAMPLE 14–19

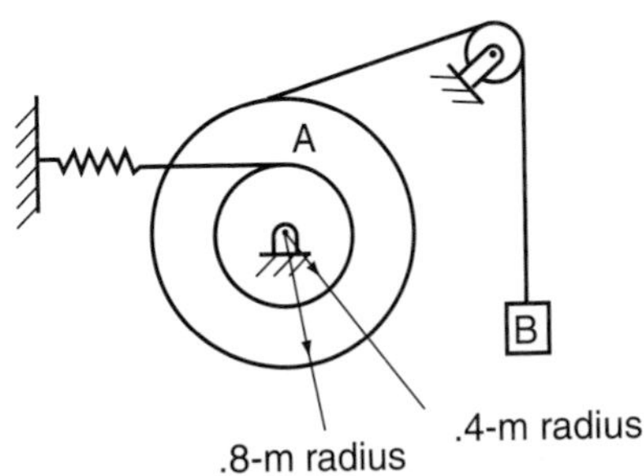

FIGURE 14–18

Body A in Figure 14–18 has a moment of inertia about its center of mass of 55 kg·m² Block B has a mass of 50 kg and a velocity of 2.7 m/s. The spring has a modulus of 400 N/m and is stretched 0.2 m at the instant shown. Determine the velocity of B after it has dropped 0.7 m.

The activating energy or the energy input to the system is the potential energy loss of B.

$$\text{PE loss of B} = \Delta\text{KE}_\text{B} + \Delta\text{KE}_\text{A} + \text{spring work}$$
$$Wh = \frac{1}{2}m(v_2^2 - v_1^2)$$
$$+ \frac{1}{2}I(\omega_2^2 - \omega_1^2) + \frac{1}{2}k(s_2^2 - s_1^2)$$

where

$$v = r\omega = (0.8 \text{ m})\omega \quad \text{or} \quad \omega = \frac{v}{0.8 \text{ m}} \text{ for A}$$

and

$$\begin{aligned} \text{final spring distance} &= \text{initial distance} \\ &\quad + \text{proportion of block B distance} \\ &= 0.2 \text{ m} + \frac{0.4}{0.8}(0.7 \text{ m}) \\ \text{final spring distance} &= 0.55 \text{ m} \end{aligned}$$

Filling known information into our energy equations, we get

$$\begin{aligned} (50 \text{ kg})(9.81 \text{ m/s}^2)(0.7 \text{ m}) &= \left[\frac{1}{2}(50 \text{ kg})(v_2^2 - (2.7 \text{ m/s})^2)\right] \\ &\quad + \frac{1}{2}(55 \text{ kg} \cdot \text{m}^2)\left[\left(\frac{v_2}{0.8 \text{ m}}\right)^2 - \left(\frac{2.7}{0.8 \text{ m}}\right)^2\right] \\ &\quad + \frac{1}{2}(400 \text{ N/m})[(0.55 \text{ m})^2 - (0.2 \text{ m})^2] \end{aligned}$$

$$\underline{v_2 = 3.4 \text{ m/s} \downarrow}$$

EXAMPLE 14–20

Wheel A weighs 40 lb (Figure 14–19) and block B weighs 25 lb. The spring has a compressive load of 20 lb. The coefficient of kinetic friction is 0.3 between the block and the wall. Assuming no slipping between wheel A and block B, determine the angular velocity of wheel A after block B drops 20 in., if it is initially at rest.

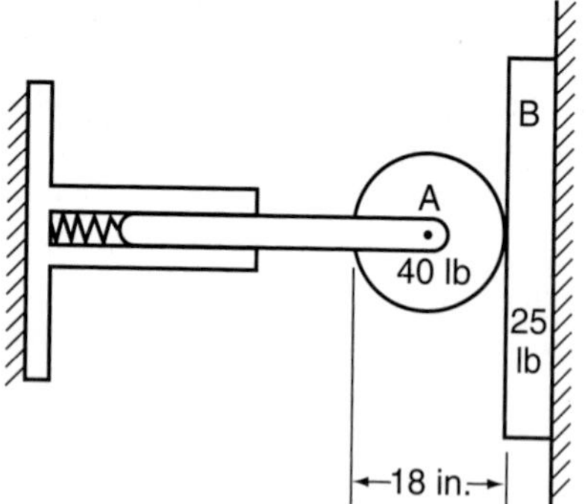

FIGURE 14–19

Since the normal force between block B and the wall is equal to the spring load of 20 lb, the friction force is found

$$\begin{aligned} F &= \mu N \\ &= .3(20 \text{ lb}) \\ F &= 6 \text{ lb} \end{aligned}$$

The mass moment of inertia for wheel A is calculated

$$I_C = \frac{1}{2}mr^2$$

$$= \frac{1}{2}\left(\frac{40 \text{ lb}}{32.2 \text{ ft/s}^2}\right)\left(\frac{9 \text{ in.}}{12 \text{ in./ft}}\right)^2$$
$$I_C = 0.349 \text{ ft-lb-s}^2$$

The velocity of block B equals the tangential velocity of the rim of wheel A.

$$v = r\omega$$
$$= \left(\tfrac{9}{12}\right)\omega$$
$$= (.75 \text{ ft})\omega$$

The potential energy loss of B is redistributed as follows

$$\text{PE}_{\text{loss of B}} = \Delta\text{KE}_\text{B} + \Delta\text{KE}_\text{A} + \text{energy lost to friction}$$
$$Wh = \frac{1}{2}mv^2 + \frac{1}{2}I\omega^2 + Fh$$
$$(25 \text{ lb})\left(\frac{20 \text{ in.}}{12 \text{ in./ft}}\right) = \frac{1}{2}\left(\frac{25 \text{ lb}}{32.2 \text{ ft/s}^2}\right)[(.75\text{ft})\omega]^2 + \frac{1}{2}(0.349 \text{ ft-lb-s}^2)\omega^2 + (6 \text{ lb})\left(\frac{20 \text{ in.}}{12 \text{ in./ft}}\right)$$
$$41.67 = 0.218\omega^2 + 0.175\omega^2 + 10$$
$$\underline{\omega = 8.98 \text{ rad/s} \circlearrowright}$$

14–8 CONSERVATION OF ENERGY: PLANE MOTION

Plane motion consists of both translational motion and rotational motion such as a rolling cylinder. The problems that we will solve are very similar to those in Section 13–6, where we had linear forces (ma) and angular torque ($I_c\alpha$). The energy method now considers both linear kinetic energy and angular kinetic energy of a rolling cylinder.

EXAMPLE 14–21

Cylinder A in Figure 14–20 weighs 200 lb and has a mass moment of inertia about its center of mass of 3 ft-lb-s^2. Block B weighs 100 lb. The weight and inertia of the pulley and cable can be neglected. If the system starts from rest, determine the velocity of B after A has rolled 10 ft on the slope.

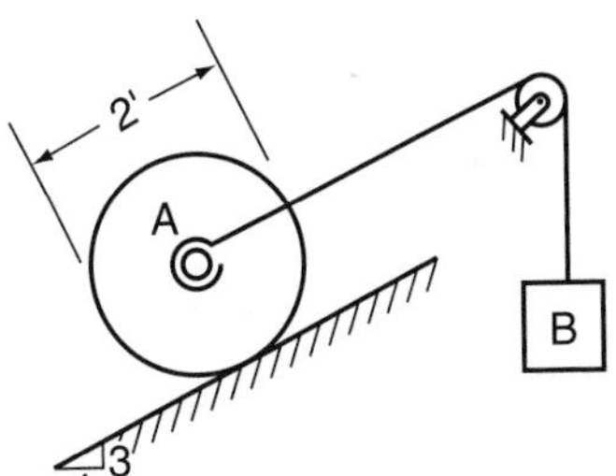

FIGURE 14–20

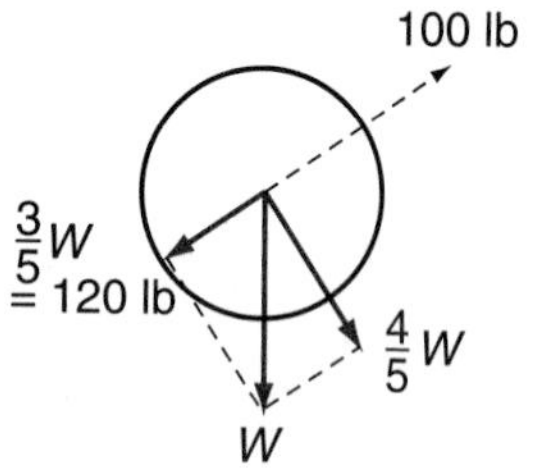

FIGURE 14–21

The first point to check is whether cylinder A rolls up or down the slope. As shown by Figure 14–21, the component ($\frac{3}{5}W$) of A is greater than the cable tension, so A rolls down the slope.

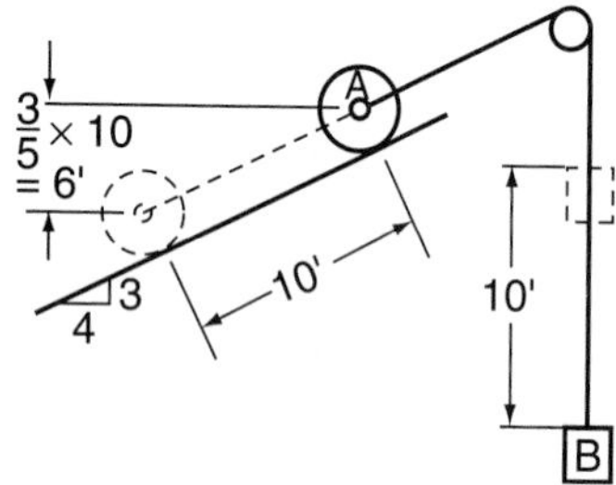

FIGURE 14–22

Now obtain the distances traveled (Figure 14–22). Equate the activating energy with all of the final forms of energy. The potential energy loss of A results in

1. KE of B
2. PE increase of B
3. KE of A (rectilinear)
4. KE of A (angular)

$$\Delta PE_A = \Delta KE_B + \Delta PE_B + \Delta KE_{A(rect.)} + \Delta KE_{A(ang.)}$$

$$W_A h_A = \left(\frac{1}{2}mv^2\right) + (W_B)(h_B) + \left(\frac{1}{2}mv^2\right) + \left(\frac{1}{2}I_C\omega^2\right)$$

where v is the rectilinear velocity of A and B.

$$\omega = \frac{v}{r} = \frac{v}{1\text{ ft}}$$

$$(200\text{ lb})(6\text{ ft}) = \left(\frac{1}{2}\right)\left(\frac{100\text{ lb}}{32.2\text{ ft/s}^2}\right)v^2 + (100\text{ lb})(10\text{ ft})$$

$$+ \left(\frac{1}{2}\right)\left(\frac{200\text{ lb}}{32.2\text{ ft/s}^2}\right)v^2 + \left(\frac{1}{2}\right)(3\text{ ft-lb-s}^2)\left(\frac{v}{1\text{ ft}}\right)^2$$

$$\underline{v = 5.7\text{ ft/s}\uparrow}$$

Example 14–22

Roller A weighs 40 lb and block B weighs 25 lb (Figure 14–23). The spring-loaded mechanism exerts a constant 20-lb load against the roller. Assuming no slipping between roller A and block B, determine the angular velocity of roller A after block B drops 20 in. from an initial position of rest.

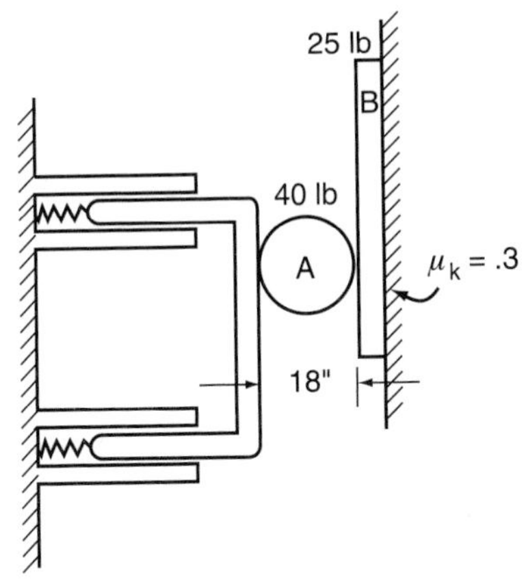

FIGURE 14–23

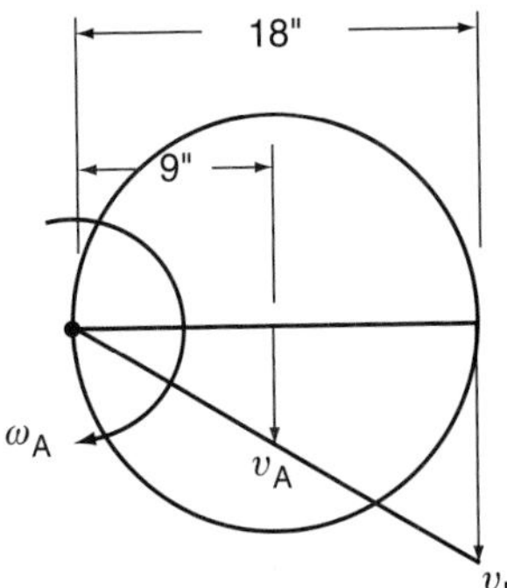

FIGURE 14–24

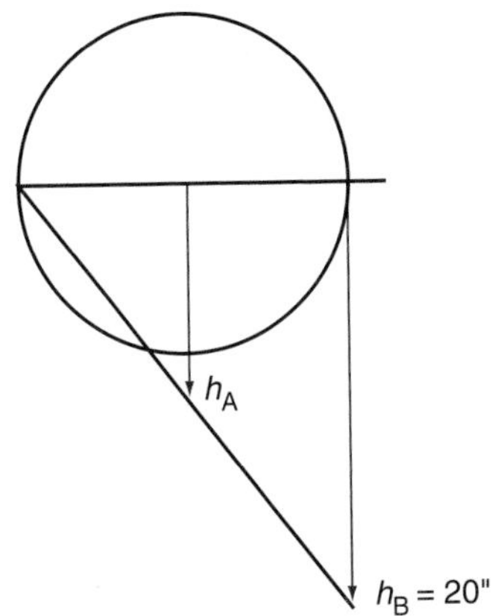

FIGURE 14–25

As in Example 14–20, the friction force on block B is

$$F = \mu N$$
$$= .3(20 \text{ lb})$$
$$F = 6 \text{ lb}$$

For roller A

$$I_C = \frac{1}{2}mr^2$$
$$= \frac{1}{2}\left(\frac{40 \text{ lb}}{32.2 \text{ ft/s}^2}\right)\left(\frac{9 \text{ in.}}{12 \text{ in./ft}}\right)^2$$
$$I_C = 0.349 \text{ ft-lb-s}^2$$

The instantaneous center of rotation for the roller is shown in Figure 14–24, where

$$v_A = r\omega_A$$
$$= \left(\frac{9 \text{ in.}}{12 \text{ in./ft}}\right)\omega_A$$
$$v_A = .75\omega_A$$

and

$$v_B = \left(\frac{18 \text{ in.}}{12 \text{ in./ft}}\right)\omega_a$$
$$v_B = 1.5\omega_A$$

Similarly, about the same instantaneous center (Figure 14–25)

$$h_A = \frac{1}{2}h_B$$
$$= \frac{1}{2}(20 \text{ in.})$$
$$h_A = 10 \text{ in.}$$

The potential energy loss of both A and B is redistributed to

1. KE of B
2. KE of A (rectilinear)

3. KE of A (angular)
4. Energy loss to friction

$$\Delta PE_A + \Delta PE_B = \Delta KE_B + \Delta KE_{A(rect.)} + \Delta KE_{A(ang.)} + \text{friction loss}$$

$$Wh_A + Wh_B = \frac{1}{2}mv_B^2 + \frac{1}{2}mv_A^2 + \frac{1}{2}I\omega_A^2 + Fh_B$$

$$(40\text{ lb})\left(\frac{10\text{ in.}}{12\text{ in./ft}}\right) + (25\text{ lb})\left(\frac{20\text{ in.}}{12\text{ in./ft}}\right) = \frac{1}{2}\left(\frac{25\text{ lb}}{32.2\text{ ft/s}^2}\right)[(1.5\text{ ft})(\omega_A)]^2 + \frac{1}{2}\left(\frac{40\text{ lb}}{32.2\text{ ft/s}^2}\right)[(.75\text{ ft})(\omega_A)]^2$$

$$+ \frac{1}{2}(.349\text{ ft-lb-s}^2)\omega_A^2 + (6\text{ lb})\left(\frac{20\text{ in.}}{12\text{ in./ft}}\right)$$

$$\omega_A = 6.82\text{ rad/s} \circlearrowleft$$

EXAMPLE 14–23

Cylinder A rolls in a slot and on a hub of 0.6-m diameter (Figure 14–26). A cable is wrapped around the cylinder's 1.2-m diameter. Cylinder A has a mass of 70 kg and a radius of gyration of 0.5 m with respect to the center of mass. Mass B is 26.5 kg and has a coefficient of kinetic friction of 0.2. Assume that there is no slippage of A and that the system is initially at rest. Determine the angular velocity of A after B has slid 2 m along the inclined plane.

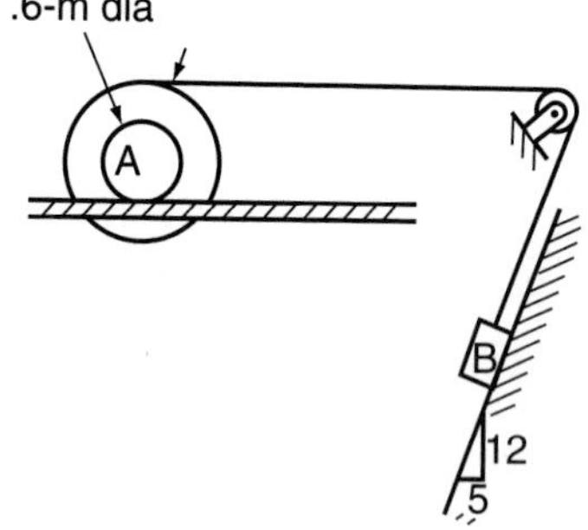

FIGURE 14–26

There is no question as to the direction of motion, but a free-body diagram of B is required in order to obtain the friction force (Figure 14–27).

$$N = 100\text{ N}$$

and

$$F = \mu N = 0.2(100\text{ N}) = 20\text{ N}$$

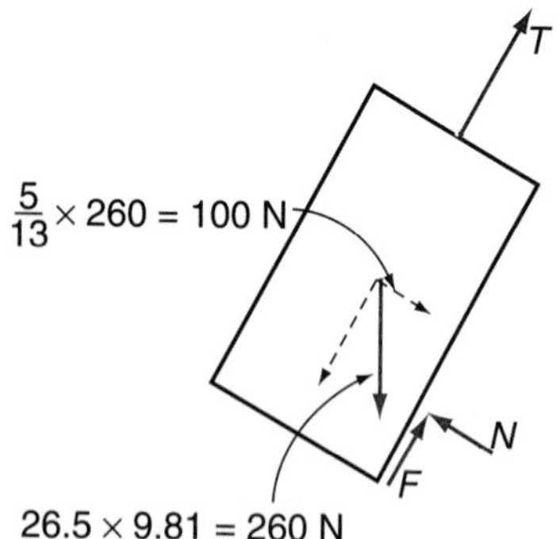

FIGURE 14–27

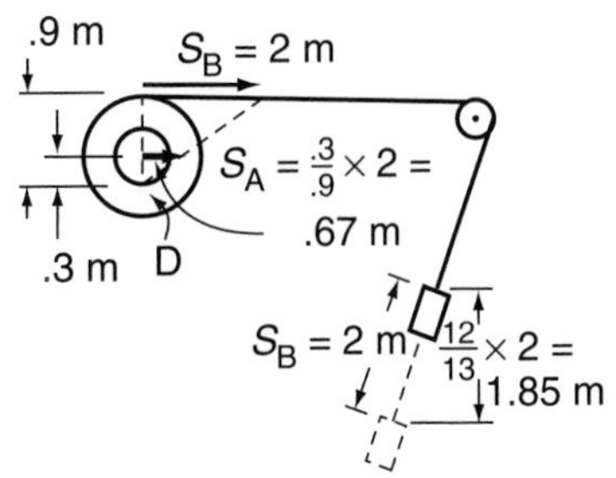

FIGURE 14–28

Next, referring to Figure 14–28, we determine some of the various distances involved. Note that D is the instantaneous center of rotation of A.

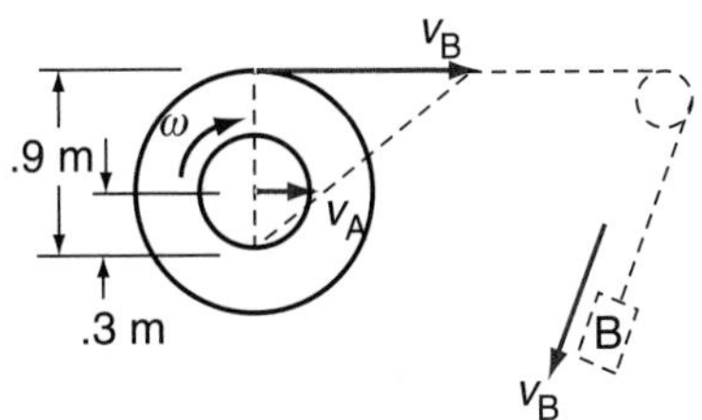

FIGURE 14–29

Since we are solving for the angular velocity of A, all velocities could be expressed in terms of ω (Figure 14–29).

$$v_A = r\omega = (0.3\ m)\omega$$
$$v_B = r\omega = (0.9\ m)\omega$$

The radius of gyration is

$$k = \sqrt{\frac{I}{m}}$$

or

$$\begin{aligned} I &= k^2 m \\ &= (0.5\ \text{m})^2(70\ \text{kg}) \\ I &= 17.5\ \text{kg} \cdot \text{m}^2 \end{aligned}$$

The potential energy loss of B results in

1. KE of B
2. friction loss at B
3. KE of A (angular)
4. KE of A (rectilinear)

$$\begin{aligned} \text{PE}_\text{B} &= \text{KE}_\text{B} + \text{friction loss} + \text{KE}_{\text{A(ang.)}} \\ &\quad + \text{KE}_{\text{A(rect.)}} \\ Wh &= \frac{1}{2}mv_B^2 + Fs + \frac{1}{2}I\omega^2 \\ &\quad + \frac{1}{2}mv_A^2 \end{aligned}$$

$$(260\text{ N})(1.85\text{ m}) = \frac{1}{2}(26.5\text{ kg})[(0.9\text{ m})(\omega)]^2 + (20\text{ N})(2\text{ m})$$

$$+ \frac{1}{2}(17.5\text{ kg}\cdot\text{m}^2)\omega^2 + \frac{1}{2}(70\text{ kg})[(0.3\text{ m})(\omega)]^2$$

$$\underline{\omega = 4.41\text{ rad/s} \circlearrowleft}$$

14–9 POWER AND EFFICIENCY

We have examined the various forms of energy—work being one of them. We will now concern ourselves not simply with the quantity of work but also with the time required to accomplish this work. The rate of doing work is *power*.

$$\text{power} = \frac{\text{work}}{\text{time}} = \frac{Fs}{t}$$

$$\text{P} = \frac{U}{t} \tag{14–8}$$

where, in the SI system,

U = work; joules (J)

t = time; seconds (s)

P = power; watts (W); since 1 watt is defined as the rate of work of 1 joule per second, $1\text{ W} = 1\text{ J/s} = \dfrac{\text{N}\cdot\text{m}}{\text{s}}$

and in common units for the U.S. Customary system,

U = work; ft-lb

t = time; seconds(s)

P = power; in ft-lb/s

but, by definition; 1 horsepower = 550 ft-lb/s = 33,000 ft-lb/min

Therefore

$$\text{power (in hp)} = \frac{Fs}{550t} \quad \text{for } t \text{ in seconds} \tag{14–9}$$

$$\text{power (in hp)} = \frac{Fs}{33{,}000\,t} \quad \text{for } t \text{ in minutes} \tag{14–10}$$

where

F = force; lb

s = distance; ft

Other ways of writing Equation (14–8) would be as follows:

$$P = \frac{Fs}{t} \quad \text{but} \quad v = \frac{s}{t}$$

Therefore,

$$P = Fv$$

For rotational motion, recall that for angular work,

$$U = \text{torque}(\theta) \qquad \text{(Equation 14–7)}$$

$$P = \frac{U}{t} = \frac{\text{torque}(\theta)}{t} \quad \text{but} \quad \omega = \frac{\theta}{t}$$

Therefore,

$$P = \text{torque}(\omega) \qquad (14\text{–}11)$$

For the SI system:

torque is N · m

ω = angular velocity, rad/s

P = power, watts (W)

and in the U.S. Customary system,

torque is lb-ft

ω = angular velocity; rad/s

P = power, ft-lb/s

We can convert between English and SI by using the conversion factor:
1 horsepower (hp) = 0.746 kW

EXAMPLE 14–24

An average force of 300 N is applied over a distance of 50 m. If the time required is 2 min, determine the work and the power.

$$\begin{aligned} \text{work} &= Fs \\ &= (300\text{ N})(50\text{ m}) \\ &= 15{,}000\text{ J} \\ \underline{\text{work}} &\underline{= 15\text{ kJ}} \end{aligned}$$

$$\text{power} = \frac{\text{work}}{\text{time}}$$

$$= \frac{15{,}000 \text{ J}}{120 \text{ s}}$$

$$\underline{\text{power} = 125 \text{ W}}$$

EXAMPLE 14–25

Determine the horsepower required to provide a force of 400 lb for a distance of 8 ft in 5 seconds.

$$\text{hp} = \frac{Fs}{550t}$$

$$= \frac{(400 \text{ lb})(8 \text{ ft})}{(550 \text{ ft-lb/s/hp})(5 \text{ s})}$$

$$\underline{\text{hp} = 1.16 \text{ hp}}$$

EXAMPLE 14–26

A pump handles 50 gal/min (1 U.S. gal = 8.33 lb) while pumping water a difference in elevation of 20 ft. What is the power input to the pump in hp?

$$\text{hp} = \frac{Fs}{33{,}000\, t}$$

$$= \frac{(50 \text{ gal})(8.33 \text{ lb/gal})(20 \text{ ft})}{(33{,}000 \text{ ft-lb/min/hp})(1 \text{ min})}$$

$$\underline{\text{hp} = 0.25}$$

EXAMPLE 14–27

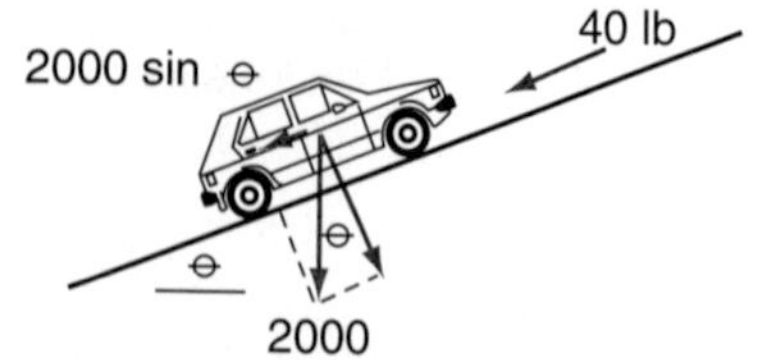

FIGURE 14–30

A 2000-lb car has 80 hp available to maintain a speed of 50 mph up a hill. How steep a hill can it climb if wind and rolling resistance forces are 40 lb?

Converting the speed (60 mph = 88 ft/s)

$$= \frac{50}{60}(88 \text{ ft/s}) = 73.3 \text{ ft/s}$$

The forces to be overcome are the 40-lb force and the component of the weight, acting down the slope, or 2000 sin θ (Figure 14–30).

$$\text{hp} = \frac{Fs}{550\ t}$$

$$80\ \text{hp} = \frac{(40\ \text{lb} + 2000\ \text{lb}\sin\theta)73.3\ \text{ft}}{(550\ \text{ft-lb/s/hp})(1\ \text{s})}$$

$$\underline{\theta = 16.3^\circ}$$

Various machines are capable of receiving power and converting it to a more useful form. An electric motor receives input energy of kilowatts and gives an output of horsepower. The efficiency at which a machine can transmit or convert this energy is described by the ratio of power output to power input. Expressed as a percentage,

$$\text{efficiency (in \%)} = \left(\frac{\text{power output}}{\text{power input}}\right)(100)$$

An interesting example of this overall *efficiency* is that of a car with an internal combustion gasoline engine. If the total heat input of the gasoline were converted at 100% efficiency to work causing motion of the car (with no friction or wind resistance), it would yield an astonishing mileage of approximately 450 miles per gallon (159 km/l).

EXAMPLE 14–28

The starting motor on a turbine applies a constant torque of 60 lb-ft while turning the turbine 250 revolutions in 15 seconds. Calculate the horsepower required.

$$U = \text{torque}(\theta) \quad \text{(Equation 14–7)}$$

$$\text{power} = \frac{U}{t} = \frac{\text{torque}(\theta)}{t}$$

since

$$1\ \text{hp} = 550\ \text{ft-lb/s}$$

$$\text{P} = \frac{\text{torque}(\theta)}{550t}$$

$$= \frac{(60\ \text{lb-ft})(250\ \text{rev})(2\pi\ \text{rad/rev})}{(550\ \text{ft-lb/s/hp})(15\ \text{s})}$$

$$\underline{\text{P} = 11.4\ \text{hp}}$$

EXAMPLE 14–29

What is the efficiency of an electric motor that supplies 8 hp while using 7.1 kW of electricity?

$$\text{efficiency} = \frac{\text{output}}{\text{input}}(100)$$
$$= \frac{(8\text{ hp})(0.746\text{ kW/hp})}{7.1\text{ kW}}(100)$$
$$\underline{\text{efficiency} = 84\%}$$

Note that if we had been using the SI metric system, we would have had the motor supplying 5.96 kW of power while using 7.1 kW of electricity. Efficiency would have been easily calculated without any cumbersome conversion; that is,

$$\text{efficiency} = \frac{5.96\text{ kW}}{7.1\text{ kW}}(100)$$
$$\underline{\text{efficiency} = 84\%}$$

EXAMPLE 14–30

A cutting tool on a lathe applies a tangential force of 3.5 kN when it is machining a bar 150 mm in diameter. If the lathe is turning at 100 rpm, what power must be supplied to the bar?

$$\text{P} = \frac{\text{torque}(\theta)}{t}$$
$$= \frac{(3500\text{ N})\left(\frac{0.15\text{ m}}{2}\right)(100\text{ rev})(2\pi\text{ rad/rev})}{60\text{ s}}$$
$$= 2750\text{ W}$$
$$\underline{\text{P} = 2.75\text{ kW}}$$

HINTS FOR PROBLEM SOLVING

1. In the equation, work = (force)(distance), the force must be in the same direction as the distance. This may require multiplying the distance by a component of a force.
2. In the English system, work has units of ft-lb, as compared to torque lb-ft.
3. A force that varies in a linear fashion produces work equal to the average force times the distance.
4. The distance that a spring is stretched or compressed is always with respect to its free length.
5. In the spring work equation $U = \frac{1}{2}k(s_2^2 - s_1^2)$, be sure to square each spring deflection term *before* subtracting.
6. The conservation of energy equation simply accounts for energy within a system and the following should be noted:
   (a) Energy is lost and gained. Look for the energy that was lost and equate it to the energy gained by other objects within the system.
   (b) We are concerned only with initial and final conditions. The intermediate conditions that may absorb and then give up energy are of no consequence.
   (c) A suggested sequence to follow would be:
   - Write the energy equation in descriptive terms.
   - Write individual equations for each block of energy.
   - Fill in each individual equation with known data.
   - Relate all unknown angular and linear velocities to the velocity for which you are solving.

   (d) The kinetic energy of plane motion, such as a rolling cylinder, is made up of linear $(\frac{1}{2}mv^2)$ and angular $(\frac{1}{2}I_C\omega^2)$ energy. Note that I_C is the mass moment of inertia about the *center* of mass.

PROBLEMS

APPLIED PROBLEMS FOR SECTIONS 14–1 AND 14–2

14–1. A man pushes a loaded cart 15 m horizontally by pushing with a horizontal force of 60 N. He then unloads 20 bags, each with a mass of 40 kg, from the cart. Each bag is lifted to a height of 0.8 m. Determine the total work done.

14–2. Using a lever and fulcrum, a worker lifts a weight of 1000 lb a height of 4 in. Determine the work done. What force did this person apply if his end of the lever traveled 3.5 ft?

14–3. Determine the work done in lifting a 250-lb refrigerator 4 ft vertically into a van. Alternatively, the refrigerator could be placed on a cart and rolled up a 10-ft-long ramp. What force, parallel to the ramp, is required to push the refrigerator along the ramp? (Neglect rolling resistance.)

14–4. Crate A is moved 8 ft to the right by the forces shown in Figure P14–4. Determine the work done.

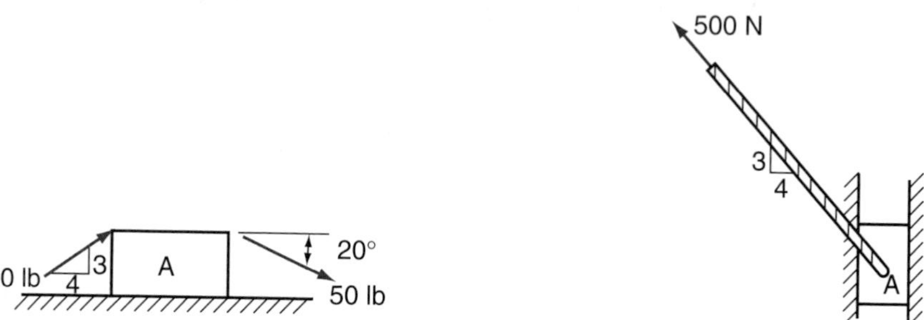

FIGURE P14–4

FIGURE P14–5

14–5. Slider A in Figure P14–5 is lifted to a vertical height of 5 m by the rope shown. The rope remains at the same angle throughout the motion. Determine the work done on A by the rope.

14–6. A construction worker must apply the horizontal and vertical forces shown in Figure P14–6 while pushing the loaded wheelbarrow 6.5 m up the slope. Determine the work done.

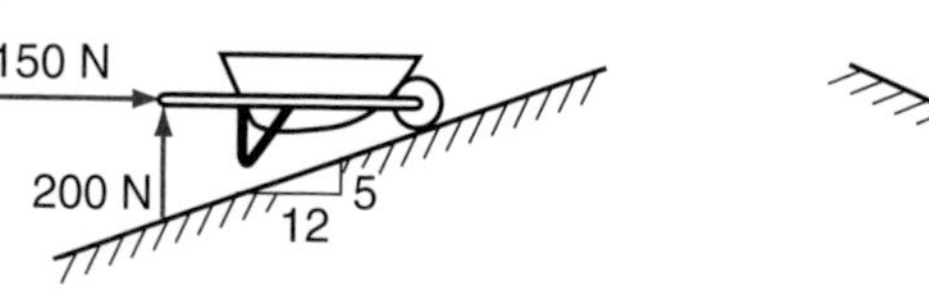

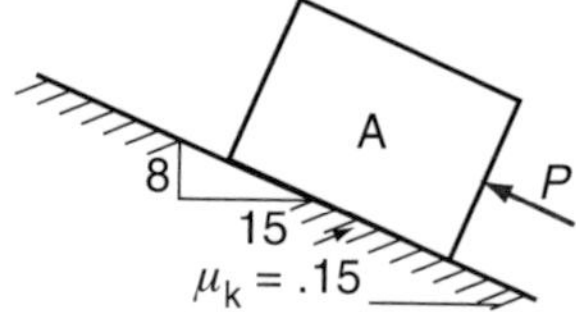

FIGURE P14–6

FIGURE P14–7

14–7. Block A in Figure P14–7 has a mass of 15 kg and is pushed 3.4 m up the slope by a force, $P = 130$ N. Determine the work done on the block by (a) force P and (b) the friction force.

14–8. Cart A is pushed 6 m along the slope by the 400-N force shown in Figure P14–8. Determine the work done on the car by the 400-N force.

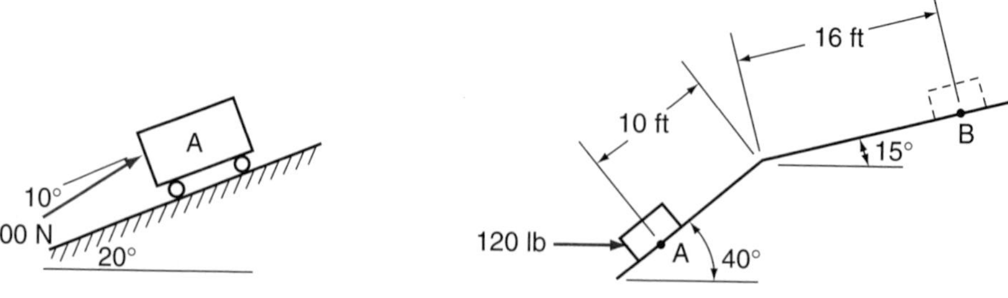

FIGURE P14–8

FIGURE P14–9

14–9. Due to different coefficients of friction for each surface, a constant horizontal force of 120 lb moves the block shown in Figure P14–9 from point A to B. Determine the work done.

14–10. The winch shown in Figure P14–10 slowly lifts a 8-kg mass a height of 2 m. Due to cable friction, the tension in the cable at the winch is 120 N.
Determine (a) the work done on the 8 kg mass, (b) the work done by the winch, and (c) the force P on the handle of the winch if nine revolutions were required.

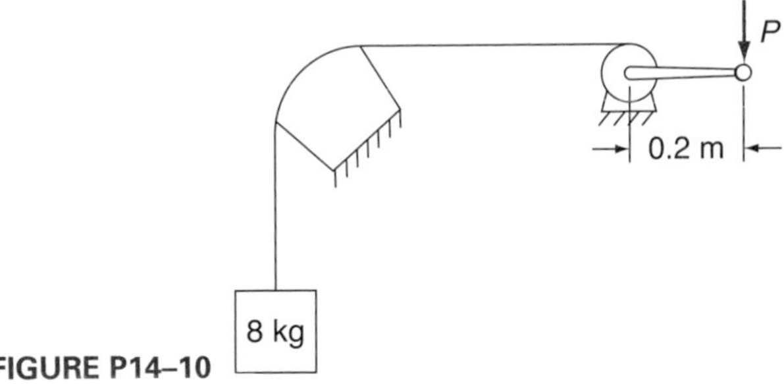

FIGURE P14–10

14–11. A 40-kg block slides 5 m down a 60° slope. If the coefficient of friction is 0.4, determine the work done on the block by the friction force.

APPLIED PROBLEMS FOR SECTION 14–3

14–12. The block shown (Figure P14–12) is pulled from A to B by means of cord tension T, which is sufficient to overcome the block friction and varies linearly from position A to B. If the coefficient of friction is 0.4, determine the work done by tension T.

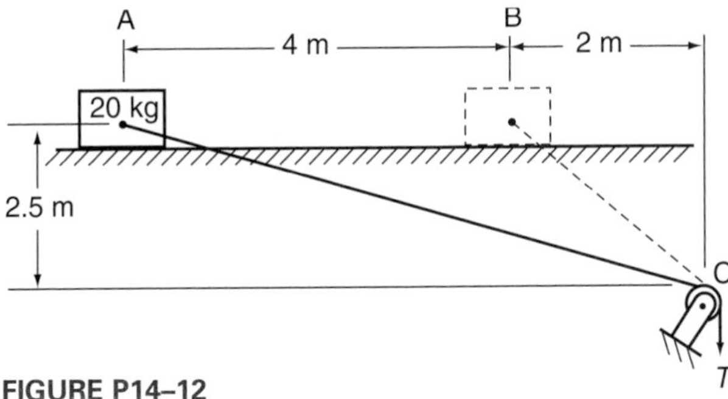

FIGURE P14–12

14–13. Determine how much a spring, with a spring constant of 3 lb/in., will be compressed by a force of 12 lb.

14–14. A spring is initially compressed 50 mm by a force, P. If P is increased by 250 N, thereby causing a total spring compression of 150 mm, what is the spring constant?

14–15. A spring is compressed 4 in. from its free length by a force of 80 lb. Determine the spring constant and the work required.

14–16. A spring is compressed 200 mm from its free length by a force of 500 N. Determine the spring constant and the work required.

14–17. A spring with a constant of 0.3 lb/in. is presently stretched 1.5 in. How much work is required to stretch it an additional 4 in.?

14–18. A shock absorber spring, initially at its free length, absorbs 1800 J of energy while deflecting 420 mm. Determine the spring constant.

14–19. A spring with a constant of 20 lb/in. is compressed 1.4 in. from free length. How much more will the spring compress when absorbing an additional 65 in.-lb of energy?

14–20. A spring is stretched 3 in. from free length by force *P*. When *P* is increased by 10 lb, the spring is stretched 8 in. from free length. Determine the work done on the spring while stretching it from 3 in. to 8 in. deflection.

14–21. A scale mechanism is loaded as shown in Figure P14–21. Assuming that the top beam was level initially, determine the angle of tilt when the 30-N load is applied.

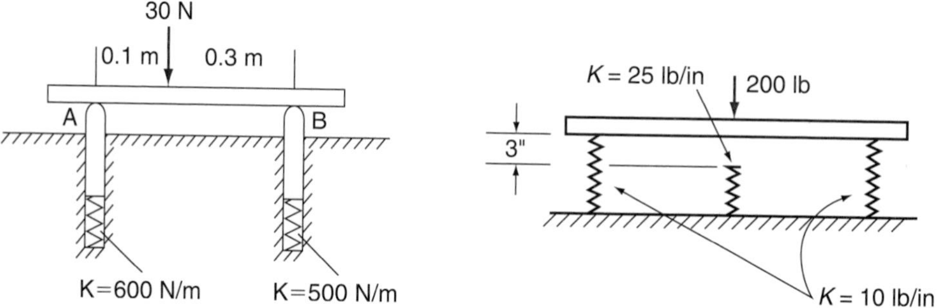

FIGURE P14–21

FIGURE P14–22

14–22. All three springs (Figure P14–22) are at their free length just before the 200-lb force is applied. Determine the deflection and work done on each spring.

14–23. The spring shown in Figure P14–23 has a free length of 0.5 m and a spring constant of 1.2 kN/m. Determine the work required to move the lever to the vertical position.

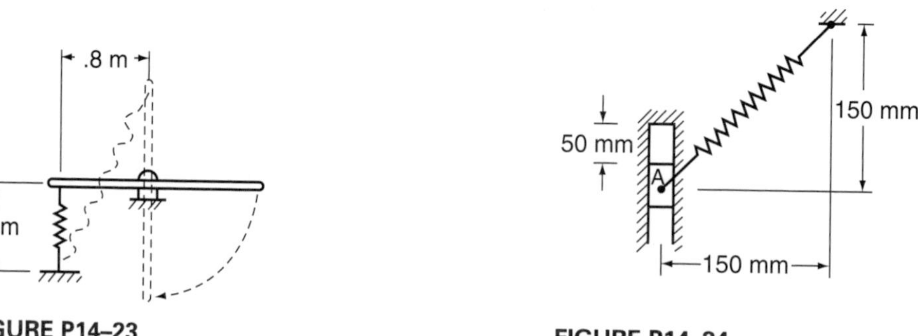

FIGURE P14–23

FIGURE P14–24

14–24. The spring shown in Figure P14–24 has a free length of 160 mm and a spring constant of 800 N/m. Neglect all friction and determine how much work the spring does in lifting block A 50 mm.

14–25. Determine the work into the spring and the work out of the spring when the lever shown in Figure P14–25 rotates from A to B. The spring has a free length of 8 in. and spring constant of K = 40 lb/in.

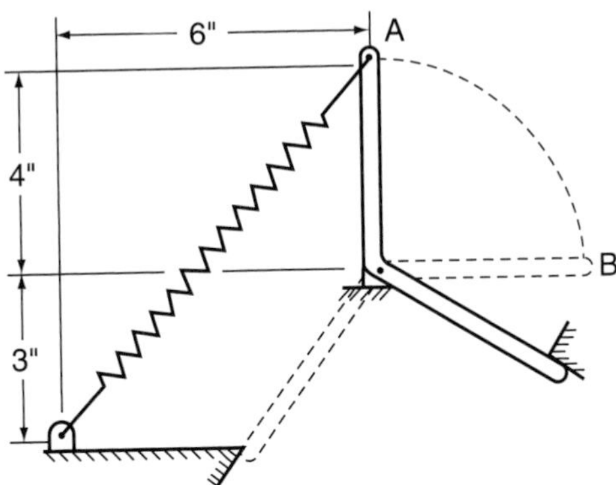

FIGURE P14–25

14–26. The 10-lb block (Figure P14–26) is slowly moved from the position shown to its equilibrium position. Determine (a) the distance it moves and (b) the work on the bottom spring.

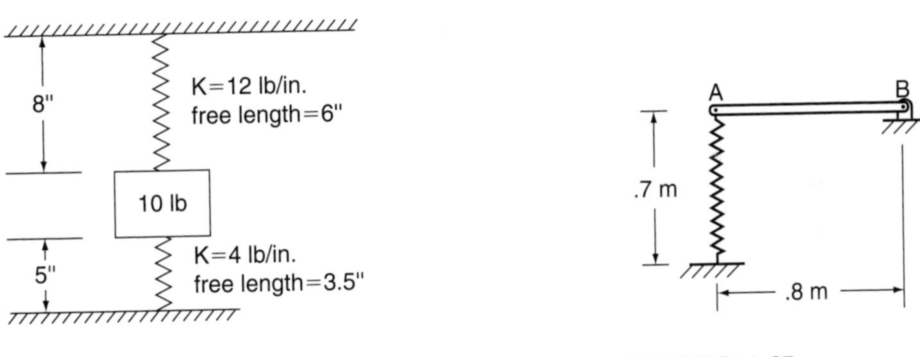

FIGURE P14–26

FIGURE P14–27

14–27. The spring in Figure P14–27 is stretched 0.1 m so that it is in the position shown. It has a spring constant of 3 kN/m. Determine the work required to rotate arm AB 90° clockwise.

14–28. Weight A in Figure P14–28 is winched to the right by a rope tension that varies linearly from 30 lb to 40 lb. Determine the work done by the rope on weight A when it is pulled 8.5 ft to the right as shown.

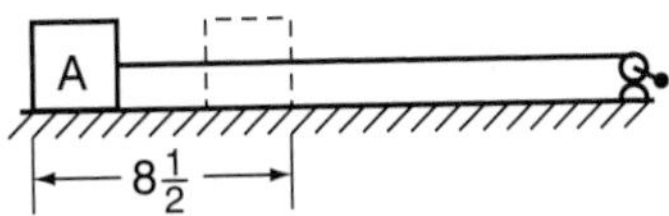

FIGURE P14–28

APPLIED PROBLEMS FOR SECTION 14–4

14–29. Determine the theoretical minimum amount of work required to excavate a hole measuring 10 ft × 8 ft × 6 ft deep. Assume that the specific weight of the soil is 30 lb/ft^3.

14–30. Determine the theoretical minimum work required to dig a trench 100 m long with a cross-sectional profile as shown in Figure P14–30. Assume that the soil weighs 4 kN/m^3.

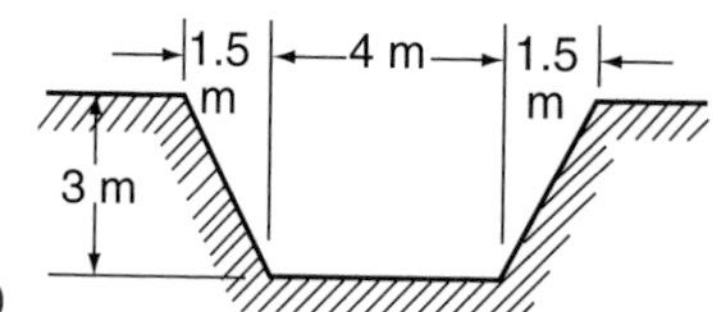

FIGURE P14–30

14–31. Determine the kinetic energy of a truck weighing 40 kN and traveling at 25 m/s. How high would the equivalent amount of energy lift the truck?

14–32. A pipe carries oil, weighing 50 lb/ft^3, at a rate of 7 ft/s. Determine the kinetic energy of each cubic foot of oil.

14–33. A construction crane drops a mass of 50 kg from a height of 4 m. Determine the kinetic energy of the mass as it strikes the ground and the velocity at which it strikes the ground.

14–34. A 32.2-lb object is dropped from a 100-ft building. Determine the velocity and kinetic energy at (a) the 50-ft level and (b) the 25-ft level above the ground.

14–35. Determine the kinetic energy of an 8-kg mass moving at 15 m/s. How many times greater is the kinetic energy if the speed is doubled?

14–36. By the time a starting pitcher is "pulled" in the eighth inning, he has thrown 350 pitches, including warmup. If the ball weighs 0.32 lb and reaches 80 ft/s on each pitch, how much work has been done on the baseball? If he weighs 180 lb, how high would he have to climb a ladder to do the equivalent amount of work?

APPLIED PROBLEMS FOR SECTION 14–5

14–37. A 1200-lb wrecking ball with a velocity of 4 ft/s at the bottom of its arc strikes a wall and comes to rest in a distance of 9 in. Neglecting any change in elevation of the ball, determine (a) the average force on the wall, (b) the deceleration rate, and (c) the time period of deceleration.

14–38. A 200-lb package travels on a conveyor as shown (Figure P14–38). The conveyor roller resistance is a constant 1.5 lb. What initial velocity at A must this package have so that it has a velocity of 2 ft/s at B?

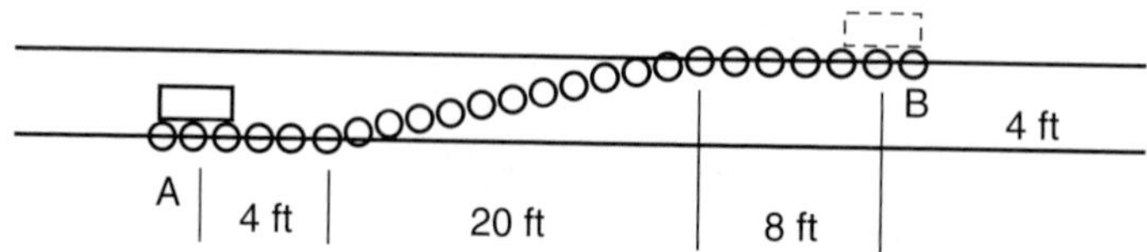

FIGURE P14–38

14–39. Weight A (Figure P14–39) drops 2 m from rest in 1.5 seconds. Determine the kinetic energy of block B at $t = 1.5$ seconds.

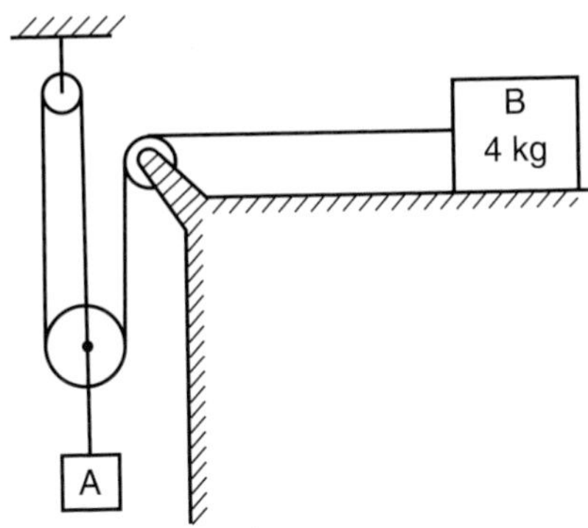

FIGURE P14–39

14–40. Mountain highways with long descents have "escape" roads with a steep incline for vehicles to take if their brakes have failed. If one of these roads is inclined at 40° above horizontal, how long must it be to stop a 10-ton truck traveling at 80 mph?

14–41. Use the work-energy method to solve for the velocity of A (Figure P14–41) if the force $P = 100$ lb moves the block 40 ft. The system is initially at rest.

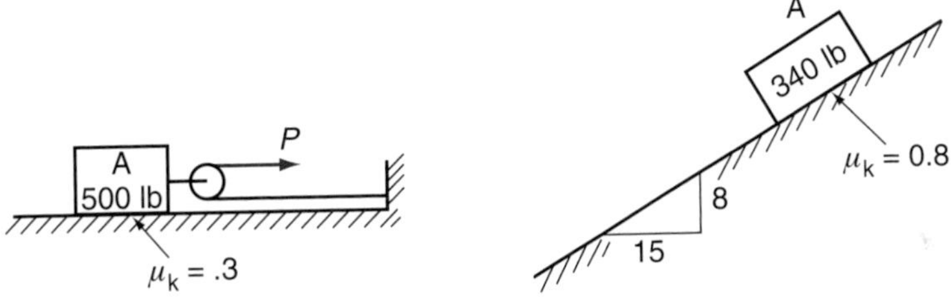

FIGURE P14–41

FIGURE P14–42

14–42. Block A of Figure P14–42 has a velocity of 15 ft/s down the slope. How far will the block slide along the slope before stopping?

14–43. The spring shown in Figure P14–43 has a spring constant of 340 N/m and is compressed 0.3 m in the position shown. If released from this position, how far down the slope will block A travel before coming to a stop? (The spring is not fastened to block A.)

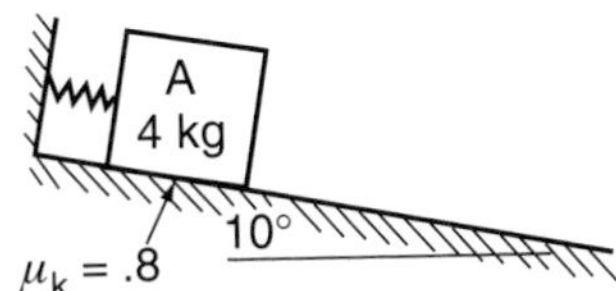

FIGURE P14–43

14–44. A 48-tonne ferry strikes a dock at 3 km/h. What average force does it exert on the dock if it is brought to rest over a distance of 0.2 m?

14–45. A parcel chute in the post office is 3 m high, 9 m long, and discharges the parcels onto a level platform. If the coefficient of kinetic friction is 0.15, determine (a) the parcel speed at the bottom of the chute and (b) the distance traveled by the parcel on the level platform. (Assume that the parcel has negligible velocity at the top of the chute.)

14–46. The Charpy testing machine shown in Figure P14–46 consists of a weight (19 lb) on the end of an arm (12 lb). Material to be tested is clamped at the bottom, and the weight is raised, released, and allowed to break the test material. Determine the energy absorbed by the test material if the initial and final positions of the arm are as shown.

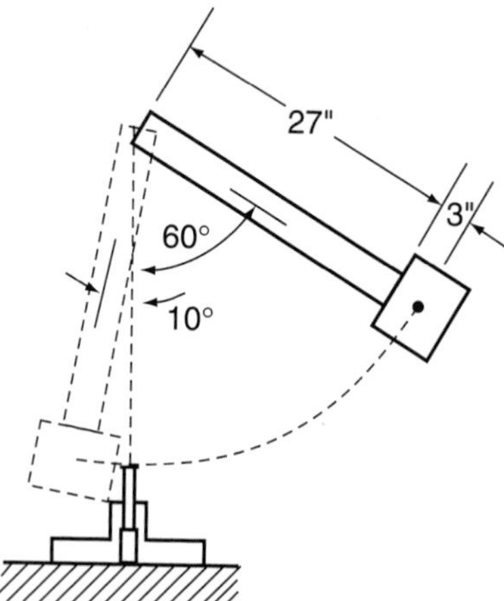

FIGURE P14–46

14–47. Door A, in Figure P14–47, weighs 50 lb. The friction and inertia of the pulleys and rollers can be neglected. If counterweight B weighs 15 lb, determine the velocity of the door if, after starting from rest, it moves 5 ft to the right.

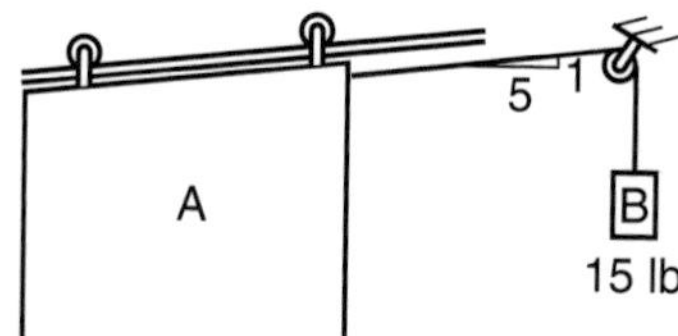

FIGURE P14–47

14–48. The spring in Figure P14–48 has a spring constant of 1.5 kN/m and a free length of 0.6 m. If the spring starts from rest at the position shown, determine the distance that A moves if it is (a) dropped and (b) lowered slowly.

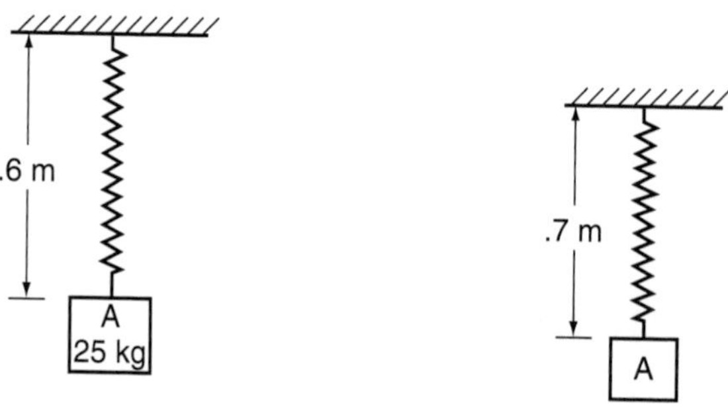

FIGURE P14–48

FIGURE P14–49

14–49. If the spring in Problem 14–48 has a 25-kg mass dropped from the position shown in Figure P14–49, determine the distance that the mass drops.

14–50. If the spring in Problem 14–48 has a 25-kg mass released from the position shown in Figure P14–50, determine the maximum distance upward that the mass can move. (No work is done in compressing the spring.)

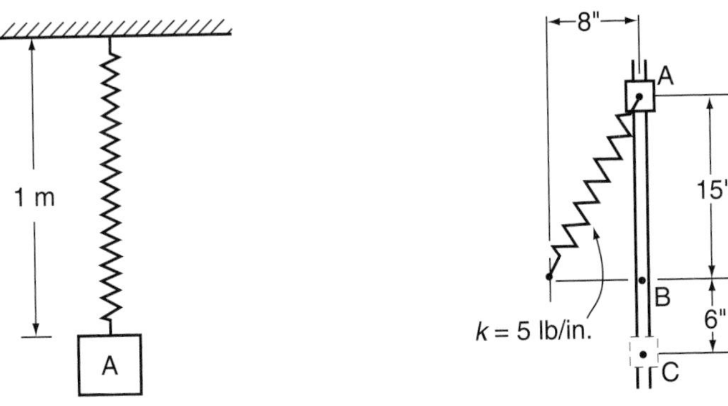

FIGURE P14–50 **FIGURE P14–51**

14–51. Collar A in Figure P14–51 weighs 10 lb and slides on a frictionless rod. The spring has a free length of 7 in. If A is released from the position shown, determine the velocity of A at point B.

14–52. Use the data in Problem 14–51 to determine the velocity of A when it reaches point C.

14–53. A steel ball with sufficient velocity will roll through the hoop as shown in Figure P14–53. Determine the velocity of the ball at A if, when it reaches point B, it is on the verge of dropping away from the track.

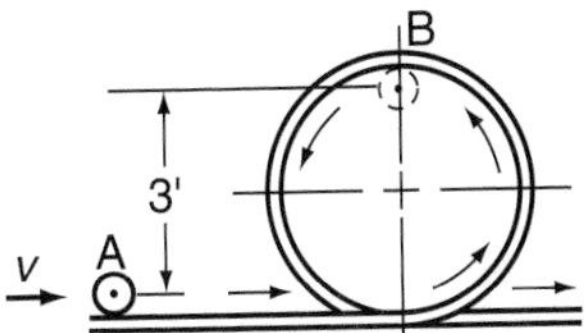

FIGURE P14–53

14–54. Spring A of Figure P14–54 is at its free length in the position shown. If released from the position shown, weight B drops 0.4 m and reverses direction. Neglecting the inertia of the cable and pulleys, determine the spring constant.

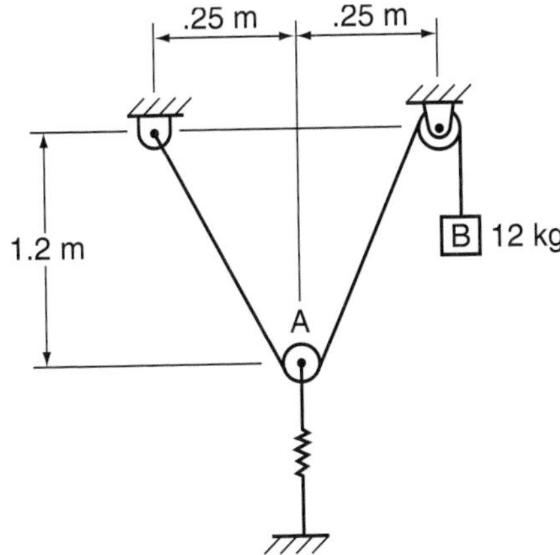

FIGURE P14–54

14–55. Springs B and C are identical springs with free lengths of 6 in. and spring constants of 25 lb/in. (Figure P14–55). Determine the velocity of A after it drops 4 in. from the position shown.

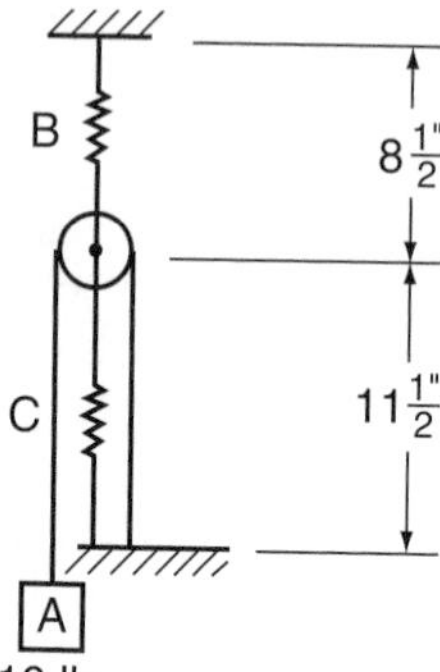

FIGURE P14–55

APPLIED PROBLEMS FOR SECTION 14–6

14–56. A gear with a mass moment of inertia of 0.45 kg·m^2 has a kinetic energy of rotation of 21 joules. What is its speed of rotation in rpm?

14–57. The front wheel of a car weighs 25 lb and has a radius of gyration of 0.5 ft. While being dynamically balanced after the installation of a tire, the wheel is rotated at 80 rpm. Determine its kinetic energy.

14–58. Determine the kinetic energy of a 2.5-m slender rod, with a mass of 10 kg, when it is rotated at 20 rpm about A in Figure P14–58.

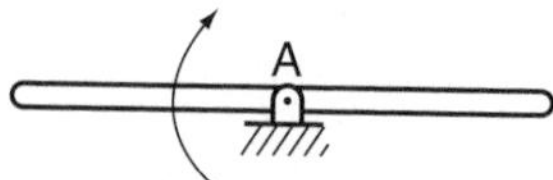

FIGURE P14–58

14–59. In order to get a weight reduction, a 4-in. solid steel shaft rotating at 120 rpm and weighing 300 lb is replaced by a hollow steel shaft with 4-in. O.D. and 3-in. I.D. (same length of shaft). Determine the kinetic energy of each of these shafts.

14–60. A nut is tightened during its last half-turn by an average torque of 20 N·m. Determine the work done.

14–61. A shaft coupling is rated for a torque of 100 N·m at 300 rpm. Determine the energy that it transmits in 1 minute.

14–62. The flywheel on a punch press turns two revolutions while supplying 3000 ft-lb of energy to punch a hole. Determine the average torque supplied during this time.

APPLIED PROBLEMS FOR SECTION 14–7

14–63. Wheel A in Figure P14–63 weighs 200 lb and has a radius of gyration of 2 ft. If the system is initially at rest, determine the angular velocity of A after B has dropped 8 ft.

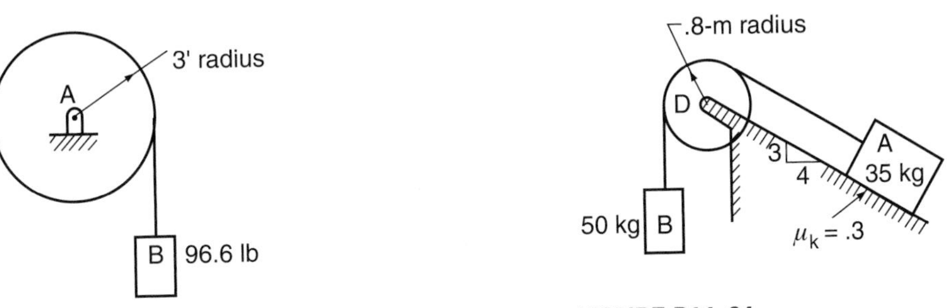

FIGURE P14–63

FIGURE P14–64

14–64. Masses A and B in Figure P14–64 are fastened together by a belt over pulley D. (Assume no slipping of the belt.) The mass moment of inertia of pulley D is 15 kg·m^2. How far does mass B drop before reaching a velocity of 2 m/s? The system is initially at rest.

14–65. Drum A in Figure P14–65 has a mass moment of inertia of 65 kg·m^2. If B has a velocity of 1.5 m/s downward, determine the force P necessary to brake drum A to a stop in one revolution.

FIGURE P14–65

FIGURE P14–66

14–66. Double-pulley D in Figure P14–66 has a mass moment of inertia of 150 ft-lb-s^2. If the system is initially at rest, determine the velocity of A just before B strikes the ground.

14–67. The system shown in Figure P14–67 is released from rest, and weight B drops 0.63 ft before reversing direction. Determine the spring constant required if the spring has an initial deflection of 0.1 ft in the position shown.

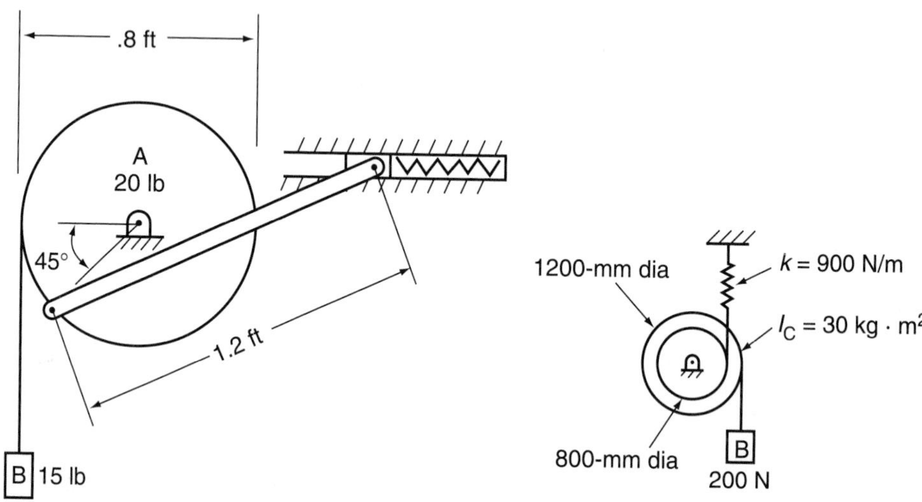

FIGURE P14–67

FIGURE P14–68

14–68. The spring in Figure P14–68 is at its free length, and the system is at rest when weight B is dropped. Determine the maximum distance that B will drop.

14–69. The wheel shown (I_C = 7 ft-lb-s^2) in Figure P14–69 has cables wound on it at the diameters shown. The spring is stretched 6 in. at the moment of release from rest. Find the velocity of A when the spring has a total deflection of 18 in.

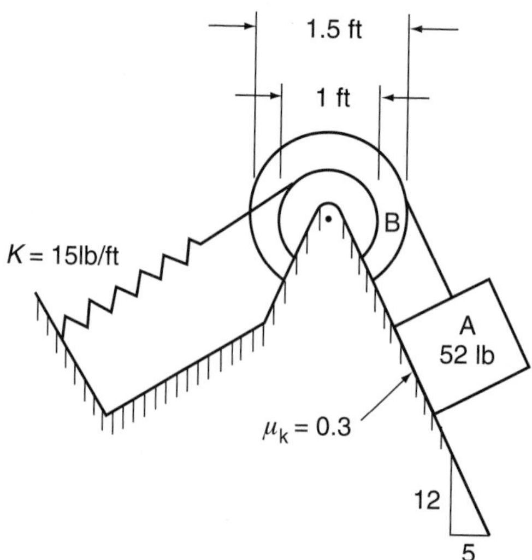

FIGURE P14–69

14–70. Pulley A in Figure P14–70 weighs 250 lb and has a mass moment of inertia of $I_C = 40$ ft-lb-s^2. If the system is initially at rest and the spring is at its free length, determine the velocity of B after it has dropped 1.2 ft.

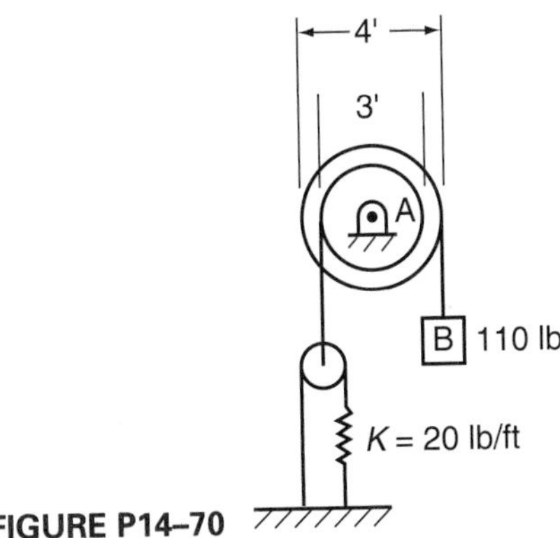

FIGURE P14–70

14–71. The wheel shown in Figure P14–71 has an initial velocity and, while coming to rest, turns 200 revolutions at constant deceleration. If the average friction force acting on piston B is 10 N, determine the initial speed of the wheel. (Neglect all other friction and the mass of the piston and member AB.)

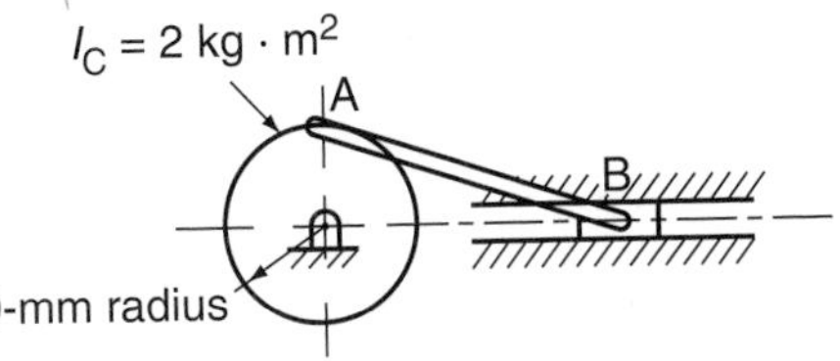

FIGURE P14–71

14–72. A drive motor is operating at 600 rpm and the load on the winch is $P = 300$ N (Figure P14–72). If power to the motor is stopped, how much cable is wound onto the drum before the system stops completely?

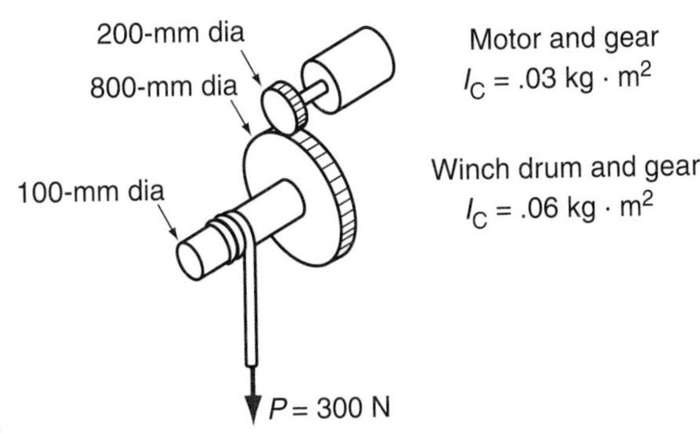

FIGURE P14–72

APPLIED PROBLEMS FOR SECTION 14–8

14–73. A log rolls without slipping down a 45° slope that is 100 ft long. The log weighs 800 lb and has an average radius of 9 in. When the log is at the bottom of the slope, determine (a) its linear kinetic energy and (b) its rotational kinetic energy.

14–74. A 50-lb drum, 3 ft in diameter, is lowered down planking by means of a rope as shown in Figure P14–74. If the rope starts to slip in the operator's hands at A and he then allows the rope to slip freely, what will be the velocity of the rope through his hands when the drum reaches the bottom of the plank?

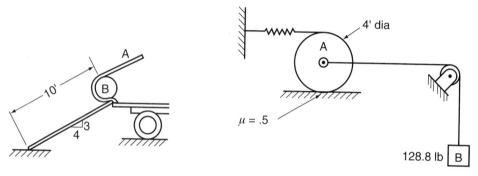

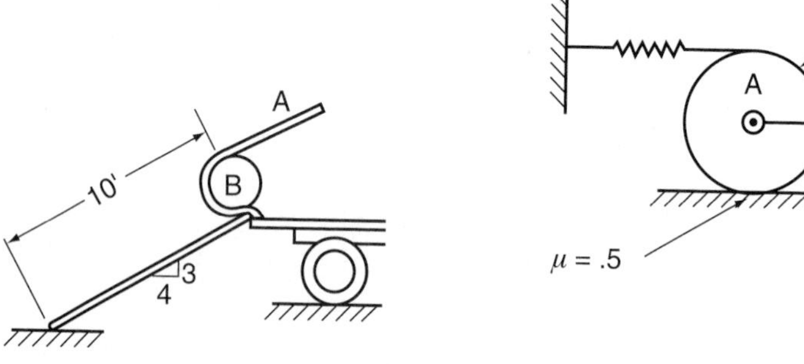

FIGURE P14–74

FIGURE P14–75

14–75. Cylinder A in Figure P14–75 weighs 322 lb. The spring is at its free length and has a spring constant of 10 lb/ft. If the system is initially at rest, determine the velocity of B after it has dropped 6 ft.

14–76. A 50-kg barrel (I_C = 2.25 kg·m^2) starts from rest and is restrained by mass B, which is supported by a cable wound around the barrel as shown (Figure P14–76). Determine the mass of B so that the barrel has a velocity of 1.2 m/s at the bottom of the slope.

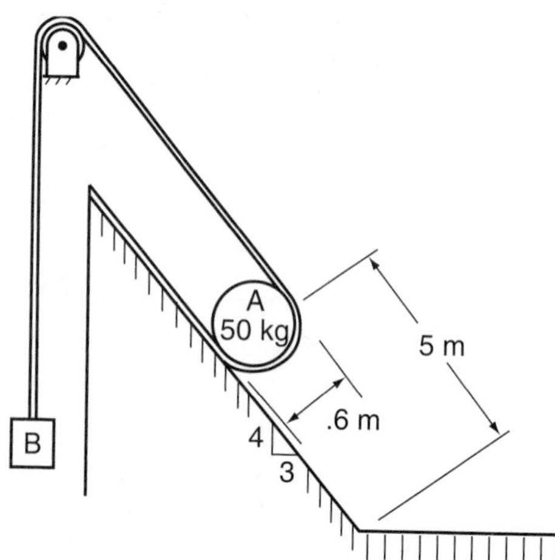

FIGURE P14–76

14–77. Cylinder A (Figure P14–77) has a mass of 20 kg ($I_C = 0.9$ kg·m^2). The spring has a free length of 0.5 m and a spring constant of 600 N/m. Assuming no slipping and neglecting the inertia of the cable and pulley, determine (a) the distance B will drop if lowered slowly and (b) the velocity of B if it is released from the position shown and has dropped 0.4 m.

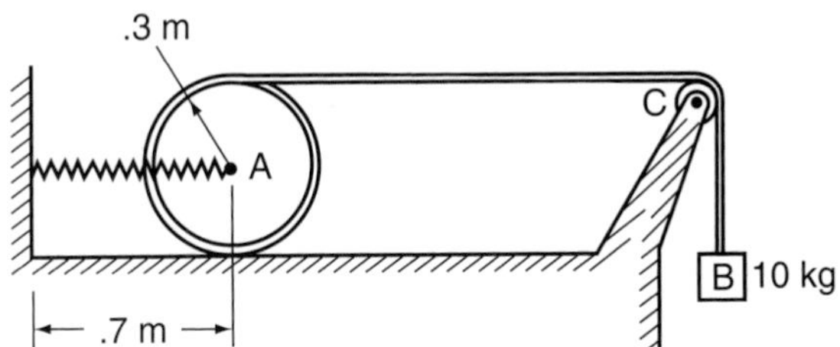

FIGURE P14–77

14–78. The cart shown in Figure P14–78 is released from rest. The spring is at its free length initially. Determine the spring constant such that the cart reverses direction after rolling 3 m down the slope. The body of the cart has a mass of 50 kg. Each wheel has a mass of 10 kg, a diameter of 0.4 m, and $I_C = 0.2$ kg·m^2.

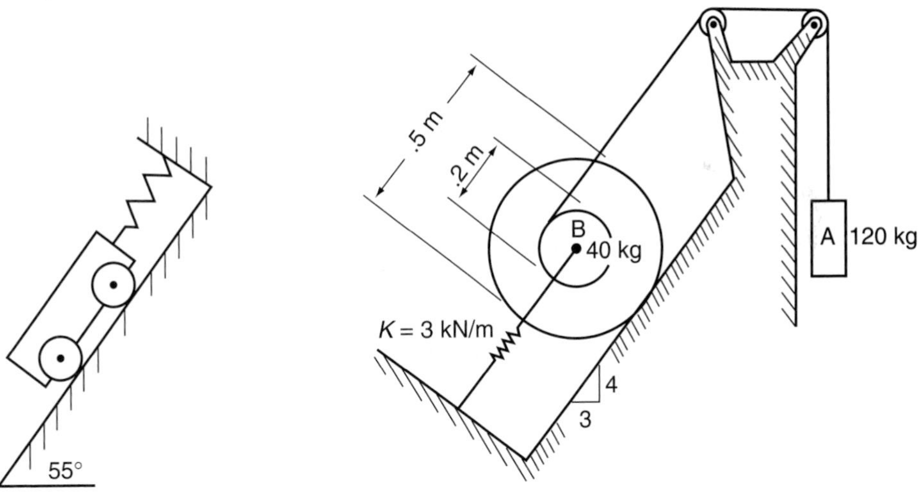

FIGURE P14–78

FIGURE P14–79

14–79. The spring is stretched 0.1 m in the position shown in Figure P14–79. Assume no slipping and neglect the weight of the pulleys and cable. Use $I_C = \frac{1}{2}mr^2$ for roller B. If the system is released from rest, determine the velocity of A after it drops 0.3 m.

14–80. The spring is stretched 0.2 m from free length in the position shown (Figure P14–80). $I_C = 1.12$ kg·m^2 for cylinder A. If there is no slipping and the system is released from rest, find the velocity of B after it has moved 0.25 m.

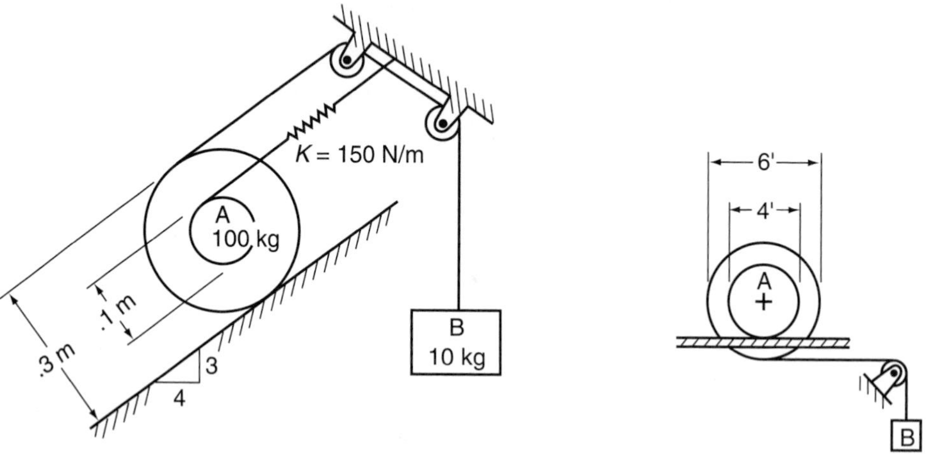

FIGURE P14–80

FIGURE P14–81

14–81. Block B in Figure P14–81 weighs 20 lb, and A weighs 32.2 lb. If they start at rest, determine the angular velocity of A when B has dropped 10 ft. (Assume no slippage of A.) The mass moment of inertia of A about its center is 2.8 ft-lb-s^2.

14–82. A belt wound around cylinder B in Figure P14–82 supports cylinder A. If the system is initially at rest, determine the angular velocity of A when it has dropped 1.5 m.

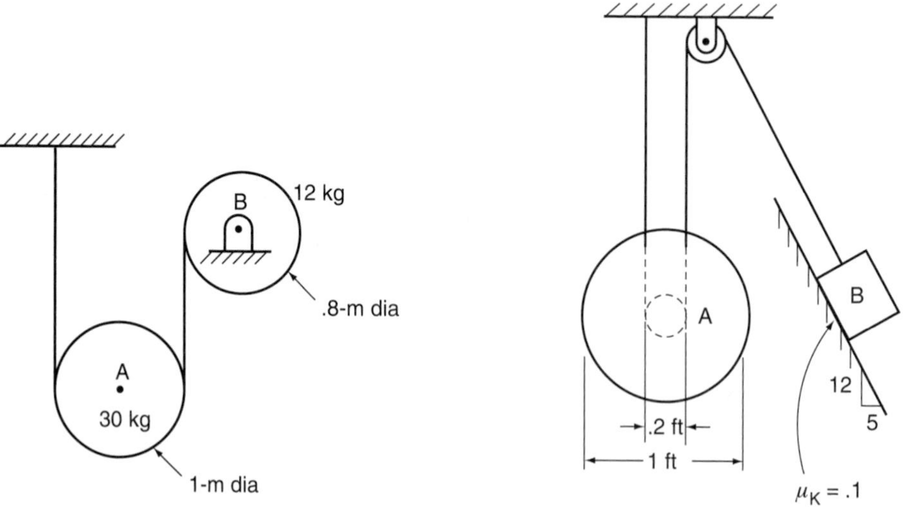

FIGURE P14–82

FIGURE P14–83

14–83. Roller A weighs 200 lb ($I_C = 0.75$ ft-lb-s^2) and block B weighs 130 lb (Figure P14–83). If the system is released from rest, determine the velocity of B after it has moved 3 ft. Neglect the weight of the cable and pulley. Assume no slipping between roller A and the cable.

14–84. Cylinder A does not slip and the spring is stretched 3 in. in the position shown (Figure P14–84). If the system is released from rest, determine the velocity of B after cylinder A has rolled 3 in. down the slope.

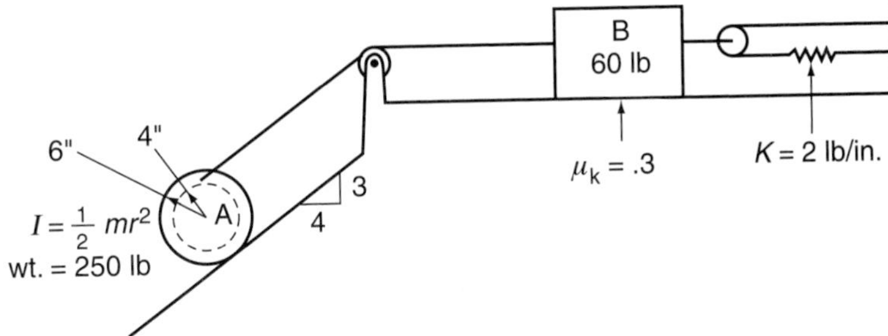

FIGURE P14–84

14–85. The system shown in Figure P14–85 is initially at rest. Roller A weighs 80 lb $(I_C = \frac{1}{2}mr^2)$. The spring has a spring constant of 1 lb/in. and is presently stretched 2 in. Block B weighs 60 lb and has a uniform friction force of 20 lb. Determine the velocity of B after it has moved 6 in. to the left.

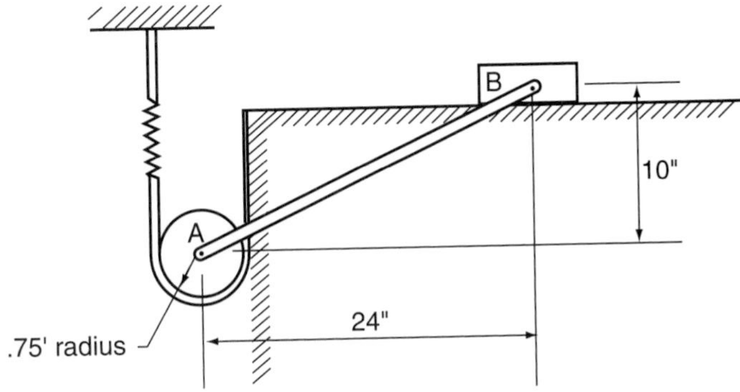

FIGURE P14–85

14–86. The 200-lb barrel shown in Figure P14–86 $(I_C = \frac{1}{2}mr^2)$ starts from rest and rolls 15 ft down the slope before contacting the spring shock absorber at A. At the instant of contact, determine the linear velocity of the barrel, the linear kinetic energy, and the angular kinetic energy.

If the shock absorber deflects 9 inches from free length in bringing the barrel to a stop, determine the spring constant. The coefficient of kinetic friction is 0.3 for all surfaces. Neglect the weight of the shock absorber.

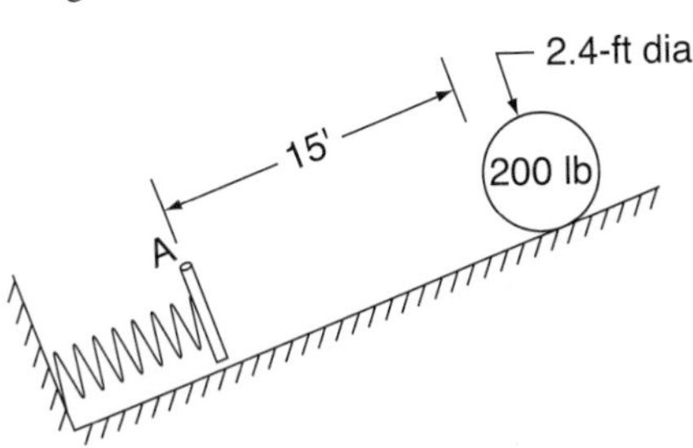

FIGURE P14–86

14–87. For the system shown (Figure P14–87), determine the velocity of A after it moves 15 in. from its initial position of rest.

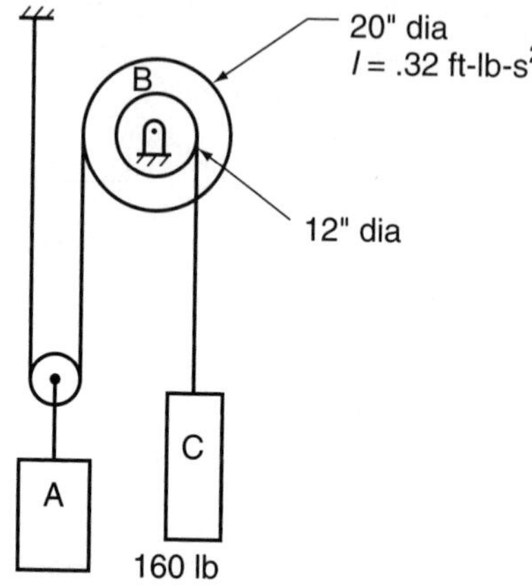

FIGURE P14–87 140 lb

14–88. If the system shown in Figure P14–88 is released from rest, determine the velocity of cart A after it moves 9 in. along the slope. Neglect the weight of the pulleys and cable.

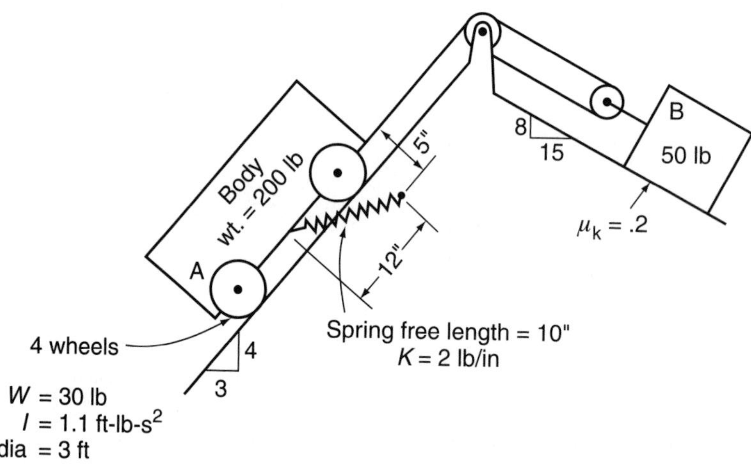

FIGURE P14–88

APPLIED PROBLEMS FOR SECTION 14–9

14–89. What thrust is being provided by the propeller of a boat if the propeller uses 22 hp to produce a boat speed of 16 mph?

14–90. A winch 2 ft in diameter lifts a 3000-lb weight at 10 ft/s. What horsepower and torque must be supplied to the winch drum?

14–91. What is the power output of a motor that provides a torque of 4 kN·m while running at 1600 rpm?

14–92. A diesel engine has a maximum rated torque of 175 lb-ft when its power output is 50 hp. At what engine rpm does this occur?

14–93. A hydraulic hoist is to lift 3 tonnes a height of 4 m in 15 seconds. To what power requirement is this equivalent (1 tonne = 1000 kg)?

14–94. Water (1000 kg/m^3) is pumped at a rate of 8 liters per second to a height of 6 meters. What is the power output of the pump motor?

14–95. Determine the hp produced by a locomotive traveling at 28 mph while providing a draw-bar pull of 23,000 lb.

14–96. A 75% efficient hoist is powered by a motor with 8 kW of output. Determine the velocity at which it could lift 1 tonne of material.

14–97. What would be the horsepower output of a 107-kW motor that is 90% efficient?

14–98. A 15-kW induction motor may be designed to run at 1165 rpm or 1750 rpm. Compare the torque developed at each speed.

14–99. A ski lift is to carry 60 riders, at 75 kg each, at 3 m/s up a 40° slope. If the lift is 250 m long, determine the power required, assuming no losses.

14–100. A driver notices that if he shifts to neutral, his 1000-kg car will slow from 70 km/hr to 60 km/hr in 4 seconds on level ground. What power will be required to maintain 65 km/hr (a) on level ground and (b) up a 6° slope?

14–101. A 160-ton string of box cars is pulled up a 2.5% slope by a 400-hp engine. If the rolling resistance is 2600 lb, determine the maximum speed of the train.

14–102. The centrifugal discharge bucket elevator in Figure P14–102 must lift 40 tons per hour to a vertical height of 80 ft. If the drive efficiency is 75%, determine the required output horsepower of the motor that drives the elevator.

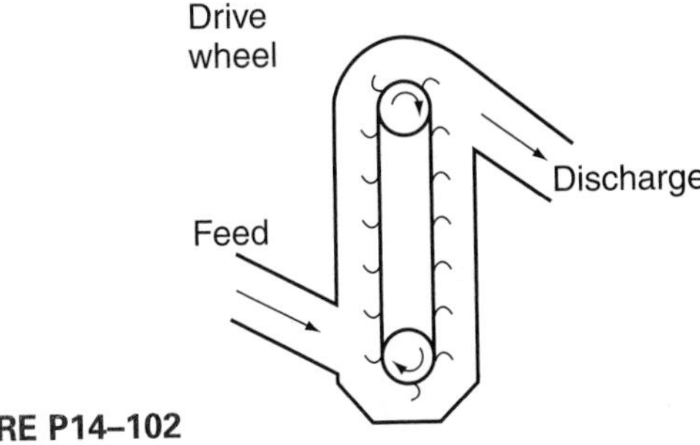

FIGURE P14–102

14–103. Determine the horsepower input to the drive pulley while rotating at 50 rpm and producing upward acceleration of the elevator car shown in Figure P10–18.

REVIEW PROBLEMS

R14–1. A 200-kg block is slowly pushed a distance of 80 m up a slope that makes an angle of 30° with the horizontal. Determine the work done by a horizontally applied force *P* if the coefficient of friction is 0.3.

R14–2. A spring is 8 in. long when loaded with 15 lb and 6.5 in. long when loaded with 25 lb. Determine the spring constant.

R14–3. A spring in the vertical position, when carrying a 30-kg load, is stretched to a total length of 150 mm. When carrying 50 kg, its length is 200 mm. Determine the work required to stretch the spring from a total length of 100 mm to 250 mm.

R14–4. A 70-kg man riding a 120-kg motorcycle starts from rest and accelerates at 5 m/s^2 for 5 seconds. Determine the total kinetic energy at $t = 5$ seconds.

R14–5. The system shown in Figure RP14–5 has zero tension in the rope at C and is released from rest. Find the linear velocity of A after it has moved 1.5 ft.

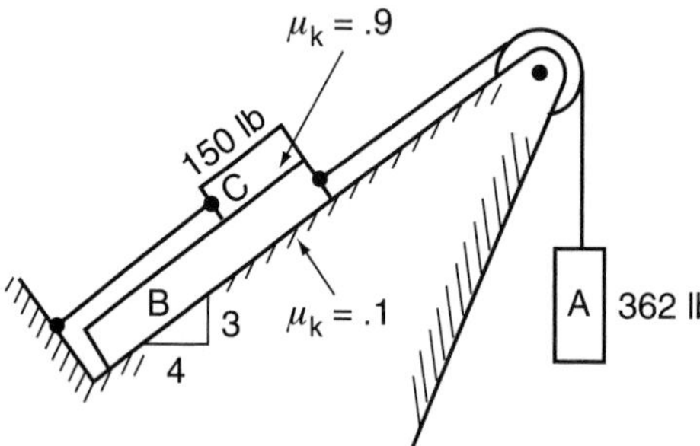

FIGURE RP14–5

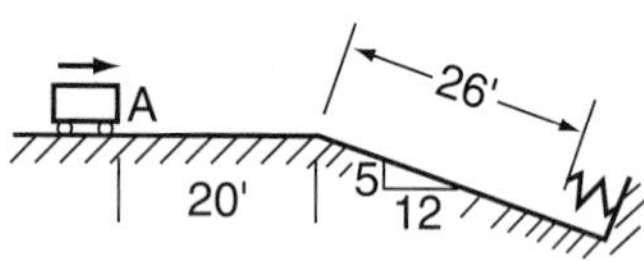

FIGURE RP14–6

R14–6. Cart A, weighing 390 lbs, has an initial velocity of 15 ft/s at the position shown in Figure RP14–6. If the cart has a constant rolling resistance of 10 lb, determine the spring deflection and the location of the cart with respect to its initial position when it comes to rest on the horizontal surface. The spring constant is 300 lb/in.

R14–7. A shaft, 5 ft long and 4 in. in diameter, weighs 220 lb. A 322-lb rotor with a radius of gyration of 15 in. is mounted on the shaft. For a speed of 1500 rpm, determine the kinetic energy of (a) the shaft, (b) the rotor, and (c) the shaft and rotor combined.

R14–8. Determine the work done in turning the pulley shown (Figure RP14–8) through two revolutions counterclockwise.

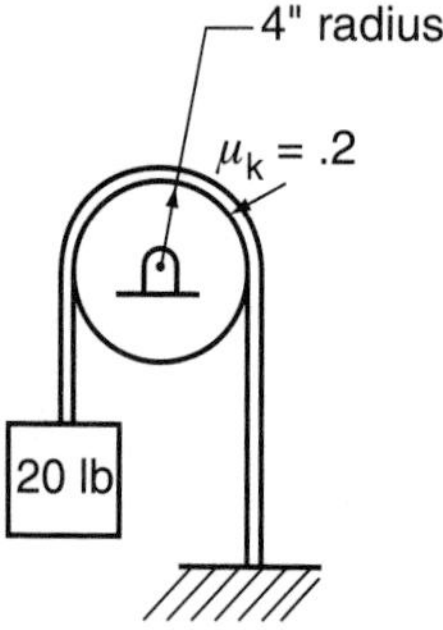

FIGURE RP14–8

R14–9. A 2500-lb car, in gear with the clutch depressed, starts from rest and accelerates down a 10° slope. After it has traveled 100 ft, the clutch is released. The engine does not start and the car comes to rest in a distance of 15 ft. Neglect the rotational kinetic energy of the wheels, and assume that both wheels drive equally and that there is no rolling resistance. Determine (a) the car velocity just before the clutch is released and (b) the average torque applied to the drive wheels during the 15-ft interval if the wheels are 20 in. in diameter.

R14–10. The combined wheel and hub in Figure RP14–10 has a mass of 25 kg and a mass moment of inertia of 27 kg·m^2. The mass of block B is 12 kg. If the center of A has a velocity of 1.5 m/s to the left at the instant shown and rolls without slipping, determine the distance that B will move before bringing the system to rest.

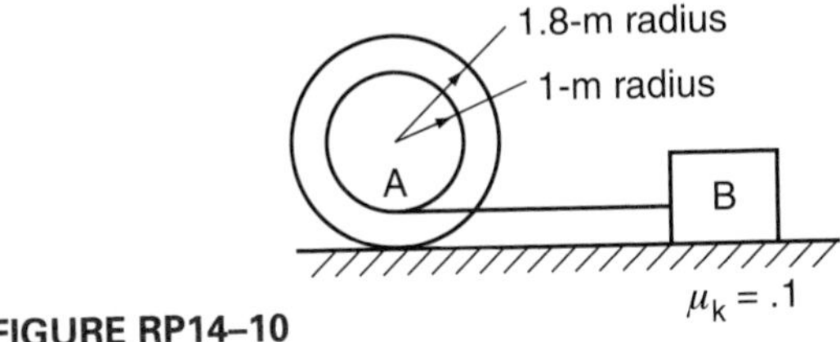

FIGURE RP14–10

R14–11. Determine the velocity of the cart (Figure RP14–11) if, starting from rest, it has moved 2 m down the slope. Neglect the weight of the pulley and cable.

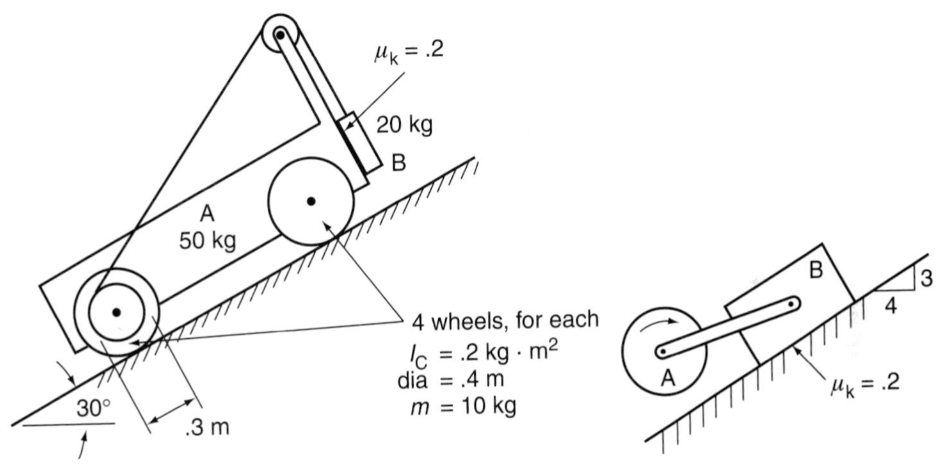

FIGURE RP14–11

FIGURE RP14–12

R14–12. Wheel A $(I_C = \frac{1}{2}mr^2)$ has a mass of 10 kg and a radius of 0.3 m (Figure RP14–12). The mass of the arm is 5 kg and block B is 15 kg. Initially, block B is at rest, and wheel A is rotating at 150 rpm clockwise. Wheel A is lowered to the sloped surface. (Assume no slipping between wheel A and the surface.) Determine (a) the velocity of block B after it has moved 0.15 m up the slope and (b) the compressive load in each arm.

R14–13. An electric traction motor provides a torque of 400 N·m while rotating through 60 revolutions in 5 seconds. If the motor is 90% efficient, determine the electrical power input.

R14–14. Driving a 2000-ft loaded horizontal conveyor at 4 ft/s requires 35 hp. What horsepower would be required if the conveyor is now raised at one end, giving a gradient of 1 ft of rise in 60 ft of horizontal run? Each foot of conveyor carries a load of 20 lb.

CHAPTER 15

Impulse and Momentum

OBJECTIVES

Upon completion of this chapter the student will be able to:

1. Use the impulse momentum method to solve problems with the following motions: linear, angular, plane, and combined linear and angular.
2. Apply the conservation of momentum method to solve for various velocities for both linear and angular motion.

15–1 LINEAR IMPULSE AND MOMENTUM

An object that is subjected to a force for a very short time is said to have received an *impulse;* impulse is equal to force times time. The movement or change in velocity of the object depends on the mass of the object and is described by the term *momentum,* which is equal to mass times velocity.

A baseball struck by a bat receives an impulse in a short period of time that results in a momentum that may by sufficient to clear the outfield fence.

The object receiving a sudden blow has in effect had a variable force applied. Considering the average value of this force, we can write

$$F = ma$$

where

$$a = \frac{v}{t}$$

Therefore,

$$F = \frac{mv}{t}$$

$$Ft = mv \qquad (15\text{–}1)$$

where Ft is *impulse* and mv is *momentum.* The change in momentum of an object is equal to the impulse applied to the object.

In the SI system,

$$\begin{aligned} F &= \text{force; N} \\ t &= \text{time; s} \\ m &= \text{mass; kg} \\ v &= \text{velocity; m/s} \end{aligned}$$

Substituting these units into Equation (15–1), we have

$$\text{N(s)} = \text{kg(m/s)}$$

but

$$1\ \text{N} = (1\ \text{kg})(1\ \text{m/s}^2)$$

or

$$1\ \text{kg} = 1\ \text{Ns}^2/\text{m}$$

Substituting this into $Ft = mv$ (Equation 15–1), we have

$$\begin{aligned} \text{N(s)} &= \frac{\text{Ns}^2}{\text{m}}\left(\frac{\text{m}}{\text{s}}\right) \\ \text{N(s)} &= \text{N(s)} \end{aligned}$$

For linear motion, the units of both impulse and momentum are newton-seconds.

The common units in the U.S. Customary system are

$$\begin{aligned} F &= \text{force; lb} \\ t &= \text{time; s} \\ m &= \text{mass; slugs} \quad \text{or} \quad \frac{\text{lb-s}^2}{\text{ft}} \\ v &= \text{velocity; ft/s} \end{aligned}$$

The units of impulse and momentum can be found by substitution of these units in Equation (15–1) ($Ft = mv$).

$$\begin{aligned} \text{lb(s)} &= \text{lb-s}^2/\text{ft(ft/s)} \\ \text{lb-s} &= \text{lb-s} \end{aligned}$$

To handle situations in which the initial velocity is not zero, Equation (15–1) can be written as

$$F\Delta t = m\Delta v$$

This equation is the basis of the impulse = momentum method referred to earlier. **Both impulse and momentum are vector quantities; therefore, direction must be taken into account when this formula is used.**

Although our initial description of impulse was that of a force applied for a short period of time, the length of time can be longer, as the following example will show.

EXAMPLE 15–1

If there is a constant towing force of 1.3 kN, how long would it take a tow truck to tow a 1500-kg car from rest to 90 km/h? (Assume a rolling resistance of 200 N.)

$$Ft = mv$$

$$(1300\text{ N} - 200\text{ N})t = (1500\text{ kg})\frac{(90{,}000\text{ m/h})}{(3600\text{ s/h})}$$

$$\underline{t = 34.1\text{ s}}$$

EXAMPLE 15–2

A 1.6-oz golf ball is struck by a club so that a force of 18 lb is applied for 0.04 second. What is the velocity of the ball due to this impulse?

$$Ft = mv$$

$$(18\text{ lb})(0.04\text{ s}) = \frac{(1.6\text{ oz})}{(16\text{ oz/lb})}\left(\frac{1}{32.2\text{ ft/s}^2}\right)v$$

$$\underline{v = 232\text{ ft/s}}$$

15–2 ANGULAR IMPULSE AND MOMENTUM

Rotating objects are also subject to impulse. The change in motion is not simply due to a force but rather to a force acting at some radius. *Angular impulse* is therefore due to a torque. From Section 13–4 we can write

$$(\text{torque})\ T = I_C\alpha \text{ where } \alpha = \frac{\omega}{t}$$

Therefore,

$$T = \frac{I_C\omega}{t}$$

$$Tt = I_C\omega \tag{15–2}$$

where, in the SI system,

T = torque; N · m
t = time; s
I_C = mass moment of inertia about the center of mass; kg · m^2
ω = angular velocity; rad/s

In the U.S. Customary system, the units are

T = torque; lb-ft
t = time; s
I_C = mass moment of inertia about the center of mass; ft-lb-s^2
ω = angular velocity; rad/s

To handle all cases of speed change in a given time period—not just those in which the object is initially at rest—Equation (15–2) could also be written as

$$T\Delta t = I_C \Delta\omega$$

This formula could be applied to pure rotation, such as occurs in a flywheel. When we come to plane motion such as with a rolling wheel, linear impulse and linear momentum must also be considered.

EXAMPLE 15–3

A torque of 30 lb-ft is applied to a turbine wheel that has a mass moment of inertia of 60 ft-lb-s^2 about its center of rotation. Neglect bearing friction and determine the angular velocity that the turbine acquires in 3 seconds from an initial state of rest.

$$Tt = I_C\omega$$

$$(30 \text{ lb-ft})(3 \text{ s}) = (60 \text{ ft-lb-s}^2)\omega$$

$$\underline{\omega = 1.5 \text{ rad/s}}$$

EXAMPLE 15–4

A belt-driven pulley is used to drive a machine that has a constant torque requirement of 60 N · m. The pulley has a mass moment of inertia, $I_C = 0.7$ kg · m^2, a diameter of 0.4 m, and a mass of 25 kg. The pulley was initially turning at 300 rpm, but the belt tensions were increased to 500 N and 150 N. Determine the new pulley speed after 3 seconds at the new tensions.

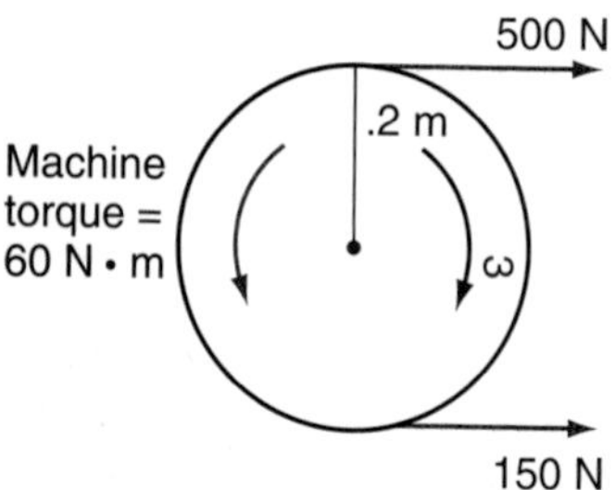

FIGURE 15–1

$$\omega_0 = \frac{(300 \text{ rev/min})(2\pi \text{ rad/rev})}{60 \text{ s/min}} = 31.4 \text{ rad/s}$$

The net torque on the pulley is input torque – output torque (Figure 15–1).

$$\text{net torque} = [(500\text{ N} - 150\text{ N})0.2\text{ m}] - 60\text{ N}\cdot\text{m}$$
$$\text{net torque} = 10\text{ N}\cdot\text{m}$$

Using Equation 15-2

$$\text{impulse change} = \text{momentum change}$$
$$\Delta Tt = \Delta I\omega$$

or

$$(\text{net torque})(t) = I(\Delta\omega)$$
$$(10\text{ m})(3\text{ s}) = (0.7\text{ kg}\cdot\text{m}^2)(\omega - 31.4\text{ rad/s})$$
$$\omega = 74.3\text{ rad/s}$$
$$\underline{\omega = 709\text{ rpm}}$$

EXAMPLE 15–5

A drum with a mass of 120 kg and 0.4 m in diameter rolls down a plane inclined 15° to the horizontal. If the drum's initial speed is 0.8 m/s and no slippage occurs, determine its speed 10 seconds later.

The mass moment of inertia is

$$I_C = \frac{1}{2}mr^2$$
$$= \frac{1}{2}(120\text{ kg})(0.2\text{ m})^2$$
$$I_C = 2.4\text{ kg}\cdot\text{m}^2$$

The initial velocities are

$$v_0 = 0.8\text{ m/s}$$
$$\omega_0 = \frac{v_0}{r} = \frac{0.8\text{ m/s}}{0.2\text{ m}} = 4\text{ rad/s}$$

The final velocities are v and ω

$$\text{where } \omega = \frac{v}{0.2}$$

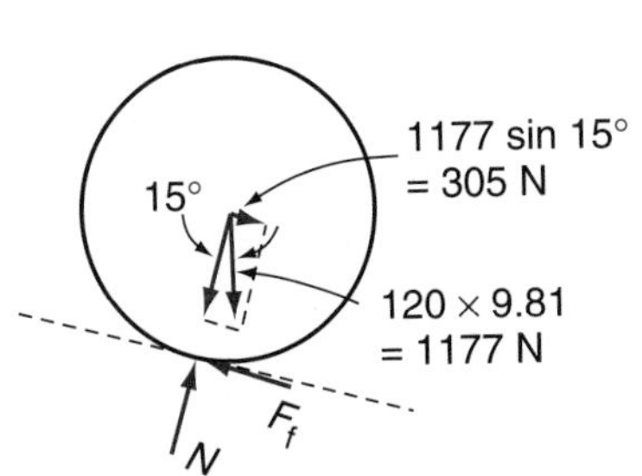

FIGURE 15–2

The free-body diagram, Figure 15–2, shows that the motion is caused by a force of 305 N and the friction force F_f. Since slipping is not impending, the friction force F_f cannot be found immediately. The impulse-momentum equation for linear motion is used first.

$$F\Delta t = m\Delta v$$
$$(305\,\text{N} - F_f)10\,\text{s} = (120\,\text{kg})\,(v - 0.8\,\text{m/s})$$
$$F_f = 315 - 12v \quad (1)$$

Writing the impulse-momentum equation for angular motion, we have

$$T\Delta t = I_C\Delta\omega$$
$$(F_f)(0.2\,\text{m})(10\,\text{s}) = (2.4\,\text{kg}\cdot\text{m}^2)\left(\frac{v}{0.2\,\text{m}} - 4\,\text{rad/s}\right)$$
$$2F_f = 12v - 9.6 \quad (2)$$

Substituting Equation (1) into Equation (2), we get

$$2(315 - 12v) = 12v - 9.6$$
$$\underline{v = 17.8\,\text{m/s}}$$

15–3 CONSERVATION OF MOMENTUM

A moving object may transfer or lose some of its momentum to another object. The momentum is transferred by means of impulse. Consider a fast-moving object that strikes and becomes attached to a slow-moving object traveling in the same direction. The fast-moving object exerts a force or impulse on the slow-moving object, thereby speeding it up. The second object at that instant exerts an equal and opposite reaction or impulse on the first object, thereby slowing it down. This action is summarized in Newton's third law, which states: For every action there is an equal and opposite reaction.

While each object has experienced a change in velocity and therefore a change in momentum, the total momentum of the system remains the same. This is, of course, known as *conservation of momentum* and applies to both linear momentum and angular momentum. In this section, we will consider only *inelastic* types of impact, that is, where the objects become attached to each other on impact. Note two other points: we are talking about vector quantities here, and there are no external forces being applied to the system.

Referring to Figure 15–3, we see that conservation of linear momentum can be written as

$$\text{initial momentum} = \text{final momentum}$$
$$(m_A v_A)_1 + (m_B v_B)_1 = (m_A v_A)_2 + (m_B v_B)_2$$

since

$$v = (v_A)_2 = (v_B)_2$$
$$m_A v_A + m_B v_B = (m_{A+B})v$$

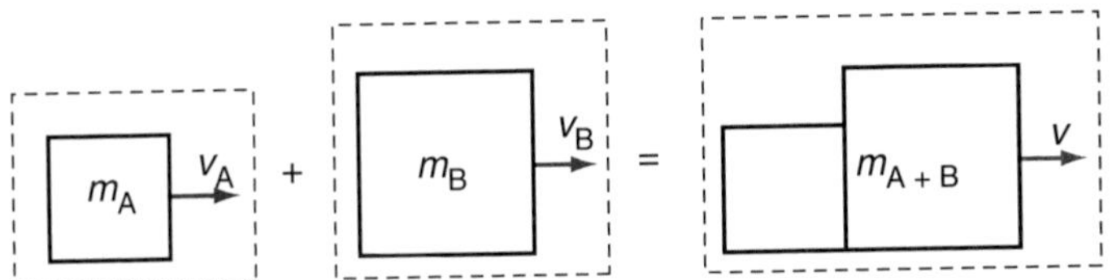

FIGURE 15–3

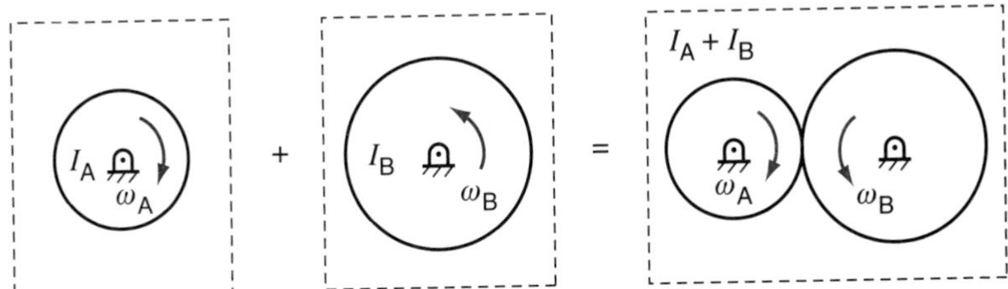

FIGURE 15–4

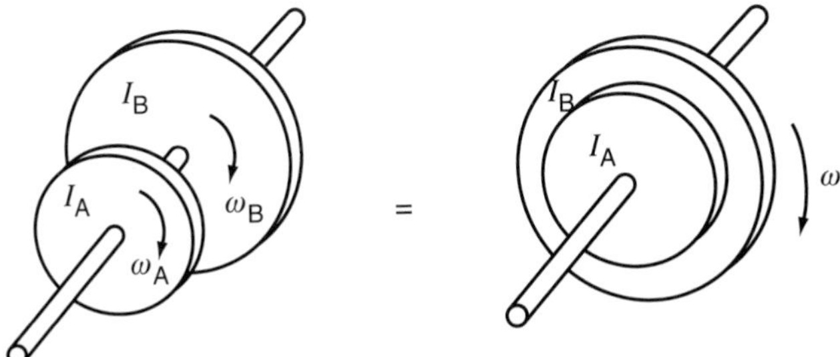

FIGURE 15–5

Angular momentum can be shown in a similar manner (Figure 15–4).

$$\text{initial momentum} = \text{final momentum}$$

$$(I_A\omega_A)_1 + (I_B\omega_B)_1 = (I_A\omega_A)_2 + (I_B\omega_B)_2 \qquad (15\text{–}3)$$

In Figure 15–5, there is a common angular velocity, ω; therefore,

$$(I_A\omega_A)_1 + (I_B\omega_B)_1 = (I_A\omega)_2 + (I_B\omega)_2$$

$$(I_A\omega_A) + (I_B\omega_B) = (I_A + I_B)_\omega \qquad (15\text{–}4)$$

In applying any of these equations, we must give velocity a negative or positive sign. Choose the direction for positive velocity and be consistent throughout.

It is not necessary to have two objects in order to apply the principle of conservation of momentum. If a single rotating object with a given initial momentum ($I\omega_0$) had its configuration changed so that its moment of inertia were reduced, there would be a corresponding increase in angular velocity. An example is a whirling figure skater who pulls her arms inward, thereby decreasing her moment of inertia and increasing her angular velocity ω.

EXAMPLE 15–6

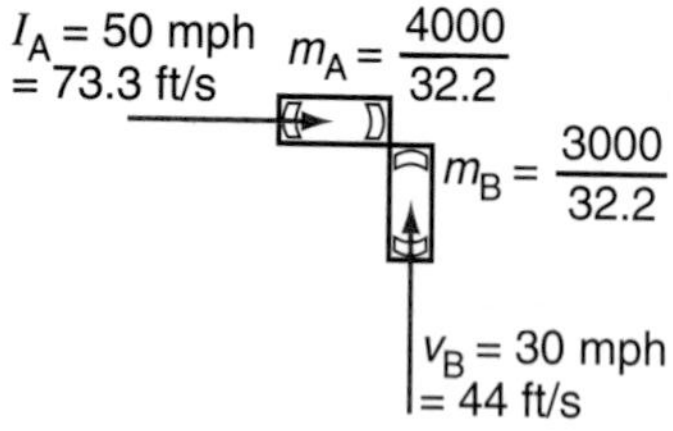

FIGURE 15–6

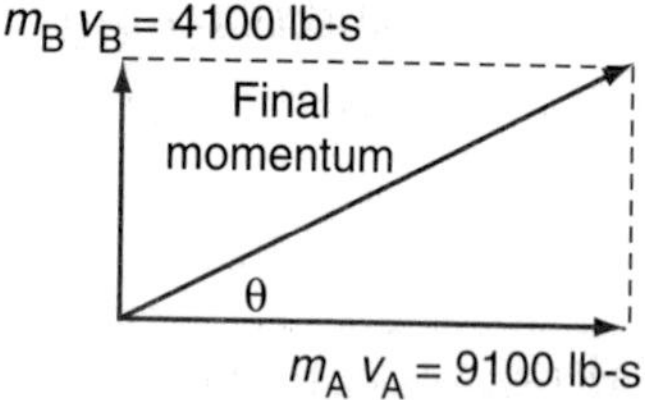

FIGURE 15–7

A 4000-lb car (A) and a 3000-lb car (B) collide on glare ice and remain together. At the time of impact, car A was traveling east at 50 mph and car B was traveling north at 30 mph. Determine their resulting velocity and direction.

The mass and velocity of each car are shown in Figure 15–6. Initial momentum equals final momentum, but the initial momentum values must be added vectorially (Figure 15–7). The direction of the final velocity will be the same as that of the final momentum.

$$m_A v_A + m_B v_B = (m_A + m_B)v$$

where

$$m_A v_A = \frac{(4000 \text{ lb})}{(32.2 \text{ ft/s}^2)}(73.3 \text{ ft/s}) = 9100 \text{ lb-s} \rightarrow$$

$$m_B v_B = \frac{(3000 \text{ lb})}{(32.2 \text{ ft/s}^2)}(44 \text{ ft/s}) = 4100 \text{ lb-s}\uparrow$$

Adding vectorially gives

$$(m_A + m_B)v = \sqrt{(9100)^2 + (4100)^2}$$

$$(m_A + m_B)v = 9980$$

$$\frac{(4000 \text{ lb} + 3000 \text{ lb})}{32.2 \text{ ft/s}^2} v = 9980$$

$$\underline{v = 45.8 \text{ ft/s} \;\angle 24.2^\circ}$$

$$\tan\theta = \frac{4100}{9100}$$

$$\underline{\theta = 24.2^\circ}$$

EXAMPLE 15–7

A whirling figure skater with her arms extended has an angular velocity of 12 rad/s. Determine her angular velocity when, by drawing her arms close to her body, she reduces her moment of inertia by 40%.

$$\text{initial momentum} = \text{final momentum}$$
$$I_0\omega_0 = I\omega$$
$$I_0(12\ \text{rad/s}) = (0.6I_0)\,\omega$$
$$\underline{\omega = 20\ \text{rad/s}}$$

EXAMPLE 15–8

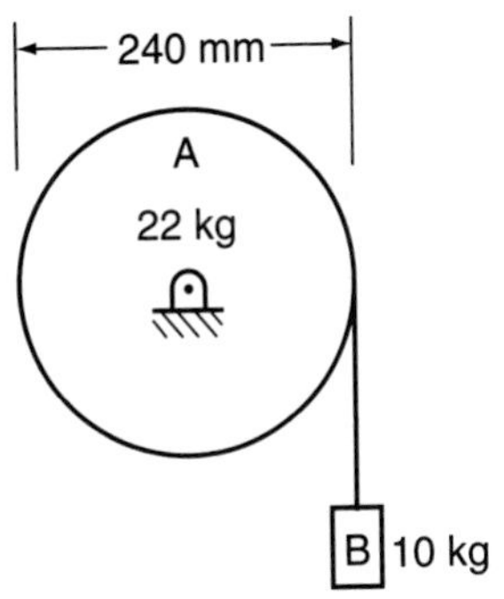

FIGURE 15–8

Cylinder A in Figure 15–8 has a radius of gyration of 85 mm. The system is initially at rest and has no bearing friction. Determine the angular velocity of A, the angular acceleration of A, and the tension in the rope at $t = 3$ seconds.

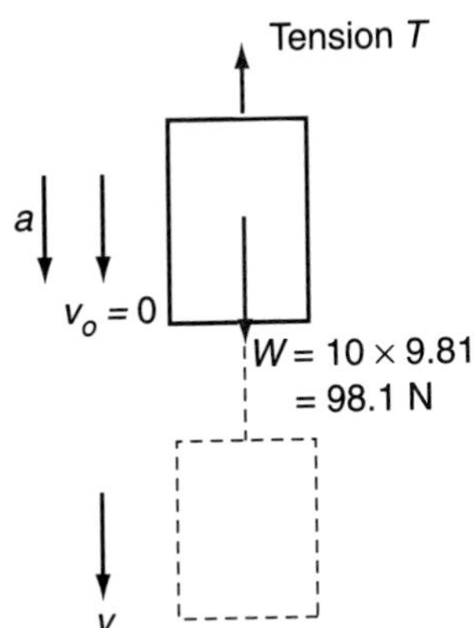

FIGURE 15–9

This example is similar to Example 13–9, which was solved by the force-inertia method. First, we will apply the impulse-momentum equation to block B (Figure 15–9).

$$\text{net impulse change} = \text{momentum change}$$
$$\Delta Ft = \Delta mv$$
$$(W - T)\,t = m(v - 0)$$
$$(98.1\ \text{N} - T)(3\text{s}) = (10\ \text{kg})v$$
$$T = 98.1 - 3.33v$$

$$\text{substitute}\quad v = r\omega = (0.12\ \text{m})\,\omega$$

$$\text{to get}\quad T = 98.1 - 0.4\omega \qquad (1)$$

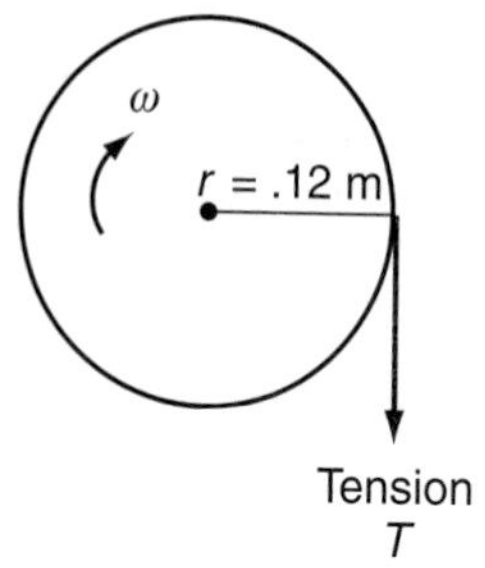

FIGURE 15–10

In preparation to use the angular impulse-momentum equation on cylinder A (Figure 15–10),

$$I_C = k^2 m$$
$$= (0.085\text{ m})^2(22\text{ kg})$$
$$I_C = 0.159\text{ kg}\cdot\text{m}^2$$

substitute into (torque)(Δt) = $I_C\omega$

$$(T)(0.12\text{ m})\,3\text{ s} = (0.159\text{ kg}\cdot\text{m}^2)\omega$$
$$0.36T = 0.159\,\omega \qquad (2)$$

Substituting Equation (1) into Equation (2), we get

$$0.36\,(98.1 - 0.4\omega) = 0.159\omega$$
$$\underline{\omega = 116\text{ rad/s}}$$

Substituting this into Equation (1), we get

$$\underline{T(\text{tension}) = 51.5\text{ N}}$$

For the cylinder,

$$\alpha = \frac{\Delta\omega}{t}$$
$$= \frac{116\text{ rad/s} - 0}{3\text{ s}}$$
$$\underline{\alpha = 38.8\text{ rad/s}^2}$$

As stated earlier, many of the problems in Chapters 13, 14, and 15 can be solved using three different methods: inertia force and inertia torque, work energy, or impulse momentum. Example 15–9 will be solved in each of these ways.

EXAMPLE 15–9

Cylinder A in Figure 15–11 rotates at 300 rpm and lifts mass B. When its source of power is removed, A and B coast to a stop in 5 seconds. Determine the mass of B.

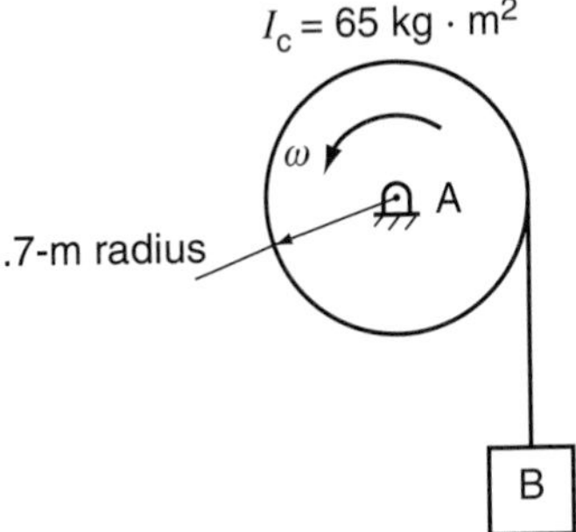

FIGURE 15–11

Method 1: Inertia Force and Inertia Torque

$$\omega_0 = 300 \text{ rpm} = (300 \text{ rev/min})\frac{(2\pi \text{ rad/rev})}{(60 \text{ s/min})} = 31.4 \text{ rad/s}$$

$$\omega = 0$$

$$t = 5 \text{ s}$$

$$\alpha = ?$$

$$\omega = \omega_0 + \alpha t$$

$$0 = (31.4 \text{ rad/s}) + (\alpha)(5 \text{ s})$$

$$\alpha = -6.28 \text{ rad/s}^2$$

(deceleration)

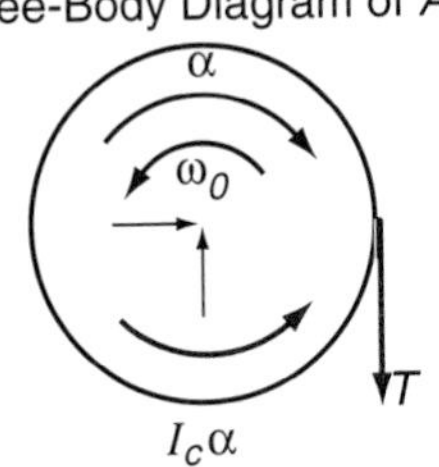

FIGURE 15–12

Take moments about the center of cylinder A (Figure 15–12).

$\Sigma M_c = 0$

$$T(\text{r}) = I_c\alpha$$

$$T(.7 \text{ m}) = (65 \text{ kg} \cdot \text{m}^2)(6.28 \text{ rad/s}^2)$$

$$T = 583 \text{ N}$$

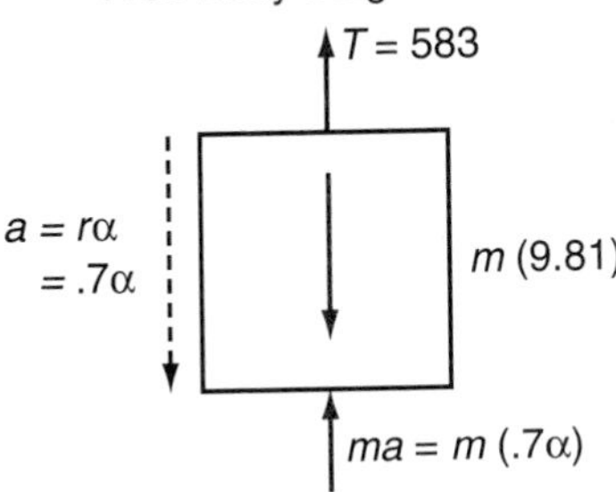

FIGURE 15–13

Summing vertical forces in Figure 15–13

$\Sigma F_y = 0$

$$583 \text{ N} + m(.7 \text{ m})(6.28 \text{ rad/s}^2) = m(9.81 \text{ m/s}^2)$$

$$\underline{m = 108 \text{ kg}}$$

Method 2: Work Energy

$\omega_0 = 31.4$ rad/s $\qquad \alpha = ?$

$\omega = 0 \qquad \theta = ?$

$t = 5$ s

$$\omega = \omega_0 + \alpha t$$
$$0 = 31.4 \text{ rad/s} + \alpha(5 \text{ s})$$
$$\alpha = -6.28 \text{ rad/s}^2 \text{ (deceleration)}$$
$$\theta = \omega_0 t + \frac{1}{2}\alpha t^2$$
$$= (31.4 \text{ rad/s})(5 \text{ s}) - \frac{1}{2}(6.28 \text{ rad/s}^2)(5 \text{ s})^2$$
$$= 78.5 \text{ rad}$$

The change in elevation for B is:

$$s = r\theta$$
$$= (0.7 \text{ m})(78.5 \text{ rad})$$
$$= 54.95 \text{ m}$$

Initial velocity of B is:

$$v = r\omega$$
$$= (0.7 \text{ m})(31.4 \text{ rad/s})$$
$$= 21.98 \text{ m/s}$$

Now equate the changes in energy

$$\Delta \text{KE}_\text{A} + \Delta \text{KE}_\text{B} = \Delta \text{PE}_\text{B}$$
$$\frac{1}{2}I\omega^2 + \frac{1}{2}mv^2 = Wh$$
$$\frac{1}{2}(65 \text{ kg} \cdot \text{m}^2)(31.4 \text{ rad/s})^2 + \frac{1}{2}m(21.98 \text{ m/s})^2 = m(9.81 \text{ m/s}^2)(54.95 \text{ m})$$
$$32043 + 241.56\, m = 539\, m$$
$$\underline{m = 108 \text{ kg}}$$

Method 3: Impulse Momentum

$$\omega_0 = 31.4 \text{ rad/s}$$
$$\omega = 0$$
$$t = 5 \text{ s}$$

Considering cylinder A alone,

$$(\text{Torque})\,t = I\Delta\omega$$
$$(\text{T})(0.7\text{ m})(5\text{ s}) = (65\text{ kg}\cdot\text{m}^2)(31.4\text{ rad/s})$$
$$T = 583\text{ N}$$

Considering mass B alone (Figure 15–14),

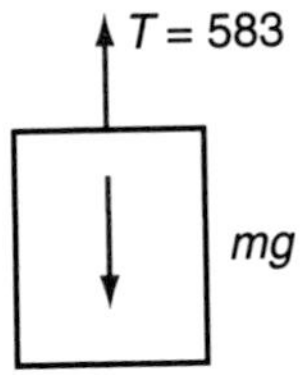

FIGURE 15–14

$$Ft = mv \quad \text{where} \quad v = r\omega = 0.7(31.4)$$
$$(mg - 583\text{ N})\,5\text{ s} = (m)(0.7\text{ m})(31.4\text{ rad/s})$$
$$(9.81\,m - 583)\,5 = 21.98m$$
$$49.05\,m - 21.98\,m = 2915$$
$$\underline{m = 108\text{ kg}}$$

Note that the length of each method of solution can vary depending upon what information is given initially.

HINTS FOR PROBLEM SOLVING

1. Problems similar to those of Chapter 15 have been solved before in two different ways.
   (a) In Chapter 13, the inertia-force or torque method was used.
   (b) In Chapter 14, the work-energy method was used.
   In Chapter 15, we are using the impulse-momentum method, where impulse is the sum of the forces or the unbalance of forces multiplied by the time.
2. Be careful and consistent with units. For the equation impulse = momentum, the units are
   (a) Linear: $Ft = mv$

   (1) U.S. Customary $$(\text{lb})(\text{s}) = \frac{\text{lb}}{\text{ft/s}^2}(\text{ft/s})$$

   or $$\text{lb-s} = \text{lb-s}$$

   (2) SI $$\text{N}\cdot\text{s} = \text{kg}\left(\frac{\text{m}}{\text{s}}\right)$$

   or $$\text{N}\cdot\text{s} = \text{N}\cdot\text{s}$$

(b) Angular: $Tt = I_C\omega$

(1) U.S. Customary $$(\text{lb-ft})(\text{s}) = \text{ft-lb-s}^2\left(\frac{\text{rad}}{\text{s}}\right)$$

(2) SI $$(\text{N}\cdot\text{m})(\text{s}) = \text{kg}\cdot\text{m}^2\left(\frac{\text{rad}}{\text{s}}\right)$$

3. Remember that you are dealing with *vector* quantities in the conservation of momentum equation.

4. For problems with plane motion, write two equations:

(a) $Ft = mv$ for linear motion

(b) $Tt = I_C\omega$ for angular motion, where T is the torque about the center of mass Then solve the two equations simultaneously.

PROBLEMS

APPLIED PROBLEMS FOR SECTION 15–1

15–1. Determine the momentum of a 90-kg motorcycle traveling at a velocity of 60 m/s.

15–2. Determine the momentum of an 8-ton truck traveling at 55 mph.

15–3. A 2-lb object on a smooth horizontal surface is subject to a horizontal force of 1 lb for 20 seconds. Determine the velocity at $t = 20$ seconds.

15–4. For how long does a 200-N force have to be applied to increase the velocity of a 30-kg mass from 80 to 150 m/s?

15–5. A soccer player kicks a 0.3-lb ball with an average force of 5 lb, giving it a velocity of 75 ft/s. For how long was the force applied?

15–6. A mass of 10 kg has its velocity increased from 3 m/s to 15 m/s in a time of 2 seconds. Determine the force required.

15–7. A 1300-kg stationary car whose brakes are not applied is subjected to a force of 5 kN for 2 seconds when struck in the rear by another car. What velocity is imparted to the car? If the car that was struck had weighed 50% less, what would have been its velocity? Assume that the 5-kN force is applied for the same length of time.

15–8. A 4000-lb car is to be towed from rest by means of a rope. A force of 800 lb is applied for 3 seconds before the rope breaks. Determine the velocity of the car.

15–9. Force P is applied as shown in Figure P15–9 for 7 seconds. Determine the velocity increase of the block at $t = 7$ seconds.

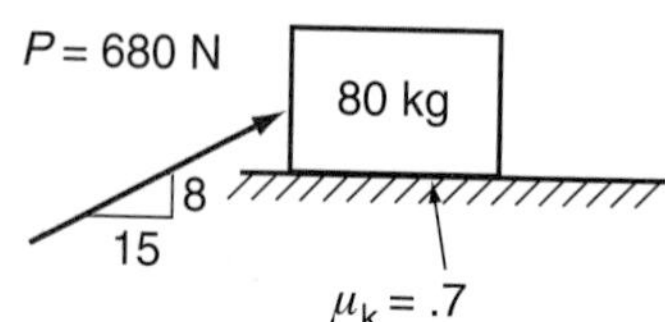

FIGURE P15–9

15–10. How long will it take block A in Figure P15–10 to come to rest ($\mu_k = 0.6$)?

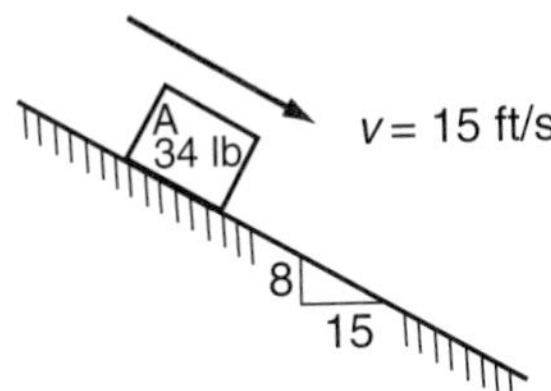

FIGURE P15–10

15–11. Determine the force required to bring an 8-ton truck traveling at 50 mph to rest in 5 seconds. How far does it travel?

15–12. A 30-g bullet is fired horizontally into a 12-kg block that is suspended on a long cord. What speed of bullet would cause the center of gravity of the block to rise 8 cm?

15–13. Determine the velocity of the blocks shown in Figure P15–13 at $t = 4$ seconds, if they are initially at rest. Neglect the weight of the connecting bar.

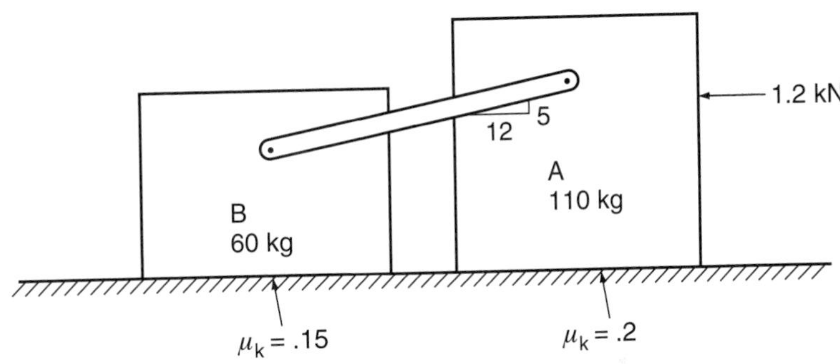

FIGURE P15–13

15–14. Solve Problem 13–10 using the impulse-momentum method. (Assume a distance of travel of 1 ft.)

15–15. Solve Problem 14–100 using the impulse-momentum method.

APPLIED PROBLEMS FOR SECTION 15–2

15–16. Calculate the angular momentum of an electric motor rotating at 1800 rpm if the rotor weighs 150 lb and has a radius of gyration of 8 in.

15–17. The rotor of an electric motor turns at 1750 rpm, weighs 150 lb, and has a radius of gyration of 9.5 in. Determine (a) the angular momentum at 1750 rpm and (b) the torque required to slow the rotor to 800 rpm in 2 seconds.

15–18. A pulp and paper machine has a roller that has a mass of 400 kg and a mass moment of inertia about its center of rotation of 30 kg · m^2 How long will it take a torque of 15 N · m to accelerate the machine from 0.8 to 2.3 revolutions per second?

15–19. A pulley weighs 200 lb and has a mass moment of inertia of 4 ft-lb-s^2. How long will it take a torque of 15 lb-in. to accelerate the pulley from 50 rpm to 150 rpm?

15–20. A flywheel (I_C=25 kg · m^2) on a baler is freewheeling at 200 rpm immediately after its shear pin fails. What tangential friction force must be applied to the flywheel (0.44 m in radius) to stop it in 3 seconds?

15–21. A small grinder consists of a grinding wheel mounted on the shaft of a 3600-rpm motor. The combined mass moment of inertia is 0.035 ft-lb-s^2. On startup it reaches 3600 rpm in 5 seconds. Determine the starting torque.

15–22. Inertia welding of two shafts end-to-end is accomplished by rotating one shaft and bringing it against the end of the other fixed shaft (shafts A and B in Figure P15–22). When sufficient heat is generated due to friction, the driving fixture on shaft A releases and shaft A comes to rest. It takes 2 seconds to weld two 20-mm-diameter, 2-m-long shafts when shaft A is initially rotating at 800 rpm. Determine the average torque contributed by shaft A alone. The density of the steel in the shafts is 0.00783 kg/cm^3.

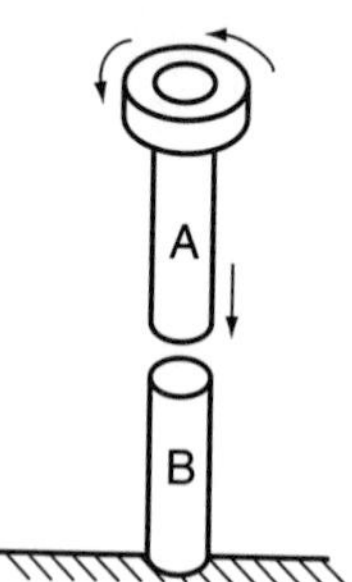

FIGURE P15–22

15–23. A helicopter rotor has a mass moment of inertia of 6500 kg·m^2. If it is initially at rest, determine its speed after a torque of 7000 N·m is applied for 45 seconds.

15–24. An 80-mm-square, 0.3-m-long wood block is mounted longitudinally in a lathe and turned at 200 rpm. It is then machined down to an 80-mm diameter. Determine the difference in angular momentum. Assume that the wood weighs 0.00084 kg/cm^3.

15–25. A steel shaft, 150 mm in diameter and 3.2 m long, transmits a constant torque of 300 N · m at 2400 rpm. Assume that the machine it is driving has no rotational inertia, and determine how long it would take the shaft to coast to a stop if its input power were removed. The steel weighs 0.00783 kg/cm^3.

15–26. Roller A in Figure P15–26 is rotating at 600 rpm as it is lowered onto plate B, which then slides to the left and causes both A and B to come to rest in 1.5 seconds. The total mass of roller A and its holder is 40 kg. Assume no slipping between roller A and plate B. Determine the mass of B.

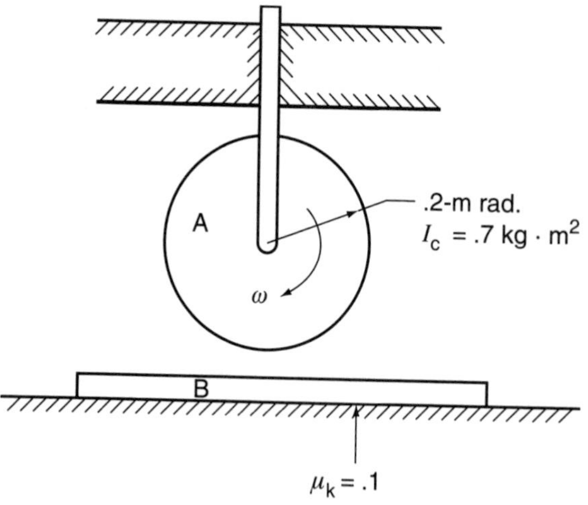

FIGURE P15–26

15–27. Determine the torque T required on gear A to accelerate gear B from rest to 200 rpm in 1.7 seconds (Figure P15–27).

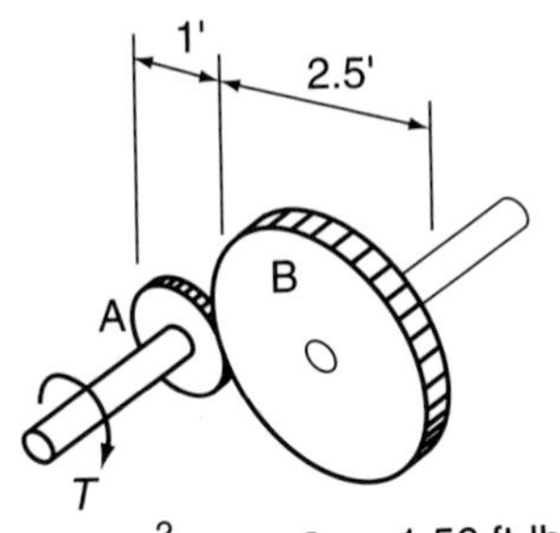

FIGURE P15–27

15–28. A torque of 40 N · m is applied to a 100-kg armature of an electric generator for 3 seconds. If the speed increase is 400 rpm, determine the radius of gyration of the armature.

15–29. Cylinder A in Figure P15–29 weighing 644 lb has a radius of gyration, $k = 0.5$ ft. Starting from rest and with no slipping, it reaches a velocity of 20 ft/s in 10 seconds. Determine the force P necessary to cause this motion.

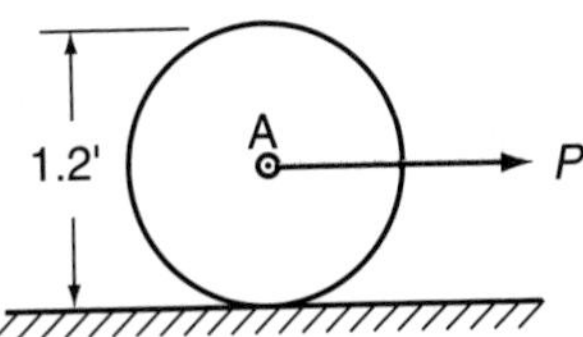

FIGURE P15–29

15–30. The cylinder and hub in Figure P15–30 weigh 322 lb and have a mass moment of inertia $I_C = 80$ ft-lb-s^2. The coefficient of static friction is 0.2. Slipping is impending as the maximum force P is applied for 4 seconds. Determine the force P and the linear velocity of the cylinder at $t = 4$ seconds.

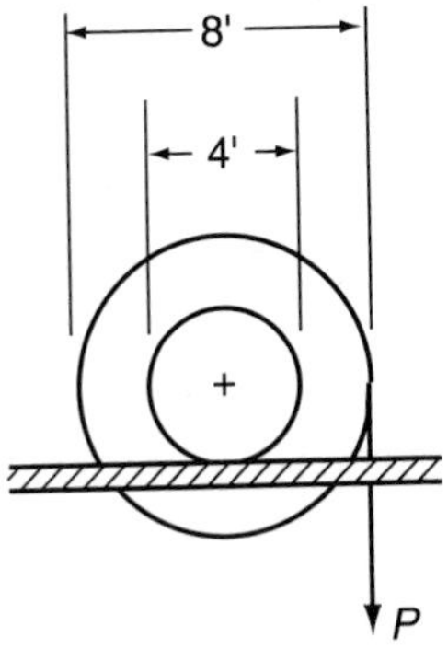

FIGURE P15–30

15–31. A pipe with a 3-in. outside diameter and a $2\frac{1}{2}$-in. inside diameter weighs 96.6 lb. It rolls from rest down a 15° slope. Assume no slipping and determine its angular velocity 5 seconds later.

APPLIED PROBLEMS FOR SECTION 15–3

15–32. A 4-kg rifle firing an 8-g bullet at 650 m/s obeys the law of conservation of momentum. Determine the recoil speed of the rifle.

15–33. A 30-ton boxcar traveling at 20 ft/s is coupled at the rear by a 50-ton boxcar traveling at 40 ft/s in the same direction. Determine the final velocity of the boxcars.

15–34. A cart on casters weighs 600 N and is traveling in a straight line at 3 m/s. A mass of 20 kg is dumped onto the cart with a horizontal velocity of 4 m/s at a right angle to the cart's direction. Determine the cart's final velocity.

15–35. A rail car with a mass of 600 kg rolls on a horizontal track at 1.1 m/s to the right. Directly above it is a conveyor that deposits 200 kg of iron ore at a velocity of 1.5 m/s to the left at an angle of 20° above horizontal. Determine the resulting velocity of the ore and car.

15–36. An 80-kg man jumps vertically downward from a bridge into a 20-kg boat that is traveling at 4 m/s. Determine the velocity of the boat after the man lands in it.

15–37. Two people, 160 lb and 120 lb, respectively, are in a boat that is moving forward at 4 ft/s. If the 160-lb person jumps directly off the back of the boat with a velocity of 7 ft/s, determine the resulting velocity of the boat and remaining person.

15–38. A can filled with sand weighs 10 lb and sits on the top of a fence post. With what velocity is the can knocked off the post when struck by a bullet weighing 0.5 oz and traveling 2000 ft/s? (The bullet remains embedded in the sand.)

15–39. A plan view of objects A and B is shown in Figure P15–39. When they collide and remain in contact, the direction of A is diverted 10° to the left. Determine (a) the mass of B and (b) the resulting velocity of A and B.

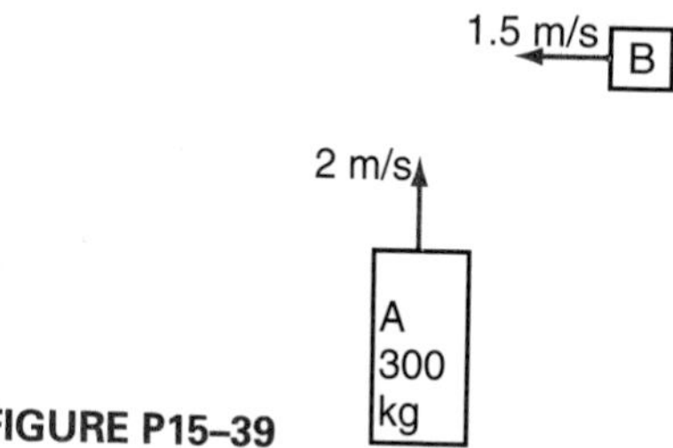

FIGURE P15–39

15–40. Block A in Figure P15–40 is released from the position shown. It has a velocity of 6 m/s when it strikes and sticks to block B. If the coefficient of friction is 0.3, how long does it take the blocks to come to rest? What is the maximum velocity of B?

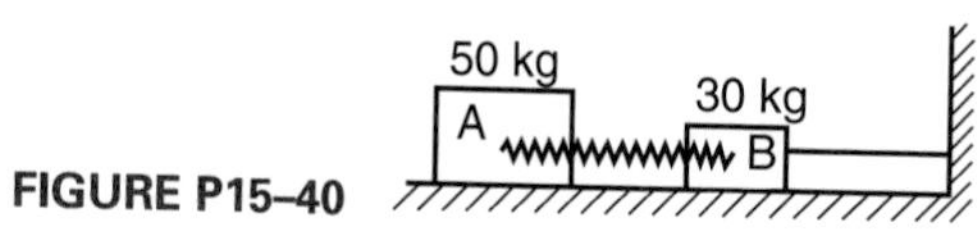

FIGURE P15–40

15–41. A filing cabinet drawer weighing 25 lb is slammed shut with a velocity of 8 ft/s. If the cabinet rests on a surface with $\mu = 0.2$ and is on the verge of sliding due to the momentum of the closed drawer, for how long was the impulse applied? (The total weight of cabinet and drawers is 60 lb.)

15–42. The spring shown brings roller A into contact with roller B and holds them together with no slipping (Figure P15–42). If the initial angular velocities are $\omega_A = 30$ rad/s clockwise and $\omega_B = 5$ rad/s counterclockwise, determine the resulting angular velocity of roller B.

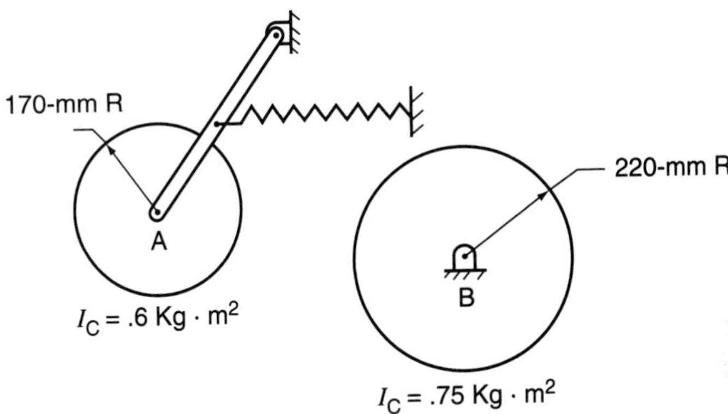

FIGURE P15–42

15–43. Disc A in Figure P15–43 ($I_C = 8$ kg · m²),while rotating at 100 rpm, is lowered onto disc B ($I_C = 10.5$ kg · m²), causing the latter to rotate from rest. Assume no slipping and determine the angular velocity of B.

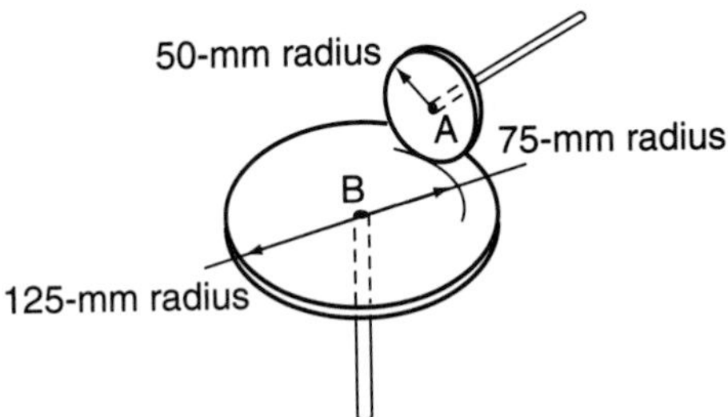

FIGURE P15–43

15–44. Wheel A is moved axially to drive wheel B in a friction drive system (Figure P15–44). Initially, wheel B is rotating at 20 rpm clockwise and wheel A is rotating at 65 rpm clockwise. Determine the resulting angular velocity of B if both wheels are freewheeling after A has contacted B.

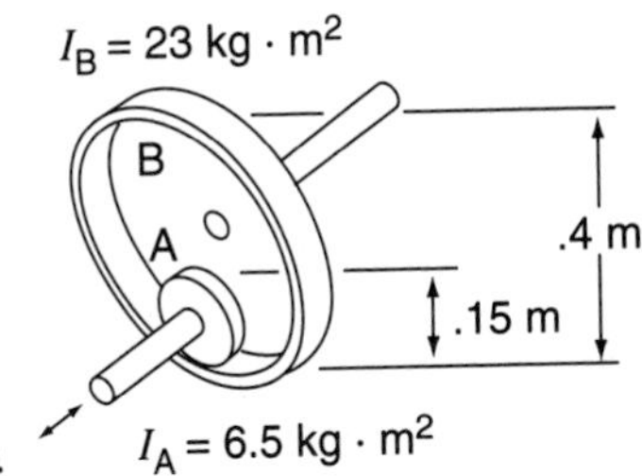

FIGURE P15–44

15–45. An 8-in.-diameter pulley A in Figure P15–45 is connected to a system with $I_C = 2.5$ ft-lb-s^2. The system is at rest and has a friction moment of 5 lb-ft. Block B weighs 200 lb. If the system is initially at rest, at $t = 4$ seconds, determine (a) the angular velocity of A, (b) the angular acceleration of A, and (c) the tension in the rope.

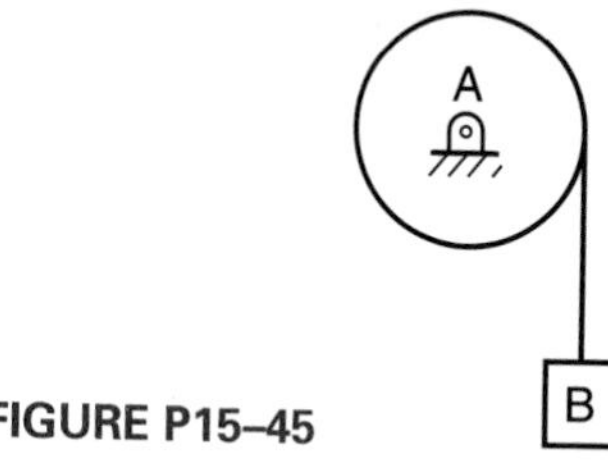

FIGURE P15–45

15–46. Solve Problem 14–65 using the impulse-momentum method.

REVIEW PROBLEMS

R15–1. The braking mechanism on an aircraft carrier can bring a 2-ton plane to a stop in a time of 1.8 seconds. The plane lands at 80 mph. Determine the force required and the deceleration rate.

R15–2. A steel shaft (specific weight = 470 lb/ft^3) is 8 ft long and is to be accelerated from rest to 400 rpm in 6 seconds by a torque of 180 lb-ft. Determine the maximum diameter of the shaft.

R15–3. Determine the tension in the cable and the velocity of weight B, 5 seconds after the system (Figure RP15–3) is released from rest. Assume no slipping and neglect the weight of the cable and pulley.

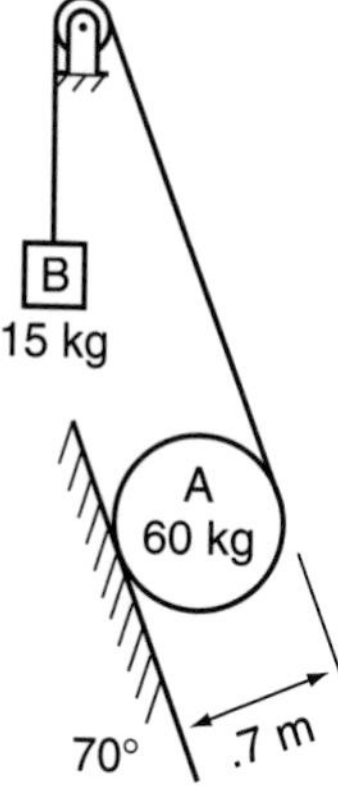

FIGURE RP15–3

R15–4. If objects A and B (Figure RP15–4) collide and remain in contact, determine the resultant velocity. Neglect friction.

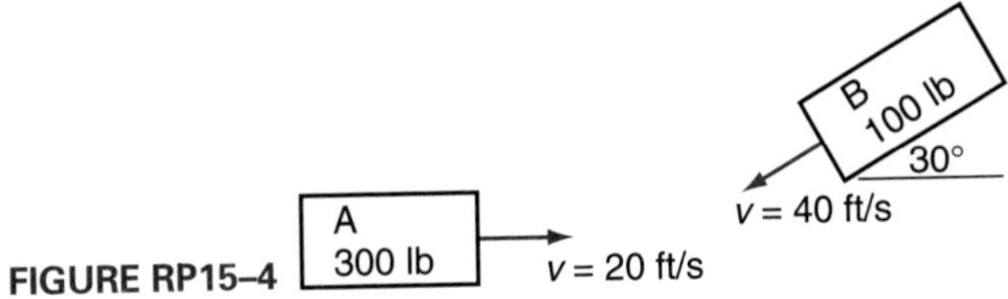

FIGURE RP15–4

R15–5. The system shown in Figure RP15–5 is accelerated from rest by force *P*. Assume no slipping of the cable on the roller. Determine the velocity of block A at $t = 0.155$ second. Confirm this velocity value by using the impulse-momentum method, the work-energy method, and the inertia method. (Block A moves 0.3 m in 0.155 second.)

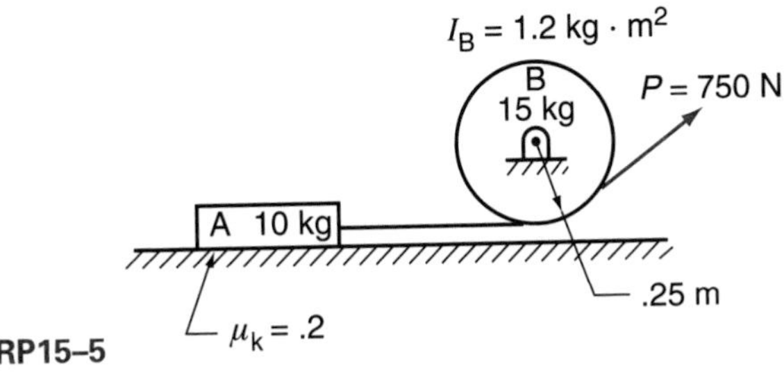

FIGURE RP15–5

Answers to Problems

CHAPTER 1

SECTIONS 1–1 TO 1–9

1–1. $x = 2$
1–2. $x = 2$
1–3. $x = 24$
1–4. $y = 1.74$
1–5. $x = 5.28$
1–6. $x = 0.865$ or -0.712
1–7. $x = 1.67$ or 1
1–8. $a = 35°$
$b = 55°$
$c = 125°$
1–9. $a = 80°$
$b = 100°$
$c = 100°$
1–10. $a = 50°$
$b = 15°$
$c = 115°$
$d = 65°$
$e = 65°$
1–11. ED = 15 in.
1–12. CE = 20 m
1–13. $A = 12.3$ m
1–14. $\theta = 66.4°$
1–15. $A = 16.5$ ft
1–16. $y = 11$ m
1–17. $\theta = 36.9°$ R = 5 in.
$\theta = 67.4°$ R = 13 in.
$\theta = 28.1°$ R = 17 in.
1–18. A = 2.18 in.
A = 3.36 ft
A = 14 in.
1–19. $\theta = 38.7°$
1–20. $y = 8.58$ mm
1–21. $\theta = 27.3°$

SECTIONS 1–10 AND 1–11

1–22. $c = 51.2$ cm
1–23. $c = 34.6$ ft
1–24. $\theta = 102.6°$
1–25. CB = 46.2 in.
1–26. $d = 12.7$ m
1–27. CD = 0.174 m
1–28. $A = 127$ in.
1–29. AC = 913 ft
AD = 865 ft
1–30. $d = 9.32$ m
1–31. $\theta = 73°$
1–32. $x = 4.1$ ft

1–33. 11.4°
1–34. $h = 3.57$ in.
1–35. $x = 1.9$ in.
$y = 2.7$ in.
1–36. 2.17 in.
1–37. $d = 6.49$ cm
1–38. $c = 2.18$ m
$\theta = 12.8°$
1–39. $h = 13.81$ m
1–40. 2.07 m
1–41. 7.41 in.
1–42. $d = 4.85$ m
$x = 1.95$ m
1–43. $\theta_1 = 21.1°$ $\theta_2 = 42.2°$
$\theta_3 = 42.2°$ $\theta_4 = 21.1°$
$\theta_5 = 68.9°$ $\theta_6 = 111.1°$

REVIEW PROBLEMS

R1–1. $x = 13.6$ m
$y = 6.34$ m
R1–2. A = 1.46
B = 4.73 m
R1–3. $\theta = 110°$
R1–4. $\phi = 53.5°$ $\theta = 86.5°$
R1–5. $\theta = 101°$
R1–6. R = 6.91 m
R1–7. $\theta = 25.3°$
R1–8. CB′ = 0.197 m
R1–9. $\phi = 14.8°$
R1–10. 0.595 in.
R1–11. 58.4°
R1–12. 3.96 in.

CHAPTER 2

SECTIONS 2–1 TO 2–3

2–1. R = 42.7 lb (8, 3)
2–2. R = 60.2 lb 74.6° R = 126 lb 71.6°
R = 1060 lb 32.6°
2–3. R = 80.8 lb 21.8° R = 539 lb 68.2°
R = 7,210 lb 33.7°
2–4. R = 15.2 kN 23.2°
R = 10.2 MN 78.7°
R = 25 N 36.9°
2–5. R = 632 N (3, 1)
2–6. R = 17 lb (15, 8)
2–7. R = 1.52 kN 29.5°
2–8. R = 4.27 kN 81.4°
2–9. R = 36.1 lb 76.3°

SECTIONS 2–4 AND 2–5

2–10. R = 330 N 80°
2–11. R = 63.6 kN 38.8°
2–12. R = 27 lb 61.8°
2–13. R = 14.1 kN 34.5°
2–14. R = 301 lb 48.7°
2–15. R = 185 lb ↓ R = 196 lb 11.3°
2–16. R = 5 kips 88.1°
2–17. $P_x = 8.55$ lb ← $P_y = 23.5$ lb ↓
$P_x = 1.29$ kips → $P_y = 1.53$ kips ↑
$P_x = 17.3$ lb ← $P_y = 10$ lb ↑
2–18. $F_x = 40$ N → $F_y = 75$ N ↑
$F_x = 28.3$ kN ← $F_y = 28.3$ kN ↓
$F_x = 96$ N → $F_y = 72$ N ↓
$F_x = 48$ kN ← $F_y = 20$ kN ↓
2–19. $F_x = 52$ lb → $F_y = 30$ lb ↑
$F_x = 27.4$ kips ← $F_y = 75.2$ kips ↓

$v_x = 300$ ft/s ← $v_y = 400$ ft/s ↑

$v_x = 30$ mph ← $v_y = 16$ mph ↓

2–20. $F_x = 158$ lb ← $F_y = 123$ lb ↑

$v_x = 11.8$ ft/s ← $v_y = 25.4$ ft/s ↓

$F_x = 169$ lb → $F_y = 86.3$ lb ↑

$F_x = 806$ lb → $F_y = 301$ lb ↓

2–21. $F_x = 0.313$ kN → $F_y = 1.77$ kN ↓

2–22. $A_x = 451$ N → $A_y = 451$ N ↓

2–23. $P_x = 76.6$ N 20°

$P_y = 64.3$ N 70°

$Q_x = 2.1$ kN 20°

$Q_y = 11.8$ kN 70°

$R_x = 38.3$ N 20°

$R_y = 11.6$ N 70°

2–24. 1.29 kN

2–25. $F_x = 10$ lb 15°

2–26. $P_x = 67.8$ N 20°

$P_y = 42.4$ N 70°

2–27. $F = 24.5$ lb 30°

2–28. $P_x = 394$ N 10°

$P_y = 69.5$ N 10° or 80°

SECTION 2–7

2–29. 62.6 N (10, 3)

2–30. R = 56.9 lb 72.6°

2–31. R = 184 N 23.7°

2–32. R = 213 lb 84.1°

2–33. R = 7.11 kN 85.8°

2–34. R = 1810 lb 57.8°

2–35. R = 5.53 kN 56.5°

2–36. R = 225 lb 23°

2–37. R = 12 lb 54.3°

REVIEW PROBLEMS

R2–1. $R_1 = 65$ N (12, 5) $R_2 = 8.54$ kN (8, 3)

$R_3 = 102$ N (15, 8)

R2–2. $R = 6.7$ kips 86.6°

R2–3. $R = 459$ N 23°

R2–4. $F_x = 20.7$ lb ← $F_y = 77.3$ lb ↓

$v_x = 15.2$ ft/s ← $v_y = 11.4$ ft/s ↑

$F_x = 1.34$ lb → $F_y = 1.49$ lb ↓

$F_x = 390$ lb ← $F_y = 156$ lb ↓

R2–5. $v_x = 3.44$ m/s ← $v_y = 4.91$ m/s ↓

$s_x = 3.13$ m → $s_y = 17.7$ m ↑

$a_x = 60$ m/s^2 ← $a_y = 32$ m/s^2 ↑

$F_x = 36.1$ N → $F_y = 54.1$ N ↑

R2–6. 2.98 kN 40° 29 kN 40°

R2–7. $R = 381$ N 49.8°

R2–8. $R = 52.5$ N 72.3°

CHAPTER 3

SECTION 3–1

3–1. $M_A = 1$ lb-ft ↺

3–2. $M_A = 95$ N · m ↻

3–3. $M_A = 2350$ lb-in. ↻

3–4. $M_A = 7.49$ N · m ↻

3–5. $M_A = 1280$ lb-ft ↻

3–6. $M_A = 2460$ N · m ↻ P = 493 N 38°

3–7. $M_A = 26{,}800$ lb-in. ↻

3–8. $M_C = 4140$ N · m ↻

$M_B = 3060$ N · m ↻

3–9. $M_A = 5940$ lb-in. ↻

3–10. $M_P = 420$ lb in. ↻

3–11. P = 315 N 75°

3–12. $M_A = 2160$ N · m ↻

3–13. $M_A = 19.8$ N · m ↻

3–14. $M_A = 60$ kN · m ↻

3–15. $M_A = 889$ lb-ft ↻

3–16. $M_B = 643$ N·m ↻ $F_S = 887$ N
3–17. $M_A = 11.2$ kip-ft ↻
$M_B = 9.92$ kip-ft ↺
3–18. 60 N
3–19. $M_B = 952$ lb-in. ↺
$M_C = 917$ lb-in. ↻
Upper arm

SECTION 3–2

3–20. $M_A = M_B = 16$ N·m ↻
3–21. $M_A = 0.612$ kN·m ↻
3–22. $M_A = 8.6$ lb-ft ↺
3–23. $M_A = 412$ lb-ft ↻
3–24. $M_A = 11.3$ N·m ↻
3–25. $M_A = 280$ lb-in. ↻
3–26. A = 13,300 N ← B = 13,300 N →
3–27. 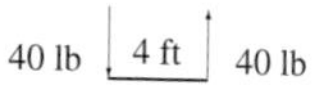 40 lb, 4 ft, 40 lb
3–28. 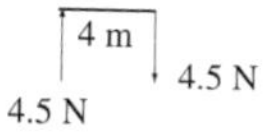 4 m, 4.5 N, 4.5 N
3–29. 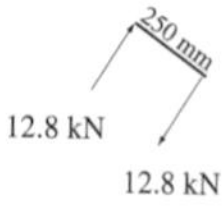 250 mm, 12.8 kN, 12.8 kN
3–30. 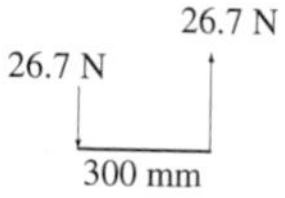 26.7 N, 26.7 N, 300 mm
3–31. 160 N, 160 N, 50 mm

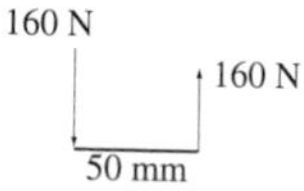

REVIEW PROBLEMS

R3–1. $M_A = 92$ N·m ↻
R3–2. $M_A = 17{,}120$ lb-ft ↻
$B = 6260$ lb ∠38°
R3–3. $M_A = 21$ N·m ↻ $M_A = 91$ N·m ↻
R3–4. $M_A = M_B = 45$ N·m ↻
R3–5. 12 lb, 12 lb, 10″

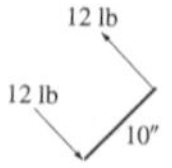

CHAPTER 4

SECTION 4–5

4–25. $T_A = 20$ lb $T_B = 20$ lb
4–26. T = 40 lb
4–27. $T_A = 60$ lb $T_B = 80$ lb
4–28. B = 75 kg
4–29. $T_A = 40$ lb $T_B = 160$ lb
4–30. T = 10 kN
4–31. T = 736 N
4–32. $T_1 = 80$ lb $T_2 = 160$ lb $T_3 = 320$ lb
4–33. T = 30 kN

SECTION 4–6

4–34. AC = 1 kN *T* BC = 800 N *C*
4–35. AB = 340 lb *C* BC = 300 lb *C*
4–36. AB = 75 lb *T* BC = 125 lb *T*
CE = 160 lb *T* CD = 241 lb *T*
4–37. AB = 750 N *T* BC = 770 N *C*
4–38. W = 156 lb
4–39. BC = 4530 lb *T* BD = 130 lb *T*
4–40. AB = 1.59 kN *C* AC = 2.19 kN *C*
4–41. AD = 160 lb
4–42. AB = 922 N *T* BC = 631 N *T*
CD = 1190 N *T* E = 126 kg
4–43. CD = 1760 N *T*
4–44. P = 38.6 N 60°
4–45. $R_A = 31.7$ lb → $R_B = 50$ lb ↑
4–46. DE = 1.79 kN *T* CE = 0.536 kN *T*
AD = 2.45 kN *T* BD = 2.2 kN *C*
4–47. T = 300 lb
4–48. M = 1.4 N·m ↺

SECTION 4–7

4–49. A = 7.75 kN ↑ B = 4.25 kN ↑

4–50. A = 19 lb ↓ B = 169 lb ↑

4–51. A = 0.85 kN ↓ B = 3.05 kN ↑

4–52. A = 42 kN ↑ B = 20 kN ↓

4–53. A = 0.89 kN ↑ B = 1.01 kN ↑

4–54. A = 6.8 kN ↑ B = 65.2 kN ↑

4–55. A = 5150 N ↑ B = 2950 N ↑

4–56. A = 86.5 kN ↑ B = 116 kN ↑

4–57. A = 5.4 kN ↑ B = 2.7 kN ↑

4–58. A = 3.28 kN ↑ B = 5.22 kN ↑

4–59. A = 800 lb ↑ B = 1000 lb ↑

4–60. A = 7.9 kips ↑ B = 9.1 kips ↑

4–61. A = 44 kN ↓ B = 100 kN ↑

4–62. B = 120 lb T

4–63. BE = 66 lb T

4–64. d = 39.7 in.

4–65. d = 26.5 in.

4–66. A = 1670 lb ↑ B = 2130 lb ↑

4–67. 10 beams

4–68. A = 4390 lb ↑ B = 3520 lb ↑
C = 3090 lb ↑

4–69. B = 3300 lb ↑ C = 736 lb ↑

4–70. C = 111 lb T

4–71. A = 334 lb ↑ B = 96 lb
C = 831 lb D = 401 lb ↑

4–72. d = 2.09 ft

4–73. d = 0.538 m

SECTION 4–8

4–74. A_x = 40 N → A_y = 600 N ↑
B = 200 N ↑

4–75. A = 58.3 kN ↓ B_x = 40 kN →
B_y = 183 kN ↑

4–76. T = 3640 N A_x = 2180 N →
A_y = 2110 N ↓

4–77. B_x = 3360 N → B_y = 600 N ↓
C = 3040 N ←

4–78. AB = 4 kN *T* AC = 7.25 kN *C*

4–79. AB = DE = 2880 lb *C*
AC = CE = 1950 lb *T*

4–80. A = 300 lb ← B_x = 300 lb →
B_y = 900 lb ↑

4–81. A = 28.6 lb → B_x = 28.6 lb ←
B_y = 400 lb ↑

4–82. A_x = 2370 lb → A_y = 2790 lb ↓
B = 5440 lb (slope 4:3)

4–83. A_x = 81.1 lb ← A_y = 148 lb ↑

4–84. A_x = 0.139 kN ← A_y = 4.45 kN ↓
C = 2.43 kN (slope 4:3)

4–85. A_x = 508 lb ← A_y = 123 lb ↑
B = 846 lb (slope 4:3)

4–86. DE = 1790 lb *T*

REVIEW PROBLEMS

R4–4. T_1 = T_3 = 981 N T_2 = 1960 N

R4–6. T = 222 N

R4–7. AB = 600 N *T* BC = 1000 N *C*
DC = 1560 N *C* CE = 2240 N *C*

R4–8. BC = 85.4 lb *T*

R4–9. Position A CB = 557 lb *T*
Position B CB = 1110 lb *T*

R4–10. T = 0.973 kN

R4–11. A = 1236 lb ↑ B = 764 lb ↑

R4–12. DC = 0.55 kN *T*

R4–13. P = 653 lb ↓

R4–14. A = B = 1160 lb *T*

CHAPTER 5

SECTION 5–1

5–1. AC = 28.6 kN *T*

5–2. AB = 5.6 kN *T* BC = 2.5 kN *C*
AC = 6.5 kN *T* DC = 6 kN *C*

5–3. BC = 56 kN *C*

5–4. CH = 7.51 kips *C*

5–5. BD = 2 kips *C*

5–6. AB = 77.8 kN *T* BC = 36 kN *T*
AC = 30 kN *C* DB = 79 kN *C*

5–7. AE = 76 kN *C* BE = 34 kN *T*
CB = 48 kN *T* CE = 67.8 kN *C*
CD = 48 kN *T* ED = 52 kN *C*

5–8. CE = 1570 lb *C* BE = 250 lb *T*

5–9. AB = 2.66 kips *T* AD = 11.4 kips *T*
AE = 3.33 kips *T* AF = *0*
FE = 16 kips *C* ED = 13.3 kips *C*
BD = 2 kips *C* BC = 3.33 kips *T*
DC = 2.66 kips *T*

5–10. AB = 5.25 kN *C* BC = 6.67 kN *C*
BE = 3.55 KN *T* AD = 5.69 kN *T*
DE = 6 kN *C* CD = 6.67 kN *C*

5–11. EB = 6.56 kips *C*, BD = 3.94 kips *T*

5–12. CD = 3000 lb *T*

5–13. CG = 3.75 kips *C*

5–14. BD = 2040 lb *T*, CA = 2670 lb *T*

5–15. AC = 810 lb *C* CE = 0

5–16. BC = 5.2 kips *T* CE = 1.99 kips *T*

5–17. AC = 6.71 kips *T*

5–18. AB = BE = 50 kN *T*
BD = BC = AC = 0
ED = DC = 56 kN *C*

5–19. AB = BC = CD = 46.6 kN *T*
HG = GE = ED = 45 kN *C*
AH = AG = BG = BE = CE = 0

5–20. BE = CE = 0 BC = 2.73 kN *C*
CD = 1.83 kN *C* AB = 2.73 kN *C*
AE = 1.77 kN *T* ED = 1.77 kN *T*

5–21. BC = 30.9 kips *T* CG = 0

5–22. GE = 0 CE = 58.7 kN *C*

5–23. BC = 17.3 kN *T*

5–24. AC = 63.8 kN *T*

SECTION 5–2

5–25. BC = 6.4 kN *T* BG = 4 kN *C*
EG = 3.2 kN *C*

5–26. CD = 3.33 kN *C* CE = 2.67 kN *C*
BD = 0.537 kN *T*

5–27. BC = 6.67 kips *C* BE = 1.94 kips *C*
BD = 0 DE = 8.33 kips *T*

5–28. BD = 0.887 kips *C* CD = 1.92 kips *C*
CE = 2.66 kips *T*

5–29. BG = 30 kN *C*

5–30. EG = 13.7 kN *T*

5–31. CD = 10.9 kips *T*

5–32. CD = 0.75 kN *C* ED = 3.36 kN *C*

5–33. BC = 2.24 kips *C* BH = 0 JH = 2 kips *T*

5–34. DE = 3.22 kN *T* DG = 32.1 kN *T*

5–35. CB = 5.2 kN *C*, BE = 3.84 kN *C*
BG = 0

5–36. CE = 567 lb *T* ED = 992 lb *T*
DG = 1550 lb *C*

5–37. BD = 1.09 kN *T* CD = 4.25 kN *T*
CE = 5.59 kN *C*

5–38. CE = 37.6 kN *C* ED = 44.7 kN *C*
BD = 9.6 kN *T*

5–39. CD = 70.7 kN *T* HG = 40.6 kN *T*
JG = 88.2 kN *C*

5–40. CD = 2.4 kips *C* CG = 3 kips *C*

5–41. BC = 91.6 kips *C* BG = 1.19 kips *C*

5–42. DE = 3.39 kN *C* JE = 0.81 kN *T*
KH = 2.25 kN *T* LM = MN = 0

SECTION 5–3

5–43. B_x = 7.5 kips B_y = 0
D_x = 7.5 kips D_y = 5 kips

5–44. D_y = 47.6 kN D_x = 16.5 kN

5–45. $B_x = 131$ lb $B_y = 0$
$C_x = 99$ lb ← $C_y = 60$ lb ↑

5–46. $B_x = 200$ lb ← $B_y = 167$ lb ↑
$C_x = 300$ lb → $C_y = 67$ lb ↓

5–47. $B_x = 3000$ lb $B_y = 4200$ lb
$C_x = 3000$ lb $C_y = 3000$ lb

5–48. AC = 36.3 kN *T*

5–49. $A_x = 128$ lb $A_y = 96$ lb
$B_x = 795$ lb $B_y = 1290$ lb

5–50. $D_x = 124$ N $D_y = 110$ N

5–51. $T = 6$ kips $B = 11.3$ kips ∠6.1°
$B_x = 11.2$ kips → $B_y = 1.2$ kips ↑

5–52. BD = 6180 N *C* $C_x = 2440$ N
$C_y = 3880$ N

5–53. $C_x = 1940$ N $C_y = 880$ N

5–54. DE = 359 N *T*

5–55. D = 7.87 kN $A_x = 2.19$ kN $A_y = 0$
$E_x = 2.19$ kN
$E_y = 4.37$ kN

5–56. $T = 131$ lb

5–57. $P = 268$ lb ←

5–58. $A_x = 62.5$ lb $A_y = 375$ lb
$D_x = 250$ lb
$D_y = 375$ lb

5–59. $C_x = 1090$ N ← $C_y = 407$ N ↓

5–60. $P = 5.21$ kN ↓ $A_x = 2$ kN →
$A_y = 6.67$ kN ↑
$B_x = 5$ kN ← $B_y = 1.46$ kN ↓

5–61. P = 12.1 kN ↑

5–62. P = 45.6 kN ← $B_x = 51.5$ kN
$B_y = 21.5$ kN

5–63. BD = 1490 lb *C*

5–64. P = 971 lb ←

5–65. $G_x = 591$ lb $G_y = 473$ lb

5–66. P = 1467 lb → T = 833 lb

5–67. 8.75 lb-in. ↺

5–68. P = 1089 N
C = D = 1045 N

5–69. GH = 44.3 lb

5–70. EH = 750 lb

5–71. AB = CD = 2.25 kN *T*

5–72. BD = 19.2 kN *T*

5–73. $D_x = 275$ N ← $D_y = 238$ N ↑
$C_x = 475$ N → $C_y = 208$ N ↓
P = 29.7 N ↓

5–74. D = 2340 lb *T* E = 8510 lb *C*

5–75. BG = 734 lb *C*

5–76. B = 146 N (slope 5/12) on BH
AC = 573 N *T*

5–77. A = 430 lb (slope 4/3) $B_x = 9.63$ lb (slope 3/4)
$B_y = 392$ lb (slope 4/3)

5–78. AC = 765 lb *C*
E = 527 lb

5–79. $D_y = 546$ lb $D_x = 495$ lb
D = 737 lb
$B_y = 1170$ lb $B_x = 539$ lb
B = 1290 lb

5–80. CD = 4310 lb $B_y = 2780$ lb ↓
$B_x = 1660$ lb ←
EH = 15,000 lb $G_y = 3060$ lb ↑
$G_x = 14{,}900$ lb ←

5–81. $A_x = 435$ lb ← $A_y = 630$ lb ↑
$B_x = 435$ lb → $B_y = 30$ lb ↑

5–82. $A_x = 3750$ lb ← $A_y = 3880$ lb ↑
$B_x = 5750$ lb → $B_y = 3380$ lb ↓

5–83. $D_x = 18.2$ kN ← $D_y = 10$ kN ↑
$B_x = 20.3$ kN → $B_y = 15.6$ kN ↓

5–84. $D_x = 725$ lb → $D_y = 300$ lb ↑

5–85. $A_x = 2.7$ kips → $A_y = 11.6$ kips ↑
$D_x = 17.3$ kips → $D_y = 13.4$ kips ↑

5–86. $A_x = 71.4$ kN ← $A_y = 75.1$ kN ↓
$C_x = 81.4$ kN → $C_y = 125$ kN ↑

REVIEW PROBLEMS

R5–1. AB = 60 kN C AC = 0
BC = 100 kN T
BD = 80 kN C CD = 14.2 kN T
CE = 99 kN T
ED = 72.1 kN C EG = 52.5 kN T

R5–2. GD = 19.5 kN T

R5–3. AB = 10 kN T AE = 6 kN C
BC = 19.2 kN T BE = 17.9 kN C
CD = 19.2 kN T ED = 20.9 kN C
EG = 0 EH = 22 kN C
HG = 0 GD = 0 CE = 0

R5–4. CD = 83.3 kN T EG = 41.7 kN C
EH = 33.3 kN C

R5–5. GE = 61.2 kips C

R5–6. CD = 7.27 kips C CH = 2.84 kips C
GH = 6.5 kips T

R5–7. $C_x = 6.37$ kN $C_y = 10.8$ kN

R5–8. $B_x = 1.43$ kN $B_y = 34.2$ kN

R5–9. $T = 138$ lb $E = 175$ lb

R5–10. $D = 204$ lb

R5–11. $BD = 480$ lb C
$C_y = 17.1$ lb $C_x = 407$ lb

R5–12. 167 lb-in. ↻ P = 145 lb

R5–13. $C_x = 675$ N → $C_y = 2630$ N ↑
$D_x = 675$ N ← $D_y = 525$ N ↓

R5–14. $A_x = 8$ kips → $A_y = 12$ kips ↑

CHAPTER 6

SECTION 6–1

6–1. $R = 50$ N ↓ $\bar{x} = 6$ m $\bar{z} = 1.8$ m

6–2. $R = 4$ kN ↓ $\bar{x} = 0.25$ m $\bar{z} = 5$ m

6–3. $R = 1$ kip ↓ $\bar{x} = -7$ ft $\bar{z} = -24$ ft

6–4. $R = 300$ lb ↓ $\bar{x} = 7$ in. $\bar{z} = 1.67$ in.

6–5. $R = 1400$ N ↑ $\bar{x} = 5.21$ m $\bar{z} = 1.21$ m

SECTION 6–2

6–6. $A = 7.5$ N ↑ $B = 1095$ N ↑ $C = 388$ N ↑

6–7. $A = 1202$ lb ↑ $D = 469$ lb ↑
$C = 929$ lb ↑

6–8. $A = 2.25$ kN ↓ $B = 4.74$ kN ↑
$C = 10.5$ kN ↑

6–9. $A = 3.55$ kN ↑ $B = 2.08$ kN ↑
$C = 0.37$ kN ↑

6–10. $A = 842$ N ↑ $B = 560$ N ↑
$C = 560$ N ↑

6–11. $A = 127$ lb ↑ $B = 130$ lb ↑
$C = 63$ lb ↑

6–12. $A = 1870$ lb ↑ $B = 1950$ lb ↑
$C = 1180$ lb ↑

SECTION 6–3

6–13. $R = 30.8$ kN (3, 2, 5)

6–14. $R = 15.3$ kips (− 5, 12, 8)

6–15. $R = 16.6$ lb (9, 5, 13)

6–16. $R = 66$ kN (5 − 10, 7)

6–17. $R_x = 13.3$ N ← $R_y = 46.4$ N ↓
$R_z = 13.3$ N ↙

6–18. $R_x = 3.1$ kips → $R_y = 6.99$ kips ↑
$R_z = 2.33$ kips ↗

6–19. $R_x = 747$ lb ← $R_y = 187$ lb ↑
$R_z = 467$ lb ↗

6–20. $R_x = 0.54$ kN ← $R_y = 2.03$ kN ↑
$R_z = 0.95$ kN ↙

6–21. $R_x = 1250$ lb → $R_y = 456$ lb ↓
$R_z = 684$ lb ↙

6–22. $R_x = 133\text{ lb} \leftarrow R_y = 665\text{ lb} \downarrow$
$R_z = 177\text{ lb} \swarrow$

6–23. $R = 250\text{ lb}\ (110, 177, 139)$

6–23. $R = 16.1\text{ kN}\ (3.93\ -13.4, 8.05)$

6–25. $R = 94.2\text{ N}\ (-24.7, -25.8, 87.2)$

SECTIONS 6–4 AND 6–5

6–26. $B_y = 152\text{ N} \uparrow \quad B_z = 60\text{ N} \nearrow$
$C_y = 25.3\text{ N} \downarrow \quad C_z = 10\text{ N} \swarrow$

6–27. $C_x = 5.9\text{ lb} \rightarrow \quad C_y = 6.1\text{ lb} \downarrow$
$D_x = 7.1\text{ lb} \rightarrow \quad D_y = 37.6\text{ lb} \uparrow$

6–28. $A = 70\text{ N} \downarrow \quad B = 10\text{ N} \downarrow$
$C = 70\text{ N} \uparrow \quad D = 10\text{ N} \uparrow$

6–29. $A_x = 16.7\text{ N} \leftarrow \quad A_y = 50\text{ N} \uparrow$
$A_z = 50\text{ N} \swarrow$
$B_x = 56.7\text{ N} \rightarrow \quad B_y = 150\text{ N} \uparrow$
$B_z = 50\text{ N} \swarrow$
$C_x = 40\text{ N} \leftarrow \quad C_y = 0 \quad C_z = 0$

6–30. $CD = 40.3\text{ lb } T$
$A_x = 7.2\text{ lb} \leftarrow \quad A_y = 40\text{ lb} \uparrow$
$A_z = 21\text{ lb} \nearrow$
$B_x = 4.8\text{ lb} \leftarrow \quad B_y = 0 \quad B_z = 21\text{ lb} \nearrow$

6–31. $CD = 2210\text{ lb } T$
$A_x = 200\text{ lb} \rightarrow \quad A_y = 11.1\text{ lb} \downarrow \quad A_z = 0$
$B_x = 1080\text{ lb} \leftarrow \quad B_y = 1080\text{ lb} \downarrow$
$B_z = 717\text{ lb} \swarrow$

6–32. $A_x = 2.14\text{ kN} \rightarrow \quad A_y = 5.71\text{ kN} \uparrow$
$A_z = 9.27\text{ kN} \nearrow \quad B_x = 0$
$B_y = 2.04\text{ kN} \downarrow \quad B_z = 2.85\text{ kN} \swarrow$
$CD = 0.611\text{ kN } C$

6–33. $A_x = 200\text{ lb} \quad A_y = 625\text{ lb}$
$A_z = 0$
$B_x = 0 \quad B_y = 125\text{ lb}$
$B_z = 200\text{ lb}$

6–34. $A_x = 6.25\text{ kN} \leftarrow \quad A_y = 14.9\text{ kN} \downarrow$
$A_z = 3.94\text{ kN} \swarrow \quad B_x = 0.25\text{ kN} \rightarrow$
$B_y = 11.4\text{ kN} \uparrow \quad B_z = 3.94\text{ kN} \swarrow$
$C_x = 10.5\text{ kN} \leftarrow \quad C_y = 10.5\text{ kN} \uparrow$
$C_z = 7.88\text{ kN} \nearrow$

6–35. $A_x = 0 \quad A_y = 4\text{ kN} \downarrow \quad A_z = 3\text{ kN} \nearrow$
$B_x = 4\text{ kN} \rightarrow \quad B_y = 6.67\text{ kN} \uparrow$
$B_z = 3\text{ kN} \nearrow$

SECTION 6–6

6–36. AC = 5.42 kips *T* BC = 5.42 kips *T*
CD = 22.3 kips *C*

6–37. AC = BC = 4.8 kips *T*
CD = 18.5 kips *C*

6–38. AC = 3.16 kN *T* CB = 2.23 kN *T*
CD = 20.6 kN *C*

6–39. AD = 1.38 kN *T* BD = 1.25 kN *T*
CD = 4.22 kN *C*

6–40. AB = 3.89 kips *T*
CB = DB = 3.11 kips *C*

6–41. AD = 52.6 kN *T* DC = 54 kN *T*

6–42. AD = 2830 kN *T* BD = 2150 kN *C*
CD = 1200 kN *T*

6–43. AD = 888 lb *C*

6–44. 30°

6–45. AB = 3.99 kips *C* BC = 3.47 kips *C*
BD = 5.93 kips *C*

6–46. AD = 220 lb *T* BD = 212 lb *C*
CD = 58.3 lb *T*

6–47. BC = 370 lb *T* DB = 295 lb *T*

6–48. BC = 370 lb *T* DB = 295 lb *T*
AB = 1590 lb *T*

6–49. AD = 16.5 kN *C* BD = 22.1 kN *T*
CD = 48.3 kN *C*

6–50. AB = 670 lb T CD = 562 lb T

6–51. AD = 1700 lb *T* CD = 2020 lb *T*

6–52. AD = 19.4 kN *C* BD = 19.8 kN *T*
CD = 13.2kN *C*

6–53. AD = 0.407 kN T BD = 6.77 kN C
CD = 7.26 kN C

REVIEW PROBLEMS

R6–1. $R = 4$ kN ↑ $\bar{x} = 0.87$ $\bar{z} = -1$ m
R6–2. $A = 1760$ lb ↑ $B = 1180$ lb ↑
$C = 2060$ lb ↑
R6–3. $R = 186$ lb $(14, -10, -7)$
R6–4. $A = 1.15$ kN ↓ $B = 1.54$ kN ↑
$C = 1.84$ kN ↑ $D = 0.333$ kN ↓
R6–5. AD = 5.56 kN C BD = 11.9 kN T
CD = 5.09 kN T
R6–6. AB = 3.35 kN C CB = 3.33 kN T
DB = 1.98 kN T
R6–7. AB = 1.72 kN T BC = 1.78 kN T
HK = 0.447 kN T $D_x = 0.16$ kN →
$D_y = 0.312$ kN ↑ $D_z = 2.57$ kN ↙

CHAPTER 7

SECTIONS 7–1 TO 7–4

7–1. 0.27
7–2. 3.82 kg
7–3. $P = 8$ lb ↗ $P = 32$ lb ↗
7–4. $P = 63.9$ lb →
7–5. $P = 17.2$ N →
7–6. $\mu = 0.466$
7–7. $P = 60$ lb ↓
7–8. $P = 37.3$ N →
7–9. $P = 33.7$ lb 40° $d = 38.2$ in
7–10. $P = 94.9$ lb 40° $d = 13.5$ in
7–11. $F = 4.14$ lb $W = 99.7$ lb
$W = 57.7$ lb
7–12. $P = 50$ lb →
7–13. $P = 273$ lb $P = 67.6$ lb
7–14. $P = 115$ lb (tipping)
7–15. tip at $\theta = 32.9°$
7–16. 3 crates
7–17. $P = 676$ N →
7–18. $P = 5.19$ kN ←
7–19. $P = 1.44$ lb
7–20. $P = 22.7$ lb ↑
7–21. $P = 4.4$ lb ↓
7–22. $P = 39.6$ N →
7–23. $T = 3.6$ lb
7–24. $P = 549$ N ←
7–25. $\mu = 0.036$
7–26. $\theta = 25°$
7–27. $N_B = 1680$ N (4, 3) $F_B = 504$ N (3, 4)
$N_A = 1410$ N → $F_A = 342$ N ↓
$\mu = 0.24$
7–28. $\theta = 16.7°$
7–29. no slipping $F = 13$ lb $\mu = 0.139$
7–30. 5270 N
7–31. $P = 209$ N ↓
7–32. $\mu = 0.454$
7–33. $P = 153$ N →
7–34. $P = 2.19$ lb ↓
7–35. T = 11.9 lb for sliding
T = 13.5 lb for tipping
d = 20 in.
7–36. slipping at A $F_B = 117$ N
$F_A = 105$ N
7–37. $\mu = 0.474$
$B_x = 121$ N $B_y = 34.3$ N

SECTION 7–5

7–38. $T_L = 1030$ lb $T_S = 243$ lb
7–39. 198 kg
7–40. $\mu = 0.368$
7–41. 1 turn
7–42. $P = 2390$ N
7–43. $B = 41.2$ lb
7–44. $B = 227$ kg
7–45. $P = 417$ lb
7–46. 165 lb-in. $\mu = 0.37$

7–47. $\theta_1 = 35°$
7–48. $\phi = 30°$
7–49. 146 N · m
7–50. 3.87 lb-ft
7–51. $B = 724$ lb
7–52. $T = 158$ N
7–53. 274 lb, 1.46 lb
7–54. $T = 248$ lb
7–55. $P = 60.2$ lb
7–56. slipping at A, $T = 110$ lb
7–57. no slipping at A or B
$F_A = 43.7$ kN $\rightarrow$
7–58. $P = 23.5$ N no slipping at A

REVIEW PROBLEMS

R7–1. no motion
R7–2. A moves at $P = 157$ N $F_B = 157$ N
R7–3. $\mu = 0.375$
R7–4. $P = 10.9$ lb $\downarrow$
R7–5. 36.9 kg
R7–6. no motion, $F_B = 52.1$ lb $10°$
R7–7. 509 N, A wears more
R7–8. slipping at B $F_c = 534$ N $\angle 10°$
R7–9. $\mu = 0.61$ 270 lb-ft
R7–10. $A = 79.8$ N $C_x = 470$ N $\leftarrow$
$C_y = 870$ N $\downarrow$ $\mu = 0.54$
R7–11. $P = 520$ N $\downarrow$
R7–12. $P = 21.1$ N $\rightarrow$
R7–13. $A = 642$ lb $15.9°$ $B = 285$ lb $76°$

CHAPTER 8

SECTIONS 8–1 TO 8–3

8–1. $\bar{y} = 4.81$ in.
8–2. $\bar{x} = 167$ mm
8–3. $\bar{x} = 4.02$ in. $\bar{y} = 5.96$ in.
8–4. $\bar{x} = 29.1$ mm $\bar{y} = 76.5$ mm
8–5. $\bar{x} = 0$ $\bar{y} = 3.57$ in.
8–6. $\bar{x} = 18.8$ mm $\bar{y} = 122$ mm
8–7. $\bar{x} = 224$ mm $\bar{y} = 50.3$ mm
8–8. $\bar{x} = 0$ $\bar{y} = 3.22$ in.
8–9. $\bar{x} = 0$ $\bar{y} = 44.7$ mm
8–10. $\bar{x} = 54.4$ mm $\bar{y} = 11.2$ mm
8–11. $\bar{x} = 48.6$ mm $\bar{y} = 180$ mm
8–12. $\bar{x} = 102$ mm $\bar{y} = 62.5$ mm
8–13. $\bar{y} = 4.92$ ft
8–14. $\bar{y} = 84.6$ mm
8–15. $\bar{y} = 112$ mm
8–16. $\bar{y} = 51.8$ mm
8–17. $\bar{y} = 8.82$ in.
8–18. $\bar{y} = 56.3$ mm above base
8–19. $\bar{x} = +0.208$ in. $\bar{y} = -0.208$ in.
from center of circle

SECTION 8–4

8–20. $\bar{x} = 5.74$ in. $\bar{y} = 1.28$ in.
8–21. $\bar{x} = 6.67$ in. $\bar{y} = 0.49$ in.
8–22. $\bar{x} = 1.02$ m $\bar{y} = 0.6$ m
8–23. $\bar{y} = 0.296$ m
8–24. $\bar{x} = 5.37$ in. $\bar{y} = 6.29$ in.
8–25. $\bar{x} = -68.2$ mm $\bar{y} = 68.2$ mm
8–26. $\bar{x} = 2.86$ in. $\bar{y} = 2.86$ in.
8–27. $\bar{y} = 75.7$ mm

REVIEW PROBLEMS

R8–1. $\bar{x} = 2$ in. $\bar{y} = 2.92$ in.
R8–2. $\bar{x} = 43$ mm $\bar{y} = 70$ mm
R8–3. $\bar{x} = 5.08$ in. $\bar{y} = 2.82$ in.

CHAPTER 9

SECTIONS 9–1 AND 9–2

9–1. $I_x = 2.13 \times 10^8\ \text{mm}^4$
9–2. $I_x = 50.9 \times 10^6\ \text{mm}^4$
9–3. $I_x = 36.9 \times 10^6\ \text{mm}^4$
9–4. $I_x = 160\ \text{in.}^4$
9–5. $I_x = 369\ \text{in.}^4$
9–6. $I_x = 154\ \text{in.}^4$
9–7. $I_{x2} = 139\ \text{in.}^4$
9–8. $b = 4$ in. $I = 1872\ \text{in.}^4$
9–31. $I_x = 99.9\ \text{in.}^4$
9–32. $I_x = 415 \times 10^6\ \text{mm}^4$
9–33. $I_x = 13.9 \times 10^6\ \text{mm}^4$
9–34. $I_x = 460\ \text{in.}^4$
9–35. $I_x = 3.57\ \text{in.}^4$
9–36. $I_x = 105\ \text{in.}^4$
9–37. $I_x = 187\ \text{in.}^4$
9–38. k = 88.3 mm
9–39. k = 51.6 mm
9–40. k = 4.32 in.
9–41. k = 0.185 m

SECTIONS 9–3 AND 9–4

9–9. $I_x = 1.14 \times 10^8\ \text{mm}^4$
9–10. $I_x = 834\ \text{in.}^4$
9–11. $I_x = 713\ \text{in.}^4$
9–12. $I_x = 95.1 \times 10^6\ \text{mm}^4$
9–13. $I_x = 58.8 \times 10^6\ \text{mm}^4$
9–14. $I_x = 7.6 \times 10^{-6}\ \text{m}^4$
9–15. k = 0.0563 m
9–16. $I_x = 241\ \text{in.}^4$
9–17. $I_x = 378 \times 10^6\ \text{mm}^4$
9–18. $I_x = 56.4 \times 10^6\ \text{mm}^4$
9–19. $I_x = 896\ \text{in.}^4$
9–20. $I_x = 53.7 \times 10^6\ \text{mm}^4$
9–21. $I_x = 87.7 \times 10^6\ \text{mm}^4$
9–22. $I_x = 2.91 \times 10^{-3}\ \text{m}^4$
9–23. $I_x = 11 \times 10^6\ \text{mm}^4$
9–24. $I_y = 535 \times 10^6\ \text{mm}^4$
9–25. $I_x = 196\ \text{in.}^4$
9–26. $I_x = 57.7\ \text{in.}^4$
9–27. $I_x = 113 \times 10^{-6}\ \text{m}^4$
9–28. k = 0.0922 m
9–29. $I_x = 1.14\ \text{in.}^4$
9–30. $I_x = 26.2\ \text{in.}^4$ $I_y = 10.9\ \text{in.}^4$

SECTIONS 9–5 TO 9–7

9–42. $I_c = 0.8$ ft-lb-s^2 k = 0.63 ft
9–43. $I_x = 0.18\ \text{kg} \cdot \text{m}^2$ k = 42.4 mm
9–44. $I_x = 18$ ft-lb-s^2
9–45. $I_x = 0.397\ \text{kg} \cdot \text{m}^2$
9–46. $I_x = 8.78\ \text{kg} \cdot \text{m}^2$
9–47. $I_c = 13.8$ ft-lb-s^2
9–48. $I_x = 0.044\ \text{kg} \cdot \text{m}^2$
9–49. k = 2.41 in.
9–50. $I = 3.76 \times 10^{-2}\ \text{kg} \cdot \text{m}^2$

REVIEW PROBLEMS

R9–1. $I_y = 2.88 \times 10^{-4}\ \text{m}^4$
R9–2. $I_{x2} = 461\ \text{in.}^4$
R9–3. 67%
R9–4. $I_x = 3.75 \times 10^{-6}\ \text{m}^4$
R9–5. k = 0.044 m
R9–6. $I = 34 \times 10^{-5}\ \text{kg} \cdot \text{m}^2$
R9–7. $I_y = 0.203$ ft-lb-s^2
R9–8. $I_c = 0.028$ ft-lb-s^2

CHAPTER 10

SECTIONS 10–1 TO 10–4

10–1. 14.8 ft 61.7°

10–2. displ. = 7.16 m 12.1° dist. = 9.5 m

10–3. dist. = 68 m displ. = 58 m 10.4°

10–4. dist. = 300 km displ. = 280 km 30.3°

10–5. displ. = 7.71 m 21.5°

10–6. displ. = 19.4 ft 59.4°

10–7. vel. = 30 ft/s

10–8. vel. = 2.49 ft/s 63.2°

10–9. $a = 1.22\ \text{m/s}^2$ 67.5°

10–10. $a = 3.8\ \text{m/s}^2$ 52.5°

10–11. $a = 5.13\ \text{m/s}^2$ 20°

10–12. $a = 11.3\ \text{ft/s}^2$ 65°

10–13. $a = 13.4\ \text{m/s}^2$ 72.5°

10–14. 2.93 hr

10–15. 2.4 m/s

10–16. clears by 15 ft

10–17. $a = 8\ \text{m/s}^2$

SECTION 10–5

10–18. $t = 4.38$ s $v = 5.48$ ft/s
total $t = 9$ s

10–19. $a = 15.1\ \text{in./s}^2$

10–20. $v_o = 15.5$ m/s $t = 2.82$ s

10–21. $d = 1220$ ft

10–22. $a_{AB} = 0.404\ \text{m/s}^2$ $t = 41.3$ s

10–23. $t = 319$ s

10–24. displ. = 461 ft uphill

10–25. $t = 3.5$ s $v = 34.3$ m/s ↓

10–26. 44.1 m

10–27. s = 30.6 m

10–28. s = 38.7 m

10–29. $v = 5.64$ ft/s $v = 13.9$ ft/s

10–30. h = 31.8 m $t = 7.32$ s
$v = 46.8$ m/s ↓
for v_o downward
$v = 46.8$ m/s ↓ $t = 2.23$ s

10–31. $v = 193$ ft/s $t = 293$ s

SECTION 10–6

10–32. $v = 37.8$ m/s 31.6°

10–33. $v = 35.1$ m/s 34.4°

10–34. d = 426 m $v_B = 163$ m/s 29.9°

10–35. horiz displ. = 47.3 m
vert displ. = 65 m

10–36. s = 138 m

10–37. d = 81.1 ft

10–38. $v_o = 40.8$ m/s 76°

10–39. d = 113 m

10–40. $v = 33.3$ m/s 82.3°

10–41. $v_1 = 153$ m/s 32.9°

10–42. 35.6°

10–43. $v = 1.43$ m/s 50°

10–44. $x = 315$ ft $v_C = 120$ ft/s 30.6°

10–45. $v_B = 14$ m/s 50°

10–46. min d = 40.2 m max d = 321 m

10–47. $v = 35.7$ ft/s 48°

10–48. d = 12,500 m $v = 399$ m/s 12.3°

10–49. d = 85 ft

10–50. B horiz. surface
$v_B = 608$ m/s 13.6°

REVIEW PROBLEMS

R10–1. 7.54 ft 30.5°

R10–2. 1.69 m 27.6° vel. = 0.242 m/s 27.6°

R10–3. $a = 10.4\ \text{ft/s}^2$

R10–4. 1.98 km

R10–5. 74.1 s

R10–6. abreast 82 ft above ground

$v_A = 208$ ft/s ↓ $v_B = 243$ ft/s ↓

R10–7. $d_1 = 953$ m

R10–8. $s = 125$ m

R10–9. $d = 609$ ft

CHAPTER 11

SECTIONS 11–1 TO 11–5

11–1. 2π rad π rad

11–2. 94.2 rad

11–3. 4.11 rev 25.8 rad

11–4. 9.42 rad/s

11–5. $\omega = 3.26$ rad/s ↻

11–6. $\omega = 4.89$ rad/s

11–7. 2100 rpm

11–8. 235 rad/min

11–9. 8.37 rad/s^2

11–10. 335 rad/s^2 60 rev

11–11. 1.13 rad/s^2

11–12. 4.36 rad/s^2 77.5 rev

11–13. 0.91 rad/s^2 33 s

11–14. 8.37 rad/s 80 rpm

11–15. 6.98 rad/s^2 2.59 rev

11–16. 2.28 min

11–17. 12 rev 14.6 rad/s

11–18. 611 rad/s^2

11–19. 31 rad/s 20.3 rev 15.5 rad/s^2

11–20. 15.7 rad/s^2 51.6 rad/s^2 220 rev

11–21. 24 s 3.48 rad/s^2

11–22. 35 rev

11–23. 0.268 rad/s^2 78.1 s 102 rpm

11–24. 4.51 s 4.18 rad/s^2 17.3 rev

11–25. 1.47 rad/s^2 1.49 rev 0.426 rev

11–26. 207 s

SECTION 11–6

11–27. 5 rad/s^2 ↺ 10 m/s ↑

11–28. 4.03 rpm

11–29. 125 m

11–30. 982 m

11–31. 3.77 min

11–32. 16.7 m/s

11–33. 27.2 m/s

11–34. 75 rad/s^2

11–35. 1270 rpm

11–36. 62.8 in./s $\omega_3 = 200$ rpm

$\omega_2 = 300$ rpm

11–37. 119 rpm

11–38. 1.2 m/s^2 ← 4.8 m/s →

11–39. 13.9 ft/s^2

11–40. 11.1 rad/s^2 ↺ 1.33 m/s^2 ↑

11–41. $\alpha_B = 3.49$ rad/s^2 ↺

$a_C = 10.5$ in./s →

11–42. $\omega_D = 3$ rad/s ↻ $v_C = 150$ mm/s →

$\omega_{AC} = 1.5$ rad/s ↻

$v_A = 180$ mm/s ←

11–43. $\theta_B = 100$ rad ↻ $\theta_C = 267$ rad ↺

$\alpha_B = 0.5$ rad/s^2 ↻ $\alpha_C = 1.33$ rad/s^2 ↺

$a_t = 0.33$ ft/s^2 ↘

11–44. $\omega_{pipe} = 3.92$ rad/s ↻ $\omega_C = 13.1$ rad/s ↺

11–45. $\omega_C = 7.85$ rad/s ↻

$v_C = 31.4$ in./s

No change

$\omega_C = 7.85$ rad/s ↺

11–46. $v_B = 320$ mm/s ←

$a_B = 53.3$ mm/s^2 ←

11–47. $s_D = 15.7$ in. ↓ $v_D = 10.5$ in./s ↓

11–48. 25.1 ft

11–49. 3.77 m/s 9.42 rad/s

11–50. 8.33%

11–51. $\omega_A = 57.3$ rpm $\alpha_A = 2$ rad/s^2

$\omega_B = 86$ rpm $\alpha_B = 3$ rad/s^2

SECTION 11–7

11–52. 166,000 ft/s^2

11–53. 163 rad/s

11–54. 220 m/s^2

11–55. 3800 rpm

11–56. 1414 rpm

11–57. 1480 ft/s^2

11–58. $a_n = 270$ m/s$^2 \rightarrow$ $a_t = 1.5$ m/s$^2 \downarrow$

11–59. 3.31 ft/s^2

11–60. 28.3 m/s

11–61. 4.27 rad/s 50.1 rad/s^2 ↻

11–62. $a_B = 87.8$ m/s^2 55.2°

11–63. 8.69 ft/s^2

11–64. $\mu = 0.81$

11–65. $\omega = 4.25$ rad/s ↻ $\alpha = 3.96$ rad/s^2 ↻

11–66. $\omega = 3.3$ rad/s ↺ $\alpha = 2.17$ rad/s^2 ↻

REVIEW PROBLEMS

R11–1. 13.9 rad/s^2 15 s

R11–2. 0.26 rad/s^2

R11–3. 0.524 rad/s^2 0.83 rev

R11–4. 1.74 rad/s 3.49 rad/s^2

R11–5. 0.433 rad/s^2 5.66 rad/s

R11–6. 3.12 rad/s ↻

R11–7. $a_B = 100$ in./s$^2 \rightarrow$

$\alpha_E = 66.7$ rad/s^2 ↺ $\theta_E = 10$ rad ↺

$\omega_c = 55$ rad/s ↻

R11–8. 52.8 mph

R11–9. $a_n = 135$ ft/s$^2 \downarrow$ $a_t = 4.5$ ft/s^2

R11–10. $a = 21.3$ ft/s^2

CHAPTER 12

SECTION 12–1

12–1. 880 ft/min 520 ft/min

12–2. 23.6°

12–3. $s_{A/B} = 6$ m $\downarrow$

12–4. $a_{A/B} = 15$ ft/s$^2 \uparrow$

12–5. $v_{t/c} = 30.8$ m/s 35.8°

12–6. $v_{A/B} = 13.5$ m/s 35°

12–7. 1.09 m/s 66.2°

12–8. $v_m = 1.98$ m/s

12–9. $S_{B/A} = 329$ ft 17°

$v_{B/A} = 65.9$ ft/s 17°

12–10. $\omega_{BC} = 0$ $v_C = 7.33$ m/s $\rightarrow$

$\omega_{BC} = 14.8$ rad/s ↺ $v_C = 1.11$ m/s $\rightarrow$

12–11. 22 in./s $\rightarrow$ 1.47 rad/s ↺

12–12. $\omega_{AB} = 2.8$ rad/s ↻

$v_D = 0.646$ m/s (12, 5)

12–13. $\omega_{DB} = 0.234$ rad/s ↻

12–14. $\omega_{AB} = 1.11$ rad/s ↻

12–15. $v_c = 9.22$ m/s 50°

$\omega_{AC} = 15.4$ rad/s ↺

$\omega_{BD} = 27.2$ rad/s ↻

12–16. $v_B = 3.72$ in./s 50°

$v_D = 4.04$ in./s $\leftarrow$

12–17. 22.5 ft/s $\rightarrow$

12–18. $v_D = 2.88$ m/s $\rightarrow$ $v_E = 4.8$ m/s $\leftarrow$

12–19. $v_{C/D} = 36$ mm/s $\rightarrow$

$v_C = 48$ mm/s $\rightarrow$

12–20. $v_C = 4.66$ m/s (5, 3) $\omega_{AC} = 10$ rad/s ↺

12–21. $v_C = 167$ mm/s (3, 4)

$\omega_{CBD} = 0.762$ rad/s ↺

$v_D = 240$ mm/s 3.41°

12–22. $\omega_{BC} = 0.4$ rad/s ↻

$v_A = 43.7$ in./s 15.9°

12–23. $v_D = 8.84$ ft/s $\angle 47.3°$

12–24. $v_D = 13.6$ m/s $80.4°$

12–25. $v_C = 15.9$ m/s $60°$

$\omega_{DBC} = 0.72$ rad/s ↻

$v_D = 19$ m/s $40.5°$

12–26. $\omega_{CBD} = 1.55$ rad/s ↻

$v_D = 60.4$ in./s $70.8°$

12–27. $v_D = 0.17$ m/s $32.9°$

$\omega_{BCD} = 0.706$ rad/s ↺

12–28. $a_C = 35.4$ in./s^2 $\angle 63.3°$

12–29. $a_C = 686$ mm/s^2 →

$\alpha_{BC} = 0.57$ rad/s^2 ↺

12–30. $a_C = 1.83$ m/s^2 $62.9°$

SECTION 12–2

12–31. 69.6 mph

12–32. $\omega = 7.5$ rad/s ↻ $v_B = 8.48$ m/s $\angle 45°$

12–33. $v_B = 8.67$ m/s (5, 12)

12–34. $v_D = 5.45$ m/s $\angle 45°$

$v_E = 7.71$ m/s →

$v_C = 3.86$ m/s → $s_C = 2.57$ m

12–35. $s_B = 10$ m ↑

12–36. $\omega_B = 10.7$ rad/s ↺ $\alpha_B = 3.55$ rad/s^2 ↺

12–37. $v_B = 4.4$ m/s ←

12–38. 4 rad/s^2 ↻

12–39. $v_B = 72$ in/s →

12–40. $v_C = 4.61$ m/s $30°$

12–41. $v_D = 3$ m/s (1, 1) $\omega_{CD} = 1.5$ rad/s ↻

12–42. $\omega_{AC} = 1.28$ rad/s ↻ $v_D = 8.4$ m/s ↑

12–43. $v_C = 73.6$ mm/s $40°$

$\omega_{BC} = 0.34$ rad/s ↻

12–44. $\omega_{BC} = 0.872$ rad/s ↺

$\omega_{CD} = 0.65$ rad/s ↻

12–45. $v_E = 1.95$ m/s →

SECTION 12–3

12–48. $v_A = 100$ in./s (3, 4)

$\omega_{AC} = 4$rad/s ↺

$v_B = 49.6$ in./s (1, 4)

12–49. $v_D = 18.9$ in./s $\angle 40°$

$\omega_{BD} = 6.9$ rad/s ↺

12–50. $v_C = 0.887$ m/s →

$\omega_{BCD} = 2.76$ rad/s ↻

$v_D = 1.04$ m/s $31.9°$

12–51. $v_A = 26$ in./s (12, 5)

$\omega_{BC} = 3$ rad/s ↺ $\omega_{AD} = 2$ rad/s ↻

12–52. $\omega_{DG} = 2.4$ rad/s ↻

$v_G = 20.9$ in./s $12.2°$

12–53. $v_A = 16.1$ ft/s ↓

$v_C = 13.3$ ft/s (17, 3)

12–54. $v_D = 3.94$ m/s (5, 12)

12–55. $v_E = 400$ mm/s $69.5°$

$\omega_{EB} = 1$ rad/s ↺ $\omega_{AB} = 1.17$ rad/s ↻

12–56. $v_C = 0.9$ m/s → $\omega_{EC} = 4$ rad/s ↺

$v_D = 4.27$ m/s $\angle 16.3°$

12–57. $v_E = 65.9$ in./s $10.5°$

12–58. $v_C = 338$ mm/s $45°$

$v_E = 272$ mm/s (8, 15)

12–59. $v_G = 314$ mm/s (5, 12)

12–60. $\omega_{CD} = 2$ rad/s ↺

$v_B = 12.2$ in./s (1, 6)

$v_A = 18$ in./s ↓

12–61. $\theta = 44°$

$v_D = 11$ in./s →

12–62. $v_C = 17.1$ m/s (3, 4)

REVIEW PROBLEMS

R12–1. $S_{B/A} = 160$ m $38.6°$

$v_{B/A} = 32$ m/s $38.6°$

$a_{B/A} = 3.2$ m/s $38.6°$

R12–2. $v_C = 48.8$ in./s 55°
$\omega_{CD} = 7.43$ rad/s ↻

R12–3. $v_C = 15.6$ ft/s (5, 6)
$v_E = 19.2$ ft/s (5, 4)
$a_{E/D} = 75.2\ \text{ft/s}^2$ (2, 25)

R12–4. $v_E = 3860$ mm/s 43.4°
$\omega_{CE} = 9.36$ rad/s ↻
$\omega_{AC} = 10$ rad/s ↻

R12–5. $v_B = 10.3$ in./s 60°
$\omega_{BD} = 0.665$ rad/s ↺
$v_E = 14.8$ in./s 59°

R12–6. $v_D = 14.8$ ft/s (15, 8)
$v_E = 20$ ft/s →
$v_C = 6.95$ ft/s →
$S_C = 3.48$ ft →

R12–7. $\omega_{BC} = 3.64$ rad/s ↻

R12–8. $\omega_{ABD} = 2.11$ rad/s ↺
$v_E = 1.45$ m/s 45°

R12–9. $v_C = 1.35$ m/s 45°

R12–10. $\omega_{ABC} = 2$ rad/s ↻
$v_A = 340$ mm/s (8, 15)
$\omega_E = 2.67$ rad/s ↻

R12–11. $v_A = 90$ in./s ←
$v_C = 238$ in./s 67.8°
$\omega_{DE} = 0.59$ rad/s ↻
$v_D = 69.9$ in./s (1, 1)

R12–12. $v_C = 1.21$ m/s →
$v_D = 1.16$ m/s 3.81°
$\omega_{BD} = 1.9$ rad/s ↺

CHAPTER 13

SECTION 13–1 TO 13–3

13–1. $a = 9.05\ \text{ft/s}^2$ →

13–2. $a = 2.96\ \text{m/s}^2$ 25°

13–3. $a = 1.81\ \text{m/s}^2$

13–4. $a = 1.97\ \text{m/s}^2$ (5, 12)

13–5. 0.53

13–6. 0.433

13–7. $a = 2.69\ \text{m/s}^2$ ↑

13–8. 264 lb

13–9. $a = 1.81\ \text{m/s}^2$ ↓

13–10. $P = 30.6$ lb →

13–11. $s = 11$ ft

13–12. $a_A = 4.9\ \text{m/s}^2$ ↑ $a_B = 2.45\ \text{m/s}^2$ ↓

13–13. $16.1\ \text{ft/s}^2$ $9.66\ \text{ft/s}^2$
weight changes are immaterial

13–14. $3.07\ \text{ft/s}^2$

13–15. 0.6

13–16. 2270 N

13–17. 157 ft

13–18. 239 N 121 N

13–19. $T = 46.5$ N $\theta = 32.5°$ AB = 1.86 m

13–20. $a_A = 0.76\ \text{m/s}^2$ $a_A = 0$

13–21. $a_A = 4.53\ \text{m/s}^2$ →

13–22. $v_B = 3.77$ m/s (12, 5)

13–23. 12.3°

13–24. 4.35 rad/s ↻

13–25. will not skid

13–26. 16.7 rpm ↻

13–27. 7.26 ft (5, 12)

13–28. $\mu = 0.285$

13–29. $d = 4.06$ m $v_A = 2.29$ m/s 15°

13–30. $a_A = 1.4\ \text{m/s}^2$ ↑ $a_B = 2.8\ \text{m/s}^2$ ↓

13–31. $s_B = 4.13$ m ↓

13–32. 1190 lb 615 lb

13–33. $a_B = 6.1\ \text{m/s}^2$ 80°

13–34. $v_A = 6.88$ ft/s ↓

SECTIONS 13–4 AND 13–5

13–35. $T = 1.38$ lb-ft
13–36. $T = 5690$ lb-ft
13–37. 95.5 rpm
13–38. 31.4 N
13–39. 18.8 N·m
13–40. 2830 lb-ft
13–41. 58.9 s
13–42. 4.52 rad/s^2 ↺
13–43. 4.69 s
13–44. 273 lb-ft ↻
13–45. $s_A = 381$ ft ↓
13–46. $\alpha = 4.95$ rad/s^2 ↻ $T = 4490$ N
13–47. $\alpha = 12.8$ rad/s^2 ↺ $T = 129$ N
13–48. $A_x = 45$ N → $A_y = 44.1$ N ↑
13–49. $A_x = 19.7$ lb ← $A_y = 96.6$ lb ↑
$\alpha = 17.7$ rad/s^2 ↺
13–50. $A_x = 71.1$ N ← $A_y = 137$ N ↑
$\alpha = 39.8$ rad/s^2 ↺

SECTION 13–6

13–51. $P = 143$ lb →
13–52. $a = 1.89$ m/s^2 →
13–53. $a = 2.9$ m/s^2 →
13–54. $a = 0.87$ m/s^2 →
13–55. $T = 141$ N $s_B = 72.6$ m ↓
13–56. $a = 0.667$ m/s^2 ← $\mu = 0.272$
13–57. $P = 251$ lb ↙
13–58. $\alpha = 4.71$ rad/s^2 ↻
13–59. $T = 474$ lb $a_B = 1.67$ ft/s^2 ↓
$\alpha_A = 0.835$ rad/s^2 ↻
13–60. $a_B = 0.402$ m/s^2 ↓
13–61. $a = 8.33$ ft/s^2 $\overline{\backslash 20°}$
13–62. $a = 12.8$ ft/s^2 →
13–63. $a_A = 0.0461$ m/s^2 (slope 12/5)
13–64. $a_B = 2.08$ m/s^2 (slope 8/15)
13–65. $a = 2.74$ ft/s^2 ←
13–66. $T = 189$ N $a_B = 0.339$ m/s^2 ↓
13–67. $v = 5.72$ m/s ↓ $v_F = 7$ m/s ↓
13–68. $a = 1.69$ m/s^2 →

REVIEW PROBLEMS

R13–1. 8.35 lb
R13–2. $r = 5.45$ ft weight has no effect
R13–3. $a_A = 13.1$ ft/s^2 ↓ $T = 45.5$ lb
R13–4. 14 N·m
R13–5. 3110 lb-ft
R13–6. $a = 0.494$ m/s^2 20° $\mu = 0.76$
R13–7. $a = 3.25$ ft/s^2 (slope 8/15) $\mu = 0.52$
R13–8. $\alpha = 11.9$ rad/s^2 ↺
R13–9. $a = 0.1$ m/s^2 ∠40°
R13–10. $a = 2.73$ m/s^2 $\overline{\backslash 20°}$

CHAPTER 14

SECTIONS 14–1 AND 14–2

14–1. 7180 J
14–2. 333 ft-lb 95.2 lb
14–3. 1000 ft-lb 100 lb
14–4. 952 ft-lb
14–5. 1500 J
14–6. 1400 J
14–7. 442 J −66.2 J
14–8. 2360 J
14–9. 2770 ft-lb
14–10. 157 J, 240 J
$F = 21.2$ lb
14–11. 392 J

SECTION 14–3

14–12. 582 J
14–13. 4 in.
14–14. $K = 2500$ N/m
14–15. $K = 20$ lb/in. 160 in.-lb
14–16. $K = 2500$ N/m 50 J
14–17. 4.2 in.-lb
14–18. $K = 20.4$ kN/m
14–19. 1.51 in.
14–20. 55 in.-lb
14–21. 3.22°
14–22. each outside spring 230 in.-lb 6.78 in.
center spring 178 in.-lb 3.78 in.
14–23. 840 J
14–24. 0.922 J
14–25. work in = 117 in. · lb
work out = 27.8 in. · lb
14–26. 0.5 in., 3.5 in.-lb
14–27. 1800 J
14–28. 297 ft-lb

SECTION 14–4

14–29. 43,200 ft-lb
14–30. 9 MJ
14–31. 1.28 MJ 31.9 m
14–32. 38 ft-lb
14–33. 1.96 kJ 8.85 m/s
14–34. 56.8 ft/s 1610 ft-lb
69.5 ft/s 2415 ft-lb
14–35. 900 J, 4 times
14–36. 8900 ft-lb 49.5 ft.

SECTION 14–5

14–37. F = 397 lb $a = -10.7$ ft/s^2
$t = 0.375$ s
14–38. $v_A = 16.7$ ft/s →
14–39. 128 J
14–40. 332 ft
14–41. $v_A = 16$ ft/s →
14–42. 14.8 ft
14–43. 0.635 m
14–44. 83.3 kN
14–45. 5.82 m/s 11.5 m
14–46. 355 in.-lb
14–47. 5.14 ft/s (slope 1:5)
14–48. 0.327 m 0.163 m
14–49. 0.127 m
14–50. 0.49 m
14–51. 14.6 ft/s ↓
14–52. 15.3 ft/s ↓
14–53. 15.5 ft/s →
14–54. $K = 2.14$ kN/m
14–55. 6.95 ft/s ↓

SECTION 14–6

14–56. 92.3 rpm
14–57. 6.8 ft-lb
14–58. 11.4 J
14–59. 34.2 ft-lb 23.5 ft-lb
14–60. 62.8 J
14–61. 188 kJ/min
14–62. 239 lb-ft

SECTION 14–7

14–63. 5.46 rad/s ↻
14–64. 1.07 m
14–65. $P = 30.3$ kN ↓
14–66. $v_A = 11.6$ ft/s ↑
14–67. 43.6 lb/ft
14–68. 1 m
14–69. $v_A = 2.61$ ft/s (slope 12:5)
14–70. $v_B = 3.85$ ft/s ↓
14–71. 31 rad/s
14–72. 0.22 m

SECTION 14–8

14–73. KE(linear) = 37,700 ft-lb
KE(rot) = 18,800 ft-lb

14–74. 32.1 ft/s

14–75. $v_B = 2.36$ ft/s ↓

14–76. $m_B = 18.9$ kg

14–77. $v_B = 0.61$ m/s ↓ $s_B = 0.254$ m ↓

14–78. 482 N/m

14–79. $v_A = 1.43$ m/s ↓

14–80. $v_B = 1.33$ m/s ↑ ↻

14–81. $\omega_A = 7.34$ rad/s ↻

14–82. $\omega_A = 7.15$ rad/s

14–83. $v_B = 1.92$ ft/s (12, 5 slope)

14–84. $v_B = 2.34$ ft/s ←

14–85. $v_B = 3.52$ ft/s ←

14–86. $v = 12.9$ ft/s
$KE_{lin} = 517$ ft-lb
$KE_{ang} = 261$ ft-lb
K = 2770 lb/ft

14–87. $v_A = 2.71$ ft/s ↑

14–88. $v = 5.25$ ft/s (4, 3 slope)

SECTION 14–9

14–89. 516 lb

14–90. 54.5 hp 3000 lb-ft

14–91. 670 kW

14–92. 1500 rpm

14–93. 7.85 kW

14–94. 471 W

14–95. 1720 hp

14–96. 0.612 m/s

14–97. 129 hp

14–98. 123 N · m 81.9 N · m

14–99. 85.1 kW

14–100. 12.3 kW 30.8 kW

14–101. 14.3 mph

14–102. 4.31 hp

14–103. 20.4 hp

REVIEW PROBLEMS

R14–1. 144 kJ

R14–2. $K = 6.67$ lb/in.

R14–3. 58.8 J

R14–4. 59.4 kJ

R14–5. $v_A = 2.35$ ft/s ↓

R14–6. 1.67 ft 40.9 ft left of initial

R14–7. 1170 ft-lb 192,000 ft-lb
193,000 ft-lb

R14–8. 73.3 ft-lb

R14–9. 33.4 ft/s 2770 lb-ft

R14–10. 3.41 m

R14–11. 2.53 m/s 30°

R14–12. 1.2 m/s 85 N

R14–13. 33.5 kW

R14–14. 39.9 hp

CHAPTER 15

SECTION 15–1

15–1. 5400 N · s

15–2. 40,100 lb · s

15–3. 322 ft/s

15–4. 10.5 s

15–5. 0.14 s

15–6. 60 N

15–7. 7.69 m/s 15.4 m/s

15–8. 19.3 ft/s

15–9. 24 m/s

15–10. 7.92 s

15–11. 7290 lb 183 ft

15–12. 502 m/s

15–13. 21.3 m/s

15–14. 30.6 lb

15–15. 12.3 kW 30.8 kW

SECTION 15–2

15–16. 420 ft-lb-s

15–17. 535 lb-s 145 lb-ft

15–18. 18.8 s

15–19. 33.5 s

15–20. 396 N

15–21. 2.64 lb-ft

15–22. 1.03×10^{-2} N · m

15–23. 462 rpm

15–24. 0.0147 kg · m^2/s

15–25. 1.04 s

15–26. 14.5 kg

15–27. 10.8 lb-ft

15–28. 0.17 m

15–29. 67.8 lb →

15–30. 138 lb ↓

15–31. 181 rad/s ↺

SECTION 15–3

15–32. 1.3 m/s

15–33. 32.5 ft/s →

15–34. 2.47 m/s ∠66.4°

15–35. 0.45 m/s →

15–36. 2.86 m/s

15–37. 8.19 ft/s →

15–38. 6.23 ft/s

15–39. 1.64 m/s 80°

15–40. 3.75 m/s → t = 1.27 s

15–41. 0.52 s

15–42. 14.2 rad/s ↺

15–43. 35.5 rpm

15–44. 21.9 rpm

15–45. ω_A = 77.2 rad/s ↻
α_A = 19.3 rad/s^2 ↻ T = 160 lb

15–46. P = 30.3 kN ↓

REVIEW PROBLEMS

R15–1. 8100 lb 65.2 ft/s^2

R15–2. 1.22 ft

R15–3. 212 N

R15–4. 8.07 ft/s 38.3°

R15–5. 3.87 m/s →

Index